PRATT

3-11-75

The Mammals of Louisiana and Its Adjacent Waters

THE MAMMALS

Published for the

LOUISIANA WILD LIFE AND FISHERIES COMMISSION

By the LOUISIANA STATE UNIVERSITY PRESS

OF LOUISIANA AND ITS ADJACENT WATERS

By GEORGE H. LOWERY, JR.

Director, Museum of Natural Science and
Boyd Professor of Zoology, Louisiana State University

Illustrated by H. DOUGLAS PRATT

ISBN 0–8071–0609–7
Library of Congress Catalog Card Number 73-89662
Copyright © 1974 by George H. Lowery, Jr.
All rights reserved
Manufactured in the United States of America

Designed by Dwight Agner. Set in Linofilm
Baskerville by Southwestern Typographics, Inc.,
Dallas, Texas. Printed and bound by
Kingsport Press, Inc., Kingsport, Tennessee.
Colorplates printed by Benson Printing Co.,
Nashville, Tennessee.

To WILLIAM HAZEN GATES
who was to me a wise and
faithful counselor, patient
teacher, and longtime friend

LOUISIANA
WILD LIFE AND FISHERIES COMMISSION

TABLE OF CONTENTS

ILLUSTRATIONS AND TABLES

DISTRIBUTIONAL MAPS

FOREWORD

IN 1928 the Louisiana Department of Conservation, as the present Wild Life and Fisheries Commission was then known, issued a book entitled *The Fur Animals of Louisiana* written by Stanley Clisby Arthur, one of Louisiana's distinguished naturalists of the time. The book has long been out of print, and the demand for a similar yet modern work has steadily mounted ever since. The fact that it has done so is not surprising. For one thing, the production of fur in Louisiana is a multi-million-dollar industry, often greater than that of all other states combined or of all Canada. Some inhabitants of our forest, such as deer, rabbits, and squirrels, help to make Louisiana the "Sportsman's Paradise" our license plates long claimed it to be. Still other species, for example the introduced house rats, do great damage to man's property and are instrumental in spreading diseases that once reached plague proportions in New Orleans. But whether a particular kind of mammal is beneficial or harmful, large or small, attractive in appearance or nondescript, it is fascinating and interesting to study. The habits of mammals, often not too well known because of their elusive ways, provide outdoorsmen with an opportunity to add to the storehouse of facts about natural history.

When the Fur Division of the Wild Life and Fisheries Commission adopted the plan of again issuing a treatise on the mammals of our state, it was only natural that Professor Lowery was asked to prepare the text. He is unquestionably the state's leading authority on the subject as a consequence of his lifetime of study of our furred animals. His book *Louisiana Birds*, also published for the Commission by the LSU Press, has been eminently successful and is now going into its third, revised edition. It was he who founded some 37 years ago what is now known as the Museum of Natural Science at Louisiana State University, and he is responsible, with the help of his associates and his students, for having brought its research collections to a position of eminence among those of the great university museums of the United States. Much of the information contained in the pages that follow is obviously the product of the museum's collections or of research by its personnel, including numerous students who over the years have studied under Professor Lowery. One of the assistant directors of the Wild Life and Fisheries Commission and five of its division chiefs, including the undersigned, are proud to be included among those who were at one time enrolled

xviii   MAMMALS OF LOUISIANA

in his "Zoology 141, Mammalogy" at Louisiana State University.

Professor Lowery, in sharing his knowledge of our mammals and their habitats with us, has created a publication that could well serve as a textbook, for it contains invaluable information of particular importance to game management students and to teachers and wildlife technicians as well.

During the preparation of the work, we were fortunate to have close at hand an artist, H. Douglas Pratt, who is a graduate student in zoology at LSU and who possesses the special talent of being able to transfer to the drawing board in lifelike manner the characteristics peculiar to each species and to render accurately other scientific illustrations to complement the author's text. I wish to thank both author and artist for their diligence and long hours of painstaking work in the preparation of this book.

As for myself, I feel that the role I may have played in guiding this project to completion climaxes a lifetime of absorption with the fur animals of Louisiana and their ecology.

TED O'NEIL
Chief, Fur Division
Wild Life and Fisheries Commission
New Orleans, Louisiana
1 July 1973

PREFACE

THE PRIMARY objectives of this work are threefold. One goal is to acquaint the people of the state with the wide array of furred animals that are denizens of our fields, forests, and marshlands. When I was first asked to undertake the writing of this book, the suggestion was made that I treat only the animals of Louisiana that produce commercial fur. Admittedly, that particular group of mammals is extremely important. Furbearers constitute a great economic asset to the state, amounting to millions of dollars each year, and therefore are of major concern to many Louisianians. But it was my belief that just as many, if not more, people in our state are interested in learning about the moles that are churning up their lawns, the bats that they may hear squeaking in their attics, or the whales and dolphins they see plowing the waters along our coast. Consequently, I argued that this work should treat all the mammals within our borders. I am grateful to the Louisiana Wild Life and Fisheries Commission for its wholehearted concurrence with this plan.

The second main objective of the book is to provide an up-to-date account of the present known status of each species and to give a summary of where each has been recorded and of its relative abundance. Many mammals are no longer found in the numbers in which our forebears encountered them when Louisiana was first settled. The Bison was long ago extirpated and several other species are now officially listed as rare and endangered and unless given protection may soon pass from the Louisiana scene. While some of our mammals are large, conspicuous, and thoroughly familiar to nearly everyone, others are tiny, extremely elusive, and often poorly known, even to those of us who have devoted a lifetime to their pursuit. The life histories and habits of some species are so well studied that entire books have been written about them, while our knowledge of some other species is distressingly incomplete.

Finally, even though several textbooks and innumerable monographs have been written covering many facets of mammalogy, creating a voluminous literature on the subject, much of this information is not readily available to the layman nor to the serious student who does not have access to a large library. For this reason, I have tried to intersperse through the following pages at least a smattering of basic facts about mammals that will,

I hope, stimulate further interest on the part of the reader, that will perhaps cause him to explore more deeply, and that may thereby lead him to discoveries filling in present gaps in our knowledge concerning the group of animals to which man himself belongs.

The study of mammals, particularly their correct identification, sometimes requires a basic knowledge of their structure, notably that of their skulls. For this reason I have provided in one of the introductory chapters instructions as to how body and skull measurements are made, drawings of typical skulls with the major bones and processes labeled, and other information that will aid the nonzoologist when he delves into the subject for the first time. I have also included explanations of how to use identification keys, and I have provided a key to the identification of the land mammals that inhabit Louisiana. Because of the immense impor-

tance of furbearers in the state's economy, I have devoted a whole chapter to the subject, and I have supplied a table showing the annual take of the various fur-producing species since 1913 and the annual monetary value of the furs to the trapper since 1928. The main body of the book is, however, made up of the accounts of the native mammals inhabiting Louisiana and its adjacent waters, species by species. The plan of these accounts is explained in the text. I have added sections that furnish a brief treatment of our domesticated mammals and a list of the fossil mammals currently known to have once occurred in the state.

GEORGE H. LOWERY, JR.
Museum of Natural Science,
Louisiana State University,
Baton Rouge, Louisiana
1 June 1973

ACKNOWLEDGMENTS

THE CHAIN of circumstances leading to this book began in 1927. That summer, as a young high school student, during a brief residence of my family in West Florida, I went camping with some companions on the mainland opposite Santa Rosa Island. One moonlit night a group of us rowed across Santa Rosa Sound to the island. On turning over a piece of driftwood on the edge of a sand dune, we were astonished to find and to catch an almost entirely white mouse. Later I learned that the little mammal was the endemic beach mouse that the distinguished mammalogist Arthur H. Howell had discovered only a few years before and had described under the name *Peromyscus leucocephalus*.

The knowledge that the nearly pure-white sands of Santa Rosa Island had played a role in the evolution, through natural selection, of a nearly pure-white mouse intrigued me no end; so I began to wonder what other kinds of exciting but elusive small mammals might be found dwelling in the fields and woodlands nearby. By setting a string of mouse traps in a marsh habitat where I often went looking for Clapper Rails and Seaside Sparrows, I caught a small mammal that subsequently proved to be a Marsh Rice Rat *(Oryzomys palustris)*. But I knew that to find out what it was I had to skin it, stuff it with cotton, and send it to some expert. The final product of my first skinning effort was rather dismal. Yet when I sent the specimen to the agency then known as the Bureau of Biological Survey, in Washington, D.C., a reply came from none other than Arthur H. Howell himself. His letter was one of the most stimulating and encouraging that I have ever received. In his typical kindly manner he not only told me what was wrong with the preparation of my specimen, but he sent me a mouse, which he himself had prepared, to show me what a small mammal study skin should look like. His letter also identified my rat, but it did more than that. It sealed irrevocably then and there the determination of a fourteen-year-old boy to pursue the study of mammals. And never once in the years that have followed has this interest waned, despite my somewhat greater devotion to the study of birds. The two endeavors go hand in hand. In my quest for some feathered rarity I have often unexpectedly come upon a furred creature from which I have derived great pleasure and excitement in making its acquaintance. The setting of a string of mouse traps in a secluded Louisiana woodland and finding the next morning a small rodent captured that has previously escaped detection can be just as electrifying to the naturalist as getting a close-up photograph of a lion or a rhinoceros in the African veld. I know, for I have experienced both kinds of rewards.

I am also indebted to the students who have taken "Zoology 141, Mammalogy" since I first offered the course in 1939. Each of these students has been required to trap or otherwise procure and to prepare as conventional museum study skins a minimum of 10 specimens of native wild mammals. Their quotas are usually met with rats and mice. The exercise provides training in making an acceptable study skin, in recording field data, in writing out labels, in measuring, weighing, and sexing specimens, and in keeping a field catalog and itinerary. Many of the specimens turned in by these students have been procured in their home areas during their Thanksgiving holidays and their Christmas vacations and have thus added valuable distributional records of species from localities not otherwise represented in museum collections, and they have helped to make the holdings of the Louisiana State University Museum of Zoology by far the largest in Louisiana, as well as in the entire southeastern United States. The interest and enthusiasm of my students have over the years been a source of great inspiration to me, and I thank each and every one of them for their individual and collective efforts.

The actual writing of this book was undertaken at the instigation of Mr. Ted O'Neil, representing the Louisiana Wild Life and Fisheries Commission as chief of its Fur Division. Once the project was underway I received from the members of the Board of Commissioners and Mr. O'Neil wholehearted support and assistance, beginning as far back as 1954, when two of my doctoral students, Stephen M. Russell and Herbert Shadowen, were hired during the summer months to trap mammals in critical areas of the state for

the purpose of filling in gaps in distributional information concerning certain species. The Commission later authorized arrangements with H. Douglas Pratt for the execution of the striking colorplates and other illustrations that enliven the pages to follow. To Mr. O'Neil and to the entire personnel of the Louisiana Wild Life and Fisheries Commission I am indebted beyond all words.

Also of special importance to me has been the assistance I have received from the curators of collections in other Louisiana colleges and universities, where highly significant regional aggregations of specimens repose. In this regard I am indebted to the following: Marshall B. Eyster, University of Southwestern Louisiana; Royal D. Suttkus and Alfred L. Gardner, Tulane University; James D. Lane, McNeese State University; D. T. Kee, Northeast Louisiana University; John W. Goertz, Louisiana Tech University; Dick T. Stalling, Northwestern State University of Louisiana; and J. Larry Crain, Southeastern Louisiana University. In the same connection I am grateful to the curators of other collections who have supplied information on Louisiana specimens in their care or who have given me advice on technical matters: F. C. Fraser, British Museum (Natural History); William J. Beecher, Chicago Academy of Science; E. Raymond Hall, Museum of Natural History, University of Kansas; Charles O. Handley, Jr., James G. Mead, Henry W. Setzer, Richard C. Banks, Clyde Jones, and Don Wilson, United States National Museum; Stephen R. Humphrey, Florida State Museum; David J. Schmidly, Texas A and M University; and Emmet T. Hooper, University of Michigan.

Grateful acknowledgment is made to the National Geographic Society for the use of the photograph on page 479; to Johns Hopkins Press for permission to reproduce the photograph on page 200; to the Boone and Crockett Club for authorization to duplicate the items that appear as Figure 234; and to E. Raymond Hall for permitting me to base, with some modifications, the small inset maps of North America, showing the overall range of each species, on maps contained in *The Mammals of North America*. Photographs are credited individually in their respective legends, but I want to pay special tribute to Joe L. Herring for the innumerable evenings that he spent photographing skulls and other materials and in helping me in a multitude of ways.

The following individuals have rendered invaluable assistance of one sort or another, and to each I express my appreciation: Randolph A. Bazet, Joseph Billiot, John Blanchard, Harvey R. Bullis, Jr., Tucker Campbell, Robert Chabreck, J. Y. Christmas, B. L. Davidge, Dan Dennett, Paul H. Duhon, Allan B. Ensminger, Alfred L. Gardner, Leslie Glasgow, Mrs. J. E. Gonzales, Claude H. Gresham, Gordon Gunter, John H. Guthans III, William G. Haag, Frank A. Hayes, Robert F. Henry, Carl L. Hubbs, David G. Huckaby, Roger Hunter, Horace H. Jeter, Ted Joanen, J. B. Kidd, G. A. Loker, John S. McIlhenny, A. Bradley McPherson, Robert Murry, Norman C. Negus, Robert W. Neuman, John D. Newsom, Robert E. Noble, A. W. Palmisano, William F. Perrin, Lester Plaisance, R. H. Potts, Jr., Stephen M. Russell, George E. Sanford, S. B. Seibenaler, Herbert Shadowen, Charles Shaw, Ned McIlhenny Simmons, Kenneth C. Smith, Stewart Springer, Philip St. Romain, G. F. Tannehill, Alexander Wetmore, and Richard K. Yancey.

Several people have read the manuscript either in part or in its entirety. David K. Caldwell generously read the section on cetaceans; John D. Newsom and Robert E. Noble, the part dealing with the artiodactyla; A. W. Palmisano, the special chapter on the Louisiana fur industry; and Kenneth C. Corkum, the paragraphs dealing with parasites. Among those who have read the entire manuscript are Robert J. Newman, Ted O'Neil, and Alfred L. Gardner. From all these individuals I have received numerous valuable suggestions for which I am grateful. But my associate Robert J. Newman deserves special mention. He has not only read every sentence critically with his usual scholarly insights, but he has helped me in a multitude of other ways. I am also indebted more than I can say to my curatorial assistant, Mrs. Peter J. Fogg, who not only plotted the localities on the range maps and compiled the lists of specimens examined, employing in both undertakings her characteristically meticulous accuracy and attention to detail, but also helped in other important ways. Mrs. Martin Richard typed most of the manuscript. Mrs. Marguerite Ponder provided indispensable liaison between my office and that of Mr. O'Neil, and my research assistant, Miss Sandra Guthans, made most of the skull measurements, performed some of the literature search, and spent untold hours in helping me proofread drafts of the manuscript.

**The Mammals of Louisiana
and Its Adjacent Waters**

INTRODUCTION

MAN IS instinctively interested in mammals, because he himself is a mammal and all the rest of the mammals of the world are his kin. He feels for them a sympathy that he feels for no other class of living things. He even credits them sometimes with his own mental traits and mistakingly tries to interpret their behavior in human terms. Their ways are indeed a source of endless fascination. But before we take up these matters we need to consider how mammals fit into the scheme of zoological classification.

Mammals belong to the phylum Chordata, which includes those members of the animal kingdom possessing a dorsal structural support to the body, the notochord, during some stage of their lives. They are in turn placed in the subphylum Vertebrata, which embraces all forms of chordates possessing a dorsal hollow nerve cord, gill pouches at least during embryonic development, and the notochord replaced by bone to form vertebrae. Within this subphylum zoologists recognize a minimum of five classes—Pisces, fishes (now separated by most experts into four, sometimes five, classes, two of which, the Placodermi and the Acanthodii, consist only of fossil forms); Amphibia, amphibians (including frogs, toads, and salamanders); Rep-

tilia, reptiles; Aves, birds; and Mammalia, mammals. The word Mammalia was derived from the Latin *mammalis*, which refers to the mammae, or breasts, at which the female suckles her young. The layman often uses the term "animal" as if it applied exclusively to mammals, but such an interpretation is incorrect. Sponges, corals, worms, insects, fishes, frogs, and birds are also animals, as well as almost all other forms of life not classified as plants. Zoologists are still debating how to classify one-celled organisms, including the protozoans and the brown and red algae, not to speak of, for example, the bacteria, the blue-green algae, and the viruses.

Only mammals possess mammary glands, and all except the Monotremata have teats, or nipples, to facilitate the transfer of the milk to the offspring. All have hair at some stage of their lives. In the whales and dolphins it may be limited to a few bristles on the snout or lacking entirely except during an embryonic stage of development, but it is nevertheless present at one time or another. This ectodermal derivative occurs for certain in no other class of vertebrates. Thus, mammae and hair are two of the most diagnostic features of the class Mammalia.

Mammals are easily distinguished from

3

other vertebrates by a combination of other salient characters. In all except the egg-laying Monotremata, the young are born alive after a period of gestation inside the mother's body. All are homoiothermic, or warm-blooded, and all are endothermic, that is, they produce their own heat, in contrast to lower vertebrates, which are ectothermic, that is, dependent upon their environment for the source of their body warmth. In some mammals, such as the Duck-billed Platypus (*Ornithorhynchus anatinus*), homoiothermy is quite imperfectly developed, the body temperature varying from 81.7° to 90.7° F. Similarly, in the slow-moving sloths, of the family Brachypodidae, the rectal temperature varies between 81.8° and 98.3° F. But in most higher mammals the body temperature is fairly constant, and in any group of individuals of a given species usually averages somewhere above 96.8° F. In man, for instance, regardless of what the ambient temperature may be, at least within certain reasonable limits, the body temperature in a healthy individual is always close to 98.6° F.

The teeth of mammals are thecodont, that is, set in sockets. This method of attachment reduces the chances of the teeth being lost when the animal is fighting or securing its prey. In most reptiles the teeth are acrodont, or merely attached to the surface of the jaw, but in crocodilians they are thecodont as in mammals. In most mammals the dentition is heterodont (differentiated into incisors, canines, premolars, and molars) in contrast to being homodont (all teeth the same), but again some reptiles share with mammals the heterodont condition. In the vast majority of mammals the dentition is diphyodont, that is, temporary milk teeth are replaced by a permanent set of teeth, instead of monophyodont. In the latter condition the milk series is retained throughout life.

Each ramus of the lower jaw is a single bone, the dentary, in contrast to the situation in reptiles, for example, in which each half of the lower jaw consists of as many as six bones

(dentary, articular, angular, surangular, coronoid, and splenial). Another osteological feature that is unique to the class Mammalia is that the lower jaw articulates directly with the cranium. In other vertebrates its suspension from the cranium is by way of the quadrate. In the evolutionary development of mammals the quadrate bone became the incus of the middle ear, one of the small bones that transmit sound waves to the inner ear. The other two of these bones, the malleus and stapes, were derived, respectively, from the articular and the hyomandibular, the latter by way of the columella of amphibians and reptiles. These three bones of the middle ear provide a classic example of evolutionary conservatism, whereby utilization is made of structures already on hand when some new structure is needed to perform a new or modified function.

The vertebrae of mammals are said to be amphiplatyan, that is, they are flat on both anterior and posterior ends of the centra, the main body of each vertebra, and they are separated by intervertebral discs.

In mammals two occipital condyles, formed by the exoccipital bones and located one on each side of the foramen magnum, through which the spinal cord passes into the braincase, serve to articulate the skull with the cervical, or neck, vertebrae. In reptiles and birds a single condyle, derived mainly from the unpaired basioccipital bone rather than the paired exoccipitals, articulates the skull to the neck vertebrae. The latter are seven in number in all but five groups of mammals. The exceptions are the manatees (*Trichechus*) and the two-toed sloths (*Choloepus*), each with six; the anteater (*Tamandua*), which has eight; and the three-toed sloths (*Bradypus*), which have nine. In cetaceans the cervical vertebrae exhibit various degrees of fusion. The vertebrae and long bones of the limbs possess epiphyses, terminal secondary centers of ossification. The phalangeal formula, that is, the number of bones in the fingers and toes, is frequently 2-3-3-3-3 but the number of toes and phalanges is highly variable from one mammal group to another.

The heart of mammals is four-chambered, as is also that of birds and basically that of crocodilians, but the dorsal aorta in mammals is derived from the left fourth aortic arch, whereas in birds the right fourth aortic arch is preserved as the dorsal aorta. In reptiles both the left and the right are utilized (Figure 1).

In mammals and crocodilians only the nasal cavity is separated from the oral cavity by a complete secondary palate. Also, in mammals a muscular diaphragm completely partitions the pericardial cavity and the two pleural cavities from the abdominal, or peritoneal, cavity. In birds and some reptiles the trans-verse and oblique septa that serve as a partition in the pleuroperitoneal cavity, dividing it into anterior and posterior compartments, is essentially nonmuscular. The red blood cells of all mammals except camels are nonnucleated when fully developed.

The class Mammalia is divided into three subclasses, Prototheria, Allotheria, and Theria. The first contains only the primitive, egg-laying mammals of the order Monotremata, comprising the spiny anteaters, or echidnas (family Tachyglossidae), and the Duck-billed Platypus (family Ornithorhynchidae), all of which are confined to the Australian Zoogeographical Realm. They possess mammary glands but no nipples. The milk oozes from pores and is lapped up by the

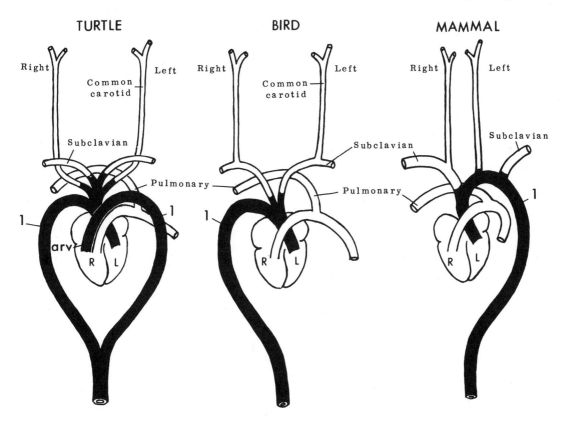

Figure 1 Fate of the right and left fourth aortic arches in a turtle, bird, and mammal. The fourth, or systemic, aortic arches (1) and their aortic trunks are shown in black. Also shown is the aortic trunk from the right ventricle (arv) of the turtle and other reptiles. (Adapted from Kent, *Comparative Anatomy of the Vertebrates,* 3rd ed., 1973, C. V. Mosby Co.)

offspring. The pectoral girdle includes interclavical and precorocoid elements that are both peculiar to the Monotremata. Epipubic bones are present and probably serve to support the marsupium, in which the echidnas place the egg for incubation. Functional teeth are absent in adults, but in the platypus they are present during embryonic development and in the newly born young. The oviducts are separate and each opens into the cloaca, a common receptacle for the urogenital system and the digestive tract. The cloaca is the basis of the name Monotremata (from the Greek *mono,* single and *trema,* hole). The penis conducts only seminal fluid containing the sperm. The eggs are leathery shelled and reptilelike. The skull differs from that of most placental mammals in lacking lacrymal and jugal bones, as well as auditory bullae. The cerebral hemispheres are smooth, but in most other respects the brain is almost typically mammalian. The laying of eggs and the structure of the pectoral girdle and of the female's reproductive tract are somewhat more reptilian than mammalian, but as Beddard (1902) long ago pointed out, the monotremes display many typically mammalian characteristics, not the least of which is the presence of hair and mammary glands. Geologically monotremes are known for certain only from the Pleistocene (or late Pliocene) to the Recent, but they are obviously the specialized descendants of an ancient line of mammals that arose independently from the Mezozoic mammallike reptiles, perhaps over 200 million years ago.

The subclass Allotheria (which Simpson, 1971, subdivides into two subclasses, Allotheria and Eotheria, the former including only the Order Multituberculata) is an ancient and aberrant group of mammals known solely as fossils that date from the Lower Jurassic to no later than the Eocene. All remaining mammals are placed in the subclass Theria, whose living representatives are viviparous and possess mammary glands with nipples and a pectoral girdle without inter-

clavical or precorocoid elements. The subclass is divided into three infraclasses, Pantotheria, Metatheria, and Eutheria. The first of these, the pantotheres, are known only as fossils and may have been the group that was ancestral to higher mammals of the subclass Theria. The infraclass Metatheria is represented by the members of one order, Marsupialia. The infraclass Eutheria comprises the remainder of both living and fossil forms, the placental mammals.

Living members of the infraclass Eutheria are placed in 17 orders, only 9 of which occur in the wild state in Louisiana: Insectivora, shrews and moles and their allies; Chiroptera, bats; Edentata, armadillos and their allies; Lagomorpha, rabbits; Rodentia, rodents; Cetacea, whales and dolphins; Carnivora, carnivores; Sirenia, manatees and their allies; Artiodactyla, hoofed mammals with an even number of toes. Characteristics of each order are given under the appropriate headings in the text that follows.

Mammals rank next to last among vertebrates in the number of known Recent species (fishes, 22,000; birds, 8,600; reptiles, 6,000; mammals, 4,060; amphibians, 3,000). Some of the 4,060 nominal species of mammals will doubtless be shown to be nothing more than geographic variants of other species; so the real number of "good" species may be somewhat less than the number cited. A total of 1,032 recognized living Recent genera currently accommodate the existing species. Only 70 species that are native, wild mammals or else have been introduced and now occur as successfully established forms living in a wholly feral state have been definitely recorded from Louisiana or its immediately adjacent waters. This tabulation excludes Man, domesticated mammals (even those occurring sometimes in a feral state), and the forms enclosed in brackets in the text (see page 54). These 70 species belong to 56 genera.

THE NAMING OF MAMMALS

Mammalogists employ a system of naming mammals known as binominal nomenclature. The same system is used by zoologists working with other groups of animals. We are indebted to the great Swedish taxonomist Carl Linnaeus (or Carolus Linnaeus, as he often called himself, and who was ennobled as Carl von Linné in 1761), for his epoch-making and revolutionary classification of animals (1758). He was responsible for the innovation of the binominal system, whereby each kind of animal has a Latin scientific name consisting of two parts, the first denoting the genus and the second the species. Each part consists of a single word. Other authors prior to Linnaeus sometimes used binominal names, and indeed so did Linnaeus himself in early editions of his *Systema Naturae,* but in none of these works was the author consistently binominal. Finally, in the 10th edition of his great opus, published in 1758, Linnaeus adopted the practice of naming all animals strictly in accordance with his binominal system. It was this edition that was later designated as the starting point of zoological nomenclature.

According to the rules, known as the Code, laid down by the International Commission on Zoological Nomenclature and now adhered to by virtually all zoologists, the first binominal name applied to an animal is the name by which it is to be known, provided certain conditions are met. One requirement is that the name must have appeared in the 10th edition of Linnaeus' *Systema Naturae* (1758) or in some work subsequent thereto. This rule and related stipulations constitute what is known as the Law of Priority. Numerous other provisions of the Code help to determine the validity and orthography of names, but none is more important than the Law of Priority. The International Code of Zoological Nomenclature (which has been often revised and amended, the latest edition being that of 1964) is indeed a remarkable document, for it insures uniform taxonomic procedure on the part of all systematists and lays down rules that prevent any two animals, whether they be a protozoan and a mammal, or a mollusk and a bird, from having the same scientific name. More than a million binominal couplets are in current use for the vast array of animal species. The fact that no two are the same speaks well for our system of binominal nomenclature and the rules governing its application.

Although many species have common or vernacular names that are sometimes used by scientists and laymen alike, these names vary from country to country just as do the languages spoken. Sometimes in one country a species may be known by several names, or the same name may be applied to several species. But the Latin name for a given species is understood by zoologists of all countries. The Japanese, for example, refer to what English-speaking people call the Giant Sperm Whale as *Physeter catodon,* even though their vernacular name for the species is *makko-kujira,* which they write as 抹香鯨..

Most of the work of the early systematists, that is, those who were engaged in classifying plants and animals, was that of pigeonholing species that they considered related by placing them in the same genus. Related genera were placed in a family, related families in an order, and so forth. Quite often decisions concerning relationships were based on superficial anatomical features. But today, in line with what Julian Huxley and others (1940) termed "the new systematics," attempts are made by systematists to apply all available evidences in arriving at true relationships. Such decisions may be based on more than external anatomy and may include internal structure, chromosome morphology, physiology, electrophoretic studies of body proteins, paleontology, zoogeographic considerations, or even behavior.

In the past 100 years considerable attention has been given to describing geographically variant populations within species, and the variations that are considered worthy of nomenclatural recognition are assigned a tri-nominal name, denoting the subspecies. Consequently, when a zoologist sees a scientific name consisting of three words, he knows at a glance that the species is polytypic, that is, it has been divided into two or more subspecies. A scientific name generally is followed by the name of a person, the one who first described and named the form. If the author's name appears in parentheses this means that the form in question was described under a different generic name from the one now being employed. For example, the name of the Common Muskrat in Louisiana is written *Ondatra zibethicus rivalicius* (Bangs) because Bangs, in describing the subspecies, placed the species in the genus *Fiber*.

WHAT IS A SPECIES?

One of the most fundamental of all biological questions is: What *is* a species? This question has long plagued zoologists, but now, at least among vertebrate zoologists, fairly close agreement prevails. The present unanimity has come about in large measure as a result of the writings of Ernst Mayr (1942, 1965) in which he admirably synthesized the wealth of information on the subject and, at the same time, injected his own perceptive insights.

None of the earlier criteria that were used in defining a species are wholly acceptable today. The contention on the part of system-atists of the past that a species is no more than what some competent worker (preferably a specialist on the group of animals in question) believes it to be, was eminently practical, at least with regard to certain difficult groups, but it implies that the species is an entirely subjective unit, which it is not. Moreover, in the pre-Darwinian era it tended to suggest that species are static and not the product of a dynamic speciation process.

The concept that species are based on the degree of morphological distinction is an old one. It is the concept on which Linnaeus founded the science of systematics in his *Systema Naturae* and is still the concept employed by workers dealing with many groups of animals. It is the only one available to the paleontologists, who have only the remains of skeletal elements with which to work. The trouble with morphology per se is that it provides no way to draw a dependable line between structural differences that are of specific magnitude and those that are only of infraspecific quality. Age and sexual differences are morphological but are certainly not species criteria. More serious still is the fact that morphological features are often subject to great geographic variation. Two forms may be morphologically very different yet clearly only subspecies. Or two forms may be morphologically extremely similar, if not superficially identical, yet on other grounds prove to be perfectly good species.

Another notion employed in the past in defining species was that they represented groups of genetically identical individuals. Such a concept is completely erroneous. Not only are subspecies genetically different from other subspecies in the same species, but, in species that consistently reproduce bisexually, all individuals, except for identical twins, are genetically different.

The matter of the fertility or sterility of offspring has also been used as a criterion in deciding what is a species. If two animals produce fertile offspring they are usually members of the same species, and, vice versa, if the offspring are sterile, the parents usually belong to two different species. But from these generalizations the conclusion does not follow that all individuals that interbreed and produce fertile offspring are members of the same species. This statement is patently untrue, there being many cases known in which two perfectly good species have interbred in the wild and produced completely fertile offspring. Conversely, all members of a single species are not always capable of producing fertile offspring. Crosses between different subspecies of a given species often result in young with varying degrees of sterility. A cross involving the male of one subspecies and the female of another may be capable of producing offspring of which only 60 percent are fertile, whereas the reciprocal cross may yield offspring of which no more than 40 percent are fertile. That two groups of animals refuse to cross in the wild when they may do so rather freely in captivity is also a fact to consider. Two kinds of animals may meet in nature and never cross, but the same two animals may interbreed freely and produce fertile offspring when brought together under the unnatural conditions of captivity, where no other choice is available to them. As Mayr describes this situation, the animals are fertile but do not cross, at least in the wild. They are reproductively isolated but not intersterile. This difference is of great biological importance, and therein lies the partial solution to the species problem.

When two forms occur side by side, that is, sympatrically, without interbreeding, they are each unquestionably "good" species. This biological criterion works well enough as long as populations are in contact. Careful field observations and the analysis of specimens taken in the zone of overlap usually disclose incontrovertibly whether interbreeding is taking place. The decision that the two populations are in fact good species is a comparatively objective one. The difficulty is that reproductive isolation cannot be tested where two populations are geographically isolated and hence do not come in contact under natural circumstances. Two forms (or species) are said to be allopatric if they do not occur together, that is, if they replace each other geographically. When this occurs the systematist is faced with the necessity of making a subjective judgment. He must ask himself what he thinks would happen if the two were to occur side by side. Would they be reproductively isolated from each other and hence good species, or would they interact as two subspecies by interbreeding in the zone of contact? Consequently, the biological spe-

cies definition has its serious flaws, but it is still our best solution to the problem. As a definition of a species Mayr (1940) proposed the following:

A species consists of a group of populations which replace each other geographically or ecologically and of which the neighboring ones intergrade or interbreed wherever they are in contact or which are potentially capable of doing so (with one or more of the populations) in those cases where contact is prevented by geographical or ecological barriers.

Or shorter: Species are groups of actually or potentially interbreeding natural populations, which are reproductively isolated from other such groups.

Space in this work does not permit me to delve into all the ramifications of the above definitions, but perhaps the direct quotation of a paragraph from Mayr (1942:120-21) will clarify several important points:

As can be seen from these definitions, it is necessary in the cases of interrupted distribution to leave it to the judgment of the individual systematist, whether or not he considers two particular forms as "potentially capable" of interbreeding—in other words, whether he considers them subspecies or species. This is necessary, because it is often impossible, for practical reasons, to test to what extent reproductive isolation exists. The life of a *Drosophila* [fruit fly] individual is very short. A population of *Drosophila melanogaster* of the year 1942 is reproductively isolated from a *Drosophila melanogaster* population of the year 1932 by a complete time barrier. Still, nobody would call them two different species. The same is true (to a slightly less absolute degree) for many geographically isolated populations. They may be reproductively isolated by a geographic barrier, but still they are not necessarily different species. Reproductive isolation is thus an immediate, practical test only for sympatric...synchronically reproducing species. The conspecifity of allopatric...and allochronic forms, which depends on their potential capacity for interbreeding, can be decided only by inference, based on a careful analysis of the morphological differences of the compared forms. This does not mean that I am retracing my steps and now propose to accept a morphological species definition; no, it means simply that we have to apply the degree of morphological difference as a yardstick in all those cases in which we cannot determine the presence of reproductive isolation. To use this method in doubtful cases is justified for the following reasons. If we examine the "good" species of a certain locality we find that the reproductive gap is associated with a certain degree of morphological difference. If we find a new group of individuals at a different locality, we use the scale of differences between the species of the familiar area to help us in determining whether the new form is a different species or not. These scales of differences are empirically reached and differ in every family and genus. An ornithologist knows by experience that the differences between good species of *Empidonax* and *Collocalia* are much slighter than the differences between male and female or between subspecies in birds of paradise or humming birds. The important point is that the biological gap between species (reproductively isolated groups) is, in general, correlated with certain morphological differences. This correlation, which naturally has exceptions like every biological phenomenon, permits the experienced taxonomist to determine in many cases whether to describe a new form as a new species or a subspecies, even though he knows nothing about its biology. When such information becomes available at a later period, it nearly always confirms the judgment of the competent taxonomist. To the adherent of a morphological species concept, any clear-cut morphological difference is a species difference. To the supporter of a biological species concept, the degree of morphological difference is simply considered as a clue to the biological distinctness and is always subordinated in importance to biological factors.

Again space here limits what can be said about the speciation process. Entire books have been devoted to the subject. But suffice it to say that among vertebrates, as well as among many invertebrate groups, the most important factors in the evolution of new species are geographic variation and spatial isolation, which can lead to reproductive isolation.

Species formation in simple form may be diagrammed as in Figure 2, which is adapted from Mayr (1942). At the start (I), species **a** occurs widely and is fairly uniform throughout. The environmental factors differ in different portions of the elongate range and through natural selection exert different influences on the evolution of the animal. Gradually the populations at the ends of the range, **a**′ and **a**‴, become recognizably distinct in appearance from each other and from **a**″, the population in the middle. The divisions between these populations are not sharply defined. They are zones of primary intergradation, in which the individuals share peculiarities of two populations.

When this condition (II) has been reached, the species is said to be polytypic, that is, divided into geographic races, or subspecies—in this case **a**′, **a**″, and **a**‴. The next development (III) is that geographical or ecological barriers completely sever the range of subspecies **a**‴ from the main range of the species. Cut off from any exchange of genes whatever with the other populations, the isolated subspecies **a**‴ develops new characteristics, intensifies old ones. In the process, it may become morphologically as distinct as a separate species.

A real test is provided if the barrier is removed and the isolated population spreads back into contact with subspecies **a**″. As Mayr pictures the possibilities, either of two outcomes eventuates. In the one situation (IV), the former isolate still interbreeds freely with subspecies **a**″ and thereby forms a zone of secondary intergradation. We then conclude that **a**′, **a**″, and **a**‴ continue to constitute one species **a**. In the other situation (V) the populations coming into secondary contact overlap but do not interbreed. We then know for certain that the former subspecies **a**‴ has differentiated into a new species, **b.**

In actual speciation, however, the results are likely to be somewhere in between the completely free interbreeding of situation IV and the complete reproductive isolation of situation V. The offspring of mixed matings may prove to be fewer and less viable than the offspring of parents of the same stock. In that event, natural selection works against mixed matings and in favor of the development of mechanisms that will minimize, and eventually prevent, such matings. In this manner, the speciation process may be completed sympatrically in a zone of secondary intergradation.

The Fox Squirrels *(Sciurus niger)* of the southeastern United States provide an interesting case that illustrates a number of points in the foregoing discussion. In the 10th edition of *Systema Naturae,* Linnaeus included descriptions of two kinds of squirrels from South Carolina and adjacent parts of the United States based on the accounts of these animals that appeared in Mark Catesby's *Natural History of Carolina, Florida, and the Bahama Islands,* published between 1731 and 1743. Even if Catesby had used binominal Latin names for the two squirrels, his names would not have been acceptable, because no name prior to Linnaeus' 10th edition is valid. In naming the two squirrels, Linnaeus gave credit to Catesby as his source of information. He called Catesby's "Black Squirrel of the

Carolinas" *Sciurus niger,* and he called Catesby's "Gray Squirrel of the Carolinas" *Sciurus cinereus.* It turned out that one was only a color phase of the other, that a single litter could contain four gray individuals, four blacks, two black and two gray, or any other combination regardless of the color of the parents. Therefore, only one species was involved, and only one of Linnaeus' names was the valid name of that species. Both had been named on the same printed page, but it so happened that the name *Sciurus niger* appeared first on the page; so that name has line priority over *Sciurus cinereus.* Consequently, the scientific name of the Fox Squirrel is today *Sciurus niger.*

The color and the morphological features of the Fox Squirrels of the Carolinas and

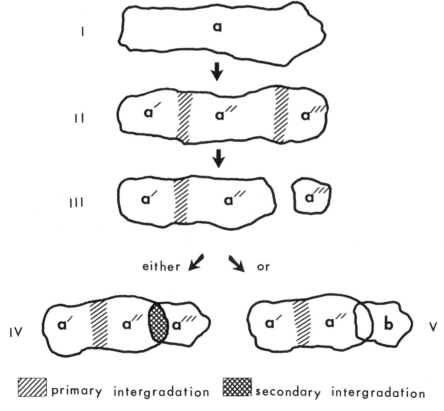

primary intergradation   secondary intergradation

Figure 2 Steps in the process of speciation explained in the text. (Adapted from Mayr, *Systematics and the Origin of Species.*)

adjacent areas are distinctive. They are large squirrels that in the gray phase are white below and gray above, interspersed with black, except on top of the head, which is solid black. All individuals, even melanistic ones, have the nose, ears, and tip of the tail extensively white. In northwestern Georgia and most of Alabama and from there to the Mississippi River, Fox Squirrels are also large, with the nose, ears, and tip of the tail white; but the gray and the remainder of the white are replaced by ferruginous tones (Plate IV). On the opposite bank of the Mississippi River the appearance of the Fox Squirrel changes abruptly. The animals constituting the population in the Atchafalaya and Mississippi bottomlands are small and never have any white on the nose, ears, and tip of the tail. Still farther westward squirrels of southwestern Louisiana and eastern Texas are extremely pale in coloration and again massive in size. Finally, in central Texas we once more encounter a population of exceedingly small squirrels, and these have the palest coloration of all.

Now, these discrete populations of Fox Squirrels are so different in color and morphology (the skulls of all are distinctly different) that they might easily be regarded as different species, were it not for the fact that where their ranges meet they interbreed freely. There is no reproductive isolation between any of them. Consequently, they must be treated as subspecies, even though some have been accorded species status in the past.

All are now known by the same species name, *Sciurus niger,* but with each a trinominal is added. The Fox Squirrel of the Carolinas, southeastern Georgia, and northern Florida is the subspecies *Sciurus niger niger* Linnaeus; the large ferruginous-colored, white-nosed form of northwestern Georgia, most of Alabama, eastern and southwestern Mississippi, and the Florida Parishes of Louisiana is *S. n. bachmani* Lowery and Davis; the small, dark squirrel of the Mississippi bottomlands and Atchafalaya Basin is *S. n. subauratus* Bachman; the large, pale squirrel of western Louisiana, southwestern Arkansas, and eastern Texas is *S. n. ludovicianus* Custis; and the small, extremely pale squirrel of central Texas is *S. n. limitis* Baird. Still other subspecies of *Sciurus niger* occur in southern Florida and the northern part of the eastern United States.

Of special importance is the fact that another kind of squirrel has its range superimposed in large part on the basic overall range of *Sciurus niger.* This is the species that we call the Gray Squirrel (*Sciurus carolinensis*). Some of the subspecies of the Fox Squirrel are as different morphologically from each other as they are from the Gray Squirrel but are still treated as subspecies, while the Gray Squirrel is regarded as a different species. The reason is simple. Both the Gray Squirrel and the Fox Squirrel occur side by side, but no case of hybridization is known. There is complete reproductive isolation between them.

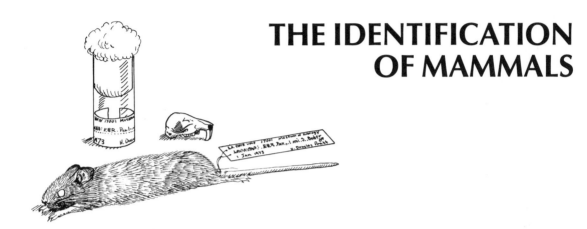

THE IDENTIFICATION OF MAMMALS

SINCE ONE of the primary objectives of this book is to acquaint Louisianians with the mammal fauna of their state, it is appropriate that I devote some attention here to methods of identifying our mammals and also that I define some of the terms used over and over again in the pages that follow. The general body form of the major groups of mammals helps us place a given mammal in its proper order, and even the uninformed layman frequently knows when he is dealing with a bat, a rodent, or a carnivore. But to determine what kind of bat, or rodent, or carnivore usually necessitates knowing key characters. Color and size may be sufficient, but more often than not the final identification to the exact species within a given group hinges on knowing a combination of more precise characters. In the case of bats, for instance, the critical diagnostic features of a certain species may be the shape of the small structure in the ear called the tragus and the number of upper incisor teeth that can be counted by lifting the lip. The identification of a mouse may depend upon the length of the hind foot, the extent to which the pinna of the ear protrudes above the pelage, or whether the incisors have a longitudinal groove on their anterior face.

The most highly diagnostic characters of mammals are generally some feature of the skull, notably the number of teeth. That is why in museum specimens the skull is whenever possible cleaned and kept with the skin. To insure that the right skull can always be associated with its skin, the catalog number of the specimen is placed on its label, and the same number appears on the skull. In the case of a skeleton, every bone large enough to accommodate the number is numbered to insure that the bones of different skeletons are never mixed up. If a mammalogist, in attempting to identify an unknown specimen, has the choice of a skin or a skull on which he must make his determination, he almost invariably prefers to have a skull. I doubt that any two of the 4,000-odd species of mammals in the world have skulls that are identical. The only problem is knowing what character to look for. The importance of the skull in the identification of mammals is the reason why in the pages that follow each species account is accompanied by a drawing or photograph of the skull.

The number of teeth in the species of a given genus is almost always the same. Mammalogists conventionally indicate the number of teeth by what is called the dental formula, wherein the number of incisors (I), canines (C), premolars (P), and molars (M) in the upper and lower jaws are written in the style of fractions. This is done in several ways. The notation sometimes shows only the number of teeth on one side of the jaw, and this number must be doubled to arrive at the total. For example, the dental formula of the Southern Flying Squirrels of the genus *Glaucomys* may be written as follows:

$$I\frac{1}{1} \ C\frac{0}{0} \ P\frac{3}{3} \ M\frac{1}{1} \times 2 = 20.$$

Or the formula may be written as I have done in the species accounts herein:

$$I\frac{1-1}{1-1} \ C\frac{0-0}{0-0} \ P\frac{3-3}{3-3} \ M\frac{1-1}{1-1} = 20.$$

Occasionally, in text, to save space, I have written the formulae in a slightly different form: I 1/1, C 0/0, P 3/3, M 1/1 × 2 = 20.

There is seldom any difficulty in distinguishing between canine, incisor, and molariform teeth, but differentiating premolars from molars is frequently impossible on superficial inspection. The upper incisors are always located on the premaxilla, the remainder of the upper teeth on the maxilla. The primitive placental mammals, such as our Insectivores, have three incisors in each half of the jaw, above and below, but in most modern forms the number of incisors is reduced. In marsupials there may be as many as five incisors on each side of the upper jaw, four on each side in the lower, and this is the complement possessed by our Virginia Opossum (*Didelphis virginiana*). Canines are generally one in number on each side, above and below, and in modern forms are simple, long, single-rooted conical teeth used for holding and piercing. The maximum number of premolars in nearly all modern eutherian mammals is four on each side, above and below, but in edentates and some cetaceans the number is greater, as it was also in many of the extinct form. They are used for shearing and cutting, and they often have one main cusp, but in advanced plant feeders the crown is

sometimes highly modified. The molars, which serve primarily for crushing, are three on each side above and below in primitive placentals. In some Mesozoic mammals the number of molars was even greater, but among modern forms the tendency is toward a reduction in the number. To distinguish molars from premolars one usually must marshal embryological data or information as to whether the teeth have milk predecessors, which never occur in the case of molars. A complete typical dental formula of a primitive placental mammal is as follows:

$$I\frac{1,2,3}{1,2,3} \ C\frac{1}{1} \ P\frac{1,2,3,4}{1,2,3,4} \ M\frac{1,2,3}{1,2,3} \times 2 = 44.$$

Among modern forms the departures from this general formula are marked by reductions. Where a tooth is lost a zero can be inserted. The complete formula for our Common Muskrat (*Ondatra zibethicus*) is as follows:

$$I\frac{1,0,0}{1,0,0} \ C\frac{0}{0} \ P\frac{0,0,0,0}{0,0,0,0} \ M\frac{1,2,3}{1,2,3} \times 2 = 16.$$

In general, the loss of premolars is from front to back, while the loss of molars is from back to front. The complete dental formula of the Domestic Cat (*Felis catus*) is:

$$I\frac{1,2,3}{1,2,3} \ C\frac{1}{1} \ P\frac{0,2,3,4}{0,0,3,4} \ M\frac{1,0,0}{1,0,0} \times 2 = 30.$$

When a skull is available, simply counting the number of teeth often leads immediately to the identification of the genus. Table 1 summarizes the dental formulae of the land mammals of Louisiana.

Dichotomous keys greatly facilitate the proper identification of an unknown specimen and, through a series of simple alternative choices, often quickly lead the user to the correct identity of the animal in question. For example, in the key that I have constructed to the land mammals of Louisiana, one sees at the outset the following couplets:

1   Female with marsupium, or pouch, in which the young are carried; hallux, or

Table 1 Dental Formulae of Louisiana Land Mammals

I	C	P	M	Total	Identities
$\frac{5-5}{4-4}$	$\frac{1-1}{1-1}$	$\frac{3-3}{3-3}$	$\frac{4-4}{4-4}$	$= 50$	*Didelphis, Marmosa*
$\frac{3-3}{3-3}$	$\frac{1-1}{1-1}$	$\frac{4-4}{4-4}$	$\frac{3-3}{3-3}$	$= 44$	*Sus*
$\frac{3-3}{3-3}$	$\frac{1-1}{1-1}$	$\frac{4-4}{4-4}$	$\frac{2-2}{3-3}$	$= 42$	*Canis, Urocyon, Vulpes, Euarctos*
$\frac{3-3}{3-3}$	$\frac{1-1}{1-1}$	$\frac{3-3}{3-3}$	$\frac{3-3}{3-3}$	$= 40$	*Equus*
$\frac{3-3}{3-3}$	$\frac{1-1}{1-1}$	$\frac{4-4}{4-4}$	$\frac{2-2}{2-2}$	$= 40$	*Bassariscus, Procyon*
$\frac{2-2}{3-3}$	$\frac{1-1}{1-1}$	$\frac{3-3}{3-3}$	$\frac{3-3}{3-3}$	$= 38$	*Myotis*
$\frac{3-3}{2-2}$	$\frac{1-1}{1-1}$	$\frac{4-4}{4-4}$	$\frac{2-2}{1-1}$	$= 36$	*Zalophus*
$\frac{3-3}{2-2}$	$\frac{1-1}{0-0}$	$\frac{3-3}{3-3}$	$\frac{3-3}{3-3}$	$= 36$	*Scalopus*
$\frac{2-2}{3-3}$	$\frac{1-1}{1-1}$	$\frac{2-2}{3-3}$	$\frac{3-3}{3-3}$	$= 36$	*Lasionycteris, Plecotus*
$\frac{3-3}{3-3}$	$\frac{1-1}{1-1}$	$\frac{4-4}{3-3}$	$\frac{1-1}{2-2}$	$= 36$	*Lutra*
$\frac{2-2}{3-3}$	$\frac{1-1}{1-1}$	$\frac{2-2}{2-2}$	$\frac{3-3}{3-3}$	$= 34$	*Pipistrellus*
$\frac{3-3}{3-3}$	$\frac{1-1}{1-1}$	$\frac{3-3}{3-3}$	$\frac{1-1}{2-2}$	$= 34$	*Mephitis, Mustela, Spilogale*
$\frac{0-0}{3-3}$	$\frac{1-1}{1-1}$	$\frac{3-3}{3-3}$	$\frac{3-3}{3-3}$	$= 34$	*Cervus*
$\frac{3-3}{1-1}$	$\frac{1-1}{1-1}$	$\frac{3-3}{1-1}$	$\frac{3-3}{3-3}$	$= 32$	*Blarina, Sorex*
$\frac{2-2}{3-3}$	$\frac{1-1}{1-1}$	$\frac{1-1}{2-2}$	$\frac{3-3}{3-3}$	$= 32$	*Eptesicus*
$\frac{2-2}{2-2}$	$\frac{1-1}{1-1}$	$\frac{2-2}{2-2}$	$\frac{3-3}{3-3}$	$= 32$	*Homo*
$\frac{1-1}{3-3}$	$\frac{1-1}{1-1}$	$\frac{2-2}{2-2}$	$\frac{3-3}{3-3}$	$= 32$	*Lasiurus borealis, L. cinereus, L. seminolus, Tadarida*
$\frac{0-0}{3-3}$	$\frac{0-0}{1-1}$	$\frac{3-3}{3-3}$	$\frac{3-3}{3-3}$	$= 32$	*Odocoileus, Bison, Bos, Capra, Ovis*
$\frac{3-3}{1-1}$	$\frac{1-1}{1-1}$	$\frac{2-2}{1-1}$	$\frac{3-3}{3-3}$	$= 30$	*Cryptotis*
$\frac{1-1}{3-3}$	$\frac{1-1}{1-1}$	$\frac{1-1}{2-2}$	$\frac{3-3}{3-3}$	$= 30$	*Lasiurus intermedius, Nycticeius*
$\frac{3-3}{3-3}$	$\frac{1-1}{1-1}$	$\frac{3-3}{2-2}$	$\frac{1-1}{1-1}$	$= 30$	*Felis*
$\frac{0-0}{0-0}$	$\frac{0-0}{0-0}$	$\frac{7\text{ to }9-7\text{ to }9}{7\text{ to }9-7\text{ to }9}$		$= 28\text{-}36$	*Dasypus*
$\frac{3-3}{3-3}$	$\frac{1-1}{1-1}$	$\frac{2-2}{2-2}$	$\frac{1-1}{1-1}$	$= 28$	*Lynx*
$\frac{2-2}{1-1}$	$\frac{0-0}{0-0}$	$\frac{3-3}{2-2}$	$\frac{3-3}{3-3}$	$= 28$	*Sylvilagus*
$\frac{1-1}{1-1}$	$\frac{0-0}{0-0}$	$\frac{2-2}{1-1}$	$\frac{3-3}{3-3}$	$= 22$	*Glaucomys, Sciurus carolinensis*
$\frac{1-1}{1-1}$	$\frac{0-0}{0-0}$	$\frac{1-1}{1-1}$	$\frac{3-3}{3-3}$	$= 20$	*Castor, Geomys, Myocastor, Perognathus, Sciurus niger, Tamias*
$\frac{1-1}{1-1}$	$\frac{0-0}{0-0}$	$\frac{0-0}{0-0}$	$\frac{3-3}{3-3}$	$= 16$	*Microtus, Mus, Neotoma, Ochrotomys, Ondatra, Oryzomys, Peromyscus, Rattus, Reithrodontomys, Sigmodon*

innermost toe of hind foot, thumblike and without nail; teeth in adults numbering 50. *Infraclass* Metatheria, *Order* Marsupialia, *Family* Didelphidae

Didelphis virginiana

1a No abdominal pouch in female; hallux, when present, not thumblike; teeth numbering 42 or less. *Infraclass* Eutheria   **2**

2 Forelimbs modified as wings. *Order* Chiroptera   **3**

2a Forelimbs not modified as wings   **13**

If the unknown animal is a mouse, it possesses no marsupium, its innermost toe of the hind foot is not thumblike, and it has considerably fewer than 42 teeth; so we are led to **2**. Since its forelimbs are obviously not modified as wings, as the forelimbs would be modified if the animal were a bat, the correct choice in the second couplet would be **2a**, which leads us to **13**, to which we can turn immediately,

skipping all the intervening couplets. We then continue to make a series of choices until we ultimately arrive at a definite species. If no erroneous choice has been made anywhere along the way, the animal can be assumed to be the species to which the key has led.

One of the best ways to learn a large number of new species is to construct a key to them yourself. Doing so helps one learn the diagnostic features, the so-called "key characters." When any of my students are preparing to leave on their first field trips to places such as Costa Rica and Peru, I have them lay out a series of museum specimens of species known to occur in the region to be visited, and then construct a key to them. In so doing they quickly equip themselves to identify what they will find when they reach their destination. I strongly commend this method to anyone having access to a museum collection. The Louisiana State University Museum of Zoology, the research division of the Museum of Natural Science, contains one such assemblage of specimens, and it maintains an open-door policy with respect to the use of its collections by interested persons. Since these collections are not available to the general public, prior notice should be given the staff when a visit is planned. The museum's bird collections house 98 percent of the species known from North America, 93 percent of those from Central America, and 76 percent of those from South America. The mammal collections, while not so nearly complete, are nevertheless highly representative of the mammalian species of the Western Hemisphere.

KEY TO THE LAND MAMMALS OF LOUISIANA

1 Female with marsupium, or pouch, in which the young are carried; hallux, or innermost toe of hind foot, thumblike and without nail; of teeth in adults numbering 50. *Infraclass* Metatheria, *Order* Marsupialia, *Family* Didelphidae
Didelphis virginiana

1a No abdominal pouch in female; hallux, when present, not thumblike; teeth numbering 42 or fewer. *Infraclass* Eutheria **2**

2 Forelimbs modified as wings. *Order* Chiroptera **3**

2a Forelimbs not modified as wings **13**

3 Tail free of interfemoral membrane for at least half of its length. *Family* Molossidae **Tadarida brasiliensis**

3a Tail not free of interfemoral membrane for more than a few millimeters, if at all. *Family* Vespertilionidae **4**

4 Ears greatly elongated, measuring an inch or more **Plecotus rafinesquii**

4a Ears not greatly elongated, measuring less than an inch **5**

5 Interfemoral membrane densely furred above **6**

5a Interfemoral membrane not densely furred above or at best only on its basal portion **9**

6 Color yellow; no white patches on shoulders or on wrists **Lasiurus intermedius**

6a Color not yellow; white patches on shoulders and wrists **7**

7 Color brick red; hairs of dorsum strongly tipped with white in female **Lasiurus borealis**

7a Color mahogany red or dark umber; no sexual dimorphism **8**

8 Size large, total length over 5 inches, wingspread more than 14 inches; chocolate brown pelage of dorsum strongly tipped with white producing hoary effect; face and cheeks yellowish; ears grayish, rimmed with black **Lasiurus cinereus**

8a Size small, total length less than 5 inches; color of dorsum dark mahogany red **Lasiurus seminolus**

9 Color sooty brown to blackish, frosted with white **Lasionycteris noctivagans**

9a Color brown or yellowish brown **10**

10 Color yellowish brown; size small, total length less than 3.5 inches of which tail is almost half **Pipistrellus subflavus**

10a Color brown **11**

11 Size large, total length 4.3 to 4.6 inches and wingspread 12 to 13 inches; ears, wings, and interfemoral membrane black or blackish **Eptesicus fuscus**

11a Size small, total length less than 4 inches and wingspread less than 11 inches **12**

12 Tragus long, slender, pointed, and straight; two incisors on each side in upper jaw **Myotis austroriparius**

12a Tragus short, broad, blunt, and bent forward; only one incisor on each side in upper jaw **Nycticeius humeralis**

13 Body with shell-like carapace hinged in the middle by nine flexible bands; dentition reduced to 7–9 peglike cheek teeth on each side of upper and lower jaws. *Order* Edentata; *Family* Dasypodidae
                                              Dasypus novemcinctus
13a Body without shell-like carapace   **14**

14 Toes provided with hoofs; no upper incisors. *Order* Artiodactyla   **15**
14a Toes with claws; upper incisors present   **17**

15 Antlers in the male, shed annually; body and legs lithe and graceful; skull with large anteorbital vacuity exposing inner bones. *Family* Cervidae   **16**
15a Solid bone horn cores attached permanently to frontal bones of the skull in both sexes; no anteorbital vacuities. *Family* Bovidae   ***Bison bison***

16 A single canine present on each side of upper and lower jaws; neck maned; white or buffy white patch on rump; antlers in male sometimes four feet in length
                                              Cervus canadensis
16a Canines normally absent; no mane; no whitish rump patch; rack relatively small compared with that of preceding species
                                              Odocoileus virginianus

17 Eyes minute or rudimentary; snout elongate; pelage short and velvety; pinna of ear absent or greatly reduced. *Order* Insectivora   **18**
17a Eyes not minute or rudimentary; snout normal; pinna present   **21**

18 Forelimbs highly modified for digging; no pinna; zygomatic process thin but complete; entirely fossorial in habits. *Family* Talpidae   ***Scalopus aquaticus***
18a Forelimbs not modified for digging; zygomatic process absent; pinna present but greatly reduced. *Family* Soricidae   **19**

19 Color reddish brown; tail more than one inch in length   ***Sorex longirostris***
19a Color gray with never more than a slight brownish tinge; tail less than one inch in length   **20**

20 Color uniformly lead gray above and below   ***Blarina brevicauda***
20a Color grayish brown above, distinctly paler below   ***Cryptotis parva***

21 Canine teeth absent; incisors long and chisel shaped, separated from cheek teeth by wide diastema   **22**
21a Canine teeth well developed, saberlike and adapted for seizing prey   **45**

22 Incisors two on each side in upper jaw; tail extremely short and powderpufflike; hind feet modified for leaping; ears as long as hind foot. *Order* Lagomorpha, *Family* Leporidae   **23**
22a Incisors one on each side; tail not short and fluffy; hind feet not extremely modified for leaping; ears not extremely long *Order* Rodentia   **24**

23 Eye-ring white; size small, total length usually under 16 inches; greatest length of skull less than 77 mm   ***Sylvilagus floridanus***
23a Eye-ring buffy; size large, total length usually over 18 inches; greatest length of skull over 80 mm   ***Sylvilagus aquaticus***

24 Total length of adults more than three feet   **25**
24a Total length of adults less than three feet   **26**

25 Tail flattened dorsoventrally and scaly. *Family* Castoridae   ***Castor canadensis***
25a Tail not flattened dorsoventrally and sparsely haired. *Family* Capromyidae
                                              Myocastor coypus

26 Tail more or less bushy, much wider than the fleshy portion. *Family* Sciuridae   **27**
26a Tail not bushy but still may be well haired   **30**

27 Tail flattened dorsoventrally; fold of skin extending as a patagium from wrist to ankle for use in gliding   ***Glaucomys volans***
27a Tail not especially flattened; no patagium   **28**

28 Body with five blackish and four pale longitudinal stripes; rump reddish; well-developed cheek pouches   ***Tamias striatus***
28a No longitudinal stripes; no cheek pouches   **29**

29 Color in normal phase basically ferruginous, brownish, or yellowish above and below; black color phase common; four cheek teeth on each side of upper jaw   ***Sciurus niger***
29a Color in normal phase grayish above, white below (sometimes may be rusty below in Atchafalaya Basin); five cheek teeth, the first small and peglike, on each side of upper jaw   ***Sciurus carolinensis***

30 Forefeet modified for digging; pinna virtually absent. *Family* Geomyidae
                                              Geomys bursarius
30a Forefeet not modified for digging; pinna present and often conspicuous   **31**

31 Tail nearly naked, the annulations clearly visible; upper cheek teeth with tubercles of crowns arranged in three longitudinal rows. *Family* Muridae   **32**

31a Tail haired, the annulations usually concealed; upper cheek teeth with tubercles of crowns arranged in two longitudinal rows or else comprised of prismatic triangles. *Family* Cricetidae **34**

32 Mouse size; upper incisors with terminal notch *Mus musculus*

32a Rat size; total length 12 to 17 inches **33**

33 Tail shorter than head and body *Rattus norvegicus*

33a Tail longer than head and body *Rattus rattus*

34 Crowns of cheek teeth in series of prismatic triangles. *Subfamily* Microtinae **35**

34a Crown of cheek teeth smoothly oval in cross section and with two longitudinal rows of tubercles or lophodont with S-shaped ridges. *Subfamily* Cricetinae **37**

35 Size large, total length 17 to 21 inches; tail flattened laterally and naked; toes partially webbed and fringed with long, stiff hairs *Ondatra zibethicus*

35a Size small, total length never more than six inches; tail less than half length of head and body; ears almost buried in pelage; eyes tiny **36**

36 Color reddish brown *Microtus pinetorum*

36a Color brownish gray interspersed with black *Microtus ochrogaster*

37 Crowns of cheek teeth with S-shaped loops *Sigmodon hispidus*

37a Crowns of cheek teeth with tubercles, not S-shaped ridges **38**

38 Size large, total length 14 inches or more; tail conspicuously furred; ears large, usually over one inch from notch to tip *Neotoma floridana*

38a Size small or medium, total length never more than 10 inches, usually much less **39**

39 Well developed fur-lined cheek pouches; auditory bullae large; color above grizzled ochraceous with a clear ochraceous line extending from the nose along sides to the tail, separating the color of dorsum from whitish underparts *Perognathus hispidus*

39a No fur-lined cheek pouches; auditory bullae normal in size; color some shade of brown, not ochraceous **40**

40 Anterior face of each upper incisor with longitudinal groove **41**

40a No groove on anterior face of incisors **42**

41 Tail three inches or more in length; rich Ochraceous-Orange or Vinaceous-Tawny line along sides of body, separating brownish upperparts from whitish underparts *Reithrodontomys fulvescens*

41a Tail 2.5 inches in length or less; color dark brown and without an ochraceous line along sides *Reithrodontomys humulis*

42 Color dark grayish, heavily interspersed with black; sides washed with Ochraceous-Buff and the underparts grayish white, not pure white; tail not distinctly bicolored *Oryzomys palustris*

42a Color of upperparts of adults some shade of reddish brown (immatures lead gray above); underparts pure white or tinged with Pale Pinkish Cinnamon; tail distinctly bicolored **43**

43 Color of upperparts rich golden brown or orange-brown; underparts pure white or tinged with Pale Pinkish Cinnamon; posterior palatine foramina closer to rear edge of hard palate than to anterior palatine foramina; front border of infraorbital plate perpendicular to horizontal axis of skull instead of bowed forward *Ochrotomys nuttalli*

43a Color of upperparts not rich golden brown but instead reddish or yellowish brown; underparts and top of feet white **44**

44 Size slightly larger; hind foot 21 mm or more; greatest length of skull 27 mm or more *Peromyscus gossypinus*

44a Size slightly smaller; hind foot 21 mm or less; greatest length of skull less than 26.5 mm *Peromyscus leucopus*

45 Feet plantigrade; larger cheek teeth with crowns relatively flat, only slightly cuspidate **46**

45a Feet essentially digitigrade; larger cheek teeth cuspidate, with pronounced cutting edges **48**

46 Size large (weight more than 100 pounds); tail short and inconspicuous; color essentially all black; skull length of adults and subadults more than 250mm; teeth 42. *Family* Ursidae *Euarctos americanus*

46a Size small or medium (weight less than 25 pounds); tail relatively long, conspicuous, and with alternating blackish and buffy or black and white rings or semirings; skull length less than 120 mm; teeth 40. *Family* Procyonidae **47**

47 Tail (except on midventral surface) with alternating black and white rings; prominent white eye-rings, no black facial mask *Bassariscus astutus*

47a Tail with alternating blackish and buffy rings; prominent black facial mask
                                                    Procyon lotor

48 Claws not retractile or only partly so, not completely concealed in hair of toes; posterior cheek teeth with crushing crowns   **49**

48a Claws completely retractile, completely concealed in hair of toes; posterior cheek teeth without crushing crowns. *Family* Felidae   **53**

49 Decidedly doglike in appearance; tail with mane of stiff hairs on basal upper surface; no specialized anal musk glands; total number of teeth 42. *Family* Canidae   **50**

49a Not especially doglike in appearance; tail without mane of stiff hairs; specialized musk glands well developed; total number of teeth 38 or less. *Family* Mustelidae **54**

50 Total length usually 42 inches or more; weight 18 pounds or more; hind foot seven inches or more; skull 6.75 inches or more; postorbital process convexed above; sagittal and lambdoidal crests prominent; in alive or freshly dead animal pupil of eye round   **51**

50a Total length usually less than 42 inches; weight usually less than 18 pounds; hind foot less than seven inches; skull 6.25 inches or less; postorbital process with shallow depression; top of cranium with prominent paired ridges beginning at postorbital process and converging posteriorly; in alive or freshly dead animal pupil of eye vertically elliptical   **52**

51 Body robust; zygomatic breadth usually more than 105 mm; rostral breadth posterior to canines nearly always more than 32 mm   *Canis rufus*

51a Body smaller; zygomatic breadth usually less than 105 mm; rostral breadth posterior to canines usually less than 32 mm
                                            Canis latrans

52 Color predominantly red or reddish yellow; tail reddish yellow mixed with black

and tipped with white; legs and feet predominantly black   *Vulpes fulva*

52a Color predominantly salt-and-pepper gray, rusty or tawny on sides of neck, lower flanks, insides of legs, and underside of tail; tail black on top and on tip
                                Urocyon cinereoargenteus

53 Size large, total length 7.5 to 8 feet, of which 35 to 40 percent is tail; color almost uniformly Cinnamon-Buff; no spots except in kittens   *Felis concolor*

53a Size small, total length 3 feet or less, of which the short tail makes up six inches or less; coloration spotted, especially on underparts and insides of legs   *Lynx rufus*

54 Size large, total length three to four feet of which the tail makes up about one-third; tail thick and muscular, tapering from base toward the tip; feet broad and large and with toes webbed; head decidedly flattened on top; semiaquatic
                                        Lutra canadensis

54a Size relatively small, never more than 27 inches in total length   **55**

55 Color all black or striped and spotted black and white   **56**

55a Color brown above and below or brown above and yellowish below   **57**

56 Color all black except for white spot on head or with white stripe on head and neck dividing into two longitudinal white stripes down the back   *Mephitis mephitis*

56a Color black heavily spotted and striped with numerous areas of white, including portions of tail   *Spilogale putorius*

57 Color dark mahogany brown above and below except for occasional white spots on the chin, throat, and chest   *Mustela vison*

57a Color light brown above, yellowish below; body greatly elongated and legs short; tail blackish on tip in contrast to basal portion and the general color of the dorsum   *Mustela frenata*

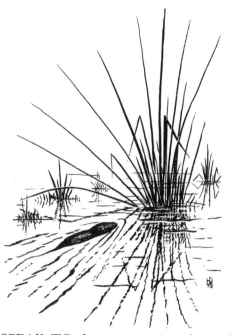

FUR IN LOUISIANA

SPEAK TO the average American of furs and fur trapping and into his mind will come thoughts of snowshoes crunching over powdered snow and leaving a trail that winds through the ice-laden hemlocks, spruces, and tamaracks of some Canadian or Alaskan wilderness. Perhaps he will think too of gray wolves and wolverines watching the trapper from concealment and stalking in his wake or of huskies and dog sleds mushing across a frozen landscape.

Such images are standard parts of the picture of the trapper's world projected and publicized by novels, movies, and television. A pirogue gliding up a bayou past trees draped with Spanish moss or a mudboat churning its way down a marshy ditch and startling Snowy Egrets into flight are no part of that picture. Yet Louisiana—the land of pirogues, Spanish moss, mudboats, and Snowy Egrets—is actually today the leading fur-producing area on the North American continent and has been so for as long as official records have been kept. In fact, in some winters the number of pelts taken here has far exceeded the combined total for all the provinces of Canada plus Alaska. And in some seasons the Louisiana catch has

amounted to as much as 65 percent of the total for all the rest of the United States combined. Though statistics such as these have been publicized many times before, they probably still come as a surprise, a nearly incredible one, to most Americans, many Louisianians included.

How Louisiana attained dominance in the trapping industry is a long story and one that has been only imperfectly recorded. The account that follows borrows heavily from five sources: O'Neil's classic treatise on muskrats (1949) and his article on the fur trade (1968); St. Amant's wildlife inventory (1959); Arthur's book on the fur animals of Louisiana (1931); and a mimeographed report on the Louisiana fur industry by Palmisano (1971).

Fur Trade in Colonial Times.—Once the clothing of cavemen, furs had become so scarce in Europe by the 15th century that they vied with rare gems and precious metals as articles of prestige, so costly that only the wealthy and the noble-born could afford to wear them. Then came the discovery of America. A whole continent teeming with furred animals was opened up to the adventurous. First to reach for these beckoning riches were the

French, spreading dominion over Canada; but soon the British by formation of the famed Hudson's Bay Company launched fierce competition.

In those days the way to obtain pelts was to barter with the Indians. The red man would happily trade away a whole year's catch for a pittance in white man's goods—beads and baubles, ribbons·and blankets, guns and gunpowder, watered-down brandy. The French supplied a superior grade of gunpowder, but they often frittered away this advantage by unfair practices. On one occasion in payment for beaver skins worth several thousand dollars, they gave a small amount of powder, which they said was a new variety that, if planted, would produce more powder—all the Indians wanted.

A difficulty was that the transport of furs from the wilderness heartland of what is now the United States to embarkation ports in Canada was an arduous undertaking. No roads had yet been built, and all the water flowed the wrong way.

Iberville, founder of the French colony that became Louisiana, perceived opportunity. Right at his doorstep was the mighty Mississippi, whose drainage system reached its fingers throughout the middle of America. Along its ready-made waterways stacks of furs could be floated on flatboats, moving with the current, down to the delta, where they could be loaded on oceangoing vessels for shipment to France. Before the Louisiana Territory was a year old, when it still had less than 400 inhabitants, the French Canadians were already complaining to the home government about encroachments on their trade by the new colony on the Gulf of Mexico.

Despite many setbacks—the spoilage of pelts in the hot climate, trouble with the Indians, the wars between England and France, the misfortunes of the Crozat regime, and the bursting of John Law's "Mississippi Bubble," Louisiana's dealings in furs prospered on the whole from their beginnings in the first year of the 18th century. Almost from the moment of its founding in 1720, New Orleans became a major fur center. Hides of buffalo and deer and pelts of bear, otter, lynx, wildcat, marten, and fox funneled down the river in great quantity for transshipment to European markets. Though all crown grants stipulated that beaver skins, then the most prized of furs, belonged to the French companies of Canada, even these found their way illicitly to the new outlet.

In 1763 two New Orleanians, Pierre Laclede and Auguste Chouteau by name, traveled upriver under a grant that gave them the right of exclusive trade with the Indians along the Missouri. On a bluff where that river joins the Mississippi to form a mighty stream, they built a small fort in which to cache beaver skins. That fort grew to be the city of St. Louis, which eventually became the "fur capital" of North America—a title it held for more than 100 years until New York City took over the fur industry for its own during the First World War. Meanwhile, as railroads began to extend out over our nation, they superseded river traffic as the chief means of transportation, the volume of New Orleans exports dwindled, and the city's former position in the fur trade was forgotten.

All during its early history, insofar as I can determine, Louisiana never was an important fur-*producing* area. New Orleans owed its position in the trade to traffic in furs that came from afar. All the time, however, there lay within the state a fabulous but largely untapped resource—3.5 million acres of coastal marsh that represented more than half of the total marshland of the 12 Atlantic and Gulf states combined. And in these Louisiana wetlands dwelt an unobtrusive little animal that was destined to revolutionize the fur industry. That animal was the Common Muskrat.

The Early History of Muskrats in Louisiana.—Muskrats have been in Louisiana for a long, long time. Fossil remains from Iberia and West Feliciana parishes (Arata, 1964;

Domning, 1969) point to their presence back in the Pleistocene, 10,000 years or more ago. In addition, archaeological explorations have revealed plenty of muskrat bones in Indian mounds and middens. The Louisiana State University Museum of Geoscience has such bones from many coastal archaeological sites ranging all the way from St. Bernard Parish in the east to Cameron Parish in the west. Far from being rare, such finds are commonplace. Some of the bones have been found on or near the surface and hence cannot be dated. The possibility has existed therefore that they are the remains of muskrats that fairly recently sought refuge from high water on the mounds and died there or of carcasses left behind by modern trappers who used the tops of middens as pelting stops or as campsites, as the Indians did of old.

Fortunately careful investigation of the internal contents of the middens has now begun. At Pecan Island, some years ago, around 1950, a road cut sliced through an archaeological deposit and exposed sherds of pottery along with muskrat bones too deeply embedded together in the shell to be of 19th- or 20th-century origin. And in 1971–1972, Robert W. Neuman, Curator of Anthropology at the Louisiana State University Museum of Geoscience, conducted a controlled dig in the Morton Shell Mound at Weeks Island. He found muskrat bones all the way from the surface down to the low-lying Tchefuncte Layer. Five acceptable radiocarbon dates of the layer ranged from 260 B.C. to A.D. 340 (Neuman, pers. comm.). Similarly, excavations in Cameron Parish at the extreme eastern end of the Little Cheniere ridge where it is cut by the Mermentau River revealed middens of the Plaquemine Culture containing muskrat bones (Springer, 1973). This controlled dig has been radiocarbon dated A.D. 1285 and 1365. So we have, as a result of the works of Neuman and Springer, firm evidence that muskrats lived in Louisiana around the time of Christ and for many centuries afterward.

A related matter needs to be noted. At our coastal Indian sites interested persons have collected numbers of arrowheads and scraping tools chipped from kinds of rock that do not occur naturally in Louisiana. These implements obviously reached the areas of discovery through travel by the Indians, barter, or both. Could muskrat remains have been brought to the coastal middens from afar in the same way? Hardly. Kitchen middens are garbage dumps, places to dispose of refuse, not repositories for valued possessions imported from a distance. And the idea of Indians carrying muskrats in the flesh hundreds of miles as provisions on an excursion into the zone of abundant seafood is barely, if at all, within the bounds of credibility.

So we can be fairly confident that the Common Muskrat flourished in our marshes up until near the threshold of historic record here. But when that record itself begins, it is almost completely silent about such an animal. The only reference I have been able to find in early historical writings dealing with Louisiana is the testimony of Jacques Gravier, S.J., whose writings appear in *The Jesuit Relations*. The Father described the Tunicas, a tribe he visited in November 1700 thus: "Most of the men have long hair and have as their only dress a wretched deerskin. Sometimes they, as well as the women, also have mantles of turkey feathers or muskrat skins well woven and well worked."

On 3 December 1700, Father Gravier wrote similarly of the Houmas, a tribe inhabiting the land that became East and West Feliciana and East and West Baton Rouge parishes: "The women wear a fringed shirt, which covers them from the waist to just below the knees. When they go out of their wigwam they cover themselves with a robe of muskrats or of turkey feathers."

Pelts, of course, do not establish the presence of a mammal with as much certitude as bones do. The muskrat skins worn by the Tunicas and the Houmas *could* have come into their possession as items of trade with tribes much farther inland or as hand-me-downs from past generations. Still, I consider an in-state source more plausible. For the making of primitive dress, muskrat pelts have no unique utility. The skins of other animals can be used instead, as, indeed, Father Gravier asserts was the case with deerskins among the Tunicas. Why should the Indians of southern Louisiana have bartered for anything so nonessential as muskrat skins when they were so acutely in need of items not available in their immediate environment—such necessities as flints, arrowheads, and other stone implements? My tentative judgment is that these Indians probably did not trade afar for muskrats, that they probably killed the animals themselves near home.

Of interest in its own right is the fact that our Gulf Coast Indians had names for the muskrat. To the Choctaw, it was *pichali;* to the Ofo, *ani okloose;* and to the Biloxi, *xanaxpe.* These words have nothing phonetic in common. They appear to be independent constructions of the sort that arise locally from direct acquaintance with a living animal.

After Father Gravier, the next unequivocal references to the Common Muskrat in our state are the mention of it by Audubon and Bachman (1846) and the description in 1895 of the Louisiana subspecies, *Ondatra zibethicus rivalicius* (Bangs). In the intervening span three historical works (Dumont de Montigny, 1753; Le Page du Pratz, 1758; and Bossu, 1768 and 1777) reported on the mammals here. None said anything about muskrats. The authors were historians primarily, naturalists secondarily. Their failure to record a species is not evidence of its absence, for they included only a small number of mammals on their Louisiana lists: Dumont de Montigny, 14; Le Page du Pratz, 18; Bossu, 12.

Audubon and Bachman, in the first comprehensive treatment of American mammals (1846), did include Louisiana in the range of the Common Muskrat. They put the matter thus: "With the exception of the alluvial lands

in Carolina, Georgia, Alabama, and Florida, it abounds in all parts of the United States north of latitude 30°." The thirtieth parallel passes through the city of New Orleans. So most of Louisiana—though not the parts where muskrats are commonest today—lies above it, in the range as delineated by Audubon and Bachman. Actually their intention may not have been to exclude our coastal marshes, as they literally did. When speaking of latitude in a general way, one commonly uses the nearest whole degree. If the range given by Audubon and Bachman is interpreted as everything north of latitude 29° 30' the greater part of our coastal marshes—the part most accessible in Audubon's day—is included. Elsewhere in the account, they specifically state that the species "extends to Louisiana."

For glimpses of the history of the Common Muskrat in Louisiana between the time of Audubon and the time of Bangs, we are indebted to the industry with which Ted O'Neil (1949) interviewed oldtime marshmen in the southern part of the state. He has reported when and where four of his informants saw their first muskrat: Henry Hillebrandt in 1880 south of Franklin, at age 22; Dave White and Ed White in 1884 at Cheniere au Tigre at age 21 and age 7, respectively; Duncan Crain in 1892 in the Little Pecan marshes at age 21. How much time these young men had spent in the marshes prior to their first meeting with muskrats is unspecified. But O'Neil does tell us that three of the four learned the identity of the mammal from their mother or father. The parents' familiarity with the species is a strong indication of the presence of muskrats in Louisiana at least a generation earlier.

As O'Neil (1949) further relates in a footnote, Mr. Frank Ritchie, engineer for the Louisiana Land and Exploration Company until his death in 1967, recalled an early surveyor's record that indicated a dense muskrat population and a possible "eat-out" in the Barataria–Lafitte area as early as

1840—back within the lifetime of Audubon. Attempts to relocate this record failed, but the field notes of land surveyor William J. Henry on file in the Louisiana Land Office describe probable muskrat eat-outs in the Turtle Bay area of Jefferson Parish in 1873.

Bangs' paper (1895) endorses the view that muskrats were not newcomers to Louisiana in the late 19th century. It states:

> The muskrat has long been known to occur in the lower Mississippi region and along the swampy coasts of the adjacent states of Alabama and Louisiana, but I know of no specimens from there having come into collections, until I sent Mr. F. L. Small, in Jan. 1895, to make a collection of specimens in this rather desirable section of the country.
>
> He reported the muskrat as being very common all through the Bayou country.

In summary, the known history of the Common Muskrat in Louisiana before 1900 is patchy—no more so, perhaps, than one should expect but enough so to invest the subject with more uncertainty than one would like. In his treatise on the species (1949), O'Neil advanced two hypotheses to account for the gaps in the record: (1) that, after inhabiting the Gulf Coast during certain periods of geological history, the animals were completely exterminated during or after the Ice Age and that they reoccupied our marshes after 1850; (2) that isolated colonies persisted all along but were so limited by the lack of a suitable food supply and by the abundance of predators that they went "unobserved by early inhabitants or explorers who studied the coastal marshes."

In the light of what has been learned since 1949, I lean strongly toward O'Neil's second explanation. Among my reasons, as detailed in the preceding pages, is an additional consideration—the subspecific distinctness of the disjunct Gulf Coast population (*O. z. rivalicius*) from muskrats in other parts of the United States. The differences in coat color, notably between *rivalicius* and *zibethicus,* the latter being the name applied to populations of the Upper Mississippi Valley and much of the remainder of eastern North America, may not be any greater than the alleged color differences between populations of *rivalicius* from various parts of southern Louisiana, but the dissimilarity of the skulls of *rivalicius* and *zibethicus* points to trenchant genetic changes of a sort that could have evolved only during a long and continuous period of isolation.

The Rise of Muskrat Trapping in Louisiana.—Hand in hand with the trapping of mink, otter, and raccoons in our marshes in the 1800s went the hunting of alligators. As these reptiles became scarcer in the southeastern part of the state, the hunters began to pole them from their holes and eventually found they could facilitate this practice by firing the vegetation. The burning created deeper peat marshes and promoted a lush growth of the subclimax plant, three-cornered sedge, which is the preferred food of muskrats. Population explosions followed.

O'Neil states that the first serious muskrat trapping in Louisiana began between 1900 and 1910, in the southeastern part of the state. Then the pelts usually sold for only 5 to 10 cents. By 1912 muskrats were so badly damaging the cattle ranges in Cameron Parish that ranchers offered a bounty of 5 cents for each animal killed, and many of them became so convinced that cattle raising in the parish was doomed that they gave up in disgust and let their marshland holdings go for taxes. Hunting of muskrats at that time and place was done mainly with pitchforks and dogs.

When Louisiana muskrat skins reached northern fur finishers in numbers, their excellence became appreciated. Why our southern subspecies should be rated highly in the trade is a question likely to perplex the general public. Almost everyone knows that animals tend to put on denser fur in colder climates. But muskrats live in water, and the temperature of water is "thermostatically" regulated in one direction by physical laws; it cannot fall below 32° F without some chemical admixture such as salt. Many of our muskrats live in brackish water, which does remain liquid below 32° F. So they have to be prepared to withstand immersion at temperatures somewhat lower actually than those endured by some northern muskrats. Perhaps, though, these considerations are not of critical importance. The animal's way of living, in burrows or in houses, largely insulates them against the cold wherever they are. In the North, muskrats often spend the winter in the dry air space beneath a protecting roof of ice.

Then too our animals have darker pelage than most of the northern ones—a feature that eliminates the need of dyeing and therefore appeals to furriers, and the leather is more durable than the northern kind. The Louisiana muskrat pelt has three distinct colors. A dark grotzen stripe runs down the center of the back, the sides are light golden, and the belly is silver. The pelt thereby provides material for three separate garments in three different colors. For instance, from 90 pelts, three separate coats can be made in three colors. In contrast, muskrats of the northern populations, even though larger, are more uniform in color. Furthermore, because of lower nap and shorter guard hair, Louisiana furs drape more easily and more gracefully when sewn together into a coat or other garment. And finally the great quantities taken in a relatively small area not only facilitate the matching of furs but also minimize the buyers' task of assembling a shipment and guarantee the industry a steady supply with uniform quality.

By 1912 fur trapping had become important enough in Louisiana for the legislature to impose a closed season on mink, otters, muskrats, and raccoons. It stipulated that these animals could be taken only by licensed trappers and only from 1 November to 1 February. Although the Act failed to set a license fee or even provide for the issuance of trapping licenses, a man was convicted under it for catching 10 mink, 6 otters, and 3 raccoons without a license. In 1914 the defect was remedied.

Meanwhile, our Department of Conservation, as the present Wild Life and Fisheries Commission was then known, began to keep statistics as to the numbers of fur animals taken. These were based on the figures obtained from raw-pelt dealers in the course of collecting a severance tax. The first report, the one for the 1913–1914 season, placed the take at more than 5 million skins, more than 4.25 million of which were muskrat pelts. A period of decline immediately set in that bottomed with a combined total for all species of 734,561 skins in 1921–1922.

The official figures from year to year do not necessarily reflect precisely the true fluctuations in the number of furbearers in our marshes, as the figures are compiled from the severance tax collections. Trapping effort varies with the prices being paid. Sometimes furs are stockpiled, payment of the severance tax is deferred, and the skins do not show up in the statistical reports until a later season. The trapper who knows he will get a good price for certain pelts will put forth greater effort to bring in a larger catch.

A graph of the total number of pelts taken, season by season, illustrates several facts. In Figure 3, muskrats are represented by a dotted line, all species together by a solid line. First, note how closely the muskrat take approached the total take winter after winter from the 1913–1914 to the 1959–1960 season. Over that 47-year period, nearly half a century, the muskrat catch never fell below 63 percent of total fur production. In the 1939–1940 season, they touched an all-time percentage high of nearly 97; and for the 47 seasons, they averaged more than 82 percent. Second, observe the striking fluctuations that have taken place over the years. Notice also how tightly the muskrat plot line is in control of the plot line for overall production up until the beginning of the 1960s. Its ups and downs so closely reflect the ups and downs of the solid line of the total catch that in shape the one plotting could almost substitute for the other. Figure 4 shows that even on the basis of dollar value the results were much the same.

Although many factors affected the figures of the period, the most deep-seated ones were the phenomenal breeding potential of muskrats and their susceptibility to disaster. Sometimes the two elements have been inter-linked. In the presence of an abundance of food, populations periodically explode. The density soars to several individuals per acre, with more being rapidly produced. Soon the muskrats eat up all the available vegetation, even digging down into the peaty floor as

much as 20 inches to devour the root systems that bind the marsh soils together. As a result, the earth disintegrates into loose muck with decaying plant remains floating in the ooze. Trappers call an area thus denuded an "eat-out" (Figures 5 and 6). It means starvation and a crash decline for the population that created it.

This sequence—low numbers of muskrats, development of a big food supply, over-population, eat-out, starvation—repeats itself over and over again, every 10 to 14 years on the better three-cornered grass marshes and at longer intervals in areas with less favored vegetation such as paille fine or cattail. The

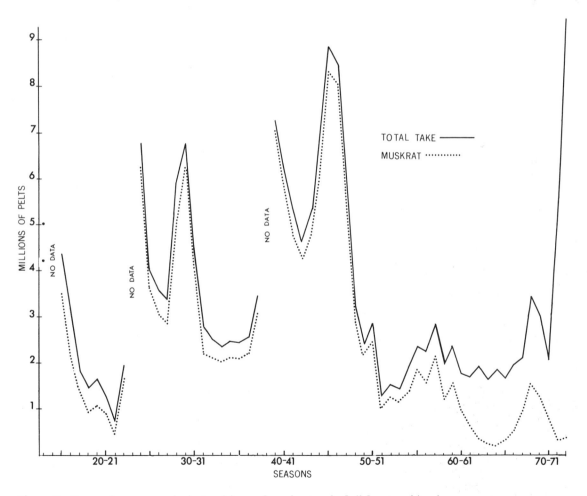

Figure 3   Season-by-season take in Louisiana of muskrat and of all furs combined.

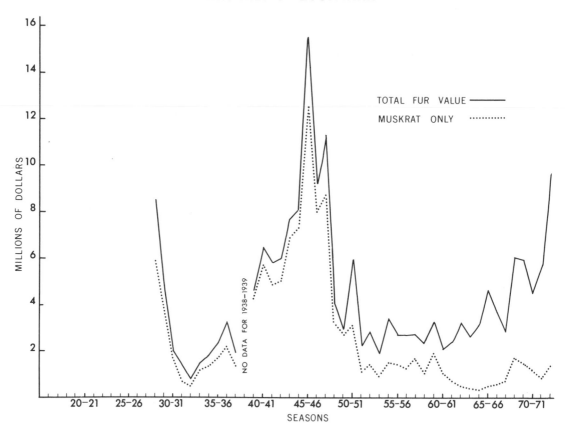

Figure 4   Season-by-season value of furs in terms of millions of dollars paid to the Louisiana trapper.

Figure 5   Photograph following a rare, light snowfall over the coastal marsh in Lake Lery area, St. Bernard Parish, February 1958, highlighting muskrat houses and severe eat-outs. (Photograph by Richard K. Yancey)

cycles that result are local and independent. There is no set trough or peak for the brackish marshes as a whole. Each area goes through its own ecological succession.

Thus, things are seldom so bad all over that statewide production of furs becomes truly poor. During the widespread hurricane of 1915, many muskrats drowned, and a succession of seven discouraging seasons followed, with the record low take of 734,561 pelts of all kinds in 1921–1922. Since then, the total production curve has never dipped below 1 million skins and only once below 1.5 million.

The Coming of the Nutria.— In 1938 occurred a momentous event that did not seem momentous at the time. The late E. A. McIlhenny purchased 14 Nutria from a dealer in Argentina. One of the females gave birth during transit, and in consequence 20 animals arrived in Louisiana. McIlhenny placed them in a 35-acre pen at Avery Island in Iberia Parish, and there they multiplied so rapidly that soon they could no longer be contained. Many escaped and in two years the species had become well established in the marshes surrounding the island.

Because Nutria fur brought a good price during the 1945–1946 season, many landowners requested breeding stock for release into their marshes. Rumor spread that the animal, which has a voracious appetite, would eat almost anything, including water hyacinths and other undesirable aquatic plants. So people over the state with weed control in mind began to purchase Nutria stock. The big rodents never lived up to their reputation as weed destroyers, but this notion helped to speed their spread over Louisiana.

By the mid-1950s these huge, ungainly, stupid-looking rodents had reached a population peak in southwestern Louisiana and had earned the thorough dislike of almost everyone. Farmers hated them because they feared for their crops. Hunters hated them because of the damage they could do to dogs.

Motorists hated them because they were dangerous obstacles at night on the highways. Everybody who waded in water hated them because they were rumored to have brought to Louisiana wetlands a parasite that penetrates clothing, tunnels under the skin, and causes a maddening affliction known as "Nutria itch." Water-control engineers hated them because the animals were forever digging tunnels through dikes and levees. But the trappers detested them more than anyone else. Muskrats were becoming scarcer as Nutria were becoming more abundant, and the trappers jumped to the conclusion that the larger animal was to blame for the decline of the smaller one. In areas where the two species came in contact, the Nutria further enraged the trappers by springing muskrat traps and then pulling free and by fighting with muskrats caught in traps and thereby damaging the fur.

Be that as it may, Nutria forced themselves increasingly into the trappers' daily operations. Some were caught as early as the 1941–1942 season, but none reached the market until the 1943–1944 and 1944–1945 seasons, when 1,338 pelts sold at the disappointingly low price of 50 cents each. Reasons for the poor return were, first, that the small quantity offered for sale could not be easily matched with other available skins and, second, that the trappers did not yet know how

Figure 6 Severe muskrat eat-out on Audubon Society property in Vermilion Parish, February 1944. (Photograph from Louisiana Wild Life and Fisheries Commission files)

to handle, stretch, and dry the pelts to produce top-quality fur. The next season brought a catch of 8,784 Nutria pelts together with a tenfold increase in price to $5.00 when several dealers in an attempt to corner the Louisiana Nutria market purchased on speculation in anticipation of an even higher price later on in Europe. They continued to pay attractive prices of $3.00 or more for five successive seasons. But meanwhile in the marshes the Nutria were increasing so rapidly that control of the market became obviously impossible. Soon the supply of Louisiana skins exceeded the then limited European demand, and in 1955 the price per pelt slumped to $1.00.

The Rise and Stabilization of the Nutria Population. — The population of Nutria in coastal Louisiana reached a peak from 1955 to 1959, when an estimated 20 million of these rodents were chewing away at the foundations of our wetlands. Freshwater marshes and marshes that had developed on a peat foundation appeared to suffer the greatest damage. The two hundred and fifty Nutria transplanted to the Mississippi River delta in 1951 increased to such a point that they had torn much of the marsh to pieces by 1957, when only the natural levees of the major channels seemed to be holding the delta together. The once dense stands of cattail there were almost completely destroyed. Similar "eat-outs" occurred throughout the coastal area.

Alligator weed and water hyacinth quickly invaded the marshes. These plants along with bulltongue, elephant's ear, and pennyworts appear to increase under heavy Nutria grazing either because the animal does not like the taste of them or, as in the case of alligator weed, because they have phenomenal ability to recover from "eat-outs."

With much of the cover vegetation removed, the stage was set for a series of happenings that eventually stabilized the Nutria population. In June 1957 Hurricane Audrey devastated the southwestern Louisiana coast. The tall growth that had in the past retarded the storm surge was gone. A great wave of seawater rolled inland from the Gulf. In spite of their aquatic adaptations, Nutria drowned by the thousands.

H. William Belknap, a graduate student of mine, who was conducting research on the White-faced Ibis at the time, later described to me the eerie desolation that greeted him when he returned to Grand Cheniere after Audrey. No humans were to be seen anywhere. No houses remained on their foundations except the east end Rockefeller Refuge quarters, where Belknap, his wife, and his two little girls had ridden out the hurricane a short time before. Great uprooted oaks, centuries old, along with other trees, lay across the cracked and gutted highway, which swarmed with snakes and other wildlife, including the surviving Nutria. And all through the day and into night, the bleats of these survivors furnished a weird vocal accompaniment to an unbelievable scene. The animals had reason to moan. Habituated to a different environment, they were not prepared to live in comfort with the increased salinity that had surrounded them.

Since Audrey, other storms have hit the coast with unusual frequency and severity. They have had a serious impact on the physical structure of the marshes, already greatly weakened by the activities of the Nutria. And five years after Audrey came a setback of another kind — the great freeze of February 1962. Temperatures in the marsh plummeted to 12° F. Extremely susceptible to frostbite and deprived of the cover that would have been furnished by the tall vegetation they themselves had razed, Nutria perished in numbers estimated in the millions. Many that lived lost tails or feet and turned up in traps in that condition for several years afterward.

Strange to say, it was near this time of direst extremity that the Nutria finally surpassed the muskrat to become the leading fur animal of the state, not only in dollar value but in number trapped. Several factors combine to explain the paradox. One of them emerges in a conversation Bill Belknap held with an old trapper at Little Cheniere just before Audrey.

"What do you use for bait?" asked Bill.

"Bait? *Mais jamais!*" said the old man incredulously. "We don't use no bait at all."

"Then how *do* you catch them?"

"We just set de trap on top of any mound of stuff dat stick out of de water. And den, sooner or later, de Nutria step into dem."

"But that way you must have to set out an awful lot of traps just to catch a few."

"No, man! No! I set out 30 trap. I catch 30 Nutria. If I set out 40 trap, I catch 40 Nutria. But 30 Nutria all I can skin in one day. So I set out only 30 trap."

The point is that the difference between superabundance, as of yesterday, and mere abundance, as of now, has had little effect on the annual take of Nutria. The limiting factor, today as before, is the number of skins a trapper can properly process. Techniques have improved. Nonetheless few present-day operators would attempt to exceed the old trapper's claim of 30 Nutria skinned per day. In the 1960s the trapper may have had to travel farther to catch his 30 Nutria than he did in the 1950s, but he still did not need to go impracticably far.

Why the Nutria Achieved Its Present Importance. — Another factor in the rising relative importance of Nutria was the decline of muskrats. The 1945–1946 season, when the Nutria made its real debut in the Louisiana market with 8,784 skins, was the zenith of muskrat production with 8,337,411 taken — nearly 1,000 times as many as Nutria! During the ensuing decade, which included the years when Nutria pelts glutted the market, they never exceeded 19.4 percent of total fur production, while muskrat pelts never fell below 72 percent.

In the face of low prices of $1.00 to $1.50,

the Nutria take continued to climb. By the 1955–1956 season, at only 17.6 percent of total fur production, it took over second place both in number of pelts and in dollar value. By the 1960–1961 season it accounted for 39.9 percent of all furs compared with the 53.6 percent contributed by muskrats. In the next winter it passed the muskrat catch with a whoosh, posting 53.1 percent of the total catch versus 36.8 percent for muskrats. Even with this impressive percentage figure, the actual Nutria catch was only 912,890 pelts; but never afterward has the take dipped below a million. The best showing yet has been the 1,754,028 pelts in 1968–1969. Figure 7 shows the comparative take of Nutria and muskrat skins from the 1945–1946 season to the 1972–1973 season in terms of numbers of pelts. Figure 8 makes the same comparison in terms of dollars paid to the trapper.

A third factor in the increasing economic importance of Nutria was the value of the meat. Back when the animal began to multiply dramatically, the Department of Home Economics at Louisiana State University experimented with dishes in which the main ingredient was Nutria and served a free meal to anyone willing to test the cuisine. They offered Nutria stew, Nutria fricassee, fried Nutria, baked Nutria, and Nutria other ways. I dropped in one noon and tested several items on the bill of fare. The meat tasted rather like rabbit, not bad, not bad at all. And, in fact, some southern Louisianians and many Europeans have come to consider it a delicacy.

Then someone got the idea that the meat might be canned as pet food. But this use never came to much either. Nutria meat did not find a real niche until fur ranchers began to use it as food for mink. In both the winters of 1962–1963 and 1963–1964, eight million pounds were sold at an average price of 10 cents a pound for a return each season of $800,000. The uppermost plot line in Figure 8 represents the total paid to trappers for Nutria including the return from both pelts and meat. As mink ranching began to slack off in the late 1960s, another market for the meat was found in the screwworm eradication program centered at Mission, Texas. This operation is the one in which millions of the flies whose larvae, known as screwworms, infest cattle are hatched in the laboratory, fed flesh during the larval stage, and raised to the adult state. Thereupon the males are sterilized and released in huge numbers. They mate with wild females, which as a result lay unfertilized eggs and produce no offspring. This outlet has consumed seven million pounds of Nutria meat annually.

Meanwhile fashions in fur were changing. Originally the only salable Nutria pelts were those from southwestern Louisiana, and the only part used was the belly fur. The fur dressers plucked the guard hair, sheared the underfur, dyed it, and made it into luxury coats that competed with mink. The process was so expensive that only the larger skins could absorb the cost and so difficult that American technicians had not yet mastered it.

Over the past decade fine furs have lost much of their glamour. Now that more and more people can afford them, fewer and fewer care about them as items of prestige. Conservationist groups have reenforced this trend by campaigns to make the wearing of fur, particularly that of rare and endangered predators such as the Leopard, seem a heartless disregard for the preservation of vanishing wildlife. One European zoo even denied admittance to ladies wearing fur.

As a result, emphasis in the industry has shifted increasingly from the making of luxury coats and stoles to the manufacture of knockabout garments from inexpensive durable fur. These products, known as "fun furs," fit in ideally with the growing interest in winter sports, and they are seen in ever greater numbers on ski slopes and on snowmobiles, at football games and around skating rinks.

From an experiment conducted by the

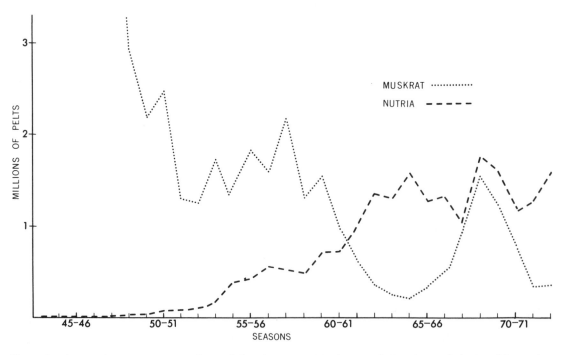

Figure 7 Season-by-season comparison of the Common Muskrat and Nutria catch in Louisiana since 1943.

Louisiana Wild Life and Fisheries Commission in cooperation with fur dressers, the long-haired, unplucked Nutria emerged. The small discarded sizes of southwestern Louisiana pelts, along with the unsalable southeastern pelts, offered an inexpensive source of material—inexpensive because for the first time the whole pelt could be used without shearing and without plucking. Also

the thin-pelted, less desirably colored Nutria meanwhile proved just what was needed for linings and trimmings, a use for which the long-haired western type was entirely too bulky. (See O'Neil, 1968.)

In just a few decades after the importation of 20 South American rodents into Louisiana, their progeny had become the basis of a multimillion dollar industry!

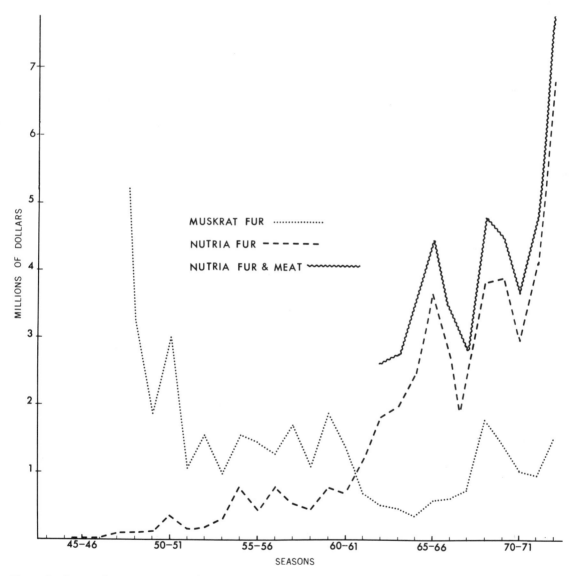

Figure 8 Season-by-season comparison of the value of the muskrat and Nutria take in terms of millions of dollars paid to the Louisiana trapper.

Other Louisiana Furs.—Up to this point, I have spoken as though muskrats and Nutria made up the entire Louisiana fur catch in modern times, or at least as though the part played by other mammals were inconsequential. The fact is that over the years from the 1928–1929 season, the first for which dollar value was reported, to the 1972–1973 season, the latest for which figures are available, the other furbearers brought our trappers a total income of nearly 46 million dollars, or an average income of slightly over a million per season.

The annual fur reports have consistently listed five species or species categories in addition to muskrat. These are mink, raccoon, otter, opossum, and skunk. In some years Spotted Skunk has been split off as a separate item, occasionally under the name of "spotted skunk" but more often as "civet cat"; in other years, it has been lumped with "cats." Fox, wolf, and bobcat (or lynx or wildcat) have appeared intermittently in the records; and since 1963 a beaver entry has been added. Space is devoted in the individual species accounts in the present book to the role of all these animals in the Louisiana fur industry; so here I shall confine attention to a few high points.

During the first three seasons for which cash values were published in Louisiana (1928–1929 to 1930–1931), raccoon was the second highest-ranking fur in dollar return, topped only by muskrat. In the winter of 1928–1929, in the days of flappers and shieks, when coonskin coats were at the height of their popularity, raccoon pelts brought an average price of $7.28 apiece, almost as much as mink. Thus the 153,194 skins taken paid the Louisiana trappers a total of more than $1,115,000. With the coming of the Great Depression, the boom collapsed, and the raccoon harvest has never exceeded a value of $1,000,000 since. Nevertheless, Louisiana has in several seasons been the leading state in the production of raccoon pelts. Prices, which bottomed at 27 cents per pelt in 1949–1950, were back to a respectable $5.50 in 1972–1973.

Wild mink took over second place in total dollar worth in the 1931–1932 season and remained there until the 1959–1960 season, when Nutria finally passed it. During this span of nearly three decades, Louisiana was several times the nation's leading producer of wild mink. The crest came in 1945–1946, with 168,598 pelts selling at an average price of $15.00 for a total remuneration in excess of $2.5 million. The catch of that winter was surpassed by the 184,552 pelts taken in 1950–1951, but by that time the price had eased off enough to make the dollar value less than in 1945–1946.

As the fur industry turned more and more to ranch mink, with its steady supply and its increasing number of glamorously named mutant colors—Tourmaline, Autumn Haze, Azurene, Black Diamond, Breath of Spring, Pink Glo, and the like—the prices paid for wild mink sank lower and lower. The $3.00 per pelt that the Louisiana trapper received in the 1967–1968 season was the poorest return ever.

In spite of having brought the trapper more money with fewer skins in some winters than all species together did in some of the other winters, mink is not Louisiana's most expensive fur. That distinction goes to otter, the most durable of all furs and one of the most beautiful, which has yielded an average return of $17.62 per pelt over the years and posted one-season averages as high as $42.00. The largest Louisiana catch so far is the 8,484 pelts in 1962–1963, which had a raw-fur value of $144,228. In that season Louisiana led the nation with 42 percent of United States otter production. In fact, it accounted for a larger share than any *seven* other states combined. In the 1972–1973 season the otter catch came to 7,668 and had a raw-fur value of $322,056. When the national figures become available they will probably show that Louisiana otter production established a new record high.

Many years ago, the lowly opossum occasionally ranked second with regard to the number of pelts harvested though never with regard to their dollar worth. The highwater mark was in the 1928–1929 season, with 518,295 skins selling for $570,125. One-third as many raccoons, by virtue of an exceptionally high price that year, represented nearly twice as much money. If one divides the 44 seasons for which fur values are available into two periods of equal length, one finds that opossums brought the trapper an average of $61,997 per season for the first 22 seasons in contrast with an average of only $3,773 for the last 22 seasons. During both periods the return per raw pelt fell as low as 10 cents. Why anyone would consider a mere dime sufficient inducement even for skinning a possum is hard to understand. Yet in the 1940–1941 season, for example, more than 131,045 opossum pelts were marketed at just such a price. The reason may be partly that when the trapper does the work he is often anticipating a higher return than he actually gets, and partly that the flesh is used for food.

Skunk and fox have been weak also-rans, bringing average annual receipts of only $5,335 and $1,236, respectively. Their record for the past 21 seasons has been even worse—only $137 and $340, respectively.

The Trapper's World—Yesterday and Today.—In years gone by, trapping was a way of life in southern Louisiana, deeply rooted in a cultural pattern that developed as a result of our rich coastal fisheries and wildlife resources. The inhabitants of the marshlands lived almost exclusively by trapping, commercial fishing, and hunting. Anyone who wanted to start harvesting fur simply bought himself some traps, sallied into any area where he thought furbearers might be found, set the traps, skinned his catch, and sold the pelts to the first visiting fur buyer.

When the value of the fur crop became evident, the owners of the marshland began to seek a share. To get it, they either trapped their property themselves or rented it at so much per acre to trappers or to leasing companies, who in turn sublet the trapping rights to trappers.

Infuriated, the so-called "free trappers" resisted attempts to deprive them of what they regarded as a God-given privilege. Here and there open warfare erupted. Then the trappers formed protective associations and engaged lawyers to challenge the legality of the trespass laws. Their efforts failed, and soon all trappers were either buying themselves trapping areas outright or leasing tracts under one of a variety of arrangements.

With the onset of winter and the beginning of the season, most professional trappers would take up residence in a cabin in a marsh, on or near their lease. With them they would bring trapping gear, wife, children, dogs, and enough staple provisions to last for an extended stay. Sometimes even pigs went along.

The usual lease covered only 100 to 300 acres. Slogging through the marsh in hip boots and making only occasional use of his pirogue, the trapper would place his traps, first working the perimeter of the tract and gradually moving in toward the center as the season progressed. In those days the law limited him to 250 sets, none of which could be nearer than 15 feet from a muskrat house and all of which he was legally required to visit daily. On a good day, in a good season, in good marsh, a skilled trapper might nevertheless catch 80 or 90 or even 100 muskrats. These he would skin on the spot, take back to the cabin, and place inside out on special drying frames (Figure 9).

Every two weeks or so, under some methods of operation, the land manager would visit the cabins in a boat. He brought perishable provisions and took away the dried pelts (Figure 10). But the big highlights of the season were the fur sales that were held at central warehouses three or four weeks apart. At the warehouse each trapper had his own stall with his name painted on the wall and there he would stack his catch (Figure 11). Two or three days before the actual sale, the fur dealers or their agents, the fur buyers, would make the rounds of the stalls to grade the fur. At a big sale as many as six or eight dealers would be represented. Each buyer would mark the price he was willing to pay for a given trapper's lot on a special form and submit it to the land manager as a sealed bid.

On the eve of the sale day, the air was charged with excitement. For the trapper it was like the night before Christmas, a time of anticipation and high hope. For the buyer it was a time of agonizing decision. If he bid too low, he might end up with no fur whatever and all his hard work of grading gone for naught. If he bid too high and overbought, he might find himself equally unpopular with his bosses. The tension produced some amazing conversations between buyers, as each grasped for clues to the others' intentions without revealing his own.

After the last sale of the season, usually sometime in February, came the balls staged for the trappers, occasions for revelry and reminiscence amid music and the flow of alcohol.

Today a great deal has changed, and the transition from muskrat to Nutria as the leading fur produced in Louisiana has been to a large extent responsible. In the two decades between 1947 and 1967, the number of fur dealers handling our pelts dropped from 40 to 21, the number of fur buyers from 263 to 128, and the number of trappers from 12,000 to only slightly more than 5,000.

The topflight trappers who have remained active have had to shift from intensive to extensive operations, from small leases of a couple of hundred acres to large leases of several thousand acres. The serious professionals cover that much ground by traveling in 16-foot inboard mudboats that hustle along plowed or dredged ditches or speed down the canals that the oil industry has provided increasingly since moving into the marsh in the 1940s. Most of the traps, which number from 300 to 400 now that legal restrictions have been removed, are set on the levees flanking the waterways.

Gone are most of the marsh cabins of earlier days. Typically, the modern trapper brings the day's catch home, where he skins the animal, then fleshes, washes, and dries the pelt. Sometimes he even makes use of his wife's clothes dryer as part of the process

Figure 9 Trapper's cabin in marsh with his stretched pelts of Nutria, muskrats, and raccoons drying in the open air. (Photograph from Louisiana Wild Life and Fisheries Commission files)

Figure 10 Counting muskrat pelts in trapper's camp on Marsh Island preparatory to transfer to New Iberia warehouse. (Photograph by Erol Barkemeyer)

before fitting the cased skin on wire stretchers in a heated drying shed.

When he has finished skinning the Nutria caught that day, the trapper hangs a red flag on his mail box. It is a signal to the roving trucks of the meat processors that he has carcasses to sell. The truckers pitchfork the Nutria bodies onto scales, dump them into the truck, and carry them to the processing plant. There they are either ground up bones and all and stuffed into 50-pound sacks or else they are laid as whole carcasses in wax-lined boxes, 50 pounds to a carton, and blast-frozen. Until recently processors have relied wholly on quick hauling to keep the meat fresh until it reaches the plant, but lately refrigerated trucks have come into use. These greatly increase the size of the area from which one processing center can gather carcasses.

The sale of the pelts is an independent transaction. Some landholders and a few state and national wildlife refuges, still arrange auctions after the old fashion. But most of our raw fur nowadays is purchased by individual buyers at the trappers' homes. Lastly, an annal Fur and Wildlife Festival, held in Cameron each February, has replaced the balls of earlier days.

What is the future of the fur industry in this state? The factors involved are so complex that no one can say for certain. With respect to the impact of one element at least — the effect of campaigns to make the wearing of fur unpopular — we are in the strongest possible position. Our trapping industry is to a large extent rodent-based. It draws upon populations of animals with extremely high breeding potential, populations that would destroy themselves agonizingly by destroying their own food supply if left alone. No rational case can be made against harvesting them.

Figure 11 Muskrat pelts stacked in Abbeville warehouse. (Photograph by Erol Barkemeyer)

Table 2　Comparative Takes of Fur Animals in Louisiana from 1914 to 1973 with Annual Prices and Receipts for the Years 1929–1973

	1913–14	1915–16	1916–17	1917–18	1918–19	1919–20	1920–21
Muskrat	4,284,000	3,500,420	2,125,000	1,387,220	963,110	1,073,200	900,170
Raccoon	401,000	532,480	678,821	220,110	258,200	252,800	160,683
Opossum	178,000	193,640	228,921	140,125	152,800	224,100	131,553
Mink	105,000	114,620	122,480	85,440	79,975	98,700	66,315
Striped Skunk	27,280	23,670	25,624	25,210	22,340	13,300	12,322
Otter	2,860	3,540	6,940	1,985	2,428	1,680	321
Foxes, Wolf, and Spotted Skunk	4,500	6,740	13,164	3,100	3,700	2,365	–
Totals	5,002,640	4,375,110	3,200,950	1,863,190	1,482,553	1,666,145	1,271,364

	1921–22	1922–23	1924–25	1925–26	1926–27	1927–28
Muskrat	465,766	1,682,571	6,236,165	3,613,765	3,036,749	2,858,834
Raccoon	119,129	140,884	145,810	128,516	127,882	103,544
Opossum	103,078	99,119	287,180	198,490	356,184	339,210
Mink	36,885	31,461	84,201	51,447	43,896	67,284
Striped Skunk	6,934	11,300	14,752	–	–	–
Otter	278	530	2,110	2,024	2,554	1,190
Foxes, Wolf, and Spotted Skunk	2,491	3,037	947	–	–	–
Skunks (unspec.)	–	–	–	30,866	27,671	22,348
Misc.	–	–	–	1,058	1,095	1,072
Totals	734,561	1,968,902	6,771,165	4,026,166	3,596,031	3,393,482

Species	Take	Approx. Price to Trapper	Value	Species	Take	Approx. Price to Trapper	Value
	1928–29 Season				*1930–31 Season*		
Muskrat	5,105,374	$ 1.15	$5,871,180.10	Muskrat	4,068,114	$ .40	$1,627,245.60
Raccoon	153,914	7.25	1,115,876.50	Raccoon	52,065	3.00	156,195.00
Opossum	518,295	1.10	570,124.50	Opossum	127,725	.40	51,090.00
Mink	99,844	8.50	848,674.00	Mink	45,390	3.00	136,170.00
Striped Skunk	31,661	1.25	39,576.25	Striped Skunk	–	–	–
Spotted Skunk	7,279	1.25	9,098.75	Spotted Skunk	–	–	–
Skunks (unspec.)	–	–	–	Skunks (unspec.)	11,487	.50	5,743.50
Otter	3,048	22.00	67,056.00	Otter	1,396	12.00	16,752.00
Foxes	1,476	2.00	2,952.00	Foxes	–	–	–
Wolf or Coyote	–	–	–	Wolf or Coyote	–	–	–
Bobcat	865	2.00	1,730.00	Bobcat	–	–	–
Cats (unspec.)	–	–	–	Cats (unspec.)	–	–	–
Nutria	–	–	–	Nutria	–	–	–
Beaver	–	–	–	Beaver	–	–	–
Misc.	314	1.50	471.00	Misc.	1,472	1.00	1,472.00
	5,922,070		$8,526,739.10		4,307,649		$1,994,668.10
	1929–30 Season				*1931–32 Season*		
Muskrat	6,269,556	$ .60	$3,761,733.60	Muskrat	2,209,382	$ .32	$ 707,002.24
Raccoon	105,381	3.50	368,833.50	Raccoon	133,470	1.50	200,205.00
Opossum	309,363	.75	232,022.25	Opossum	197,712	.15	29,656.80
Mink	69,680	5.00	348,400.00	Mink	127,245	3.50	445,357.50
Striped Skunk	–	–	–	Striped Skunk	–	–	–
Spotted Skunk	–	–	–	Spotted Skunk	–	–	–
Skunks (unspec.)	27,034	.60	16,220.40	Skunks (unspec.)	122,679	.30	36,803.70
Otter	1,447	11.50	16,640.50	Otter	1,664	8.00	13,312.00
Foxes	–	–	–	Foxes	–	–	–
Wolf or Coyote	–	–	–	Wolf or Coyote	–	–	–
Bobcat	–	–	–	Bobcat	–	–	–
Cats (unspec.)	–	–	–	Cats (unspec.)	–	–	–
Nutria	–	–	–	Nutria	–	–	–
Beaver	–	–	–	Beaver	–	–	–
Misc.	1,877	1.00	1,877.00	Misc.	3,117	.50	1,558.50
	6,784,338		$4,747,727.25		2,795,269		$1,433,895.74

Table 2 Continued

Species	Take	Approx. Price to Trapper	Value	Species	Take	Approx. Price to Trapper	Value
	1932–33 Season				*1935–36 Season*		
Muskrat	2,150,915	$ .26	$ 559,237.90	Muskrat	2,100,550	$ .80	$1,680,440.00
Raccoon	87,674	.80	70,139.20	Raccoon	84,321	2.70	227,666.70
Opossum	127,890	.15	19,183.50	Opossum	125,830	.25	31,457.50
Mink	82,364	1.50	123,546.00	Mink	82,465	5.00	412,325.00
Striped Skunk	—	—	—	Striped Skunk	—	—	—
Spotted Skunk	—	—	—	Spotted Skunk	—	—	—
Skunks (unspec.)	52,349	.08	4,187.92	Skunks (unspec.)	45,947	.25	11,486.75
Otter	870	5.00	4,350.00	Otter	650	7.00	4,550.00
Foxes	—	—	—	Foxes	—	—	—
Wolf or Coyote	—	—	—	Wolf or Coyote	—	—	—
Bobcat	—	—	—	Bobcat	—	—	—
Cats (unspec.)	—	—	—	Cats (unspec.)	—	—	—
Nutria	—	—	—	Nutria	—	—	—
Beaver	—	—	—	Beaver	—	—	—
Misc.	2,159	.50	1,079.50	Misc.	2,503	.50	1,251.50
	2,504,221		$ 781,724.02		2,442,264		$2,369,177.45
	1933–34 Season				*1936–37 Season*		
Muskrat	2,025,915	$ .58	$1,175,030.70	Muskrat	2,200,520	$ 1.00	$2,200,520.00
Raccoon	85,764	1.25	107,205.00	Raccoon	86,500	3.00	259,500.00
Opossum	123,540	.30	37,062.00	Opossum	128,000	.28	35,840.00
Mink	81,463	2.25	183,291.75	Mink	81,530	10.00	815,300.00
Striped Skunk	—	—	—	Striped Skunk	—	—	—
Spotted Skunk	—	—	—	Spotted Skunk	—	—	—
Skunks (unspec.)	50,959	.20	10,191.80	Skunks (unspec.)	46,890	.20	9,378.00
Otter	780	7.00	5,460.00	Otter	780	9.00	7,020.00
Foxes	—	—	—	Foxes	—	—	—
Wolf or Coyote	—	—	—	Wolf or Coyote	—	—	—
Bobcat	—	—	—	Bobcat	—	—	—
Cats (unspec.)	—	—	—	Cats (unspec.)	—	—	—
Nutria	—	—	—	Nutria	—	—	—
Beaver	—	—	—	Beaver	—	—	—
Misc.	2,557	.60	1,534.20	Misc.	2,600	.50	1,300.00
	2,370,978		$1,519,775.45		2,546,820		$3,328,858.00
	1934–35 Season				*1937–38 Season*		
Muskrat	2,125,720	$ .65	$1,381,718.00	Muskrat	3,110,540	$ .45	$1,399,743.00
Raccoon	83,467	1.85	154,413.95	Raccoon	87,300	1.25	109,125.00
Opossum	126,000	.35	44,100.00	Opossum	131,000	.12	15,720.00
Mink	80,364	3.25	261,183.00	Mink	82,480	5.25	433,020.00
Striped Skunk	—	—	—	Striped Skunk	—	—	—
Spotted Skunk	—	—	—	Spotted Skunk	—	—	—
Skunks (unspec.)	47,955	.22	10,550.10	Skunks (unspec.)	48,500	.12	5,820.00
Otter	789	10.00	7,890.00	Otter	920	7.00	6,440.00
Foxes	—	—	—	Foxes	—	—	—
Wolf or Coyote	—	—	—	Wolf or Coyote	—	—	—
Bobcat	—	—	—	Bobcat	—	—	—
Cats (unspec.)	—	—	—	Cats (unspec.)	—	—	—
Nutria	—	—	—	Nutria	—	—	—
Beaver	—	—	—	Beaver	—	—	—
Misc.	2,415	.70	1,690.50	Misc.	2,700	.15	405.00
	2,466,710		$1,861,545.55		3,463,440		$1,970,273.00

Table 2 Continued

No audit for 1938–39 Season

1939–40 Season

Species	Take	Approx. Price to Trapper	Value
Muskrat	7,034,661	$ .60	$4,220,796.60
Raccoon	77,446	1.40	108,424.40
Opossum	54,354	.10	5,435.40
Mink	92,061	3.00	276,183.00
Striped Skunk	5,739	.10	573.90
Spotted Skunk	144	.50	72.00
Skunks (unspec.)	–	–	–
Otter	1,392	6.60	9,187.20
Foxes	1,870	.50	935.00
Wolf or Coyote	–	–	–
Bobcat	–	–	–
Cats (unspec.)	–	–	–
Nutria	–	–	–
Beaver	–	–	–
Misc.	235	.50	117.50
	7,267,902		$4,621,725.00

1940–41 Season

Species	Take	Approx. Price to Trapper	Value
Muskrat	5,778,750	$ 1.00	$5,778,750.00
Raccoon	169,531	1.00	169,531.00
Opossum	131,045	.10	13,104.50
Mink	114,323	4.00	457,292.00
Striped Skunk	15,499	.10	1,549.90
Spotted Skunk	227	.50	113.50
Skunks (unspec.)	–	–	–
Otter	1,728	5.00	8,640.00
Foxes	3,768	.50	1,884.00
Wolf or Coyote	–	–	–
Bobcat	–	–	–
Cats (unspec.)	–	–	–
Nutria	–	–	–
Beaver	–	–	–
Misc.	1,715	.50	857.50
	6,216,586		$6,431,722.40

1941–42 Season

Species	Take	Approx. Price to Trapper	Value
Muskrat	4,846,492	$ 1.00	$4,846,492.00
Raccoon	181,309	1.50	271,963.50
Opossum	128,027	.30	38,408.10
Mink	155,391	4.00	621,564.00
Striped Skunk	20,880	.40	8,352.00
Spotted Skunk	399	.50	199.50
Skunks (unspec.)	–	–	–
Otter	1,760	6.00	10,560.00
Foxes	4,446	.50	2,223.00
Wolf or Coyote	–	–	–
Bobcat	–	–	–
Cats (unspec.)	–	–	–
Nutria	–	–	–
Beaver	–	–	–
Misc.	4,027	.50	2,013.50
	5,342,731		$5,801,775.60

1942–43 Season

Species	Take	Approx. Price to Trapper	Value
Muskrat	4,253,397	$ 1.18	$5,019,008.46
Raccoon	164,695	2.00	329,390.00
Opossum	78,738	.35	27,558.30
Mink	128,311	4.00	513,244.00
Striped Skunk	17,349	.75	13,011.75
Spotted Skunk	231	.15	34.65
Skunks (unspec.)	–	–	–
Otter	1,399	7.50	10,492.50
Foxes	3,727	1.25	4,658.75
Wolf or Coyote	–	–	–
Bobcat	–	–	–
Cats (unspec.)	–	–	–
Nutria	–	–	–
Beaver	–	–	–
Misc.	48	.50	24.00
	4,647,895		$5,917,422.41

1943–44 Season

Species	Take	Approx. Price to Trapper	Value
Muskrat	4,743,790	$ 1.44¾	$6,866,636.03
Raccoon	208,010	1.15	239,211.50
Opossum	91,033	.50	45,516.50
Mink	144,582	4.05	585,557.10
Striped Skunk	–	–	–
Spotted Skunk	–	–	–
Skunks (unspec.)	28,037	.50	14,018.50
Otter	2,404	6.00	14,424.00
Foxes	8,534	.50	4,267.00
Wolf or Coyote	–	–	–
Bobcat	–	–	–
Cats (unspec.)	918	.50	459.00
Nutria	436	.50	218.00
Beaver	–	–	–
Misc.	1,146	.50	573.00
	5,228,890		$7,770,880.63

1944–45 Season

Species	Take	Approx. Price to Trapper	Value
Muskrat	5,963,156	$ 1.23	$ 7,334,681.88
Raccoon	176,911	1.15	203,447.65
Opossum	66,000	.50	33,000.00
Mink	132,821	4.05	537,925.05
Striped Skunk	–	–	–
Spotted Skunk	–	–	–
Skunks (unspec.)	11,155	.50	5,577.50
Otter	1,912	6.00	11,472.00
Foxes	5,124	.50	2,562.00
Wolf or Coyote	–	–	–
Bobcat	–	–	–
Cats (unspec.)	628	.50	314.00
Nutria	902	.50	451.00
Beaver	–	–	–
Misc.	86	.50	43.00
	6,358,695		$ 8,129,474.08

Table 2 Continued

Species	Take	Approx. Price to Trapper	Value	Species	Take	Approx. Price to Trapper	Value
	1945–46 Season				*1948–49 Season*		
Muskrat	8,337,411	$ 1.50	$12,506,116.50	Muskrat	2,948,281	$ 1.10	$ 3,243,109.10
Raccoon	244,502	1.50	366,753.00	Raccoon	104,034	.70	72,823.80
Opossum	90,433	.50	45,216.50	Opossum	23,713	.50	11,856.50
Mink	168,598	15.00	2,528,970.00	Mink	91,541	8.00	732,328.00
Striped Skunk	—	—	—	Striped Skunk	—	—	—
Spotted Skunk	—	—	—	Spotted Skunk	—	—	—
Skunks (unspec.)	12,224	1.00	12,224.00	Skunks (unspec.)	1,144	.50	572.00
Otter	2,367	20.00	47,340.00	Otter	2,222	12.00	26,664.00
Foxes	3,823	.50	1,911.50	Foxes	1,727	.50	863.50
Wolf or Coyote	—	—	—	Wolf or Coyote	6	.50	3.00
Bobcat	—	—	—	Bobcat	—	—	—
Cats (unspec.)	1,430	.50	715.00	Cats (unspec.)	222	.50	111.00
Nutria	8,784	5.00	43,920.00	Nutria	26,738	3.50	93,583.00
Beaver	—	—	—	Beaver	—	—	—
Misc.	37	.50	18.50	Misc.	759	.50	379.50
	8,869,609		$15,553,185.00		3,200,387		$ 4,182,293.40
	1946–47 Season				*1949–50 Season*		
Muskrat	8,029,764	$ 1.00	$ 8,029,764.00	Muskrat	2,195,324	$ .77	$ 1,690,399.48
Raccoon	186,750	1.50	280,125.00	Raccoon	73,035	.27	19,719.45
Opossum	77,264	.60	46,358.40	Opossum	16,361	.20	3,272.20
Mink	153,027	5.00	765,135.00	Mink	123,850	8.90	1,102,265.00
Striped Skunk	—	—	—	Striped Skunk	—	—	—
Spotted Skunk	—	—	—	Spotted Skunk	—	—	—
Skunks (unspec.)	4,830	.50	2,415.00	Skunks (unspec.)	302	.15	45.30
Otter	2,832	12.50	35,400.00	Otter	2,968	8.00	23,744.00
Foxes	2,645	.50	1,322.50	Foxes	226	.15	33.90
Wolf or Coyote	—	—	—	Wolf or Coyote	—	—	—
Bobcat	—	—	—	Bobcat	—	—	—
Cats (unspec.)	383	.50	191.50	Cats (unspec.)	87	.15	13.05
Nutria	18,015	3.00	54,045.00	Nutria	38,988	3.00	116,946.00
Beaver	—	—	—	Beaver	—	—	—
Misc.	53	.50	26.50	Misc.	142	.15	21.30
	8,475,563		$ 9,214,782.90		2,451,283		$ 2,956,477.68
	1947–48 Season				*1950–51 Season*		
Muskrat	5,794,317	$ 1.50	$ 8,691,475.50	Muskrat	2,477,464	$ 1.25	$3,096,830.00
Raccoon	126,933	1.00	126,933.00	Raccoon	104,420	.43	44,900.60
Opossum	31,744	.70	22,220.80	Opossum	27,308	.21	5,734.68
Mink	153,120	15.00	2,296,800.00	Mink	184,552	12.93	2,386,257.30
Striped Skunk	—	—	—	Striped Skunk	—	—	—
Spotted Skunk	—	—	—	Spotted Skunk	—	—	—
Skunks (unspec.)	1,415	.70	990.50	Skunks (unspec.)	1,006	.15	150.90
Otter	5,078	18.00	91,404.00	Otter	4,801	12.15	58,332.15
Foxes	2,528	.50	1,264.00	Foxes	1,250	.25	312.50
Wolf or Coyote	—	—	—	Wolf or Coyote	—	—	—
Bobcat	—	—	—	Bobcat	—	—	—
Cats (unspec.)	70	.50	35.00	Cats (unspec.)	58	1.05	60.90
Nutria	28,176	3.00	84,528.00	Nutria	78,422	4.65	364,662.30
Beaver	—	—	—	Beaver	—	—	—
Misc.	679	.50	339.50	Misc.	41	.15	6.15
	6,144,060		$11,315,990.30		2,879,322		$5,957,247.54

Table 2 Continued

Species	Take	Approx. Price to Trapper	Value	Species	Take	Approx. Price to Trapper	Value
	1951–52 Season				*1954–55 Season*		
Muskrat	1,011,202	$ 1.09	$1,102,210.18	Muskrat	1,383,114	$ 1.10	$1,521,425.40
Raccoon	89,353	.55	49,144.15	Raccoon	84,223	.75	63,167.25
Opossum	18,017	.24	4,324.08	Opossum	7,497	.35	2,623.95
Mink	95,784	9.39	899,411.76	Mink	65,170	15.00	977,550.00
Striped Skunk	—	—	—	Striped Skunk	—	—	—
Spotted Skunk	—	—	—	Spotted Skunk	—	—	—
Skunks (unspec.)	1,062	.34	361.08	Skunks (unspec.)	120	.50	60.00
Otter	4,849	11.00	53,581.45	Otter	5,407	16.00	86,512.00
Foxes	1,181	.45	531.45	Foxes	31	.50	15.50
Wolf or Coyote	—	—	—	Wolf or Coyote	—	—	—
Bobcat	—	—	—	Bobcat	—	—	—
Cats (unspec.)	35	.25	8.75	Cats (unspec.)	542	.40	216.80
Nutria	77,966	2.00	155,932.00	Nutria	374,199	2.05	767,107.95
Beaver	—	—	—	Beaver	—	—	—
Misc.	33	.28	9.24	Misc.	—	—	—
	1,299,482		$2,265,514.14		1,920,303		$3,418,678.85
	1952–53 Season				*1955–56 Season*		
Muskrat	1,247,705	$ 1.15	$1,434,860.75	Muskrat	1,821,840	$ .80	$1,457,472.00
Raccoon	63,429	.50	31,714.50	Raccoon	61,520	.90	55,368.00
Opossum	12,114	.20	2,422.80	Opossum	6,211	.15	931.65
Mink	113,073	10.00	1,130,730.00	Mink	57,142	11.00	628,562.00
Striped Skunk	—	—	—	Striped Skunk	—	—	—
Spotted Skunk	—	—	—	Spotted Skunk	—	—	—
Skunks (unspec.)	982	.50	491.00	Skunks (unspec.)	273	.20	54.60
Otter	4,198	11.00	46,178.00	Otter	4,653	18.00	83,754.00
Foxes	377	.25	94.25	Foxes	185	.25	46.25
Wolf or Coyote	—	—	—	Wolf or Coyote	—	—	—
Bobcat	—	—	—	Bobcat	—	—	—
Cats (unspec.)	21	.50	10.50	Cats (unspec.)	102	.25	25.50
Nutria	89,526	2.00	179,052.00	Nutria	418,772	1.00	418,772.00
Beaver	—	—	—	Beaver	—	—	—
Misc.	2	.15	.30	Misc.	—	—	—
	1,531,427		$2,825,554.10		2,370,698		$2,644,986.00
	1953–54 Season				*1956–57 Season*		
Muskrat	1,173,209	$ .80	$ 938,567.20	Muskrat	1,589,433	$ .80	$1,271,546.40
Raccoon	49,857	.50	24,928.50	Raccoon	72,522	1.00	72,522.00
Opossum	6,344	.25	1,586.00	Opossum	5,056	.15	758.40
Mink	58,706	8.50	499,001.00	Mink	62,872	7.00	440,104.00
Striped Skunk	—	—	—	Striped Skunk	—	—	—
Spotted Skunk	—	—	—	Spotted Skunk	—	—	—
Skunks (unspec.)	311	.25	77.75	Skunks (unspec.)	526	.20	105.20
Otter	3,884	13.00	50,492.00	Otter	5,261	16.00	84,176.00
Foxes	86	.25	21.50	Foxes	76	.25	19.00
Wolf or Coyote	—	—	—	Wolf or Coyote	—	—	—
Bobcat	—	—	—	Bobcat	—	—	—
Cats (unspec.)	5	.25	1.25	Cats (unspec.)	31	.25	7.75
Nutria	160,654	2.00	321,308.00	Nutria	543,160	1.50	814,740.00
Beaver	—	—	—	Beaver	—	—	—
Misc.	—	—	—	Misc.	2	.25	.50
	1,453,056		$1,835,983.20		2,278,939		$2,683,979.25

Table 2 Continued

Species	Take	Approx. Price to Trapper	Value	Species	Take	Approx. Price to Trapper	Value
	1957–58 Season				*1960–61 Season*		
Muskrat	2,165,723	$ .80	$1,732,578.40	Muskrat	961,287	$ 1.10	$1,057,415.70
Raccoon	75,048	.90	67,543.20	Raccoon	65,588	1.50	98,382.00
Opossum	11,194	.15	1,679.10	Opossum	11,450	.40	4,580.00
Mink	61,377	6.00	368,262.00	Mink	32,272	7.00	225,904.00
Striped Skunk	—	—	—	Striped Skunk	—	—	—
Spotted Skunk	—	—	—	Spotted Skunk	—	—	—
Skunks (unspec.)	478	.20	95.60	Skunks (unspec.)	388	.40	155.20
Otter	4,382	14.00	61,348.00	Otter	3,602	17.00	61,234.00
Foxes	25	.25	6.25	Foxes	165	.50	82.50
Wolf or Coyote	—	—	—	Wolf or Coyote	—	—	—
Bobcat	—	—	—	Bobcat	—	—	—
Cats (unspec.)	200	.25	50.00	Cats (unspec.)	39	.50	19.50
Nutria	510,679	1.00	510,679.00	Nutria	716,435	1.00	716,435.00
Beaver	—	—	—	Beaver	—	—	—
Misc.	2	.20	.40	Misc.	—	—	—
	2,829,108		$2,742,241.95		1,791,226		$2,164,207.90
	1958–59 Season				*1961–62 Season*		
Muskrat	1,302,606	$ .80	$1,042,084.80	Muskrat	632,558	$ 1.20	$ 759,069.60
Raccoon	68,139	1.00	68,139.00	Raccoon	109,470	1.50	164,205.00
Opossum	3,660	.10	366.00	Opossum	16,662	.40	6,664.80
Mink	88,365	8.00	706,920.00	Mink	41,399	7.00	289,793.00
Striped Skunk	—	—	—	Striped Skunk	—	—	—
Spotted Skunk	—	—	—	Spotted Skunk	—	—	—
Skunks (unspec.)	623	.15	93.45	Skunks (unspec.)	464	.40	185.60
Otter	5,166	14.00	72,324.00	Otter	4,195	16.00	67,120.00
Foxes	33	.25	8.25	Foxes	192	.50	96.00
Wolf or Coyote	—	—	—	Wolf or Coyote	—	—	—
Bobcat	—	—	—	Bobcat	—	—	—
Cats (unspec.)	126	.25	31.50	Cats (unspec.)	41	.50	20.50
Nutria	461,311	1.00	461,311.00	Nutria	912,890	1.25	1,141,112.50
Beaver	—	—	—	Beaver	—	—	—
Misc.	—	—	—	Misc.	—	—	—
	1,930,029		$2,351,278.00		1,717,871		$2,428,267.00
	1959–60 Season				*1962–63 Season*		
Muskrat	1,531,788	$ 1.20	$1,838,145.60	Muskrat	346,647	$ 1.60	$ 554,635.20
Raccoon	80,814	1.50	121,221.00	Raccoon	160,218	1.80	288,392.40
Opossum	9,498	.25	2,374.50	Opossum	20,157	.50	10,078.50
Mink	58,838	8.00	470,704.00	Mink	51,316	7.50	384,870.00
Striped Skunk	—	—	—	Striped Skunk	—	—	—
Spotted Skunk	—	—	—	Spotted Skunk	—	—	—
Skunks (unspec.)	524	.25	131.00	Skunks (unspec.)	465	.50	232.50
Otter	5,559	18.00	100,062.00	Otter	8,484	17.00	144,228.00
Foxes	123	.50	61.50	Foxes	553	1.00	553.00
Wolf or Coyote	—	—	—	Wolf or Coyote	—	—	—
Bobcat	—	—	—	Bobcat	110	2.00	220.00
Cats (unspec.)	252	.50	126.00	Cats (unspec.)	—	—	—
Nutria	694,110	1.10	763,521.00	Nutria	1,357,806	1.35	1,833,038.10
Beaver	—	—	—	Beaver	—	—	—
Misc.	—	—	—	Misc.	—	—	—
	2,381,506		$3,296,346.60		1,945,756		$3,216,247.70

Table 2 Continued

Species	Take	Approx. Price to Trapper	Value	Species	Take	Approx. Price to Trapper	Value
	1963–64 Season				*1966–67 Season*		
Muskrat	240,079	$ 1.75	$ 420,138.25	Muskrat	529,438	$ 1.10	$ 582,381.80
Raccoon	58,424	1.25	73,030.00	Raccoon	83,876	1.50	125,814.00
Opossum	7,593	.25	1,898.25	Opossum	6,553	.30	1,965.90
Mink	18,805	6.00	112,830.00	Mink	62,150	4.00	248,600.00
Striped Skunk	—	—	—	Striped Skunk	260	.50	130.00
Spotted Skunk	—	—	—	Spotted Skunk	23	1.00	23.00
Skunks (unspec.)	301	.25	75.25	Skunks (unspec.)	—	—	—
Otter	4,274	18.00	76,932.00	Otter	4,118	18.00	74,124.00
Foxes	139	1.00	139.00	Foxes	303	2.00	606.00
Wolf or Coyote	—	—	—	Wolf or Coyote	—	—	—
Bobcat	26	1.00	26.00	Bobcat	10	2.00	20.00
Cats (unspec.)	—	—	—	Cats (unspec.)	—	—	—
Nutria	1,304,267	1.50	1,956,400.50	Nutria	1,307,121	2.00	2,614,242.00
Beaver	25	5.00	125.00	Beaver	21	5.00	105.00
Misc.	—	—	—	Misc.	—	—	—
	1,633,933		$2,641,594.25		1,993,873		$3,648,011.70
	1964–65 Season				*1967–68 Season*		
Muskrat	201,510	$ 1.75	$ 352,642.50	Muskrat	929,964	$ .75	$ 697,473.00
Raccoon	68,627	1.25	85,783.75	Raccoon	50,790	1.50	76,185.00
Opossum	7,047	.25	1,761.75	Opossum	3,009	.30	902.70
Mink	23,311	6.00	139,866.00	Mink	27,498	3.00	82,494.00
Striped Skunk	—	—	—	Striped Skunk	—	—	—
Spotted Skunk	—	—	—	Spotted Skunk	—	—	—
Skunks (unspec.)	296	.40	118.40	Skunks (unspec.)	136	.70	95.20
Otter	3,288	25.00	82,200.00	Otter	3,466	14.00	48,524.00
Foxes	145	1.00	145.00	Foxes	95	2.00	190.00
Wolf or Coyote	—	—	—	Wolf or Coyote	—	—	—
Bobcat	31	2.00	62.00	Bobcat	16	3.00	48.00
Cats (unspec.)	—	—	—	Cats (unspec.)	—	—	—
Nutria	1,568,233	1.60	2,509,172.80	Nutria	1,115,410	1.75	1,951,967.50
Beaver	57	5.00	285.00	Beaver	89	5.00	445.00
Misc.	—	—	—	Misc.	—	—	—
	1,872,545		$3,172,037.00		2,130,473		$2,858,324.40
	1965–66 Season				*1968–69 Season*		
Muskrat	324,204	$ 1.75	$ 567,357.00	Muskrat	1,556,764	$ 1.10	$1,712,440.40
Raccoon	78,348	2.00	156,696.00	Raccoon	95,654	2.50	239,135.00
Opossum	7,137	.30	2,141.10	Opossum	9,496	.50	4,748.00
Mink	28,216	6.00	169,296.00	Mink	46,918	5.00	234,590.00
Striped Skunk	—	—	—	Striped Skunk	—	—	—
Spotted Skunk	—	—	—	Spotted Skunk	—	—	—
Skunks (unspec.)	161	.50	80.50	Skunks (unspec.)	250	.50	125.00
Otter	3,588	20.00	71,760.00	Otter	5,426	20.00	108,520.00
Foxes	264	2.00	528.00	Foxes	324	3.00	972.00
Wolf or Coyote	—	—	—	Wolf or Coyote	—	—	—
Bobcat	13	2.00	26.00	Bobcat	56	5.00	280.00
Cats (unspec.)	—	—	—	Cats (unspec.)	—	—	—
Nutria	1,257,385	2.90	3,646,416.50	Nutria	1,754,028	2.14	3,762,084.00
Beaver	14	5.00	70.00	Beaver	124	5.00	620.00
Misc.	—	—	—	Misc.	—	—	—
	1,699,330		$4,614,371.10		3,469,040		$6,063,514.40

Table 2   Continued

Species	Take	Approx. Price to Trapper	Value	Species	Take	Approx. Price to Trapper	Value
	1969–70 Season				*1971–72 Season*		
Muskrat	1,232,052	$ 1.23	$1,515,424.00	Muskrat	326,513	$ 3.05	$ 995,795.50
Raccoon	103,725	2.25	233,381.25	Raccoon	80,632	2.94	237,212.00
Opossum	7,648	.50	3,824.00	Opossum	8,310	.50	4,155.00
Mink	46,294	5.00	231,470.00	Mink	24,299	4.00	97,196.00
Striped Skunk	—	—	—	Striped Skunk	—	—	—
Spotted Skunk	—	—	—	Spotted Skunk	3	2.00	6.00
Skunks (unspec.)	108	.50	54.00	Skunks (unspec.)	114	.50	57.00
Otter	6,632	23.00	152,536.00	Otter	5,440	38.00	206,720.00
Foxes	636	3.00	1,908.00	Foxes	476	4.50	2,142.00
Wolf or Coyote	3	5.00	15.00	Wolf or Coyote	11	5.00	55.00
Bobcat	110	5.00	550.00	Bobcat	136	6.00	816.00
Cats (unspec.)	—	—	—	Cats (unspec.)	—	—	—
Nutria	1,604,175	2.39	3,826,680.00	Nutria	1,286,622	3.22	4,146,488.00
Beaver	646	5.00	3,230.00	Beaver	126	6.00	756.00
Misc.	—	—	—	Misc.	—	—	—
	3,002,043		$5,965,700.25		1,732,682		$5,691,398.50
	1970–71 Season				*1972–73 Season*		
Muskrat	777,960	$ 1.58	$1,230,246.00	Muskrat	346,787	$ 4.12	$1,430,497.00
Raccoon	55,726	1.28	71,089.00	Raccoon	149,274	5.50	821,733.00
Opossum	3,563	.50	1,781.50	Opossum	17,065	1.25	21,331.25
Mink	21,648	5.00	108,240.00	Mink	44,062	6.00	264,372.00
Striped Skunk	—	—	—	Striped Skunk	—	—	—
Spotted Skunk	—	—	—	Spotted Skunk	—	—	—
Skunks (unspec.)	6	.50	3.00	Skunks (unspec.)	405	1.00	405.00
Otter	4,808	25.00	120,200.00	Otter	7,668	42.00	322,056.00
Foxes	242	3.50	847.00	Foxes	1,899	10.00	18,990.00
Wolf or Coyote	—	—	—	Wolf or Coyote	112	10.00	1,120.00
Bobcat	55	5.00	275.00	Bobcat	481	12.00	5,772.00
Cats (unspec.)	—	—	—	Cats (unspec.)	—	—	—
Nutria	1,226,739	2.43	2,980,217.00	Nutria	1,611,623	4.18	6,737,774.75
Beaver	14	5.00	70.00	Beaver	956	5.00	4,780.00
Misc.	—	—	—	Misc.	—	—	—
	2,090,761		$4,512,968.50		2,180,332		$9,628,831.00

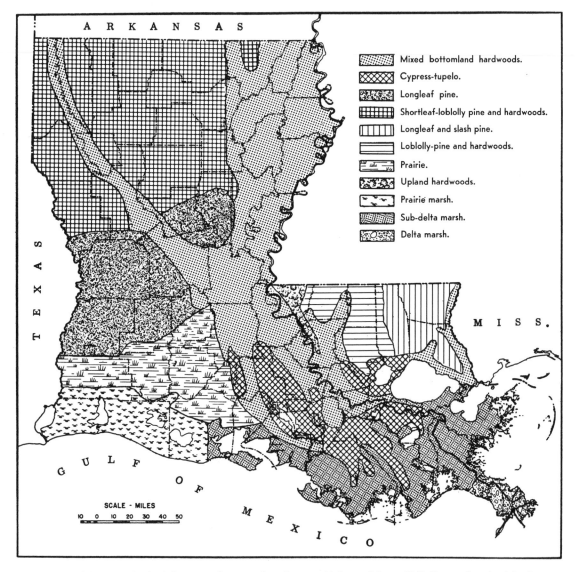

Map 1 Louisiana's principal forest and vegetational types. (Adapted from U.S. Forest Service Map)

FOREST AND VEGETATIONAL TYPES IN THE STATE

ELEVEN FAIRLY distinct forest and vegetational types can be recognized in the state, and these are roughly delineated in Map 1. None of these regions, with the possible exception of Mixed Bottomland Hardwoods and Upland Hardwoods, represent pure stands of their respective floras. There is much interdigitation of forest and vegetational types. For instance, an extensive area in southwestern Louisiana is shown on the map as Prairie, but it is by no means one uniform, short-grass, treeless expanse. Anywhere one stands on the Prairie there are trees visible on the horizon, outlining the course of streams that lace the terrain. Not unexpectedly forest-dwelling mammals follow these streams and the associated trees, and when the locality records of the mammals in question are plotted they appear to occur throughout the Prairie when in fact such is not the case.

While no species of mammal can be said to be truly endemic to any one vegetational type, the Eastern Chipmunk *(Tamias striatus)* comes close to qualifying. It is found almost exclusively on the Pleistocene terraces identified on the accompanying map as the area of Upland Hardwoods. More often species of mammals appear to be associated with two or three closely related vegetational types. The Hispid Pocket Mouse *(Perognathus hispidus)* is known only from the Longleaf Pine and the mixed Shortleaf and Loblolly Pine-Hardwood forest of western Louisiana, and the Plains Pocket Gopher *(Geomys bursarius)* is almost exclusively an animal of the same vegetational types. The range of the gopher is closely tied to the presence of certain soil types, which in turn dictate the kind of plants to be found in an area. The presence of a disjunct population of pocket gophers in Morehouse Parish highlights the close ecological relationship that exists between soil and vegetation on the one hand and where an animal is to be found on the other. What would appear to be excellent pocket gopher habitat in the Florida Parishes is, however, devoid of gophers because they have been prevented from reaching this area by the intervening bottomlands in the floodplains along the Mississippi and Atchafalaya rivers in the central part of the state.

PLAN OF THE ACCOUNTS

AT THE BEGINNING of each order of mammals a synopsis is given of the distinguishing features of the group, mentioning mainly those characters that are applicable to all members of the order. In addition I have tried to tell something about the diversity found in the taxon and about its worldwide distribution. The same sort of information is furnished for each family. For each genus I have given only the dental formula or a brief description of the dentition.

The species accounts are uniformly divided into a series of captioned sections. The first of these deals with *Vernacular and other names*. There I have listed the common names by which the species is known, particularly in Louisiana, including those used for the species by our French-speaking inhabitants in the southern part of the state, and I have given the derivation of the words making up the scientific and vernacular names. In a few instances I have felt compelled to coin an official name, at least to the extent of supplying an adjectival modifier where one is needed. In most cases I have been able to choose a name that has already been employed in some standard reference work. I abhor the practice of some mammalogists of

using an unmodified group name for a particular species of the group. For example, we should not call *Procyon lotor* "the Raccoon," inasmuch as there are several other species of raccoons, such as the Crab-eating Raccoon, *Procyon cancrivorus*, of Central and South America. To do so would be no different from our omitting the word Fox from the name Fox Squirrel and calling the species "the Squirrel," while, at the same time, supplying adjectival modifiers to other species, such as the Gray Squirrel, the Tassel-eared Squirrel, the Fire-bellied Squirrel, and so forth. The modifier "Fox" is absolutely essential in the vernacular name of *Sciurus niger*. And so is also a modifier in the vernacular name of *Procyon lotor*. I have called it the Northern Raccoon because it is the most northern ranging of the several species.

Mammalogists will observe that I have followed the practice that has been adopted by most ornithologists of capitalizing official vernacular names. The failure to capitalize these names can lead to serious confusion on the part of the layman, and even in some cases, professional mammalogists as well. Some classic examples among birds are the Lovely Cotinga, the Beautiful Bunting, and the Little Gull, but equally difficult combinations are found among mammals—the Big Brown Bat, the Vagrant Shrew, and the Black Rat. If one were to write, "I saw a big brown bat," how would the reader have any way of knowing that the adjectives were not simply descriptive rather than the name of a particular species.

The next paragraph, *Distribution,* gives the overall range of the species and its distribution in the state. Accompanying nearly every species account is a map of Louisiana showing the localities from which specimens are available in one or more museum collections. A single dot covers an area approximately 5.5 miles in diameter. The plotting of localities

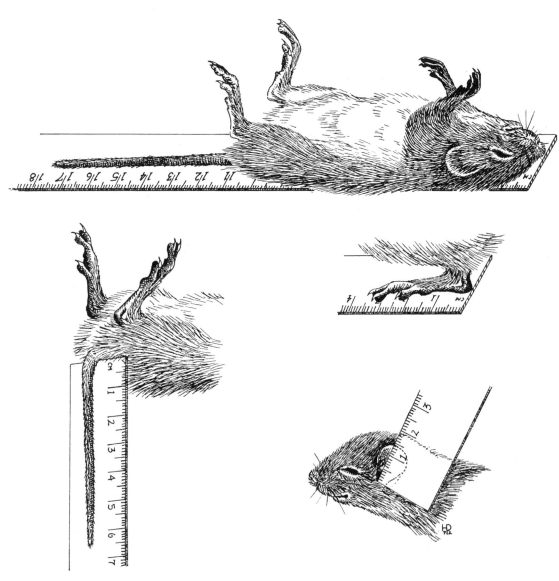

Figure 12   Illustrations showing how the standard measurements—total length, and the lengths of the tail, hind foot, and ear—are made.

that would result in broadly overlapping dots has been avoided by consolidation of localities lying in close proximity to one another. In a few instances I have used an open circle to show the location of an important sight record, one not supported by a preserved specimen. Where a species is represented in the state by two or more subspecies, their respective ranges are distinguished by differential shading. Small inset maps delineate the range of the species in North America.

In the paragraph captioned *External appearance,* I have attempted to highlight diagnostic features. Of special help to the users of this book will be the colorplates, black and white drawings, and photographs, which depict in one form or another all the mammals of the state. Under the heading *Color phases* I have described such color variants as are known for the species in question.

Measurements and weights are in nearly all instances based on specimens in the Louisiana State University Museum of Zoology. Unless otherwise specified the measurements are given in millimeters and weights are in grams. In many instances, particularly in the case of large mammals, I have also given measurements in inches or feet and weights in pounds and ounces. The following list of equivalents may help the reader to visualize various measurements:

1 millimeter = 0.04 inch
1 inch = 25.40 millimeters
1 meter = 39.37 inches
1 mile = 1.61 kilometers
1 kilometer = 0.62 mile
1 kilometer = 3,281 feet
1 gram = 0.04 ounce
1 ounce = 28.35 grams
1 kilogram = 2.20 pounds
1 acre = 4,047 square meters
1 acre = 43,560 square feet
1 hectare = 2.47 acres

The body measurements, taken directly from specimen labels, are the standard ones employed by mammalogists: total length, tail, hind foot, and ear. The manner in which these measurements are made is shown in

Figure 12. Skull measurements, except of exceedingly large skulls, were made with dial calipers, usually to the closest tenth of a millimeter. The main standard skull measurements (Glass, 1969) used herein, but not necessarily with respect to every species, are shown in Figure 13, as follows:

greatest length, A–A'
condylobasal length — distance from anterior border of median incisive alveoli to plane of posterior border of occipital condyles, B–B'
basilar length — distance from posterior border of median incisive alveoli to midventral border of foramen magnum, C–C'
cranial breadth — the greatest width of the cranium, D–D'
greatest height — the overall greatest height of the skull measured with skull lying on a flat surface to the plane of the topmost elevation (used in Cetacea only)
cranial height — depending on the kind of skull being measured, this dimension is either from the plane of the basisphenoid-basioccipital bones to the plane of the topmost elevation of the braincase or from the plane of the palatines to the plane of topmost elevation of the frontals (Felidae only), E–E'
postorbital breadth — the least width of the skull immediately posterior to the postorbital processes, between the temporal fossae, F–F'
zygomatic breadth — the greatest width across the zygoma, G–G'
interorbital breadth — the least distance between the orbits, H–H'
length of nasals, I–I'
length of diastema — distance from posterior border of alveolus of the most lateral incisor to anterior border of alveolus of first molariform tooth on the same side, J–J'
length of maxillary toothrow — distance from anterior border of the alveolus of the first tooth on the maxilla (the canine when present) to the posterior border of the last

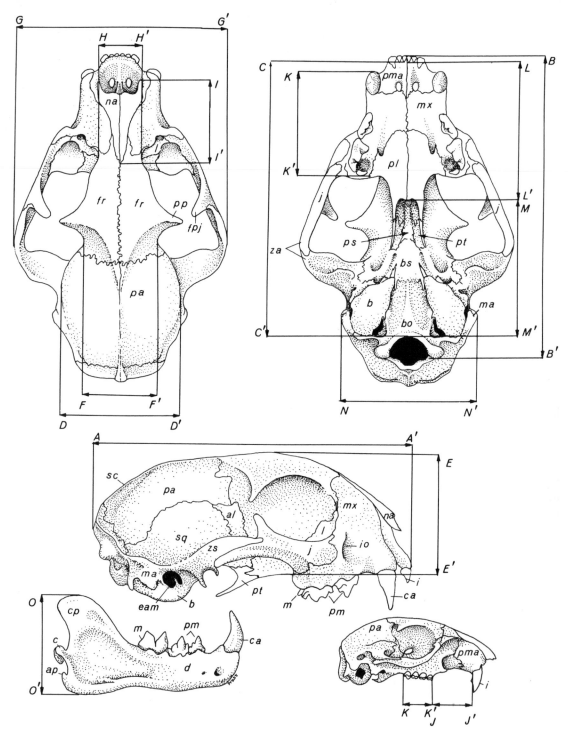

Figure 13 Cranium and mandible of the Bobcat *(Lynx rufus)* and cranium of the Fox Squirrel *(Sciurus niger),* showing the standard measurements and the names of the bones, foramina, and parts referred to frequently in the text: *A-A',* greatest length; *B-B',* condylobasal length; *C-C',* basilar length; *D-D',* cranial breadth; *E-E',* cranial height; *F-F',* postorbital breadth; *G-G',* zygomatic breadth; *H-H',* interorbital breadth; *I-I',* length of nasals; *J-J',* length of diastema; *K-K',* length of maxillary toothrow; *L-L',* palatilar length; *M-M',* postpalatal length; *N-N',* mastoid breadth; *O-O',* coronoid height; *al,* alisphenoid; *ap,* angular process; *b,* auditory bulla; *bo,* basioccipital; *bs,* basisphenoid; *c,* condyle; *ca,* canines; *cp,* coronoid process; *d,* dentary; *eam,* external auditory meatus; *fr,* frontal; *fpj,* frontal process of jugal; *i,* incisors; *io,* infraorbital foramen; *j,* jugal; *l,* lacrimal; *m,* molars; *ma,* mastoid; *mx,* maxillary; *na,* nasal; *oc,* occipital; *occ,* occipital condyle; *pa,* parietal; *pl,* palatine; *pp,* postorbital process; *pm,* premolars; *pma,* premaxillary; *ps,* presphenoid; *pt,* pterygoid; *sc,* sagittal crest; *sq,* squamosal; *za,* zygomatic arch; *zs,* zygomatic process of squamosal.

molariform tooth, except that in rodents this measurement is from the anterior border of the alveolus of the first molariform tooth to the posterior border of the alveolus of the last tooth in the series, K–K′

palatilar length—distance from posterior border of median incisive alveoli to the posterior border of the palate, not including the spines, L–L′

postpalatal length—distance from posterior border of palate, not including the spines, to the midventral border of the foramen magnum, M–M′

mastoid breadth—the greatest width across the mastoid processes, measured at right angles to the long axis of the skull, N–N′

coronoid height—height of lower jaw from dorsalmost edge of coronoid process to the ventral edge of lower jaw, O–O′

Under *Cranial and other skeletal characters* I have tried to summarize the salient features of the skull of each species, and in species where a penis bone is present, to describe its shape and dimensions. The names of the principal bones, foramina, and parts of the skull referred to in the text are also shown in Figure 13. Under *Sexual characters and age criteria* an attempt has been made to tell how the sexes differ and to state, where means are known, how to age individuals of the species of mammal.

The main subdivision of the species accounts is the one captioned *Status and habits*. Here I have first given the present-day relative abundance of the species on a statewide basis and discussed any local or sectional variations in status. In the general consideration of habits I have tried to give the reader some of the outstanding features of the animal's life history, including the season of its mating, the time when its young are born, the number of young per litter, the number of litters per year, the habitat where the species lives, its food, and its behavior. In many instances no detailed studies have been made that are based on the populations of a species inhabiting Louisiana, and I have had to rely on investigations made in other states. In the case of some species the lack of information available from any source is distressing, thus highlighting the need for further research. To facilitate finding certain kinds of life history information, I have used separate headings for *Predators* or *Predators and other decimating factors,* and I have segregated information on *Parasites* or *Parasites and diseases.* Here I have made no definitive search of the voluminous parasitological literature, particularly with regard to some species, but I have tried to list the main parasites known from a given species or to give bibliographic references to published summaries where they exist.

In the case of polytypic species the final topical heading is **Subspecies**. Here I have given the primary citations of the subspecies known to occur in Louisiana and the citations of the first use of the combinations employed in this work. In the case of monotypic species the same has been done for the binominal names employed, and the heading has been omitted. When a species or subspecies has been recorded from the state under a different name, I have often also included that name and the bibliographic reference in the synonymy. Under each species or subspecies I have listed the specimens examined with their precise localities except in the case of a few common forms for which only one subspecies occurs in the state and for which the number of specimens is excessive and represents nearly all parishes. In the listing of *Specimens examined from Louisiana,* of which I have personally examined nearly every one, the entries are arranged alphabetically by parish and by locality within each parish. When a specimen is available from a parish but the label on the specimen gives no exact locality within the parish, it is listed under the known parish as "locality unspecified." In listing the specimens, I have indicated the museum collection in which each reposes by one of the following abbreviations:

AMNH   American Museum of Natural History, New York

ANSP   Academy of Natural Sciences of Philadelphia

CAS   Chicago Academy of Science

CMNH   Colorado Museum of Natural History, Denver

FM   Field Museum of Natural History, Chicago

LSUMZ   Louisiana State University Museum of Zoology, Baton Rouge

LSUS   Louisiana State University at Shreveport

LTU   Louisiana Tech University, Ruston

MCZ   Museum of Comparative Zoology, Harvard

MSU   McNeese State University, Lake Charles

MVZ   Museum of Vertebrate Zoology, University of California, Berkeley

NLU   Northeast Louisiana University, Monroe

NSU   Northwestern State University of Louisiana, Natchitoches

SLU   Southeastern Louisiana University, Hammond

TCWC   Texas Cooperative Wildlife Collections, Texas A and M University, College Station

TU   Tulane University, New Orleans

USL   University of Southwestern Louisiana, Lafayette

USNM   United States National Museum, including Biological Survey Collections, Washington, D.C. This institution is now officially called the National Museum of Natural History.

I have presented accounts for a few species whose occurrence in Louisiana is supported by evidence somewhat less than completely satisfactory. Following my policy in *Louisiana Birds,* I have inserted such accounts in their proper place in the main taxonomic sequence but have enclosed the species names in brackets to distinguish these mammals from those whose right to be on the state list is beyond challenge. In the case of cetaceans, I have treated, although sometimes only briefly, all species known from the northern Gulf Coast but have clearly indicated which species have actually been seen from our shores or stranded on one of our beaches and which have not. Unfortunately, we have few stretches of beach along which we can drive for any appreciable distance. Consequently, our chances of finding stranded cetaceans are rather limited. I fervently hope that anyone who does find a dead whale or porpoise along our shores will contact me immediately in order that its skull or perhaps its complete skeleton may be salvaged. Some of the cetaceans are among the rarest animals in the world insofar as museum collections are concerned. For this reason the identification and preservation of every stranded individual is vitally important if our quest for a better understanding of the worldwide distribution of these interesting mammals is to be successful.

Capitalized color names are from Ridgway (1912). The sequence of genera for the most part follows Walker et al. (1968), but in some instances I have been influenced by the sequence employed by Anderson and Jones (1967) and their collaborators.

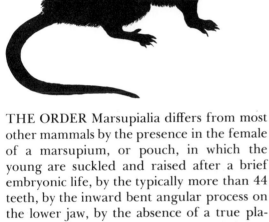

POUCHED MAMMALS
Order Marsupialia

THE ORDER Marsupialia differs from most other mammals by the presence in the female of a marsupium, or pouch, in which the young are suckled and raised after a brief embryonic life, by the typically more than 44 teeth, by the inward bent angular process on the lower jaw, by the absence of a true placenta, and by the presence of epipubic bones (Figure 14) in the pelvic girdle that serve to support the pouch (Figure 15). The order Marsupialia includes many bizarre forms such as kangaroos, wallabies, the Koala, the Tasmanian Devil, bandicoots, and the Marsupial Mole, as well as our own Virginia Opossum. The marsupials occur principally in the Aus-

tralian Realm, but are also represented in tropical America by numerous species of the families Didelphidae and the Caenolestidae. Only the Virginia Opossum is normally found north of Mexico.

Figure 15 The genitalia and associated structures of the Virginia Opossum *(Didelphis virginiana):* (A) male, showing the location of the penis behind the scrotum; (B) female, showing the marsupium, cloaca, and anal orifice.

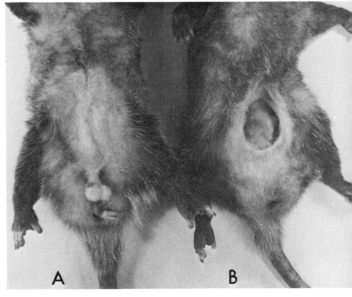

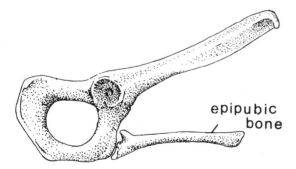

Figure 14 Right pelvic bones and the epipubic bone of the Virginia Opossum *(Didelphis virginiana).*

The American Opossum Family
Didelphidae

THE FAMILY Didelphidae ranges throughout most of tropical and temperate regions of North, Central, and South America. It contains 12 Recent genera and about 65 species. Some of the members are small, others medium in size. Most are terrestrial or arboreal, but one genus, *Chironectes* of southern Mexico and Central and South America, is semi-aquatic with webbed toes.

Genus *Didelphis* Linnaeus

Dental formula:

$$I\frac{5-5}{4-4} \ C\frac{1-1}{1-1} \ P\frac{3-3}{3-3} \ M\frac{4-4}{4-4} = 50$$

VIRGINIA OPOSSUM   Plate I
(*Didelphis virginiana*)

Vernacular and other names. — This marsupial is commonly called the opossum or simply possum, names derived from the Algonquian Indian name *apasum*. The French-speaking people of the southern part of the state refer to it as *rat de bois*, meaning rat of the woods. But one also sometimes hears in St. Bernard and Plaquemines parishes the name *topo* or *raibua*. According to Arthur (1928), the Ofo Indians named it *fesca tic-nki* or little pig, while the Biloxi called it *kcicka* or hog. The first part of the scientific name, *Didelphis*, is of Greek origin and means double womb (*di*, double and *delphys*, womb). The combination by an early zoological scholar may have been

in reference to the paired uteri or it may have been an allusion to the marsupium, or pouch, where the young are carried while they complete their development. The specific name *virginiana* refers, of course, to the state of

Plate I
1. Virginia Opossum (*Didelphis virginiana*): **a,** melanistic phase; **b,** normal phase.
2. Swamp Rabbit *(Sylvilagus aquaticus)*.
3. Eastern Cottontail *(Sylvilagus floridanus)*.
4. Nine-banded Armadillo *(Dasypus novemcinctus)*.

PRATT
1971

Virginia, wherefrom the animal first became known to science.

Distribution.—The Virginia Opossum is common throughout Louisiana, occurring in nearly all wooded areas and in the coastal marshes. The species ranges throughout most of the eastern United States and extreme southern Canada but is not common north of Illinois and Pennsylvania or west of central Kansas. It has been introduced into the Pacific Coast states. Within the United States it is most abundant in the southeastern quarter of the country from southern Texas and Missouri to the Atlantic Coast. The species ranges southward over most of Mexico, except for parts of the Central Plateau, from central Sonora and Tamaulipas to Chiapas and Yucatan thence to southern Nicaragua. It overlaps in range with *D. marsupialis* in southern Tamaulipas, eastern San Luis Potosí, Veracruz, northern and eastern Hidalgo, Oaxaca, Chiapas, Campeche, and southern Yucatan south through Guatemala, British Honduras, El Salvador, and parts of Honduras and Nicaragua (Gardner, 1973). In Louisiana the species is statewide in its occurrence. (Map 2)

External appearance.—Adults are about as large as the Domestic Cat but have shorter legs and a heavier body. Some of the major characters are: an external, fur-lined abdominal pouch, or marsupium, in the female for carrying the young; body fur long and dense, white basally and black terminally, strongly interspersed with exceptionally long white guard hairs; general color grayish white above with the fore and hind quarters darker and the belly much lighter; head conical with a pointed snout; face white except for black eye spots and a narrow line of black down the center of the crown; legs, feet, and basal portions of the toes black, the remainder of the toes white; ears leathery, naked, and black, often with a terminal spot of white; tail dirty white and scantily haired except at the

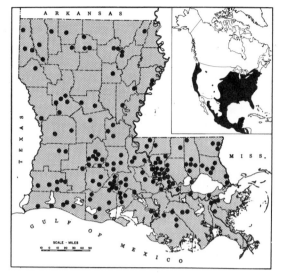

Map 2 Distribution of the Virginia Opossum (*Didelphis virginiana*). *Shaded area,* known or presumed range within the state; •, localities from which museum specimens have been examined; *inset,* overall range of the species.

base, which is black and heavily furred; the front foot with five claw-bearing toes; the hind feet with four claw-bearing toes and a large first toe that is opposable or thumblike and clawless (Figure 16); eyes blue-black; no seasonal variation in the pelage not attributable to excessive wear.

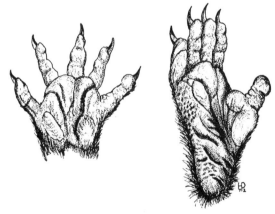

Figure 16 Right front and right hind foot of the Virginia Opossum (*Didelphis virginiana*), the latter showing the opposable inner toe, which is clawless. Approx. ⁴/₅ ×.

Color phases. — A black color phase is not uncommon in Louisiana, although encountered much less frequently than the normal, gray phase. In the dark individuals, the white underfur is strongly tipped with black, and the long guard hairs of the upperparts are usually black instead of gray (Plate I). Albinos occur rarely, there being two examples from the state in the LSUMZ.

Measurements and weights. — Twenty-two adult males in the LSUMZ, in which all teeth have erupted and which come from widespread localities in the state, averaged as follows: total length, 821 (751–940); tail, 310 (223–380); hind foot, 62 (50–85); ear, 51 (40–60); greatest length of skull, 115.3 (98.3–129.3); zygomatic breadth, 60.3 (59.0–73.1); interorbital breadth, 31.0 (22.7–38.1); palatilar length, 63.0 (52.9–71.9); postpalatal length, 41.6 (35.0–51.9). Weights ranged from 2,740 to 4,672 grams (6.0–10.3 pounds). Twenty-two adult females (also in the LSUMZ) averaged as follows: total length, 732 (613–900); tail, 302 (260–329); hind foot, 60 (38–75); ear, 50 (42–58); greatest length of skull, 103.9 (94.0–128.0); zygomatic breadth, 53.6 (47.9–63.1); interorbital breadth, 24.7 (20.9–28.7); palatilar length, 59.0 (51.7–69.4); postpalatal length, 37.0 (32.0–48.3). Weights ranged from 1,159 to 2,180 grams (2.5–4.8 pounds), but these figures are probably well below the average for adult females.

Cranial and other skeletal characters. — The adult Virginia Opossum possesses 50 teeth, more than any other Louisiana land mammal. The dentition is complete, that is, incisors, canines, premolars, and molars are present. The teeth are, for the size of the animal, small and rather weak except for the canines, which are long, somewhat curved, and sharp. The skull is easily recognized by the great number of teeth, by the wide posterior expansion of the nasal bones where the latter meet the frontals, by the small braincase, and

by the inwardly turned angle of the lower jaw (Figure 17). The pelvic, or hip, girdle possesses epipubic bones that project forward more than an inch in both males and females and in the latter serve to support the pouch (Figure 15).

Sexual characters and age criteria. — Males average slightly larger than females. The latter are readily distinguished by the presence of the pouch, the males by the scrotum, which lies in front of the penis. Petrides (1949) compiled information based on his data and that of Hartman (1928), Moore and Bodian (1940), and Reynolds (1945) that shows the growth rate of pouch young (Table 3). Since considerable variation occurs in the size of pouch mates, any age calculation should be based on the average measurements for the litter.

The sex organs of pouch young become discernible between the 11th and 17th day of age. The first tooth to emerge, the third premolar, does so on the 60th day. The second tooth, the second premolar, follows 15 days later. Petrides states that by the 80th day the teeth may consist of 4 incisors, 1 canine, and 3 premolars, above and below, with a lower molar also erupted, for a total of 34 teeth. At four months, opossums have a total of 40 teeth; at 5 to 8½ months, 44; and at 7 to 11 months, 48. But anytime after 10

Table 3 Average Weights and Snout-Anus Lengths of Pouch Young in *Didelphis virginiana* in Relation to Age[1]

Age (days)	Snout–Anus Length (in mm)	Weight (grams)
1	13	0.13
5	17	0.4
10	24	0.9
20	33	1.7
30	45	3.9
40	60	7.0
50	75	13.0
60	100	25.0
70	125	45.0
80	150	80.0
100	180	125.0

[1]After Petrides, 1949.

months the full complement of 50 teeth may be found. There are no data on the longevity of opossums in Louisiana, but judging from statements in the literature pertaining to other parts of the United States they can be presumed to live as long as four or five years. Probably, though, few individuals reach that age.

Status and habits. — The opossum is probably one of the best known of all our mammals. Despite its seeming stupidity it is eminently successful, as indicated by its widespread abundance. There are, indeed, few wooded areas in the state in which it does not reside, and it is frequently found in residential sections of our towns and cities. Its habits, how-

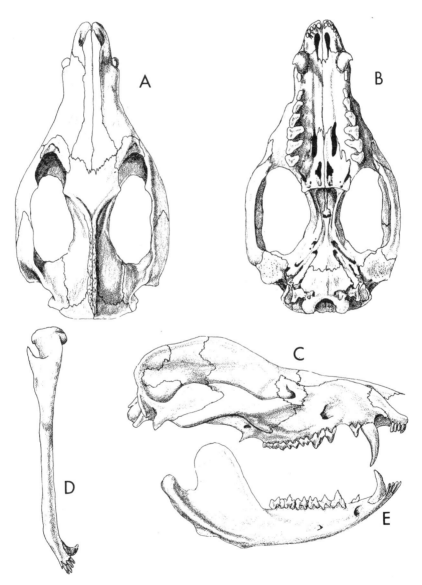

Figure 17 Skull of Virginia Opossum *(Didelphis virginiana pigra)* ♂, Vowells Mill, Natchitoches Parish, Louisiana, LSUMZ 1345. (A) Dorsal view of cranium; (B) ventral view of cranium; (C) lateral view of cranium; (D) ventral view of right ramus of lower jaw; (E) lateral view of lower jaw. Approx. ³/₄ ×.

Tracks of Virginia Opossum

Opossums were among the strange and bizarre forms of life that greeted the early colonists of Virginia as they began to carve their new home out of the wilderness. From the outset the animal was shrouded in myths and misconceptions, some of which persist even today. Captain John Smith, in his *The Generall Historie of Virginia* ... (1624:27), includes an excerpt from one of his earlier works (*A True Relation*, 1608) in which he describes the strange new animal the settlers found on their arrival. He said, "An Opassom hath a head like a Swine, and a taile like a Rat, and is of the bignesse of a Cat. Under her belly shee hath a bagge, wherein she lodgeth, carrieth, and suckleth her young."

Mating occurs in late January or early February. Gestation, or pregnancy, lasts only 12½ to 13 days. Although, according to Hartman (1952), an average of 22 eggs may be shed by the ovaries at each estrous, or heat, period, not all are fertilized and even fewer develop. The number of young that survive is limited by the number of teats available. Most females possess 13 teats, all inside the marsupial pouch. Twelve are typically arranged in a horseshoe design with the remaining teat in the center between the third pair from the rear (Figures 15 and 18). The young at birth are extremely tiny and incompletely developed, as one would expect after only 12 to 13 days of embryonic life. A litter of 20 at the time of their birth can be accommodated in a teaspoon. They are blind and are nothing more than glorified embryos, but, strangely enough, the claws on the forefeet are well developed and therefore serve a useful purpose in helping the young scramble into the pouch. Their length is about 10 mm (0.4 inch) and their weight approximately 0.13 gram (0.0046 ounce). It would take, in other words, 217 of them to weigh an ounce.

Among the many misconceptions surrounding the opossum is the still persistent belief on the part of the lay public that mating takes place through the nose. This notion probably stems from two facts, one being that

ever, are almost entirely nocturnal. It spends the daylight hours "holed up" in a hollow tree stump, rotten log, or culvert, from which it emerges at nightfall to begin its forays for food. Perhaps one of the reasons it is so successful is the fact that it eats virtually anything. Insects, such as grasshoppers, crickets, ants, and beetles, are devoured, as are also many fruits and berries and even corn. Birds and their eggs are taken occasionally. But when insects and desired vegetable matter are scarce the opossum readily becomes a scavenger. Around our homes it frequently overturns our garbage cans, and it is not averse to carrion of any sort. Although I have long wanted myself to eat some proverbial possum cooked with sweet potatoes, my enthusiasm for doing so diminished considerably after I poked a carcass of a dead horse that I chanced upon in a wood and was startled when a possum emerged from inside.

the male's penis is forked. Since the only visible paired openings in the female are her nostrils, the fallacious assumption is that mating takes place through them. Moreover, when parturition is about to happen, the female licks the interior of the pouch and the fur between it and the vulva. This behavior gives rise to the notion that she is blowing the embryos from her nostrils into the pouch. Nothing, of course, could be further from the truth. The forked penis serves to deliver spermatozoa to the paired uteri of the female, which are not visible externally. The licking of the fur-lined pouch and lower abdomen by the female prior to parturition encourages another erroneous belief—that she assists the young in their movements towards the pouch. Careful observations of the associated events have now revealed that the young get to the pouch entirely under their

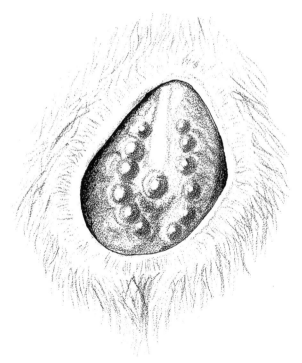

Figure 18 Marsupium, or pouch, of Virginia Opossum *(Didelphis virginiana),* showing six pairs of teats arranged in a horseshoe pattern with thirteenth teat in the middle.

own power, without any help from the mother. They have been recorded making the journey in less than 17 seconds, but they usually take longer (Burns and Burns, 1956).

Dr. Carl G. Hartman, who carried on a lifetime of research on the embryology, birth, and development of the opossum in his laboratory at the University of Texas, described the birth of the young in words (Hartman, 1920) that deserve direct quotation:

... the animal showed signs of restlessness and soon began cleaning out the pouch, which she did about four times. Then began a short series of spasmodic contractions of the abdominal wall, after which she came to a sitting posture with legs extended. ...

After assuming the sitting posture, our specimen bent her body forward and licked the vulva; however, her position at this time was such that we could not see the embryos, which very likely passed into the pouch with the first licking of the genital opening. Hence we went to the outside where we could plainly hear her lap up the chorionic fluid; then suddenly a tiny bit of flesh appeared at the vulva and scampered up over the entanglement of hair into the pouch to join the other foetuses, which now could be seen to have made the trip without our having observed them. Unerringly the embryo traveled by its own efforts; without any assistance on the mother's part, other than to free it of liquid on its first emergence into the world, this ten-day-old embryo, in appearance more like a worm than a mammal, is able, immediately upon release from its liquid medium, to crawl a full three inches over a difficult terrain. Indeed, it can do more: after it has arrived at the pouch it is able to find the nipple amid a forest of hair. This it must find—or perish. ...

The pouch contained a squirming mass of eighteen red embryos of which twelve were attached, though thirteen might have been accommodated. The remainder were, of course, doomed to starvation. Even some of these unfortunates, however, held on with their mouths to a flap of skin or to the tip of a minute tail, while several continued to move about.

With the mother under the influence of ether, we now gently pulled off a number of embryos from the teats in order to test their reactions. The teats had already been drawn out from about a millimeter in height to double that length, doubtless by the traction of the embryo itself, for the bottom of the pouch certainly presented a busy scene with each member of the close-pressed litter engaged in very active breathing and sucking movements.

One detached young, placed near the vulva, crawled readily back into the pouch. Two or three others regained the teats after some delay, and one wanderer, which lost out in the first scramble, found a vacated teat and attached itself even after twenty minutes' delay, showing that the instinct to find the teat persists for some time. If the skin be tilted, the embryos can be made to travel upward and even away from the pouch, for they are negatively geotropic.

For locomotion the embryo employs a kind of "overhand stroke," as if swimming, the head swaying as far as possible to the side opposite the hand which is taking the propelling stroke. With each turn of the head the snout is touched to the mother's skin as if to test it out, and if the teat is touched, the embryo stops and at once takes hold.

It is thus apparent that the opossum embryo at birth possesses not only fairly well-developed respiratory and digestive systems, but that it has attained a neuromuscular development sufficient to enable it to find its place in the pouch where food and shelter await it.

After successfully reaching the pouch and finding a "filling station," the young animal begins sucking immediately. Within an hour, as a result of the traction exerted by the young, the nipple is greatly extended in length. It serves not only to transport milk but gradually becomes longer and longer (up to 1½ inches) and acts as a tether that permits some freedom of movement to the young. The young can be sexed externally by the 10th day in the pouch, the males showing a distinct scrotum, the females a fold in the abdomen that will later become the pouch. Even though the young are destined to be warm blooded, that is, able to maintain a fairly constant body temperature, they are at the outset not yet equipped to do so. The portions of the brain that serve to regulate this physiological function are not yet developed. The baby opossum is kept warm solely by heat from its mother's body. However, the young develop rapidly. Within two months they are the size of a House Mouse and the eyes are open. They then release their hold on the nipples for the first time, but they can

return to them at will. Final weaning does not take place until the young are between 75 and 80 days of age. They remain with the mother for three or four months. After the growing litter mates reach a size where all will not fit in the pouch, some individuals ride on the back of the mother, clinging to her fur, while the others remain in the pouch (Figure 19).

In the opossum the paired uteri are connected to the common urogenital opening or cloaca by two lateral vaginal canals. When birth occurs, however, the young do not retrace the circuitous route taken by the sperm in reaching the uteri but instead break through to the cloaca by way of a new, temporary passage that is more direct. This new birth passage characterizes all marsupials (Hartman, 1952).

The opossum has an estrous cycle of about 28 days. After raising an initial litter in late winter, the female often brings forth a second litter in early summer. Allegedly a third litter may be produced, but in a two-year study of 117 individual opossums on an 86-acre plot near Beaumont, Texas, Daniel W. Lay (1942) found no support for the idea. He found

Figure 19 Virginia Opossum (*Didelphis virginiana*) with its young, some of which are suckling while partially concealed in the pouch. (Photograph by A. B. McPherson)

Figure 20   The Virginia Opossum *(Didelphis virginiana)* can indeed hang by its tail. (Photograph by Joe L. Herring)

peaks of breeding in February and June and a low in April. None of the 92 adult females examined between late July and the end of December carried young in the pouch. The sex ratio among the 117 opossums trapped was 57 males to 43 females. McKeever (1958), in a study of reproduction in the opossum in southwestern Georgia and northwestern Florida, obtàined results similar to those of Lay. He found two distinct breeding periods. The first began in the second week of January, reached a peak in the second week of February, and ended in the first week of March. The second season began in the first week of April, reached a peak in the first and second weeks of May, and ended in the second week of June. Burns and Burns (1956) state that females can reproduce at an age of eight months, when they weigh just over two pounds, but Hartman (1922) records a pregnant female that weighed only 651 grams (1.4 pounds).

The nest of the opossum is usually in a hollow tree or fallen log or in a secluded cranny in an outbuilding. Rarely it is in the crotch of a tree. It consists of a fairly large quantity of dead leaves, which the opossum carries to the site in its tail.

The ambling gait of the opossum is distinctive. It runs with the tail straight and generally held above the ground. It climbs with slow, measured dexterity, and often uses its tail as a "fifth leg." Not infrequently it hangs from a limb by its tail (Figure 20).

When frightened the animal bares its teeth and drools saliva from its mouth. Not infrequently it enacts the behavioral pattern for which it is famous—that of "playing possum." It rolls over, closes its eyes, allows the tongue to hang out of its mouth, and appears dead. If, however, the animal is turned back to a normal position, it sometimes immediately rolls over again, thereby revealing that the behavior is in part only an act. Interestingly enough, though, physiological changes do take place in the perpetrator. The rate of heartbeat slows and other responses occur that are seemingly involuntary and thus analogous to fainting in humans.

In the folklore of Indians, Negroes, and whites of the southern United States, the opossum has played a prominent role. One legend tells how the opossum lost the hair on its tail when it was burned over a fire. Another story ascribes the loss of the hair to an encounter with a ghost when the opossum was caught stealing corn near a graveyard. The grinning mouth, according to a Choctaw Indian legend in what is now St. Tammany Parish, resulted from a joke played by the opossum on a deer. As the story goes the deer inquired why the possum was so fat when the other animals of the forest were suffering from the lack of food during a severe dry season. The opossum responded that he was feasting on persimmons that he shook from the high branches by running downhill head on into the tree. When the deer tried to duplicate this feat it struck the tree with such great force that every bone in its body was broken. The opossum was convulsed with laughter at the plight of the deer, and it laughed so hard that it stretched its mouth, which remains large even to this day (Bushnell, 1909).

The night hunting of the possum used to be a great pastime among our rural inhabitants and is still practiced to a large extent in some areas. It is done with headlight and dogs. No special breed of dog is required. A mutt or run-of-the-mill cur serves well enough. When pursued the possum will rather quickly "tree," usually in a small sapling from which it can be dislodged by shaking (Figure 21).

To many of our rural inhabitants, possum is a delicacy, especially when baked and served with sweet potatoes. Possum hunters generally like to take their animals alive so that they can be kept for several weeks and fattened. One recipe calls for dressing the animal by immersing it in very hot (not boiling) water for one minute. Next remove the hair by scraping (like scraping a hog) with a dull knife or hog scraper, so that the skin is not cut. Then slit the underside from neck to hind legs and remove the entrails, washing inside and out with hot water. Cover with cold water, to which has been added one cup of salt, and allow to stand overnight. The following morning drain off the salt water and rinse again, with boiling water. Make a stuffing as follows: brown one large, finely chopped onion in one tablespoon of butter, add chopped opossum liver and heart and cook until tender; add two cups of browned bread crumbs, a chopped bay leaf, a chopped red pepper, a finely chopped hard-boiled egg, a generous dash of Tabasco sauce, salt and pepper to taste, and moisten with three or four tablespoons of water; stuff opossum with the above mixture and sew up opening with needle and heavy thread; roast in oven at 350° until tender and richly brown; skim fat from gravy and serve with baked yams, grits, and salad. Of course, anything garnished with Tabasco sauce is bound to be good—even opossum.

Predators.—Although dogs, and occasionally foxes, Coyotes, Bobcats, and Great Horned Owls include opossums in their diet, the greatest decimating factor to our opossum population is the automobile on our high-

Figure 21   Virginia Opossum *(Didelphis virginiana),* Morehouse Parish, 1968. (Photograph by Joe L. Herring)

ways. One can seldom drive for more than a few miles along a well-traveled highway without seeing one or more dead opossums that were killed the night before. The numbers thus killed are possibly exceeded only by the roadside mortality to the armadillo.

Parasites and diseases. — Professor Harry J. Bennett of Louisiana State University in Baton Rouge has furnished me (pers. comm.) with the following information concerning the parasites of the opossum in Louisiana (see also Lumsden and Zischke, 1961, and Lumsden and Winkler, 1962). They include various species of protozoans, trematodes, cestodes, and nematodes, many of which frequently occur in large numbers in individual hosts. The most common of the trematodes are *Rhopalias macracanthus* and *Proalaria variabilis,* both of which may be present by the hundreds. The cestodes *Mesocestoides latus* and *M. variabilis* are also common. The most prevalent nematodes are *Aspidodera harwoodi, Physaloptera turgida* and *Gnathostoma didelphis.* Acanthocephala have been reported in the opossum but are infrequently found and usually occur in small numbers. The most prevalent acanthocephalan is *Travassosia tumida.* The opossum has been reported infected with tularemia (Mease, 1929).

Subspecies

Didelphis virginiana pigra Bangs

Didelphis virginiana pigra Bangs, Proc. Boston Soc. Nat. Hist., 28, 1898:172; type locality, Oak Lodge, on east peninsula opposite Micco, Brevard County, Florida.
Didelphis virginiana virginiana, Lowery, Occas. Papers Mus. Zool., Louisiana State Univ., 13, 1943:216; name applied to certain Louisiana specimens now treated as *D. v. pigra.*

Although I previously (1943) assigned specimens from Monroe, Ruston, and Tallulah to *D. v. virginiana,* I now regard all Louisiana specimens of this species referable to the race *pigra* in which the ears and toes are generally all-black and the overall coloration is darker. In *virginiana* the tail is shorter, the

ears and toes usually possess some white, and the general coloration is grayer, less blackish. I am following Gardner (1973) in considering *D. virginiana* Kerr specifically distinct from *D. marsupialis* Linnaeus, with which it had been combined by nearly all recent taxonomists. I think that he has demonstrated convincingly that two species are clearly involved and that they overlap in a zone of sympatry in eastern and southern Mexico, Yucatán, and in Guatemala, British Honduras, Honduras, and Nicaragua. Aside from possessing distinctive karyotypes, the two species can be distinguished on conventional morphological grounds, including slight but diagnostic skull differences.

Specimens examined from Louisiana. — Total 324, representing all parishes except Allen, Assumption, Concordia, East Carroll, Franklin, Jackson, Jefferson, La Salle, Madison, Morehouse, Orleans, Red River, Union, and Winn (137 LSUMZ; 81 USL; 18 MSU; 16 NLU; 16 SLU; 16 TU; 13 LTU; 11 USNM; 6 CAS; 5 MCZ; 3 FM; 2 NSU).

[Genus *Marmosa* Gray]

Dental formula:

$$I\frac{5-5}{4-4} \ C\frac{2-2}{2-2} \ P\frac{3-3}{2-2} \ M\frac{3-3}{3-3} = 48$$

[ALSTON'S MOUSE OPOSSUM]   Figure 22
(*Marmosa alstoni*)

Vernacular and other names. — The small opossums of this segment of the family Didelphidae are called Mouse, or Murine, Opossums. They have numerous vernacular names in their native haunts in Central and South America. The generic name *Marmosa* is the Latinized French word *marmose,* which is of undetermined implication in the present connection. The specific name *alstoni* honors Edward R. Alston, the distinguished British zoologist (1845–1881), who contributed the section on Mammalia in the multivolume work *Biologia Centrali-Americana.*

Figure 22 Alston's Mouse Opossum *(Marmosa alstoni).*

Distribution. — Alston's Mouse Opossum is normally found in tropical and subtropical forests from southern British Honduras to central and eastern Nicaragua and Costa Rica. The two records of its occurrence in Louisiana are unquestionably the result of accidental introductions from Latin America. (Map 3)

External appearance. — *Marmosa alstoni* is a small, exceedingly long-tailed opossum with upperparts grayish brown and of woolly texture. The tail is much longer than the head and body. The first inch or slightly more of the tail is densely furred, the remainder scaly and bare. The basal three-fifths of the scaly portion is blackish, the remainder yellowish white. The underparts are grayish, sometimes strongly tinged with buff. The eye is ringed with black.

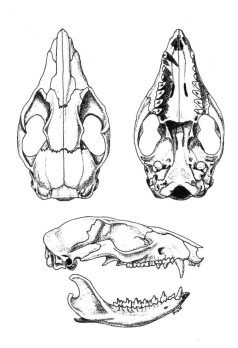

Map 3 Distribution of the Alston's Mouse Opossum *(Marmosa alstoni).* •, locality where introduced into Louisiana.

Figure 23 Skull of Alston's Mouse Opossum *(Marmosa a. alstoni),* ♀, ca. 2 km. NW Santa Ana, Prov. San José, Costa Rica, LSUMZ 12637. Approx. natural size.

Measurements and weights. — The LSUMZ specimen (a skin only) from the state measured as follows: total length, 445; tail, 263; hind foot, 25. Two other adult specimens in the LSUMZ from Costa Rica averaged as follows: total length, 454; tail, 266; hind foot, 31; weight, 113.8 grams (approx. 0.25 lb.).

Cranial characters. — This species differs from other members of the genus by its larger size — greatest length of skull 44 mm or more instead of less than 44 mm. For other features of the skull see Figure 23.

Status and habits. — The inclusion of this species in the present work hinges solely on two specimens captured in or near New Orleans. One is specimen no. 222887 in the United States National Museum caught on 3 January 1917. No other details are available. The second specimen was obtained in an uninhabited expanse of marsh outside of New Orleans on 15 October 1935 by Allan G. Watkins and is now in the LSUMZ. Both were unquestionably introduced from somewhere in Central America, possibly via a banana boat. But such

is the way other species, as the House Mouse and the various forms of the genus *Rattus,* gained access to the United States to become established as an integral part of our mammalian fauna. Consequently, there is potential historical importance in the fact that at least two individuals of the Alston's Mouse Opossum found their way to our shores.

The mammalogist who obtained the 1935 specimen is reported to have set a line of traps in a marsh outside of New Orleans, hoping to obtain specimens of the common native Marsh Rice Rat *(Oryzomys palustris),* and was surprised when he found the Alston's Mouse Opossum in one of his traps. A matter of interest is that it was not caught near one of the wharves or even in the city at all but in a marsh some distance from human habitation. The other specimen, the one taken on 3 January 1917, was probably obtained on a boat at one of the docks in the downtown part of the city, since on the same date a specimen of the Mexican Mouse Opossum, *Marmosa mexicana,* was also taken. I have not included either species on the official state list, and to indicate their equivocal status, I have placed

Figure 24 Mexican Mouse Opossum *(Marmosa mexicana).*

the names in the species headings in brackets (see page 54 for an explanation of the policy employed in this work with regard to such species). But the circumstances surrounding the 1935 specimen of *M. alstoni* seem to me to entitle it to more than casual consideration.

The tail in Alston's Mouse Opossum is strongly prehensile. The species of its genus lack a marsupial pouch. In their native haunts they frequent banana plantations, forests of small trees, and tangles of vines. Their diet is said to include insects and fruits, as well as small rodents, lizards, and bird eggs. In the higher, cooler portions of their range they are alleged to breed one to three times a year but in the tropical regions they give birth to young throughout the year.

Subspecies

Marmosa alstoni nicaraguae Thomas

Marmosa alstoni nicaraguae Thomas, Ann. Mag. Nat. Hist., ser. 7, 16, 1905:313; type locality, Bluefields, sea level, Nicaragua.

The LSUMZ specimen is referable to *nicaraguae*, and Tate (1933) assigned the USNM specimen also to that race.

Specimens examined from Louisiana.—Total 2, as follows: *Orleans Par.:* New Orleans, 2 (1 USNM; 1 LSUMZ).

[MEXICAN MOUSE OPOSSUM]   Figure 24
(Marmosa mexicana)

The only known occurrence of this species in the state is substantiated by a specimen collected in New Orleans on 3 January 1917. It is now number 222888 in the U.S. National Museum. The label on the specimen does not give the name of collector nor any details as to the circumstances under which it was obtained. Since a specimen in the same museum of *Marmosa alstoni* was also procured in New Orleans on the same date, I would presume that both were probably found on a boat at one of the unloading docks, possibly in a cargo of bananas.

Marmosa mexicana differs from *M. alstoni* in

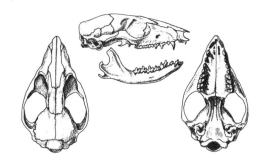

Figure 25 Skull of Mexican Mouse Opossum (*Marmosa m. mexicana*), ♀, New Orleans, Orleans Parish, USNM 222888. Approx. natural size.

being Cinnamon Brown above instead of grayish. It is also decidedly smaller. The greatest length of skull is less than 40 mm instead of over 40 mm (Figure 25). In contrast to the condition in *M. alstoni*, the base of the tail is unfurred and the pelage is not woolly. In the latter species the total length is more

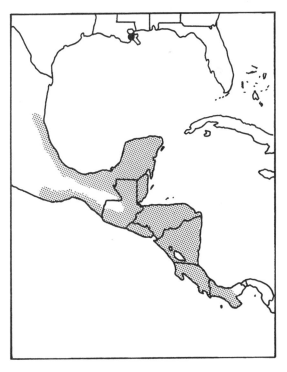

Map 4 Distribution of the Mexican Mouse Opossum (*Marmosa mexicana*). •, locality where introduced into Louisiana.

than 385 mm and the tail over 195 mm, whereas in *M. mexicana* the total length is less than 385 mm and the tail is less than 190 mm.

Subspecies

Marmosa mexicana mexicana Merriam

Marmosa murina mexicana Merriam, Proc. Biol. Soc. Washington, 11, 1897:44; type locality, Jaquila, 1,500 m, Oaxaca.

Marmosa mexicana, Bangs, Bull. Mus. Comp. Zool., 39, 1902:19.

Marmosa mexicana mexicana, Tate, Bull. Amer. Mus. Nat. Hist., 66, 1933:133.

The New Orleans specimen measures as follows: total length, 280; tail, 160; greatest length of skull, 33.8; zygomatic breadth, 19.8. It appears referable to the race *mexicana*, and therefore probably originated from southern Mexico, Guatemala, Honduras, or northern Nicaragua (see Map 4).

Specimens examined from Louisiana.—Total 1, as follows: *Orleans Par.:* New Orleans (USNM).

SHREWS AND MOLES
Order Insectivora

MEMBERS OF this order are among the most primitive of the living mammals of the world, and they are generally regarded as the evolutionary stem line that gave rise to numerous more advanced groups, including the Primates. The tree shrews of the family Tupaiidae of southeastern Asia, Borneo, and the Philippine Islands were once considered members of this order, but are now treated by most mammalogists as primitive Primates.

The name Insectivora, a combination of two Latin words, means insect-eating, as indeed is appropriate for most members of the group. Some, however, are flesh eaters. Most of them are small. The smallest mammal in the world and the smallest mammal in Louisiana are both insectivores. The snout is long and narrow and each foot bears five clawed digits. Neither the thumb nor the big toe is opposable. The eyes are small and in the case of moles lack external openings. The pinna, the cartilaginous projection of the external ear, is greatly reduced or absent, and the short, dense fur consists of only one kind of hair. In other words, it lacks the long guard hairs characteristic of most mammals. In the Insectivores inhabiting Louisiana the testes do not descend into a scrotum. The skull is low and flat, the teeth are primitive, and the cerebral hemispheres are smooth, lack fissures, and do not extend backward over the cerebellum. A thin zygomatic arch is present in the moles but absent in the shrews.

This order occurs throughout the world except for Australia, the southern portion of South America, and Antarctica. The recent members of this group are divided among eight families, some 63 genera, and close to 300 species. Included are the Eurasian and African hedgehogs, the golden moles and elephant shrews of Africa, the solenodons of the West Indies, and the tenrecs of Madagascar, as well as other bizarre forms.

The Shrew Family
Soricidae

THE FAMILY name is derived from the Latin word *sorex,* which was applied to the Common Shrew of Europe by the Romans. The word shrew probably stems from the Anglo-Saxon *screawa,* meaning a wicked or dangerous thing or person, and was applied to this little animal because it was thought, and correctly so, to be poisonous (see account of Short-tailed Shrew).

The family is represented in Louisiana by three species. In all three the first pair of upper incisors are exceptionally large and bicuspid, thus modified as a pair of pincers for grasping food (Figure 26). The next four or five teeth on each side of the upper jaw are unicuspid, the exact number depending on the genus involved. The first two upper molars have their cusps arranged in a W-shaped pattern. The first lower incisor is unicuspid and projects forward and slightly upward. All the teeth have the apices of the cusps strongly pigmented with chestnut, especially so in the genus *Blarina.* The zygomatic arch is lacking. All shrews are plantigrade, that is, they walk on the entire foot with the heel touching the ground. The long snout, which contains the nostrils at the tip and extends beyond the mouth, is flexible and highly sensitive. The external ears barely extend beyond the tips of the fur, and the eyes though extremely small do actually open to the exterior. The middle ear is surrounded by an open ring of bone. The dense fur is not differentiated into underfur overlaid with guard hairs but consists of hairs of short uniform length that can be stroked equally well in any direction.

Our three Louisiana members of the family are terrestrial, but *Sorex palustris* of the northern and western United States, for example, has its hind toes fringed with hairs and is semiaquatic. It has been observed running on water, a feat that it accomplishes by the support it receives from surface tension and the bouyancy of the air trapped in the hairs of its feet.

Genus *Sorex* Linnaeus

Dental formula:

$$I\frac{3-3}{1-1} \ C\frac{1-1}{1-1} \ P\frac{3-3}{1-1} \ M\frac{3-3}{3-3} = 32$$

SOUTHEASTERN SHREW   Plate II
(*Sorex longirostris*)

Vernacular and other names. — Because this species is rare in the state, it has probably been seen by few of our residents and therefore has no local names. If encountered it would likely be called a shrew-mouse or misidentified as a baby mole. The generic name *Sorex* is the Latin word for shrew, while the specific name *longirostris* means long snout or long-nosed and refers, of course, to a distinctive structural feature that is characteristic of all shrews.

Distribution. — There are only nine records of this shrew from the state, all in the Florida Parishes. The species occurs mainly in the southeastern United States·from Maryland,

72

Kentucky, and Illinois south to Florida and southeastern Louisiana. (Map 5)

External appearance. — This species of the genus *Sorex* is distinguished from its congeners by its third unicuspid being smaller than the fourth. The brown color above readily separates it from the Short-tailed Shrew (*Blarina brevicauda*) and the Least Shrew (*Cryptotis parva*), which are decidedly grayish even though sometimes with a brownish tinge. Also, the tail is relatively much longer in this species. The coloration is decidedly paler below than above.

Measurements and weights. — Eight of the nine specimens from Louisiana averaged as follows: total length, 81.8 (78–85); tail, 29.9 (25–33); hind foot, 10.2 (9–12). The ninth specimen, probably an immature individual, measured, respectively, 72, 27, 10. Only two of the eight specimens that are adults show weights, one being 2.2 grams and the other 3.0 grams. In other words, *Sorex longirostris* weighs approximately 0.10 of an ounce or approximately the same as a penny.

Cranial characters. — The diminutive skull of the species, when compared with that of its only rival, *Cryptotis parva*, clearly indicates that it is indeed the smallest mammal in the state. The skulls of five of the nine specimens known from Louisiana measured as follows:

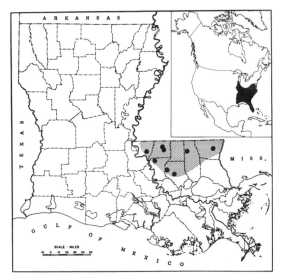

Map 5 Distribution of the Southeastern Shrew (*Sorex longirostris*). *Shaded area,* known or presumed range within the state; •, localities from which museum specimens have been examined; *inset,* overall range of the species.

greatest length, 14.6 (14.1–15.3); cranial breadth, 7.2 (7.0–7.3); interorbital breadth, 3.2 (2.9–3.9); palatilar length, 5.8 (5.0–6.1); postpalatal length, 6.7 (6.5–7.2); anteorbital breadth, 4.0 (4.0–4.1). The fact that the third unicuspid tooth of the upper jaw is smaller than the fourth is diagnostic. (Figure 27)

Status and habits. — Credit for the discovery of this shrew in Louisiana goes to the distinguished biologist, Mrs. Frances Hamerstrom. She obtained the first Louisiana specimen during her brief residence in the state when she and her equally distinguished biologist husband were stationed at Baton Rouge during World War II. The shrew was trapped a few miles north of the city near Plains. Subsequently I saturated the wood with dozens of traps but failed to obtain another specimen. As a matter of fact, I set out a score of traps ringing the tree at the base of which Mrs. Hamerstrom caught her shrew, all to no avail. My futile efforts, though, are not without parallel. The first record of the species in

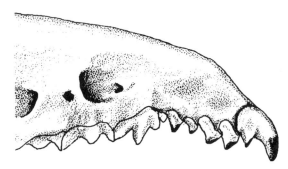

Figure 26 Rostrum of Southeastern Shrew (*Sorex l. longirostris*), sex ?, ca. 1 mi. SE Clinton, East Feliciana Parish, LSUMZ 11848. Approx. 13 ×.

Alabama was a specimen found by Arthur H. Howell in the stomach of a Barred Owl *(Strix varia)* obtained on the border of Bear Swamp, near Autaugaville. He later set numerous traps for several days in the vicinity of the place where the owl was shot but never succeeded in securing there another *Sorex longirostris* (Howell, 1921).

Three of the eight specimens obtained in the nearly 30 years since Mrs. Hamerstrom's initial discovery have been caught under somewhat fortuitous circumstances. LSU Forestry students captured two alive on separate occasions, when carrying on controlled experimental burning of pine habitats in Livingston and East Feliciana parishes. Another was found by Gayle Strickland when he turned over a large piece of cardboard near a refuse dump three miles west of Fluker in Tangipahoa Parish. Five were obtained by LSU mammalogy students by conventional trapping procedures, one each in East Feliciana, Livingston, and Washington parishes and two in West Feliciana Parish.

The apparent rarity of this species reflects, I am sure, nothing more than its elusive habits and the difficulties associated with trapping it. Shrews are sometimes caught in traps baited with peanut butter and rolled oats or with crushed pecans combined with pieces of liver. Also, cans set into the earth with the open end level with the forest floor serve as pitfalls, and are often a rewarding means of capturing shrews, including *Sorex longirostris* (Tuttle, 1964).

Little is known about the life history and habits of *Sorex longirostris*. The specimen obtained by Strickland from beneath a piece of cardboard at a refuse dump in a wood near Fluker was one of several that scurried in every direction when he lifted the cover. He said (pers. comm.) that they came from a fairly bulky nest composed of fine grasses. Negus and Dundee (1965) report finding a nest of this species in the Homochitto National Forest in Mississippi, not far from the Louisiana–Mississippi state line. It was located inside a rotten log that lay across a small spring-fed creek in a shaded ravine. When the log was broken open an adult *Sorex* escaped into the ground cover, but four half-grown young were captured within the nest. The same workers found a second nest containing two young at the same locality in a fallen pine tree on a well-drained hillside. The nest consisted of a mass of cut leaves located in a cavity just beneath the bark on the upper surface of the log.

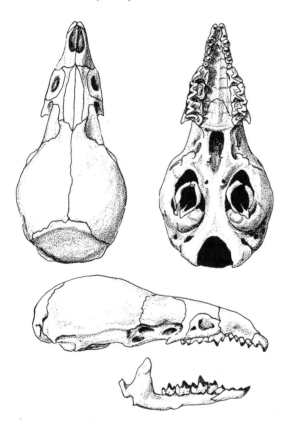

Figure 27 Skull of Southeastern Shrew *(Sorex longirostris)*, sex ?, ca. 1 mi. SE Clinton, East Feliciana Parish, Louisiana, LSUMZ 11848. Approx. 4½ ×.

Subspecies

Sorex longirostris longirostris Bachman

Sorex longirostris Bachman, Jour. Acad. Nat. Sci. Philadelphia, ser. 1, 7, 1837:370; type locality, Hume Plantation, swamps of the Santee River [=Cat Island, mouth of the Santee River], South Carolina.

The Louisiana specimens are referred provisionally to the nominate race. They are really too few to permit a judgment regarding any geographical differences that may exist between the South Carolina population and the one here at the western edge of the species' range.

Specimens examined from Louisiana.—Total 9 (all LSUMZ), as follows: *East Baton Rouge Par.:* Plains, 1. *East Feliciana Par.:* 1 mi. SE Clinton, 1; 5 mi. SE Clinton, 1. *Livingston Par.:* ca. 5 mi. SE Walker (Ward 7), 1; 3 mi. NNE Denham Springs, 1. *Tangipahoa Par.:* 3 mi. W Fluker, 1. *Washington Par.:* 10 mi. E Franklinton, 1. *West Feliciana Par.:* 5 mi. NE St. Francisville, 2.

Genus *Blarina* Gray

Dental formula:

$$I\frac{3-3}{1-1} \ C\frac{1-1}{1-1} \ P\frac{3-3}{1-1} \ M\frac{3-3}{3-3} = 32$$

SHORT-TAILED SHREW   Plate II
(Blarina brevicauda)

Vernacular and other names.—This species is sometimes called the mole shrew or shrewmouse. The basis of the generic name *Blarina* is not known, for Gray, the zoologist who coined the name, did not explain its etymology. The word *brevicauda* is, of course, the Latin for short tail.

Distribution.—The Short-tailed Shrew occurs throughout the wooded portions of the state except in the coastal tier of parishes, where it appears to be absent. The species as a whole ranges from southeastern Saskatchewan across southern Canada to Nova Scotia and Newfoundland and south to extreme eastern Colorado, eastern Oklahoma, and southeastern Texas to southern Florida. (Map 6)

External appearance.—This small molelike mammal is dark slate gray above and below. The underparts are only very slightly if at all lighter colored than the upperparts. The lack of contrast readily distinguishes this species

from the Least Shrew, in which the underparts are decidedly paler. Sometimes the pelage has a slight brownish tinge. The velvety fur is of nearly equal length throughout. The short tail, long snout, tiny eyes, and ears concealed in the hair readily distinguish it from any mouse. Winter pelage is slightly darker than that of summer, which is browner, less slaty gray.

Color phases.—No true color phases occur in the Short-tailed Shrew, and albinos are extremely rare (Hamilton, 1939; Ulmer, 1940; Jackson, 1961).

Measurements and weights.—Seventy-three adults in the LSUMZ from various localities in the state averaged as follows: total length, 85.0 (72.0–95.0); tail, 17.8 (13.0–22.6); hind foot, 12.0 (10.0–14.0). The sixteen adults with data on weight averaged 8.9 (6.5–10.5) grams. Cranial measurements of 75 adults in the LSUMZ from various localities in the state average as follows: greatest length, 18.4

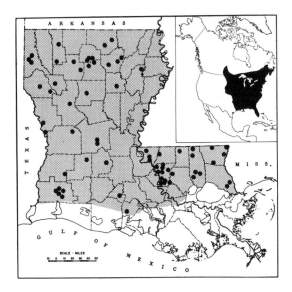

Map 6 Distribution of the Short-tailed Shrew (Blarina brevicauda). *Shaded area,* known or presumed range within the state; •, localities from which museum specimens have been examined; *inset,* overall range of the species.

(16.9–19.2); cranial breadth, 9.7 (9.1–10.6); interorbital breadth, 5.1 (4.7–5.8); palatilar length, 7.6 (7.1–8.7); postpalatal length, 8.2 (7.4–8.8); greatest anteorbital breadth, 6.4 (5.5–6.9).

Cranial characters. — With respect to the skull, this species is readily distinguished from *Sorex longirostris* and *Cryptotis parva* by its large size, especially that of the braincase, by its more bulky rostrum, and by the more extensive reddish brown or chestnut tipping to the crowns of its teeth. Both *Sorex longirostris* and *Blarina brevicauda* have 32 teeth, but *Cryptotis parva* has only 30, since it lacks a premolar on each side of the upper jaw. (Figure 28)

Sexual characters and age criteria. — Males are only a trifle, if any larger than females. Since in males the testes do not descend into a scrotum, Short-tailed Shrews are difficult to

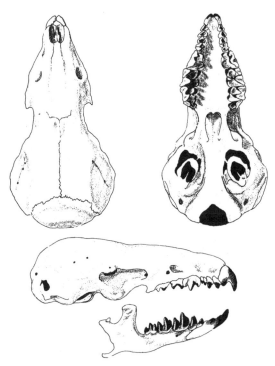

Figure 28 Skull of Short-tailed Shrew *(Blarina brevicauda minima)*, ♀, 3 mi. S, 0.5 mi. E University, East Baton Rouge Parish, LSUMZ 15164. Approx. 3×.

sex externally. Males and females differ little if at all in color. Two molts take place, one in the fall, the other in spring. Glands on the flanks are well developed in males but small and indistinct in females. Their secretions probably serve in the role of sex recognition. The broad head of old adults reflects the more massive skull. Few individuals probably survive past their first year and those that live a year and a half have probably attained a ripe old age.

Status and habits. — The Short-tailed Shrew is an abundant species in the state. It is probably more numerous than the Least Shrew because of its somewhat greater diversity of habitat. Although it is most common in wooded areas, it is also occasionally found in brushy thickets adjacent to forests.

Ounce for ounce this denizen of our woodlands is one of the most pugnacious of all mammals (Figure 29). It will attack and kill adversaries much larger and seemingly more powerful than itself, and it has no compunction against cannibalism. Its appetite is so ravenous that it has been alleged to consume its own weight in food every twenty-four hours. W. J. Hamilton, Jr. (1930) asserts that half that much food is sufficient, but even this amount is prodigious. Not only does it have an insatiable lust for food, but it can digest and assimilate it rapidly. All shrews are high-strung and of nervous temperament. They often die from sheer fright when handled or subjected to other stress situations.

President Theodore Roosevelt, who was an accomplished naturalist, in 1888 sent his observations on the food habits of *Blarina* to Dr. C. Hart Merriam, then head of the Division of Economic Ornithology and Mammalogy, the governmental agency that eventually became the U.S. Fish and Wildlife Service. Roosevelt's notes *(fide* Jackson, 1961) are worthy of quotation in full.

Of course its food consists mainly of insects, such as ants, ground beetles, caterpillars, or earthworms, and of these it consumes enormous quantities, for it has a most insatiable appetite. But any

Figure 29 Agonistic posture of Short-tailed Shrew *(Blarina brevicauda).* (After Olsen, 1969)

small or weak vertebrate will be taken quite as readily. Once, while walking through the woods in the late evening, my attention was attracted by the cries of a small pair of Ovenbirds ... and on approaching I found them hovering about their domed nest in great distress. As I stopped to look into it a small animal jumped out; I caught it at once, and it proved to be a Mole Shrew [=Blarina *brevicauda*], while inside the nest the torn and bloody remains of three poor little nestlings explained only too clearly the alarm of the parent birds.

I once kept one of these stout little Shrews for some weeks in a wire cage, and certainly a more bloodthirsty animal of its size I never saw. In the day it seemed rather confused by the glare, and spent most of its time in a small heap of dried grass, but with the approach of dark it became restless, running round and round the cage, and occasionally uttering a low, fine squeak. Its eyes did not seem to be very good, but its sense of smell was evidently excellent, and it seemed to rely mainly on its nose in discovering its prey. For the first few days I fed it only on insects, which it devoured greedily, holding them down firmly with its fore paws. Then one day I placed a nearly full grown Pine Mouse in the cage. The Shrew smelt it at once and rushed out. The mouse was hopping about, raising itself on its hind legs and examining the walls of its prison, when it was suddenly seized from behind by the bloodthirsty little Shrew. It squeaked piteously, and tried to use its long incisors on its antagonist; indeed I think it did inflict one bite, but in spite of this, and of its struggling and scratching, it was very soon killed and partially eaten. Two days afterward I introduced a young garter-snake, about seven inches long, into the cage. The little snake at first moved slowly about, and then coiled itself up on a piece of flannel. The Shrew had come out from its nest, but did not seem to see the snake, and returned to it. Soon

afterward it came out again and quartered across the cage. While doing so it evidently struck the scent (the snake all the time was in plain sight), raised its nose, turned sharply around and ran rapidly up to the flannel. It did not attack at once, as with the mouse, but cautiously smelt its foe, while the little snake moved its head uneasily and hissed slightly, then with a jump the Shrew seized it low down, quite near the tail. The snake at once twisted itself right across the Shrew's head and under one paw, upsetting him; but he recovered himself at once, and before the snake could escape flung himself on it again, and this time seized it by the back of the neck, placing one paw against the head and the other on the neck, and pushing the body from him while he tore it with his sharp teeth. The snake writhed and twisted, but it was of no use, for his neck was very soon more than half eaten through, and during the next twenty-four hours he was entirely devoured.

Merriam had himself earlier (1884) made observations on the Short-tailed Shrew as a mouser that, among other things, demonstrated the rapidity with which it assimilates its food.

... having caught a vigorous, though undersized Shrew, I put him in a large wooden box and provided him with an ample supply of beechnuts, which he ate eagerly. He was also furnished with a saucer of water, from which he frequently drank. After he had remained two days in these quarters, I placed in the box with him an uninjured and very active white-footed mouse. The Shrew at the time weighed 11.20 grammes, while the mouse, which was a large adult male, weighed just 17 grammes. No sooner did the Shrew become aware of the presence of the mouse than he gave chase. The mouse, though much larger than the Shrew, showed no disposition to fight, and his superior agility enabled him, for a long time, easily to evade his pursuer, for at a single leap he would pass over the latter's head and to a considerable distance beyond. The Shrew labored at great disadvantage, not only from his inability to keep pace with the mouse, but also, and to a still greater extent, from his defective eyesight. He frequently passed within two inches (31 mm) of the mouse without knowing of his whereabouts. But he was persistent, and explored over and over again every part of the box, constantly putting the mouse to flight. Indeed, it was by sheer perseverance that he so harassed the mouse, that the latter, fatigued by almost continuous exertion, and also probably weakened by fright, was no longer able to escape. He was first caught by the tail; this proved a temporary stimulant, and he bounded several times across the box, dragging his adversary after him. The Shrew did not seem in the least disconcerted at being thus harshly jerked about his domicil, but continued the pursuit with great determination. He next seized the mouse in its side, which resulted in a rough and tumble, the two rolling over and over and biting each other with much energy. The mouse freed himself, but was so exhausted that the Shrew had no difficulty in keeping alongside, and soon had him by the ear. The mouse rolled and kicked and scratched and bit, but to no avail. The Shrew was evidently much pleased and forthwith began to devour the ear. When he had it about half eaten-off the mouse again tore himself free; but his inveterate little foe did not suffer him to escape. This time the Shrew clamored up over his back and was soon at work consuming the remainder of the ear. This being satisfactorily accomplished, he continued to push on in the same direction till he had cut through the skull and eaten the brains, together with the whole side of the head and part of the shoulder. This completed his first meal, which occupied not quite fifteen minutes after the death of the mouse. As soon as he had finished eating I again placed him on the scales and found that he weighed exactly 12 grammes—an increase of .80 gramme.

When trapping for small mammals in Louisiana one frequently finds in the trapline the remains of a Hispid Cotton Rat (*Sigmodon hispidus*) or a Marsh Rice Rat (*Oryzomys palustris*), as well as other rodents and even shrews that were in all probability devoured either by a Short-tailed or a Least Shrew after being caught.

Shrews eat a variety of animals and even ingest some plant materials. No detailed studies have been undertaken with regard to their food habits in Louisiana, but Hamilton (1930) found on examination of the stomachs of 244 shrews that insects made up nearly one-half of the food, while next in abundance were vegetable matter, annelids, crustacea, mollusks, vertebrates, centipedes, inorganic matter, arachnids, and millipedes in the order named.

The Short-tailed Shrew is active the year around. Not much is known of its breeding habits in the state, but a considerable amount of information has been published concerning it from elsewhere in eastern North

America. The gestation period is 21 to 22 days (Hamilton, 1929). The nest is usually placed under a fallen tree or rotten stump and may be 12 to 15 inches below the surface with several tunnels leading to it. The average number of young appears to be 6 or 7, but 10 have been found. The male possesses a gland on the side of the body, located near the forelimbs, that becomes prominent during the breeding season. It is present in the female but is less noticeable. The secretion of the gland is oily and is the source of the pungent, musty odor given off by these animals, particularly by the male. The secretion probably plays a role in species recognition and as part of the shrew's sex life. It doubtless rubs off on the sides of the burrows and on the pathways followed and thereby leaves a scent trail for other individuals of the species. According to Gould (1969) the male exudes great amounts of the odoriferus substance during courtship just prior to copulation.

The toxic effects of the bite of shrews have long been known in Europe. A 17th century naturalist by the name of Reverend Topsell had this to say in his book entitled *History of the Four-footed Beasts and Serpents,* published in London in 1658: "It is a ravening beast, feigning itself gentle and tame, but, being touched, it biteth deep, and poysoneth deadly." In this country an account by C. J. Maynard (1889) describes the effects of bites that a shrew inflicted on his hands when he picked it up.

All this occupied perhaps thirty seconds, when I began to experience a burning sensation in the first two bites, followed by a peculiar sensation . . . in the right hand. I walked to the house, only a few hundred yards away, but by this time, the pain which had been rapidly increasing, had become quite severe, and by the time I had placed the shrew in an improvised cage, I was suffering acutely.

The burning sensation, first observed, predominated in the immediate vicinity of the wounds, but was now greatly intensified, accompanied by shooting pains, radiating in all directions from the punctures but more especially running along the arm, and in half an hour, they had reached as high as the elbow. All this time, the parts in the immediate vicinity of the wounds, were swelling, and around the punctures the flesh had become whitish.

I bathed the wounds in alcohol and in a kind of liniment, but with little effect. The pain and swelling reached its maximum development in about an hour, but I could not use my left hand without suffering great pain for three days, nor did the swelling abate much before that time. At its greatest development, the swelling on the left hand caused that member to be nearly twice its ordinary thickness at the wound, but appeared to be confined to the immediate vicinity of the bites, and was not as prominent on the right hand; in fact, the first wound was by far the most severe.

The burning sensation disappeared that night, but the shooting pains were felt, with less and less severity, upon exertion of the hand, from the elbow downward, for a week, and did not entirely disappear until the total abatement of the swelling, which occurred in about a fortnight.

It remained, however, for Dr. Oliver P. Pearson (1942, 1950) to work out the source and experimental effects of the poison on various animals. Among his important conclusions were the following:

1. A poisonous extract prepared from the submaxillary salivary glands of the shrew, *Blarina brevicauda,* contains toxic material also found in the saliva, which may be introduced into wounds made by the teeth.
2. The extract causes a local reaction, the lowering of the blood pressure, slowing of heart beat, and inhibition of respiration.
3. A lethal dose for a 20-gram mouse is provided by the extract of 5.7 mg of fresh submaxillary tissue when injected intraperitoneally, or of 0.4 mg by intravenous injection. The lethal dose for rabbits injected intravenously is extract of approximately 7 mg per kilogram of body weight.
4. The toxic material is soluble in water and salt solution, but not in acetone. Heating to 100° C in neutral or acid solution does not affect toxicity. Similar heating in basic solution destroys toxicity.
5. Extracts of the parotid salivary glands, unlike those of the submaxillary salivary glands, produce no serious effects on mice, cats, or rabbits, even when injected in large doses.

When attacking frogs, crickets, and grasshoppers, shrews have been observed to bite the joints of the largest legs of their victim until they have succeeded in "hamstringing" it and thereby rendering it helpless (Hatt, 1938). The poison may paralyze the legs of the shrew's prey, thereby making it easier to subdue.

Short-tailed Shrews in captivity emit a series of clicks and twitters that, according to Gould et al. (1964), may serve as part of a system of echolocation. In a series of experiments with shrews of the genus *Sorex,* the researchers demonstrated that under the test conditions performance was greatly impaired in subjects in which the ears had been plugged. Similar but less conclusive results were obtained with regard to *Blarina,* but they were convinced that the sounds emitted were enabling the shrews to ascertain in darkness the position of objects and other space relationships.

Predators.—Except for certain birds, notably owls, few animals feed on shrews. Cats, and sometimes dogs, capture them but seldom eat them. As a matter of fact, the chief way by which these little animals become known to most people is by their cats bringing them in and depositing them on doorsteps.

Parasites.—Nothing is known about the parasites of the Short-tailed Shrew in Louisiana, but Jameson (1950) has made a study of external parasites found on 405 specimens that he examined from Ontario and New York. He found 14 species of fleas, of which 10 are more or less restricted to *Blarina.* Internal parasites include an acanthocephalan in the mesenteries, trematodes in the stomach and intestines, cestodes in the small intestines, and quantities of nematodes in the lungs, esophagus, stomach, duodenum, and small intestine (Oswald, 1958; Jackson, 1961).

Subspecies

Blarina brevicauda minima Lowery

Blarina brevicauda minima Lowery, Occas. Papers Mus. Zool., Louisiana State Univ., 13, 1943:218; type locality, Comite River, 13 mi. NE Baton Rouge, East Baton Rouge Parish, Louisiana.

Through its wide range in eastern North America the Short-tailed Shrew is subject to great geographical variation in size, cranial characters, and the shade of its grayish coat color. Specimens in Louisiana, southeastern Texas, and western Mississippi are smaller than those found elsewhere in the southeastern United States, and they are decidedly smaller than representatives of any northern population. Figure 30 shows a series of adults from Louisiana in comparison with a series from West Virginia (*B. b. kirtlandi*). The geographical race that occurs throughout Louisiana has been designated *B. b. minima,* a name that calls attention to the population's diminutive body size.

Genoways and Choate (1972) have shown that the large and the small *Blarina* come together in southeastern Nebraska with little or no evidence of intergradation. Of 66 specimens taken in or near the zone of overlap between the two types of shrews, only one proved to be a probable hybrid. The large and small individuals appear therefore to be acting as if they represent two species, rather than subspecies, with isolating mechanisms actively preventing, or at least restricting, hybridization, thereby resulting in negligible introgression. Genoways and Choate discuss other possibilities that could account for the situation in Nebraska, such as the large and small forms being in that area the ends of a circle of races. Should further studies show that the large northern and the small southern *Blarina* occur together in sympatry at other places where their ranges meet, the two would unequivocally require treatment as two separate species. The *Blarina* in Louisiana, which are extremely small, would then become *Blarina carolinensis minima.* The various races of the large northern form would go under the name *Blarina brevicauda.*

Specimens examined from Louisiana.—Total 201, as follows: *Allen Par.:* 0.25 mi. NW Oakdale, 1 (USL). *Beauregard Par.:* 1.5 mi. E Edith, 2 (LSUMZ). *Bienville Par.:* Bryceland, 1 (LSUMZ); 6 mi. N Mt. Olive, 2 (LSUMZ). *Caddo Par.:* 7 mi. WSW Shreveport, 1 (LSUMZ); 3 mi. S, 2 mi. W Blanchard, 3 (LSUMZ); 8 mi. NW Shreveport, 1 (LTU); 7 mi. NW Shreveport, 1 (LTU). *Calcasieu Par.:* Lake Charles, 1 (LSUMZ); 7 mi. W Lake Charles, 2 (LSUMZ); Sam Houston, 1 (LSUMZ); 2 mi. W Sulphur, 1 (LSUMZ); Moss Bluff, 2 (MSU). *Caldwell Par.:* Columbia, 14 (FM). *De Soto Par.:* Mansfield, 1 (USNM); 6.5 mi. N, 4.5 mi. E Pelican, 1 (LSUMZ). *East Baton Rouge Par.:* various localities, 75 (73 LSUMZ; 2 USNM). *East Feliciana Par.:* 3 mi. S Jackson, 2 (LSUMZ). *Evangeline Par.:* 4 mi, NNW Ville Platte, 1 (LSUMZ). *Franklin Par.:* 9 mi. S Delhi, 2 (NLU). *Iberia Par.:* Avery Island, 1 (CAS). *Jackson Par.:* 6 mi. SE Ruston, 1 (LTU); Clay, 5 (LTU). *Lafayette Par.:* 6.3 mi. S Lafayette city limits, 2 (USL); 2 mi. S Lafayette city limits, 1 (USL). *Lincoln Par.:* Ruston, 8 (1 LSUMZ; 7 LTU); 1 mi. S Ruston, 6 (LTU); 2 mi. W Ruston, 1 (LTU); 6 mi. SW Ruston, 1 (LTU). *Livingston Par.:* 3 mi. NNE Denham Springs, 2 (LSUMZ). *Natchitoches Par.:* Natchitoches, 9 (6 NSU; 3 USNM). *Ouachita Par.:* Monroe, 5 (3 NLU; 2 LSUMZ); 3 mi. N Monroe, 1 (LSUMZ); Swartz, 2 (NLU); Cheniere Lake, 1 (NLU). *Rapides Par.:* 18 mi. S Alexandria, 1 (USL); 15 mi. S Alexandria, 1 (LTU). *Red River Par.:* 5 mi. N Coushatta, 1 (LSUMZ). *Sabine Par.:* Bayou Negreet at Sabine River, 1 (LSUMZ); 7.5 mi. SW Toro, 1 (TU). *St. Helena Par.:* 5.5 mi. NNW Chipola, 1 (LSUMZ). *St. Tammany Par.:* 1 mi. E Pearl River, 1 (SLU); 1 mi. N Slidell, 1 (TU). *Tangipahoa Par.:* 4 mi. NE Kentwood, 1 (LSUMZ); 6 mi. N Robert, 1 (LSUMZ); Hammond, 2 (LSUMZ); 3 mi. W Fluker, 1 (LSUMZ); 3 mi. W Ponchatoula, 2 (SLU); 2 mi. W Ponchatoula, 2 (SLU); 3.25 mi. E Hammond, 1 (SLU). *Union Par.:* 2 mi. SE Farmerville, 1 (LTU); Haile, 3 (NLU). *Vernon Par.:* 0.75 mi. SE Simpson, 2 (LSUMZ). *Washington Par.:* 2 mi. S Enon, 3 (LSUMZ); 2 mi. N Angie, 1 (LSUMZ); 2 mi. NE Varnado, 1 (SLU); Hackley, 1 (FM). *Webster Par.:* Evergreen, 1 (LTU); 1 mi. SE Lake Bistaneau State Park, 2 (LSUS). *West Feliciana Par.:* Bains, 2 (LSUMZ); ca. 2.5 mi. N. 1 mi. E Weyanoke, 3 (LSUMZ); St. Francisville, 1 (LSUMZ). *Winn Par.:* 2 mi. W Winnfield, 1 (NSU).

Genus *Cryptotis* Pomel

Dental formula:

$$I\frac{3-3}{1-1}\ C\frac{1-1}{1-1}\ P\frac{2-2}{1-1}\ M\frac{3-3}{3-3}=30$$

LEAST SHREW   Plate II
(*Cryptotis parva*)

Vernacular and other names.—Local names of this species include little shrew, little short-tailed shrew, little mole-shrew, small shrew, bee shrew, cryptotis, and mole-mouse. The generic name *Cryptotis* comes from the Latin

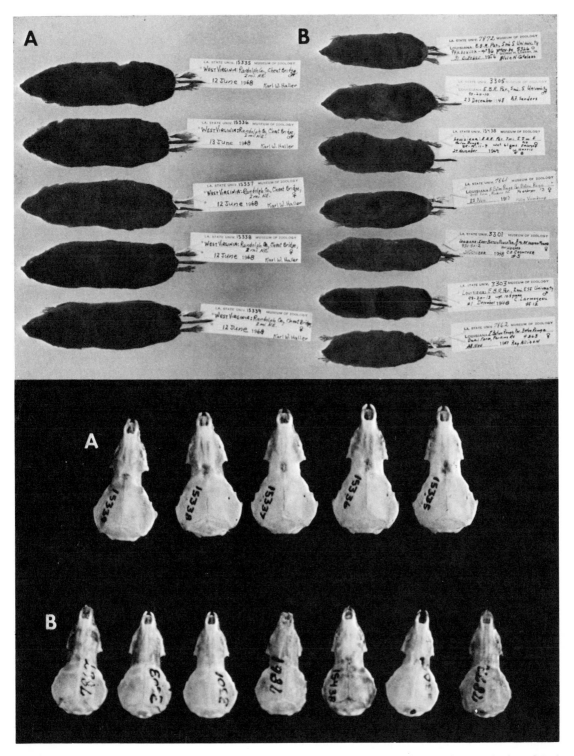

Figure 30 Geographical variation in size of Short-tailed Shrews *(Blarina brevicauda)*. (A) Series of *B. b. minima* from East Baton Rouge Parish, Louisiana; (B) series of *B. b. kirtlandi* from Randolph County, West Virginia. (Photograph by Joe L. Herring)

and translates as hidden ear. The specific name *parva* is the Latin for small.

Distribution. — The Least Shrew is widespread throughout most of Louisiana wherever there are grassy fields or thickets along the edges of woodlands. It probably occurs in every parish, but as yet no specimens have been taken in a considerable number of parishes. The species ranges from South Dakota across southern Minnesota, Wisconsin, Michigan, Ontario, and New York to Connecticut and from there south to northeastern Colorado, northern, eastern, and southern Texas, northeastern Mexico, Louisiana, Mississippi, Alabama, and southern Florida. (Map 7)

External appearance. — This species is the second smallest mammal in the state. It can be confused only with the Short-tailed Shrew, from which it is readily distinguished by its much smaller size, by the generally browner (less slaty gray) coloration of its upperparts, and by the fact that it is decidedly paler below

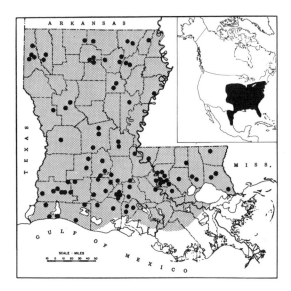

Map 7 Distribution of the Least Shrew (*Cryptotis parva*). *Shaded area,* known or presumed range within the state; •, localities from which museum specimens have been examined; *inset,* overall range of the species.

instead of nearly uniform gray both above and below. Not only is the body of the animal smaller, but the tail is shorter and the feet smaller. The color above is Deep Plumbeous to Mouse Gray and the underparts vary from Pale Olive-Gray to Light Olive-Gray sometimes with a decidedly brownish tinge. The tail is almost imperceptibly bicolored. The ear openings are large but there is no pinna extending above the fur. The eyes are extremely minute, black, and beady. The snout, as in other shrews in the state, is long and pointed. The fur is fine and rather short but not nearly as velvety as that of the Short-tailed Shrew.

Color phases. — Some specimens, apparently without regard to sex, age, season, or locality, are much grayer than average, both above and below, than is usually the case. Whether this color type actually represents a color phase or is only one extreme in a wide range of individual variation is not known. Many of the older museum specimens appear to have undergone postmortem color change to the extent of being somewhat browner than fresh material.

Measurements and weights. — One hundred and thirty adults in the LSUMZ from localities throughout the state (exclusive of Cameron, Calcasieu, and Jefferson Davis parishes), but mainly from East Baton Rouge Parish, averaged as follows: total length, 70.6 (59–90); tail, 15.1 (13–20); hind foot, 9.6 (7–12). The weight of 26 of these averaged 4.5 (3.3–6.5) grams. Cranial measurements were as follows: greatest length, 15.2 (14.2–16.7); anteorbital breadth, 4.9 (4.2–5.6); interorbital breadth, 3.6 (3.2–4.9); cranial breadth, 7.4 (6.7–8.1); palatilar length, 5.9 (5.1–6.9); postpalatal length, 7.2 (6.4–7.9).

Cranial characters. — The cranial character that immediately sets *Cryptotis* apart from both *Sorex* and *Blarina* is the fact that it possesses only 30 instead of 32 teeth. It lacks one of the

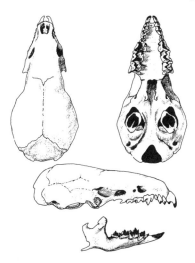

Figure 31 Skull of Least Shrew *(Cryptotis p. parva)*, ♂, 6 mi. S Baton Rouge, East Baton Rouge Parish, LSUMZ 13407. Approx. 3 ×.

premolars on each side of the upper jaw. Consequently, instead of having five unicuspid teeth, it has only four. The minute last unicuspid is smaller than the third, is wedged between the third unicuspid and the first molariform tooth, and is visible only from the occlusal or ventral view. (Figure 31)

Sexual characters and age criteria.—Specimens of this species, like other shrews, are difficult to sex externally because the testes do not descend into a scrotum. In color the sexes are alike. There are two molts a year, one in spring, the other in early fall. Molting usually starts on the face and proceeds backward, but this sequence may vary considerably. The young are blind and naked at birth.

Status and habits.—The Least Shrew is one of the commonest mammals in the state. It abounds in grassy fields, stands of broomsedge *(Andropogon)*, railway rights-of-way, hedgerows, and briar thickets and other tangles of vegetation adjacent to wooded areas. It is also one of our state's smallest mammals (Figure 32). A Least Shrew has a body not much, if any, larger than some bumblebees,

and weighs approximately 4.5 (3.3–6.5) grams or only 0.12 to 0.23 of an ounce. The Southeastern Shrew *(Sorex longirostris)* weighs slightly less, but its slimmer body and longer tail make it appear somewhat larger. In any event, the Least Shrew surely qualifies as one of our two most diminutive mammals, each weighing considerably less than a nickle. The largest terrestrial mammal, the elephant, weighs up to 14,500 pounds, and the largest of all mammals, the Blue Whale, weighs up to 236,000 pounds. Consequently, 1,461,535 Least Shrews would be required to equal one elephant, and 23,765,528 to equal one Blue Whale. Such figures serve to illustrate the wide range of variation in size and weight among mammals as a group.

This tiny midget is almost never seen. Consequently most of what is known about its behavior and habits is based on observations of it in captivity. Like other shrews it is extremely active and high strung, but it does not appear to be as prone to collapse from nervous shock as some of its relatives. The muzzle is kept constantly twitching from one side to the other as the little animal searches for its adversary or prey. The small eyes seem to be of little use, as is also its sense of smell. Its hearing of low frequency sounds is likewise not especially acute. It cannot detect an insect at a distance of more than a few inches, and it pays little or no attention to the clanging of a bell or clapping of the hands as long

Figure 32 Least Shrew *(Cryptotis parva)*, Owensville, Clermont County, Ohio, 1 October 1956. (Photograph by Woodrow Goodpaster)

as no air movements reach it. Perhaps it can hear supersonic sounds beyond the range of the human ear. Despite these apparent deficiencies it possesses highly sensitive tactile organs, such as the vibrissae and the hairs of the head and back, that enable the animal to detect objects readily. It is itself noisy, uttering a high-pitched birdlike twitter or chatter, especially when in the presence of others of its kind.

The Least Shrew occurs in great numbers in grassy situations, but we know this mainly through indirect evidence rather than direct observation. A line of traps set in the proper habitat almost always results in a few being captured. Traps set for shrews are rather difficult to bait because the little insectivores feed so largely on animal matter. Also they are so small that when they sit on the trap near the treadle feeding on the bait the striking bar is likely to clear the intended victim when the trap is sprung. The most efficient "collectors" of shrews are, however, not mammalogists but the various species of owls whose keen eyesight enables them to see these little insectivores even at night. I recently examined approximately 125 Barn Owl pellets from Gum Cove in Cameron Parish and found therein no fewer than 49 skulls of *Cryptotis parva*. When an owl eats a small mammal it swallows it whole and in the stomach the hair and bones are compressed into a pellet that is then regurgitated. Since all mammal skulls possess diagnostic characteristics that distinguish them as to species, the identification of the owl's prey is comparatively easy. To catch 49 Least Shrews in the vicinity of the Barn Owl roost at Gum Cove a mammalogist would probably have to set more than 100 traps a night over a period of several weeks.

The reproductive behavior of the Least Shrew is poorly known. Little or no information on the animal based on observations in Louisiana is available. The breeding season in the northern part of the range extends from March until November, but at our latitude the species is probably capable of producing young almost any month of the year. The gestation period is stated by Conaway (1958) to be 21 to 23 days, but Walker (*in* Walker et al., 1968) asserts that it is 15 days or less. The number of young ranges from two to six, but five appears to be the usual number. At birth the young are naked with the eyes and ears closed and with no teeth visible. The toes are separate and equipped with distinct claws. Within a week baby Least Shrews are furred above, and by the end of their third week, when weaning occurs, they look exactly like adults except for their slightly smaller size. The male helps the female in the care of the young, and both are highly solicitous of the welfare of their offspring. To the female, though, goes the main credit when a brood is successfully reared. Hamilton (1944) tells of a parent shrew that he had under observation, which, weighing but 5.4 grams (0.19 ounce), supplied sufficient milk to nourish and develop five young whose combined weight just prior to weaning exceeded that of their mother by slightly more than three times.

The Least Shrew commonly uses the runways of Hispid Cotton Rats, with whom it shares its grassy habitats. The nests are often in subterranean excavations that lie some 8 inches below the surface and are no more than an inch in diameter. A tunnel described by Davis and Joeris (1945) in clay loam soil in a Bermuda grass pasture near College Station, Texas, was at least five feet long and 8 inches below the surface at the deepest point excavated. It had a short side branch that led to a rather large nest chamber. The nest itself is a globular mass of dry grass lined with finely shredded material. McCarley (1959) describes an unusually large, communal nest that he found beneath a log just east of Nacogdoches, Texas, not far from our border. When the dead tree was gently rolled over, the nest was revealed in the center of the area where the log had lain. It was about 2 inches deep and 4 × 6 inches in lateral measurements, was composed of leaves and grass, and was more or less closed over at the top. When the nest was gently prodded shrews began to run out and no fewer than 31 were counted. For about two weeks prior to the finding of the nest the temperatures had been somewhat colder than normal, dropping to 32° F at night. McCarley thinks that this factor may have been responsible for the formation of so large an aggregation, which permitted the little animals to concentrate and share their body heat and thereby counteract the effect of the cool temperatures.

This shrew is cleanly and sanitary in its habits and always deposits its feces at a distance from the nest. Its natural food consists of beetles, bugs, crickets, grasshoppers, earthworms, millipedes, sowbugs, snails, and often salamanders and small frogs. It also has a reputation for entering beehives, sometimes building its nest therein, and feeding on the larvae and pupae. This habit results in its being called the bee shrew in some quarters. There is, however, another species, *Notiosorex crawfordi*, or the Desert Shrew, that also frequents beehives. It has not yet been taken within our state, but it occurs as close as extreme southeastern Oklahoma, less than 150 miles from the northwestern corner of Louisiana. A collection of owl pellets from anywhere in extreme northwestern Louisiana would be extremely valuable for they could reveal the presence of *Notiosorex*, as they have done elsewhere. Indeed, some of the comparatively few localities from which this rare little animal is known are based solely on remains found in owl pellets. It is easily identified by having only 3 unicuspid teeth on each side of the upper jaw, giving it a total of only 28 teeth, 2 less than *Cryptotis* and 4 less than *Sorex* and *Blarina* (Figure 26). The 3 unicuspids form a graded series, the 3rd being more than half as large as the 2nd and hence not minute. The ears are conspicuous, extending well above the fur, and they alone would separate *Notiosorex* from *Cryptotis parva*, with which it agrees in size and often in color.

Predators.—Cats, and sometimes dogs, catch Least Shrews, as in the case of the previous species, but do not often eat them. Instead they are simply deposited on our steps. Their remains, as previously noted, are frequently found in owl pellets.

Parasites.—Little or nothing is known about the internal and external parasites of *Cryptotis parva* in Louisiana.

Subspecies

Cryptotis parva parva (Say)

Sorex parvus Say, *in* Long, Account of an exped. . . . to the Rocky Mts., 1, 1823:163; type locality, west bank of Missouri River, near Blair, formerly Engineer Cantonment, Washington County, Nebraska.
Cryptotis parva, Miller, Bull. U.S. Nat. Mus., 79, 1912:24.
Cryptotis parva parva, Hall and Kelson, Mammals N. Amer., 1959:56–58.

Only one subspecies of *Cryptotis parva*, the nominate race *parva*, occurs in Louisiana. Eighteen specimens from Cameron, Calcasieu, and Jefferson Davis parishes are slightly larger than specimens from elsewhere in the state. They averaged as follows: total length, 76.6 (71–92); tail, 16.0 (13–21); hind foot, 10.5 (9–13); greatest length of skull, 16.3 (15.4–17.4); anteorbital breadth, 5.1 (4.7–5.6); interorbital breadth, 3.8 (3.7–4.1); cranial breadth, 7.6 (6.7–8.1); palatilar length, 6.7 (5.1–6.9); postpalatal length, 7.5 (6.9–7.8). Of these 18 specimens several are exceedingly grayish in color, but others in the series are indistinguishable from specimens from elsewhere in the state. Consequently, all Louisiana populations are assigned to the subspecies *parva*.

Specimens examined from Louisiana.—Total 354, as follows: *Acadia Par.:* 7 mi. SW Eunice, 1 (LSUMZ); 1 mi. NE Church Point, 2 (LSUMZ); 2 mi. N, 1 mi. W Crowley, 1 (LSUMZ); 1 mi. S Eunice, 1 (USL); Mermentau, 16 (FM). *Allen Par.:* 4.5 mi. NW Oberlin, W. Bay Game Mgt. Area, 2 (LSUMZ); just W of Oakdale city limits, 1 (TU). *Ascension Par.:* Gonzales, 1 (LSUMZ). *Avoyelles Par.:* 2 mi. E Marksville, 1 (LSUMZ); 2 mi. E Cottonport, 1 (LSUMZ). *Bossier Par.:* 3 mi. N Red Point, 1 (LSUMZ); Bossier City, 2 (USL). *Caddo Par.:* Belcher, 1 (USNM); 8.25 mi. NW Shreveport, 1 (LTU); 7.75 mi. NW Shreveport, 1 (LTU); 3 mi. S, 2 mi. W Blanchard, 1 (LSUMZ); 3 mi. NW

Shreveport P.O., 1 (LSUMZ). *Calcasieu Par.:* Lake Charles, 3 (1 LSUMZ; 2 MSU); 4 mi. E Lake Charles, 1 (LSUMZ); 3 mi. N Sulphur, 2 (LSUMZ); Sulphur, 2 (LSUMZ); 1 mi E Iowa, 1 (LSUMZ); 3.3 mi. W Ararat, 0.2 mi. W Indian Bayou ent. Sam Houston St. Pk., 1 (LSUMZ); 7.5 mi. N Toomey, 1 (LSUMZ); Sam Houston St. Pk., 1 (TU). *Caldwell Par.:* 1 mi. N Clarks, 3 (LSUMZ); Columbia, 6 (FM). *Cameron Par.:* Lowry, 1 (LSUMZ); Grand Cheniere, 1 (LSUMZ); 12 mi. S Vinton, Cameron Farms, 4 (LSUMZ); 1.5 mi. N Cameron, 1 (LSUMZ); 2 mi. NE Cameron, 1 (LSUMZ); Holly Beach, 1 (MSU). *Catahoula Par.:* 6 mi. N Harrisonburg, 5 (LSUMZ). *Claiborne Par.:* 6 mi. W Bernice, 1 (LSUMZ). *East Baton Rouge Par.:* various localities, 128 (127 LSUMZ; 1 USNM). *Evangeline Par.:* 8 mi. NW Ville Platte, 1 (USL); 1.5 mi. N Chataignier, 1 (USL). *Iberia Par.:* 3 mi. NW Jeanerette, 3 (USL); Jeanerette, 1 (USL); 3 mi. SE New Iberia, 1 (USL); Avery Island, 3 (TU); locality unspecified, 1 (USNM). *Iberville Par.:* 0.5 mi. N Plaquemine, 1 (LSUMZ). *Jackson Par.:* 6 mi. S Ruston on US Hwy. 167, 1 (LSUMZ). *Jefferson Davis Par.:* 3 mi. N, 1 mi. W Fenton, 2 (LSUMZ); 4.6 mi. E, 3.5 mi. N Gillis, 1 (LSUMZ). *Lafayette Par.:* various localities, 33 (USL). *Lincoln Par.:* 5 mi. NE Ruston, 1 (LSUMZ); 1 mi. S Ruston, 3 (LTU). *Livingston Par.:* locality unspecified, 1 (SLU). *Morehouse Par.:* 2 mi. N Bastrop, 1 (NLU); 1 mi. E Mer Rouge, 1 (NLU). *Natchitoches Par.:* Provencal, 4 (LSUMZ); Natchitoches, 3 (NSU). *Ouachita Par.:* Monroe, 28 (8 LSUMZ; 20 NLU); 1 mi. SE Monroe, 2 (NLU); 3.5 mi. E Monroe, 1 (NLU); Swartz, 1 (NLU); 4 mi. NE Monroe, 1 (NLU); 1 mi. S Intersection Hwys. La. 165 and 15, 2 (NLU); 0.25 mi. W Intersection of Hwys. La. 80 and I-20, Calhoun, 1 (NLU). *Pointe Coupee Par.:* 5.5 mi. N Innis, 1 (LSUMZ). *Rapiaes Par.:* 2.5 mi. W Pineville, 1 (LSUMZ); Alexandria, 1 (LSUMZ); 5 mi. S Alexandria, 1 (USL); Glenmora, 2 (USNM). *Richland Par.:* 4.9 mi. S Start, 1 (LSUMZ); 6 mi. S Rayville, 1 (NLU). *Sabine Par.:* 2 mi. SE Fort Jesup, 1 (LSUMZ). *St. Helena Par.:* 5 mi. NE Montpelier, 1 (LSUMZ). *St. Landry Par.:* 7 mi. N Henderson, 4 (USL); 6 mi. S Opelousas courthouse, 3 (USL); 7 mi. W Opelousas, 1 (USL); Eunice, 1 (USL). *St. Martin Par.:* 1 mi. W St. Martinville, 7 (USL); 1 mi. S Henderson, 1 (USL); Henderson, 3 (USL). *St. Tammany Par.:* 4 mi. NE Covington, 1 (LSUMZ); 10 mi. NE Covington, 1 (LSUMZ); 1 mi. N Slidell, 2 (TU). *Tangipahoa Par.:* 3 mi. SE Ponchatoula, 1 (LSUMZ); 3 mi. E Ponchatoula, 1 (SLU); 1.5 mi. W Hammond, 1 (SLU); 3.5 mi. E Hammond, 1 (SLU). *Tensas Par.:* 3 mi. S St. Joseph, 1 (LTU); *Union Par.:* Bernice, 1 (LTU). *Vermilion Par.:* 3 mi. N Abbeville, 1 (USL); 3 mi. NE Gueydan, 1 (USL); Abbeville, 2 (USL). *Vernon Par.:* New Llano, 3 (USL). *Washington Par.:* Hackley, 3 (FM). *West Baton Rouge Par.:* 0.5 mi. S Port Allen, 1 (LSUMZ); ca. 1 mi. S Port Allen, 1 (LSUMZ). *West Carroll Par.:* 4 mi. W Oak Grove, 3 (LSUMZ). *West Feliciana Par.:* St. Francisville, 1 (MSU).

The Mole Family
Talpidae

THE FAMILY comprises 12 living genera, of which 8 are monotypic, that is, they contain but a single species. The some 20 species in the family are found in Eurpoe and Asia south to the Mediterranean and the Himalayas and in eastern North America from southern Canada south to extreme northern Mexico and southern Florida. The one genus occurring in Louisiana is represented here by a single species, *Scalopus aquaticus*.

Moles are among the most highly specialized of all terrestrial mammals. They are indeed a marvelous bundle of adaptations to their particular way of life, which involves spending probably 99 percent of their time in their subterranean passages. The front legs are greatly foreshortened to provide increased leverage in digging, and the forefeet have the palms turned permanently outward and are in other ways modified as digging tools (Plate II and Figure 33). But the creature's specializations do not end there. The humerus or upper arm bone is without parallel in the way in which it is not only shortened to increase leverage but flattened as well to provide ample surface for the attachment of powerful muscles. In turn the sternum or breastbone extends far forward and has a prominent keel for the attachment of the strong pectoral muscles. The neck is short and muscular. Both the forefeet and hind feet have five toes that are webbed to make them more effective dirt movers. The head is abruptly conical with the nose elongated into a snout that is bare of hair to a line far back of the nostrils, which open upward.

Many of the attributes of animals living on the surface have, in the mole, been modified or eliminated through the process of natural selection. For example, there are no external ears. The ear openings are tiny holes buried in the fur. The eyes are not much larger than the head of a pin, lack external openings, and are degenerate to the point of serving little use. The fur is soft and velvety and will lie in any direction, thereby not hindering the mole from going forward or backward in its small burrows. The tail, which in surface-dwelling terrestrial mammals is usually long and gen-

erally serves as a balancing structure or, in arboreal forms, as a "fifth leg," is in the mole shortened to less than one-fourth of the animal's total length and is nearly devoid of hair.

Genus *Scalopus* E. Geoffroy Saint-Hilaire

Dental formula:

$$I\frac{3-3}{2-2} \quad C\frac{1-1}{0-0} \quad P\frac{3-3}{3-3} \quad M\frac{3-3}{3-3} = 36$$

EASTERN MOLE   Plate II
(*Scalopus aquaticus*)

Vernacular and other names.—In most of Louisiana the species is commonly referred to simply as the mole, but the French-speaking people of Pointe Coupee Parish call it *taupe*. Other names include garden mole and ground mole. The word "mole" comes from the Middle English word *molle*, which is related to another Middle English word, *moldwarpe*, meaning earth-thrower. The generic name, *Scalopus*, comes from the Greek word *skalops* meaning to dig and refers to the tunnels that the animal excavates with its large front feet. The specific name, *aquaticus*, is a Latin word meaning found in water. This name may have been based on the misconception that the mole's webbed front feet were used in swimming, though, of course, they are not. The name Eastern Mole serves to differentiate the moles of the eastern United States from the Western Mole of British Columbia, Washington, Oregon, Idaho, California, and Nevada. The latter belongs to the genus *Scapanus*.

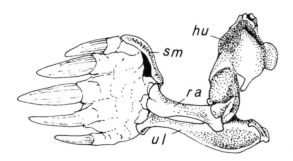

Figure 33 Bones of the forelimb of the Eastern Mole (*Scalopus aquaticus*), LSUMZ 15958: *ra*, radius; *ul*, ulna; *hu*, humerus; *sm*, sesamoid. The foreshortening of the elements provides greater leverage and the broad humerus allows for the attachment of powerful muscles. All these features and the specialized foot, with its sesamoid bone, are adaptations for digging.

Distribution.—The Eastern Mole is found throughout the uplands of the state. It does not occur regularly in coastal situations even on higher ground, which is isolated by broad marshes where, as expected, moles are absent. The species occurs throughout the eastern half of the United States from southeastern South Dakota, central Minnesota, the southern peninsula of Michigan, eastern Pennsylvania, and southern New England south to extreme northern Mexico, the Gulf Coast, and the southern tip of Florida. (Map 8)

External appearance. — Moles are about the size of small rats but bear no further resemblance (Plate II and Figure 34). The head and body are usually no more than six inches in length, and the tail is seldom more than an inch and a quarter long. The color is Slate-Gray above with a silvery sheen when viewed from certain angles and only very slightly, if at all, paler below. The forefeet, which are highly modified for digging, the velvety pelage, the small eyes and absence of external ears, and the numerous skeletal characters prevent confusion of the mole with any other mammal. See the general comments in the introduction to the family Talpidae for other characters.

Color phases. — No color phases occur in the members of this family, and albinos are rare. A specimen in the U.S. National Museum from St. Francisville is all white and is probably a true albino rather than a so-called "black-eyed white," which would be difficult to discern in the absence of eye color.

Measurements and weights. — See discussions under subspecies at the end of this account.

Figure 34 Eastern Mole *(Scalopus aquaticus).*

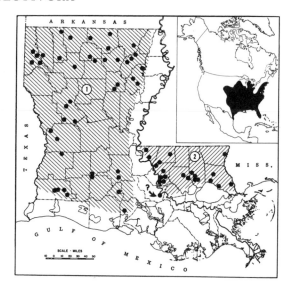

Map 8 Distribution of the Eastern Mole *(Scalopus aquaticus).* Shaded area, known or presumed range within the state; •, localities from which museum specimens have been examined; *inset,* overall range of the species. 1. *Scalopus aquaticus pulcher;* 2. *Scalopus aquaticus howelli.*

Cranial and other skeletal characters. — The skull of our mole is unique among the mammals of the state. It is conoidal with a relatively broad, flattened braincase. The mastoids are rather heavy and prominent, the frontal region is flat, the frontal sinuses are swollen, the anterior ends of the premaxillary bones are thickened and extend beyond the nasals to form an acute notch in front of the nasals, the auditory bullae are complete, and zygomatic arches are present although comparatively thin and not much curved. The first incisor is long and broad, but the second and third are tiny. The canine is about two-thirds as large as the first incisor. The first two molars are almost equal, but the third is decidedly smaller. Also, the molars possess three cusps and the crowns are W-shaped. For cranial measurements see the remarks on subspecies at the end of this account, and for other skeletal characters see the remarks pertaining to the family Talpidae. (Figure 35)

Sexual characters and age criteria.—The scrotum of the male is a slight swelling beneath the skin. The penis, which serves both urinary and genital functions, is directed backward. In the female there are two openings in addition to the anus. The most anterior one is the urinary opening situated in a small urinary papilla; the middle one is the vagina; and the most posterior opening is the anus. The female has three pairs of teats, one on the sides of the pectoral region, one on the sides of the abdomen, and the last inguinal. Males are slightly larger than females, and the young tend to be lighter colored than adults.

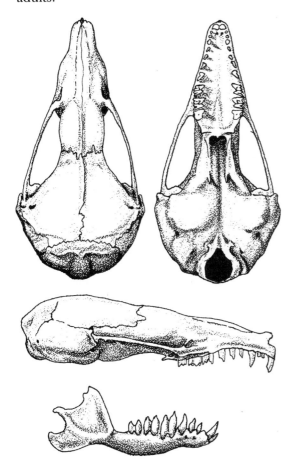

Figure 35 Skull of Eastern Mole *(Scalopus aquaticus aereus)*, ♂, Lafayette, Lafayette Parish, LSUMZ 15959. Approx. ½ ×.

Status and habits.—The mole prefers loose sandy soils or those consisting of light loams, in either of which it finds burrowing comparatively easy. Grassy prairies, meadows, pastures, gardens, cemeteries, lawns, golf courses, and wooded areas are all preferred habitat, particularly if the soil is moist but not wet. From most points of view the mole is very beneficial because it tills the soil, permitting air and moisture to penetrate deeper, and it eats a prodigious number of destructive insects. Its penchant for earthworms might seem to act at cross purposes to man's best interest, for these invertebrates are very beneficial as soil builders. But, like all predator-prey relationships, the mole's activities merely serve to maintain the balance of nature. The mole becomes really objectionable to some people when its burrows disfigure lawns and golf courses or when it inadvertently uproots rows of vegetables that we have planted in our gardens.

The tunnels are of two quite different types. One is an inch or so beneath the surface of the ground. In constructing this type of burrow the mole simply raises a ridge of earth that marks its pathway below the surface (Figure 36a). The other type is 6 to 10 inches underground and is more permanent. As a mole excavates its deep passageways it opens up at intervals vertical shafts through which it pushes up dirt into mounds several inches in height and 8 to 10 inches in diameter (Figure 36 a, b, c, and d).

For a long time mammalogists thought that the mole used a "swimming" motion in excavating its tunnels, employing a sort of breast stroke. However, a careful study by Hisaw (1923) demonstrated that such is not the case. He found that when excavating surface runways, forming the conspicuous ridges with which nearly everyone is familiar, the mole employs a lateral motion that involves only one of the forefeet at a time. When the earth is about to be pushed up by the left foot the mole rotates the front part of the body about 45 degrees to the right, brings

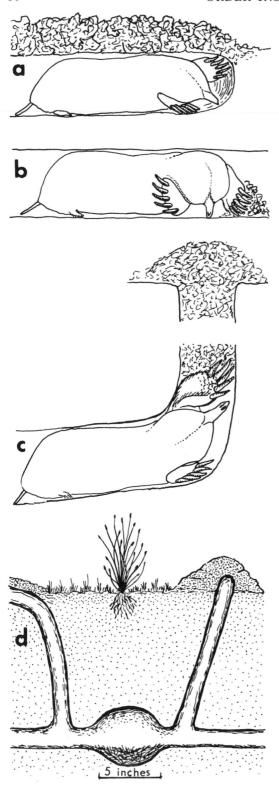

both feet together at the lower right side of the burrow, and, after a few rapid thrusting motions to place the claws, pushes the left leg vigorously upward (Figure 36 a). If the right foot is used in throwing up the soil, the reverse is true. In both cases, the leg that is used to push the earth upward, thereby forming the surface ridge, is aided by the extension of the other, which is held against the bottom of the burrow (Figure 36 a and c). During the entire process the nose is employed in making rapid exploratory examinations. It is in no way used for loosening the soil but seems to function entirely as a tactile organ for directing the powerful claws. Hisaw further discovered that when moles deepen surface burrows in periods of drought they drive a new tunnel beneath the old one and dispose of the excavated dirt by pushing it into the burrow being vacated, as well as by bringing it to the surface and piling it in mounds.

The nest of the mole is usually in one of the deep tunnels and is from four to six inches in diameter and several inches high. It is made of fine grasses and leaves. Two authorities (Arlton, 1936; Jackson, 1961), give the gestation period as 42 and 45 days, respectively, but another author (Conaway, 1959) has given the time as "four weeks or less." Obviously, the matter needs clarification. A baby mole plowed up in a field in Natchitoches

Figure 36 Tunnels of the Eastern Mole *(Scalopus aquaticus)*. (a) Construction of a tunnel that appears as a surface ridge. When the left foot is being employed the body is turned about 45 degrees to the right and the left leg is extended vigorously upward raising the earth above. (b) The mole transferring dirt in a deep tunnel. The head and shoulders are turned abruptly to the right or left, and the loose earth is pushed forward by one of the forepaws. (c) The mole constructing a mound of dirt from one of its deep tunnels. The body is turned at an angle, and the earth is pushed to the surface by extending the foreleg upwardly and laterally while the other leg and foot are used as a brace. (d) Deep burrow, nest, surface ridge, and surface mound of dirt from deep burrow of the Eastern Mole. (Diagrams a, b, and c redrawn from Prince, *in* Hisaw, 1923; diagram d, redrawn from Jackson, 1961)

Parish on 23 March 1939 and prepared as a specimen by Horace A. Hays, was probably a week old when found. On the basis of a 45-day gestation period this would have put the mating of its parents sometime in the first 10 days of February.

At birth the baby mole is nearly two inches in total length, exceptionally large for the size of the mother. It is hairless, although a few tiny vibrissae soon appear on the lips. It is mostly lead gray in color except that the snout, legs and feet, tail, and underparts are pink. Body hair does not begin to show until the animal is a week or 10 days old. Only one litter a year is produced and the number of young per litter is usually four but varies from two to five (Conaway, 1959).

Food of the mole consists of any kind of insect, both in the adult and larval stages, that it encounters, as well as great quantities of earthworms, sowbugs, millipedes, centipedes, and slugs. It has been known to eat frogs and occasionally to catch and eat one of the other small mammals that utilize its tunnels.

When moles become a nuisance, as in our lawns and gardens, control methods may be warranted. But they are not easy to capture or otherwise eliminate. Rat traps set in their runways are completely ineffective. The mole simply burrows beneath the obstruction in its tunnel, lifting it to the surface. Three special mole traps have been designed, the choker-loop type, the scissor-jaw type, and the harpoon type. Of these the last two are the most effective, but even they work well only in loose, sandy, loamy soil. These traps operate on the principle that a mole loses no time in repairing a damaged tunnel. So, when one tramps the ridge of earth and sets the treadle of the trap flat on the pressed earth above the tunnel, the mole springs the trap by lifting the soil against the trigger that releases the harpoons or scissor-jaws.

Sometimes after a rain a mole can be seen excavating a new tunnel. It will progress as much as 18 feet in an hour. If a person approaches the site of the operation quietly, he can often shove a spade or hoe into the earth behind the digger and heave the mole to the surface. Where a hydrant is available, a hose can sometimes be used successfully to drown the mole in its burrow, but this method is ineffective if the soil is too loose and sandy. Still other methods of control are to place paradichlorobenzene or napthalene in the tunnels or to insert a hose from the exhaust pipe of an automobile into a burrow and to kill the mole by carbon monoxide. Finally, the claim has been made that empty soft drink bottles can be buried at an angle to the surface of the ground with the bottoms in the mole's runway and the necks sticking out. Allegedly, when the wind blows a piping sound is made that so disconcerts the occupant of the tunnel that it deserts the area almost overnight.

As for me, I rather delight in seeing the workings of a mole in my yard. I cannot help admiring the industrious little creatures that work so laboriously utilizing the superb structural modifications with which they have been provided to help them escape the intense competition for survival on the crowded surface of the earth.

Predators. — Moles have comparatively few enemies. Perhaps their musty odor makes them unpalatable to the animals that might succeed in catching them. Cats, to be sure, do sometimes manage to capture moles but do not often eat them. Foxes, skunks, snakes, and owls may sometimes obtain one, and the shrews that use their tunnels conceivably eat baby moles that they chance upon in a nest, but the flooding of bottomland is probably the greatest decimating factor affecting the mole populations in low lying areas.

Parasites. — Moles have few external parasites but are not infrequently infested with a louse, *Euhaematopinus abnormis,* as well as by a small parasitic beetle, and several fleas. Internally they harbor nematode worms, often in great quantities, as well as no less than five species of cestodes (Stiles and Stanley, 1932). The acanthocephalan *Moniliformis clarki* has also been reported (Olive, 1950).

Subspecies

Scalopus aquaticus howelli Jackson

Scalopus aquaticus howelli Jackson, Proc. Biol. Soc. Washington, 27, 1914:19; type locality, Autaugaville, Autauga County, Alabama.

This subspecies occurs east of the Mississippi River and north of Lake Pontchartrain (Map 8). It is characterized by its smaller size and by certain cranial differences.

Measurements and weights. — Twenty-three males from the Florida Parishes averaged as follows: total length, 146.4 (128.0–172.0); tail, 19.5 (15.0–26.5); hind foot, 17.9 (15.0–23.0); greatest length of skull, 34.2 (33.1–35.0); mastoidal breadth, 17.6 (16.8–18.5); interorbital breadth, 7.9 (7.4–8.4); palatilar length, 13.8 (13.0–15.0); postpalatal length, 13.9 (13.3–14.9); maxillary molariform toothrow, 10.7 (10.1–11.5). Five males weighed 53.9, 56.5, 65.8, 79.0, and 87.0 grams. Fourteen females from the Florida Parishes averaged, respectively, as follows: 141.6 (120.0–157.0); 21.5 (18.0–24.0); 17.8 (16.0–21.0); 33.3 (30.7–34.4); 17.3 (16.1–18.7); 7.8 (7.4–8.5); 13.4 (12.9–14.1); 13.4 (12.4–13.9); 10.4 (9.7–10.7). Two female specimens weighed 47.0 and 64.3 grams.

Specimens examined from Louisiana. — Total 75, as follows: *East Baton Rouge Par.:* various localities, 50 (LSUMZ). *East Feliciana Par.:* 3 mi. N Zachary, 1 (LSUMZ); 2 mi. SE Clinton, 1 (LSUMZ). *Livingston Par.:* 0.5 mi. N Holden, 1 (SLU); 2.5 mi. SE Albany, 1 (SLU). *St. Tammany Par.:* 4 mi. NE Covington, 1 (LSUMZ); 3 mi. N Lacombe, 1 (LSUMZ); St. Joe, 1 (LSUMZ); Madisonville, 2 (TU); locality unspecified, 1 (AMNH). *Tangipahoa Par.:* Hammond, 6 (3 SLU; 3 LSUMZ); 2 mi. E Ponchatoula, 1 (SLU); 2 mi. W Natalbany, 2 (SLU); locality unspecified, 1 (SLU); 2 mi. W Ponchatoula, 1 (SLU). *Washington Par.:* Varnado, 2 (1 LSUMZ; 1 SLU). *West Feliciana Par.:* 38 mi. NNW Baton Rouge, 1 (LSUMZ); St. Francisville, 1 (USNM).

Scalopus aquaticus aereus (Bangs)

Scalops texanus aereus Bangs, Proc. Biol. Soc. Washington, 10, 1896:138; type locality, Stillwell, Adair County, Oklahoma.

Scalopus aquaticus aereus, Miller, Bull. U.S. Nat. Mus., 79, 1912: 8. Also see Hall and Kelson, Univ. Kansas Publ., Mus. Nat. Hist., 5, 1952:326.

Scalopus aquaticus pulcher, Lowery, Occas. Papers Mus. Zool., Louisiana State Univ., 13, 1943:217; name applied to population in Louisiana now assigned to *S. a. aereus.*

The subspecies *aereus* inhabits that part of the state lying west of the Mississippi River. It is distinguished from *S. a. howelli* by its larger size and the tendency for many individuals to be suffused over the entire body with a rich golden or coppery brown. I have followed Hall and Kelson (1952) in synonymizing *S. a. pulcher* Jackson with *S. a. aereus* (Bangs). The latter was originally described on the basis of a single specimen from Stillwell, Oklahoma, that was then considered unique in possessing the coppery color. But specimens from throughout the range of so-called *pulcher* show this feature, and since *aereus* has priority over *S. a. pulcher,* it takes precedence as the name for the mole occurring in parts of eastern Oklahoma, the southern two-thirds of Arkansas, northeastern and extreme southeastern Texas, and western Louisiana.

Measurements and weights. — Twelve males from localities west of the Mississippi River averaged as follows: total length, 172.5 (150.0–190.0); tail, 26.2 (20.4–35.0); hind foot, 20.4 (17.5–23.0); greatest length of skull, 35.9 (33.9–37.9); mastoidal breadth, 18.1 (17.0–19.3); interorbital breadth, 7.8 (7.5–8.1); palatilar length, 14.7 (13.0–17.5); postpalatal length, 14.3 (13.9–15.1); maxillary molariform toothrow, 11.2 (10.4–12.1). Two male specimens weighed 79 and 87 grams. Seven females from the same area averaged, respectively, as follows: 159.8 (146.0–175.0); 26.7 (21.0–36.0); 19.8 (18.0–21.0); 35.9 (34.0–38.1); 17.8 (16.4–19.1); 7.5 (7.1–8.9); 13.9 (11.0–17.0); 14.2 (13.8–14.9); 10.9 (10.3–11.4). Two female specimens weighed 63 and 65 grams.

Specimens examined from Louisiana. — Total 130, as follows: *Acadia Par.:* few miles S Basile, 1 (LSUMZ). *Beauregard Par.:* 5 mi. NE De Ridder, 2 (LSUMZ). *Bienville Par.:* Bienville, 2 (LSUMZ). *Bossier Par.:* Bossier City, 1 (LSUMZ). *Caddo Par.:* Blanchard, 1 (LSUMZ); Shreveport, 5 (3 USNM; 2 LSUS); 8.25 mi. NW Shreveport, 1 (LTU); 5 mi. E Greenwood, 1 (LSUS); 8.7 mi. S, 2.7 mi. E

LSUS Campus, 1 (LSUS). *Calcasieu Par.:* 6 mi. E Lake Charles, 1 (LSUMZ); Holmwood, 1 (MSU); Lake Charles, 5 (MSU); Maplewood, 1 (MSU). *Caldwell Par.:* Columbia, 1 (FM); Clarks, 1 (USNM). *Claiborne Par.:* Camp, 1 (NLU). *Evangeline Par.:* Basile, 1 (LSUMZ). *Franklin Par.:* Winnsboro, 1 (LTU); 2 mi. N Gilbert, 1 (NLU); 9 mi. S Delhi, 1 (NLU). *Iberia Par.:* Avery Island, 2 (CAS). *Iberville Par.:* White Castle, 1 (LSUMZ; immature and actually not identifiable to subspecies except on basis of geographical probability). *Jackson Par.:* 6 mi. SE Ruston, 1 (LTU). *Lafayette Par.:* Lafayette, 19 (5 LSUMZ; 14 USL). *Lincoln Par.:* Ruston, 16 (1 LSUMZ; 15 LTU). *Morehouse Par.:* Mer Rouge, 23 (5 MCZ; 18 USNM).

Natchitoches Par.: Provencal, 12 (LSUMZ); Kisatchie, 2 (LSUMZ); Mora, 1 (LSUMZ); Natchitoches, 1 (USNM). *Ouachita Par.:* Monroe, 2 (NLU); 3 mi. W Monroe, 1 (NLU); 17 mi. W Monroe, 1 (NLU); 16 mi. W Monroe, 1 (NLU); Calhoun, 2 (NLU). *Richland Par.:* 3.5 mi. NW Rayville, 1 (NLU); 5 mi. S Start, 1 (NLU); 5 mi. E Oak Ridge, 1 (NLU); Clear Lake, 1 (NLU). *St. Landry Par.:* Opelousas, 2 (USL); 0.5 mi. W Sunset, 1 (USL); Grand Coteau, 1 (USNM). *Union Par.:* Bernice, 1 (LTU); 5 mi. N Marion, 1 (LTU); Point, 1 (LTU). *Vernon Par.:* Leesville, 1 (NLU); 5 mi. SW Hornbeck, 1 (NSU). *Webster Par.:* Shongaloo, 1 (LSUMZ); 5 mi. S Sibley, 1 (LTU).

BATS
Order Chiroptera

MAMMALS of this order are the only ones that have the forelimbs modified as wings and hence that are capable of true flight. Other mammals, such as the so-called flying lemurs and the flying squirrels, glide but do not fly. Bats are so highly specialized as flying machines that locomotion by other means is accomplished with difficulty. The hind limbs are relatively small and posteriorly directed. The upper arm consists of the humerus, the forearm consisting of a long somewhat curved radius and a vestigial ulna, and the wrist of six small bones. The thumb is exceedingly short. The second, third, fourth, and fifth digits and their metacarpals are greatly elongated and connected in between by a leathery membrane that is anchored at the heel of the hind foot. The mouselike tail is often incorporated in what is known as the interfemoral membrane, which may extend from the tip of the tail to the ankles of the hind feet. An accessory ankle bone, the calcar, is frequently highly developed and serves to reinforce the attachment of the inter-femoral membrane to the ankle. (Figure 37)

Bats are worldwide in their distribution in the temperate and tropical regions of the world and sometimes occur on remote oce-anic islands that are uninhabited by other native mammals. On New Zealand, a pair of ancient continental islands, that is, islands once long ago connected with a mainland land mass, two bats are the only mammals occurring naturally, everything else having been introduced by man. The order is divided into two suborders, the Megachiroptera and the Microchiroptera. The former contains only one family, the fruit-eating Pteropodidae of the Old World, that is distinguished by the fact that most of its genera have a claw on the second digit, whereas in the Microchiroptera only the thumb bears a claw. The subordinal names are misleading because some of the Megachiroptera are smaller than some of the Microchiroptera. The latter suborder includes 16 families, only 2 of which occur in Louisiana. Our bats are almost exclusively insect eaters, but elsewhere in the world there are bats whose principal diet is pollen and nectar of flowers, fruit and vegetable matter, fish and other vertebrates such as small mammals, or blood obtained from a living victim. The last group consists of the vampires of the American tropics that make a small incision in the skin, usually on the neck and shoulders or on the ears or toes

95

often while their prey is asleep, and then lap the blood that flows from the wound.

Many superstitions prevail regarding bats, one being that they carry bedbugs. Such is, however, not true. Some of their ectoparasites may be true bugs of one sort or another and even members of the same genus as the bedbug, but not the same species that afflicts man. Another myth is that a bat in a room is sure to become inextricably entangled in the hair of any woman present. I am aware of no case of this happening.

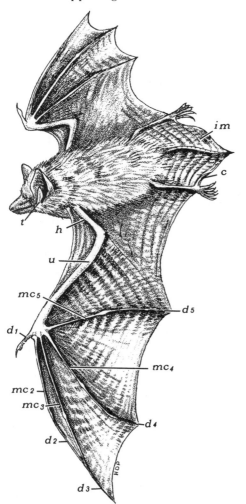

Figure 37 Anatomical features of a bat: *c*, calcar; *d*, digits; *h*, humerus; *im*, interfemoral membrane; *mc*, metacarpals; *t*, tragus; *u*, ulna, or forearm.

The eyes of bats are small and beady and probably not too good. But the highly developed ability of echolocation more than compensates for any deficiency in this regard. Bats emit a series of supersonic sounds that strike objects ahead; and, by means of their acute sense of hearing, they are able to convert the echos into information that enables them to avoid obstacles and to detect flying insects even in total darkness. The generally well-developed pinnae of the external ears help in this regard, as does possibly also the tragus, the shape of which is highly diagnostic in some species. (Figure 38)

Bats seek shelter in daytime in caves, crevices, buildings, abandoned houses, and in the case of the lasiurines and a few others, in Spanish moss and the thick foliage of trees. However, there is not to my knowledge a single cave in Louisiana large enough to be suitable for bats and consequently we lack several species that habitually roost in underground passages. Most species of the northern United States either hibernate or migrate to southern climes. But here in Louisiana on warm nights in winter we often find bats on the wing in search of insects. In their daytime roosts sleeping bats sometimes become semitorpid and their physiological processes slow down appreciably. Oxygen consumption may be one-tenth of normal, active rates. At rest during the day bats generally hang head downward by their hind feet. Colonial species aggregate loosely in attics or lofts of buildings under warm ambient temperatures, but when the temperature drops to near freezing or lower they assemble in a closely compact mass in some favored corner of the attic and thereby conserve their body heat. Bats live longer than most small mammals and even many large ones. The Zoological Society of London kept a fruit bat for 17 years; in nature banded individuals have been known to live for 21 years.

A peculiarity of the reproductive cycle of certain bats, particularly vespertilionids of the northern states in which hibernation takes place in winter, is the fact that copulation may occur both in the fall and again in spring. The sperm from fall matings are stored in the uterus or the Fallopian tubes of the female until the following spring. When ovulation occurs, the sperm that are already present can then fertilize the ova. The delayed implantation of the fertilized ova many months after copulation is a marked deviation from the usual course of events in the reproductive cycle of mammals.

Bats are of great economic importance. That they are held in such low esteem by most of our citizens is therefore unfortunate. Not only do bats consume great quantities of insects, but they are aesthetically worthy of protection. Few other mammals except some of the large game species such as deer, rabbits, and squirrels are as readily observed. Anyone who fails to derive some pleasure from watching one of these marvelously agile creatures feeding at twilight must indeed be a callous individual unperceptive of the beauties of the world of nature that surround us. Here are creatures that through their powers of flight are superbly adapted to fill a niche

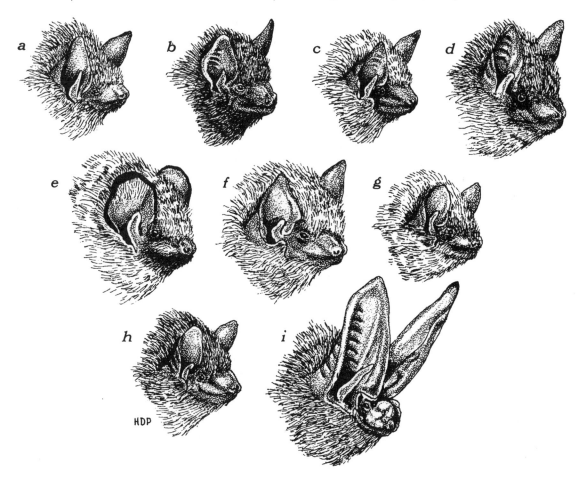

Figure 38   Heads of certain Louisiana bats, showing especially the shape of the ears and the tragus: (a) *Myotis austroriparius;* (b) *Lasionycteris noctivagans;* (c) *Pipistrellus subflavus;* (d) *Eptesicus fuscus;* (e) *Lasiurus cinereus;* (f) *Lasiurus intermedius;* (g) *Lasiurus borealis;* (h) *Nycticeius humeralis;* (i) *Plecotus rafinesquii.* Approx. natural size.

occupied by no other mammal. They have been present in North America in almost unchanged form since the Middle Eocene—for over 50 million years—hence eons before Man first came into being.

One species, the Brazilian Free-tailed Bat, admittedly is obnoxious because its pungent, musty smell spreads through our homes when it takes up residence there, but it can be forced to move its domicile elsewhere without resorting to its wholesale destruction. Moreover, some bats are now known to carry rabies and hence their bites are dangerous to humans and livestock. But no one would dare to recommend that all dogs be eliminated just because a few are known to be rabid. Parents should warn their children against handling bats, but they should likewise admonish them against familiarity with strange curs. Bats are

remarkable creatures and therefore should be appreciated as an interesting component of our wildlife community. Every form of life on our planet serves some role in the scheme of things around us. Man should not assume the right of deciding from his own selfish point of view what is worth preserving and what should be eliminated.

Among our French-speaking inhabitants all bats go by the name *souris-chaude*, which is sometimes written *chauve-souris* or *souris-chauve*. *Souris-chaude* is said (Read, 1931 and 1963) to have originated among Canadian woodcutters and was doubtless brought to Louisiana by Acadian immigrants. Meaning hot mouse, it stems from the fact that bats sometimes sleep in holes in trees during the winter and remain warm enough to defy the cold.

The Vespertilionid Bat Family
Vespertilionidae

THIS LARGE family of 38 living genera and approximately 275 species ranges widely throughout the temperate and tropical regions of the world. All the species of bats in Louisiana except the Brazilian Free-tailed Bat belong to this family. The dental formula varies from I 1/2, C 1/1, P 1/2, M 3/3 × 2 = 28 to I 2/3, C 1/1, P 3/3, M 3/3 × 2 = 38. Most vespertilionid bats are cave dwellers, but in Louisiana, where suitable caves are lacking, they roost in hollow trees, in clusters of moss and leaves, in attics of buildings and old abandoned houses, and in culverts and well shafts. The tail extends to the tip of the interfemoral membrane or a few millimeters beyond. In some species the interfemoral membrane is heavily furred, in others almost naked. The ears are usually large and the tragus is well developed, its shape varying according to the species (Figure 38). The skull is relatively broad, the braincase inflated, and the zygomata complete. The cusps on the molariform teeth form a W-shaped pattern.

Genus *Myotis* Kaup
Mouse-eared Bats

Dental formula:

$$I\frac{2-2}{3-3} \quad C\frac{1-1}{1-1} \quad P\frac{3-3}{3-3} \quad M\frac{3-3}{3-3} = 38$$

SOUTHEASTERN MYOTIS   Plate II
(Myotis austroriparius)

Vernacular and other names. — This bat is so poorly known in the state that it has been given no truly vernacular appellations by our residents. The word "bat" comes from the Middle English *bakke,* which in turn is derived from an old Norse word *blaka* meaning to flutter. The first part of the scientific name is from two Greek words, *mys,* meaning mouse, and *otis,* meaning ear. The specific name *austroriparius* is from two Latin words, *austro,* meaning southern, and *riparius,* which means frequenting banks of streams. The species is indeed mostly a bat of the southern states, and, like many of its relatives, it feeds over ponds, lakes, and rivers.

Distribution. — In Louisiana this bat is now known from nearly all parts of the state, the exception being the southwestern quarter. The species ranges from extreme southeastern Oklahoma and extreme northeastern Texas across southern Arkansas to western Tennessee, western Kentucky, southern Illinois, and central-southern Indiana south over most of Louisiana, Mississippi, southern Alabama, most of Georgia, and northern and central Florida as far as the vicinity of Tampa Bay. (Map 9)

External appearance. — LaVal (1970) has made a comprehensive analysis of the individual and possible geographic variation represented in his study of over 800 specimens from all parts of the range of this species. He set up four classes to describe the range of color of the dorsum: (1) Ochraceous-Tawny; (2) Snuff Brown to Prout's Brown; (3) Sac-

cardo's Umber; and (4) Mouse Gray. For the venter he also set up four color classes: (1) Yellow Ochre; (2) Cream Buff to Cartridge Buff; (3) Ivory Yellow; and (4) White. Expressed in more general, less technical terms, these colors of the upperparts might be described, respectively, as a bright orange-brown, a duller brown, a gray-brown, and gray, while the colors of the underparts might be described as slightly brownish yellow, cream, ivory, and white. Obviously, with such a broad range of colors, the Southeastern Myotis must be described as a highly variable species. LaVal further pointed out that 16 color categories were possible, expressing a combination of one of the four colors of the upperparts with one of the four colors of the underparts but that only 12 were actually represented in the material he examined. He sorted these combinations into four groups: A (1–1, 1–2, 1–3); B (2–2, 2–3); C (3–2, 3–3, 3–4); and D (4–1, 4–2, 4–3, 4–4). LaVal discovered that when 49 specimens from southern Louisiana were assigned to

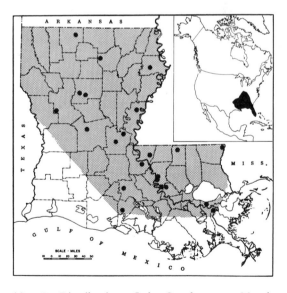

Map 9 Distribution of the Southeastern Myotis *(Myotis austroriparius). Shaded area,* known or presumed range within the state; •, localities from which museum specimens have been examined; *inset,* overall range of the species.

one of the 12 combinations they were fairly evenly distributed among his four color groups, as follows: A, 16; B, 10; C, 14; D, 9. These results again emphasize the great variability in color of our bats of this species.

Confusing species.—The Southeastern Myotis is small in size, although not as small as the Eastern Pipistrelle. In its brownish form it is most easily confused with the Evening Bat, which it resembles closely in color. It is, however, readily distinguished from the Evening Bat by its long, pointed, and straight—instead of short, blunt, and curved—tragus and by its four, instead of two, upper incisors, which are plainly visible in the living animal when one lifts its upper lip. (Figure 39)

Color phases.—Although the individuals that are bright orange-brown above and brownish yellow below differ markedly from individuals that are gray or gray-brown above and white below, I interpret these not as color phases but rather as color forms in a broad continuum of individual variation. In Florida, however, the reddish form is far less prevalent (6 out of 78 from Alachua County, and 1 out of 499 from Citrus County, *fide* LaVal, *loc. cit.*).

Measurements and weights.—Eighteen males from Louisiana in the LSUMZ averaged as follows: total length, 83.7 (77.0–89.0); tail, 36.8 (26.0–44.0); hind foot, 9.5 (7.0–11.0); ear, 12.0 (11.0–16.0); forearm, 36.0 (33.0–39.0); third metacarpal, 31.1 (29.5–33.8). Weights of 9 males averaged 5.9 (5.1–6.8) grams. Twenty-nine females averaged, respectively, as follows: 87.2 (80.0–97.0); 38.0 (29.0–42.0); 10.0 (8.0–12.0); 13.3 (9.0–15.0); 38.6 (33.5–40.0); 32.7 (30.5–36.0). Weights of 19 females averaged 6.9 (5.2–8.1) grams.

Cranial and other skeletal characters.—The Southeastern Myotis is the only bat so far found in the state that possesses as many as 38 teeth. But there are one or two other species of this genus that will, I feel certain, sooner or later be recorded here. The skull of *Myotis austroriparius* in most cases shows a definite and sharp-edged sagittal crest. Although a few of the skulls examined lack this feature, the character is nevertheless often useful in separating the species from other bats presently known from Louisiana. The calcar is not keeled. For a description of the baculum, see Davis and Rippy, 1968. (Figure 40)

Sexual characters and age criteria.—Females tend to be brighter in color than males and are on the average slightly larger. A young female that I obtained at Baton Rouge on 22 September 1940 was at the time in full molt, but I can detect among adults in the LSUMZ series of specimens little evidence of exten-

Figure 39 Southeastern Myotis *(Myotis austroriparius),* ♀, Dutch Town, Ascension Parish. (Photograph by Richard K. LaVal)

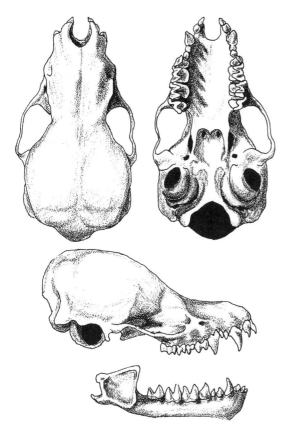

Figure 40 Skull of Southeastern Myotis *(Myotis austroriparius)*, ♀, 1 mi. W Dutch Town, Ascension Parish, LSUMZ 11305. Approx. 4 ×.

sive molt at any time of the year. Almost certainly, though, molt does take place in late summer or early fall.

Status and habits.—The Southeastern Myotis was once regarded as one of the rarest bats in the state and indeed everywhere (Miller and Allen, 1928). Its original discovery in Louisiana was made by Professor Horace A. Hays at Vowells Mill, in Natchitoches Parish on 23 November 1938, when he caught one in a house. The next 23 years produced only 8 additional specimens, most of which were obtained in buildings on the LSU campus in Baton Rouge. Then several of my students, notably Richard K. LaVal, began the systematic use of Japanese mist nets set across ponds and small streams, which resulted in the capture in only a few years of an additional 31 specimens, involving numerous additional localities in the state. Drs. Clyde Jones and Royal D. Suttkus and their students at Tulane University found a colony and obtained 27 specimens in the old ammunition storage bunkers on Tulane's Hebert Center campus near Belle Chase. Jones also obtained one at Avery Island. Dr. Marshall B. Eyster and his students collected a specimen at Breaux Bridge, in St. Martin Parish, and another 10 miles SW Tallulah, in Madison Parish. B. A. Heck, a student of Dr. D. T. Kee at Northeast Louisiana University, obtained one at the LSU Experiment Station at Calhoun, in Ouachita Parish, and A. W. Bagur, a student of Dr. John Goertz at Louisiana Tech University, obtained a specimen near Winnfield, in Winn Parish. Thus the species instead of being extremely rare, as originally thought, is probably fairly common over most of the state.

For our present better understanding of the status of this bat and others we can be thankful for the role that Japanese mist nets have played in enabling mammalogists to capture specimens. The nets come in various lengths, but the ones that are often used are 30 feet long and approximately 5 feet high when in position (Figure 41). The fine mesh is similar to that of a lady's hair net and is hence almost invisible when stretched across a clearing or small pond or stream. If by its sonar the bat detects the net in its flight path it usually does so too late to swerve and avoid collision. Hence it strikes the net and immediately becomes inextricably entangled in the mesh. Not only do these nets enable the mammalogist to obtain specimens of bats that would be extremely difficult to shoot on the wing in the fading light of evening, but also the specimens obtained are undamaged.

Capturing bats in mist nets has other advantages over shooting them. If one finds in his nets a species that is not needed for a specimen, the animal can be extracted and released unharmed. In addition to being difficult to shoot on the wing, bats that are hit are sometimes almost impossible to find in the semidarkness of early evening. Consequently, valuable specimens are often lost. Then too some species of bats do not emerge from their daytime retreats until after complete darkness has settled over the landscape;

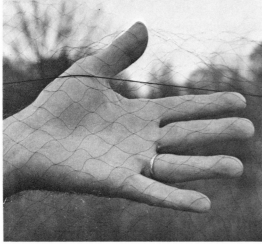

Figure 41   Mist net in position.

hence the collector never has the opportunity to shoot one of these late flyers. Some of our LSU Museum's field parties in the American tropics have more than once obtained species of bats in their nets that they have never otherwise seen.

Little is known about the life history of the Southeastern Myotis in Louisiana. To my knowledge only one communal roost has ever been found in the state, that being in the old ammunition storage bunkers at the Tulane University Hebert Center at Belle Chasse. Fortunately, however, the excellent studies of the species in Florida by H. B. Sherman (1930) and Dale W. Rice (1957) have provided valuable details that are probably applicable in large measure to our own populations. Much of the following information is derived from these two sources.

There are, though, basic differences between most populations of *Myotis austroriparius* and those inhabiting Louisiana. In Illinois, Indiana, west Florida, and probably also Kentucky, Tennessee, and Mississippi, *Myotis austroriparius* is a cave bat, and its life history is essentially the same as that of most other cave-inhabiting species of *Myotis*. There it hibernates for four to seven months of the year, from September or October to late February or early March. The bats remain torpid for most if not all of this period, as there is no indication that they leave the caves to feed during the cold months. In Louisiana and peninsular Florida they remain active throughout much of the winter and feed regularly during this time. The LSUMZ collections show representations in every month of the year except June. In Louisiana they roost in such places as the attics of buildings, in abandoned houses, in culverts, and possibly also in outdoor sites such as hollow trees. When the temperature drops below 40° F they apparently become inactive and go into a semitorpid condition, but when the days become warm again, they awaken and resume their nightly activity.

As previously noted, one of the peculiarities of northern, hibernating bats is that copulation takes place in the fall, but fertilization does not occur until after ovulation the following spring. The sperm are stored in the genital tract of the female during the winter. But I doubt that matings occur exclusively in the fall in bats that remain active throughout most of the winter. In fact, the suggestion has been made that the only reason the spermatozoa from fall matings are able to survive for such a prolonged period is that the body temperature of the hibernating female has dropped to a low level during the stage of winter dormancy. Even though matings do occur in fall, we cannot safely assume that other matings fail to take place in spring. Indeed, females less than one year old become pregnant for the first time in the spring following their birth, and they do so as a result of matings that take place at that time, for they are not believed to copulate during their first autumn. In peninsular Florida, among noncave populations, Rice found that virtually all males have enlarged epididymides, suggesting active spermatogenesis, from 12 February to 16 April. A similar situation is likely among outdoor roosting populations in Louisiana.

In northern Florida, Rice found that in spring and summer practically all females, accompanied by some of the males, congregate in certain caves to bear their young. They begin to arrive in these maternity caves about the second week of March, and number anywhere from 2,000 to 90,000 adult bats in a single cave. The majority of the males and the nonbreeding females remain apart from the maternity colonies, roosting singly or in small groups outside caves. The mystery is what the pregnant females of our Louisiana *Myotis* do about maternity wards in the absence of any caves. Do they aggregate at all when ready to bear their young? Paradoxically, I am informed by Dr. Royal D. Suttkus of Tulane University that the females in the small colony of *Myotis austroriparius* in the old ammunition storage bunkers at Belle Chasse do not form maternity colonies nor give birth to young in these bunkers, but the species is present there every month of the year. Obviously, our knowledge of the habits of bats of this species in Louisiana is sadly lacking.

Parturition in peninsular Florida extends from the end of April until late May, with the peak occurring about the second week of May. The earliest date that Rice detected young bats was 30 April, and no pregnant females were found after 17 May. He further learned that the young are able to fly five or six weeks after birth.

Sherman (1930) described parturition in this species in peninsular Florida. When the embryo is about to appear, the mother assumes a position in which the interfemoral and alar membranes form a receptacle where the young are received and can crawl about as soon as born. At birth the young are naked except for a few vibrissae and a few hairs on the toes and about the knees. When the placenta appears several hours after birth it is eaten by the mother, who also severs the umbilical cord with her teeth. In Florida *Myotis austroriparius* normally gives birth to two young. It is the only member of the genus that ever produces more than one young at a time. No dates or other information are available on the birth of Louisiana bats of this species.

Predators.—In Florida caves the Virginia Opossum *(Didelphis virginiana)* and several species of snakes were observed eating the Southeastern Myotis. Cockroaches *(Periplaneta americana* and *P. australasiae)* were detected preying on newborn bats. The Rat and Corn Snakes *(Elaphe obsoleta* and *E. guttata)* were both present in the bat caves and, being excellent climbers, probably account for the loss of some bats. Owls were also discovered to be a major predator.

Parasites.—The Southeastern Myotis in Florida was found by Rice (1957) to be heavily infested with mites *(Spinturnix* sp. and *Ichoronyssus quadridentatus)* and parasitic flies *(Basilia boardmani* and *Thichobius major).*

Subspecies

Myotis austroriparius (Rhoads)

Vespertilio lucifugus austroriparius Rhoads, Proc. Acad. Nat. Sci. Philadelphia, 49, 1897:227; type locality, Tarpon Springs, Pinellas County, Florida.

Myotis austroriparius, Miller and G. M. Allen, Bull U.S. Nat. Mus., 144, 1928:76.

Myotis austroriparius gatesi Lowery, Occas. Papers Mus. Zool., Louisiana State Univ., 13, 1943:219; type locality, Louisiana State University Campus, Baton Rouge, East Baton Rouge Parish, Louisiana.

Myotis austroriparius mumfordi Rice, Quart. Jour. Florida Acad. Sci., 18, 1955:67; type locality, Bronson's Cave, Spring Hill State Park, 3 mi. E Mitchell, Lawrence County, Indiana.–Crain and Packard, Jour. Mamm., 47, 1966:323.

In 1943, when I described *M. a. gatesi,* only five specimens of the species were available from Louisiana and, except for a series from several localities in peninsular Florida, there were less than two dozen specimens extant from the entire range of the species as then known. The original five specimens from the state were so strikingly different from material from Florida and elsewhere that at least two authorities on bats expressed the opinion that I should have accorded *gatesi* full species rank instead of describing it as a subspecies of *austroriparius.* Now, though, in the light of all the new material that is available, I am following the recommendations of one of my former students, Dr. Richard K. LaVal, in not recognizing either *M. a. gatesi* or *M. a. mumfordi.* In other words, I am treating *Myotis austroriparius* as a monotypic species. LaVal has demonstrated (1970), in my opinion conclusively, that the array of color categories exhibited by the species throughout its range are not sufficiently correlated with the geography of the populations sampled to justify the recognition of subspecies. In one part of his study, LaVal assigned each of 824 specimens, representing all parts of the geographical range of the species, to one of his subjective 16 primary color classes that comprised 4 secondary color categories. In some instances the overwhelming majority of specimens from one locality fell in one of the 4 categories, but every locality or aggregation of localities was represented by at least 2 color categories; and of his 12 locality groupings no less than 8 included specimens assignable to 3 of the 4 color categories. As previously noted, of 49 specimens from Louisiana, 16 were in Category A, 10 in B, 14 in C, and 9 in D. He found some geographical variation in the means of certain mensural characters, but the range of variation in any one character from one particular locality invariably broadly overlapped the range of variation in that character from other localities. Crain and Packard (1966) refer five specimens (LSUMZ 8918–8922) obtained by Crain in Washington Parish to *M. a. mumfordi.*

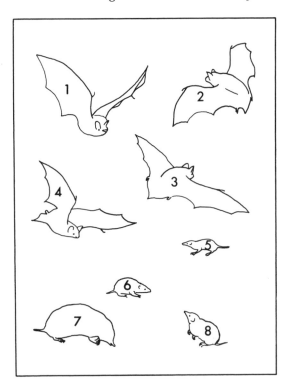

Plate II
1. Southeastern Myotis *(Myotis austroriparius).*
2. Evening Bat *(Nycticeius humeralis).*
3. Silver-haired Bat *(Lasionycteris noctivagans).*
4. Eastern Pipistrelle *(Pipistrellus subflavus).*
5. Southeastern Shrew *(Sorex longirostris).*
6. Least Shrew *(Cryptotis parva).*
7. Eastern Mole *(Scalopus aquaticus).*
8. Short-tailed Shrew *(Blarina brevicauda).*

As noted above, however, I am following LaVal (1970) in synonymizing this race along with *M. a. gatesi* and therefore treating *Myotis austroriparius* as monotypic.

Specimens examined from Louisiana.—Total 74, as follows: *Ascension Par.:* 1 mi. W Dutch Town, 10 (LSUMZ). *Avoyelles Par.:* 14 mi. ENE Marksville, 4 (LSUMZ); 3.5 mi. ENE Mansura, 1 (USL). *Claiborne Par.:* 4.5 mi. SW Summerfield, 1 (LSUMZ). *Concordia Par.:* 4 mi. W Ferriday, 3 (LSUMZ). *East Baton Rouge Par.:* University, 9 (LSUMZ); Baton Rouge, 1 (LSUMZ). *Iberia Par.:* Avery Island, 1 (TU). *Iberville Par.:* Spanish Lake Swamp, 1 (LSUMZ). *Jefferson Par.:* Metairie, 1 (LSUMZ); *Madison Par.:* 10 mi. SW Tallulah, 1 (USL). *Natchitoches Par.:* Vowells Mill, 1 (LSUMZ). *Ouachita Par.:* Calhoun, 1 (NLU). *Plaquemines Par.:* Tulane Univ. Riverside Campus, Belle Chasse, 27 (14 LSUMZ; 13 TU). *Pointe Coupee Par.:* New Roads, 1 (LSUMZ). *Rapides Par.:* 0.5 mi. SE Kincaid, 1 (LSUMZ). *St. Helena Par.:* 5.4 mi. NNE Chipola, 1 (LSUMZ). *St. Martin Par.:* Breaux Bridge, 1 (USL). *Washington Par.:* 3 mi. E Angie, 2 (LSUMZ); 2.75 mi. E, 0.25 mi. N Angie, 3 (LSUMZ). *West Feliciana Par.:* 4.2 mi. W ENE St. Francisville, 1 (LSUMZ). *Winn Par.:* 8 mi. W Winnfield, 1 (LSUMZ); 2 mi. S Winnfield, 1 (LTU).

Genus *Lasionycteris* Peters

Dental formula:

$$I\frac{2-2}{3-3} \ C\frac{1-1}{1-1} \ P\frac{2-2}{3-3} \ M\frac{3-3}{3-3} = 36$$

SILVER-HAIRED BAT   Plate II
(*Lasionycteris noctivagans*)

Vernacular and other names.—This species is sometimes called the black bat, silvery bat, or silver-black bat. The latter two names refer to the silvery tips of the hairs on the back. The generic name comes from two Greek words, *lasios,* meaning hairy, and *nykteris,* meaning bat, and alludes to the dense fur of the upperparts, including the basal portion of the tail membrane. The specific name is from two Latin words, *nox* for night and *vagans* for wandering, hence night wandering.

Distribution.—The species is widespread from the Pacific Coast of British Columbia across southern Canada to New Brunswick and Nova Scotia south to central California,

southern Arizona, central-southern Texas, central Alabama, and South Carolina. There is only one record for the occurrence of the species in Louisiana. (Map 10)

External appearance.—The Silver-haired Bat is a small to medium-sized bat that is larger than the Eastern Pipistrelle and Southeastern Myotis but smaller than the Red and Seminole Bats and considerably smaller than the Big Brown Bat. It is easily distinguished by its color alone, being a dark sooty brown or black with the tips of the hairs, particularly on the back, flanks, chest, and abdomen tipped with ashy white, producing a frosted appearance. The basal half of the interfemoral membrane is sparsely furred. The ears are fairly large but relatively short and broad, not pointed. The tragus is short and broad, rather truncate. The tail is short, being less than 40 percent of the total length.

Color phases.—No color phases nor color abnormalities have been described.

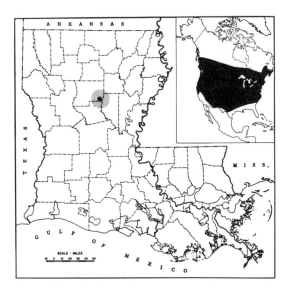

Map 10 Distribution of the Silver-haired Bat (*Lasionycteris noctivagans*). •, the only locality from which a specimen has been examined and the only place where the species is known to have occurred in the state; *inset,* overall range of the species.

Measurements and weights. — Total length, 95–115; tail, 35–45; hind foot, 8.5–10.5; ear, 15–17; forearm, 40–42; weight, 7–10 grams (Jackson, 1961).

Cranial and other skeletal characters. — The skull is considerably flattened and the rostrum is broad, almost as wide as the braincase when the skull is viewed from above. The upper surface of the rostrum is distinctly concave on each side between the lacrimal region and the premaxillary notch. The calcar is not keeled. The skull is 16–16.9 mm in length and 9.5–10.3 mm in greatest width. (Figure 42)

Sexual characters and age criteria. — The sexes are alike, without seasonal variation. In the young the silvery tipping of the hairs is sometimes much more pronounced than in adults.

Status and habits. — The only record of the Silver-haired Bat in Louisiana is a female shot three miles west of Tullos, in Winn Parish, on 22 March 1958 by B. J. Pendarvis. Through the courtesy of Dr. Herbert E. Shadowen the specimen is now in the Louisiana State University collections (LSUMZ 8567).

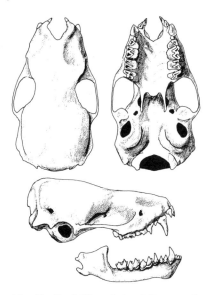

Figure 42 Skull of Silver-haired Bat *(Lasionycteris noctivagans)*, ♂, Glenwood Fish Hatchery, Catron County, New Mexico, TU 2208. Approx. 3 ×.

Since this species is well known for its migratory habits, the Tullos record could very well be a migrant that had moved into the area during the previous fall. Jackson (1961) states that the species leaves Wisconsin about the first of October and returns in April and May. It has been observed migrating at sea along the Maine coast, 15 miles from the nearest land, and, still more remarkably, makes regular visits to the Bermuda Islands, more than 700 miles from the coast of the United States. Migrations take place not only at night but in daytime as well. On a cloudy and mild day, 28 September 1907, at Washington, D.C., many bats, including this species, were seen migrating toward the southwest (Howell, 1908).

On the other hand, the Tullos record may indicate that the species is an exceedingly rare permanent resident in the northern part of the state. The bat's solitary habits render it difficult to locate except by chance encounter. The species seldom aggregates in numbers, and it roosts in hollow trees or clings behind the loose bark of trees or in clumps of leaves. Occasionally it roosts in nooks and crannies in outbuildings or enters attics.

According to Jackson (1961), the Silver-haired Bat in Wisconsin is a late flier, taking wing after most other bats have commenced their evening forays in search of their diet of insects. Contrariwise, Schwartz and Schwartz (1959) describe it as an early flier in Missouri, leaving its daytime roost in late afternoon before sunset. The flight is said to have a fluttering quality that is distinctive. The one or two young are born from late June to early July and are black and wrinkled. The female carries them on her search for food until they are too heavy for her to transport, but the young are able to fly themselves by the time they are three to four weeks old.

Lasionycteris noctivagans (Le Conte)

V[espertilio]. noctivagans Le Conte, in McMurtrie, The animal kingdom…by the Baron Cuvier…, 1, 1831:431; type locality, eastern United States.
Lasionycteris noctivagans, H. Allen, Bull. U.S. Nat. Mus., 43, 1894:105; LaVal, Jour. Mamm., 48, 1967:647.

Lasionycteris noctivagans is monotypic, that is, it shows no recognizable geographic variation anywhere within its range.

Specimens examined from Louisiana. — Total 1, as follows: *Winn Par.*: 3 mi. W Tullos, 22 March 1958 (LSUMZ).

Genus *Pipistrellus* Kaup

Dental formula:

$$I\frac{2-2}{3-3} \ C\frac{1-1}{1-1} \ P\frac{2-2}{2-2} \ M\frac{3-3}{3-3} = 34$$

EASTERN PIPISTRELLE   Plate II and
(Pipistrellus subflavus)   Figure 43

Vernacular and other names. — This species has long been known in the literature as the Georgia bat, Georgian bat, or Georgian pipistrelle. It is also referred to simply as the

Figure 43 Eastern Pipistrelle *(Pipistrellus subflavus)*, Wayne County, Mississippi. (Photograph by Richard K. LaVal)

pipistrelle or pipistrel, or as the yellowish-brown bat. The most appropriate name is Eastern Pipistrelle, which serves to distinguish it from its close relative the Western Pipistrelle *(Pipistrellus hesperus)*, which occurs no further east than western Oklahoma and western Texas. The generic name *Pipistrellus* is the Latinized form of the Italian word *pipistrello*, meaning bat. The specific name *subflavus* is from the Latin *sub*, meaning below, and *flavus*, meaning yellowish, and hence signifies yellowish bellied.

Distribution. — The Eastern Pipistrelle ranges widely in eastern North America from Maine, southern Quebec, and Ontario to Ohio, southern Indiana, northern Illinois, northern Wisconsin, and southeastern Minnesota south through the eastern Great Plains and the entire eastern United States to central Florida and to eastern Mexico, Yucatan, and northern Honduras. In Louisiana it occurs widely over the state, but, for some unexplained reason, has not yet been taken anywhere in the southwestern parishes. (Map 11)

External appearance. — This small bat is easily distinguished by its somewhat yellowish brown coloration. More precisely, the color of the upperparts varies between Snuff Brown and Sayal Brown. When the main hairs of the back are separated by blowing into the fur, they are seen to be dark gray at the base, then broadly banded with yellowish brown, and tipped with a darker brown. The scattered long guard hairs are entirely yellowish. The underparts are also yellowish brown, but the bases of the hairs are, as on the dorsum, grayish in color. The ears are moderately large and when laid forward extend just beyond the tip of the nose. The tragus is fairly long and narrow and nearly straight (Figure 38). The membranes are blackish, but the skin over the forearm is slightly reddish in color, and the membrane between the second and third metacarpals is unpigmented. The interfemoral membrane is scantily furred near its base.

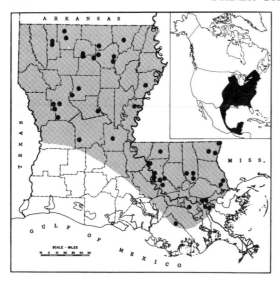

Map 11 Distribution of the Eastern Pipistrelle *(Pipistrellus subflavus). Shaded area,* known or presumed range within the state; •, localities from which museum specimens have been examined; *inset,* overall range of the species.

Color phases. — Specimens showing erythristic and melanistic tendencies have been reported from New York, and a partially white individual was found in Texas. A wide range of variation in the shade of the brownish color of the dorsum is evident in the LSUMZ series, but there is no indication that two color phases are involved. See comments beyond under *Status and habits.*

Measurements and weights. — Thirty-three adults from Louisiana in the LSUMZ collections averaged as follows: total length, 79.8 (71–95); tail, 37.4 (30–46); hind foot, 7.9 (6–9); ear, 10.9 (10–15); forearm, 33.6 (29.9–36.0); third metacarpal, 29.2 (25.2–32.0); greatest length of skull, 12.7 (12.2–13.5); cranial height, 4.7 (4.3–5.1); cranial breadth, 6.8 (6.5–7.2); zygomatic breadth, 7.9 (7.7–8.3); interorbital breadth, 3.6 (3.2–3.9); palatal breadth, 5.3 (4.8–5.3); palatilar length, 4.7 (4.2–5.2); postpalatal length, 4.4 (4.3–4.8); maxillary toothrow, 4.1 (3.8–4.4). Thirty specimens weighed an average of 5.4 (4.7–6.3) grams.

Cranial and other skeletal characters. — The Eastern Pipistrelle has a small and lightly built skull, but the braincase is somewhat inflated. It is the only bat in the state with a total of 34 teeth. The palate extends only a short distance behind the last molars. The first incisor on each side of the upper jaw (which is actually I 2) is unique in being bicuspidate. In the closely related *P. hesperus,* which is not yet known from this state, I 2 is unicuspidate and the first premolar (P 1) is concealed in its labial view by the large canine (C 1) and the second premolar (P 4), which is thus in contact with the canine. In *P. subflavus* P 4 is not in contact with the canine, being clearly separated from it by P 1. The calcar is short and not keeled. In *P. subflavus* the thumb is decidedly longer than in *P. hesperus.* The mean thumb length in 33 specimens of *P. subflavus* from Louisiana was 5.4 mm and the extremes ranged from 4.5 to 6.0 mm. Hall and Dalquest (1950) give the corresponding measurements of the thumb of 10 specimens of *P. hesperus maximus* as 3.9, 3.6, and 4.3. (Figure 44)

Sexual characters and age criteria. — The sexes are superficially alike. The young are darker than adults. Pipistrelles that have been banded are known to have lived up to 10 years in the wild.

Status and habits. — The Eastern Pipistrelle is one of the commonest bats in the state. It has been collected in numerous localities in the northern part of Louisiana, as well as in the southeastern section. For some reason it has not yet been obtained in the southwestern quadrant of the state, although it almost certainly occurs in wooded areas of that section.

In most places where the species occurs it is cave inhabiting, although females regularly establish nursery colonies outside of caves. At the Tulane University Hebert Center, near Belle Chasse, pipistrelles inhabit the cavelike spaces under the old ammunition storage bunkers during the winter, and tagging has demonstrated that at least certain individuals are there throughout the year. But Jones and Pagels (1968) found no females with young among the pipistrelles present at any time and assumed that they must go elsewhere to bear their offspring.

The possibility exists that some of the bats in this colony migrate north in spring and thereby are responsible for the diminished size of the colony in summer, but I doubt that such is the case. To be sure, some Louisiana specimens are suggestive of the northern subspecies, *P. s. obscurus,* in that they are grayer, less yellowish, than most examples of the southern population. But I do not believe that these individuals represent migrants from Wisconsin, Indiana, Ohio, Ontario, or other states in the northeastern part of the United States where the subspecies *obscurus* occurs. I think instead that these grayish individuals are only color variants of the race *subflavus.* They are represented among speci-

mens from nearly every month of the year, as they would not be if they were migrants from the north. Moreover, one such specimen is labeled as a postlactating female.

Jones and Pagels found more females than males in the Tulane Hebert Center roosts in fall and winter, but in summer the reverse was true. They suggested also that females may depart for more northern regions to bear their young. I think instead that the females merely take up residence in other situations, such as some nearby outdoor retreat, to bear their young. None of the Louisiana specimens of this species is what might be called an extremely young individual only recently on the wing. However, the labels on some of the females in the collection bear the notation that they were lactating at the time of their capture, and other specimens show immaturity to the extent that the epiphyses of the wing bones are not fully ossified.

The Eastern Pipistrelle in Louisiana has been found in hollow stumps, in clusters of moss, in culverts, and in the attics of buildings. Jones and Pagels (1968) report that *Pipistrellus subflavus* exhibits daily torpidity in its roosts under the old ammunition storage bunkers on the Tulane University Hebert Center campus, but their data indicate that the bats frequently move during the winter from one bunker to another. The pipistrelle begins its feeding forays at sundown or shortly thereafter. Its flight is rather weak, erratic, and fluttering. To some it suggests the flight of a butterfly. Like other small bats, the pipistrelle feeds on tiny flying insects. Studies elsewhere show that the young number either one or two and are born in May and June.

Subspecies

Pipistrellus subflavus subflavus (F. Cuvier)

V[espertilio]. subflavus F. Cuvier, Nouv. Ann. Mus. Hist. Nat. Paris, 1, 1832:17; type locality, eastern United States, probably Georgia.
Pipistrellus subflavus, Miller, N. Amer. Fauna, 13, 1897:90.

Specimens from Louisiana are all referable to *P. s. subflavus.*

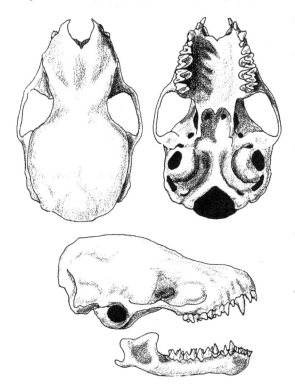

Figure 44 Skull of Eastern Pipistrelle *(Pipistrellus s. subflavus),* ♀, 2.5 mi. SE University, East Baton Rouge Parish, LSUMZ 11138. Approx. 4 ×.

Specimens examined from Louisiana.—Total 129, as follows: *Ascension Par.:* 1 mi. W Dutch Town, 7 (LSUMZ). *Avoyelles Par.:* 2.5 mi. ENE Mansura, 2 (USL). *Claiborne Par.:* Homer, 1 (LTU); 3 mi. E Haynesville, 3 (LTU). *Concordia Par.:* 4 mi. W Ferriday, 3 (LSUMZ). *East Baton Rouge Par.:* various localities, 19 (LSUMZ). *East Feliciana Par.:* Clinton, 1 (LSUMZ). *Grant Par.:* Fishville, 1 (LSUMZ). *Jackson Par.:* Indian Village, 1 (NLU). *Jefferson Par.:* Metairie, 1 (LSUMZ); Marrero, 1 (LSUMZ). *Lincoln Par.:* Ruston, 2 (LSUMZ); 4 mi. S Ruston, 1 (LTU). *Morehouse Par.:* Mer Rouge, 10 (USNM). *Natchitoches Par.:* Kisatchie, 1 (LSUMZ); Vowells Mill, 1 (LSUMZ); Provencal, 3 (LSUMZ); 5 mi. S Robeline, 1 (TU); in cave 2 mi. N Kisatchie, 8 (NSU). *Orleans Par.:* New Orleans, 7 (1 USNM; 6 LSUMZ). *Ouachita Par.:* Bayou Lafourche on St. Hwy. 15, 1 (LSUMZ); 5 mi. SE Monroe, 1 (LSUMZ); Calhoun, 2 (NLU). *Plaquemines Par.:* Tulane Univ. Riverside Campus, Belle Chasse, 20 (TU). *Rapides Par.:* 11 mi. W Forest Hill on Calcasieu River, 1 (LSUMZ). *St. Charles Par.:* Paradis, 1 (LSUMZ). *St. Tammany Par.:* locality unspecified, 3 (2 AMNH; 1 SLU); 1 mi. E Pearl River, 2 (SLU); Slidell, 1 (LSUMZ). *Tangipahoa Par.:* 5 mi. E Ponchatoula, 1 (SLU); 2 mi. S Lee's Landing (10 mi. SE Ponchatoula), 1 (SLU); 7 mi. E Ponchatoula, 1 (SLU); 7 mi. W Ponchatoula, 4 (SLU); 2 mi. E Ponchatoula, 1 (SLU); 2 mi. S Ponchatoula, 1 (SLU). *Tensas Par.:* 0.75 mi. NW Saranac on Clark Bayou, 1 (LSUMZ). *Terrebonne Par.:* Houma, 2 (USNM). *Union Par.:* 13 mi. W Sterlington, 1 (NLU). *Washington Par.:* 10 mi. W Bogalusa, 1 (LSUMZ); locality unspecified, 1 (SLU); 2.75 mi. E, 0.25 mi. N Angie, 1 (SLU); 3 mi. E, 0.25 mi. N Angie, 1 (SLU); Bogalusa, 1 (LSUMZ). *Webster Par.:* 3 mi. N Caney Lake, 2 (LSUMZ); 9 mi. N Minden, 1 (LSUMZ). *Winn Par.:* 2 mi. S Winnfield, 1 (LTU); 5 mi. E Clarence, 1 (NSU).

Figure 45 Big Brown Bat *(Eptesicus fuscus)* in flight, Hamilton County, Ohio, July 1961. Note the numbered aluminum band on the bat's left forearm. (Photograph by Woodrow Goodpaster)

Genus *Eptesicus* Rafinesque

Dental formula:

$$\text{I}\frac{2\text{-}2}{3\text{-}3}\ \text{C}\frac{1\text{-}1}{1\text{-}1}\ \text{P}\frac{1\text{-}1}{2\text{-}2}\ \text{M}\frac{3\text{-}3}{3\text{-}3} = 32$$

BIG BROWN BAT   Plate III and
(Eptesicus fuscus)   Figure 45

Vernacular and other names.—The Big Brown Bat sometimes goes by the names Carolina bat, brown bat, dusky bat, house bat, large brown bat, and serotine bat. The last of these names is derived from the designation of the European species *E. serotinus* and alludes to the fact that the bat is so often seen in late summer. The generic name is of Greek origin and is probably a corruption of the *ptetikos,* meaning able to fly. The specific name *fuscus* comes from the Latin and means brown, which is the bat's basic color. The addition of the word Big serves to distinguish this large species from the brown bats that are small.

Distribution.—This species is not especially widespread in Louisiana, for it has been taken in only 12 parishes. Three of these parishes are in the central-western part of the state, five are in north central and northeastern Louisiana, and four are in the southeastern section. It has not yet been recorded in the center of the state or anywhere in extreme southern Louisiana except at New Orleans. The species ranges across southern Canada from New Brunswick and Quebec to British Columbia south to Baja California, over most of Mexico and Central America to Panama, and eastward to Florida and the Atlantic Coast. It also occurs widely in the Greater Antilles. (Map 12)

External appearance.—Big Brown Bats, as the name implies, are large and brown, and by these characters alone can be easily distinguished from any other species in Louisiana. The color above in different individuals

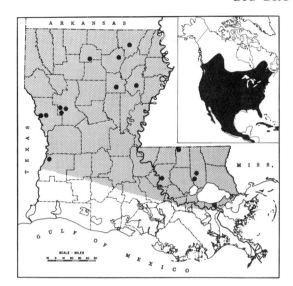

Map 12 Distribution of the Big Brown Bat (*Eptesicus fuscus*). *Shaded area,* known or presumed range within the state; •, localities from which museum specimens have been examined; *inset,* overall range of the species.

varies from a rich, almost golden Sepia or Amber Brown to a much duller Bister or a shade darker. The underparts are Pinkish Cinnamon or Light Vinaceous-Cinnamon to a Pale Olive-Buff. The fur is soft and lax and is about a half inch or slightly less in length in the middle of the back. It is Dark Plumbeous basally and brown for the terminal half. The thick, naked, black or blackish ears are rather short and rounded and, when laid forward, barely reach the nostrils. The membranes are blackish, those of the wing being blacker than the interfemoral membrane, which is naked. The tragus is medium in size, narrowing terminally and having a slight bend forward at the tip (Figure 38). The interfemoral membrane is naked except for a sprinkling of hairs on the basal one-fourth.

Color phases. — An adult male in the Northeast Louisiana University collection at Monroe is unique among the specimens I have examined. It is almost melanistic, being between Fuscous-Black and Sooty Black above and

Brownish Drab below. Its heavily worn dentition excludes the possibility that it might be a young animal.

Measurements and weights. — Nine specimens in the LSUMZ collection averaged as follows: total length, 113.5 (104–125); tail, 45.1 (37–48); hind foot, 9.8 (8.0–9.8); ear, 14.6 (14–15); forearm, 45.3 (44.0–46.9); greatest length of skull, 18.4 (17.5–20.0); cranial height, 6.0 (5.6–6.6); cranial breadth, 9.3 (8.8–10.6); zygomatic breadth, 12.7 (12.1–13.7); interorbital breadth, 4.4 (4.2–4.8); palatal breadth, 8.2 (7.4–9.4); palatilar length, 7.9 (7.3–9.4); postpalatal length, 6.5 (6.2–7.3); maxillary toothrow, 6.9 (6.5–9.4). The only two specimens for which weights are available are two males, one weighing 12.5, the other 15.8 grams.

Cranial and other skeletal characters. — The skull of *Eptesicus fuscus* is large and massive, and its length is greater than that of either *Lasiurus cinereus* or *L. intermedius.* It is not nearly so truncate, and the braincase is much less expanded than it is in either of these two species. The zygomata converge inwardly and form a nearly straight line with the convergence of the maxillae anteriorly. Both *Eptesicus* and *Lasiurus (sensu stricto)* possess 32 teeth, but *Eptesicus* has 2 upper incisors and 1 upper premolar on each side, whereas *Lasiurus* has only 1 upper incisor and usually 2 premolars on each side. The first premolar (P 2) when present is minute and peglike. It is absent in *Lasiurus intermedius* and the other members of the subgenus *Dasypterus.* The upper incisors are well developed, the inner one being distinctly larger than the outer one and having a secondary cusp. The calcar is keeled. (Figure 46)

Sexual characters and age criteria. — The sexes are alike in size and color, but the young differ from the adults by being darker and duller in coloration. The young at birth are naked. The eyes are closed but are said to open on the second day. Walker (*in* Walker et al., 1968) tells of keeping a Big Brown Bat in captivity, which at the time when he was writing had been a pet for 12½ years.

Status and habits. — Though nowhere common in Louisiana, the Big Brown Bat is fairly widespread in the state. Instead, it appears singly or in small groups and always unexpectedly. A cluster of a few individuals may be found in an attic or in some outbuilding or storm sewer. I once discovered in January 1932 several semitorpid individuals behind a shutter on a window of a dormitory on the Louisiana Tech University campus. Occasionally one is captured in a mist net set across a pond or small stream. Big Brown Bats leave their daytime roost in search of food well after sundown and continue on the wing until at least 10 o'clock, at which time I have caught them in nets.

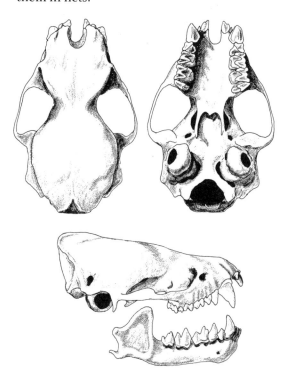

Figure 46　Skull of Big Brown Bat *(Eptesicus f. fuscus)*, ♂, 5 mi. NNE Chipola, St. Helena Parish, LSUMZ 11604. Approx. 2⅗ ×.

The number of young produced by *Eptesicus fuscus* is probably only two (Figure 47), although four are alleged to occur occasionally. Gates (1937), studying a small colony in an old barn near Hammond, in Tangipahoa Parish, wrote as follows:

Until 9 P.M. there was no evidence of any bats. At 10 P.M. there they were, a small colony of them, actively squeaking and jostling each other. They remained until some time between 2 and 3 A.M., for they were gone at the latter hour. The next night, at about 10 o'clock, the whole colony was captured. It consisted of 10 females of *Eptesicus f. fuscus.* All of these, and also the one captured on April 21, obviously were in advanced stages of pregnancy.

It has been noted before that apparently the environment may cause a hibernating bat to ovulate. Especially is this true if the bat be taken from hibernating quarters after February 1 and placed in a warm laboratory (see Guthrie, Jour. Mamm., vol. 14, pp. 199–216, 1933). If this be so, then it is natural to suppose that most of the females of any one bat colony would ovulate at nearly the same time in the spring. This would lead to parturition by the females of any one colony within a relatively short period. In this particular case, all the females of the colony gave birth to young within a period of 48 hours. One female gave birth to one young before 6 A.M. on Saturday, May 16. An hour or so later a second young was born to the same female. By 6 o'clock that evening 4 of the 11 bats had produced young. By 6 o'clock the following morning 5 more had given birth, and by evening all but one had parturiated. The remaining bat gave birth before the next morning, May 18. The 11 bats had 23 young, and later examination showed these young to comprise 10 females and 13 males. Just how many young were born to each female could not be ascertained in every instance, for the majority of the bats were confined together. At least one female had a single young, and 3 other females had 2 apiece. The remaining 7 females produced 16 young.

This record furnishes evidence that all the females in this colony must have ovulated within a short time of each other, probably not more than 48 hours apart. It is hardly likely that any causative agent other than the environment would have induced ovulation almost simultaneously among all these females. This record also tends to support the theory that spring copulations are not necessary for fertilization, for it is rather beyond the range of possibility that all these bats would have

Figure 47 Big Brown Bat *(Eptesicus fuscus)* with suckling young, Clermont County, Ohio, May 1958. (Photograph by Woodrow Goodpaster)

copulated within such a short time. The most plausible explanation seems to be that as a result of fall copulations sperm had been stored in the uterus and oviduct, and that upon favorable spring conditions, ovulation had taken place with resulting fertilization, the same conditions affecting all females in the locality similarly, and eventually producing more or less simultaneous parturition.

The sex ratio at birth appears to be close to 1:1. Further details on the life history and behavior of *Eptesicus fuscus* have been supplied by Phillips (1966) for northeastern Kansas and by Davis et al. (1968) for Kentucky, and some of the facts in these papers may be applicable to Louisiana populations. Much still needs to be learned about Louisiana bats of this species. Consequently, anyone locating a colony of Big Brown Bats should communicate immediately with the mammalogists at the closest state college or university.

Predators.—No information is available regarding predators on Big Brown Bats in Louisiana, but owls and any mammalian carnivores that by chance happen upon a member of this species are almost certain to capture and eat it.

Parasites.—The parasites of the species in this state constitute a subject about which nothing is known; but, in Kansas, Phillips (1966) records a tapeworm *Hymenolepis roudabushi* and a fluke *Plagiorchis micracanthos,* and seven kinds of ectoparasites, including one species of flea, one bat bug, four mites, and the larvae of a tick.

Subspecies

Eptesicus fuscus fuscus (Palisot de Beauvois)

Vespertilio fuscus Palisot de Beauvois, Catalogue raisonné du muséum de Mr. C. W. Peale, Philadelphia, 1796:18 (p. 14 of English edition by Peale and Beauvois); type locality, Philadelphia, Pennsylvania.
Eptesicus fuscus, Méhely. Magyarország denevéreinek monographiája (Monographia Chiropterorum Hungariae), 1900:206 and 338.

The only subspecies in the state is *E. f. fuscus.* There are, however, aberrant specimens that I cannot satisfactorily interpret other than as mere color or age variants. One such specimen in the LSUMZ collections from Sabine Parish is remarkably close to *E. f. pallidus* Young of the Great Plains and the southwestern United States, but it was taken on 8 July and could hardly be a migrant. It therefore probably represents nothing more than a color variant.

Specimens examined from Louisiana.—Total 32, as follows: *Beauregard Par.:* Merryville, 3 (LSUMZ). *Caldwell Par.:* Columbia, 1 (FM). *East Baton Rouge Par.:* Baton Rouge, 2 (LSUMZ). *Franklin Par.:* 1.5 mi. NW Gilbert, 2 (NLU); Gilbert, 1 (NLU). *Lincoln Par.:* Ruston, 4 (LSUMZ). *Morehouse Par.:* Mer Rouge, 1 (USNM). *Natchitoches Par.:* Provencal, 2 (LSUMZ); Vowells Mill, 1 (LSUMZ); Robeline, 5 (NSU). *Orleans Par.:* New Orleans, 1 (USNM). *Ouachita Par.:* Monroe, 1 (NLU). *Sabine Par.:* 7 mi. W Many, 1 (LSUMZ); 15 mi. W Many, 1 (LSUMZ). *St. Helena Par.:* 5 mi. NNE Chipola, 1 (LSUMZ). *Tangipahoa Par.:* Hammond, 4 (3 LSUMZ; 1 SLU); 2 mi. W Ponchatoula, 1 (SLU).

Genus *Lasiurus* Gray
Hairy-tailed Bats

Dental formula:

$$\text{I}\frac{1-1}{3-3}\ \text{C}\frac{1-1}{1-1}\ \text{P}\frac{2-2}{2-2}\ \text{M}\frac{3-3}{3-3}=32$$

or

$$\text{I}\frac{1-1}{3-3}\ \text{C}\frac{1-1}{1-1}\ \text{P}\frac{1-1}{2-2}\ \text{M}\frac{3-3}{3-3}=30$$

RED BAT   Plate III
(Lasiurus borealis)

Vernacular and other names. — The Red Bat is well named, for its color is indeed red. The bats of this genus are often referred to as the Hairy-tailed Bats because their interfemoral membranes are heavily furred. Other names include tree bat and red tree bat. The name *Lasiurus* is derived from the Greek words *lasios* meaning hairy, woolly, or shaggy, and *oura*, the tail. The specific name *borealis* is the Latin word for northern.

Distribution. — The species occurs almost throughout Louisiana. It ranges widely in North America from Nova Scotia and New Brunswick west across southern Canada to southern British Columbia and south to Panama and beyond, as well as east across the Gulf states to northern Florida and the Atlantic Coast but is absent within this area from the Rocky Mountains and the central plateau of Mexico. Both closely related nominal species and subspecies occur in the Greater Antilles and in South America. (Map 13)

External appearance. — This is one of the few species of bats with marked sexual dimorphism. The upperparts of the male are a brick red, whereas the females are a rather purplish red with most of the hairs so strongly washed with white that they produce a frosted appearance. There is a yellowish white patch on each shoulder and likewise thin patches of yellowish hair at the base of the thumb and of the third, fourth, and fifth metacarpals. The interfemoral membrane is

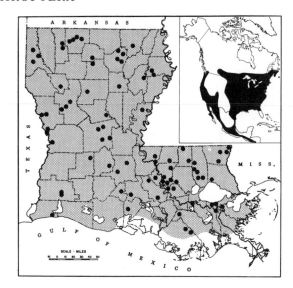

Map 13 Distribution of the Red Bat *(Lasiurus borealis). Shaded area,* known or presumed range within the state; •, localities from which museum specimens have been examined; *inset,* overall range of the species.

heavily furred all the way to the posterior edge, as is also the under surface of the wing along the humerus and radius and the basal portions of the metacarpals. The ears are short and rounded and not furred inside except near the tip. When laid forward they do not reach the nostrils. The tragus is short and broad with a slight forward bend at the top. (Figure 38)

Color phases. — Although considerable variation occurs in the intensity of the red coloration and the amount of frosting on the upperparts, none of these differences appear to qualify as color phases.

Measurements and weights. — Twenty-nine males in the LSUMZ from various localities in Louisiana averaged as follows: total length, 97.5 (81–107); tail, 47.4 (40–54); hind foot, 7.2 (6–11); ear, 10.6 (8.0–13.0); forearm, 38.9 (36.2–46.0); third metacarpal, 38.4 (36.0–40.7); condylobasal length of skull, 12.1 (11.9–13.6); cranial height, 5.6 (4.1–6.1);

cranial breadth, 7.4 (6.8–8.2); zygomatic breadth, 9.0 (8.4–10.3); interorbital breadth, 4.2 (4.0–4.5); palatal breadth, 6.1 (5.9–6.9); palatilar length, 4.1 (3.5–4.7); postpalatal length, 5.0 (4.7–5.4); maxillary toothrow, 4.2 (4.0–4.6). Sixteen males in the LSUMZ from various localities in Louisiana averaged 8.3 (6.5–10.7) grams. Fifty females from the same source averaged as follows: total length, 105.4 (95.0–116.0); tail, 51.8 (42.6–56.0); hind foot, 7.7 (7.0–9.6); ear, 11.1 (8.9–13.0); forearm, 41.2 (36.3–43.4); third metacarpal, 41.4 (38.0–43.6); condylobasal length, 12.9 (12.0–13.8); cranial height, 5.8 (5.4–6.2); cranial breadth, 7.7 (7.1–8.3); zygomatic breadth, 9.8 (9.4–10.5); interorbital breadth, 4.2 (4.0–4.5); palatal breadth, 6.5 (5.8–7.0); palatilar length, 4.8 (4.0–5.0); postpalatal length, 5.3 (4.9–6.0); maxillary toothrow, 4.5 (4.3–5.0). The weights of 38 females (ex-cluding those with well-developed embryos) from the same source averaged 11.8 (8.5–15.2) grams.

Cranial and other skeletal characters. — The skull of *Lasiurus borealis* is broad, short, and deep. The rostrum is nearly as broad as the brain-case. A tiny, peglike first premolar is present (Figure 48 and 49). A cranial difference be-tween *L. borealis* and *L. seminolus* is discussed in the account of the latter species on page 119.

Sexual characters and age criteria. — See the pre-ceding paragraph on external appearance for color differences between the sexes. The suckling young also show the same sexual dimorphism. Females average slightly larger than males, but there is a broad overlap in the measurements of each.

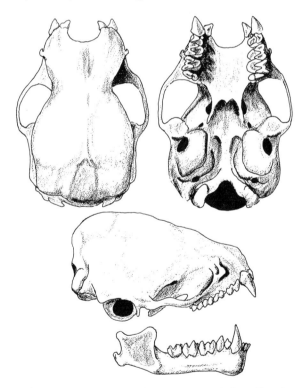

Figure 48 Skull of Red Bat *(Lasiurus b. borealis)*, ♂, 1 mi. W. Dutch Town, Ascension Parish, LSUMZ 11147. Approx. 3½×.

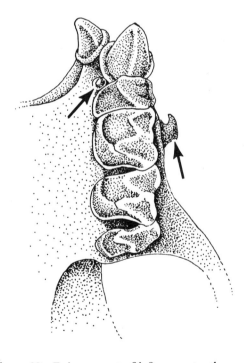

Figure 49 Enlargement of left upper toothrow of the Red Bat *(Lasiurus borealis)*, emphasizing the peglike first premolar and the prominent lacrimal shelf. Approx. 12×.

Status and habits. — This beautiful little bat is apparently one of the more common and widespread species in the state. Like other lasiurine bats, it is solitary in habits and for this reason may not appear to be as common as some of the communal roosters, which are often encountered in large aggregations and leave the impression that they are more abundant. Red Bats roost individually in trees, frequently in clumps of Spanish moss, but also by hanging head downward from twigs by one or both of their hind legs (Figure 50). When at rest one is so well camouflaged that it appears to be a leaf. Not infrequently one is found clinging to the rafter of an open porch or in some other more or less exposed situation around our homes.

In the northern part of its range the Red Bat is at least partially migratory (Davis and Lidicker, 1956), but whether our population

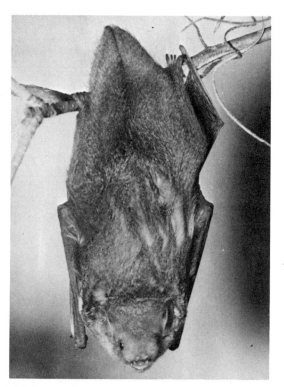

Figure 50 Red Bat *(Lasiurus borealis),* Dutch Town, Ascension Parish. (Photograph by Richard K. LaVal)

in winter is substantially increased by an influx of these migrants is not known. According to Thomas (1921), a migrating flock of approximately 100 individuals of this species and the Silver-haired Bat was recorded in southward flight 20 miles off the coast of North Carolina on 3 September 1920, and the Red Bat is known to have reached the Bermuda Islands, over 700 miles from the nearest land along the Atlantic Coast. Obviously, it is a strong flier and does quite likely filter into Louisiana in winter from the northern part of the United States and southern Canada. Indeed, a specimen picked up dead on 21 November 1953 by Dr. Marshall B. Eyster on the edge of the surf at Holly Beach, which is located on a long stretch of treeless coastline in Cameron Parish, was possibly such a migrant from the north.

For daytime roosts the Red Bat generally selects a site with a southern and western exposure and usually with a clearing beneath into which it can drop when taking flight. Although it sometimes feeds over the tops of trees, it is more often seen pursuing insects in clearings at heights of 10 to 30 feet above the ground. It begins its forays shortly after sunset and is often active up to midnight. Sometimes it seems to follow a somewhat territorial course by returning night after night to the same clearing. One of its favorite places to search for food is around streetlights, and it will sometimes alight on the pole itself to capture its prey. It also will alight on vegetation to capture an insect, and it must also sometimes go close to ground level for the flightless crickets that have been found among the contents of its stomach. (McClure, 1942; Constantine, 1966)

Little or nothing is known about the breeding behavior of Red Bats in Louisiana. In the northern United States mating is said to occur in late summer and early fall (Dearborne, 1946) with ovulation and implantation not taking place until the following spring. But our resident population of this species, which is often active during warm weather throughout the winter, may engage in matings in early spring. In any event the young are born from late May to mid-June, after a gestation period reported to be approximately 80 to 90 days. The females have four mammae and the number of young is one to four, rarely five (Allen, 1939; McClure, 1942; Constantine, 1966; Barbour and Davis, 1969; Hamilton and Stalling, 1972). No known chiropteran except this species and the following one normally gives birth to more than two young. Interestingly enough, the Red Bat is one of the few species that is sexually dimorphic. The color differences between males and females are readily apparent in young that are only a week or so old. A mother will sometimes carry her three or four babies with her on a feeding foray even when the combined weights of the young exceed her own weight. Generally, though, when the young reach this stage of development they are left behind until the mother returns. A female that I obtained at Baton Rouge on 23 June 1971 had four babies no more than a week old. She weighed 10.6 grams, and the young weighed 3.6, 3.3, 3.1, and 3.1 grams, or in aggregate, 13.1 grams.

Gates (1936 and 1938) demonstrated his ability in keeping certain species of eastern bats in captivity and, in the case of this species, of actually rearing individuals from an early age on an artificial diet.

Predators and other decimating factors.—Because the Red Bat roosts in exposed places it is often subjected to a wider range of temperature than are most bats. Consequently, severe freezes and inclement weather, such as cold winter storms, undoubtedly take a considerable toll. Its roosting habits also probably account for the fact that Blue Jays, hawks, and owls are often listed as the main predators on the Red Bat. It has been discovered in the stomach of the Roadrunner (Wilks and Laughlin, 1961) and the Virginia Opossum. And it has been found impaled on thorn bushes and on barbs of a barbwire fence.

Collision with the guy wires of TV towers results in the death of an occasional Red Bat. This type of mortality surprises me because I have not heard of other species of bats hitting these obstacles. In southern Louisiana the collection of Spanish moss for commercial purposes probably results in some diminution of their numbers.

Parasites.—Fleas of the genus *Eptescopsalla* and a bat bug, *Cimex pilosellus,* have been found on the Red Bat. It is also known to harbor trematodes of the genus *Lecithodendrium* and cestodes of the genus *Taenia.*

Subspecies

Lasiurus borealis borealis (Müller)

Vespertilio borealis Müller, Des Ritters Carl von Linné ... vollstandiges Natursystem..., Suppl., 1776:20; type locality, New York.
Lasiurus borealis, Miller, N. Amer. Fauna, 13, 1897:105.

The area comprising the entire eastern two-thirds of the United States and southern Canada is occupied by the nominate race. Other subspecies occur in extreme western United States, southern Mexico, and Central and South America.

Specimens examined from Louisiana.—Total 255, as follows: *Allen Par.:* Oakdale, 1 (LSUMZ). *Ascension Par.:* 1 mi. W Dutch Town, 4 (LSUMZ). *Caddo Par.:* Shreveport, 2 (1 LSUMZ; 1 USNM); 1 mi. N Hwy.1, 5 mi. N Blanchard, 1 (LSUMZ); Shreveport, 7mi. SW, 1 mi. N I-20, 1 (LSUMZ). *Calcasieu Par.:* Lake Charles, 6 (2 LSUMZ; 4 MSU); McNeese State College, Lake Charles, 1 (MSU). *Caldwell Par.:* Columbia, 3 (FM). *Cameron Par.:* Holly Beach, 1 (USL). *Claiborne Par.:* 7.5 mi. NNE Homer, 2 (LSUMZ); 3.5 mi. N Homer, 5 (LSUMZ); 4.5 mi. SW Summerfield, 3 (LSUMZ); 3.5 mi. W Homer, 2 (LSUMZ). *Concordia Par.:* 4 mi. W Ferriday, 1 (LSUMZ). *East Baton Rouge Par.:* various localities, 33 (LSUMZ). *East Feliciana Par.:* Reiley, 1 (LSUMZ). *Evangeline Par.:* 3 mi. S Ville Platte, 1 (USL). *Franklin Par.:* 13.5 mi. NE Gilbert, 2 (LSUMZ). *Grant Par.:* Fishville, 1 (LSUMZ). *Iberia Par.:* Avery Island, 1 (TU). *Iberville Par.:* St. Gabriel, 5 (LSUMZ); Grosse Tete, 1 (LSUMZ); Pigeon, 1 (LSUMZ). *Jackson Par.:* 6 mi. S Ruston, 1 (LSUMZ). *Jefferson Par.:* Gretna, 1 (LSUMZ); Metairie, 3 (LSUMZ); Marrero, 1 (LSUMZ); Harvey, 1 (LSUMZ). *Lafayette Par.:* Lafayette, 13 (12 USL; 1 USNM). *Lafourche Par.:* 2.5 mi. W, 1.3 mi. N Thibodaux, 1 (LSUMZ). *Lincoln Par.:* Ruston, 9 (3 LSUMZ; 6 LTU). *Livingston Par.:* Harrell's Ferry, 1 (LSUMZ). *Madison Par.:* McGill Bend of Tensas River, 2

(LSUMZ); 11 mi. SSW Tendal on Tensas River, 1 (LSUMZ). *Morehouse Par.:* Mer Rouge, 1 (LTU). *Natchitoches Par.:* Provencal, 2 (LSUMZ); 2 mi. W Natchitoches, 2 (LSUMZ); 5 mi. S Robeline, 1 (TU); Natchitoches, 5 (NSU). *Orleans Par.:* New Orleans, 16 (13 LSUMZ; 1 USNM; 2 ANSP). *Ouachita Par.:* 5 mi. SE Monroe, 17 (LSUMZ); 6 mi. SE Fairbanks, 1 (NLU); Sterlington, 4 (NLU); Monroe, 5 (2 NLU; 3 LSUMZ). *Rapides Par.:* Deville, 20 mi. SE Pineville, 1 (LSUMZ); 1.5 mi. WSW Lecompte, 4 (USL); Alexandria, 2 (1 USNM; 1 LTU); Pineville, 1 (USNM). *Red River Par.:* Coushatta, 5 (NSU). *Sabine Par.:* 7 mi. W Many, 4 (LSUMZ); 15 mi. W Many, 7 (LSUMZ); 7.5 mi. SW Toro, 1 (TU); Anthony's Ferry, 8 mi. SW Toro, 1 (TU); 2 mi. S Toro, 1 (TU). *St. Bernard Par.:* Chalmette, 1 (LSUMZ). *St. Helena Par.:* 0.25 mi. W Hillsdale, 1 (LSUMZ); 5 mi. NNE Chipola, 3 (LSUMZ). *St. John the Baptist Par.:* Edgard, 1 (LSUMZ). *St. Landry Par.:* Palmetto, 1 (LSUMZ); 2 mi. N Palmetto, 1 (LSUMZ); Opelousas, 1 (USL). *St. Martin Par.:* St. Martinville, 1 (USL). *St. Tammany Par.:* locality unspecified, 1 (AMNH); Madisonville, 1 (LSUMZ). *Tangipahoa Par.:* Hammond, 6 (SLU); 3.5 mi. E Hammond, 1 (SLU); 3.75 mi. E. Hammond, 2 (SLU); 7 mi. E Ponchatoula, 1 (SLU); 2 mi. W Natalbany, 1 (SLU); Ponchatoula, 5 (LSUMZ); Amite, 1 (LSUMZ). *Terrebonne Par.:* Houma, 1 (LSUMZ); Bourg, 1 (LSUMZ). *Vernon Par.:* 4 mi. N Burr Ferry, 2 (LSUMZ). *Washington Par.:* 4 mi. E Enon, 1 (LSUMZ); LSU Forestry Camp, Sheridan, 1 (LSUMZ); 2.75 mi. E, 0.25 mi. N Angie, 3 (LSUMZ); 1 mi. N Angie, 1 (SLU); 3 mi. E Angie, 2 (SLU); locality unspecified, 1 (SLU); Varnado, 1 (TU). *Webster Par.:* 5 mi. N Minden, 1 (LSUMZ); 9 mi. N Minden, 3 (LSUMZ). *West Baton Rouge Par.:* locality unspecified, 2 (USNM). *West Feliciana Par.:* 6 mi. SE St. Francisville, 2 (LSUMZ); 4.2 mi. ENE St. Francisville, 9 (LSUMZ). *Winn Par.:* 8 mi. W Winnfield, 1 (LSUMZ).

SEMINOLE BAT   Plate III
(Lasiurus seminolus)

Vernacular and other names. — This species is also sometimes called the mahogany bat. The specific name *seminolus* refers to the region inhabited by the Seminole Indians from which the bat was first known.

Distribution. — The Seminole Bat has been recorded from numerous places in Louisiana but not yet from all sections of the state. In summer it extends from the coastal and Piedmont regions of North and South Carolina south to southern Florida and west across Georgia, Alabama, Mississippi, Louisiana, southern Arkansas, and southern Texas into northeastern Mexico. In late summer it apparently wanders northward, for it has been

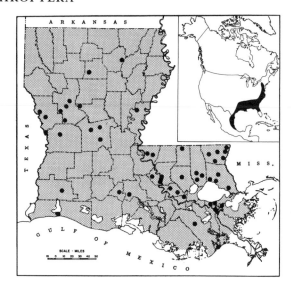

Map 14 Distribution of the Seminole Bat *(Lasiurus seminolus). Shaded area,* known or presumed range within the state; •, localities from which museum specimens have been examined; *inset,* overall range of the species.

taken at that time in Pennsylvania and southern New York. (Map 14)

External appearance. — The species is very similar to the Red Bat, from which it differs only in color and by one minor cranial character. The color, instead of being red or reddish, is a deep mahogany, with a light frosting above produced by the white tips of the hairs of the dorsum. Like the Red Bat it has small, rounded ears, a heavily furred interfemoral membrane, patches of hair at the base of the thumb and the third, fourth, and fifth fingers, hair along the under surface of the upper arm and forearm, and a white patch on each shoulder.

Color phases. — None.

Measurements and weights. — Twelve males in the LSUMZ from various localities in Louisiana averaged as follows: total length, 97.7 (89–114); tail, 39.7 (35–45); hind foot, 8.3 (6–11); ear, 9.0 (7–13); forearm, 39.7

(37.2–43.0); third metacarpal, 40.7 (39.5–43.5); condylobasal length, 12.3 (11.9–12.9); cranial height, 5.7 (5.5–6.1); cranial breadth, 7.3 (7.1–7.6); zygomatic breadth, 9.2 (9.0–9.4); interorbital breadth, 4.2 (4.0–4.5); palatal breadth, 6.2 (6.0–6.4); palatilar length, 4.0 (3.8–4.5); postpalatal length, 5.1 (4.8–5.5); maxillary toothrow, 4.1 (4.0–4.4). One male weighed 9.3 grams. Twelve females, also from various localities in Louisiana, averaged as follows: total length, 103.5 (90–113); tail, 45.5 (40–50); hind foot, 8.2 (6–9); ear, 11.1 (9–13); forearm, 40.7 (38.2–42.0); third metacarpal, 42.0 (41.2–43.4); condylobasal length, 12.9 (12.3–13.9); cranial height, 5.7 (5.4–6.1); cranial breadth, 7.7 (7.4–8.1); zygomatic breadth, 9.6 (9.4–10.3); interorbital breadth, 4.2 (4.1–4.3); palatal breadth, 6.5 (6.0–6.9); palatilar length, 4.3 (4.1–4.9); postpalatal length, 5.4 (5.3–5.5); maxillary toothrow, 4.4 (3.9–4.9). Weights of five females, none with well-developed embryos, averaged 10.7 (8.6–12.5) grams.

Cranial characters.—The skull is almost identical with that of the Red Bat. One minor difference, however, is the presence in the latter of a more pronounced ridge above the lacrimals (Handley, 1960; Hall and Jones, 1961). This character can be observed best in dorsal view and is highly diagnostic in separating the two species. It thereby contradicts the frequently repeated statement in the literature that the two species differ only in color. (Figures 51 and 52)

Sexual characters and age criteria.—Unlike the Red Bat, this species shows no appreciable color differences between the sexes. Young only two weeks of age are fully furred and appear almost identical in color to the mother.

Status and habits.—The Seminole Bat is a common species in certain parts of the state (see Map 14). Its apparent absence in some sec-

tions may reflect nothing more than fortuitous circumstances or insufficient effort in searching for it. Like the Red Bat, it is solitary in habits and roosts in clumps of Spanish moss. Although active throughout the year, it becomes torpid during periods of cold weather. Hays (1941) reported finding two "hibernating" in dense clumps of Spanish moss collected from oak trees in Natchitoches Parish in winter. And Constantine (1958a) noted in southwestern Georgia that Seminole Bats do not fly when the temperature drops below 68°F, but that at 70° F they launch immediately into flight when disturbed.

Also like the Red Bat, this species sometimes gives birth to as many as four young. There is little precise information on the breeding of the species in the state, but suckling young have been taken with a female on

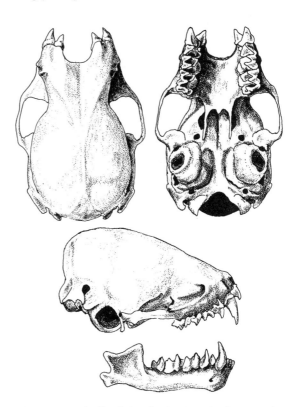

Figure 51 Skull of Seminole Bat *(Lasiurus seminolus)*, ♂, Baton Rouge, East Baton Rouge Parish, LSUMZ 9301. Approx. 3½ ×.

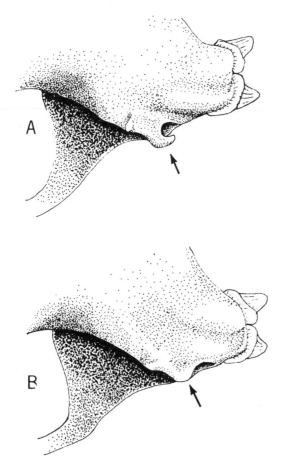

Figure 52 Lacrimal shelf of (a) *Lasiurus borealis* (LSUMZ 11724) and (b) *Lasiurus seminolus* (LSUMZ 11613).

20 June at Baton Rouge. A female obtained on 2 June at Pineville, in Rapides Parish, had three embryos of unspecified size, and one of two females collected on 3 June a few miles from St. Francisville had a single embryo measuring 16 mm in crown-rump length and the other had two embryos measuring 18 mm. Still another female, obtained on 6 July not far outside of Natchitoches, was labeled as postlactating. I strongly suspect that most females give birth to their young by the end of the second week of June.

Various authors, including myself, have suggested the possibility that the Seminole Bat may be nothing more than a color phase of the Red Bat. I am now convinced that such is not the case. Their ranges broadly overlap in the southeastern United States and no intermediates or mixed litters have yet been found. Despite their morphological similarity, the Seminole Bat does possess one minor but significant cranial character that serves to identify it (Figure 52). Moreover, there is a possibility that the two possess slight behavioral differences. Barbour and Davis in their superb treatise *Bats of America* (1969) expressed the belief that the Seminole Bat seldom flies close to the ground, as does the Red Bat, because they were unable to catch it in their nets except when the nets were set above roof level. My students and I, however, have occasionally netted the species a few feet above ground level.

Predators. — Because of its close similarity to the Red Bat in its manner of roosting in rather exposed situations, this species is doubtless subjected to much the same decimating factors as the Red Bat. Except for predation by birds, notably Blue Jays, and losses resulting from adverse weather, I suspect that commercial moss collecting destroys more Seminole Bats than any other factor.

Parasites. — Little or nothing is known about the internal and external parasites of this species in Louisiana.

Lasiurus seminolus (Rhoads)

Atalpha borealis seminola Rhoads, Proc. Acad. Nat. Sci. Philadelphia, 47, 1895:32; type locality, Tarpon Springs, Pinellas County, Florida.
Lasiurus borealis seminolus, Miller, N. Amer. Fauna, 13, 1897:109.
Lasiurus seminolus, Poole, Jour. Mamm., 13, 1932:162.
Lasiurus seminola, Lowery, Proc. Louisiana Acad. Sci., 3, 1936:17.

This species is monotypic.

Specimens examined from Louisiana. — Total 93, as follows: *Ascension Par.:* Sorrento, 1 (LSUMZ); 1 mi. W Dutch Town, 1 (LSUMZ). *Calcasieu Par.:* Westlake, 1 (LSUMZ); Lake Charles, 2 (MSU). *Cameron Par.:* Cameron, 1 (LSUMZ). *Concordia Par.:* Frogmore, 1 (LSUMZ). *East*

Baton Rouge Par.: various localities, 17 (LSUMZ). *Grant Par.:* 2.5 mi. N Montgomery, 1 (LSUMZ). *Jackson Par.:* Jonesboro, 1 (LSUMZ). *Jefferson Par.:* Grand Isle, 1 (LSUMZ); Metairie, 1 (LSUMZ); Gretna, 1 (LSUMZ). *Lafayette Par.:* Lafayette, 2 (USL). *Lafourche Par.:* 7 mi. SE Raceland, 1 (USL); Lockport, 1 (LSUMZ). *Livingston Par.:* 2.5 mi. S Livingston, 3 (TU). *Natchitoches Par.:* Provencal, 1 (LSUMZ); Kisatchie, 1 (LSUMZ); Vowells Mill, 1 (LSUMZ); 2 mi. W Natchitoches, 1 (LSUMZ). *Orleans Par.:* New Orleans, 2 (1 LSUMZ; 1 USNM). *Ouachita Par.:* Russell Sage Game Mgt. Area, Monroe, 1 (NLU). *Plaquemines Par.:* Belle Chasse, 1 (LSUMZ). *Rapides Par.:* 0.5 mi. SE Kincaid, 1 (LSUMZ); Pineville, 2 (LSUMZ); 1.5 mi. WSW Lecompte, 1 (USL). *Sabine Par.:* 15 mi. W Many, 1 (LSUMZ); 7.5 mi. SW Toro, 1 (TU); 8 mi. S Toro, 3 (TU); Wall Creek, 2 mi. S Negreet, 1 (TU). *St. Bernard Par.:* Chalmette, 2 (LSUMZ). *St. Helena Par.:* 5.4 mi. NNE Chipola, 1 (LSUMZ). *St. James Par.:* Blind River, 2 (LSUMZ). *St. Martin Par.:* St. Martinville, 3 (USL). *St. Tammany Par.:* Abita Springs, 1 (SLU); Madisonville, 1 (USNM); Mandeville, 1 (USNM); Slidell, 1 (LSUMZ). *Tangipahoa Par.:* 3 mi. SE Ponchatoula, 1 (LSUMZ); 7 mi. E Ponchatoula, 1 (SLU); 1 mi. N Hammond, 2 (SLU); 3.75 mi. E Hammond, 3 (SLU); Hammond, 1 (SLU). *Vernon Par.:* 2 mi. W Leesville, 2 (TU). *Washington Par.:* 1 mi. NW Franklinton, 1 (LSUMZ); LSU Forestry Camp, Sheridan, 1 (LSUMZ); 1 mi. S Enon, 1 (LSUMZ); Bogalusa, 2 (LSUMZ); 2.75 mi. E, 0.25 mi. N Angie, 1 (LSUMZ); 3 mi. E Angie, 1 (SLU); 3.25 mi. E, 0.25 mi. N Angie, 1 (SLU); Varnado, 1 (LSUMZ). *West Baton Rouge Par.:* 10 mi. W Port Allen, 1 (LSUMZ); locality unspecified, 3 (USNM). *West Feliciana Par.:* 4.2 mi. ENE St. Francisville, 3 (LSUMZ); Hwy. 66 on west side of Bains, 1 (TU).

HOARY BAT   Plate III
(Lasiurus cinereus)

Vernacular and other names. — This bat derives its vernacular name from the frosting of white on its fur. The specific name *cinereus* comes from the Latin and means ash colored and again refers to the white-tipped hairs of its pelage.

Distribution. — Hoary Bats range widely in North America from Nova Scotia north to the Northwest Territories and MacKenzie south to southern British Columbia and over the entire length and breadth of the United States except southern Florida, thence over most of Mexico. In Louisiana the species has been recorded in only 12 parishes. (Map 15)

External appearance. — This large, attractive bat is easily recognized by its size and distinctive coloration. It is a dark brown or umber with the hairs both on the dorsum and the belly strongly tipped with white. When the fur is separated by blowing into it the hairs are found to be strongly banded. Basally they are blackish, followed by a wide yellowish band, then a wide dark brown band, and finally a white tip. The face and chin are yellow, and the short, round ears are furred inside except on the black rim. The interfemoral membrane is densely furred above except on the lateral corners near the feet and along the posterior borders. Beneath, the membrane is bare of fur except near the base of the tail. Dense patches of yellowish fur extend along the median ventral border of the wings adjacent to the body and along the upper arm and forearm. The wrist and the membrane at the base of fifth metacarpal show small patches of yellowish-white hair and there are often small patches of the same color near the elbow. The species cannot be confused with any other. It is the largest bat in Louisiana and in size is approximated only by the next

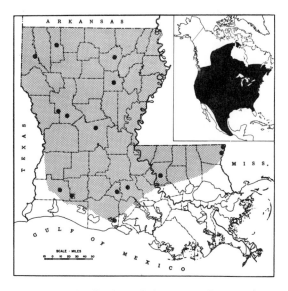

Map 15 Distribution of the Hoary Bat *(Lasiurus cinereus)*. *Shaded area,* known or presumed range within the state; •, localities from which museum specimens have been examined; *inset,* overall range of the species.

species, the Northern Yellow Bat. Its unique coloration renders it easily distinguishable from that species.

Measurements and weights. — Twelve specimens in the LSUMZ from various localities in North America (only eight or nine specimens involved in skull measurements) averaged as follows: total length, 131.4 (122–144); tail, 56 (50–67); hind foot, 10.8 (9–12); ear, 17.0 (15–19); forearm, 53.1 (48.4–57.0); third metacarpal, 57.7 (48.9–62.8); greatest length of skull, 16.8 (16.0–17.7); cranial height, 6.9 (6.6–7.2); cranial breadth, 9.6 (9.1–10.1); zygomatic breadth, 12.1 (11.6–12.8); interorbital breadth, 5.2 (5.0–5.7); palatal breadth, 8.6 (8.3–9.1); palatilar length, 5.1 (4.5–5.7); postpalatal length, 7.2 (6.9–7.8); maxillary toothrow, 5.7 (5.4–6.3). Two males weighed 16.7 and 17.7 grams, and two females, 21.5 and 34.5 grams.

Cranial and other skeletal characters. — The skull is quite similar to that of the two preceding species but, because the animal is much larger in all dimensions, the skull is likewise considerably larger. It is relatively heavy and broad, and decidedly truncate. The skull is definitely the largest of any bat in Louisiana. The calcar is moderately keeled. (Figure 53)

Sexual characters and age criteria. — The sexes are alike except that females average slightly larger. I have not seen the young of this species, but they are said to be a pale gray.

Status and habits. — Only 17 specimens of this handsome bat have been taken in the state. Its overall range is wide, but nowhere does it appear to be common. The Hoary Bat is, however, highly migratory and occasionally appears in numbers along the pathways between its summer and winter abodes. For example, Barbour and Davis (1969) tell of netting 59 near Portal, Arizona, on 18 June 1966, and Findley and Jones (1964) noted migratory waves in May and June in New Mexico. They captured 48 near Glenwood on

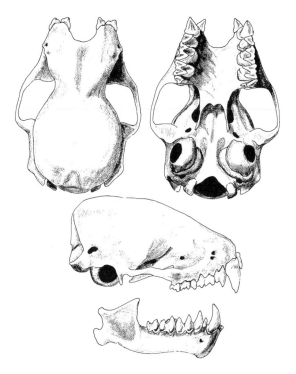

Figure 53 Skull of Hoary Bat (*Lasiurus cinereus*), ♂, 2.75 mi. E, 0.25 mi. N Angie, Washington Parish, LSUMZ 9233. Approx. 3×.

the nights of 13 and 14 May 1960 and even larger numbers during June. During August and September very few were taken despite intensive netting operations.

Of the 17 Louisiana specimens 2 were obtained in January, 4 in February, 4 in March, 2 in May, 3 in June, 1 in July, and 1 in October. Findley and Jones (1964) discounted the importance of the southeastern states as a wintering area for the Hoary Bat, but the Louisiana data seem not to support this conclusion, for 10 of the 17 records from the state were in the winter or early spring months. That the species breeds in the state is almost unquestionable. A specimen taken by Richard K. LaVal in Webster Parish on 15 July 1966 that was labeled as a postlactating female initially suggested that it might do so, but on 11 June 1973, Horace H. Jeter obtained a dead juvenile that had been found in a yard in Shreveport. It was too young to fly and had evidently fallen from its mother.

Strong presumptive evidence that the species breeds in Arkansas is available (Baker and Ward, 1967).

Hoary Bats are strong fliers. They normally appear well after sunset and they fly with the tail and interfemoral membrane curved forward under the body, in contrast to the Red and Seminole Bats, which fly with their tails extended except when using the membranes to assist in the capture of an insect. In many parts of its range the Hoary Bat seems to be associated with coniferous forests. Most, but not all, specimens obtained in Louisiana were taken in forested areas in which pine predominated.

Predators and other mortality factors.—Some of these bats are perhaps lost in the course of their long migrations, and almost certainly they are preyed upon by hawks and owls.

Subspecies

Lasiurus cinereus cinereus (Palisot de Beauvois)

Vespertilio cinereus Palisot de Beauvois, Catalogue raisonné du muséum de Mr. C. W. Peale, Philadelphia, 1796:18; type locality, Philadelphia, Pennsylvania.
Lasiurus cinereus, H. Allen, Smiths, Misc. Coll., 7 (publ. 165), 1864:21.—Lowery, Occas. Papers Mus. Zool., Louisiana State Univ., 13, 1943:223.
Lasiurus cinereus cinereus, Crain and Packard, Jour. Mamm., 47, 1966:324.

The species is polytypic with a subspecies in South America and another on the Hawaiian Islands. All North American populations belong to the nominate race.

Specimens examined from Louisiana.—Total 17, as follows: *Caddo Par.:* Shreveport, 1 (LSUMZ). *Calcasieu Par.:* Lake Charles, 1 (MSU); Bell City, 1 (MSU). *Caldwell Par.:* Columbia, 1 (USL). *East Baton Rouge Par.:* 4 mi. E Baton Rouge, 1 (LSUMZ). *Lafayette Par.:* Lafayette, 1 (LSUMZ). *Natchitoches Par.:* Vowells Mill, 1 (LSUMZ); Long Leaf Vista, Kisatchie National Forest, 1 (USL). *Ouachita Par.:* Monroe, 2 (1 NLU; 1 LSUMZ). *Rapides Par.:* Pineville, 1 (USNM; wet preserved, no. 52968). *St. Martin Par.:* 4 mi. E Breaux Bridge, 1 (USL). *Vermilion Par.:* 7 mi. SE Intracoastal City at Red Fish Point, 1 (USL). *Washington Par.:* 2.75 mi. E, 0.25 mi. N Angie, 2 (1 LSUMZ; 1 SLU); Varnado, 1 (LSUMZ). *Webster Par.:* 9 mi. N Minden, 1 (LSUMZ).

NORTHERN YELLOW BAT   Plate III
(Lasiurus intermedius)

Vernacular and other names.—In Louisiana this species has been called the Florida bat and Florida yellow bat. Harrison Allen proposed the specific name *intermedius* in 1894 because he thought the new species he was naming was intermediate in size between what he believed to be two species of Hoary Bats, the North American *L. cinereus* and a closely allied form from South America that is now treated as a subspecies, *L. c. villosissimus.* His misconception is of no great consequence, and his name *intermedius* for this species stands as the earliest name applied to it. The word Northern must be added to the name to distinguish this species from its tropical relative, the Southern Yellow Bat, *Lasiurus ega.*

Distribution.—The Northern Yellow Bat occurs on the Atlantic coastal plain from extreme southeastern Virginia, North and South Carolina, southern Georgia, and over all of Florida, thence west along the Gulf Coast of southern Alabama, Mississippi, and Louisiana to southwestern Texas and from there southward along the lowlands of eastern and southern Mexico and across the Yucatan Peninsula; it also occurs in Cuba. In Louisiana it is found in the southern part of the state with specimen records as far north as Winn and Concordia parishes. (Map 16)

External appearance.—This is the second largest bat in the state. It has long wings, short ears, and an interfemoral membrane that is furred on the anterior half of the dorsal surface. The color is usually yellowish above and below, but variants are sometimes grayish or even slightly brownish. Its overall yellowish color clearly separates it from any other species in the state except possibly from some specimens of *Pipistrellus subflavus,* from which it differs in being decidedly larger.

Measurements and weights.—Sixteen specimens (12 with body measurements) in the LSUMZ

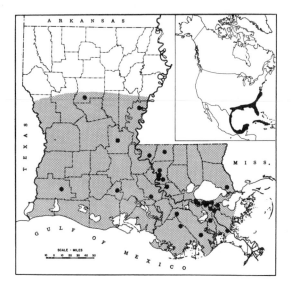

Map 16 Distribution of the Northern Yellow Bat (*Lasiurus intermedius*). *Shaded area,* known or presumed range within the state; •, localities from which museum specimens have been examined; *inset,* overall range of the species.

collections from Lousiana averaged as follows: total length, 123.5 (113.0–132.7); tail, 56.2 (50.0–62.0); hind foot, 9.5 (7.6–11.0); ear, 14.6 (9.0–16.0); forearm, 49.2 (47.0–54.0); third metacarpal, 52.9 (49.2–56.7); condylocanine length of skull, 17.5 (17.0–18.4); cranial height, 7.3 (6.6–7.6); cranial breadth, 9.8 (9.3–10.5); zygomatic breadth, 13.0 (12.1–13.9); interorbital breadth, 4.9 (4.5–5.3); palatal breadth, 8.9 (8.3–8.9); palatilar length, 5.9 (5.5–6.4); post-palatal length, 7.0 (6.5–8.3); maxillary tooth-row, 6.2 (5.9–6.6). Four males averaged 17.0 (14.0–20.0) grams. A female with three embryos 15 mm in crown-rump length weighed 31.2 grams.

Cranial and other skeletal characters. — The skull of *Lasiurus intermedius* is large and robust and decidedly truncate although not as much so as the skull of *L. cinereus.* The tiny, peglike premolar that in the Red, Seminole, and Hoary Bats is located on the lingual edge of the toothrow at the junction of the canine and

the large premolar, is absent in the Northern Yellow Bat. The calcar is only slightly keeled. (Figure 54)

Sexual characters and age criteria. — The sexes are similar, if not identical, insofar as color is concerned. The young resemble the adults.

Status and habits. — The Northern Yellow Bat is a fairly common resident in the southern part of the state. It has been taken at all times of the year, and, although females with extremely small young have not yet been found, partially grown young and females with embryos have been secured in late May and June. At Baton Rouge on 22 May, Michel Buquoi collected a female with four embryos, and on 3 June, near St. Francisville, Richard K. LaVal netted a female with three embryos that measured 15 mm in crown-rump length. Mating, as in many vespertilionid bats, prob-

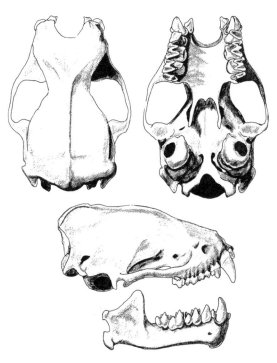

Figure 54 Skull of Northern Yellow Bat (*Lasiurus intermedius*), ♀, 4.2 mi. ENE St. Francisville, West Feliciana Parish, LSUMZ 11742. Approx. 2½ ×.

ably takes place in the fall, though the ova are not shed and fertilization does not take place until the following spring. All young are probably on the wing by the end of June or the first week of July.

Yellow bats begin to leave their solitary roosts in the leaves of trees and in Spanish moss well before dark, often as early as sunset or shortly thereafter. They are strong, direct fliers, and are usually readily recognized by their large size and strong wing beats. In this regard they could be confused only with the rare Hoary Bat or perhaps the Big Brown Bat. I used to watch them regularly in summer as they left their roost somewhere in the Collegetown subdivision south of the LSU campus and commenced a flight high above the trees and cane fields in the direction of the Mississippi River, a mile and a half to the west. There they fed over the river itself or along the battures.

Predators and other mortality factors. — All tree-roosting species of bats may be presumed to have much the same predators and to be subjected to the same causes of death, such as inclement weather.

Parasites. – I have not located a reference to any internal or external parasites having been found in the Northern Yellow Bat.

Subspecies

Lasiurus intermedius floridanus (Miller)

Dasypterus floridanus Miller, Proc. Acad. Nat. Sci., Philadelphia, 54, 1902:392; type locality, Lake Kissimmee, Osceola County, Florida. — Lowery, Proc. Louisiana Acad. Sci., 3, 1936:17; Occas. Papers Mus. Zool., Louisiana State Univ., 13, 1943:223.

Lasiurus intermedius floridanus, Hall and Jones, Univ. Kansas Publ. Mus. Nat. Hist., 14, 1961:84. — Barbour and Davis, Univ. Press Kentucky, 1969:149. — Handley, Proc. U.S. Nat. Mus., 112, 1960:478.

Louisiana populations of this species are somewhat intermediate in measurements between *L. i. intermedius* of Texas and *L. i. floridanus* of Florida but are closer to the latter.

The following taxonomic comments pertain to the question of the generic allocation of the yellow bats to *Lasiurus* instead of a separate genus *Dasypterus*. The species *intermedius* was originally described in 1862 as a member of the genus *Lasiurus,* but later, in 1871, Peters erected the genus *Dasypterus* for it, mainly because it has only one premolar on each side of the upper jaw. Subsequently, several other species were named in the genus *Dasypterus* or transferred to it, including *D. floridanus* Miller from Florida. The current trend, however, is to treat *Dasypterus* as congeneric with *Lasiurus.* I am not wholly convinced that this course is wise even though I am aware that *Lasiurus* and *Dasypterus* have certain features in common. Their differences are many. In addition to the loss of the small, peglike premolar, *Dasypterus* lacks a completely furred dorsal surface of the interfemoral membrane, possesses a bold sagittal crest on the skull, and has decidedly more pointed ears. Its members are yellowish in color instead of some shade of red, and are without prominent white patches on the shoulders and wrists. *Lasiurus (sensu stricto),* on the other hand, consists of species in which the peglike P 2 is present, the interfemoral membrane is heavily furred above, a bold saggital crest is lacking, the ears are short and rounded, the color is some shade of red instead of yellow, and the wrists and shoulders show prominent white or whitish patches. Consequently, the Red, Seminole, and Hoary Bats, along with their close relatives in South America, comprise a compact group of species that I myself would prefer to retain in *Lasiurus* while setting the Northern Yellow Bat, Southern Yellow Bat, and their closely related species in South America apart in the genus *Dasypterus.* However, I have here subscribed to the treatment followed by most present-day workers by synonymizing *Dasypterus* with *Lasiurus.* A compromise treatment would be to accord subgeneric rank to *Dasypterus,* thereby showing that the species included therein are more like each other than they are like any of the species in the sub-genus *Lasiurus*.

Specimens examined from Lousiana. — Total 112, as follows: *Ascension Par.:* 1 mi. W Dutch Town, 1 (LSUMZ). *Avoyelles Par.:* 3.5 mi. ENE Mansura, 1 (USL). *Calcasieu Par.:* Lake Charles, 2 (1 LSUMZ; 1 MSU). *Concordia Par.:* 4 mi. W Ferriday, 2 (LSUMZ). *East Baton Rouge Par.:* various localities, 40 (LSUMZ). *East Feliciana Par.:* Clinton, 1 (LSUMZ). *Jefferson Par.:* Moisant Airport, Kenner, 2 (TU); Marrero, 10 (LSUMZ); Kenner, 3 (LSUMZ); Harahan, 1 (LSUMZ); Gretna, 2 (LSUMZ); Metairie, 3 (LSUMZ); Lafitte, 1 (LSUMZ); Harvey, 1 (LSUMZ); Westwego, 1 (LSUMZ). *Iberville Par.:* North Island, Grand Lake, 1 (LSUMZ). *Lafayette Par.:* Lafayette, 2 (USNM). *Lafourche Par.:* Galliano, 4 (LSUMZ); near Thibodaux, 2 (LSUMZ). *Orleans Par.:* New Orleans, 19 (18 LSUMZ; 1 SLU). *Plaquemines Par.:* 2 mi. N Port Sulphur, 1 (SLU); Belle Chasse, 1 (LSUMZ). *St. Bernard Par.:* Chalmette, 1 (LSUMZ). *St. Charles Par.:* Sarpy Swamp, 2 mi. S Norco, 1 (TU); 2 mi. E Norco, 1 (TU). *St. Tammany Par.:* Slidell, 1 (LSUMZ). *Terrebonne Par.:* Houma, 3 (USNM). *West Feliciana Par.:* 4.2 mi. ENE St. Francisville, 3 (LSUMZ). *Winn Par.:* 18 mi. E Clarence, 1 (NSU).

Genus *Nycticeius* Rafinesque

Dental formula:

$$I\frac{1-1}{3-3} \ C\frac{1-1}{1-1} \ P\frac{1-1}{2-2} \ M\frac{3-3}{3-3} = 30$$

EVENING BAT   Plate II
(*Nycticeius humeralis*)

Vernacular and other names. — This species is sometimes called the twilight bat. The generic name is derived from the Greek *nyktos*, meaning night, and the Latin *eius* denoting belonging to, hence belonging to the night. The specific name is from the Latin *humerus*, referring to the upper arm bone, and *alis*, meaning pertaining to.

Distribution. — The Evening Bat occurs in the southeastern quadrant of North America from northeastern Nebraska, northern Illinois, southern Michigan, and central Pennsylvania south to northeastern Mexico, the northern Gulf Coast, Florida, and Cuba. It probably is present in every parish of Louisiana, but a few remain from which it has not yet been recorded. (Map 17)

External appearance. — The species is a small to medium-sized bat with short, narrow wings and is dark brown above and yellowish brown below. The body fur is rather dense and long and when parted is revealed to be blackish on the basal half and then brown. An occasional individual shows a faint grayish cast and the venter is sometimes dark grayish with no tinge of yellow. The ears are small, rather thick, and rounded at the tips. The tragus is short and blunt with a slight bend forward at the tip (Figure 38). The interfemoral membrane is virtually hairless and the wing membranes reach to the toes. The Evening Bat can be confused only with *Myotis austroriparius*, from which it is distinguished at a glance by its short, blunt tragus and by the presence of only one incisor on each side of the upper jaw. The Southeastern Myotis has a long, pointed tragus and two incisors on each side of the upper jaw.

Color phases. — No color phases are known to occur in this species.

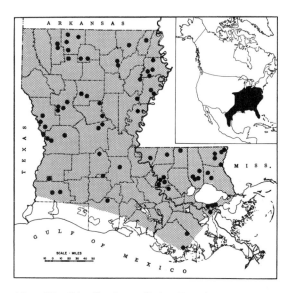

Map 17 Distribution of the Evening Bat (*Nycticeius humeralis*). Shaded area, known or presumed range within the state; •, localities from which museum specimens have been examined; *inset*, overall range of the species.

Measurements and weights.—Sixty-one specimens in the LSUMZ from various localities in the state averaged as follows: total length, 89.0 (81–98); tail, 38.7 (31.5–45.0); hind foot, 7.0 (6–10); ear, 12.0 (8–14); forearm, 35.0 (33–38); third metacarpal, 32.2 (29.0–39.8); condylobasal length, 13.1 (12.7–14.2); cranial height, 5.1 (4.6–5.6); cranial breadth, 7.9 (7.5–8.5); zygomatic breadth, 9.2 (9.1–10.5); interorbital breadth, 4.0 (3.7–4.3); palatal breadth, 6.4 (6.1–6.8); palatilar length, 5.4 (5.1–6.6); postpalatal length, 4.6 (4.2–5.3); maxillary toothrow, 4.8 (4.5–5.2). Weights of 21 males from the same collection averaged 9.2 (6.8–14.0) grams, while weights of 28 females averaged 10.0 (7.6–13.0) grams.

Cranial and other skeletal characters.—The Evening Bat is the only bat in Louisiana with as few as 30 teeth. The skull is short and relatively broad and flat. There is only one incisor on each side of the upper jaw. This feature and the fact that the braincase is only slightly elevated above the rostrum separates this bat at a glance from the similarly colored *Myotis austroriparius.* The calcar is only slightly, if at all, keeled. (Figure 55)

Sexual characters and age criteria.—The sexes are indistinguishable except by inspection of the external sex organs. There seems to be no seasonal variation, but the immatures are characterized by their blacker and more scanty pelage. By late fall they appear identical with the adults.

Status and habits.—This species is one of the commonest bats in the state. We have records for every month of the year except February and December. I suspect that the absence of specimens from these months is nothing more than an accident of sampling, especially since it has been taken in January.

The Evening Bat is a communal rooster in hollow trees, in attics in residential areas, and in old abandoned houses, barns, outbuildings, warehouses, cisterns, and similar

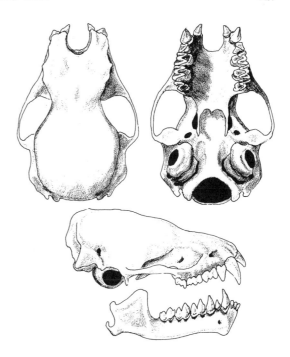

Figure 55 Skull of Evening Bat *(Nycticeius humeralis),* ♀, Covington, St. Tammany Parish, LSUMZ 11831. Approx. 3³/₁₀ ×.

sites. Professor Horace A. Hays found a colony of not less than 500 in the belfry of a church in Robeline in Natchitoches Parish. He reported that the bats had been known to descend in numbers into the church when disturbed. Nets set over farm ponds almost anywhere in the state on any occasion will very likely yield several of these bats. At a pond a few miles west of Ferriday on 8 October Richard K. LaVal caught the unusual number of 58 between 6:30 and 8:30 P.M. The species sometimes is found in cohabitation with the Brazilian Free-tailed Bat in hollow trees, as well as in old buildings. Sometimes, but not always, they tend to segregate within the roost.

In my experience the species does not begin to fly until late twilight. At first it flies above the trees, but as darkness falls it descends lower. Perhaps this habit is the reason why it seldom is caught in a net until complete darkness has descended over the landscape. I have never caught one after midnight, but perhaps my sampling in the later hours of the night has not been sufficiently extensive.

The young are born in late May and June. The number is usually two, but occasionally there is only one. Three young are exceedingly rare. A female shot in Alabama was reported to contain four embryos. I doubt, though, that all four would have been delivered. Wimsatt (1945) has shown that in many vespertilionids the number of ova shed normally exceeds the number of young brought to term. Although all the ova that are shed may succeed in implanting, the excess number is usually subsequently resorbed.

Gates (1941), in a study of 31 Evening Bats that he obtained alive from Natchitoches Parish in May 1940, pointed out that they comprised only females and all were pregnant. Of those that he was able to keep alive, all gave birth to their young almost simultaneously, within a period of 48 hours, thereby suggesting that ovulation had occurred within an equally short interval.

For an informative account of the growth and development of the Evening Bat we are indebted to Dr. Clyde Jones (1967). He studied 28 young of 14 litters born in captivity to females netted in Franklin County, Mississippi, not far from the Louisiana–Mississippi state line. He found that 14 adult females gave birth to 16 males and 12 females on dates between 25 May and 4 June. All births observed were by breech presentation and required from 3 to 114 minutes. The females ate the placenta and umbilical cords and licked the young thoroughly soon after birth. In every case observed the young found and grasped the teats within a short time, usually five to eight minutes following birth. The young bats, in the hour after birth, were pink with smooth, soft skin. Only a few hairs were present on the feet, shoulders, and the top of the head. By the 18th hour, pigmentation appeared on the venter and within 24 hours it was extensive. After five days soft and gray

fur was present on the dorsum across the scapular region to the flanks and extended onto the rump. By the ninth day of age the young were furred completely. At birth the eyes were sealed shut, but the line of fusion was distinctly evident. At 18 to 24 hours following birth, the eyes opened. Initially the pinnae were folded over, but when the young were between 24 and 36 hours old the ears unfolded and became erect. For the first 10 days the young almost constantly uttered weak squeaks and chirps but after this time made vocal sounds only when disturbed or handled.

Of special interest in the study by Jones were his observations concerning the behavior of the young and adults. His comments in this connection deserve quotation here (Jones, 1967:9–15).

The young evening bats seemed weak and uncoordinated at birth. Although the babies attached themselves firmly to the parent, young 1 day of age seemed rather helpless when separated from the adult. At 1 day of age, the young could crawl about only feebly and were unable to right themselves when placed on the dorsum on a flat surface.

By 3 days of age the young could crawl about very well and could right themselves quickly. In part, these abilities may be a reflection of the aforementioned unfolding of the pinnae at this time.

For nearly the first 2 weeks of age the young were attached to the nipples of the adults almost constantly and remained enveloped by the membranes of the adults. With the exception of a few occasions when a young bat was observed uncovered, the young did not leave the close association with the parent and move about in the containers until about 3 weeks of age. After this time, young scampered frequently about the cages, but hung adjacent to the female when at rest.

. . . the young were highly vocal for the first 10 days following birth, but then made vocal sounds only when disturbed or handled. During these observations, the adults emitted sounds only when disturbed, handled, or sometimes when offered food.

Some definite specificities of adults for their young were noted. As long as the young bats were returned to the same nipples from which they were taken, no female refused to accept the young after they had been removed from the mother and measured or handled otherwise. On several occasions attempts were made to get adult females to accept nursing young from other females; all such efforts failed. The adults would bite and move away from the strange young. One young bat that was allowed to become attached to the nipple of a restrained female was attacked and thrown from her when the adult was released. This same adult accepted her own young a few minutes later. Litters and females could not be mixed successfully until nursing ceased. The refusal of adult females to accept other young may be a reflection of the manner in which the animals were maintained in relative isolation from other young and adults. Gates (1941) reported that he detected no specificity with regard to nursing young and adult females when the young and adults no longer remained together during periods of feeding.

The young first showed an interest in food and water at approximately 3 weeks of age, when they appeared to smell and lick items of food (portions of mealworm larvae) held before them. Early interests of young in water included considerable licking of the end of a water-filled dropper. At the age of 4 weeks, young bats were taking water from a dropper and eating small mealworms that were presented with forceps.

Throughout the course of this study, the adults were given mealworms from forceps and water was administered from a dropper. In only two cases did individuals become accustomed to picking up and eating mealworms that were not presented by hand. In general, the bats made little effort to fly or move about while being fed and it was possible to feed four to six animals at one time. One adult crawled about almost continually while being fed. The young reared by this female behaved in similar fashion during the periods of feeding.

Some animals began eating immediately when food was offered; others simply held a food item in the mouth for a short time. During this time the animals exhibited considerable shivering, presumably while the body temperature was increased.

Management or control.—Since this species often enters human dwellings and establishments, control measures may be required when its presence becomes obnoxious. The simplest procedure is to find the place where the bats are entering and to close it up. This may be only a loose piece of siding or a broken window that the bats are using to gain entrance to their daytime attic retreat. Actually their presence in the attic or loft of a human dwelling is not altogether bad. Personally I wish that I did have a colony of bats in my own attic.

Subspecies

Nycticeius humeralis humeralis (Rafinesque)

Vespertilio humeralis Rafinesque, Amer. Monthly Mag., 3, 1818:445; type locality, Kentucky.
N[ycticeius]. *humeralis* (Rafinesque), Jour. Phys. Chim. Hist. Nat. et Arts, Paris, 88, 1819:414.

The Evening Bat is polytypic with recognized races in extreme southern Florida *(subtropicalis)* and in northeastern Mexico *(mexicanus),* but all other populations, including those in Louisiana, are referable to the nominate race, *humeralis.*

Specimens examined from Louisiana.—Total 287, as follows: *Ascension Par.:* 1 mi. W Dutch Town, 6 (LSUMZ). *Bienville Par.:* near Taylor, 3 (LTU). *Calcasieu Par.:* Lake Charles, 1 (LSUMZ); DeQuincy, 1 (MSU); Sulphur, 1 (LSUMZ). *Catahoula Par.:* 4 mi. NNE Foules, 2 (LSUMZ). *Claiborne Par.:* Summerfield, 5 (LSUMZ); 3.5 mi. W Homer, 2 (LSUMZ); 3 mi. E Haynesville, 7 (LTU); 10 mi. E Minden, 1 (LTU). *Concordia Par.:* 4 mi. W Ferriday, 29 (LSUMZ). *East Baton Rouge Par.:* Baton Rouge, 8 (LSUMZ); University, 4 (LSUMZ); 5 mi. S University, 3 (LSUMZ); 12.2 mi. ESE University Campus, 2 (LSUMZ); 10.3 mi. ENE LSU, 14 (LSUMZ). *East Feliciana Par.:* Clinton, 5 (LSUMZ). *Evangeline Par.:* Turkey Creek area, 2 (LSUMZ). *Franklin Par.:* 13.5 mi. NE Gilbert, 10 (LSUMZ). *Grant Par.:* Fishville, 5 (LSUMZ); 0.25 mi. W Georgetown, 1 (LTU). *Jefferson Par.:* Metairie, 1 (LSUMZ); Marrero, 1 (LSUMZ). *Lincoln Par.:* Ruston, 3 (l LSUMZ; 2 LTU); 10 mi. W Ruston, 1 (LTU). *Madison Par.:* 11 mi. SW Tendal on Tensas River, 7 (LSUMZ); McGill bend of Tensas River, 2 (LSUMZ); 10 mi. SW Tallulah, 1 (USL); 8 mi. NW Tallulah, 1 (USNM). *Morehouse Par.:* Mer Rouge, 18 (USNM). *Natchitoches Par.:* Robeline, 6 (LSUMZ); Provencal, 7 (LSUMZ); 2 mi. W Natchitoches, 1 (LSUMZ); 5 mi. S Robeline, 1 (TU). *Orleans Par.:* New Orleans, 2 (1 TU; 1 LSUMZ); 1 mi. E Lakefront Airport at Lake Pontchartrain, 12 (TU). *Ouachita Par.:* Monroe, 15 (LSUMZ); 5 mi. SE Monroe, 1 (LSUMZ). *Rapides Par.:* Pineville, 2 (1 LSUMZ; 1 USNM); 7 mi. NE Pineville, 2 (LSUMZ). *Sabine Par.:* Bayou Negreet at Sabine River, 4 (LSUMZ); 7 mi. W Many, 4 (LSUMZ); 7.5 mi. S Toro, 1 (TU); 5 mi. N Toro, 1 (TU). *St. Helena Par.:* 5 mi. NNE Chipola, 5 (LSUMZ). *St. Landry Par.:* 7.5 mi. N, 2.75 mi. E Washington, Thistlethwaite GMA, 10 (LSUMZ); 1 mi. E Eunice, 1 (USL). *St. Tammany Par.:* Covington, 1 (LSUMZ); 6 mi. S Folsom,

1 (SLU). *Tangipahoa Par.:* 5.5 mi. E Ponchatoula, 1 (SLU); 5 mi. N Robert, 1 (SLU); locality unspecified, 2 (SLU); 2 mi. S Hammond, 2 (SLU); 2 mi. E Ponchatoula, 3 (SLU); 7 mi. E Ponchatoula, 1 (SLU); 1 mi. N Hammond, 1 (SLU); 2 mi. S Ponchatoula, 2 (SLU); 3.75 mi. E Hammond, 3 (SLU); 1.5 mi. S Ponchatoula, 1 (SLU); 3 mi. S Hammond, 2 (SLU). *Tensas Par.:* 0.75 mi. NW Saranac on Clark Bayou, 7 (LSUMZ); St. Joseph, 3 (LSUMZ). *Terrebonne Par.:* 1 mi. NE Montegut, 1 (LSUMZ); Cocodrie, 1 (LSUMZ). *Vermilion Par.:* Indian Bayou, 1 (USL). *Vernon Par.:* 4 mi. N Burr Ferry, 1 (LSUMZ); Slagle, 1 (LSUMZ); 7 mi. W Leesville, 2 (LTU). *Washington Par.:* 10 mi. S Bogalusa, 1 (LSUMZ); 10 mi. W Bogalusa, 1 (LSUMZ); Sheridan, LSU Forestry Camp, 1 (LSUMZ); Pine, 1 (SLU); locality unspecified, 1 (SLU); 2 mi. W Varnado, 2 (SLU); Varnado, 1 (SLU); 2.75 mi. E, 0.25 mi. N Angie, 1 (SLU). *Webster Par.:* Caney Lake, 1 (LSUMZ); 9 mi. N Minden, 5 (LSUMZ). *West Carroll Par.:* Oak Grove, 5 (NLU). *West Feliciana Par.:* 4.2 mi. ENE St. Francisville, 3 (LSUMZ). *Winn Par.:* 8 mi. W Winnfield, 3 (LSUMZ); 2 mi. S Winnfield, 1 (LTU).

Genus *Plecotus* E. Geoffroy Saint-Hilaire
Big-eared Bats

Dental formula:

$$I\frac{2-2}{3-3} \ C\frac{1-1}{1-1} \ P\frac{2-2}{3-3} \ M\frac{3-3}{3-3} = 36$$

RAFINESQUE'S BIG-EARED BAT   Plate III
(*Plecotus rafinesquii*)

Vernacular and other names. — This bat has often gone by the name of eastern lump-nosed bat or eastern big-eared or long-eared bat. The name Rafinesque's Big-eared Bat serves to distinguish this species from its close relative the Western, or Townsend's, Big-eared Bat (*Plecotus townsendii*). The generic name is derived from the Greek word *pleko,* meaning to twist and the Latin word *otus,* referring to the ear, and hence means twisted ear, and alludes probably to the manner in which the large ears are sometimes coiled when the bat is at rest or in a torpid state.

Distribution. — The species occurs from southern Virginia west across southern Ohio, southern Indiana, southern Illinois, southeastern Missouri, and northwestern Arkansas south to the northern Gulf Coast and to central Florida. In Louisiana records of its occurrence are fairly widespread but are still lacking from numerous parishes. (Map 18)

External appearance. — This bat cannot be confused with any other species presently known from the state. The immense ears that are more than an inch in length and the prominent lumps on the nose render it unmistakable (Figure 56). The color is dark brown above and grayish white below. When the fur is parted by blowing into it the hairs are seen to be strongly bicolored. On the back they are blackish basally and brown terminally. On the belly they are again blackish basally but grayish white terminally. The strongly bicolored hair serves to distinguish the species from its western relative, *Plecotus townsendii,* in which the hair shows little or no contrast between the basal and terminal portions. Another important difference between the two species is

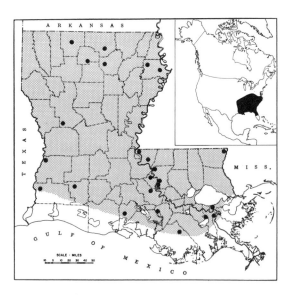

Map 18 Distribution of the Rafinesque's Big-eared Bat (*Plecotus rafinesquii*). *Shaded area,* known or presumed range within the state; •, localities from which museum specimens have been examined; *inset,* overall range of the species.

grams, while 12 females in the LSUMZ and the Tulane collections averaged 9.1 (7.9–10.2) grams.

Cranial and other skeletal characters.—The skull of *Plecotus rafinesquii* is readily identifiable by its highly arched or vaulted braincase that slopes evenly from the rostrum without forming an angle where the nasals meet the frontals. The first incisor is definitely bicuspid, whereas in *P. townsendii* it is unicuspid. The calcar is not keeled. (Figure 57)

Sexual characters and age criteria.—The sexes are alike in color and there is little or no seasonal variation. The young are nearly black and the fur is shorter and thinner than

Figure 56 Rafinesque's Big-eared Bat *(Plecotus rafinesquii)*, McClellanville, South Carolina. (Photograph by Richard K. LaVal)

that the hairs on the feet of *P. rafinesquii* are so long that they project beyond the end of the toes. The tragus in all members of the genus *Plecotus* is straight, extremely long, and somewhat pointed (Figure 38).

Measurements and weights.—Thirty-four specimens in the LSUMZ from various localities in Louisiana averaged as follows: total length, 96 (80–102); tail, 42 (33–51); hind foot, 10 (8–13); ear, 30 (27–34); forearm, 41.5 (38.8–43.5); third metacarpal, 37.4 (35.7–39.5); greatest length of skull, 14.0 (13.2–15.1); cranial height, 5.9 (5.6–6.4); cranial breadth, 8.8 (8.2–9.1); zygomatic breadth, 8.5 (8.2–8.9); interorbital breadth, 3.6 (3.5–4.0); palatal breadth, 6.0 (5.8–6.4); palatilar length, 5.1 (4.9–5.4); maxillary toothrow, 4.9 (4.7–5.4). Three males in the Tulane collection weighed 7.9, 8.5, and 8.0

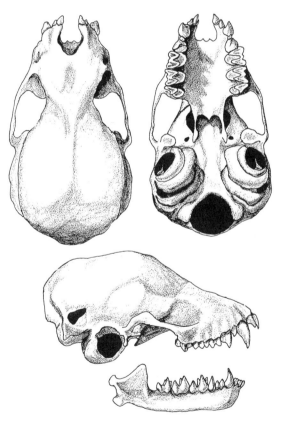

Figure 57 Skull of Rafinesque's Big-eared Bat *(Plecotus rafinesquii)*, ♀, 7 mi. NW Plaquemine and Grosse Tete Bayou on Hwy. 77, Iberville Parish, LSUMZ 9304. Approx. 3⁸/₁₀ ×.

that of the adults, but by late fall they are indistinguishable from their parents. Adult males in the fall and winter possess greatly enlarged testes, because the epididymides (tubes containing sperm) are at their maximum stage of development. By spring the epididymides have dwindled so much in size that they are no larger than those of young males.

Status and habits. — Rafinesque's Big-eared Bat is common in the state. I am confident that the absence of records of it from many parishes is attributable solely to the fact that no search for it has been made in those places. Favorite roosting sites include attics, lofts of barns and other outbuildings, open cisterns, culverts, and particularly old, dilapidated, and abandoned houses in rural areas.

In the northern part of its range the species hibernates in caves, cisterns, and similar places, but even there it appears to be active off and on throughout the cold months, as is certainly the case here in Louisiana. At rest the bat hangs by its feet, head downward and coils its long ears in a spiral over its neck. When disturbed while at rest, it erects the ears and waves them in the direction of the intruder as if trying to keep track of him by echolocation or by the sounds he produces. Barbour and Davis (1969) state that under these circumstances the bat gives the impres-

sion that it is peering intently at the observer. Hall (1963) describes its utterances when disturbed as a "low, hoarse bark, sounding rather like a small dog."

Rafinesque's Big-eared Bat is seldom if ever seen on the wing (an exception is provided by Figure 58), for it emerges only after dark. Barbour and Davis (1969) found no record of its foraging during twilight, and my own observations agree.

Nursery colonies, consisting of a dozen or more females, are found in spring. The young are apparently born in May and very early June in Louisiana. Vernon Bailey's field notes tell of collecting females with single well-developed embryos in the attic of a large plantation house near Houma, in Terrebonne Parish, on 12 May. And I have seen numerous specimens from Louisiana collected in July that seem to represent young only recently on the wing. Some still retained their milk dentition. Females normally bear only one offspring at a time.

So little is known about this interesting bat that its life history warrants much intensive study. Anyone finding a group of them, which would probably be a nursery colony, should report their whereabouts to one of the biologists at the closest state college or university.

Subspecies

Plecotus rafinesquii macrotis Le Conte

Plec[otis]. macrotis Le Conte, *in* McMurtrie, The animal kingdom ... by the Baron Cuvier ..., 1, 1831:431; type locality, none designated but subsequently fixed by Miller (1897:51) as the Le Conte plantation, 5 miles south of Riceboro, Liberty County, Georgia. See also Handley, 1959:160.
Plecotis rafinesquii macrotis, Handley, Proc. U.S. Nat. Mus., 110, 1959:161.
Corynorhinus macrotis, Lowery, Occas. Papers Mus. Zool., Louisiana State Univ., 13, 1943:224.

Handley (1959) recognized two races of this species. He assigned specimens from Illinois, Indiana, Ohio, West Virginia, Kentucky, and Tennessee to *P. r. rafinesquii* Lesson, while those from southern Louisiana, southern Mississippi, southern Alabama,

Figure 58 Rafinesque's Big-eared Bat *(Plecotus rafinesquii)* in flight, Reelfoot Lake, Tennessee, July 1961. Note the numbered aluminum band on forearm. (Photograph by Woodrow Goodpaster)

southern Georgia, Florida, South Carolina, eastern North Carolina, and extreme southeastern Virginia he called *P. r. macrotis* Le Conte. He considered as intergrades specimens from northern and central Louisiana, extreme eastern Oklahoma, Arkansas, northern Alabama, northern Georgia, and western North Carolina. However, he believed those specimens from northern and central Louisiana were nearer to *macrotis* than to the nominate race. In this treatment I definitely concur despite the fact that a small series I have seen from Natchitoches Parish (LSUMZ) is not as dark brown above as specimens from the southern part of the state. Indeed, they are so light colored above and show so much less contrast between the basal and terminal bands of color of the hairs that they suggest the possibility of their being *Plecotus townsendii*. But the two cusps on their first upper incisors leave no doubt about their identity as *P. rafinesquii*.

Specimens examined from Louisiana.—Total 105, as follows: *Assumption Par.:* 1 mi. S swing bridge at Belle River (12 mi. N Morgan City), 1 (USL). *Beauregard Par.:* 6.5 mi. N Merryville, 2 (MSU). *Calcasieu Par.:* Starks Road, 1 (MSU). *Claiborne Par.:* locality unspecified, 1 (LTU). *East Baton Rouge Par.:* various localities, 42 (LSUMZ). *Iberia Par.:* Avery Island, 2 (TU). *Iberville Par.:* 7 mi. NW Plaquemine Bayou and Grosse Tete Bayou on Hwy. 77, 2 (LSUMZ). *Jefferson Par.:* Couba Island, 1 (LSUMZ). *Jefferson Davis Par.:* Fenton, 1 (MSU). *Lincoln Par.:* Ruston, 4 (3 LSUMZ; 1 LTU). *Madison Par.:* Waverly, 1 (LSUMZ); Tallulah, 2 (USNM). *Natchitoches Par.:* Kisatchie, 5 (LSUMZ). *Orleans Par.:* New Orleans, 1 (LSUMZ). *Ouachita Par.:* 10 mi. W Monroe, 1 (LTU). *Plaquemines Par.:* Tulane Univ. Riverside Campus, Belle Chasse, 27 (26 TU; 1 LSUMZ). *Terrebonne Par.:* Houma, 4 (USNM). *Union Par.:* 13 mi. W Sterlington, La. Hwy. 2, 1 (NLU). *Washington Par.:* 3 mi. E Angie, 1 (SLU); 4 mi. NE Angie, 1 (LTU). *West Baton Rouge Par.:* Lobdell, 1 (USNM). *West Feliciana Par.:* 14 mi. NW Bains, 1 (LSUMZ); St. Francisville, 1 (LSUMZ); 2 mi. N St. Francisville, 1 (LSUMZ).

The Free-tailed Bat Family
Molossidae

THE FREE-TAILED BATS, as their name implies, have a portion of the tail extending free beyond the interfemoral membrane (Figure 59). The short hair is velvety to the touch. The wings are exceptionally narrow, and the thick, leathery ears, which tend to project forward, are broad basally, relatively short, and somewhat triangular in shape. In many species in the family the ears meet on top of the head, but in our species they are an eighth of an inch apart. The face is sparsely furred and has stiff, erectile hairs. Especially characteristic of our native molossid bat are the long, stiff tactile hairs on its toes.

Living members of the family comprise 12 genera and approximately 80 species. The group ranges widely throughout the warmer regions of the world from southern Europe and southern Asia south through Africa, India, and Malaysia and east to the Solomon Islands in the Old World, and from the southern United States, through Mexico,

Figure 59 Brazilian Free-tailed Bat *(Tadarida brasiliensis cynocephala)*, ♂, St. Amant, Ascension Parish, LSUMZ 11841. (Photograph by Gary Eberle)

Central America, and the West Indies to southern South America in the New World.

Molossid bats are not particularly attractive in appearance, for their features are rather coarse and their facial expressions ugly in the extreme. They are the one bat in Louisiana that sometimes makes a nuisance of itself by inhabiting old business establishments in our towns and cities. Then their deposits of feces and urine, called guano, create an unpleasant odor that smells strongly of ammonia. Moreover, free-tails have facial sebaceous glands (Werner et al., 1950) that exude an oily substance on the muzzle and beneath the lower jaw. Its strong, musty, and rather unpleasant

odor will pervade a building in which they roost. Under such conditions control measures are sometimes required. Interestingly enough, the large accumulations of guano produced by these bats were once of great economic importance. During the War Between the States the Confederacy used guano as a source of nitrates for making gun powder and other explosives. A close relative of our local free-tailed bat is the principal species among the millions of bats inhabiting caves in Texas and New Mexico, including Carlsbad Caverns. Commercial quantities of guano were once removed from the latter cave at the rate of one to three carloads daily, each carload weighing about 40 tons. According to Allen (1939), the price obtained for the product was said to range from \$20 to \$80 a ton.

Genus *Tadarida* Rafinesque

Dental formula:

$$I\frac{1-1}{2-2} \text{ or } I\frac{1-1}{3-3} \ C\frac{1-1}{1-1} \ P\frac{2-2}{2-2} \ M\frac{3-3}{3-3} = 30 \text{ or } 32$$

BRAZILIAN FREE-TAILED BAT   Plate III
(Tadarida brasiliensis)

Vernacular and other names.—When the free-tailed bats of Louisiana were considered a distinct species under the name *Tadarida cynocephala* (Le Conte), they were called the Le Conte's Free-tailed Bat. But the form *cynocephala* is here being treated as a subspecies of *T. brasiliensis* (see **Subspecies** at the end of this account). Consequently, I am employing the common name Brazilian Free-tailed Bat. It has also been called the guano bat. The generic name *Tadarida* was coined by Rafinesque, who gave no clue as to its etymology. The specific name *brasiliensis* is a combination of the name of the country Brazil and the Latin word *ensis,* meaning belonging to. The species was originally named from Brazil.

Distribution. — The Brazilian Free-tailed Bat ranges across the entire southern part of the United States south through Mexico, the West Indies, and Central America, as well as over most of South America. In Louisiana it is virtually statewide in its occurrence. (Map 19)

External appearance. — This bat is brownish black above, near Bister, and grayish brown below, close to Saccardo's Umber. An occasional individual is faintly tinged with a grayish above. The velvety texture to the fur is highly diagnostic, as are also the leathery ears, the tail that extends beyond the interfemoral membrane for at least one-half its length, the long hairs on the toes (Figure 60), and the black, spaniellike muzzle with its vertically arranged wrinkles on the upper lip. The tragus is short and truncate.

Color phases. — Bailey (1951), in the course of handling thousands of individuals of this species during his study of a colony in Baton Rouge, noted that about 10 percent show an

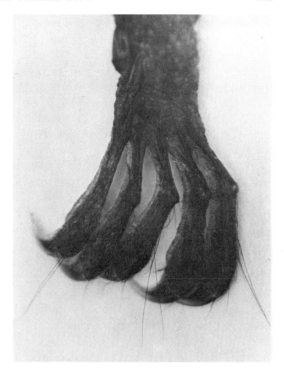

Figure 60 Enlargement of foot of Brazilian Free-tailed Bat *(Tadarida brasiliensis cynocephala)*, LSUMZ 11841, showing the exceptionally long hairs that are characteristic of the toes of this species. Approx. 11 ×. (Photograph by Gary Eberle)

aberrant flecking of white, usually on the back but sometimes on the underparts as well. One individual (LSUMZ 6163) is extremely reddish, near Russet above and between Sayal Brown and Cinnamon beneath. Numerous specimens appear to be browner, less blackish, than the average, but some of this brownness may be attributable to discoloration by the ammonia fumes that pervade the bats' daytime retreats (Constantine, 1958b). Both true albinos (individuals lacking pigment in the hair and possessing pink irises) and normal-eyed, white-coated individuals are known for this species (McCoy, 1960).

Measurements and weights. — The skins of 23 males in the LSUMZ, from various localities in the state, averaged as follows: total length,

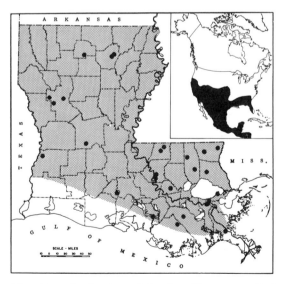

Map 19 Distribution of the Brazilian Free-tailed Bat *(Tadarida brasiliensis)*. *Shaded area,* known or presumed range within the state; •, localities from which museum specimens have been examined; *inset,* overall range of the species.

102 (90–109); tail, 35 (29–44); hind foot, 8 (7–14); ear, 15 (10–20); forearm, 42.2 (40.0–45.0); third metacarpal, 42.1 (40.2–44.9). Forty-six females averaged, respectively, as follows: 95 (90–100); 33 (29–42); 8 (7–12); 16 (14–19); 42.8 (38.2–45.6); 41.6 (39.1–43.5). The skulls of 32 males averaged as follows: greatest length, 16.9 (16.5–17.4); cranial height, 6.2 (5.7–6.8); cranial breadth, 9.7 (9.3–10.1); zygomatic breadth, 10.2 (9.3–10.1); interorbital breadth, 4.1 (4.0–4.5); palatal breadth, 7.1 (6.8–7.5); palatilar length, 5.7 (5.4–6.1); postpalatal length, 6.8 (5.7–7.3); maxillary toothrow, 6.0 (5.6–6.6). The skulls of 49 females averaged as follows: greatest length, 16.6 (16.0–17.5); cranial height, 6.2 (5.9–6.9); cranial breadth, 9.5 (9.1–9.9); zygomatic breadth, 10.0 (9.7–10.8); interorbital breadth, 4.1 (3.9–4.3); palatal breadth, 7.1 (6.9–7.7); palatilar length, 5.4 (5.1–6.0); postpalatal length, 6.6 (6.3–7.2); maxillary toothrow, 5.5 (5.2–6.3). The weights of 10 males averaged 11.7 (9.6–14.4) grams and 16 females averaged 12.9 (11.5–14.7) grams.

Cranial and other skeletal characters.—The skull readily separates *Tadarida brasiliensis* from any of the vespertilionid bats in Louisiana. It is relatively flat in profile with a depression in the center of the crown followed by a weak sagittal crest posteriorly. There is a prominent notch in the front end of the hard palate, but, because the single incisors on each side project inwardly, the notch appears rather round instead of U-shaped as in most vespertilionids. The crown of the last molar is Z-shaped. The calcar is not keeled. (Figure 61)

Sexual characters and age criteria.—Color in this species does not differ between the sexes. In vespertilionid bats the females generally average larger than the males, but the opposite is alleged to be true of the Brazilian Free-tailed Bat. But my own measurements of 23 males and 46 females in the LSUMZ collections fail to support this contention. The means of 4

out of 15 measurements are the same in the two sexes, and in the eight measurements in which the males exceed the females, the differences lack statistical significance. However, according to Herreid (1959), the teeth of *Tadarida brasiliensis* reflect a disparity in favor of larger males. He showed that the canine of the male free-tailed bat is larger than that of the female. Young free-tailed bats do not differ appreciably from the adults except for their smaller size and shorter ears.

Status and habits.—This species is probably the commonest bat in Louisiana. Surely few towns in the state are without at least one

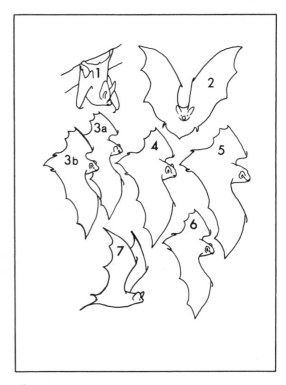

Plate III
1. Rafinesque's Big-eared Bat *(Plecotus rafinesquii)*.
2. Big Brown Bat *(Eptesicus fuscus)*.
3. Red Bat *(Lasiurus borealis):* **a,** female; **b,** male.
4. Northern Yellow Bat *(Lasiurus intermedius)*.
5. Hoary Bat *(Lasiurus cinereus)*.
6. Seminole Bat *(Lasiurus seminolus)*.
7. Brazilian Free-tailed Bat *(Tadarida brasiliensis)*.

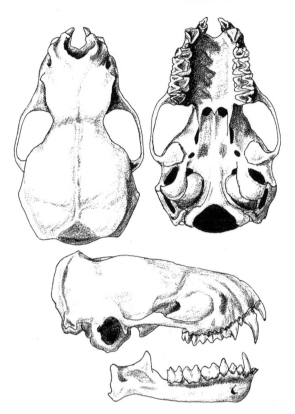

Figure 61 Skull of Brazilian Free-tailed Bat *(Tadarida brasiliensis cynocephala),* ♂, Clinton, East Feliciana Parish, LSUMZ 13455. Approx. 3³/₁₀ ×.

colony of free-tailed bats in one of their buildings. Cities such as New Orleans, Baton Rouge, Monroe, and Shreveport doubtless house many large colonies. A study conducted in such a colony in Baton Rouge in 1951 by Floyd L. Bailey was located in the attic of an old building in the downtown part of the city, less than 200 yards from the Mississippi River. The colony held an estimated 20,000 bats and had been in existence for more than 35 years. Bailey's study involved almost daily visits to the colony between 20 September and the end of December. During this time he kept careful notes on the behavior of the bats, fluctuations in their weights and the amount of fat that they deposited, and the correlation of their daily activities and movements with tempera-

ture. Among the many interesting facts he uncovered was that the sexes tended to segregate by occupying different parts of the attic.

Bailey also found that the bats always began to leave the building after sunset, though the actual time varied greatly. In September it was from 7 to 15 minutes following sunset, in October from 26 to 59 minutes, and in November from 3 to 44 minutes. Most of the bats returned to the attic between 3:00 A.M. and 4:30 A.M., but when the temperature outside approached freezing they returned much earlier. On 11 November, for instance, when the temperature outside was low, many more bats were returning at 8:00 P.M. than were leaving. On cold nights not all the bats would leave the building, and those that did leave would return before 10:00 P.M., usually around 8:00 P.M.

Bailey found that in fall the bats deposit a quantity of fat beneath the skin, and perhaps elsewhere on the body. He weighed 100 males and 100 females on 29 October. The mean weight of the males was 11.76 grams, that of the females 13.01 grams. But in the first week of December 500 males and 700 females had mean weights of 13.87 grams and 14.09 grams, respectively. In other words, in slightly over one month the males had increased 2.12 grams while the females had increased only 1.08 grams. Although the females had started out heavier, the males increased in weight more rapidly and by the first week in December weighed only .92 grams less, on the average, than did the females.

The most startling discovery made by Bailey was, however, the events that took place in early December. The numbers of bats present had fluctuated in late October and early November, with a general exodus between 29 October and 2 November, but afterwards the numbers again built up. On 29 November there was a small group of about 200 bats in one corner of the attic. The numbers rose to about 1,000 individuals during the next six days and remained at this level for the following five days. But on 15 December only about 50 scattered bats could be found. Two days later, on 17 December, a careful search revealed not a single bat. Thus between 10 December and 17 December there was an exodus of the entire colony. No bats were again present in the building until the following March.

The question is: Did Bailey's colony migrate or simply move to some other retreat? Colonies of free-tailed bats in Texas, Oklahoma, New Mexico, and Arizona definitely do migrate. Almost the entire population of literally millions of these bats that spend the summer in this area move into Mexico in winter. Thousands of free-tails dwelling in caves in the Southwest have been banded, and sufficient recoveries have resulted to enable workers to piece together their movements (Villa, 1956; Glass, 1958 and 1959; Short et al., 1960; Davis et al., 1962; and Villa and Cockrum, 1962). But it has often been said that populations east of Texas, in the southeastern states, do not leave in winter. Capture of free-tails in Louisiana in December, January, and February are not infrequent. Consequently, by no means all of them leave. Yet the question still remains as to what happened to Bailey's Baton Rouge colony when it completely deserted its attic roost in the third week of December. In the event they did not migrate and instead merely moved to another nearby location, one wonders what features the alternate roost provided that the old one did not. Unfortunately, the building in which Bailey's colony was located is no longer in existence, and his study cannot be resumed.

Free-tailed bats prefer the attics of old buildings but are not averse to roosting in hollow trees, where they sometimes share quarters with the Evening Bat. According to Richard K. LaVal (MS), free-tails that he banded in houses in Clinton in East Feliciana Parish (Figure 62) were later found in nearby Jackson and at Amite, Baton Rouge, and New Orleans. The Baton Rouge recoveries were among free-tails found in company with Evening Bats in a hollow tree. Free-tails that he banded in Baton Rouge in October were found in Clinton the following spring and early summer. Some of LaVal's data are summarized in Table 4. His study has disclosed other interesting facts about the species in Louisiana.

Sherman (1937) found that, in a building in Florida housing more than 10,000 free-tailed bats from spring until early autumn, the population decreased as the weather became colder, although several hundred were usually present throughout the winter months. Few or no bats were present for about a week in March, and he thought that copulation occurred at this time. When the bats returned, nearly all the females were pregnant. He further found that parturition

Table 4 "Foreign" Recoveries of *Tadarida brasiliensis cynocephala* Banded by Richard K. LaVal in Louisiana[1]

Age	Sex	Louisiana Banding Site	Date	Louisiana Recovery Site	Date	Air Miles and Direction from Banding Point
Adult	♀	Clinton	29 June 1965	Jackson	8 Oct. 1970	12 W
Young	♂	Clinton	14 July 1965	New Orleans	23 Oct. 1967	84 SE
Young	♀	Clinton	20 July 1965	Baton Rouge	22 Oct. 1965	29 S
Adult	♀	Clinton	24 July 1965	Baton Rouge	22 Oct. 1965	29 S
Adult	♀	Baton Rouge	22 Oct. 1965	Clinton	6 May 1966	29 N
Adult	♂	Baton Rouge	22 Oct. 1965	Clinton	4 June 1966	29 N
Adult	♀	Clinton	12 Nov. 1965	New Orleans	13 June 1966	84 SE
Adult	♀	Clinton	12 Nov. 1965	Amite	June 1970	32 ESE

[1]The directions are from the point of banding and are only approximate.

occurred no earlier than 31 May but sometimes it was much later. For instance, in one year he could find no young in the colony on 10 June, but three nights later, on 13 June, he counted 350 young, on 15 June there were 567, and on 17 June, the number had increased to 1,459. On 28 June he examined several hundred females before finding a single pregnant female, and on the following day he could not locate even one. The gravid female found on 28 June gave birth on 30 June, the latest date for which he had evidence of parturition in this species.

All the evidence adduced by Sherman and by other investigators seems to indicate that free-tailed bats are reproductively quite unlike the vespertilionid bats, which mate in the late summer and fall months, with the sperm being stored in the tract of the female until the following spring, when ovulation and fertilization takes place. In the molossid free-tail, mating is in the spring in a span of a few days and is followed immediately by ovulation and fertilization (Cockrum, 1955). Nearly all the young are then born in a period of two weeks. Barbour and Davis (1969) cite the case of a Texas colony in which two-thirds of all births took place within 5 days, and 90 percent within a 15-day period.

The single baby is blind and naked at birth. As with most species of bats, the young are usually born by breech presentation. Sherman (1937) watched a captive female give birth. With the female hanging head down, the young bat emerged completely within two minutes. After all but the head was free, the female scratched the amnion to shreds with her feet. After birth the baby, while still suspended by the umbilical cord, soon found a nipple and began to nurse. The placenta did not emerge until two hours later and was not eaten by the mother; it remained attached to the youngster for two days by the cord, which then dried and broke close to the body.

The female produces a copious amount of milk and will accept and nurse any baby that attaches itself to one of her nipples (Davis et al., 1962). When a lactating female is held in a cluster of young she is swarmed over by them and makes no attempt to reject the first two that discover her mammae. The free-tail female differs in this respect from vespertilionid bats, which normally accept only their own young.

Growth in free-tailed bats is rapid, but because their wings are narrow and pointed they do not make their first flight until they are as much as five weeks of age, some two weeks later than is usual for vespertilionids, which have much broader flight membranes (Barbour and Davis, 1969).

Ingeniously devised photographic techniques (Edgerton et al., 1966), by which pictures were made of free-tailed bats leaving Carlsbad Caverns, showed that in flight the bat is able to extend the membrane almost to the end of the tail by slipping it along the vertebrae. These investigators suggest that the abbreviation of the membrane while the bat is at rest may be an adaptation that serves to reduce water loss by evaporation from the membrane's surface, but when the full membrane is needed to aid in catching an insect it can be extended.

The species is certainly one of the most abundant bats in Louisiana, but its population in this state is probably but a fraction of its numbers in single caves in New Mexico, Texas, and Oklahoma. The estimates of populations of some of the caves, although in-

Figure 62 A cluster of Brazilian Free-tailed Bats in the attic of an old house in Clinton, East Feliciana Parish. (Photograph by Richard K. LaVal)

credibly large, appear highly conservative. Davis et al. (1962) give the total number for nine caves in Texas in 1957 as 100 million, distributed in millions as follows: Bracken, 20; Nye, 10; James, 6; Davis, 4; Goodrich, 14–18; Racker, 12–14; Frio, 10–14; Fern, 8–12; and Devil's Sink Hole, 6–10. At Carlsbad Caverns in New Mexico in 1936 Allison (1937) estimated the population as being as high as 8,700,000, but in 1966 the number was said to be as low as 250,000 (Edgerton et al., 1966).

Many visitors to Carlsbad Caverns have witnessed the mass exodus of bats at nightfall and found it enthralling. It is a truly spectacular sight. The diminution in the number of bats in Carlsbad Caverns is indeed regrettable. Many biologists are now much concerned about what appears to be a drastic decrease in bat populations in numerous areas, but the environmental factor causing the reduction is unknown. Bats are important economically, for they consume tens of thousands of tons of insects in the course of a year. Davis et al. (1962) estimated that each free-tailed bat captures a nightly average of one gram of insects. They computed that the species destroys about 6,600 tons of insects a year in Texas alone. Assuming a more realistic nightly consumption of 3 grams of insects per bat, Barbour and Davis (1969) think that the annual take would at least triple the value computed by R. B. Davis and his collaborators. In any event bats are important enough to man to deserve his protection.

Predators. — Free-tailed bats in the attic of a building probably have few enemies. Rat Snakes *(Elaphe obsoleta)* and Corn Snakes *(E. guttata)* are excellent climbers and sometimes find their way into other daytime retreats of free-tails to prey upon them. In roosts in hollow trees the bats are probably captured by owls and occasionally by raccoons.

Parasites and Disease. — Bailey (1951) found few external parasites on the free-tailed bats in the colony he studied at Baton Rouge. A small unidentified species of mite and a species of bat flea were the only ones he detected. But in southwestern caves Eads et al. (1957), Jameson (1959), George and Strandtmann (1960), and Strandtmann (1962) found *Tadarida brasiliensis* infected with seven species of mites, two kinds of fleas, the Giant Bat Bug *(Primicimex cavernis)*, and a parasitic fly *(Trichobias major)*. Internal parasites include various species of cestodes, trematodes, and nematodes (Cain, 1966).

When Bailey carried on his study of the free-tailed bats in the colony in downtown Baton Rouge in 1950, no one yet knew that bats commonly carry rabies. If I had been aware of this danger, I would not have suggested and encouraged his undertaking, for prophylactic immunization against rabies was not yet generally available. Pasteur treatment, however, could then be taken after having been actually bitten by a rabid animal. But now anyone planning to handle numbers of bats or to work in the confined quarters of their roosts should first be injected with a preventive serum that has since become available. Apparently one can contract the disease without being bitten; evidence indicates that the causative agent, a virus, may be airborne. In recent years a bat researcher and another individual died of rabies after entering one of the Texas bat caves, although nothing suggested that either had actually been bitten.

Constantine (1962) conducted experiments in Frio Cave in Texas with Coyotes, Gray Foxes, Ringtails, and other carnivores placed as test animals in four types of cages. Those in Group I were in cages with 1-inch wire mesh. Thus they had no assurance against contact with bats, other vertebrates, arthropods, bat excreta, or cave atmosphere. Experimental animals in Group II were in cages within larger boxlike enclosures covered with 1/4-inch wire mesh and were hence protected against bats and other vertebrates but were exposed to arthropods, bat excreta, and the cave atmosphere. The animals of Group III were in cages with 1/18-inch mesh and were hence exposed only to small arthropods (such as mites), bat excreta, and the cave air. Finally, the cages for Group IV were constructed of 1/26-inch dacron mesh, reinforced with 1/18-inch plastic mesh, and provided with a moat designed to prevent the entry of tiny arthropods such as bat mites. The experimental animals in these cages were exposed only to the air of the cave! After exposure of 24 to 30 days, experimental animals in all four groups developed rabies within 28 to 109 days. Prior to placement in the cave, all but two of the experimental animals had been caged in isolation from 6 to 20 months. The exceptions were a Coyote in Group I and a fox in Group IV that were trapped only four months previously. These findings support the idea that the rabies virus can be transmitted directly through the air. Bats apparently contract rabies and carry it, but they themselves survive the disease.

Subspecies

Tadarida brasiliensis cynocephala (Le Conte)

Nyct[icea]. cynocephala Le Conte, in McMurtrie, The animal kingdom . . . by the Baron Cuvier . . . , 1, 1831:432; type locality, Georgia, probably in the neighborhood of the Le Conte plantation, Liberty County.

Tadarida cynocephala, Shamel, Proc. U.S. Nat. Mus., 78, 1931:7.—Lowery, Occas. Papers Mus. Zool., Louisiana State Univ., 13, 1943:224.

Tadarida brasiliensis cynocephala, Schwartz, Jour. Mamm., 36, 1955:108.—Hall and Kelson, Mammals N. Amer., 1959:205.

The Brazilian Free-tailed Bats inhabiting the southeastern part of the United States east of Texas have frequently been accorded full species rank under the name *Tadarida cynocephala*, as have also the populations of Texas and the southwestern and western parts of the United States under the name *Tadarida mexicana*. However, the morphological differences between the two are slight. Alleged skull differences broadly overlap. The main distinctions are ecological differences. The free-tailed bat in Texas and thence westward is a cave bat, but *cynocephala* is mainly a dweller in buildings. The former is highly migratory, but the latter is not yet proven to be migratory at all. I personally regard the two forms as nothing more than geographical representatives of the wide-ranging *Tadarida brasiliensis*. All free-tails taken in Louisiana belong to the race *cynocephala*, although an occasional *T. b. mexicana* may conceivably wander into the state from Texas.

Specimens examined from Louisiana.—Total 487, as follows: *Ascension Par.:* St. Amant, 3 (LSUMZ). *Beauregard Par.:* Merryville, 19 (4 LSUMZ; 15 MSU). *East Baton Rouge Par.:* Baton Rouge, 123 (121 LSUMZ; 2 USNM); University, 5 (LSUMZ); locality unspecified, 1 (LSUMZ). *East Feliciana Par.:* 9 mi. E Norwood, 5 (LSUMZ); Clinton, 40 (LSUMZ). *Lafayette Par.:* Broussard, 1 (USL). *Lafourche Par.:* 2 mi. S Raceland, 1 (LSUMZ). *Lincoln Par.:* Ruston, 1 (LSUMZ). *Natchitoches Par.:* Robeline, 5 (LSUMZ); Marthaville, 1 (NSU); Natchitoches, 1 (NSU). *Orleans Par.:* New Orleans, 77 (11 LSUMZ; 16 USL; 14 TU; 36 USNM); south shore of Lake Pontchartrain, 30 (TU). *Ouachita Par.:* West Monroe, 59 (LSUMZ); Monroe, 6 (LSUMZ); NLSC [NLU] Campus, Monroe, 1 (NLU). *Plaquemines Par.:* Port Sulphur, 2 (SLU). *Rapides Par.:* Glenmora, 4 (LSUMZ). *St. Mary Par.:* Morgan City, 4 (USNM). *St. Tammany Par.:* Covington, 5 (LSUMZ). *Tangipahoa Par.:* 5 mi. E Hammond, 1 (LSUMZ); Amite, 1 (SLU). *Terrebonne Par.:* Houma, 79 (USNM). *Washington Par.:* 2 mi. W Varnado, 11 (SLU); Franklinton, 1 (LSUMZ).

TREE SHREWS, LEMURS, LORISES, POTTOS, GALAGOS, LANGURS, MONKEYS, GIBBONS, ORANGUTANS, GORILLAS, AND MAN

Order Primates

THE ORDER contains 11 families, 62 genera, and some 192 living species. Primates are characterized by their manner of walking on the sole of the foot (plantigrade), by being pentadactylous (that is, five fingered and five toed); by having the eyes directed forward and their sockets separated from the temporal fossa by a bony bar or partition; and by possessing a well-developed clavicle. An even more notable feature of the primates is the great development of the cerebral hemispheres of the brain, a condition that achieves its maximum expression in *Homo*, the genus to which modern Man belongs. The first digit of the hind foot is opposable in all primates except man, but the pollex, or thumb, is also opposable in many forms, as it is in Man, and provides a useful aid in climbing and gives man a dexterity without which he probably could never have developed his complicated present-day culture and civilization.

The Man Family

Hominidae

THE FAMILY contains the single genus *Homo* and its one living species, *H. sapiens.* Because of Man's superior intellect and his great adaptability, he has been able to alter his environment or adjust to its extremes. He withstands low temperatures through the use of fire and clothing, and he regulates high temperatures with his air-conditioning equipment. He survives in some of the coldest regions of the world and in the middle of its hottest deserts. By his ingenious machines he has acquired great mobility that has taken him to all quarters of the earth (and even beyond to the moon), and his versatility has permitted him permanently to inhabit all the land areas of the world, even Antarctica on a limited basis. Through his development of agriculture and his domestication of other

143

Figure 63   American Indian (*Homo sapiens americanus*). (Portrait by H. Douglas Pratt)

animals he has increased his available food supply, and by his medical research and sanitation he has improved his survival rate to the point that his numbers now exceed three billion individuals.

Genus *Homo* Linnaeus

Dental formula:

$$I\frac{2-2}{2-2}\ C\frac{1-1}{1-1}\ P\frac{2-2}{2-2}\ M\frac{3-3}{3-3} = 32$$

MAN   Figure 63
(Homo sapiens)

The order Primates, the only representative of which in Louisiana is Man, is included here briefly for the sake of completeness.

The genus *Homo* and its one living species, *H. sapiens,* is distinguished by having the hind limbs longer than the forelimbs and modified for an erect, upright stance; the femur longer than the humerus; the foramen magnum located near the center of the vertical face of the skull (Figure 64); a diastema absent between the second upper incisor and the canine; tendencies present toward (1) the formation of a pronounced chin, (2) the loss of the supraorbital prominences, (3) a reduction in prognathism, or the degree to which the lower jaw projects beyond the plane of the face, (4) a reduction in the length of the

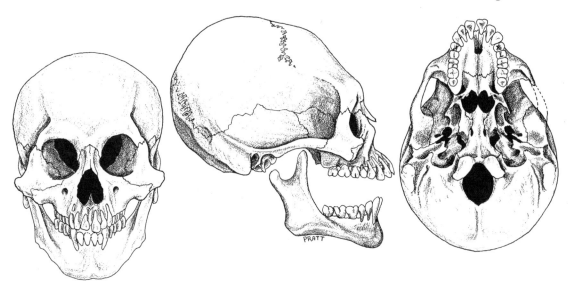

Figure 64   Skull of a Louisiana Indian *(Homo sapiens)* from a prehistoric site in Assumption Parish (LSU Museum of Geoscience no. 16AS00). Approx. ¼×.

canines, (5) an increase in the absolute and relative size of the brain, and (6) a decrease in the thickness of the cranial bones (Hall and Kelson, 1959).

Subspecies

Although some students of modern man consider all living forms of *Homo sapiens* to be members of a single race, other experts would recognize from four to as many as forty races, the figure depending on which authority is being consulted. For the four alleged primary races, conventional subspecific names are available, as follows: Caucasian, *H. s. sapiens*, of Europe, western and southern Asia, and northern Africa; American Indian, *H. s. americanus*, of North, Central, and South America (now almost universally regarded as the product of an early invasion of mongoloids by way of the Bering Bridge and possibly other routes as well); Negro, *H. s. afer*, originally of Africa south of the Sahara; and Mongolian, *H. s. asiaticus*, of northern and eastern Asia.

The races *afer, asiaticus,* and *sapiens* are now introduced and widely established in nearly all parts of North America, including Louisiana.

[*Homo*] *sapiens* Linnaeus, Syst. nat., ed. 10, 1, 1758:20; type locality, Sweden.

[*Homo sapiens*] *americanus* Linnaeus, Syst. nat., ed. 10, 1, 1758:22; type locality, eastern North America.

[*Homo sapiens*] *afer* Linnaeus, Syst. nat., ed. 10, 1, 1758:22; type locality, Africa.

[*Homo sapiens*] *asiaticus* Linnaeus, Syst. nat., ed. 10, 1, 1758:21; type locality, Asia.

SLOTHS, ANTEATERS, AND ARMADILLOS

Order Edentata

THIS ORDER is made up of three living families, all that remain of the ten that once reigned in the Western Hemisphere. The distribution of both the extinct and the living forms is exclusively New World. The seven extinct families include such spectacular creatures as the giant ground sloths and the glyptodonts. Some of the ground sloths attained immense size, as large as a rhinoceros. The glyptodonts were remarkable for the extreme development of their defensive armor. The top of the head was protected by a thick, long shield or casque; the body and limbs were enclosed in an immense carapace; and, in some forms, the end of the tail was equipped with a huge, club-shaped, horny mass that may have been employed against its attackers in the same manner as a knight's mace.

A few of the ground sloths existed in the West Indies and in South America until Recent times. Indeed, they may have been seen by the early Europeans that visited Hispaniola and Puerto Rico in the early part of the 16th century, for their bones have been found in caves in association with fragments of pottery and with bones of man and the domestic pig. In 1888, near the southern end of the South American continent, a piece of hide, covered with long, reddish hair, was found in direct association with human remains, tools, and pieces of fur of the Guanaco (*Lama guanaco*), one of the four camels endemic to the Neotropical region. The remains of the sloth hide were studded with highly diagnostic dermal bones, and the hair was covered with a dried substance that on later chemical analysis proved to be serum. Some of the skin itself was boiled and found to contain glue, showing that the collagen and gelatinous substances were perfectly preserved. All evidences point to the animal having been alive not too long before its remains were found. Legends prevail among the indigenous Indian inhabitants of Patagonia that their ancestors knew a strange, huge, ugly monster, which had its abode in the cordillera of the Andes south of latitude 37°. There is even some evidence that it was kept in captivity by man, for some of its bones have been found in what appear to have been man-made stone corrals (Beddard, 1902).

The three present-day families of the order Edentata are the tree sloths, anteaters, and armadillos, which bear little external resemblance to each other. The fact that they are related at all is revealed only by skeletal similarities. The word Edentata comes from the Latin and means without teeth, which is true of the highly specialized anteaters but not of the tree sloths and armadillos. Indeed, the Giant Armadillo *(Priodontes giganteus)* of the Amazon Basin possesses as many as 100 small teeth, more than any other mammal except certain toothed whales.

The anteaters, which make up the family Myrmecophagidae, comprise three genera and three species. The largest is the appropriately named Giant Anteater of the humid tropical forest and savannahs of Central and South America. It attains a length of six feet, part of which is taken up by the huge, bushy tail. Its greatly elongated, tapering snout is equipped with a tongue well over a foot long and covered with a sticky substance from the salivary glands that aids in picking up ants and termites and their eggs, larvae, and pupae. The heavily clawed forefeet serve to tear apart ant and termite mounds. The much smaller Collared Anteater or Tamandua and

the rather diminutive Two-toed Anteater extend as far north as central and southern Mexico, respectively, but the Giant Anteater reaches only to Guatemala and southern British Honduras.

The Tree Sloths, which constitute the Family Bradypodidae, are a bizarre group consisting of two genera and some seven species that inhabit the tropical forest from Honduras south to Argentina, Paraguay, and Brazil. They derive the name sloth from their sluggish movements. Their forefeet and hind feet are equipped with long claws that hook over horizontal branches from which they normally suspend themselves in an upside down position. On the ground they are unable to walk, but they will extend the forelegs and claws and pull themselves forward. The diet is entirely vegetarian. The coloration is sometimes greenish in hue because of the presence of algae in the hair.

Armadillos, the only family of edentates with a Louisiana representative, will be the subject of the remainder of this account.

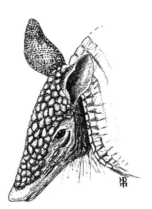

The Armadillo Family
Dasypodidae

THE FAMILY contains nine genera and about 20 species and ranges from the south-central and southeastern United States through Mexico and Central America to Chile and Argentina. The word armadillo is of Spanish origin and refers to the armorlike covering of its head, body, and tail. Only the Naked-

tailed Armadillo *(Cabassous centralis),* which does not range north of Guatemala, and the Nine-banded Armadillo *(Dasypus novemcinctus)* reach this continent. The remaining genera and species are confined to South America.

The South American three-banded armadillos *(Tolypeutes)* can roll themselves into an almost completely armored ball. Armadillos range in size from the extremely small, seven-inch Fairy Armadillos *(Chlamyphorus)* of central western Argentina to the enormously large Giant Armadillo *(Priodontes giganteus)* of Amazonia, which weighs up to 120 pounds. All species possess some hair, chiefly on the underparts but also in between the scutes and the bands on the dorsal surface. *Dasypus pilosus,* a rare Andean species, is unique among armadillos in having almost as much dorsal hair as a Common Muskrat. The hair entirely conceals the dorsal plates. The LSUMZ collections contain the only two specimens of this species in the museums of North America. In fact the only other specimens in the Western Hemisphere are two mounted examples in the Museo de Historia Natural "Javier Prado" in Lima, Peru.

Genus *Dasypus* Linnaeus

Dental formula:

$$I\frac{0-0}{0-0} \ C\frac{0-0}{0-0} \ P\frac{0-0}{0-0} \ M\frac{7 \text{ to } 9-7 \text{ to } 9}{7 \text{ to } 9-7 \text{ to } 9} = 28 \text{ to } 36$$

NINE-BANDED ARMADILLO   Plate I and
(Dasypus novemcinctus)   Figure 65

Vernacular and other names. — The word Nine-banded refers to the nine flexible bands on the carapace between the anterior scapular shield and the posterior pelvic shield. The species is sometimes called the "grave digger," but the reason is obscure because it does not habitually excavate its holes in cemeteries. The generic name *Dasypus* comes from a combination of two Greek words generally taken to mean hairy-footed, which makes no sense at all. Perhaps one of the alternative meanings, thick-footed or rough-footed, was the one intended by Linnaeus, who coined the word. The specific name *novemcinctus* is derived from two Latin words, *novem,* meaning nine, and *cinctus,* meaning banded or girdled.

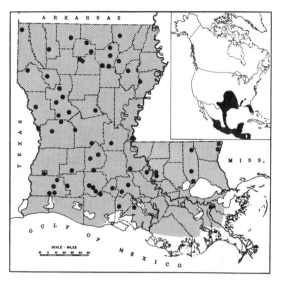

Map 20 Distribution of the Nine-banded Armadillo *(Dasypus novemcinctus). Shaded area,* known or presumed range within the state; •, localities from which museum specimens have been examined; *inset,* overall range of the species.

Distribution. — The species ranges from southeastern Oklahoma and southwestern Arkansas south and east through the Gulf Coast states to southern Florida and south across most of Texas and over most of Mexico, except parts of the central plateau, through Central America to southern South America. In Louisiana it is now virtually statewide in its occurrence. (Map 20)

External appearance. — The Nine-banded Armadillo is about the size of an opossum but heavier. Talmage and Buchanan (1954) aptly remark that perhaps the most expressive, if not the most scientific, description of the armadillo is that given to Strecker (1926) by a Negro resident of Louisiana who called it a "dry-lan varmint, wid er turpin [=terrapin] shell." Most of the animal is encased in a bony carapace beginning with an anterior scapular shield or buckler, three or four inches wide, followed by 9 flexible bands of more than 60 alternating triangular scutes, and terminating with a two- to four-inch posterior, pelvic shield, which, like the anterior buckler is comprised of 18 to 20 rows of small octagonal scutes surrounded by a multitude of minute peripheral plaques. The tail is encased in a series of bony rings, and the top of the head is covered with a plate of scutes that extends from the nose, over the eyes, to the ears. The thick, leathery ears are hairless, densely reticulated with scutes, and almost joined medially at their base. The face, chin, throat, neck, abdomen, and upperparts of the legs are studded with scalelike dermal nuclei that each give rise to 6 to 10 white hairs. The feet and legs are covered with scutes. There are four toes on the forefeet, five on the hind feet, and all toes are heavily clawed. The two middle claws on the forefeet and the claw on the middle toe on the hind foot are the largest. Most of the major dorsal scutes are spotted with ivory yellow, especially along the sides, whereas on the flexible bands each posterior triangular plate is terminally marked with an ivory spot. The tail is predominantly ivory colored although the anterior dorsal scutes and the terminal fourth of the tail are blackish or horn colored. Each posterior scute of the flexible bands is triangular and has four fairly long white hairs arising from beneath its posterior edge.

Measurements and weights. — Two females and a male in the LSUMZ from the western part of the state measured, respectively, as follows: total length, 730, 763, 730; tail, 350, 373, 310; hind foot, 65, 70, 65; ear, 32, 35, 38. The skulls of these three specimens plus that of a male from West Baton Rouge Parish, averaged as follows: greatest length, 96.5 (93.7–100.0); interorbital breadth, 24.1 (22.7–25.9); zygomatic breadth, 37.4 (35.6–38.5); cranial height, 26.1 (25.6–27.5); palatilar length, 62.5 (59.5–65.9); postpalatal length, 15.1 (14.1–16.4). Dr. Alfred L. Gardner informs me that six males and six females

Figure 65 Nine-banded Armadillo *(Dasypus novemcinctus mexicanus).* (Photograph by Joe L. Herring)

from the Hebert Center of Tulane University at Belle Chasse, Plaquemines Parish, averaged, respectively, 3.3 (2.4–4.2) and 4.5 (3.1–5.7) kilograms, or 7.3 and 9.9 pounds.

Cranial and other skeletal characters. — The dorsal aspect of the skull of *Dasypus novemcinctus* is somewhat bottle shaped in that the cranium is inflated but narrows down abruptly into a rather long rostrum. The teeth are a series of peglike structures that are flattened laterally. They vary in number from seven to nine on each side of both the upper and lower jaws. There are no incisors or canines (Figure 66).

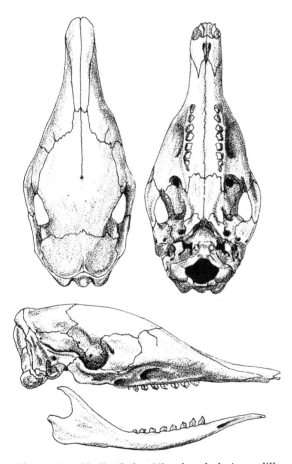

Figure 66 Skull of the Nine-banded Armadillo (*Dasypus novemcinctus mexicanus*), ♂, Provencal, Natchitoches Parish, LSUMZ 1349. Approx. ⅔ natural size.

The skeletal system is closely associated with the dermal plates at the top of the skull, and all the major processes of the vertebrae of the tail are in contact with the bony rings.

Sexual characters and age criteria. — This species shows no secondary sexual characters other than the somewhat larger average size of the males. The testes do not descend into a scrotum. The young are like the adults except for size. Sexual maturity is attained at one year of age.

Status and habits. — The Nine-banded Armadillo has been gradually expanding its range from southern Texas for the past 100 years. It appeared in numbers in Louisiana for the first time not long prior to 1925 but is alleged to have been in De Soto Parish as early as 1917. By 1935 it was already abundant in the pine woodlands north of Lake Charles, as well as in various localities in northern and central Louisiana (Lowery, 1936). Within another decade it was widely distributed over the prairie region of southwestern Louisiana and in pinelands and hardwood uplands of the entire western two-thirds of the state, but it was still not completely established east of the Mississippi River in the Florida Parishes (Lowery, 1943). Now, however, the armadillo seems to be everywhere. There is not a parish in the state from which it has not been reported, and it is abundant in many localities, including the Florida Parishes, as well as in the parishes lying south of Lake Pontchartrain.

There is little doubt that numerous armadillos have been transported from one place to another by man, but, at the same time, the species has unquestionably spread in major part by its own volition. No other mammal, except possibly the opossum, is more frequently seen dead along our highways. Indeed, the high mortality it suffers from being struck by automobiles makes one wonder how it can sustain its numbers, much less continue to increase. A peculiar habit of the species is to bounce upward when startled and this doubtless results in many being killed by cars that might otherwise pass over them without inflicting injury.

The armadillo is generally looked upon with disfavor. It digs burrows into which horses and cattle sometimes step and sustain injury. Fox hunters dislike this animal because they claim that their best-trained dogs will, more frequently than not, desert a "hot" fox trail to pursue an armadillo that happens to cross their paths. And the armadillo, in its quest for insects, will often dig up a garden row. Quail hunters claim that the armadillo eats the eggs of the Bobwhite. But it does so only on extremely rare occasions (Kalmbach, 1944; Fitch et al., 1952). On the contrary, the armadillo is very beneficial. It consumes great quantities of noxious insects, and it is said also to eat poisonous snakes. Its burrows provide retreats for rabbits, skunks, and other forms of wildlife, some of which are valuable fur bearers.

In a year-around study of 104 stomachs, mostly from the longleaf pine belt of the Kisatchie National Forest of western Louisiana, Fitch et al. (1952) found that armadillos consumed mostly beetles and their larvae (44.6 percent of the total diet), followed by plant food (fruits, berries, seeds, mushrooms, 9.8 percent). Other items eaten were Orthoptera (8.5 percent); millipedes and centipedes (8.5 percent); snails, slugs, and earthworms combined (6.4 percent); Lepidoptera (4.7 percent); amphibians and reptiles (4.5 percent); Diptera (4.2 percent); Hymenoptera, mainly ants (3.9 percent); Arachnida (2.1 percent); Hemiptera (1.6 percent); Isoptera (1.2 percent).

Armadillos are predominantly nocturnal in habits, but they also are frequently seen feeding in broad daylight. They become so preoccupied in their search for food among leaves and debris that one can often approach extremely close to them. I have even had one walk within a few feet of my legs as I stood quietly watching a group of birds in a tree. When finally detecting one's presence they amble off at a rapid gait into some nearby thicket or else disappear into one of their holes. They do not roll themselves into a ball as they are often alleged to do. Sometimes, however, when overtaken they will curl up as much as possible, so that their armor protects their soft underparts. But since they run with considerable speed and employ diversionary tactics of darting first one way and then another, they often elude their pursuer.

The nest of leaves and grasses sometimes amounts to a half bushel in volume. Home burrows often have two or more tunnels leading to the outside, though only one is regularly used (Talmage and Buchanan, 1954). They are sometimes, but not always, in the side of a bank. I have them in my yard beneath azalea plants and other shrubbery. If the soil is loose an armadillo can dig a hole and disappear from view in a matter of a few minutes, so effective are its claws for this purpose.

Breeding occurs in July or August. The copulatory position is unusual among quadrupeds, for the female assumes a position on her back. Implantation does not occur immediately; indeed, it is delayed to November. Then follows a gestation period of 120 days. The oddest feature of their reproduction is the fact that the complement of four young

Tracks of Nine-banded Armadillo

is always made up of identical quadruplets, hence members of the same sex. The reason is that all come from only a single fertilized ovum, which divides and then divides again to establish four embryonic growth centers, from which further development into separate embryos proceeds. Since only one ovum, fertilized by a single sperm, is involved, all the young have exactly the same genetic makeup and are consequently identical.

The female has four mammae, two thoracic and two abdominal. The young are born in February in one of the underground excavations. They are well developed at birth and are able to walk when only a few hours old. Their main structural difference from the adults is their more pliable carapace. Their eyes are open and they appear to be miniature replicas of their parents. The life span of the Nine-banded Armadillo is said to be about four years.

The body temperature of the Nine-banded Armadillo is relatively low for a mammal; according to Talmage and Buchanan (1954), it is, at the same time, fairly constant at 32° C (89.6° F) even when the ambient temperature is varying between 16° and 28° C (60.8°–82.4° F). For the same environmental range Wislocki (1933) reported a more variable body temperature of 32° to 35° C (89.6°–95.0° F). At very low ambient temperatures of 11° C (51.8° F), the body temperature of the armadillo has been recorded as low as 29° C (84.2° F). Since the animal does not hibernate, it cannot remain in its burrows for the duration of long spells of cold weather, and this factor alone probably limits the further extension of its range northward. Even in Louisiana individuals often succumb during periods of intense and prolonged cold weather. The severe cold weather of January and February 1948 in western Louisiana resulted in the death of numerous armadillos and left many others in poor physical condition. The overall reduction in the population may have been close to 80 percent (Fitch et al., 1952).

The only sound I have heard from the Nine-banded Armadillo is a wheezy grunt that it emits almost constantly while feeding, but my associate Robert J. Newman tells me that he has heard a piglike squeal from frustrated or frightened individuals. Their hearing is reputed to be inferior, but their sense of smell is probably excellent, at least in sniffing out their prey of earthworms and arthropods. Not infrequently when rooting in the leaves for food the animal will rise on its haunches and sniff the air and then immediately resume its explorations. In some way it can detect the presence of an earthworm or insect several inches beneath the surface. It then proceeds to dig after the prospective morsel with powerful motions of the claws on the forefeet, and it often buries its snout and eyes in the dirt.

Armadillos are said to cross streams by walking underwater on the bottom. As unlikely as this ability may seem to be, Kalmbach (1944) and Taber (1945) experimentally demonstrated that the animal can actually do so. But the procedure is not necessarily normal. To be sure, the armadillo is aided by its dense body, which gives it a specific gravity greater than that of water (1.06 according to Kalmbach, 1944). Moreover, as shown by Scholander et al. (1944), the armadillo can build up a considerable oxygen debt and can struggle violently without access to oxygen for periods of as much as six minutes. This ability enables it to burrow beneath the surface in a matter of minutes, during which time the nostrils and the mouth are for the most part covered with dirt or debris. It also can thereby remain under water for surprisingly long periods. Hardberger (1950) reports that a Louisiana schoolboy claimed to have caught one in a seine in six feet of water. But to just what extent armadillos habitually cross bodies of water by walking on the bottom requires much additional study.

Predators and other decimating factors.—Bobcats, wolves, and Coyotes doubtless take their toll of armadillos, but one of their worst natural enemies is the dog that happens to catch one of them too far away from its hole. But, despite these occasional depredations, the greatest factor decimating their population is the automobile on our highways. Some losses are sustained from their being captured for the conversion of their shells into baskets and other curios.

Parasites. — Literature on the parasites of armadillos is summarized by Chandler, in Talmage and Buchanan (1954). This species is strikingly free of parasites. It certainly has few ectoparasites, a fact that is not surprising because the bony carapace excludes fleas, lice, ticks, and mites from the dorsal surfaces, and the scantily haired underparts discourage parasites from infesting those areas. According to Chandler, internal parasites are also limited, possibly because of the low body temperature of the armadillo, which remains fairly constant at 32° C (89.6° F). This factor would, however, not appear to be important, for cold-blooded vertebrates, such as reptiles and amphibians, are often loaded with internal parasites. Chandler lists only two protozoan parasites from *Dasypus novemcinctus*, one of which, *Schizotrypanum cruzi*, is the cause of American human trypanosomiasis, or Chagas' disease. But the infective organism is not unduly host specific, being found in dogs, cats, opossums, and several species of rodents, besides man and armadillos. Chandler noted the occurrence of remarkably few helminths in *Dasypus novemcinctus*, nematodes being by far the most prevalent.

Subspecies

Dasypus novemcinctus mexicanus Peters

Dasypus novemcinctus var. *mexicanus* Peters, Monatsb. preuss. Akad. Wiss., Berlin, 1864:180; type locality, Matamoros, Tamaulipas. See Hollister, Jour. Mamm., 6, 1925:9.

Tatu novemcinctum texanum Bailey, N. Amer. Fauna, 25, 1905:52.

Dasypus novemcinctus texanus, Lowery, Proc. Louisiana Acad. Sci., 3, 1936:32–34.

The species is divisible into several well-marked races in its wide range from the United States to southern South America, but all populations in this country, including those from Louisiana, are referable to *D. n. mexicanus*. The names *D. n. texanum* and *D. n. texanus* are now included in the synonymy of *D. n. mexicanus*.

Specimens examined from Louisiana. — Total 73, as follows: *Acadia Par.:* Egan, 1 (LSUMZ); 1 mi. W Crowley, 1 (USL); 2 mi. S Ebenezer, 1 (MSU). *Allen Par.:* Kinder, 2 (MSU). *Avoyelles Par.:* 1 mi. S Cottonport, 1 (USL). *Caddo Par.:* Hwy. 1, 7.5 mi. N Vivian, 1 (LSUMZ). *Calcasieu Par.:* Lake Charles, 1 (LSUMZ); 2.5 mi. E, 2.5 mi. S Gillis, 1 (MSU); 6 mi. S McNeese State University, 1 (MSU); south of Moss Lake, 1 (MSU); De Quincy, 1 (MSU). *Claiborne Par.:* 8 mi. E Homer, 1 (LTU). *Concordia Par.:* Point Breeze, 1 (LSUMZ). *De Soto Par.:* 4 mi. N Gloster, 1 (LTU). *East Baton Rouge Par.:* Baton Rouge, 1 (LSUMZ); 3.1 mi. E University, 1 (LSUMZ). *Evangeline Par.:* Chicot State Park, 1 (USL); 1.7 mi. W Easton, 1 (USL); Cocodrie Lake, 2 (USL); 4 mi. W Easton, 2 (USL). *Franklin Par.:* brushy bank of Big Creek near Baskin, 1 (LTU). *Grant Par.:* Pollock, 1 (LSUMZ). *Iberia Par.:* Avery Island, 1 (TU). *Jackson Par.:* NE edge of Jackson–Bienville Refuge, 2 (LTU); Hwy. 167 south of Clay, 1 (LTU). *Jefferson Davis Par.:* 1.5 mi. E Iowa, 1 (MSU). *Lafayette Par.:* Lafayette, 3 (USL). *Lincoln Par.:* Choudrant, 1 (LTU); 6 mi. NE Unionville, 1 (LTU). *Morehouse Par.:* 5 mi. N Bastrop, 1 (NLU). *Natchitoches Par.:* Kisatchie, 1 (LSUMZ); Provencal, 1 (LSUMZ); 4 mi. S Derry, 2 (USL); Coulee Bayou, 1 (NLU); 1 mi. N Natchitoches, 1 (NSU). *Ouachita Par.:* Monroe, 1 (NLU); 15 mi. S West Monroe, 1(NLU); 15 mi. SW West Monroe, 1 (NLU); Old Sterlington Road, 1 (NLU); 5 mi. S Luna, 1 (NLU). *Plaquemines Par.:* Tulane Univ. Riverside campus, Belle Chasse, 2 (TU). *Pointe Coupee Par.:* 4.5 mi. SE Krotz Springs, 1 (LSUMZ). *Rapides Par.:* 3 mi. S Sieper, 1 (LSUMZ). *Red River Par.:* 2 mi. SE of East Point, 1 (LTU). *Richland Par.:* 5 mi. NW Alto, 1 (NLU). *Sabine Par.:* 10 mi. W Many, 1 (LSUMZ); 7.5 mi. SW Toro, 1 (TU); 7.5 mi. S Toro, 1 (TU). *St. Landry Par.:* 3 mi. W Eunice, 1 (USL). *St. Martin Par.:* Butte La Rose, 1 (USL); St. Martinville, 1 (USL). *St. Mary Par.:* 3.4 mi. E Iberia Parish line on Hwy. 90, 1 (USL); 3 mi. E Iberia Parish line on Hwy. 90, 1 (USL); 3.1 mi. E Iberia Parish line on Hwy. 90, 1 (USL). *Tangipahoa Par.:* Manchac, 1 (SLU); 13 mi. W Franklinton, 1 (SLU); 5 mi. E Hammond, 1 (SLU). *Tensas Par.:* 3 mi. down Tensas River from Newlight Landing, 1 (LTU). *Vermilion Par.:* 3 mi. NW Indian Bayou, 1 (USL). *Vernon Par.:* Hutton, 1 (LSUMZ); Leesville, 1 (USNM). *Washington Par.:* 10 mi. S Angie, 1 (SLU); 0.5 mi. S, 1.5 mi. E Angie, 1 (SLU). *Webster Par.:* La. Army Ammo Plant, W of Minden, 1 (NLU). *West Baton Rouge Par.:* ca. 6 mi. E Rosedale, 1 (LSUMZ).

PIKAS, RABBITS, AND HARES

Order Lagomorpha

THIS ORDER was once regarded as a suborder of Rodentia, but its members have little in common with rats and mice, and few modern zoologists treat them as rodents. Lagomorphs differ from rodents in a number of salient features. For one thing, they possess a pair of incisors, one behind the other, on each side of the upper jaw (Figures 68 and 70), while rodents have only one incisor on each side (Figure 13). Lagomorphs lack a baculum, or penis bone, which is almost always present in rodents. The testes are in a scrotum in front of the penis, as in the marsu-

Figure 67 Eastern Cottontail (*Sylvilagus floridanus*), Hamilton County, Ohio, July 1962. (Photograph by Woodrow Goodpaster)

pials, rather than behind the penis as in rodents and most other mammals.

Rabbits and their allies, which are wholly vegetarians, have a peculiar, perhaps wholly unique, way of obtaining the maximum amount of nutrients from the food they eat. Two types of pellets are expelled from the digestive tract. The dry, brownish pellets are the true feces; they represent the remains of completely digested food from which virtually all nutrients have been extracted. The moist, greenish pellets consist of food that has been only partially digested; they are re-eaten to pass again through the digestive tract (Southern, 1942; Kirkpatrick, 1956). The process is functionally analogous to the method employed by ruminants, such as the cow, which swallow their food into one of the chambers of the stomach and then regurgitate it to the mouth to be chewed again at leisure and reswallowed (Watson and Taylor, 1955).

The order contains only two families, the Ochotonidae, the pikas, or conies, and the Leporidae, the rabbits and hares. Twelve species of pikas inhabit Russia and Siberia and two species occupy the mountains of western North America. They differ from rabbits in being smaller and in having shorter ears and limbs. Anyone who has traveled through the mountainous regions of our western states has almost certainly seen these interesting little animals in the talus formations at high elevations.

Since lagomorphs are strictly vegetarians, they sometimes do damage to crops and girdle the trees of orchards. But, on the other hand, their thin skin and dense fur have served many uses, including that of providing clothing, both for the early aborigine and for modern man. Their succulent meat is important as food and the hunting of them is a major recreational activity in many parts of the United States. When English immigrants to Australia did not find rabbits to hunt, they imported them for this purpose. Unfortunately, in doing so, they took the rabbits away from their natural enemies and other factors that had served to keep their numbers in check, such as their parasites and diseases. Consequently, the introduced rabbits multiplied rapidly and soon overran their new domain, destroying the vegetation and sometimes doing irreparable damage to the habitat. Millions of dollars have been expended in trying to bring their numbers under control, sometimes to no avail.

PRATT
1973

The Hare and Rabbit Family
Leporidae

THIS FAMILY consists of nine living genera and about 50 species and is almost worldwide in distribution, occurring naturally everywhere except in Antarctica, parts of the Middle East, Madagascar, Australia, New Zealand, southern South America, and on oceanic islands. They have, however, been widely introduced in Australia and New Zealand and on many islands, usually to the detriment of the native birds and mammals.

Although often confused, hares and rabbits differ in a number of important details. The young of rabbits are born blind and helpless in a fur-lined nest especially constructed for them by the adults. At birth, they are virtually naked in most species but have a scant covering of hair in at least one. Young hares, on the other hand, are born with a heavy coat of hair, with the eyes open, and with the ability to run about within a few minutes after birth. They do not have a nest prepared for them. Some species of hares of the genus *Lepus* turn white in winter, especially those that live in a snowy winter climate, but none of the rabbits do so. The various jack rabbits of our western United States and the Snowshoe Rabbits of Arctic regions are hares, as are all members of the genus *Lepus*. The so-called Belgian Hare, a strain of the widespread European *Oryctolagus cuniculus*, is, however, a true rabbit. The only members of the family inhabiting Louisiana belong to the genus *Sylvilagus* and are rabbits.

Genus *Sylvilagus* Gray

Dental formula:

$$I\frac{2-2}{1-1} \ C\frac{0-0}{0-0} \ P\frac{3-3}{2-2} \ M\frac{3-3}{3-3} = 28$$

EASTERN COTTONTAIL   Plate I
(Sylvilagus floridanus)

Vernacular and other names. — The Eastern Cottontail is sometimes called the cottontail rabbit or simply cottontail. The addition of the word Eastern serves to distinguish it from various species in other parts of the United States and Mexico, such as the New England, Pygmy, Forest, Brush, Nuttall's, Desert, and Mexican Cottontails. Cottontail refers, of course, to the white puff of fur on the underside of the tail, which it flashes when running.

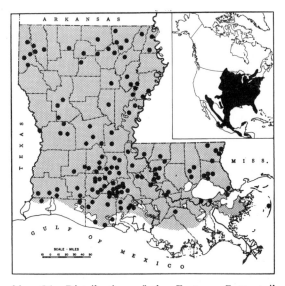

Map 21 Distribution of the Eastern Cottontail *(Sylvilagus floridanus)*. *Shaded area,* known or presumed range within the state; •, localities from which museum specimens have been examined; *inset,* overall range of the species.

The word rabbit comes from the Middle English *rabet*. Among the French-speaking people of southern Louisiana, all rabbits are called *lapin,* there being no distinction made between this species and the Swamp Rabbit *(S. aquaticus)*. The generic name *Sylvilagus* is of Latin and Greek derivation and translates as wood hare *(sylva,* a wood, and *lagos,* hare). The specific name *floridanus* means of Florida and refers to the state from which the species was first collected and first recognized, by J. A. Allen in 1890.

Distribution. — The species occurs throughout most of the eastern two-thirds of the United States and extreme southern Canada south through eastern and much of southern Mexico with more or less disjunct populations in the southwestern United States, western Mexico, Yucatan, and parts of Central America. It ranges widely over most of Louisiana, being absent only from sections of the coastal marshes. (Map 21)

External appearance. — The Eastern Cottontail is a typical rabbit with long ears, large hind legs and feet, short front legs and feet, a soft pelage, and a short, fluffy tail that is white beneath (Figure 67). The upperparts vary from grayish brown to reddish brown except for the nape, which is decidedly rusty, and the face and flanks, which are gray. The tops of the front and hind feet are white or whitish, and a pale cream-colored eye-ring surrounds the eye. The chin and upper throat are white, as are also the belly and undersides of the legs, but the throat is brownish.

Color phases. — No color phases as such are known, but the LSUMZ collections contain a perfect albino that was obtained near Farmerville, in Union Parish.

Measurements and weights. — Specimens in the LSUMZ, from numerous localities in Louisiana, averaged as follows (because not all body measurements were recorded on every label

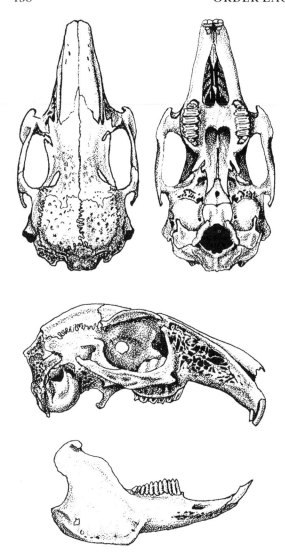

Figure 68 Skull of Eastern Cottontail *(Sylvilagus floridanus alacer)*, ♀, 1 mi. N Glenmora, Rapides Parish, LSUMZ 2433. Approx. natural size.

and because some skulls were partially defective, the number of specimens involved in each measurement is given in parentheses after the name of the measurement): total length (28), 411 (385–480); tail (21), 52 (40–62); hind foot (32), 89 (80–111); ear (33), 58 (47–71); greatest length of skull (28), 73 (69.0–76.8); cranial breadth (28), 26.4 (24.5–28.5); zygomatic breadth (27), 34.3

(32.1–36.6); interorbital breadth (37), 17.2 (15.4–20.0); breadth of rostrum at P 1 (37), 19.0 (17.3–22.4); breadth of rostrum at anterior end (38), 9.4 (7.3–11.2); palatal breadth (40), 27.7 (25.3–29.9); palatilar length (32), 29.6 (25.7–31.6); postpalatal length (31), 29.5 (25.7–38.8); maxillary toothrow (34), 13.3 (12.0–14.8). The average weight of six specimens from the same material was 957.6 (784–1,244) grams or 2.1 (range, 1.60–2.75) pounds. The average weight of 123 cottontails obtained by Hastings (1954) in southern Louisiana was 2.54 (1.3–3.5) pounds.

Cranial characters.—The skull of a rabbit is easily distinguished from that of all other mammals of Louisiana by the prolonged diastema, two pairs of upper incisors, and the bony network or fenestration of the sides of the long rostrum (Figure 68). The hindmost pair of incisors are tiny and almost cylindrical without much of a cutting edge. Their function is not known. From the Swamp Rabbit *(S. aquaticus)*, the Eastern Cottontail differs by its much smaller skull, by the lack of obvious sutures where the interparietal bone meets the parietals and the supraoccipital, and by the absence of fusion of the posterior extension of the supraorbital process where it touches the skull.

Sexual characters and age criteria.—The sexes are superficially identical. Females average only slightly larger than males. The young are quite like the adults in color, but the pelage is softer and more woolly until they are about half grown. Juveniles can be distinguished from adults up to approximately nine months of age by the stage of development of the proximal end of the upper arm, or humerus. In the young the ball-like cap by which the humerus fits into the socket of the shoulder is separated from the main shaft of the bone by a groove filled with cartilage. In an individual over nine months of age the cap is fused with the shaft into one solid bone (Figure 69).

Status and habits. — The Eastern Cottontail is one of the commonest mammals in Louisiana. Its total population in the state may run as high as several million and, in good years, may approach 10,000,000 (St. Amant, 1959). Consequently, few of our native mammals are better known. None perhaps is as dear to the hearts of people of all ages. The child first learns of it while in the cradle by one of the lullabies that it hears sung while being rocked to sleep. And afterwards come the stories associated with the Easter Bunny, Aesop's fable of the race between the tortoise and the hare, the immortal Uncle Remus tales, and countless other myths in which the rabbit is the key figure. But to the man of the family and to innumerable sportsmen throughout the land the rabbit is more than a symbol. It is a major source of food for the table, and it provides inestimable hunting pleasure as the country's number one game species. More hunters spend more time and more money in pursuit of rabbits than all other game combined. No figures are available on the annual kill of cottontails in Louisiana, but, according to Rue (1965), 6,018,914 were killed in Missouri alone in 1958, and hunters of that state figured that they had left 10,000,000 for breeding purposes.

The Eastern Cottontail is most frequently

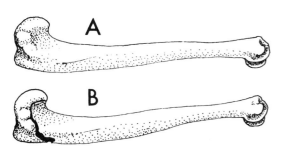

Figure 69 Humerus of (A) an adult Eastern Cottontail and of (B) an immature Eastern Cottontail less than nine months old. In the latter a groove, indicating the presence of cartilage, remains between the epiphysis and the main shaft of all the long bones. In the adult the cartilage is replaced by bone and the epiphysis is ankylosed to the main shaft.

found in fairly open country, pastures, and grassy areas adjacent to croplands. The Swamp Rabbit, on the other hand, more often frequents woodlands, swamps, and marshes. The cottontail is generally hunted with dogs, and when flushed it takes off at a rapid clip, zigzagging first one way and then another, with its tail flashing white. As every rabbit hunter knows, hitting a cottontail on the run is no easy feat.

When feeding, rabbits sometimes walk, moving the left front leg in unison with the right hind leg, then advancing the right front and the left rear leg, just as do most four-legged animals. But, when startled, they explode into a series of bounding hops. The two forefeet come down one after the other and the two hind feet are then brought up and hit the ground ahead of the forefeet. The next time the forefeet hit the ground they may be anywhere between one and ten feet ahead of the last tracks made by the hind limbs.

One of the main reasons why rabbits can easily withstand heavy hunting pressure is their great fecundity. The breeding season extends almost throughout the year, although the main period is from February to mid-October. The litter size in Louisiana ranges from 1 to 7 with an average of 3.7 young per litter (Hastings, 1954). The gestation period is 28 to 32 days, and a female may produce as many as 6 litters a year.

The nest of the cottontail is in a saucerlike depression in the earth in which the female places soft dead grass mixed with hair that she pulls from her breast with her teeth. In addition to lining the nest with hair, the female also makes a blanket of grass and hair with which to cover the young, especially when she is not in attendance. The nest is usually in a dense stand of grass, often beneath a stand of taller vegetation such as a hedgerow or briar thicket.

As soon as the young are born the female is again in estrus. She continues in heat for about one hour and at this time she is vigorously pursued by one or more males. After giving birth to her young the female remains away from the nest until near nightfall, but she is not far away, usually close enough to watch the location of the nest and to distract any intruder should the need arise. Late on the first day she returns to the nest, removes the cover, and proceeds to suckle the young from her eight nipples. She may return to nurse them again in the middle of the night and always does so at dawn. She then retires to her "form," as the resting place of a rabbit is called. Growth of the young is rapid, for the milk of the mother is exceptionally rich. Rue (1965) states that 12 percent of it is protein and 13 percent butterfat. This author points out that milk with a 13 percent butter-fat content is about four times as rich as the average cow's milk that we ourselves drink.

At birth the young cottontail is about four inches in length and weighs approximately one ounce. The eyes are sealed shut and the little rabbits are quite helpless. But within two weeks their weight is close to a quarter of a pound and they are able to make short forays from the nest and to carry on many playful activities, by which they gain strength and coordination. At the end of three to four weeks they are weaned and on their own. Should the female have bred following parturition she is about ready to deliver another litter. Consequently, she wants nothing to do with the last brood. She is too busy preparing a nest and getting ready for her new off-spring.

The food habits of the Eastern Cottontail in Louisiana have been the subject of several investigations by game management students at Louisiana State University, Baton Rouge (Lapham, 1950; Bryant, 1954; Croft, 1961; Richardson, 1963). Their studies indicate that rabbits eat a wide variety of grasses and that their preference for certain species varies in the course of the year. Much depends on the seasonal availability of the plants. Some cultivated foods are relished such as rye grasses, vetch, chufa, oats, soy beans, and various truck crops.

The cottontail, when engaged in eating, does not use the forefeet to grasp or hold the food, for it possesses no dexterity with its paws. The upper lip is deeply cleft and is the basis of the expression "harelipped" that describes a human abnormality. Since in rabbits the maxillary toothrows are much farther apart than are the mandibular toothrows, only the teeth on one side occlude at one time. Consequently, the animal chews with a transverse motion of its jaws, causing a movement of the nose that makes it appear to be twitching.

Cottontails are usually silent but when frightened by being captured by a predator or when fighting among themselves they emit a shrill high-pitched scream. The sound, when once heard, is never to be forgotten. It is recognized by other animals of field and forest and causes an instantaneous response. Some freeze in their tracks, but others, such as competing predators, race to the source of

Tracks of Eastern Cottontail

the sound as if in hope of joining the feast. Predator calls are manufactured that imitate the sound and, when correctly employed, usually bring up predators to the waiting hunter. The female gives a soft grunting sound while nursing her young, and she also has a call that she makes to assemble her offspring. Also, the large hind feet may be thumped on the ground as a danger signal.

The fact that rabbits are sometimes infected with tularemia discourages many people from handling or eating them. To avoid contamination, hunting of rabbits should be confined to the cold, winter months, when the danger of infection is less. In preparing a rabbit anyone with open cuts or abrasions on the hands should use rubber gloves, and under all circumstances the meat should be thoroughly cooked, for then it is perfectly safe. Certain mycin-type drugs are now a specific cure for tularemia, and for this reason the disease is no longer so dreaded as it was formerly. Any rabbit, however, that appears sick should be avoided. Once in my early days when I was still a student I came to the laboratory and found a beautiful specimen of an Eastern Cottontail on my table with no explanation as to its source or the identity of its collector. Presuming that sooner or later I would be advised, and having free time available, I immediately set about preparing it as a museum specimen. After I had completed the task one of my classmates did arrive to tell me that the animal had been easily caught by hand! The thought immediately occurred to me that surely no one could pick up a live rabbit unless it were ill or injured and that perhaps this particular animal had been sick with tularemia. Although the move was probably futile at that stage, I quickly rewashed my hands and bathed my arms to the shoulder with the strongest possible disinfectant.

Predators and other adverse factors. — Few rabbits ever reach a "ripe old age," because of the great pressures constantly exerted on the population by a wide variety of factors.

Among the principal enemies of rabbits are the Great Horned Owl, our two species of foxes, the Coyote, and the Bobcat. Mink, weasels, and Striped Skunks likewise kill them, and crows destroy many young cottontails in the nest, as do also dogs, cats, and snakes. Mowing and plowing operations and the burning of fields and wastelands take a further heavy toll of young in the nest. Hunting is perhaps one of the greatest decimating factors, but the experience of game managers and other students of animal populations has demonstrated unequivocally that hunters are taking only a harvestable portion of the overall population, that if hunting losses did not occur an equal number would be eliminated by some other factor such as a natural predator, adverse weather, or disease. Heavy rains often wash out nests or cause the death of young by chilling them. Periodic cyclic outbursts of tularemia sometimes reduce the rabbit populations drastically. Indeed, the life of the cottontail is fraught with one hazard after another, but despite all the dangers with which it copes, its numbers remain remarkably high.

Parasites. — I am aware of no specific studies of the parasites of the Eastern Cottontail in Louisiana, but studies elsewhere within the range of the species show it to be infected with the usual array of internal parasites. Harkema (1936) listed the parasites then known to occur in the cottontail, citing six species of protozoans, three trematodes, seven cestodes, eight nematodes, five mites, eight fleas, and two flies. Rabbits are an important link in the life cycle of the dog tapeworm (*Taenia pisiformis*). The eggs pass out of the dog in its feces onto the grass, which is later eaten by the rabbit. The eggs hatch in the rabbit's digestive tract and then spread to the liver, where they develop into cysts that produce white, watery pustules that are sometimes confused with the smaller spots characteristic of tularemia infections. Rabbits are also subject to myxomatosis, a deadly disease that has been used in various places, such as in Australia, to try to reduce the rabbit population by introducing individuals infected with the disease. Generally, however, such attempts are unsuccessful, for a certain percentage of the population will prove either to be immune or else will soon develop an immunity to the disease. This ability on the part of certain individuals to resist the disease is usually relayed to their offspring and in time a whole population is able to escape infection. Numerous kinds of ticks, lice, and fleas infest rabbits, but only in exceptional cases do they weaken the animal to an appreciable extent, in which case the host generally falls victim to some predator. Rarely, excessive numbers of ticks may induce an anemia that results in weakness leading to death. A hunter who kills a rabbit with fleas on it should not throw away the animal, for if the fleas should get on a human they get off at the earliest opportunity in search of another rabbit. Likewise, rabbits with botfly larvae, or warbles, are still perfectly fit for human consumption, since the insects are only just beneath the skin and do not affect the underlying meat.

Subspecies

Sylvilagus floridanus alacer (Bangs)

Lepus sylvaticus alacer Bangs, Proc. Biol. Soc. Washington, 10, 1896:136; type locality, Stilwell, Boston Mountains, Adair County. Oklahoma.
Sylvilagus floridanus alacer, Lyon, Smiths. Misc. Coll., 45, 1904:336.

All Eastern Cottontails in this state are referable to the subspecies *alacer,* which occupies the lower Mississippi Valley from southeastern Kansas, southern Missouri, and extreme southern Illinois south through eastern Oklahoma, Arkansas, eastern Texas, Louisiana, and Mississippi.

Specimens examined from Louisiana.—Total 259, as follows: *Acadia Par.:* 3 mi. E Church Point, 1 (LSUMZ); 11 mi. N Crowley, 1 (USL); 5 mi. S Crowley, 1 (USL); 3 mi. SE Egan, 1 (USL); 4.5 mi. S Crowley, 2 (USL); 3 mi. SE Rayne, 1 (USL); 3 mi. SW Duson, 1 (USL); Rayne, 1 (AMNH); Pointe aux Loups Springs, 2 (MCZ); Cartville, 6 (3 MCZ; 3 USNM). *Allen Par.:* Kinder, 2 (MSU); 4.5 mi. E Hwy. 26 on Hwy. 104, 1 (TU). *Ascension Par.:* 3 mi. NE Donaldsonville, 1 (LSUMZ); Burnside, 1 (LSUMZ). *Avoyelles Par.:* 6 mi. N Marksville, 1 (LSUMZ); 1 mi. W Cottonport, 1 (USL); 2 mi. W Cottonport, 1 (USL). *Bossier Par.:* 2 mi. E Bossier City, 1 (LSUMZ); Bossier Air Force Base, 1 (LSUMZ); Haughton, 1 (USNM); 5 mi. N Bossier, 1 (LTU). *Caddo Par.:* 3 mi. S, 1 mi. W Blanchard, 1 (LSUMZ); 5 mi. N Shreveport P.O., 1 (LSUMZ); 15 mi. S Shreveport, 1 (NLU); Belcher, 1 (USNM); Foster, 5 mi. E Shreveport, 1 (USNM). *Calcasieu Par.:* 7 mi. S Sulphur, 1 (LSUMZ); 2 mi. S Lake Charles, 1 (MSU); 7.8 mi. S Lake Charles, 1 (MSU); Ged, 2 (CAS). *Caldwell Par.:* Columbia, 5 (FM). *Cameron Par.:* 1 mi. S Lowry, 1 (LSUMZ); Gum Cove Ferry Rd., 2 (MSU). *Catahoula Par.:* 4 mi. NNE Foules, 1 (LSUMZ). *Claiborne Par.:* 5 mi. N Arcadia, 1 (LTU). *Concordia Par.:* Ferriday, 1 (LTU); 4 mi. N Monterey, 1 (NLU). *East Baton Rouge Par.:* various localities, 18 (17 LSUMZ: 1 USL). *East Carroll Par.:* 5 mi. W Sonofemier, 1 (LTU); Lake Providence, 1 (NLU). *East Feliciana Par.:* 9 mi. SSW Clinton, 1 (LSUMZ). *Evangeline Par.:* 4 mi. N Ville Platte, 1 (LSUMZ); 1.5 mi. N Chataignier, 2 (USL); 1 mi. W Chataignier, 1 (USL); 4 mi. N Basile, 1 (USL); 5 mi. N Basile, 1 (USL); 3 mi. W Ville Platte, 1 (USL); 3.5 mi. W Ville Platte, 1 (USL). *Franklin Par.:* 5.5 mi. E Bayou Macow bridge on Hwy., 1 (LSUMZ); 15 mi. SE Winnsboro, 2 (NLU); 2 mi. S Winnsboro, 2 (NLU); Wisner, 2 (NLU). *Iberia Par.:* Jeanerette, 1 (USL); 1 mi. N Jeanerette, 1 (USL); 3 mi. S Jeanerette, 1 (USL); 5 mi. W New Iberia, 1 (USL); 1 mi. S Patoutville, 1 (USL); Avery Island, 9 (1 USNM; 8 CAS). *Iberville Par.:* 0.5 mi. NW Plaquemine, 1 (LSUMZ); 2 mi. S Grosse Tete, 1 (LSUMZ); Ramah levee, 1 (LSUMZ). *Jackson Par.:* Jackson–Bienville Game Reserve, 1 (LTU). *Jefferson Par.:* Kenner, 1 (SLU). *Jefferson Davis Par.:* Welsh, 1 (MSU); 3 mi. W Lake Arthur on Morgan Plantation, 1 (LSUMZ). *Lafayette Par.:* various localities, 23 (USL). *Lafourche Par.:* 5 mi. ENE Thibodaux, 1 (LSUMZ). *La Salle Par.:* Jena, 2 (LTU). *Lincoln Par.:* Ruston, 2 (LTU). *Livingston Par.:* locality unspecified, 1 (SLU). *Morehouse Par.:* 0.5 mi. E Vaughn, 1 (LSUMZ); Wham Break, 1.5 mi. S Hwy. 134, 2 (NLU); 4 mi. W Oak Ridge, 2 (NLU); Perryville, 1 (NLU); Mer Rouge, 3 (USNM). *Natchitoches Par.:* Provencal, 3 (LSUMZ); Kisatchie, 9 (LSUMZ); Natchitoches, 1 (USNM). *Orleans Par.:* New Orleans, 1 (LSUMZ); 3 mi. E New Orleans, 1 (USL); Lake Catherine, 1 (AMNH). *Ouachita Par.:* 1.5 mi. SE Monroe, 1 (NLU); 2 mi. NW Calhoun, 1 (NLU); Monroe, 2 (NLU); Old Sterlington Rd., 1 (NLU). *Pointe Coupee Par.:* 5 mi. N Lottie, 1 (LSUMZ); Innis, 1 (LSUMZ). *Rapides Par.:* 1 mi. N Glenmora, 1 (LSUMZ); Alexandria, 1 (USNM); Lecompte, 1 (USNM); 3 mi. S Cheneyville, 1 (LTU). *Richland Par.:* Rayville, 1 (NLU). *Sabine Par.:* 7.5 mi. SW Toro, 1 (TU). *St. Bernard Par.:* Chandeleur Islands (opp. North Island), 2 (USL). *St. Landry Par.:* 2 mi. N Port Barre, 1 (LSUMZ); 1 mi. SW Palmetto, 1 (LSUMZ); 6 mi. S Opelousas courthouse, 1 (USL); 7 mi. W Opelousas, 8 (USL); 1 mi. E Eunice, 2 (USL); Rosa, 1 (USL); 1 mi. S Washington, 1 (USL); 4 mi. S Opelousas, 1 (USL); Leonville, 1 (USL). *St. Mary Par.:* 3 mi. E Iberia Parish line on Hwy. 90, 1 (USL). *St. Martin Par.:* 14 mi. E Lafayette, 1 (LSUMZ); 2 mi. N St. Martinville, 1 (USL); St. Martinville, 2 (USL); Pointe Claire, 1 (USL); on levee near Catahoula, 1 (USL); 1 mi. S Parks, 7

(USL); 1 mi. SE Parks, 1 (USL); 8 mi. SW Cecelia, 1 (USL); 2 mi. N Parks, 1 (USL); 6 mi. W St. Martinville, 1 (USL); Lake Martin, 3 (USL). *St. Tammany Par.:* Salt Bayou, 1 (LSUMZ); 6 mi. S Covington, 1 (LSUMZ); 6 mi. E Covington, 1 (LSUMZ); Covington, 1 (SLU); Madisonville, 1 (AMNH). *Tangipahoa Par.:* locality unspecified, 1 (SLU); 3 mi. S Hammond, 1 (SLU); 2 mi. S Ponchatoula, 1 (SLU); 7 mi. NE Hammond, 1 (SLU). *Tensas Par.:* 1 mi. S St. Joseph, 1 (LSUMZ); 1 mi. S Waterproof, 1 (LSUMZ). *Union Par.:* near Farmerville, 1 (LSUMZ). *Vermilion Par.:* 2 mi. SE Gueydan, 1 (LSUMZ); 7 mi. S Klondike, 1 (LSUMZ); Pecan Island, 2 (USL); Red Fish Point, 1 (USL); Maurice, 1 (USL); Perry, 3 (USNM); Cheniere au Tigre, 5 (ANSP). *Vernon Par.:* Hutton, 1 (LSUMZ); 3 mi. E Slagle, 1 (USL); 8 mi. SE Simpson, 1 (USL). *Washington Par.:* Bogalusa, 1 (LSUMZ); 10 mi. SW Bogalusa, 1 (LSUMZ); 2 mi. E Varnado, 1 (LSUMZ); 4 mi. NW Varnado, 1 (LSUMZ); Varnado, 1 (TU); Hackley, 8 (FM). *West Carroll Par.:* Oak Grove, 1 (NLU). *West Feliciana Par.:* 2 mi. N St. Francisville, 1 (LSUMZ).

SWAMP RABBIT   Plate I
(Sylvilagus aquaticus)

Vernacular and other names. — The Swamp Rabbit goes also by the name of marsh rabbit, particularly in the coastal parishes, where marshes are its main habitat and where the word swamp connotes, as it does elsewhere, a wet, wooded area. The name marsh rabbit is unfortunate because *Sylvilagus palustris,* of extreme southeastern United States, officially goes by that name. In the interior of the state the species is sometimes called the "cane cutter" or "cane jake." The second part of the scientific name, *aquaticus,* comes from the Latin and means found in water. It refers to the animal's preference for wet situations and its complete lack of hesitation in taking to water and swimming.

Distribution. — The species occurs from extreme southeastern Kansas, southern Missouri, southern Illinois, and extreme southwestern Indiana, south through eastern Oklahoma, eastern Texas, Arkansas, Louisiana, Mississippi, Alabama (except the southeastern corner), northwestern Georgia, and extreme western South Carolina. In Louisiana it is statewide. The range of *S. palustris* extends westward along the northern Gulf Coast to Mobile Bay, but *S. aquaticus* occurs

west of Mobile Bay thence westward along the coast. (Map 22)

External appearance. — The Swamp Rabbit resembles the Eastern Cottontail *(S. floridanus)* but is appreciably larger and darker in coloration. The top of the head, the cheeks, and the entire dorsum are much browner, with the center of the back showing a more extensive interspersion of black. The flanks, legs, rump, and the top of the tail are decidedly browner (less grayish), as are also the tops of the feet. The eye-ring in *S. floridanus* is whitish or at most a Light Buff, but in *S. aquaticus* it is a Pinkish Cinnamon. I find that I can distinguish *S. aquaticus* at a glance by the color of the eye-ring and by the overall brownish, instead of grayish, coloration, particularly on the sides, flanks, rump, tail, and feet.

Measurements and weights. — Specimens in the LSUMZ, from numerous localities in Louisiana, averaged as follows (because not all body measurements were recorded on every label,

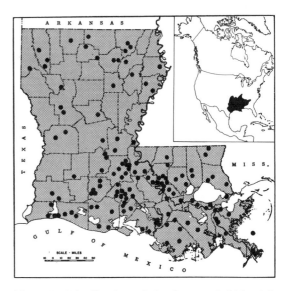

Map 22 Distribution of the Swamp Rabbit *(Sylvilagus aquaticus). Shaded area,* known or presumed range within the state; •, localities from which museum specimens have been examined; *inset,* overall range of the species.

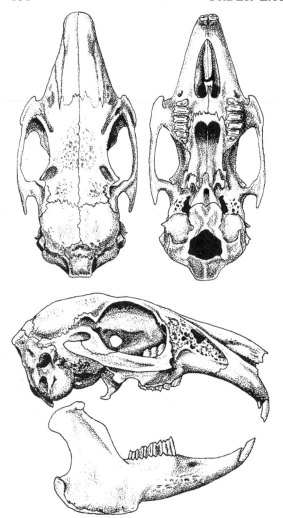

23.2 (21.4–26.2); breadth of rostrum at anterior end (28), 11.9 (10.5–13.7); palatal breadth (36), 35.5 (32.6–38.2); palatilar length (34), 33.2 (31.3–35.7); postpalatal length (33), 35.5 (32.6–38.2); maxillary toothrow (39), 15.4 (14.0–16.5). The average weight of seven adults in the same series was 1,698 (1,280–2,019.4) grams, or 3.74 (2.6–4.4) pounds. The average weight of 70 Swamp Rabbits obtained by Hastings in southern Louisiana was 4.39 (3.0–6.5) pounds.

Cranial characters. — The skull of *S. aquaticus* resembles that of *S. floridanus* but is larger and much more massive. Some of the more striking differences are its flatter profile (which results from the much deeper rostrum), the longer and the more robust premaxilla, and the larger and longer incisors. The rostrum is much wider basally and tapers less sharply, so that its width at the anterior end of the nasals is greater and hence the rostrum is more truncate. The supraorbital process is also much heavier than it is in *S. floridanus,* and it is much more extensively fused posteriorly with the frontals. The palatal shelf is wider, the palatal and postpalatal lengths are greater, and the maxillary toothrow is longer than in *S. floridanus.* (Figures 68 and 70)

Figure 70 Skull of Swamp Rabbit *(Sylvilagus aquaticus),* ♂, Hico, Lincoln Parish, LSUMZ 2452. Approx. ⁴⁄₅ natural size.

and because some skulls were partially defective, the number of specimens involved in each measurement is given in parentheses after the name of the measurement): total length (30), 501 (452–552); tail (27), 59 (50–74); hind foot (28), 101 (90–113); ear (31), 70 (60–80); greatest length of skull (34), 87.5 (83.7–99.6); cranial breadth (24), 29.1 (27.9–30.5); zygomatic breadth (27), 42.0 (39.1–49.8); interorbital breadth (34), 20.0 (17.5–29.7); breadth of rostrum at P^1 (30),

Tracks of Swamp Rabbit

Sexual characters and age criteria.—The sexes, as in the Eastern Cottontail, are superficially similar. Half-grown young are almost identical with the adults, but, as is the case among cottontails, they lack, up to about nine months of age, fusion of the proximal epiphyses of the humerus to the main shaft of the bone (Figure 69).

Status and habits.—This species is abundant throughout the state, occurring mainly in heavily wooded areas and in our coastal marshes. It is generally replaced by the cottontail in dry, upland cultivated areas and pastures, but itself largely replaces the cottontail in our hardwood bottomland swamps and coastal marshes. In the latter habitat the species is particularly numerous where canal banks and chenieres, or wooded ridges, provide an abundance of excellent cover and where its resting places, or forms, remain fairly dry and its runways easy to travel along. But marsh-dwelling Swamp Rabbits are perfectly at home in the dense stands of giant roseau cane, bulrush, three-cornered grass, as well as in paille fine marshes.

Swamp Rabbits, when pursued by dogs, take readily to water but swim rather slowly, for a dog can easily overtake one of them. But, in eluding a dog, they will often lie still in the water, usually under a bush or amid a drift of debris, with only the nose visible.

The young are born in almost any month of the year, but by far the greatest reproductive activity is in the period from late January to the end of September, with a peak between February and May, when an abundance of green vegetation is available. The number of young varies from one to six. Hastings (1954), working with populations in Ascension and East Feliciana parishes, found the average number to be 2.84, but Svihla (1929), in her study of Swamp Rabbits in the marshes below Houma, in Terrebonne Parish, reported the average to be 3.7. The gestation period is 39 to 40 days (Hunt, 1959).

The nest, like that of the cottontail, is a slight depression in the earth that is filled with grasses mixed with rabbit hair. Svihla (1929) describes a nest that she found below Houma located on the ground at the base of an old cypress stump in the crotch between two "knees." It was made of Spanish moss and rabbit fur, with which the young were covered.

This docile-appearing rabbit is not always meek and unobtrusive. Rival males often engage in ferocious encounters, meaning literally that they "meet with hostile intent" as the word encounter, in one of its earlier usages, is

Figure 71 Swamp Rabbit *(Sylvilagus aquaticus).* (Photograph by Lloyd Poissenot)

defined by Webster. When fighting among themselves males stand on their hind feet face to face and free to use their teeth and the claws on their forefeet to inflict wounds on each other that not infrequently result in the death of one or the other of the combatants. Often they jump off the ground to strike with the claws of the hind feet, which are so sharp that they sometimes tear great gashes in the skin of their opponent.

Cane cutters eat emergent aquatic vegetation and succulent herbaceous vegetation, such as grasses, sedges, and cane. They feed mostly at night, but in my yard, where I have them in numbers, I often see them abroad in the daylight hours, particularly in early morning and late afternoon. After rains, it is not at all unusual to see them feeding at any hour of the day.

Predators and other decimating factors. — Swamp Rabbits are eaten by much the same predators as the cottontail. Their numbers are likewise reduced by hunting pressure and disease and by the flooding of their habitat by excessive rains, which often take a heavy toll of young still in the nest. Hurricanes sometimes eliminate Swamp Rabbits entirely from wide expanses of coastal marshes. Hurricane Camille of August 1969 is said to have wiped out the Swamp Rabbit population on Grand Isle and in the marshes surrounding Barataria Bay. Such was surely also the case in much of Cameron Parish following Hurricane Audrey of June 1957. Under such circumstances no place is available for the rabbits to seek refuge, for even the chenieres, the only high ground to which they might retreat, are often themselves completely inundated. Hurricane Audrey, during which over 500 human lives were lost, covered the main chenieres of Cameron Parish with water to a depth of nine feet or more.

Parasites. — Harkema (1936), in listing the parasites then known from the Swamp Rabbit, cited only two protozoans, one cestode,

three nematodes, four mites, and one flea. But Hunt (1959) added at least one parasitic mite, one tick, one flea, and one fly. Lumsden and Zischke (1961) reported the occurrence of the trematode *Hasstilesia texensis* from the Swamp Rabbit in Louisiana.

Subspecies

Sylvilagus aquaticus (Bachman)

Lepus aquaticus Bachman, Jour. Acad. Nat. Sci. Philadelphia, 7, 1837:319; type from western Alabama.
Sylvilagus aquaticus, Nelson, N. Amer. Fauna, 29, 1909:270.
Sylvilagus aquaticus aquaticus, Lowery, Proc. Louisiana Acad. Sci., 3, 1936:31; Occas. Papers Mus. Zool., Louisiana State Univ., 13, 1943:251. — Hall and Kelson, Mammals N. Amer., 1959:269.
Sylvilagus aquaticus littoralis Nelson, N. Amer. Fauna, 29, 1909:273; type from Houma, Terrebonne Parish, Louisiana. — Lowery, Proc. Louisiana Acad. Sci., 3, 1936:32; Occas. Papers Mus. Zool., Louisiana State Univ., 13, 1943: 251. — Hall and Kelson, Mammals N. Amer., 1959:269.

The Swamp Rabbits of our tidal marshes have long been recognized as a valid subspecies under the name *Sylvilagus aquaticus littoralis.* The race is allegedly characterized by being much darker and more reddish brown than *aquaticus,* especially on rump, hind legs, and tops of all the feet. In my opinion, however, the subspecies is not worthy of recognition. Although the series that I have studied from tidal marshes comprises individuals that are, as a whole, somewhat darker than most specimens from the interior swamps of the state, each specimen can nevertheless be matched almost exactly with one or more of the specimens from the inland, nontidal swamps. Some of the excessive, reddish brown coloration is, in my opinion, attributable to a ferruginous stain that the marsh-dwelling rabbits pick up from the terrain on which they live. An exactly parallel situation is the brown stain that often discolors the plumages of marsh-inhabiting birds, such as the Snow Goose in both its white and blue phases or morphs, as well as various ducks. Experience has shown that the ferruginous stain is easily removed by applying a weak acid, such as oxalic acid. To test my hypothesis with regard to marsh rabbits, I selected a specimen of Swamp Rabbit from Yscloskey, St. Bernard Parish, that showed exceptionally brown feet, and I proceeded to wash those on one side, leaving the feet on the other side untreated as a control. The results were dramatic. Nearly all the reddish brown coloration washed off and even the pads of the feet came out no darker than a pale yellowish buff. This experiment was all the evidence that I needed to convince me that the alleged redder brown coloration of the feet of "*littoralis*" is purely adventitious. As noted above, specimens from the tidal marshes show an interspersion of a considerable amount of black on the upperparts, but specimens with an equal amount of black are represented in the material from freshwater swamps even in the northern part of the state.

I have not undertaken a comparison of Swamp Rabbits from Louisiana with specimens from other parts of the range of the species, but this should be done to determine if significant geographic variation does exist. I have noted, for example, that specimens from Caddo and Bossier parishes are much lighter in coloration than are specimens from elsewhere in the state. But at present not enough material is available to warrant rendering a judgment in this regard. At any rate, for the time being, the relegation of *littoralis* to the synonymy of *aquaticus* makes the species monotypic.

Specimens examined from Louisiana.—Total 255, as follows: *Acadia Par.:* Egan, 1 (LSUMZ); 3 mi. SE Egan, 1 (USL); Mermentau, 1 (FM); Cartville, 4 (MCZ); Pointe aux Loups Springs, 2 (MCZ). *Allen Par.:* 4 mi. S LeBlanc, 1 (USL). *Ascension Par.:* 0.75 mi. E Sorrento, 1 (LSUMZ); 1.5 mi. SE Donaldsonville, 1 (LSUMZ); 0.75 mi. E Gonzales, 1 (SLU). *Assumption Par.:* 3 mi. W Labadieville, 1 (LSUMZ). *Avoyelles Par.:* 3 mi. E Marksville, 1 (LSUMZ); 5 mi. S Cottonport, 1 (USL). *Bossier Par.:* 11 mi. N Bossier, 1 (LSUMZ); 2 mi. N Taylortown, 1 (LSUMZ); Haughton, 1 (USNM). *Caddo Par.:* 15 mi. S Shreveport, 1 (NLU). *Calcasieu Par.:* 1 mi. S Maplewood, 7 mi. W Lake Charles, 1 (LSUMZ); 7.8 mi. S Lake Charles, 1 (MSU). *Caldwell Par.:* Clarks, 1 (USNM). *Cameron Par.:* 1 mi. NE Holly Beach, 1 (LSUMZ); Willow Island, 1 (LSUMZ); Creole, 1 (LSUMZ); Cameron, 2 (LSUMZ); 1 mi. E, 3 mi.

N Creole, 5 (MSU); Little Cheniere Rd., 4 (MSU); 15 mi. E Creole, 1 (MSU); 1.5 mi. S Creole, 1 (MSU); 2 mi. N Little Cheniere on Little Cheniere Rd., 1 (MSU); Hackberry, 3 (MCZ). *Claiborne Par.:* locality unspecified, 1 (LTU). *Concordia Par.:* Three Rivers Mgt. Area, 1 (LSUMZ). *East Baton Rouge Par.:* various localities, 15 (LSUMZ). *East Feliciana Par.:* 4 mi. S, 2 mi. E Clinton, 1 (LSUMZ). *Evangeline Par.:* 12 mi. N Ville Platte, 1 (USL); 12 mi. NE Ville Platte, 1 (USL); 1.5 mi. N Chataignier, 3 (USL). *Franklin Par.:* near Baskin, 1 (LTU). *Grant Par.:* 1.5 mi. NW Pollock, 1 (USL). *Iberia Par.:* 6 mi. W New Iberia, 1 (USL); 1 mi. SW Spanish Lake, 4 (USL); Avery Island, 8 (CAS); Lydia, 3 (CMNH). *Iberville Par.:* 0.5 mi. S St. Gabriel, 1 (LSUMZ); Spanish Lake, 1 (LSUMZ); 5 mi. SE Krotz Springs, 1 (USL). *Jefferson Par.:* Grand Terre, 1 (LSUMZ); east end Greater New Orleans Miss. River bridge, 1 (USL). *Lafayette Par.:* Lafayette, 2 (1 LSUMZ; 1 USL); 2.5 mi. SE Lafayette courthouse, 1 (USL); 3 mi. E Broussard, 1 (USL); 13 mi. N Lafayette, 2 (USL); 4 mi. N Milton, 10 (USL); 4 mi. SSW Lafayette courthouse, 1 (USL); 1 mi. S Lafayette, 1 (USL); 3 mi. S Lafayette courthouse, 1 (USL); 2 mi. SSW Lafayette courthouse, 1 (USL). *Lafourche Par.:* 5 mi. NE Mathews, 1 (LSUMZ); 3 mi. E Larose, 1 (USL). *Lincoln Par.:* Hico, 1 (LSUMZ); 2.5 mi. E Tremont, 1 (LSUMZ). *Livingston Par.:* 2 mi. S Satsuma, 1 (LSUMZ); 10 mi. S Montpelier, 1 (LSUMZ); 8 mi. S Springfield, 1 (SLU). *Madison Par.:* 4 mi. E Tallulah, 1 (LSUMZ). *Morehouse Par.:* Horseshoe Lake, 1 (NLU); 4 mi. W Oak Ridge, 1 (NLU); Mer Rouge, 2 (USNM); locality unspecified, 1 (USNM). *Natchitoches Par.:* Provencal, 3 (LSUMZ); 3.25 mi. SW Flora, 1 (LSUMZ). *Orleans Par.:* New Orleans, 3 (1 LSUMZ; 1 SLU; 1 USNM); Lake Catherine, 4 (1 TU; 3 AMNH); 3 mi. E Gentilly, 1 (TU). *Ouachita Par.:* Monroe, 2 (NLU); 7 mi. E Monroe, 1 (NLU). *Plaquemines Par.:* Pilottown, 1 (LSUMZ); 0.5 mi. W Port Sulphur, 1 (LSUMZ); 2.5 mi. S, 0.5 mi. W Port Sulphur, 1 (LSUMZ); Port Sulphur, 1 (SLU); Tulane Riverside campus, Belle Chasse, 10 (TU); Breton Island, Breton Sound, near La. coast, 2 (TU); Belair, 1 (USNM); Pass a Loutre, 1 (USNM); Burbridge, 4 (MCZ). *Pointe Coupee Par.:* Lottie, 1 (LSUMZ); Lakeland, 1 (LSUMZ); Morganza, 1 (SLU). *Rapides Par.:* 15 mi. SE Alexandria, 1 (LSUMZ). *Red River Par.:* 10 mi. E Coushatta, 1 (NSU). *Richland Par.:* 4 mi. W Rayville, 1 (NLU). *St. Bernard Par.:* Yscloskey, 1 (LSUMZ); Violet, 1 (SLU). *St. Charles Par.:* Sarpy, 1 (TU). *St. James Par.:* 2 mi. NE Gramercy, 1 (LSUMZ). *St. John the Baptist Par.:* Laplace, 1 (LSUMZ). *St. Landry Par.:* 4 mi. NE Palmetto, 1 (LSUMZ); Melville Hwy., 0.25 mi. W, 3 mi. N Krotz Springs, 1 (LSUMZ); 9 mi. NNE Opelousas, 1 (LSUMZ); 7.25 mi. N, 2.75 mi. E Washington, TGMA, 1 (LSUMZ); Thistlethwaite Game Mgt. Area, 1 (LSUMZ); Melville, 1 (USL); 2 mi. E Leonville, 1 (USL); 5 mi. W Melville, 1 (USL); 6 mi. S Opelousas courthouse, 2 (USL); 3 mi. E Port Barre, 1 (USL). *St. Martin Par.:* 1.4 mi. E Coteau Holmes, 1 (LSUMZ); 2 mi. S St. Martinville, 1 (USL); west side Lost Lake Island, 1 (USL); Cypress Island, 1 (USL); 8 mi. SW Cecelia, 1 (USL). *St. Mary Par.:* 0.25 mi. S Morgan City, Bateman Island, 3 (USL); 2.5 mi. E Morgan City courthouse, 1 (USL); 15 mi. S Centerville on Bayou Sale Rd., 1 (USL); Avoca Island, 1 (USL); Morgan City, 18 (USNM); Belle

Isle, 1 (USNM). *St. Tammany Par.:* Pearl River, 1 (LSUMZ); 6 mi. E Covington, 1 (LSUMZ); 2 mi. SE Abita Springs, 1 (SLU); Madisonville, 1 (SLU); Covington, 3 (USNM). *Tangipahoa Par.:* 4 mi. E Tickfaw, 1 (SLU); 2 mi. SE Hammond, 1 (SLU). *Tensas Par.:* St. Joseph, 1 (LTU); Newlight, 1 (USNM). *Terrebonne Par.:* 16 mi. N Houma, 1 (LSUMZ); 5 mi. S Dulac, 1 (LSUMZ); 30 mi. SW Houma, 1 (LSUMZ); Houma, 3 (USNM); Gibson, 4 (MCZ). *Union Par.:* 6 mi. N Bernice, 1 (LTU). *Vermilion Par.:* 4.5 mi. ESE Pecan Island, 1 (LSUMZ); 2 mi. SE Gueydan, 1 (LSUMZ); 5 mi. SW Gueydan, 1 (USL); Pecan Island, 2 (USL).: 4 mi. S Abbeville, 1 (USL); Perry, 1 (USNM); Cheniere au Tigre, 8 (7 USNM; 1 CMNH). *Vernon Par.:* Leesville, 1 (LSUMZ); 3 mi. E Slagle, 1 (USL). *Washington Par.:* 1 mi. NW Franklinton, 1 (LSUMZ); Franklinton, 1 (SLU); Hackley, 1 (FM). *West Baton Rouge Par.:* Cinclare, 1 (LSUMZ); 7 mi. SE Rosedale, 1 (LSUMZ).

RODENTS OR GNAWING MAMMALS
Order Rodentia

THE RODENTS constitute by far the largest order of mammals from the standpoint of the number of species and the number of individuals as well. The order comprises 40 families, of which 32 survive today. The latter are presently divided into 358 living genera and approximately 2,000 extant species. Hence rodents are almost equal in number of species to all other kinds of mammals combined. The group as a whole is so highly adaptable that it is represented in practically all land habitats on every continent except Antarctica. Rodents occupy the world's driest deserts and its wettest rain forest, and they range from areas below sea level to the bleak paramos on some of the highest mountains. The breeding potential of certain species is so great that one pair could produce more than a million progeny in one year, provided all offspring survived. Fortunately, though, many factors, including adverse weather, predation, disease, parasites, and competition within species and among them for living space and available food, serve in most instances to keep the prodigious numbers in check.

Most rodents are terrestrial, many are arboreal, and a number live out their lives in tunnels beneath the surface. Some move about by running, others by leaping kangaroo fashion, still others by climbing, and a few by gliding. Some are great swimmers and are otherwise adapted to a semiaquatic existence. But, despite the diversity of the members of the order in their ways of life and the habitats they occupy, they are nevertheless remarkably uniform in their basic structure.

All rodents possess chisel-shaped incisors that are used in gnawing, one pair in the upper jaw and another pair in the lower jaw. The incisors grow throughout the animal's life. The tip of each incisor is normally gradually worn away in the course of cutting hard materials, but growth at the base continues to extend the tooth outward. If for one reason or another the teeth are not worn down, the tips may grow past each other. Since the teeth curve like segments of a circle, those of the upper jaw sometimes continue growing downward, backward, and then upward to penetrate the roof of the mouth, while those of the lower jaw grow upward in front of the nose. A wide gap, or diastema, intervenes between the incisors and the first cheek teeth. Canines are absent. As a further adaptation for gnawing, the masseter muscle of the jaw is massive and powerful.

169

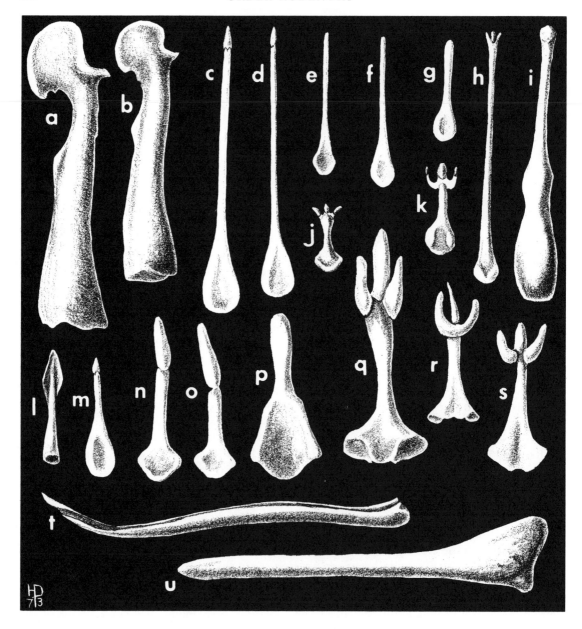

Figure 72   The bacula, or penis bones, of certain rodents: (a) Fox Squirrel; (b) Gray Squirrel; (c) Cotton Mouse; (d) White-footed Mouse; (e) Eastern Harvest Mouse; (f) Fulvous Harvest Mouse; (g) Golden Mouse; (h) Hispid Pocket Mouse; (i) Plains Pocket Gopher; (j) Woodland Vole; (k) Prairie Vole; (l) Eastern Chipmunk; (m) House Mouse; (n) Roof Rat; (o) Norway Rat; (p) Eastern Wood Rat; (q) Common Muskrat; (r) Marsh Rice Rat; (s) Hispid Cotton Rat; (t) Southern Flying Squirrel; (u) American Beaver. Scale: a–s, 6 ×; t and u, 3 ×. (i, after Kennerley, 1958; h, after Burt, 1936; e, j, k, l, and u, after Burt, 1960; others from specimens in the LSUMZ.)

Most rodents are plantigrade, that is, they walk on the soles of their feet, with the heel and the wrist touching the ground. The radius and ulna are unfused, and the elbow joint permits a free semirotary motion of the forearm. A baculum, or penis bone, is usually present (Figure 72). Its absence is generally regarded to be of considerable taxonomic significance in defining the higher taxa, notably the squirrels (Wade and Gilbert, 1940). Some rodents possess internal or external cheek pouches, which open near the corners of the mouth. The external pouches are fur-lined.

The order Rodentia is divisible into three well-defined suborders (some mammalogists, for example, Wood, 1955, recognize seven): Sciuromorpha, the squirrellike rodents; Myomorpha, the ratlike rodents; and Hystricomorpha, the porcupinelike rodents. In the first of these suborders, the infraorbital canal is tiny, never slitlike as in the Myomorpha (see Figure 73B), nor greatly enlarged as in the Hystricomorpha. In the hystricomorph Nutria, for example, the infraorbital canal is large enough to admit the end of one's little finger (see Figure 73C). But in the American Beaver, a sciuromorph, the infraorbital canal is not much larger than the lead of a pencil (see Figure 73A). The extremely large canal in the hystricomorphs serves not only as a passageway for part of the fifth cranial nerve and blood vessels but also for an extension of a portion of the masseter muscle. In the Hystricomorpha the zygomatic arch is robust and in some forms massive, and is supported anteriorly by two struts that arise from the maxilla and constitute the ventral and lateral borders of the infraorbital canal. Other cranial and postcranial features serve to differentiate the three suborders.

Rodents are of great economic importance to man. Some of them destroy injurious insects and the seeds of noxious weeds. Others, such as the American Beaver, Common Muskrat, and Nutria, annually provide furs worth millions of dollars. Squirrels furnish inesti-

mable sporting pleasure to outdoorsmen. A few rodents such as rats, mice, and guineapigs are valuable as laboratory animals and are used in medical research on human ailments and studies relating to basic biology. Unfortunately, though most rodents are scrupulously clean, a few carry diseases and parasites to which man is susceptible. The commensal, introduced Norway Rat, which

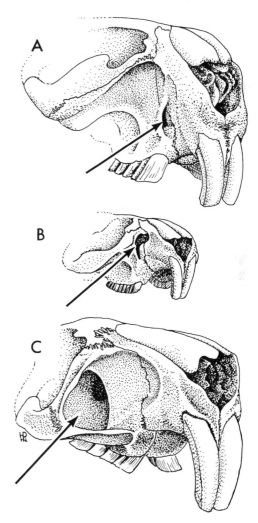

Figure 73 Facial view of three rodents representing three suborders of Rodentia: (A) American Beaver (Sciuromorpha); (B) Common Muskrat (Myomorpha); (C) Nutria (Hystricomorpha). Note the great difference in the size and shape of the infraorbital foramina.

often invades human dwellings, can spread bubonic plague and typhus by the fleas and other external parasites that it harbors. The allegation has been made that man's history has been altered more extensively by rat-borne typhus than by any single person who ever lived, however great or powerful, and that rat-borne diseases have taken more human lives than all the wars and revolutions in which man has ever fought. Outbreaks of plague have been known to reduce the population of western Europe by one-fourth or more on numerous occasions. Explosive irruptions of mouse populations sometimes result in the wholesale destruction of crops and stored grains. One such irruption occurred in Kern County, California, in 1926 and 1927. In the dry bed of Buena Vista Lake the number of mice in places was estimated to be as high as 82,800 per acre. In one grain bin that was two-thirds full of stacked barley, 3,520 mice were in sight at one time, and, at another grainery, two tons of mice were killed in one day (Hall, 1927).

The Squirrel Family
Sciuridae

SQUIRRELS ARE widespread over most of the world, being absent only from Australia, Madagascar, southern South America, many oceanic islands, and parts of Arabia and northeastern Africa. The family contains about 52 genera, but only 3 of these occur regularly in Louisiana—*Sciurus, Tamias,* and *Glaucomys.* All squirrels have a well-developed postorbital process, a character peculiar to the members of this family. The number of cheek teeth is either 5/4 or 4/4.

Many of the 93 species of sciurids in North America are well known to the general public, at least in the regions where they occur. Among them are the tree squirrels that scamper about our city parks and lawns, the prairie-dogs that establish "towns" on the outskirts of our own western cities, the attractive little chipmunks and ground squirrels that occur in one form or another over most of the United States, and the Woodchuck, or groundhog, about which a superstition has arisen with regard to what happens if it sees its shadow on the second day of February when, according to the story, it emerges from hibernation. Our familiarity with the members of this family comes about as a result of their conspicuous diurnal habits, which sometimes bring them into close association with man.

Genus *Sciurus* Linnaeus

Dental formula:

$$\mathrm{I}\frac{1-1}{1-1}\ \mathrm{C}\frac{0-0}{0-0}\ \mathrm{P}\frac{1-1}{1-1}\ \text{or}\ \frac{2-2}{1-1}\ \mathrm{M}\frac{3-3}{3-3} = 20\ \text{or}\ 22$$

GRAY SQUIRREL Plate IV
(*Sciurus carolinensis*)

Vernacular and other names. — In Louisiana, the Gray Squirrel is almost always referred to as the cat squirrel, but the name fluffy-tail is also heard applied to it. The French-speaking inhabitants of the southern part of the state call it écureuil gris. The generic name *Sciurus* comes from the Greek word for squirrel, *skiouros* (*skia,* shadow; *oura,* tail—that is, "the creature that sits in the shadow of its tail"). The specific name *carolinensis* is a Latinized adjective that means belonging to Carolina, and refers to the part of the country from which the species was first named.

Distribution. — The species ranges from Maine, southern Quebec, southern Ontario, northern Michigan, northern Minnesota, and southern Manitoba south throughout the eastern United States to southern Florida, the northern Gulf Coast, and eastern Texas. In Louisiana it is statewide wherever there are trees except for the chenieres of the coastal marshes. (Map 23)

External appearance. — The general form of a squirrel is so well known that it hardly requires description. The Gray Squirrel, as the name implies, is usually grayish or grayish brown above, sometimes with a yellowish cast. It is white or grayish white below. The vast majority of the individual hairs on the top of the head and neck, along the middle of the dorsum, and on the sides of the belly are gray basally and then successively banded with yellowish brown, dusky, and finally yellowish brown or sometimes pure gray. (In squirrels from northwestern Louisiana, the flanks and lower back are decidedly more grayish than in specimens from the other areas of the state because the tips of the individual hairs of these parts are predominantly grayish.) Black or grayish black guard hairs are interspersed throughout the upperparts and contribute to the overall grizzly appearance of the dorsum. Each hair of the bushy tail is yellowish brown basally with a broad subterminal band of black and with a white tip. The feet vary from dark rusty brown, through yellowish brown, to almost pure gray, and the scantily haired ears are brown except for a prominent patch of white hairs that is often present at the base of the medial edge. The Gray Squirrel is readily distinguished from the Fox Squirrel by its smaller size, by its basically grayish overall coloration, and by the fact that its underparts are almost always white (Figure 74), not ferruginous. The tips of the hairs of the tail are always grayish or white, never yellow or yellowish as in the Fox Squirrel.

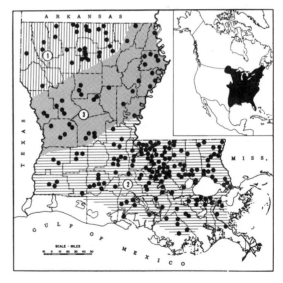

Map 23 Distribution of the Gray Squirrel (*Sciurus carolinensis*). *Shaded areas,* known or presumed range within the state; •, localities from which museum specimens have been examined; *inset,* overall range of the species. 1. *S. c. carolinensis;* 2. *S. c. fuliginosus;* 3. zone of intergradation between *S. c. carolinensis* and *S. c. fuliginosus.*

Figure 74 Gray Squirrel *(Sciurus carolinensis)*, Baton Rouge, East Baton Rouge Parish. (Photograph by Joe L. Herring)

Color phases.—Albinos are fairly frequent in this species, as are also color variants that are referred to as "blonds." The latter have the dorsum pale yellowish, an abnormality possibly attributable to what some geneticists call the dilution factor. The eyes of these individuals are sometimes pink, as in true albinos, but such is not always the case. In one of the residential subdivisions of Baton Rouge a number of these abnormal "blond" Gray Squirrels have been observed regularly for years. In the lower Atchafalaya Basin a dark, brown-bellied variant (illustrated in Plate IV) is frequently encountered. True melanistic individuals of the species that are solid black are extremely rare. Indeed, I have seen only one such specimen, a skin in the Tulane collections, which was obtained in Sarpy Swamp, near Norco, in St. Charles Parish. Still another color variation that is occasion-

ally seen is an erythristic phase, in which the pelage is decidedly reddish. I have examined a few specimens with only the tail abnormally reddish.

Measurements and weights.—One hundred and twenty-nine specimens in the LSUMZ from various localities in the state, averaged as follows: total length, 455.8 (395–600); tail 215.4 (160–300); hind foot, 56.5 (40–65); ear, 26.3 (18–33). The skulls of 105 specimens from the same source averaged as follows: greatest length, 57.4 (53.8–60.9); zygomatic breadth, 31.4 (23.6–33.6); cranial breadth, 23.3 (21.7–29.5); interorbital breadth, 17.0 (14.8–19.4); length of nasals, 19.2 (17.2–26.5); palatilar length, 25.4 (23.8–27.7); postpalatal length, 18.4 (17.0–20.1); maxillary toothrow, 10.2 (9.0–11.5). The weights of 13 adult males averaged 419.7 (356–650) grams, those of 14 adult females 401.8 (317–460.0) grams.

Cranial and other skeletal characters.—The bones of the skull and other skeletal parts of the Gray Squirrel are creamy white, whereas they are reddish in the Fox Squirrel. In the adult Gray Squirrel a small, extra, peglike premolar is present, giving it five cheek teeth on each side of the upper jaw, one more than in *Sciurus niger* (Figures 75 and 77). The os penis of this species is shown in Figure 72. It possesses a stout base, slightly ellipsoidal in cross section, that tapers gradually toward the distal end, where it becomes flattened to form a shallow disclike scoop with the edges slightly curled. The dorsal and ventral edges of the scoop each possess a prominent projection, the ventral one being spurlike and more prominent. The shaft exhibits a slight twist to the right. Its length is approximately 10.6 mm and its diameter near the middle is approximately 1.6 mm (Fairchild, 1950).

Sexual characters and age criteria.—The sexes are alike in color and do not appear to differ appreciably from each other in weight. Im-

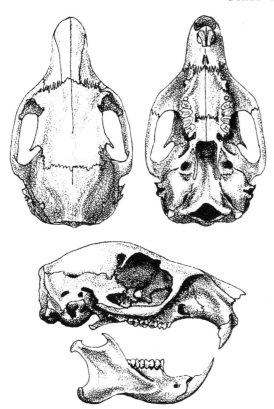

Figure 75 Skull of Gray Squirrel *(Sciurus carolinensis),* ♀, 3 mi. S University, East Baton Rouge Parish, LSUMZ 3386. Approx. natural size.

matures can be recognized by their small size, by the fluffy texture of their pelage, especially that of the underparts, and by their incomplete dentition. In the juvenile squirrel only four cheek teeth are present on each side of the upper jaw, the first of these being a deciduous premolar. At the time when it is replaced by a permanent tooth, the tiny, peglike additional premolar, which is diagnostic of the adult Gray Squirrel, also erupts. I am uncertain as to the precise age when this event occurs, but I have seen no Gray Squirrel weighing 300 grams (or approximately ⅔ lb.) or more that lacked the full complement of teeth. By the time the squirrel season opens in early fall, young males born during the previous summer can be recognized by their light-colored, undeveloped scrotum,

into which the testes have not descended. Juvenile females can be identified by their small, inconspicuous teats as compared with the much larger and darker mammae of adult females that have already given birth to young. J. B. Kidd (pers. comm.) informs me that a Gray Squirrel that he tagged as an adult on the Thistlethwaite Wildlife Management Area in St. Landry Parish on 18 May 1962 was killed by a hunter nearly seven and one-half years later on 8 October 1969. I doubt, though, that many individuals survive to this age.

Status and habits.—The Gray Squirrel is an abundant species in virtually all wooded areas of the state. The most recent Small Game Survey by the Louisiana Wild Life and Fisheries Commission, covering the 1966–1967 hunting season, indicates that the squirrel population in the state was at that time between 12 and 15 million, of which approximately 60 percent were stated to be Gray Squirrels. The total squirrel kill in the same period was estimated at 3 million, of which 1.8 million were Gray Squirrels. Figures such as these provide ample evidence of the general abundance of the species in the state and its importance as a game animal.

Sportsmen regard the Gray Squirrel as one of their favorites because they must pit their stealth and cunning against the sly wariness and elusiveness of their quarry. Perhaps Seton (1929) summed it up best for sportsman and naturalist alike when he wrote about the Gray Squirrel in these words: "It supplies to the city man, the farm boy, and the sportsman the most alluring motive for a glorious tramp in the ever-inspiring woods, with a sufficiency of material reward, an amplitude of present joy, and a heritage of delightful memories."

Three methods are employed in hunting squirrels. One is with dogs and is especially effective at a time of the year when squirrels are feeding on the ground on mast. Owners of squirrel dogs probably derive as much gratification from the actions and vocalizations of their hounds as they do from the game they are able to bag. Other methods employed — and from which most hunters take greatest delight — are either to station one's self at daybreak at the base of a large tree in a heavily wooded area and to wait for a squirrel to appear, or to walk slowly and quietly through a wooded area watching for some movement along one of the limbs and listening for the barking chatter of a squirrel by which its presence is often revealed.

The Gray Squirrel, perhaps more so than the Fox Squirrel, has a nervous disposition, which adds to its attractiveness as a game species. At the sudden approach of danger it will generally freeze, but not infrequently it sidles out of sight behind a branch or the trunk of the tree. But once it is routed it will run rapidly through the tree branches or jump directly to the ground in an effort to escape. Sometimes it will retreat into a hollow in a tree trunk or seek refuge in one of its nests of twigs and leaves.

Both this species and the Fox Squirrel build two types of nests. One is a den in a tree cavity that may be at ground level or as high up as 40 feet. In most cases it is approximately 14 inches in depth and 7 to 8 inches in diameter. The opening is often one made previously by a woodpecker or it may have been formed when a branch broke off the trunk. The other type of nest is made of leaves and twigs. Such nests are often built by young squirrels that cannot find suitable tree dens or by males that are chased out of the tree dens when the females are ready to give birth to their young. Both sexes will resort to leaf-twig nests when the dens become heavily infested with fleas (Trippensee, 1948; Fitzwater and Frank, 1955). The nest of leaves and twigs is a waterproof hollow ball or dome, the cavity of which is lined with mosses and fibers. It is placed either on a platform in the fork of a tree or intertwined among the terminal parts of the branches. The hollow ball nests at the ends of branches are most often used for escape and

resting, although they are sometimes employed for reproductive purposes.

In his study of the leaf composition of nests examined in the vicinity of Baton Rouge, T. C. Stewart (pers. comm.) found, among other things, that in one sample plot 79 percent of the trees were willow oak (*Quercus phellos*) and only 8 percent were post oak (*Q. stellata*). But a Gray Squirrel nest in a post oak was made of 78 percent post oak leaves. In other words, the squirrel used the leaves nearest at hand.

The Gray Squirrel is active in all the hours of daylight but most so from the break of dawn to about 9 o'clock in the morning and then again in late afternoon. In summer and fall it will sometimes come out to feed on moonlit nights. Breeding activity is also bimodal. Males are in rut from late December to late February, and again from late May to the middle of August. The gestation period has an average duration of 44 days; hence the first litter is usually born between mid-February and mid-April, and a second litter may appear anytime from June through September, rarely as late as October. Young from a spring litter may produce young the following spring, and young of a fall litter, the following fall. Goodrum (1940), on the basis of his study of the species in eastern Texas, states that he found squirrels every month of the year that were either in rut, pregnant, or suckling young. And Kidd, on the basis of his studies on the Thistlethwaite Game Management Area in St. Landry Parish, reports (pers. comm.) that pregnant Gray Squirrels were actually found in February, March, April, May, June, July, August, and October, but he also believes that pregnant animals could be found every month of the year.

The number of young in a litter is variable but probably averages somewhere close to three. As many as eight young to a litter have been recorded in the wild state (Barkalow, 1967), and in captivity nine have been noted (Uhlig, 1955). Seton (1929) expressed the opinion that among northern populations three to five is the normal number and that one, two, and six are rarities. The young are born hairless, with the eyes and ears tightly closed. By the 21st to 30th day the ears open and by the 37th day the eyes are likewise open. The little squirrels are not well furred until approximately nine weeks of age. The mother continues to suckle the young for a period of three to four months, at which time she is usually ready to bring forth another litter. She will then often leave the den to the young and seek out a new site in which to bear her second litter.

This species is one of the noisiest of the tree squirrels. Its frequent barking often discloses its presence to the hunter and perhaps to the predator. The voice is alleged by some sportsmen to have a more musical quality during the mating season. Seton (1929) describes one vocalization as a sort of song, a "loud *Quack-quack-quack quaaaaaaaaaaaaaaa, Quack-quack-quack-quack-quack quaaaaaaaaaaaaaaa.*" The latter part of the series is said to quicken into a harsh rattle. Goodrum (1940) states that the most common and characteristic sound is a loud and emphatic *"cherk, cherk, cherk,"* repeated rapidly and followed by a heavy buzzing, often interpolated with a grunt and accompanied by a vigorous wagging of the tail.

A popular belief in Louisiana and elsewhere, especially widespread among hunters, is that Gray Squirrels emasculate their own young, as well as adult Fox Squirrels. In other parts of the country other species of squirrels are credited with performing the same feat. In Michigan, for example, I have been told that Red Squirrels emasculate both Gray and Fox Squirrels. Thus, an interesting feature of the misconception is that when any two species are involved, it is more often than not the smaller of the two squirrels that is credited with performing the job of castration on the larger. A possible basis for the erroneous belief that adults castrate their young is the fact that up to the time a squirrel is three or four months of age the testes are undevel-

oped and not yet descended from the abdominal cavity into the scrotum. The apparent absence of the testes may therefore suggest that castration has been performed. Another factor, however, may be involved in the absence of testes, even in adults. When, as often happens, a botfly lays an egg in the skin of the scrotal region, the resulting warble may cause one or both of the testes to atrophy. In any event, no evidence supports the notion that any species of squirrel purposefully castrates another species or its own young.

Another widespread misconception among sportsmen, particularly in the southern part of the state, is that Gray and Fox Squirrels occasionally hybridize. Proven instances of such crosses are unknown. The notion stems, I believe, from the occurrence of a color phase in the Gray Squirrels of the Atchafalaya Basin in which the underparts are not white but instead dark brown or rust colored, suggesting the characteristic color of the belly of the Fox Squirrel. These dark-bellied Gray Squirrels belong to the subspecies *S. c. fuliginosus* (see page 179) and are not hybrids.

The use of the tail by tree squirrels for a variety of purposes is well known. It flicks as a signal of recognition when one individual approaches another, and it helps to maintain balance while the animal is running rapidly along the limbs of trees. And when occasionally a squirrel loses its footing and falls, it spreads the tail as a "parachute" so that it hits the ground with less force and is able to bound away without injury. I am often amused by the manner in which the tail is also used as an umbrella. Gray and Fox Squirrels sit on the feeders outside my home when it is raining heavily, but they bring the tail over the back with the hairs fluffed to shed the falling raindrops.

Much has been written about squirrels burying great quantities of nuts and acorns at times when these items are plentiful. How many of these nuggets are later remembered and dug up is not known, but I suspect only a small percentage are salvaged. One thing is certain—the instinct serves a useful purpose insofar as future generations of squirrels are concerned. Many of the nuts doubtless germinate and grow into trees that provide habitat for squirrels of the years ahead to enjoy.

Students of wildlife management are in general agreement that the most practicable method of increasing squirrel populations is by regulating the annual kill. Overhunting, depletion of habitat by excessive logging, and competition from cattle and hogs unquestionably cause declines in squirrel populations. Special care must be exercised in setting hunting seasons so that they do not conflict with the two distinct peaks in breeding activity of squirrels. Numbers fluctuate considerably from year to year depending on the condition of the mast crop. In areas of mast failure squirrels decline drastically either by moving out in search of better conditions elsewhere or by marked reductions in their breeding productivity. Biologists have long been intrigued by what appears to be an ability on the part of squirrels to anticipate years of heavy mast production by coming up with a bumper crop of young even before the mast has formed and by producing few young in advance of a poor mast yield.

Predators and other causes of mortality.—Bobcats, foxes, weasels, snakes, hawks, and owls are all enemies of the Gray Squirrel in Louisiana, but this predation is probably not serious. Where good cover in the form of tree cavities, Spanish moss, vines, and leaves is available, the squirrel has an excellent chance of effecting an escape from one of these adversaries. Its greatest peril comes at the time it is feeding extensively on the ground on mast, but even then it usually manages to scurry to a tree at the first hint of danger. In spring both Gray and Fox Squirrels feed on the new buds and leaves and are then sometimes easy prey for Red-shouldered Hawks with whom they share the forest. Audubon and Bachman (1851–1854) describe a technique sometimes employed by Red-tailed Hawks of working in pairs to catch a Gray Squirrel. I believe the passage is worth quoting here.

The Red-tailed hawk seems to regard it [the Gray Squirrel] as his natural and lawful prey. It is amusing to see the skill and dexterity exercised by the hawk in the attack, and by the Squirrel in attempting to escape. When the hawk is unaccompanied by his mate, he finds it no easy matter to secure the little animal; unless the latter be pounced upon whilst upon the ground, he is enabled by dodging and twisting round a branch to evade the attacks of the hawk for an hour or more, and frequently worries him into a reluctant retreat.

But the Red-tails learn by experience that they are most certain of this prey when hunting in couples. The male is frequently accompanied by his mate, especially in the breeding season, and in this case the Squirrel is soon captured. The hawks course rapidly in opposite directions, above and below the branch; the attention of the Squirrel is thus divided and distracted and before he is aware of it, the talons of one of the hawks are in his back, and with a shriek of triumph, the rapacious birds bear him off, either to the eyrie in which their young are deposited, to some low branch of a tree, or to a shelter situation on the ground, where, with a suspicious glance towards each other, occasionally hissing and grumbling for the choice parts, the hawks devour their prey.

Man is, of course, the worst enemy of the Gray Squirrel. Fortunately, though, its reproductive potential is so great that the animal can withstand a reasonable harvest of its numbers by man for recreational and culinary purposes. Around our towns and cities, where Gray and Fox Squirrels are sometimes exceedingly abundant, both species suffer considerable mortality by being run over by automobiles. Gray Squirrels are especially stupid in their reactions to an approaching car. They will often seem to stop to allow a car to pass only at the last moment to dart across the road in front of the approaching wheels.

Parasites.—The parasites of the Gray Squirrel are summarized by Doran (1954 a and b, 1955 a and b), Clark (1959), and Parker (1968), who list approximately 48 species of ectoparasites and over 30 species of endoparasites. Of these only two sucking lice, one flea, five arachnids, two protozoans, and two nematodes are considered by Clark to be species that are found frequently on, or in, the Gray Squirrel. Goodrum (1940) found the chigger or redbug *(Trombicula tropica)* the most serious external parasite of the Gray Squirrel in eastern Texas. Seventy-five percent of the squirrels examined in summer were infested with this pest, which sometimes produced a severe skin irritation. In many instances the squirrels had scratched the skin raw. Goodrum observed that fleas were most numerous in winter, probably in correlation with the habits that keep squirrels confined more closely in their nests in tree cavities in winter. The American dog tick *(Dermacentor variabilis)* is occasionally found on the Gray Squirrel and is capable of transmitting tularemia and Rocky Mountain spotted fever, but I am not aware of any instance of its having done so in Louisiana.

Subspecies

Sciurus carolinensis carolinensis Gmelin

[*Sciurus*] *carolinensis* Gmelin, Syst. nat., ed. 13, 1, 1788:148; type locality, Carolina.
Sciurus carolinensis var. *carolinensis*, Allen, Monog. N. Amer. Sciuridae, 1877:704.

The nominate race of the species occurs from southeastern Virginia south through the Piedmont region of North Carolina and in Georgia and Alabama to central Florida; west of the Allegheny Mountains (which are included in a southward extension of the range of *S. c. pennsylvanicus*) it occurs from southern Ohio, southern Indiana, southern Illinois, and extreme eastern Nebraska south to the Gulf Coast except for the area in the southern portion of Louisiana occupied by *S. c. fuliginosus*. The only specimens from Louisiana that are clearly referable to *carolinensis* are those from the northwestern parishes. See the following subspecies for a summary of the differences between *carolinensis* and *fuliginosus*.

Specimens examined from Louisiana.—Total 41, as follows: *Bienville Par.:* locality unspecified, 1 (LSUMZ); Gibsland, 1 (LSUMZ); Bienville, 2 (1 LSUMZ; 1 LTU). *Bossier Par.:* Bodcau Reservation, 1 (LSUMZ); 2 mi. E Midway, 1 (LSUMZ); Midway, 1 (LTU). *Caddo Par.:* 3 mi. S, 1 mi. W Blanchard, 1 (LSUMZ); 3 mi. S, 1.5 mi. W Blanchard, 2 (LSUMZ). *Claiborne Par.:* 4 mi. S Summerfield, 1 (LSUMZ); Corney Creek, 3 mi. N Summerfield, 1 (LTU). *De Soto Par.:* 15 mi. SE Mansfield, 1 (LSUMZ). *Jackson Par.:* Vernon, 1 (LTU); 10 mi. S Ruston, 1 (LTU); Jackson–Bienville Game Reserve, 1 (LTU); between Clay and Ansley, 1 (LTU). *Lincoln Par.:* 5 mi. N Ruston, off Hwy. 167, 1 (LTU); Ruston, 2 (LTU); 5 mi. S Ruston, 2 (LTU); 4 mi. W Vienna, 1 (LTU). *Ouachita Par.:* Choudrant Creek Bottom, near Calhoun, 1 (LTU); 6 mi. SSW Calhoun, 1 (LTU); 2 mi. NW Calhoun, 1 (NLU); Moon Lake [R4E, T19N, Sec. 30], 1 (NLU); 1 mi. N Calhoun, 1 (NLU); 5 mi. SSW Calhoun, 1 (NLU); 1.5 mi. SSW Calhoun, 1 (NLU). *Union Par.:* 4 mi. N Marion, 2 (LSUMZ); 1.5 mi. S Lillie, 1 (LSUMZ); 1.5 mi. S Spencer, 1 (NLU); 5 mi. NE Bernice, 1 (LTU); 3 mi. N Lillie, 1 (LTU); Linville, 1 (LTU); 3 mi. N Bernice, 1 (LTU); 2 mi. [E] Truxno, 1 (LTU). *Webster Par.:* Shongaloo, 25 mi. N Minden, 1 (LSUMZ); 10 mi. N Minden, 1 (LTU).

Sciurus carolinensis fuliginosus Bachman

Sciurus fuliginosus Bachman, Proc. Zool. Soc. London [for] 1838 [=1839]:97; type locality, near New Orleans, Louisiana.
Sciurus carolinensis fuliginosus, Bangs, Proc. Boston Soc. Nat. Hist., 26, 1895:543.

Although Bachman (1839) and later Bangs (1895) diagnosed this squirrel incorrectly, it is nevertheless a subspecies of *Sciurus carolinensis* that is worthy of recognition. These two authors each described *fuliginosus* as being large—according to them almost as large as the northern squirrel that we now call *S. c. pennsylvanicus*. But this alleged largeness is definitely erroneous. Indeed, it is actually much smaller than *pennsylvanicus*, and I can detect no significant mensural differences between it and *carolinensis*. Both Bachman and Bangs asserted that the color of the underparts in *fuliginosus* is never pure white and that the line of demarcation between the colors of the underparts and the upperparts is never distinct. In so describing *fuliginosus* they were doubtless referring to what is now known to be an extremely dark color phase of the population in southern Louisiana, in which the venter is a rich Cinnamon-Rufous, the feet are Ochraceous-Tawny, and the dor-

sal surface and the tail are exceedingly blackish (see Plate IV herein and Plate 149, Figure
2, in Audubon and Bachman, 1851–1854).
These extremely dark (but *not* melanistic) individuals make up only a small percentage of
the population in southern Louisiana, but
nearly all the members of the population are
nevertheless darker than the darkest examples of *carolinensis*. Consequently, I consider
the race *fuliginosus* valid. As noted previously,
solid black Gray Squirrels are almost nonex-
istent in the Louisiana population, as evidenced by the fact that I have seen only one
such individual. But, as is well known, melanism is common in northern populations that
belong to the race *pennsylvanicus*.

The race *fuliginosus* occurs throughout
southern Louisiana, including the southern
part of the Florida Parishes from southern
East Baton Rouge Parish east through southern Livingston, southern Tangipahoa, and
extreme southern St. Tammany parishes.
Specimens from elsewhere in the Florida Parishes are intermediate toward *carolinensis*, although occasional specimens from the
northern parts of this area are fairly close in
color to *carolinensis*. Specimens from most of
central Louisiana and from the northeastern
corner of the state are also intermediate between *carolinensis* and *fuliginosus*. Only those
taken in extreme northwestern Louisiana are
typical *carolinensis*. (Map 23)

Specimens examined from Louisiana.—Total 384, as follows:
Acadia Par.: 9 mi. W Crowley, 1 (LSUMZ); 2 mi. SE Egan,
1 (LSUMZ); 2 mi. N, 1 mi. W Crowley, 1 (LSUMZ);
Pointe aux Loups Springs, 20 (MCZ); Cartville, 3 (MCZ);
3 mi. SE Egan, 1 (USL); 2 mi. N Crowley, 1 (USL); 4.5
mi. SE Crowley, 1 (USL); Rayne, 3 (AMNH). *Allen Par.:*
4.5 mi. NW Oberlin, West Bay Game Mgt. Area, 2
(LSUMZ); 5 mi. SW Oakdale, 1 (LSUMZ); West Bay
GMA, 2 (MSU). *Ascension Par.:* 2 mi. S Donaldsonville, 1
(LSUMZ); 2.5 mi. S Donaldsonville, 1 (LSUMZ); 2 mi. W
Geismar, 1 (LSUMZ); Donaldsonville, 3 (LSUMZ). *As-
sumption Par.:* 22 mi. N Morgan City, 1 (LSUMZ); 10 mi.
SW Napoleonville, 1 (LSUMZ). *Avoyelles Par.:* 15 mi. N
Marksville, 1 (LSUMZ); 5 mi. S Dupont, 1 (USL); 5 mi. S
Cottonport, 1 (USL); near Coco Lake, Moreauville, 1
(LTU). *Beauregard Par.:* Longville, 1 (MSU); 4 mi. W, 2
mi. N Longville, 1 (MSU). *Calcasieu Par.:* Lake Charles,
13 (1 LSUMZ; 11 MSU; 1 USNM); 7 mi. NE Lake

Charles, 1 (LSUMZ); Sulphur, 1 (MSU); Interstate 10, W
of Sulphur, 5 (MSU); locality unspecified, 1 (MSU). *East
Baton Rouge Par.:* various localities, 64 (59 LSUMZ; 5
USL). *Iberia Par.:* 3 mi. N Jeanerette, 1 (USL); Avery
Island, 24 (1 TU; 4 USNM; 19 CAS); Lydia, 1 (CMNH).
Iberville Par.: Indian Village, 5 (LSUMZ); Bayou Sorrel, 2
(LSUMZ); 5 mi. SW Grosse Tete, 1 (LSUMZ); 4 mi. SW
Rosedale, 1 (LSUMZ); 3 mi. SW Indian Village, 1
(LSUMZ); 1 mi. SW Indian Village, 1 (LSUMZ); Spanish
Lake Swamp, 1 (LSUMZ); 3 mi. S Ramah, 2 (LSUMZ).
Jefferson Par.: 1.5 mi. E of river near H.P. Long Bridge, 1
(LSUMZ); 5 mi. S Kenner, 1 (LSUMZ); 2–3 mi. N Ken-
ner, 1 (LSUMZ); 1.5 mi. E Lake Pontchartrain Causeway
along levee, 1 (USL); 1 mi. W New Orleans, 2 (USL).
Jefferson Davis Par.: Lake Arthur, 1 (MSU). *Lafayette Par.:*
3 mi. SE Lafayette courthouse, 2 (USL); 1 mi. NE
Lafayette airport, 2 (USL); 2 mi. SW Lafayette court-
house, 1 (USL). *Lafourche Par.:* 6 mi. ENE Thibodaux, 3
(LSUMZ); 4 mi. SE Raceland, 1 (USL); Raceland, 1 (TU).
Livingston Par.: 1 mi. W Springville, 1 (LSUMZ); 6 mi. N
Maurepas, 1 (LSUMZ); 0.5 mi. NE Hoo Shoo Too Rd., 2
(LSUMZ); 2 mi. E, 1 mi. S Denham Springs, 1 (LSUMZ);
7 mi. SW Walker, 1 (LSUMZ); 2 mi. S Springfield, 1
(SLU). *Orleans Par.:* 4 mi. E New Orleans, 1 (LSUMZ); 3
mi. SW Algiers, 1 (LSUMZ); New Orleans, 20 (2 LSUMZ;
13 TU; 1 USNM; 4 AMNH). *Plaquemines Par.:* Fanny, 7
(LSUMZ); Port Sulphur, 1 (SLU). *Pointe Coupee Par.:*
Lottie, 7 (LSUMZ); locality unspecified, 1 (LSUMZ); 5
mi. N Fordoche, 3 (LSUMZ); 4 mi. W Morganza, 1
(LSUMZ); 5 mi. E Batchelor, 1 (LSUMZ); 2 mi. N Lottie,
2 (LSUMZ). *St. Bernard Par.:* Toca Village, 34 (LSUMZ);
Violet, 1 (SLU). *St. Charles Par.:* 6 mi. S Edgard, 1
(LSUMZ); Boutte, 1 (TU); Sarpy Swamp, 4 (TU). *St.
James Par.:* 10 mi. NW Vacherie, 1 (LSUMZ). *St. John the
Baptist Par.:* Manchac, 1 (LSUMZ); W shore of Lake
Maurepas at confluence of Blind River and Dutch
Bayou, 1 (LSUMZ). *St. Landry Par.:* 5 mi. N Palmetto, 1
(LSUMZ); 0.25 mi. N Palmetto, 1 (LSUMZ); 5 mi. W
Beggs, 1 (LSUMZ); 7 mi. W Port Barre, 1 (LSUMZ); 2.25
mi. E Port Barre, 1 (LSUMZ); 10 mi. SW Lebeau, 1
(LSUMZ); 10 mi. S Krotz Springs, 1 (LSUMZ); Grand
Coteau, 4 (1 MCZ; 3 USNM); Thistlethwaite Game Mgt.
Area, 10 (USL); 9 mi. N Opelousas, 1 (USL); 5 mi. W
Melville, 1 (USL). *St. Mary Par.:* 4 mi. W Morgan City, 1
(USL); Cypremort Point, 1 (USL); 2.5 mi. E Jeanerette, 4
(LSUMZ). *St. Tammany Par.:* 1 mi. NW Pearl River, 1
(LSUMZ); 3 mi. NE Lacombe, 1 (LSUMZ); 5 mi. N
Slidell, 1 (LSUMZ); 6 mi. S Gainesville, Miss., on same
lat. as Slidell on Pearl River, 1 (LSUMZ); 2.1 mi. S, 0.8
mi. E Pearl River, 1 (LSUMZ); 1.5 mi. SW Pearl River, 3
(LSUMZ); Pearl River, 2 (1 LSUMZ; 1 USNM); 1.3 mi. E,
1.2 mi. S Pearl River, 1 (LSUMZ); 6 mi. S Covington, 2
(LSUMZ); 2 mi. S Covington, 2 (SLU); Madisonville, 1
(SLU); 3 mi. N Slidell, 1 (SLU); 3 mi. N Lacombe, 1
(SLU); Slidell, 1 (USNM); Mandeville, 1 (AMNH). *Tangipahoa Par.:* Ponchatoula, 3 (LSUMZ); 6 mi. W Ponchatoula, 1 (LSUMZ); 1 mi. S Ponchatoula, 1 (LSUMZ); 2
mi. N Ponchatoula, 1 (SLU); Manchac, 1 (SLU). *Terrebonne Par.:* Grand Caillou, 1 (LSUMZ); 1 mi. NE Montegut, 1 (LSUMZ); 4 mi. SSE Gray, 1 (LSUMZ); Gibson, 13
(MCZ); Houma, 5 (USNM). *West Baton Rouge Par.:* Cinclare, 9 (LSUMZ); 6 mi. S Port Allen, 1 (LSUMZ);

Erwinville, 1 (LSUMZ); 4 mi. SW Port Allen, 2 (LSUMZ); 2 mi. W Port Allen, 3 (LSUMZ); 3 mi. NW Port Allen, 1 (LSUMZ); 2 mi. SE Erwinville, 1 (USL).

Specimens examined from Louisiana showing varying degrees of intermediacy between S. C. CAROLINENSIS *and* S. C. FULIGINOSUS. — Total 199, as follows: *Catahoula Par.:* Jonesville, 1 (LSUMZ); 2 mi. S Sicily Island, 1 (NLU). *Concordia Par.:* Deer Park, 1 (LTU). *East Baton Rouge Par.:* various localities, 22 (LSUMZ). *East Carroll Par.:* Lake Providence, 1 (LTU). *East Feliciana Par.:* various localities, 23 (LSUMZ). *Franklin Par.:* 20 mi. SE Winnsboro, 1 (LSUMZ). *Grant Par.:* 4 mi. SW Georgetown, 1 (USL); 10 mi. SW Georgetown, 1 (LTU); Georgetown, 1 (LTU). *La Salle Par.:* 4.5 mi. NE Olla, 2 (LSUMZ). *Livingston Par.:* 4 mi. N Holden, 1 (LSUMZ); 7 mi. N Holden, 1 (LSUMZ); 5 mi. N Holden, 3 (LSUMZ); 8 mi. E Watson, 1 (LSUMZ); 4 mi. N Denham Springs, 1 (LSUMZ); 3 mi. NNE Denham Springs, 1 (LSUMZ). *Madison Par.:* 11 mi. SW Tendall on Tensas River, 2 (LSUMZ); Tallulah, 2 (1 LSUMZ, 1 NLU); 8 mi. NW Tallulah, 2 (USNM); 5 mi. E Delhi, 1 (NLU); 12 mi. E Delhi, 1 (LTU). *Morehouse Par.:* 10 mi. NE Jones, 1 (LSUMZ); Mer Rouge, 8 (1 MCZ; 7 USNM). *Natchitoches Par.:* Provencal, 3 (LSUMZ); Lotus, 3 (LSUMZ); Vowells Mill, 1 (LSUMZ); Bellwood, 1 (LSUMZ); 2.5 mi. E, 2 mi. N Bellwood, 1 (LSUMZ). *Ouachita Par.:* 5 mi. N Monroe, 1 (LTU). *Rapides Par.:* 15 mi. S Alexandria, 4 mi. NE Forest Hill, 1 (LSUMZ); 18 mi. S Alexandria, 1 (USL); 8 mi. W Lecompte, 6 (USL); 5 mi. W Glenmora, 1 (USL). *Sabine Par.:* locality unspecified, 1 (LTU); 3 mi. S Florien, 1 (NSU); 2 mi. W Florien, 1 (NSU). *St. Helena Par.:* Grangeville, 2 (LSUMZ); Chipola, 4 (LSUMZ); 5 mi. N Chipola, 1 (LSUMZ); Greensburg, 3 (LSUMZ); 2 mi. S Greensburg, 2 (LSUMZ). *St. Tammany Par.:* 8 mi. N Covington, 2 (1 USL; 1 SLU); 15 mi. N Covington, Hwy. 437, 2 (TU); Covington, 2 (SLU); 7 mi. W Covington, 4 (SLU); 3 mi. NE Covington, 1 (SLU). *Tangipahoa Par.:* Amite, 2 (LSUMZ); 2 mi. S Robert on Hwy. 190, 1 (LSUMZ); Hammond, 3 (SLU); 4 mi. E Hammond, 1 (SLU). *Tensas Par.:* 1 mi. S St. Joseph, 1 (LSUMZ); 4 mi. S, 2 mi. E Newellton, 1 (LSUMZ); 20 mi. S, 4 mi. W Tallulah, 2 (LSUMZ); 20 mi. S, 5 mi. W Tallulah, 1 (LSUMZ). *Vernon Par.:* Anacoco, 1 (LSUMZ); 5 mi. SE Simpson, 1 (LSUMZ); 9 mi. SE Simpson, 1 (LSUMZ); Fort Polk Game Reservation, 1 (USL); 2 mi. S Fort Polk, 1 (LTU). *Washington Par.:* Bogalusa, 1 (LSUMZ); 15 mi. SW Bogalusa, 1 (LSUMZ); 5 mi. N, 1 mi. W Bogalusa, 1 (LSUMZ); 0.5 mi. N Angie, 2 (LSUMZ); 5 mi. S Bogalusa, 1 (LSUMZ); 10 mi. NW Bogalusa, 1 (LSUMZ); 7 mi. SW Bogalusa, 3 (LSUMZ); 22 mi. N Covington along Hwy. 450, 1 (TU); 1 mi. W Warnerton, 4 (TU); Angie, 1 (SLU); 3 mi. E Angie, 1 (SLU). *West Feliciana Par.:* various localities, 35 (25 LSUMZ; 10TU). *Winn Par.:* 7 mi. WSW Winnfield, 2 (LSUMZ); Atlanta, 1 (USL); 3 mi. N Atlanta, 1 (NLU); 2 mi. W Winnfield, 1 (NSU).

FOX SQUIRREL Plate IV
(Sciurus niger)

Vernacular and other names. — In Louisiana this species is sometimes called the red squirrel. Because it prefers a pine habitat in certain parts of the state, it is known there as the piney-woods squirrel. Other names sometimes heard are big head, chucklehead, and bushy-tail. The Latin name *niger* refers to the black color phase. It so happened that when Linnaeus, the great Swedish naturalist and originator of our system of binominal nomenclature, named this animal he thought the black individuals were a different species from the normal-colored ones. Even though we now know that black squirrels are only a color phase, the name *niger* still stands. For other names applied to tree squirrels in general, see the Gray Squirrel account.

Distribution. — The Fox Squirrel occurs from northern North Dakota and Minnesota, the Upper Peninsula of Michigan, extreme western New York, and southern Pennsylvania south over the eastern half of the United States to southern Florida and extreme northern Mexico. In Louisiana it is virtually statewide, being absent only in the coastal marshes, on coastal islands, and on some isolated chenieres. In some of these places, however, it has been introduced, as, for example, in the town of Cameron and on Grand Cheniere in Cameron Parish. On Cheniere au Tigre and Pecan Island, in Vermilion Parish, Fox Squirrels have been present as far back as any of the elderly residents can recall. (Map 24)

External appearance. — The Louisiana races of this species are large-sized tree squirrels that are basically ferruginous or rust colored, particularly on the underparts. In most of the bottomlands of the state, the solid black color phase is common, and in some of these areas it is the predominant form present. Geographically, the species varies drastically in

Figure 76 Fox Squirrel *(Sciurus niger),* Hamilton County, Ohio, September 1961. (Photograph by Woodrow Goodpaster)

color, and three well-marked subspecies occur in the state. For a detailed characterization of the three forms, see the descriptions of them under the different subspecies headings at the end of this account.

Color phases. — As noted above, the melanistic color phase is common in certain parts of the state, rare in others. In the subspecies *S. n. bachmani* of the Florida Parishes and in *S. n. ludovicianus* of the western part of the state, black individuals are encountered infrequently, but in the subspecies *S. n. subauratus* of the bottomlands along the Tensas and Mississippi rivers south through the Atchafalaya Basin, the black color phase is common. In some parts of this area it is more numerous than are normal-colored individuals. J. B. Kidd (pers. comm.) informs me that while engaged in a trapping and marking operation in the so-called "Dismal Swamp" region of southern Catahoula Parish, he obtained 36 Fox Squirrels and every one of them was black. Yet in checks of more than 4,000 speci-

mens in hunters' bags on the Thistlethwaite Wildlife Management Area in the western portion of St. Landry Parish only one black squirrel was found in a period of over 10 years. Also, my personal impression in the 1930s in Madison and Tensas parishes was that the proportion of blacks to reds was close to 50:50. On the other hand, I recall that a black squirrel was so unusual in Lincoln Parish in the early 1930s that a local newspaper published a picture of a hunter holding one that he had shot in that area.

Measurements and weights. — See the measurements for the various subspecies at the end of this account.

Cranial and other skeletal characters. — The skull of *S. niger* is similar to that of *S. carolinensis* but is somewhat more massive, and the bones are reddish instead of creamy white. One of the main differences, however, is that *S. niger* possesses only four cheek teeth on each side of the upper jaw, whereas in adult *S.*

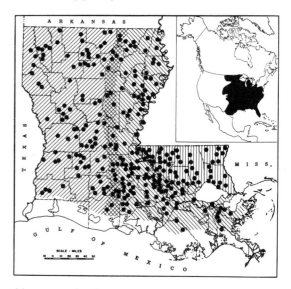

Map 24 Distribution of the Fox Squirrel *(Sciurus niger). Shaded areas,* known or presumed range within the state; •, localities from which museum specimens have been examined; *inset,* overall range of the species. 1. *S. n. ludovicianus;* 2. *S. n. subauratus;* 3. *S. n. bachmani.*

carolinensis a small, peglike premolar is added. The os penis is likewise similar to that of *S. carolinensis* but is longer and its perceptible twist gives it a slightly sigmoid shape. The scoop-shaped distal end is more expanded and flared than in *S. carolinensis* (Fairchild, 1950). (Figures 72 and 77)

Sexual characters and age criteria.—The sexes are superficially similar. As in the Gray Squirrel, immatures can be recognized by their small size and the fluffy texture of their pelage, especially that of the posterior underparts. In young males born during the previous summer, the scrotum is light colored instead of black, and the testes are undescended. The teats of young females are

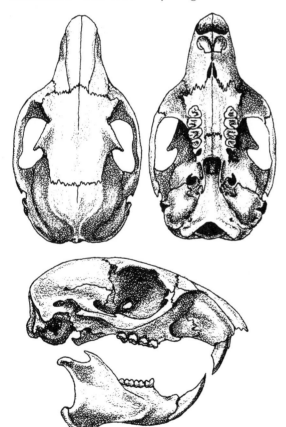

Figure 77 Skull of Fox Squirrel *(Sciurus niger)*, 1 mi. E Shongaloo, Webster Parish, LSUMZ 2495. Approx. natural size.

small and often difficult to locate in the hair, whereas those of adult females are decidedly larger and darker. Moore (1957), on the basis of his studies of the subspecies *S. n. shermani* in Florida, which is probably comparable at least to the race *S. n. bachmani* in our Florida Parishes, recognized four age classes: (1) nestlings, still feeding primarily on milk, weighing less than 550 grams, and being less than 85 days old; (2) immatures, now feeding on solid food, retaining fuzzy pelage, and weighing between 500 and 750 grams; (3) subadults, having glossy pelage like adults but weighing less than 900 grams and lacking fully erupted premolars; and (4) adults, having fully erupted complete dentition and weighing over 900 grams.

Status and habits.—The Fox Squirrel is common to abundant throughout nearly all the wooded portions of the state. Its habitat preference is for rather open situations in hardwood forests or in tracts of mixed hardwoods and pines. Such is at least the case with respect to populations of the subspecies *bachmani* in the Florida Parishes and those of *ludovicianus* in western Louisiana. On the other hand, the subspecies *subauratus* of the Atchafalaya Basin and adjacent areas in the southern part of the state is found in deep swamps in stands of cypress, tupelo, bitter pecan, and other hardwoods.

Most hunters agree that compared with the previous species Fox Squirrels are fat, logy beasts. They do sometimes appear to be lazy creatures, for I often see them spraddled out on the limbs of trees in my yard with their bellies flat on a branch and with their legs and tail dangling over the side. However, when we watch several of them scampering through the treetops and frolicking on our lawns, there appears then to be no evidence of indolence on their part. Fox Squirrels are sometimes said to be less gregarious than the Grays, to be instead shy and retiring. But such is not always the case, just as Kennicott (1857) noted over a hundred years ago. He

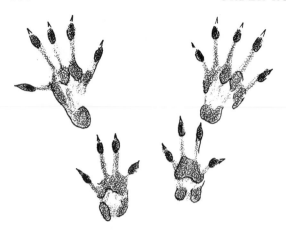

Tracks of Fox Squirrel

watched from concealment while four or five of them disported themselves overhead in a hickory tree. "They seldom remained quiet for a minute at a time. If not eating, they would scamper up and down, sometimes springing from limb to limb, with no other apparent object than fun. Their long bushy tails were never kept still, but continually jerked up and down. Occasionally, one would stop for an instant, and utter a short bark, or chatter. One kept aloof from the rest, and when approached by one of them in their play, chased them off with an angry guttural snarl."

Where Fox and Gray Squirrels come together and compete for the same habitat, the Gray Squirrel often appears to be the more successful and to replace its larger relative. The habitat requirements of the two species are remarkably similar. Consequently, it is not surprising that they are unable to coexist side by side without conflicts of interest.

The breeding biology of the Fox Squirrel is very much like that of the Gray Squirrel. Here in Louisiana there are two major periods, one in spring and one in summer, when the young are born, although pregnant females can be found every month of the year (Kidd, 1954). Breeding begins in late December and early January, reaches a peak in the third week of January, but declines to a low

level by late February. In May and June it reaches another peak, which tapers off in July. After a female is two years old, she is capable of producing two litters a year. Spring-born females will bear an early spring litter the following year but will skip the second litter in the summer, and females born in the summer will not produce a litter until the following summer. The average number of young to a litter is three, but as few as one or as many as six occur with some regularity. I know of one case of a Fox Squirrel that was

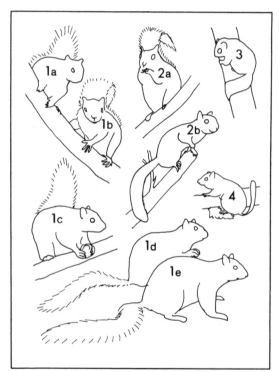

Plate IV

1. Fox Squirrel *(Sciurus niger):* **a,** melanistic phase and **b,** normal phase, of *S. n. subauratus;* **c,** *S. n. ludovicianus;* **d,** melanistic phase and **e,** normal phase, of *S. n. bachmani.*
2. Gray Squirrel *(Sciurus carolinensis):* **a,** *S. c. fuliginosus,* based on LSUMZ 132 from Atchafalaya Basin population; **b,** *S. c. carolinensis,* based on LSUMZ 15203 from Claiborne Parish population.
3. Southern Flying Squirrel *(Glaucomys volans).*
4. Eastern Chipmunk *(Tamias striatus).*

Pratt
1972

carrying seven fetuses, none of which showed any sign of being reabsorbed (Hoover, 1954).

The gestation period is 45 days in length. As in the case of the Gray Squirrel, the young are born naked, except for a few vibrissae on the chin and nose, and the eyes and ears are tightly closed. A new-born baby weighs about half an ounce. By the end of the third week its weight has increased to about two ounces, but the eyes and ears do not open until the end of the fourth week (Brown and Yeager, 1945). By this time the little squirrels are almost completely furred. The young are under parental care for about three months. They do not begin to venture forth from the nesting hollow into daylight until their seventh or eighth week or to forage on the ground until the eleventh or twelfth week.

Fox Squirrels differ somewhat from Gray Squirrels in the periods of the day in which they are most active. The latter are earlier risers, often being abroad at the break of dawn, but Fox Squirrels are seldom moving about until at least one hour after sunrise and are most active between 8 and 10 A.M. with other peaks around midday and in late afternoon. Adverse weather of any kind, high wind especially, tends to decrease activity. The heat of midsummer reduces midday movements during July, August, and early September.

Practically every kind of vegetable food that grows, whether beneath the soil or in the treetops, is consumed by squirrels. In spring they eat buried nuts, acorns, the new buds of trees, beetles, other insects, tubers, bulbs, roots, eggs of birds, and seeds of spring-fruiting trees. In summer they consume berries, fruits, nuts, corn, grains, and a wide variety of animal matter such as the larvae, pupae, and adults of insects. In the fall and winter the supply of mast produced by oaks, beeches, magnolias, gums, dogwoods, hickories, pecans, and other trees of our forest is more than sufficient for squirrels and their competitors. There are, however, areas in the state in which the carrying capacity of the forest is marginal and hence the number of squirrels present is low (St. Amant, 1959). Young forest and areas in which lumbering has been exhaustive both produce a smaller volume of food and, therefore, fewer squirrels. Best mast production occurs in middle-aged to old stands of forest. St. Amant considers that the prime squirrel habitat in the state, if indeed not in the entire country, is the approximately 5.5 million acres along the Upper and Lower Mississippi Alluvial Plain. Here large blocks of mixed hardwoods furnish excellent food and range conditions for the Gray Squirrel and one subspecies (subauratus) of the Fox Squirrel. St. Amant estimates an average of 0.7 squirrels per acre in this area. He also regards an additional 4 million acres of mixed loblolly pine and hardwoods and of creek-bottom hardwoods as excellent habitat that produces an average of 0.37 squirrels per acre.

Despite what in most years appear to be optimal squirrel populations, the fact remains that in many areas of the state the number of squirrels is often far below the level that could be achieved by improved management practices. The loss of forest as more and more land is being converted to agriculture and the elimination in the remaining forest of wolf trees in the interest of forestry practices designed to create better stands of merchantable timber are factors contributing to a diminution of our squirrel populations. Illegal preseason hunting must be stopped and hunting pressure should be better controlled to reduce the kill in critical areas. And the legal hunting season should be adjusted to protect females still with young. Range conditions could be improved by preventing damage by timber operations and practices. Private owners who want more game could be taught how to save den trees and to rotate their timber cutting operations so as to leave a greater number of old mast-producing trees. On our national and state forests, which are considered recreational areas, the improvement of game conditions should be given equal status with good forestry practices. There will never be maximum squirrel populations in areas where free-ranging hogs and even cattle cause a loss of mast on the ground. Admittedly, some conflict will persist between the end results sought by agricultural, forest, and livestock interests and those of sportsmen and biologists. Fortunately, though, the conflict can be minimized if all parties are open minded in seeking a solution.

Predators.—Animals that feed occasionally upon the Gray Squirrel also prey to some extent on the present species. Foxes, Bobcats, weasels, or owls would all relish a meal of a bushy-tail that they might succeed in catching. Fortunately, none of these potential predators are normally abroad when sunlight bathes the forest and when the Fox Squirrel is active. They are creatures of the night as much as the Fox Squirrel is a creature of the day. Nevertheless, some predation does occur, notably by hawks.

Parasites.—Although I have made no exhaustive search of the parasitological literature, I find that 1 protozoan, 15 cestodes, 3 acanthocephalan worms, over 10 nematodes, 4 ticks, 3 lice, 8 mites (including 2 chiggers), and 2 fleas have been reported from the Fox Squirrel (Chandler, 1942a; Lucker, 1943; Harwood, 1949; Morlan, 1952; Doran, 1954a and b, 1955a and b; Self and Esslinger, 1955; Moore, 1957; Jackson, 1961). However, these records are based on squirrels from Wisconsin, Georgia, and Florida and may not all represent forms that occur in or on squirrels inhabiting Louisiana.

Subspecies

Sciurus niger bachmani Lowery and Davis

Sciurus niger bachmani Lowery and Davis, Occas. Papers Mus. Zool., Louisiana State Univ., 9, 1942:156; type locality, 10 mi. NW Enon, Washington Parish, Louisiana.

This well-marked race of the Fox Squirrel is characterized by its large size, ferruginous underparts, and the presence always of a patch of white on the nose and varying amounts of white on the ears, toes, and tip of the tail (Plate IV). It is found in the mixed pine and hardwood habitat of the Florida Parishes (Map 24). Intergradation takes place with *S. n. subauratus* where the two come together on the edge of the bottomlands east of the Mississippi River in the western parts of West Feliciana and East Baton Rouge parishes. For instance, squirrels on the floodplain just west and south of the Louisiana State University campus at Baton Rouge are fairly typical *subauratus,* but those on the campus itself and on the high ground along Highland Road average larger and often show some white on their noses, an evidence of gene flow between this population and that of typical *bachmani* from farther east in the parish. In the mixed pine and hardwood habitat, beginning east of the Baton Rouge Municipal Airport, typical *bachmani* is encountered, which then ranges eastward in the Flordia Parishes, north of the bottomlands adjacent to Lake Maurepas and Lake Pontchartrain, through most of Mississippi, western Florida, and Alabama. In southeastern Alabama and Georgia it meets the range of *S. n. niger,* from which it differs by its general ferruginous instead of gray coloration.

Melanistic individuals of *bachmani* occur occasionally. They are frequently solid black except for white on the nose, ears, toes, and tip of the tail (Plate IV), but sometimes the black is confined mainly to the underparts. Sometimes the black of the upperparts is in the form of an agouti pattern.

Measurements.—Eleven skins and 22 skulls from the Florida Parishes measured as follows: total length, 570 (530–620); tail, 258 (178–295); hind foot, 66 (52–75); ear, 26 (19–30); greatest length of skull, 64.0 (59.1–67.5); cranial breadth, 25.9 (24.7–29.2); zygomatic breadth, 36.1 (33.9–38.8); interorbital breadth, 19.1 (16.9–20.7); length of nasals, 22.0

(20.1–23.0); palatilar length, 28.8 (26.4–30.4); postpalatal length, 22.1 (20.0–24.0); maxillary toothrow, 11.1 (10.3–11.8).

Specimens examined from Louisiana. — Total 122, as follows: *East Baton Rouge Par.:* Pride, 2 (LSUMZ); 1 mi. N Pride, 2 (LSUMZ). *East Feliciana Par.:* various localities, 30 (LSUMZ). *Livingston Par.:* 2 mi. SE Frost, 1 (LSUMZ); 1 mi. NE Killan, 1 (LSUMZ); 2.5 mi. E Watson, 1 (LSUMZ); 8 mi. SE Killan, 1 (LSUMZ); 5 mi. N Holden, 1 (LSUMZ); 8 mi. S Livingston, 1 (LSUMZ); 7 mi. S Walker, 1 (LSUMZ); 8 mi. E Watson, 1 (LSUMZ); 1 mi. W Springfield, 1 (LSUMZ); 2 mi. N Watson, 1 (LSUMZ). *St. Helena Par.:* Chipola, 1 (LSUMZ); 2.75 mi. N Pine Grove, 1 (LSUMZ); 8 mi. W Greensburg, 1 (LSUMZ). *St. Tammany Par.:* various localities, 41 (4 LSUMZ; 3 USNM; 13 TU; 17 AMNH; 4 SLU). *Tangipahoa Par.:* Ponchatoula, 9 (6 USNM; 1 TU; 2 AMNH); Independence, 3 (2 TU; 1 SLU); Amite City, 3 (TU); Tangipahoa, 2 (TU); Arcola, 1 (AMNH); 7 mi. SE Ponchatoula, 1 (SLU); 5 mi. E Tickfaw, 2 (SLU). *Washington Par.:* Enon, 1 (LSUMZ); 12 mi. E Franklinton, 1 (LSUMZ); 5 mi. N, 1 mi. W Bogalusa, 1 (LSUMZ); 2.75 mi. W Angie, 1 (LSUMZ); 5 mi. E, 2 mi. S Mt. Herman, 1 (LSUMZ); 4 mi. SW Bogalusa, 1 (LSUMZ); 2.5 mi. W Varnado, 2 (SLU); 2 mi. W Rio, 1 (SLU). *West Feliciana Par.:* 2 mi. S, 2 mi. W Jackson, 1 (LSUMZ); 1 mi. W Laurel, 1 (LSUMZ); 9 mi. E St. Francisville, 1 (LSUMZ); 2.5 mi. NE Weyanoke, 1 (LSUMZ).

Sciurus niger subauratus Bachman

Sciurus subauratus Bachman, Proc. Zool. Soc. London [for] 1838 [=1839]:87; type locality, Louisiana, restricted to Iberville Parish by Lowery and Davis, Occas. Papers Mus. Zool, Louisiana State Univ., 9, 1942:166.

Sciurus niger subauratus, Lowery and Davis, Occas. Papers Mus. Zool., Louisiana State Univ., 9, 1942:166.

Sciurus auduboni Bachman, *op. cit.,* 1839:97; type obtained in market in New Orleans, Louisiana; melanistic example of *S. n. subauratus.*

Another of the clearly defined races of the Fox Squirrel inhabiting Louisiana is this small, dark form that occurs in the bottomland forest of the Tensas, Mississippi, and Atchafalaya floodplains in the eastern and central-southern parts of the state (Map 24). It differs markedly from *S. n. bachmani* to the east by its much smaller size and darker coloration and by invariably lacking any white on the nose, ears, feet, and tail. From *S. n. ludovicianus* in the western part of the state, it differs again by its much smaller size and darker coloration (Plate IV). As noted in the general account on page 182, in certain localities melanistic individuals sometimes equal or even outnumber the normal color phase.

Measurements. — Thirty skins and 41 skulls from central Louisiana measured as follows: total length, 531 (490–590); tail, 252 (217–305); hind foot, 61 (51–68); ear, 27 (21–30); greatest length of skull, 60.2 (57.2–64.9); cranial breadth, 24.7 (23.6–25.9); zygomatic breadth, 33.3 (32.4–37.2); interorbital breadth, 18.2 (16.7–20.0); length of nasals, 20.3 (17.6–22.9); palatilar length, 26.5 (24.7–29.8); postpalatal length, 20.5 (19.2–22.4); maxillary toothrow, 10.7 (9.6–11.9).

Specimens examined from Louisiana. — Total 218, as follows: *Ascension Par.:* 4 mi. SE St. Gabriel, 1 (LSUMZ); Sorrento, 1 (LSUMZ); 3 mi. SE Burnside, 1 (LSUMZ); 3 mi. S Burnside, 1 (LSUMZ); 2 mi. SW Donaldsonville, 1 (LSUMZ); 2 mi. N Prairieville, 1 (LSUMZ); Gonzales, 1 (LSUMZ). *Assumption Par.:* Labadieville, 1 (LSUMZ); 7 mi. SW Napoleonville, 1 (LSUMZ); 9 mi. S Donaldsonville, 1 (LSUMZ). *Avoyelles Par.:* 15 mi. N Marksville, 1 (LSUMZ); Hamburg levee, 1 (USL); near Coco Lake, Moreauville, 1 (LTU). *Caldwell Par.:* 12 mi. SE Grayson, 1 (LSUMZ). *Catahoula Par.:* 10 mi. S Jonesville, 1 (LSUMZ); 6 mi. SW Jonesville, 1 (LSUMZ); 2 mi. S Sicily Island, 3 (1 LSUMZ; 2 NLU); 4 mi. NNE Foules, 2 (LSUMZ); Sicily Island, 1 (NLU). *Concordia Par.:* 11 mi. N Ferriday, 1 (LSUMZ); 10 mi. S Vidalia, 1 (USL); 7 mi. NE Monterey, 2 (NLU); locality unspecified, 1 (LTU); Cross Bayou near Jonesville, 1 (MSU); Ferriday, 1 (LTU); 1 mi. W Shaw, 3 (LSUMZ). *East Baton Rouge Par.:* University, 7 (LSUMZ). *East Carroll Par.:* Transylvania, 2 (LTU); Lake Providence, 2 (LTU). *Franklin Par.:* 20 mi. SE Winnsboro, 1 (LSUMZ); 2 mi. N Gilbert, 3 (NLU); 7 mi. NE Baskin, 1 (LTU); 5 mi. E Baskin, 1 (LTU). *Iberia Par.:* New Iberia, 6 (2 LSUMZ; 3 USL; 1 USNM); 3 mi. W New Iberia, 1 (USL); near Weeks Island, 1 (USL); 10 mi. E New Iberia (Lake Fausse Point), 1 (USL); 5 mi. N New Iberia, 2 mi. S La. Hwy. 14, 3 (USL); Avery Island, 1 (CAS). *Iberville Par.:* various localities, 21 (LSUMZ). *Jefferson Par.:* 2.5 mi. S Waggaman, 1 (LSUMZ). *Livingston Par.:* 4 mi. S Whitehall, 1 (LSUMZ). *Madison Par.:* Tallulah, 1 (LSUMZ); Eldorado, 1 (USNM); 8 mi. NW Tallulah, 3 (USNM); mouth of Fools River, Tensas River, 1 (LTU). *Morehouse Par.:* 9 mi. SE Bastrop, 2 (LSUMZ); Mer Rouge, 13 (8 USNM; 5 MCZ); 3 mi. SE Bastrop, 1 (NLU); 4 mi. W Oak Ridge, 1 (NLU); 10 mi. E Mer Rouge, 1 (LTU). *Orleans Par.:* New Orleans, 1 (USNM). *Ouachita Par.:* Russell Sage Game Mgt. Area, 3 (NLU); Swartz, 2 (NLU). *Plaquemines Par.:* Fanny, 3 (LSUMZ);

English Turn, 1 (LSUMZ); Dalcour, 1 (LSUMZ); Lake Hermitage, 1 (SLU). *Pointe Coupee Par.:* Raccourci, 2 (LSUMZ); Lottie, 4 (LSUMZ); Lakeland, 3 (LSUMZ); 8 mi. NW Lottie, 5 (LSUMZ); 5 mi. N Fordoche, 4 (LSUMZ); 0.25 mi. W Old River, 1 (LSUMZ); Krotz Springs, E Atchafalaya River, 3 mi. S Hwy. 90, 1 (LSUMZ); 0.5 mi. S Atchafalaya River bridge and 0.25 mi. E of river, 1 (LSUMZ); Innis, 1 (LSUMZ); 4.5 mi. SE Krotz Springs, 1 (LSUMZ); Morganza, 2 (SLU). *Richland Par.:* 2 mi. S Richland–Ouachita parish line on Hwy. 15, 1 (NLU). *St. Bernard Par.:* Toca Village, 2 (LSUMZ); Violet, 3 (LSUMZ); Yscloskey, 1 (USL). *St. James Par.:* 10 mi. NW Vacherie, 1 (LSUMZ); Old River Road at Convent, 1 (TU). *St. John the Baptist Par.:* 2 mi. NE Laplace, 1 (LSUMZ). *St. Landry Par.:* 5 mi. N Port Barre, 1 (LSUMZ); Pearsonville, 1 (LSUMZ); Melville, 4 (1 LSUMZ; 1 USL; 2 USNM); 5 mi. N Lebeau, 1 (LSUMZ); 5 mi. N Palmetto, 1 (LSUMZ); 4 mi. N Palmetto, 1 (LSUMZ); 3 mi. S Palmetto, 1 (LSUMZ); 0.25 mi. S Palmetto, 1 (LSUMZ); Courtableau Bayou, 5 mi. S Hwy. 190, 1 (USL); 2 mi. E Arnaudville, 1 (USL); Lake Fordoche, 6 mi. N Henderson, 1 (USL); 4 mi. E Port Barre, 1 (USL). *St. Martin Par.:* 4 mi. E Parks, 1 (LSUMZ); St. Martinville, just S Evangeline State Park, 1 (USL); Butte La Rose, 1 (USL); Cade, 2 (USL); St. Martinville, 1 (USL); Catahoula, 2 (USL); 2 mi. SE St. Martinville, 1 (USL); 1 mi. E St. Martinville, 1 (USL); 2 mi. E St. Martinville, 2 (USL); Lake Martin, 2 (USL); 8 mi. SW Cecelia, 1 (USL); 4 mi. W Cecelia, 1 (USL); 8 mi. N St. Martinville, 1 (USL); 1 mi. W Breaux Bridge, 1 (LTU). *Tensas Par.:* 12 mi. W Waterproof, 1 (LSUMZ); 20 mi. NW St. Joseph, 2 (LSUMZ); 1 mi. S St. Joseph, 1 (LSUMZ); 13 mi. W St. Joseph, 1 (LSUMZ); 12 mi. W St. Joseph, 2 (LSUMZ); St. Joseph, 1 (LSUMZ); 10 mi. NW Newellton, 1 (NLU); 6 mi. E Newellton, 1 (LTU); near Newellton, 1 (LTU); Waterproof, 1 (LTU); Tensas River, Cooter Point, 1 (LTU). *Terrebonne Par.:* 0.5 mi. E Schriever, 1 (LSUMZ); 3 mi. SE Houma, 1 (LSUMZ). *West Baton Rouge Par.:* 10 mi. W Miss. River Bridge on US Hwy. 190, 1 (LSUMZ); Cinclare, 3 (LSUMZ); Erwinville, 1 (LSUMZ); 10 mi. S Baton Rouge, Hwy. 1, 1 (LSUMZ); 10 mi. W Port Allen, 1 (LSUMZ); 4 mi. SW Port Allen, 2 (LSUMZ); 3 mi. NW Port Allen, 1 (LSUMZ). *West Carroll Par.:* Oak Grove, 1 (LSUMZ); 7 mi. E Epps, 1 (LTU). *West Feliciana Par.:* Tunica Island, 12 mi. NW St. Francisville, 2 (LSUMZ).

Sciurus niger ludovicianus Custis

Sciurus niger ludovicianus Custis, Philadelphia Med. Phys. Jour., 2, 1806:47; type locality, Red River of Louisiana; restricted to Natchitoches Parish by Lowery and Davis, Occas. Papers Mus. Zool., Louisiana State Univ., 9, 1942:164.

Sciurus texianus Bachman, Proc. Zool. Soc. London [for] 1838 [=1839]:86.

[*Sciurus niger*] var. *ludovicianus*, J. A. Allen, *in* Coues and Allen, Monogr. N. Amer. Rodentia, 1877:718.

Sciurus niger ludovicianus, Lowery and Davis, Occas. Papers Mus. Zool., Louisiana State Univ., 9, 1942:164.

This third and final subspecies of the Fox Squirrel that occurs in Louisiana occupies the western one-third of the state (Map 24). It is a large race with a massive skull. The coloration, although similar to that of *S. n. subauratus,* is decidedly paler with the tips of hairs of the dorsum being pale yellow and with the color of the underparts and the light areas of the tail being a pale yellow.

Measurements. — Twenty-seven skins and skulls from western Louisiana measured as follows: total length, 562 (490–616); tail, 268 (230–341); hind foot, 61 (53–70); ear, 25 (18–31); greatest length of skull, 64.6 (62.1–67.0); cranial breadth, 25.9 (24.4–27.2); zygomatic breadth, 36.5 (35.3–38.0); interorbital breadth, 19.3 (17.7–20.6); length of nasals, 22.1 (20.4–23.9); palatilar length, 29.5 (27.2–31.3); postpalatal length, 22.1 (20.9–23.4); maxillary toothrow, 11.3 (9.2–11.9).

Specimens examined from Louisiana. — Total 154, as follows: *Acadia Par.:* Crowley, 1 (LSUMZ); Egan, 2 (LSUMZ); 1.5 mi. SW Egan, 1 (LSUMZ); 2 mi. S Egan, 1 (LSUMZ); Bayou des Cannes, 1 (LSUMZ); 1 mi. S Eunice, 2 (USL); 3 mi. SE Egan, 1 (USL); 2 mi. N Crowley, 1 (USL); 4.5 mi. SE Crowley, 1 (USL); Rayne, 4 (TU); Pointe aux Loups Springs, 16 (MCZ). *Allen Par.:* 4 mi. S Elizabeth, 1 (LSUMZ); 2 mi. W Oakdale, 1 (LSUMZ); 10 mi. W Oberlin, Whiskey Chitto Cr., 5 (LSUMZ); Oberlin, 2 (LSUMZ); 7 mi. W Elizabeth, 1 (LSUMZ); 12 mi. W Mamou (Castor Creek), 1 (USL); West Bay GMA, 1 (MSU); Kinder, 1 (LTU). *Beauregard Par.:* 13 mi. S De Ridder, 1 (LSUMZ); 4 mi. S Sugartown, 1 (LSUMZ); 10 mi. S, 3 mi. E De Ridder, 1 (LSUMZ); 3 mi. NE De Ridder, 1 (LSUMZ); 7 mi. SW Merryville, 1 (MSU). *Bienville Par.:* Bienville, 2 (LTU); Jackson–Bienville Game Res., 2 (LTU); 5 mi. W Gibsland, 1 (LTU). *Bossier Par.:* Bossier Air Force Base Reservation, 2 (LSUMZ); 4 mi. N Princeton, 5 (LSUMZ); 1 mi. NE Red Point, 1 (LSUMZ); Redchute Bayou and U.S. Hwy. 80, 1 (LSUMZ); 2 mi. E Midway, 1 (LSUMZ); Midway community, 1 (LTU); Benton, 2 (LTU). *Caddo Par.:* 2.5 mi. NW Shreveport, 1 (LSUMZ); 1 mi. N Hwy. 1, 5 mi. N Blanchard, 4 (LSUMZ); 10 mi. S Shreveport on La. Hwy. 1, 1 (USL). *Calcasieu Par.:* 10 mi. NE Lake Charles, 1 (LSUMZ); 10 mi. W Westlake on Hickory Creek, 1 (LSUMZ); Toomey, 3 (1 TU; 2 AMNH); 4 mi. W Sulphur on I-10, 3 (MSU). *Claiborne Par.:* 4 mi. E Lisbon, 1 (LSUMZ); Summerfield, 1 (LTU). *De Soto Par.:* 4 mi. SE Mansfield, 1 (LSUMZ); 4.5 mi. E Grand Cane, 1 (LSUMZ); 2 mi. W Stonewall, 1 (NLU). *Evangeline Par.:* 9 mi. N Ville Platte, 1 (USL); 1.5 mi. N Chataignier, 4 (USL); 2 mi. S Easton, 1 (USL). *Grant Par.:* 4 mi. SW Georgetown, 1 (USL); 1.5 mi. NW Pollock, 1 (USL);

Georgetown, 1 (LTU). *Jackson Par.:* 0.75 mi. E Ansley, 1 (LTU); Vernon, 1 (LTU); Jackson–Bienville Game Reserve, 1 (LTU); East of Ansley in Cypress Creek, 1 (LTU); 9 mi. SE Chatham, 1 (LTU). *Lincoln Par.:* Ruston, 1 (LSUMZ); 0.5 mi. N Tremont, 1 (LSUMZ); 3 mi. S Ruston, 1 (LTU); 5 mi. N Ruston, 1 (LTU); 9 mi. N Ruston on Big Creek, 1 (LTU); 7 mi. NW Ruston, 1 (LTU); bottomland south of Ruston, 1 (LTU). *Natchitoches Par.:* 2.4 mi. N, 3.5 mi. E Kisatchie, 1 (LSUMZ); Cypress, 5 (LSUMZ); Kisatchie, 1 (LSUMZ); Provencal, 1 (LSUMZ); Bellwood, 1 (LSUMZ); Lotus, 1 (LSUMZ); Ashland, 2 (LSUMZ); 1 mi. W Vowells Mill, 1 (LSUMZ). *Ouachita Par.:* 1 mi. N Calhoun, 1 (NLU); Choudrant Creek bottom, Calhoun, 1 (LTU). *Rapides Par.:* Lake Valentine, 2 (LSUMZ); 15 mi. S Alexandria, 4 mi. NE Forest Hill, 1 (LSUMZ); Glenmora, 1 (LSUMZ); 8 mi. N Alexandria, 1 (LSUMZ); 25 mi. SW Alexandria, 1 (LSUMZ); 3 mi. S Boyce, 1 (LSUMZ); 3 mi. S Cheneyville in Lone Pine Woods, 1 (LSUMZ); 18 mi. S Alexandria, 1 (USL); Bayou Boeuf Rd., 5 mi. off Lecompte-Forest Hill Hwy., 1 (USL); 5 mi. W Glenmora, 1 (USL); Westport, 1 (USL); 3 mi. SW Mangham, 1 (NLU); 3 mi. S Alexandria, 1 (NLU). *Sabine Par.:* 13 mi. S Many, Hwy. 171, 1 (LSUMZ). *Union Par.:* 5 mi. N Marion, 1 (LTU); 8 mi. S Farmerville, 1 (LTU); 4 mi. SE Farmerville, 1 (LTU); 2 mi. SE Farmerville, 1 (LTU); Litroe, 1 (MSU). *Vernon Par.:* 3 mi. E Simpson, 1 (LSUMZ); 5 mi. E Simpson, 1 (LSUMZ); Fort Polk Game Reservation, 1 (USL); 3 mi. S Leesville, Hwy. 171, 1 (MSU); 0.25 mi. E Lacamp, 1 (LSUMZ). *Webster Par.:* 1 mi. E Shongaloo, 1 (LSUMZ); Shongaloo, 25 mi. N Minden, 1 (LSUMZ); 10 mi. N Minden, 2 (LTU). *Winn Par.:* Atlanta, 2 (USL); 3 mi. N Atlanta, 1 (NLU).

Specimens examined from Louisiana showing varying degrees of intermediacy between S. N. BACHMANI *and* S. N. SUBAURATUS.—Total 82, as follows: *East Baton Rouge Par.:* various localities, 73 (70 LSUMZ; 2 USL; 1 SLU). *Livingston Par.:* 0.5 mi. N Hoo Shoo Too Rd., 1 (LSUMZ). *West Feliciana Par.:* 5 mi. E Angola, 1 (LSUMZ); 5 mi. NW St. Francisville, 2 (LSUMZ); 3 mi. S, 2 mi. E St. Francisville, 1 (LSUMZ); 5 mi. W St. Francisville, 1 (LSUMZ); 38 mi. NNW Baton Rouge, 1 (LSUMZ); Bains, 1 (TU); 2 mi. SSE St. Francisville, 1 (LSUMZ).

Specimens examined from Louisiana showing varying degrees of intermediacy between S. N. LUDOVICIANUS *and* S. N. SUBAURATUS. Total 101, as follows: *Avoyelles Par.:* 4 mi. N Bunkie, 1 (LSUMZ); 3 mi. W Marksville, 2 (LSUMZ); 5 mi. S Cottonport, 1 (USL); Cottonport, 5 (USL). *Lafayette Par.:* various localities, 50 (5 LSUMZ; 43 USL; 2 USNM). *La Salle Par.:* Jena, 1 (LSUMZ); 4.5 mi. NE Olla, 1 (LSUMZ); Olla, Hwy. 165, 1 (NLU). *Morehouse Par.:* 7 mi. W Beekman, 1 (NLU); Georgia Pacific GMA, 1 (LTU). *Ouachita Par.:* West Monroe, 1 (NLU); 12 mi. SW Monroe, 2 (NLU); 5 mi. S Monroe, 1 (NLU); 7 mi. SE Monroe, 1 (NLU); 15 mi. SW West Monroe, 1 (NLU); between Sterlington and Monroe, 2 (NLU); 5 mi. N Monroe, 1 (LTU). *Rapides Par.:* 7 mi. NE Pineville, 1 (LSUMZ). *St. Landry Par.:* 3 mi. N Opelousas, 1 (LSUMZ); 2 mi. S Opelousas, 8 (USL); Thistlethwaite Game Refuge, 1 (USL); 1 mi. N Sunset, 1 (USL); Opelousas, 7 (USL); 3 mi. NW Washington, 2 (USL); Grand Coteau, 2 (USNM). *Vermilion Par.:* 5 mi. N Abbeville, 1 (USL); 2 mi. W Abbeville, 1 (USL); Perry, 3 (USNM).

[Genus *Marmota* Blumenbach]

Dental formula:

$$I\frac{1-1}{1-1} \quad C\frac{0-0}{0-0} \quad P\frac{2-2}{1-1} \quad M\frac{3-3}{3-3} = 22$$

[WOODCHUCK] Figure 78
(*Marmota monax*)

Vernacular and other names.—The Woodchuck is also called the groundhog, whistling pig, whistler, and eastern marmot. The generic name is the Latin word for marmot, the term applied to relatives of the Woodchuck in Europe and to certain North American representatives of the group. The last part of the scientific name, *monax,* is an American Indian name for the animal and means "the digger."

Figure 78 Woodchuck (*Marmota monax*).

The name groundhog clearly refers to the animal's habit of living in the ground, its squatty shape, and its waddling gait. The exact basis of the name Woodchuck is not known, but some writers have expressed the opinion that it was derived by folk etymology from one of the northern Indian names such as *wejack, otchig,* or *otcheck.*

Distribution. — The species occurs from Labrador and Nova Scotia across the southern half of Canada to southeastern Alaska south over most of the eastern half of the United States as far as western North and South Carolina, northern Georgia, central Alabama, southern Arkansas, and western Kansas. Only one record is available that points to its occurrence in Louisiana (see Map 25 and the discussion beyond).

External appearance. — The Woodchuck is the largest member of the squirrel family and among Louisiana rodents is exceeded in size and weight only by the American Beaver and the Nutria. An adult usually measures 20 to

26 inches in total length, but the length of the short, bushy tail seldom exceeds more than 4 or 5 inches. The body of the Woodchuck is relatively broad and somewhat flat and the legs are short. The color above is dark brown, strongly grizzled by the yellowish tips of the guard hairs and by the yellowish terminal portion of the dense underfur that shows through from below. The hair on the top of the head, nape, tail, and feet is black or brownish black, and that of the legs is heavily overlaid and mixed with burnt sienna, producing a pronounced reddish brown effect, especially on the forelegs. The color of the underparts is strongly reddish brown. That of the nose, face, and chin is whitish or whitish buff.

Color phases. — Albinism appears to be rare. Melanism is much more frequent. The LSUMZ collections contain a solid black individual from Pennsylvania.

Measurements and weights. — Jackson (1961) gives the following figures for adults: total length, 550–640 (21.6 to 25.2 in.); tail, 135–160 (5.3 to 6.3 in.); hind foot, 82–92 (3.2 to 3.6 in.). The species is said to weigh from about 5 pounds when emerging from hibernation in spring to about 12 pounds when prepared for hibernation in the fall. The skull has a total length of 94 to 102 mm, a zygomatic breadth of 60 to 69 mm.

Cranial and other skeletal characters. — The skull of *Marmota monax* is highly diagnostic and therefore easily distinguished from that of any other mammal in the state. It is massive in structure and, when viewed in profile, decidedly flattened. The postorbital processes extend out at right angles to the long axis of the skull, and between them a marked depression in the frontal bones is evident (Figure 79). The front feet possess four toes not counting the thumb, which is vestigial yet retains a small nail; the hind feet have five toes bearing slightly curved claws. The os

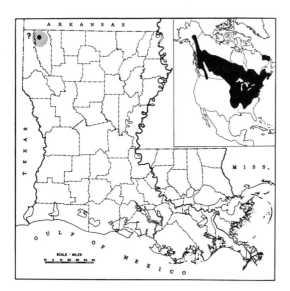

Map 25 Distribution of the Woodchuck *(Marmota monax).* •, the only locality where the species is alleged to have occurred in the state (see text); *inset,* overall range of the species.

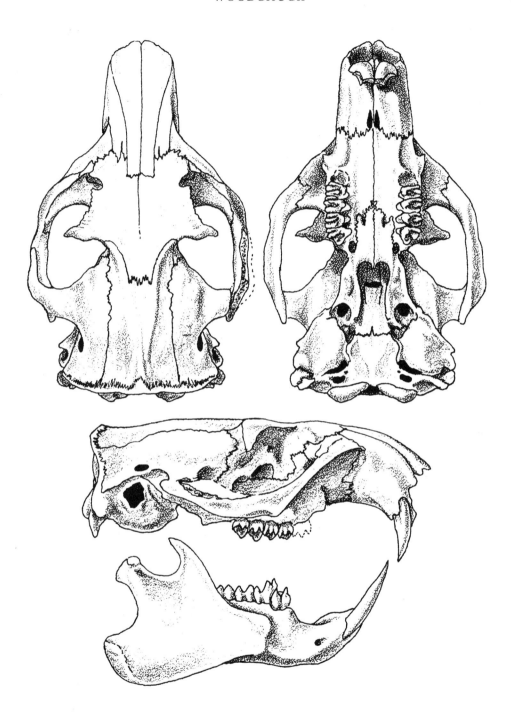

Figure 79 Skull of Woodchuck *(Marmota monax monax)*, ♂, Benton County, Arkansas, LSUMZ 6802. Approx. natural size.

penis, which is approximately 4 mm in length, is broad basally in both dorsal and lateral aspects. It is curved slightly upward and has a slight cup on the distal end.

Sexual characters and age criteria. — Males average slightly heavier than females, but the sexes are otherwise indistinguishable except by inspection of the external sex organs. The young resemble the adults but tend to be darker in coloration. By the time they are one year old they have acquired the rich colors of the legs and underparts that are generally characteristic of adults.

Status and habits. — The only record of this species in Louisiana is the capture of an adult approximately three miles north of Plain Dealing, in Bossier Parish, on or about 12 February 1952. The Shreveport *Journal* for 13 February carried the following story with the photograph that is also reproduced here (Figure 80):

WHAT IS IT? A strange animal (pictured above) has citizens of Plain Dealing puzzled as to its identity. The "thing" was brought to Plain Dealing by Bossier Hamiter, farmer, after it had been found by his bulldog near his home. Displayed on the streets, it aroused considerable speculation among the citizenry as to just what it was. Some said it was a badger, others a muskrat, prairie-dog, beaver, a cross between a squirrel and a coon [!], or just a groundhog that had made a belated appearance. Old timers like Dude McKeller and Jim Chamblee, now in their eighties, opined that they had never seen anything like it. The animal weighs 10 lbs., is 2 feet long, and 9 inches across, and has the appearance of being a large flying squirrel [!]. Its color is black and brownish gray.

The animal was, of course, a Woodchuck. The possibility exists that it was brought into the state and released, but the circumstances surrounding its capture in a somewhat remote wooded area not far from the Louisiana–Arkansas state line, militates against this explanation. Unfortunately, though, the record is rendered suspect on other grounds. A roadside zoo, now long defunct, located between Minden and Bossier City, is said to have owned at about the same time a Woodchuck that escaped. For the Plain Dealing animal to have been the same individual, it would have had to travel over 30 miles and to have crossed the bottomlands along Bayou Bodcau, a feat that I do not regard as highly plausible on the part of an escapee. Since the species is definitely known to occur in nearby Hempstead County in Arkansas (Sealander, 1956), I believe the Plain Dealing animal just as likely represented a southward range extension. However, because of the one equivocal element associated with the record, I have placed the name of the species in brackets (see statement of policy on page 54) and have not counted it as an indisputable member of our state's mammal fauna.

One of the first clues pointing to the presence of the Woodchuck, other than actually seeing the animal, is to find the entrance to one of its underground burrows. The opening can be confused with that of an armadillo's tunnel but not with the opening of

Figure 80 Photograph of Woodchuck *(Marmota monax)* killed by a dog near Plain Dealing, Bossier Parish, on or about 12 February 1952. (Photograph courtesy of Harbuck & Womack, Inc., Shreveport)

the tunnel made by any other Louisiana mammal. The entrance is usually 12 inches or more in diameter and is often located in the side of a bank or near the base of a large tree. The dirt is removed by the animal kicking vigorously with the hind feet and by pushing it out of the hole with the face and chest. No attempt is made to conceal the entrance to the hole, and the animal is often seen perched on the mound of refuse dirt in an upright position, surveying the landscape for a possible source of danger. A side entrance to the underground passage and nest chamber is often constructed, as well as blind offshoots, where the occupant sometimes buries its feces.

The Woodchuck runs with a waddling gait, and, although it spends most of its time on the ground, it will sometimes climb trees in search of fruit or to survey the landscape. The call is a shrill whistle that is often followed by several short chuckling notes. Once the sounds are heard they are thereafter easily recognized.

In the northern parts of its range the species hibernates, and presumably it does so in its southern domain. In northern Louisiana it might emerge as early as Groundhog Day, on 2 February, but whenever it ventures forth, whether it sees its shadow or not would make no difference at all to the Woodchuck. The popular fable has absolutely no factual basis.

Residents of the northern part of the state, especially of the northwestern corner, should be on the alert for additional evidence of the presence of the Woodchuck in that section.

Subspecies

Marmota monax monax (Linnaeus)

[Mus] Monax Linnaeus, Syst. nat., ed. 10, 1, 1758:60; type locality, Maryland.

Although the subspecies of the animal captured at Plain Dealing cannot be determined in the absence of a preserved specimen, it was, if a wanderer from Arkansas, almost certainly, on the basis of geographical probability, an example of monax, the race that

occurs throughout the southern half of the range of the species in the eastern United States and, of course, the race that inhabits adjacent Arkansas.

Genus *Tamias* Illiger

Dental formula:

$$I\frac{1-1}{1-1} \quad C\frac{0-0}{0-0} \quad P\frac{1-1}{1-1} \quad M\frac{3-3}{3-3} = 20$$

EASTERN CHIPMUNK   Plate IV
(Tamias striatus)

Vernacular and other names.—This species is also called chipmuck, striped ground squirrel, chippy, hackee or hackey, chipping squirrel, and piping squirrel. Although the modern name chipmunk is generally assumed to have originated from the animal's "chip, chip" call notes, Henisch and Henisch (1970) contend that the word is an anglicized corruption of an Indian word that was sometimes written down by white settlers as *atchitamon* or *chetamon,* from which "chipmunk" developed, by way of "chitmunk." In early days the animal was often called chitmunk. The word *atchitamon* is said to mean headfirst, referring to the manner in which squirrels often come down trees. The generic name *Tamias* is the Greek word for steward or "one who stores and looks after provisions," an allusion to the use by the chippie of its cheek pouches in collecting and carrying food. The second part of the scientific name, *striatus,* is the Latin for striped. The subspecific name of the chipmunk of Louisiana, *pipilans,* is of Latin derivation and refers to the piping or chirping notes made by the little animal.

Distribution.—The Eastern Chipmunk ranges from the shores of Lake Winnipeg, across southern Canada to southern Labrador, south through the eastern half of the United States to eastern Oklahoma and south-central Arkansas, central Mississippi, the northwestern corner of the Florida Parishes of

Louisiana, southern Alabama and adjacent parts of central-western Florida (Stevenson, 1962), central Georgia, extreme western South Carolina, central North Carolina, and southeastern Virginia. It enters Louisiana from Mississippi only in the region of the Tunica Hills in East and West Feliciana parishes and immediately adjacent areas. I have seen it in, and examined specimens from, northeastern East Baton Rouge Parish as far south as a wooded area just north of the brick factory immediately above the industrial complex on the north side of Baton Rouge, and it occurs in extreme western East Feliciana Parish. Otherwise, it is present only in West Feliciana Parish. I strongly suspect that the species once ranged southward along Highland Road on the south side of the present campus of Louisiana State University. With the growth of Baton Rouge and the great industrial developments on the north side of that city, this possible southern extension of the species' range has now been cut off from a source of population replenishment. Hence the chipmunk is now absent in the suitable habitat on the south side of the city. (Map 26)

External appearances. — This small squirrel, in which the total length is 10 to 11 inches and the tail is only 3.6 to 4.5 inches long, is extremely attractive in appearance (Plate IV and Figure 81). The head and flanks are

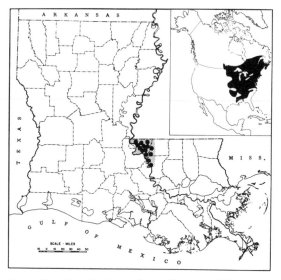

Map 26 Distribution of the Eastern Chipmunk *(Tamias striatus). Shaded area,* known or presumed range within the state; •, localities from which museum specimens have been examined; *inset,* overall range of the species.

strongly tinged with Ferruginous or Orange-Rufous, which changes to a rich Russet or Mahogany Red on the rump. On its back it has five dark stripes. The central one begins on the nape as an extension of the Vinaceous-Rufous of the crown but on the shoulders merges into Liver Brown or Black that in turn fades into the Russet of the rump. The central dark stripe is bordered on each side by grizzled grayish brown that likewise merges into the rich Russet or Mahogany Red of the rump. The other four dark stripes are in pairs on each side of the body, from the shoulder to the hip, separated by a band of Pale Yellow-Orange to Light Orange-Yellow. The flattened tail, viewed from above, is fuscous with some brown showing through from below and with the tips of the hairs gray. The tail, when viewed from the underside, is Ferruginous bordered on the edges with fuscous and then gray. The root of the tail, the thighs, and the hind feet are between Tawny and Ferruginous. The underparts are mostly pale Cream Color but often a Light Ochraceous-Salmon posteriorly. On the

Figure 81 Eastern Chipmunk *(Tamias striatus),* Hamilton County, Ohio, October 1960. (Photograph by Woodrow Goodpaster)

cheeks a Russet line passes through the eye, and an Amber Brown line lies beneath the eye, separated from each other by Ferruginous or Orange-Rufous. Two rather spacious cheek pouches open between the lips and the molars and extend backward along the cheeks and sides of the neck beneath the outer skin. The inner surfaces of the pouches are naked of hair.

Color phases.—No color phases as such are known in this species, but albinos are of rare occurrence.

Measurements and weights.—Twenty-one skins and 12 skulls from Louisiana in the LSUMZ averaged as follows: total length, 255 (228–285); tail, 94.5 (82–115); hind foot, 34.7 (31–39); ear, 19.3 (15–22); greatest length of skull, 43.8 (41.2–49.6); cranial breadth, 18.0 (17.0–20.8); zygomatic breadth, 23.7 (21.9–26.4); interorbital breadth, 11.7 (10.9–12.8); length of nasals 14.7 (13.4–16.1); palatilar length, 20.2 (18.9–24.1); postpalatal length, 14.8 (14.1–15.9); length of maxillary toothrow, 6.8 (6.3–8.6). The weights of three females and one male were, respectively, 117.0, 120.0, 165.2, and 121.0 grams, or approximately one-quarter to slightly less than one-third of a pound.

Cranial and other skeletal characters.—The skull of *Tamias striatus* is narrow but relatively broad interorbitally; the rostrum is moderately elongate, attenuated, and slightly depressed anteriorly; the infraorbital canal is large, both relatively and actually larger than in other Louisiana squirrels; only four cheek teeth are present on each side in the upper jaw. The os penis is slender, 4.5–5 mm in length, and nearly straight but upturned at the tip and slightly expanded distally into a narrow scoop with a median ridge on the the undersurface (Howell, 1929). (Figure 82)

Sexual characters and age criteria.—No significant differences between the sexes are evident in size or color. Half-grown young are like the adults in external markings. Even in newborn babies, which are hairless, pigmented lines are discernible by the sixth day, indicating the position of the future blackish hairs that will form the longitudinal stripes. Hair begins to appear by the 12th day, but not until the little squirrels are 30 days old do the eyes open (Seton, 1929; Allen, 1938). Weaning is said to take place after about five weeks. The mammae are eight in number, two being pectoral, four abdominal, and two inguinal.

Status and habits.—Unfortunately, this delightful little squirrel is found in only one limited section of the state embracing West Feliciana

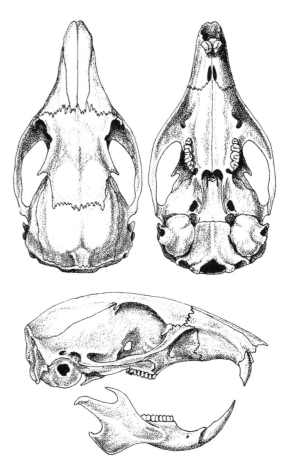

Figure 82 Skull of Eastern Chipmunk *(Tamias striatus pipilans)*, ♂, 1 mi. W St. Francisville, West Feliciana Parish, LSUMZ 16177. Approx 1½ ×.

Parish, the extreme western portion of East Feliciana Parish, and the northwestern part of East Baton Rouge Parish. There it is common, especially in the Tunica Hills north and northwest of St. Francisville. It frequents banks of streams and gullies clad with beech, magnolia, and maidenhair fern in this botanically distinctive forest. This was the setting in which it attracted the attention of John James Audubon over a hundred years ago, and even before Audubon and since, it has provided amusement and entertainment to the residents of this part of Louisiana. I regret that it is not widespread in our state, for it is indeed an attractive little animal to have around.

The chippy is not a gregarious animal. Aside from its immediate family it does not have much to do with others of its kind. Even the female chases off her young when they are no more than eight weeks old. Each individual jealously hoards its own cache of food and otherwise goes about its own business except during the mating season when the males vigorously pursue the females in rollicking chases. Despite its general lack of sociability with its own kind, the chipmunk sometimes displays tolerance of man while accepting his handouts of nuts and grain. But it shows little inclination to become tame except when taken in the nest and raised in captivity. The species is always alert and active and gives one the impression that it is exceptionally bright, for which, however, there is probably no tangible supporting evidence.

Tracks of Eastern Chipmunk

The vernacular names chippy, chipping squirrel, and piping squirrel all stem from the vocalizations rendered by the species. The most common notes or songs emitted are: a *chuck, chuck, chuck* or *chock, chock, chock* that is uttered slowly, possibly as a mild warning to its associates that danger may be near; a slightly higher-pitched *kuk, kuk, kuk* or *check, check, check* often continued for several minutes, which seems to be a kind of song that is uttered when its domain is quiet and serene; and a high-pitched musical whistle that is employed when the animal is aroused and running from danger toward its burrow with its tail held upright and nervously flipping (Seton, 1929; Jackson, 1961).

The burrows are often complex and extensive. Sometimes in excavating the tunnels, the chipmunk brings dirt out at one place but finally plugs the opening and makes a new opening from the inside some distance away (Figure 83). Consequently, the door to the tunnel is not always disclosed by a pile of excavated dirt. It is often not seen unless one happens to observe a chippy entering the retreat. More often than not the opening is under the edge of a root or by a fallen log. The nest itself is an enlarged chamber, often 10 inches or more in diameter, usually containing dried leaves but rarely grasses as well. Special chambers are provided for the storage of nuts, woody plants, seeds of weeds and grasses, and grains such as corn and oats. According to Thomas (1973), the foods most commonly stored by chipmunks in Louisiana are acorns and hickory nuts. The food of the Eastern Chipmunk also includes insects, particularly cicadas and grasshoppers, and it is known to eat birds' eggs, young mice, frogs, salamanders, and small snakes, as well as fruits and berries and occasionally mushrooms.

Food for storage is carried to the underground chamber in the cheek pouches, which the little squirrel sometimes crams so full that its face bulges on each side as shown in Figure 84. In addition to the food stashed away safely in a burrow, the chippy also caches food in shallow excavations on the surface of the forest floor in a manner similar to workings of Gray and Fox Squirrels.

Chipmunks in the north become dormant in winter during periods of severe, inclement weather, but they do not hibernate as do ground squirrels and Woodchucks. The state of torpidity in chipmunks seldom extends for a period in excess of 24 hours, and when subjected to low temperatures under laboratory conditions, the chipmunk actually steps

Figure 84 "I can't believe I ate the w-h-o-l-e thing." Eastern Chipmunk with its cheek pouches stuffed with food. Drawn from an actual photo.

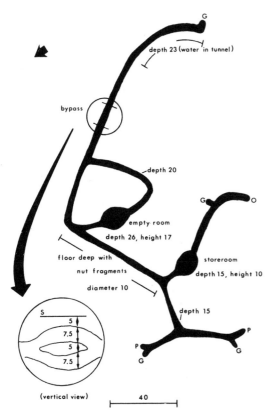

Figure 83 Top view of an extensive burrow system in West Feliciana Parish. Abbreviations are: O, open entrance to burrow; P, plugged entrance; G, gallery; S, surface of ground. All measurements are given in centimeters (2.54 cm = 1 inch). Small arrow points north. Tunnel width is not in proportion but is generally 4 to 8 cm. Depth is from the surface of the ground to the floor of the burrow. Height of rooms is from the floor to the highest point of the ceiling. (After Thomas, 1973)

up its activity. At our latitude the periods of inactivity in winter are of much shorter duration than they are farther north. Specimens in the LSUMZ were obtained throughout November and December and in the first part of January. I am certain that the species can be found abroad on any of the warm, balmy days in winter. By the end of February the chippy is extremely active and in rut. Like its relatives the tree squirrels, it has two seasonal peaks in breeding, one in early spring and the other in midsummer. The gestation period is approximately 31 days. Two to seven young, usually four or five, are born in each litter. During periods of intense heat in late summer, the Eastern Chipmunk has been said to aestivate, that is, enter into a period of summer dormancy. But Mrs. Thomas, in her year-long study of Louisiana chipmunks, concluded that only a lull in activity occurs in late summer (see also Dunford, 1972) and that at this time a few chipmunks are always abroad.

Predators. — One of the worst enemies of chipmunks anywhere is the Long-tailed Weasel, which has the ability to follow the chippy wherever it might go. Fortunately, though, weasels are rare in the region where chipmunks are found in Louisiana and are probably not a serious threat to them. Red-shouldered, Broad-winged, and Cooper's Hawks are, however, all fairly common in the Tunica Hills and therefore doubtless take a considerable toll of chipmunks, as do also foxes, Domestic Cats, and snakes of one sort or another.

Parasites.—No studies have been conducted of the parasites of the Eastern Chipmunk in Louisiana, but elsewhere it has been found to harbor the louse *Hoplopleura erratica* and five species of fleas. Internal parasites include the trematode *Scaphiostomum pancreaticum,* the cestodes *Hymenolepis diminuta* and *Taenia crassiceps,* eight species of nematodes, and two acanthocephalans, *Moniliformis clarki* and *Macrocanthorhynchus hirundinaceus* (Doran, 1955a; Freeman, 1954; Jackson, 1961).

Subspecies

Tamias striatus pipilans Lowery

Tamias striatus pipilans Lowery, Occas. Papers Mus. Zool., Louisiana State Univ., 13, 1943:235; type locality, 5 mi. S [= SE] Tunica, West Feliciana Parish, Louisiana.

This subspecies occurs in the region of Louisiana outlined above thence northward and eastward through central Mississippi to central northern Alabama. It is the largest and most richly colored of all the races of *Tamias striatus.* From its closest ally, *T. s. striatus* of southern Illinois, Kentucky, Tennessee, and extreme western Virginia, central South Carolina, western South Carolina, and northern Georgia, it differs by way of its larger size and the much lighter coloration of its upperparts. The light coloration of the cheeks, sides, and flanks of *pipilans* immediately separates it from *striatus,* in which these parts are Russet, Bay, or Auburn. The top of the head and the ears are paler than in *striatus,* as are also the hind feet and tail. The underparts of *striatus* are usually creamy white, whereas in *pipilans* these parts are strongly tinged, especially posteriorly, with Light Ochraceous-Salmon.

Specimens examined from Louisiana.—Total 64, as follows: *East Baton Rouge Par.:* 2 mi. W Baton Rouge city limits on Miss. River, 1 (LSUMZ); 3.5 mi. S Port Hudson, 1 (LSUMZ). *East Feliciana Par.:* 27 mi. N Baton Rouge, 1 (LSUMZ); 3 mi. NW Port Hudson, 1 (LSUMZ). *West Feliciana Par.:* Cornor, 12 (LSUMZ); Bains, 9 (2 LSUMZ; 7 TU); Tunica, 2 (LSUMZ); 5 mi. S [= SE] Tunica, 11 (LSUMZ); 9 mi. NW St. Francisville, 1 (LSUMZ); 6 mi. N St. Francisville, 2 (LSUMZ); 1 mi. W Laurel, 1 (LSUMZ); 5 mi. NW St. Francisville, 1 (LSUMZ); 10 mi. NE St. Francisville, 1 (LSUMZ); 38 mi. NNW Baton Rouge, 1 (LSUMZ); St. Francisville, 4 (3 LSUMZ; 1 LTU); 1 mi. E St. Francisville, 1 (LSUMZ); 2 mi. E St. Francisville, 1 (LSUMZ); 3 mi. S, 2 mi. E St. Francisville, 1 (LSUMZ); 5.5 mi. ENE St. Francisville, 1 (LSUMZ); 5 mi. ENE St. Francisville, 1 (LSUMZ); 2 mi. W St. Francisville, 1 (LSUMZ); 1 mi. W St. Francisville, 1 (LSUMZ); 5.6 mi. ENE St. Francisville, 1 (LSUMZ); Little Bayou Sara, La. Hwy. 66, 1 (NLU); 5 mi. NW Bains on La. Hwy. 66, 1 (USL); 3 mi. NE Tunica on La. Hwy. 66, 1 (USL); Plettenberg, 1 (USL); Hollywood, 3 (LTU).

Genus *Glaucomys* Thomas

Dental formula:

$$I\frac{1-1}{1-1} \ C\frac{0-0}{0-0} \ P\frac{2-2}{1-1} \ M\frac{3-3}{3-3} = 22$$

SOUTHERN FLYING SQUIRREL   Plate IV
(*Glaucomys volans*)

Vernacular and other names.—In Louisiana I have heard no other name than flying squirrel applied to this species. Even among our French-speaking inhabitants in the southern part of the state it is referred to as *l'éscureuil volant.* The modifier "Southern" is necessary to differentiate this species from the large, northern *G. sabrinus.* The generic name *Glaucomys* is derived from the Greek words *glaukos,* meaning gray, and *mys* for mouse, hence gray mouse. The name *volans* is the Latin word for flying. The subspecific names of the two races that occur in Louisiana, *texensis* and *saturatus,* mean, respectively, of Texas and saturated, the latter referring to the dark color of the population in question.

Distribution.—The species occurs from southern Maine, southern Ontario, the Upper Peninsula of Michigan, and central Minnesota south to southern Florida, the northern Gulf Coast, and eastern Texas. Disjunct populations are found in Mexico south to Nicaragua. The species is statewide in Louisiana in forested areas except for the wooded chenieres. (Map 27)

External appearance. — The flying squirrel is characterized by the fold of skin, the so-called flying membrane, that extends along the sides of the body from the wrists of the forelegs to the ankles of the hind legs (Figure 85). The body is rather flat and the tail is especially so. The fur is dense, extremely soft, and plain colored, that is, without spots or stripes. The eyes are black and exceptionally large and the lids are rimmed with black. The total length is from about 8.5 to 10 inches, of which approximately 3.5 to 4 inches is the flat, softly furred tail. The upperparts are Drab Snuff Brown as a result of the color of the tips of the hairs, which are dark gray or slate colored basally. The hair of the sides and of the folds of skin making up the flying membranes lacks for the most part any tipping of brown and hence is darker than the color of the upperparts. The tail from above is similar in color to the back but paler. The face and sides of the neck are grayish, but the underparts from the chin to belly, including the undersides of the legs, are creamy white, sometimes strongly tinged with Pinkish Cinnamon, especially on the bottom of the tail.

Color phases. — Although an occasional individual is abnormally gray, no true color phases are evident in this species. One such gray color mutant is a female in the LSUMZ from Bryceland, in Bienville Parish. All-white individuals are on record, but I do not know if they represented true albinos (that is, ones with pink eyes) or only so-called "black-eyed whites," which are genetically not the same thing at all.

Measurements and weights. — See diagnoses of subspecies at the end of this account.

Cranial and other skeletal characters. — The skull of *Glaucomys* cannot be confused with that of any other native mammal except possibly *Tamias striatus*, from which it differs in possessing five cheek teeth and being decidedly smaller with the rostrum only one-third as long and more nearly parallel-sided. The infraorbital canal is relatively much smaller but the braincase is relatively larger. The membrane extending along the side of the animal from the wrist to the ankle is supported by a cartilaginous extension of the pisiform bone of the wrist. The penis bone is highly specialized and distinctive. It is long and slender, with the tip slightly expanded into a knob, which, when viewed from the end is seen to be bilobed. A narrow and shallow furrow extends down the full length of the shaft and onto the more proximal lobe of the knobbed end (Wade and Gilbert, 1940). The total length and greatest breadth of two bacula in the LSUMZ measured, respectively, 11.5 × 1.0 and 11.6 × 1.0 mm. (Figures 72 and 86)

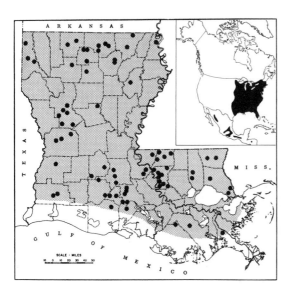

Map 27 Distribution of the Southern Flying Squirrel *(Glaucomys volans). Shaded area,* known or presumed range within the state; •, localities from which museum specimens have been examined; *inset,* overall range of the species.

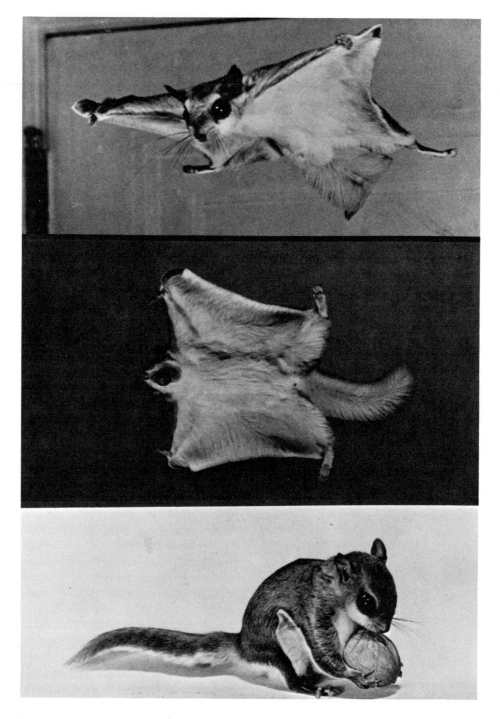

Figure 85   Southern Flying Squirrel *(Glaucomys volans)*. (Photographs by Ernest P. Walker, courtesy of John L. Paradiso and The Johns Hopkins Press)

Sexual characters and age criteria. — Superficially the sexes are identical. By the time of weaning, at about six weeks of age, the young are like the adults except for their smaller size. According to Sollberger (1943), the weight in grams at birth averages 3.63; at the end of one week, 7.3; at two weeks, 13.2; at three weeks, 19.2; at four weeks, 26.2; at six weeks, 36.8; and at eight weeks, 44.5.

Status and habits. — The Southern Flying Squirrel is common throughout most of the state's forested areas. Because of its nocturnal habits, it is seldom seen but is nevertheless reasonably well known, especially to schoolboys and woodsmen in rural areas. Many a youngster has smuggled his pet flying squirrel to school in the pocket of his jacket, allowing the little animal to sleep there docilely through the day.

Anyone who is familiar with the chippering vocalization of a flying squirrel is likely to discover its presence in numbers close at hand, such as in the trees of residential sections of our towns and cities. Not infrequently it enters attics, where at night we hear it scampering about. One of the easiest ways to locate the species is to rap on the trunks of trees containing cavities such as woodpecker holes and to watch for a head to peek out in search of the cause for the disturbance. If a person is stationed at the base of each nearby tree while the trunk containing the family of squirrels is pounded heavily, the little animals will sometimes volplane out of the hole toward the base of one of the surrounding trees and can often be caught in the hands like a ball by the waiting fielder.

Although we call *Glaucomys volans* a flying squirrel, it does not fly. It simply climbs to a high perch, leaps out into space with its feet spread, thereby extending the folds of skin along the sides of its body, and glides in a descending arc toward the base of another tree (Figure 85). Just before alighting it swerves upward and lands gently on the new trunk. If it wishes to continue its journey, it climbs again to the top of the tree and launches forth on another glide. I have always been intrigued by the slowness with which it seems to float through the air in one of these "flights." Perhaps the flat tail serves as a rudder or as a sort of steering device, for an abrupt midcourse alteration in direction can be effected. The distance covered in a glide may be of any length from a few feet to 30 yards or more.

The large eyes are indicative of its nocturnal habits. They glow red in the beam of a headlight. As soon as night falls, the little squirrels emerge from their daytime retreats and begin to frolic and engage in all sorts of playful antics. Flying squirrels are much more gregarious than their relatives the tree squirrels. Not only do as many as seven or more sometimes roost together in a single cavity, but aggregations of them often as-

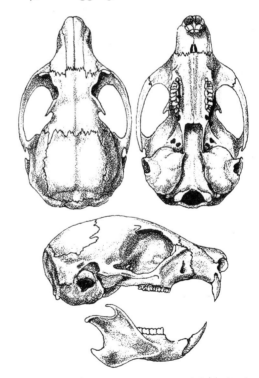

Figure 86 Skull of Southern Flying Squirrel (*Glaucomys volans saturatus*), ♂, 18 mi. NW Baton Rouge, Pointe Coupee Parish, LSUMZ 16022. Approx. 1½×.

semble to partake in their nightly gambols. I once watched a group in the moonlight outside a bedroom window of my home as they appeared to chase each other through a wide vertical and circular pathway between two large trees. The chase seemed to continue interminably but without exhaustion overtaking any of the participants.

Although the nests of flying squirrels are most frequently placed in tree cavities, occasionally they are located in an outside situation, where, like those of the tree squirrels, they are composed of leaves and small twigs with the leaves attached. A cavity nest and the lining of an outside nest are made of a dense mass of shredded bark. No one, to my knowledge, has yet observed a flying squirrel in the process of constructing one of the outside nests, but the fact that it takes over the old nests of Gray Squirrels and even old bird nests is unquestioned. I was once shown a Purple Martin house that at the time contained a half-dozen flying squirrels, as well as several bird occupants, one of which was a flicker.

The calls of our flying squirrel include a chucking note not too different from that made by its larger relatives, the tree squirrels. But the distinctive note, which may be likened to a song, is a clear, musical chirp that is birdlike in quality and often repeated. Although it is heard throughout the year, I have the impression that at certain seasons the little animal is much more vociferous than at others. Unfortunately, my field notes are rather vague on this point.

Mating apparently takes place twice a year with litters being produced in spring and again in early fall. Proof of fall litters in Louisiana is provided by Svihla (1930b), who obtained a female at Slidell on 4 October that gave birth to four young four days later, and by Goertz (1965b), who reported the birth of two young at Ruston on a date between 5 and 10 October.

The gestation period is 39 to 40 days in duration and the number of young in a litter varies from two to four. At birth the babies are naked except for having both lips sparsely mystacial. The color is pinkish and the patagium, or flight membrane, is clearly evident between the wrists and ankles. The ears are closed and the pinnae are folded over. The scrotum is clearly evident in the newborn males. By the end of two weeks the baby is clothed in hair, that of the underparts being whitish and that of the dorsum brown, as in the adults. The eyes open on about the 28th day, the ears somewhat earlier. At the end of the sixth week the young are able to take care of themselves to a large degree, and at this time the female ceases to nurse them but permits them to remain in her presence.

As pointed out by Sollberger (1943), estrus in the flying squirrel can be easily recognized by the changes that take place in the external genitalia. For most of the year the vagina is closed by a membrane. Only a faint line marks the suture that will open when the animal is in heat. As estrus approaches, the lips of the vulva swell to approximately five times their normal size and become pinkish, instead of flaccid and flesh colored. Within eight days after mating, the enlarged condition of the vulva gradually disappears. In males the testes are descended or partially descended throughout most of the year, the exception being in late fall and early winter. In captivity the male has been observed to respond quickly to the introduction of a female in heat by rapidly vibrating his hindquarters up and down and kicking backward with his feet. It then approaches the female but is not always immediately accepted. Generally, though, when left together overnight, they mate.

Food of the Southern Flying Squirrel consists mainly of vegetable substances—especially seeds, fruit, berries, acorns, nuts, corn, and grain—but it also has a strong predilection for meat. The late Herbert L. Stoddard, who was considered by many to be without a peer among the Gulf Coast naturalists of his day, wrote (1920) the following:

On April 6, 1914, an adult female flying squirrel (*Glaucomys volans*) was captured with her two young and placed in a roomy cage in the workshop with a section of tree trunk containing a flicker's hole as a nest. Two or three days later a fine male yellow-bellied sapsucker was captured unhurt, and placed in the same cage where he made himself at home on the stump. I was greatly surprised the next morning to find his bones on the bottom of the cage, picked clean. This strong, hardy woodpecker in perfect health had been killed and eaten during the few hours of darkness, by the old mother flying squirrel, though she had other food in abundance. While pondering the tragedy visions of the many holes in the woods that had been found containing feathers and other remains of small birds came to mind, and I wondered if the beautiful and apparently inoffensive flying squirrels were responsible.

The carnivorous inclinations of some flying squirrels are also documented by Audubon and Bachman (1851–1854), who tell of setting traps for ermine baited only with meat and catching flying squirrels instead.

Nevertheless, despite any acts, which to some people might appear villainous, our flying squirrel remains as one of the gentlest of all rodents. It can be handled with ease and without much fear of its biting, although it may occasionally do so. In many respects it makes a delightful pet except for its proclivity to sleep all day and to come out to play only at night. But a young boy who has never had the experience of keeping a pet flying squirrel has missed something in life.

Predators. — Doubtless because of its nocturnal and arboreal habits, this little mammal enjoys partial immunity from predation. The Domestic Cat is perhaps its most serious intentional enemy, and owls are known to catch a few. Man probably does it the most harm without realizing he is doing so by cutting down and clearing away dead trees with their suitable nesting cavities. And, of course, forest fires take their toll, not only of flying squirrels but of most denizens of the woods.

Parasites. — External parasites known to infest *Glaucomys volans* include three lice (*Neo-haematopinus sciuropteri*, *Hoplopleura trispinosa*, *Enderleinellus replicatus*); some eight or ten kinds of fleas, including *Orchopeas howardii*, *Conorhinopsylla stanfordi*, and *Leptopsylla segnis*; three arachnids (*Haemolaelaps megaventralis*, *Euhaemogamasus ambulans*, *Trombicula microti*). Among the internal parasites are three protozoans (*Eimeria glaucomydis*, *E. sciurorum*, and *Trypanosoma denysi*) and four roundworms (*Capillaria americana*, *Citellinema bifurcatum*, *Enterobius sciuri*, and *Syphacia thompsoni*), and the acanthocephalan *Moniliformis clarki* (Doran, 1954a; Jackson, 1961).

Subspecies

Glaucomys volans saturatus Howell

Glaucomys volans saturatus A. H. Howell, Proc. Biol. Soc. Washington, 28, 1915:110; type locality, Dothan, Houston County, Alabama.

This subspecies of *Glaucomys volans* occurs in the southeastern United States (excepting peninsular Florida) from South Carolina and western North Carolina west across parts of Tennessee and Georgia, Alabama, and Mississippi to the eastern two-thirds of Louisiana and the extreme eastern part of Oklahoma. It differs from *volans* to the north in having darker upperparts and no appreciable amount of white on the toes in winter. The skull is similar to that of *volans* but smaller. It is practically the same size as that of the race *querceti* of the Florida peninsula, but the bullae average smaller. From *texensis* of eastern Texas, *saturatus* is said to differ by having a skull that averages slightly longer and broader and by being somewhat darker in color. The race *texensis* may not be worthy of recognition.

Measurements and weights. — Twenty-four specimens in the LSUMZ from numerous localities in the state (exclusive of parishes in the western part) averaged as follows: total length, 232 (210–251); tail, 100 (84–117); hind foot, 27.5 (22–31); greatest length of skull, 34.1 (32.7–35.9); mastoid breadth, 18.1 (17.3–19.4); zygomatic breadth, 20.5

(19.4–21.5); interorbital breadth, 7.0 (6.3–8.3); length of nasals, 9.7 (8.6–11.6); palatilar length, 15.1 (14.6–15.5); postpalatal length, 11.8 (10.8–12.7); alveolar length, 6.1 (6.0–7.0). The weights of three adults averaged 63.0 (62.6–63.5) grams.

Specimens examined from Louisiana.—Total 103, as follows: *Acadia Par.*: Eunice, 1 (MSU); 3 mi. S Rayne, 1 (USL). *Bienville Par.*: Bryceland, 4 (LSUMZ). *Claiborne Par.*: 3 mi. S Summerfield, 1 (LSUMZ); 5 mi. N Homer, 1 (NLU). *East Baton Rouge Par.*: various localities, 22 (20 LSUMZ; 2 USL). *East Carroll Par.*: Lake Providence, 1 (LTU). *East Feliciana Par.*: Idlewild Exp. Sta., 1 (LSUMZ); 2.5 mi. SW Jackson, 1 (LSUMZ); 8 mi. N Jackson, 2 (LSUMZ); 4 mi. E, 1 mi. N Woodland, 1 (LSUMZ). *Evangeline Par.*: Ville Platte, 2 (USL); 4 mi. N Basile, 1 (USL). *Franklin Par.*: Winnsboro, 1 (NLU). *Grant Par.*: Georgetown, 1 (LTU). *Iberia Par.*: New Iberia, 3 (USL). *Jackson Par.*: 1 mi. E Ansley, 1 (LTU). *Lafayette Par.*: 2 mi. W Lafayette courthouse, 1 (USL); 6 mi. W Lafayette, 1 (USL); 1.3 mi. S Lafayette courthouse, 1 (USL); 2 mi. S Lafayette courthouse, 2 (USL). *Lafourche Par.*: 5 mi. NE Mathews, 1 (LSUMZ). *Lincoln Par.*: Hico, 1 (LSUMZ); 1 mi. S Ruston, 1 (LTU); 15 mi. N Ruston, 1 (LTU). *Morehouse Par.*: 2 mi. N Jones, 1 (NLU); Mer Rouge, 3 (USNM). *Ouachita Par.*: Monroe, 6 (1 LSUMZ; 5 NLU); 4.8 mi. N Monroe, 1 (NLU). *Plaquemines Par.*: Lake Hermitage, 1 (TU). *Pointe Coupee Par.*: Grand Swamp, lower end of False River, 1 (LSUMZ); 18 mi. NW Baton Rouge, 1 (LSUMZ). *St. Landry Par.*: Opelousas, 2 (1 LSUMZ; 1 USL); 5 mi. W Arnaudville, 1 (USL); 6 mi. S Opelousas courthouse, Opelousas, 4 (USL); 3 mi. NW Washington, 1 (USL). *St. Martin Par.*: 1 mi. S Parks, 1 (LSUMZ); 8 mi. SW St. Martinville, 1 (USL); Nina, 1 (USL); 4 mi. E Breaux Bridge, 2 (USL); 4 mi. S St. Martinville, 1 (USL). *St. Tammany Par.*: Covington, 2 (LSUMZ); 0.3 mi. S, 0.8 mi. E Pearl River, 1 (LSUMZ); 0.2 mi. S, 0.6 mi. E Pearl River, 1 (LSUMZ); Porters Island, 1 (LSUMZ); locality unspecified, 1 (SLU); Slidell, 3 (USNM). *Tangipahoa Par.*: 3.5 mi. E Tickfaw, 1 (LSUMZ); 2.5 mi. SW Hammond, 1 (LSUMZ); Hammond, 1 (SLU). *Terrebonne Par.*: 0.5 mi. E Schriever, 1 (LSUMZ). *Union Par.*: 15 mi. E Farmerville, 1 (LTU); 3 mi. S Linville, 1 (LTU); 8 mi. S Farmerville, 1 (LTU); Lake D'Arbonne, 1 (LTU). *Vermilion Par.*: 3 mi. N Abbeville, 1 (USL). *Washington Par.*: 12 mi. E Franklinton, 1 (LSUMZ); 1 mi. E Franklinton, 1 (LSUMZ). *West Feliciana Par.*: 6 mi. NW Jackson, 1 (LSUMZ).

Glaucomys volans texensis Howell

Glaucomys volans texensis A. H. Howell, Proc. Biol. Soc., Washington, 28, 1915:110; type locality, 7 mi. NE Sour Lake, Hardin County, Texas.

The subspecies *texensis* was originally described by Howell on the basis of material from eastern Texas and was alleged to differ from *saturatus* on the basis of its slightly paler coloration above and in having a shorter and broader skull. I have previously (1943) assigned specimens from Vernon and Natchitoches parishes to *texensis*, but I now entertain considerable doubts concerning the validity of the race. In my recent examination of nine specimens from eastern Texas (Cooke, Brazos, Trinity, and Newton counties) and additional material from the western parishes of Louisiana, I was unable to detect any skull differences between *texensis* and *saturatus*. Admittedly, colorwise this material does appear to average slightly paler, that is, lighter brown, than the series of *saturatus*. Because of this slight color difference I hesitate to place *texensis* in the synonymy of *saturatus* without first examining additional material.

Measurements.—Six specimens from Natchitoches, Vernon, Beauregard, and Calcasieu parishes measured as follows: total length, 227 (210–240); tail, 101.5 (85–110); hind foot, 27.0 (23–29); ear, 15.8 (15–19); greatest length of skull, 33.9 (33.2–35.1); mastoid breadth, 17.7 (16.5–18.7); zygomatic breadth, 20.4 (19.6–20.9); interorbital breadth, 7.1 (6.6–7.5); length of nasals, 10.0 (9.4–11.2); palatilar length, 15.3 (14.9–16.1); postpalatal length, 12.0 (11.5–12.7); alveolar length, 6.3 (6.1–6.8). Certain skull measurements of eight specimens from eastern Texas (Cooke, Brazos, Trinity, and Newton counties) were as follows: greatest length, 34.3 (33.3–35.6); mastoid breadth, 16.9 (16.0–17.8); zygomatic breadth, 20.5 (18.5–21.5).

Specimens examined from Louisiana.—Total 24, as follows: *Beauregard Par.*: 10 mi. S De Ridder, 2 (LSUMZ). *Caddo Par.*: Shreveport, 1 (NLU); 1 mi. N Vivian, 1 (NSU); 3 mi. S, 1.5 mi. W Blanchard, 1 (LSUMZ). *Calcasieu Par.*: Maplewood, 1 (LSUMZ); 7 mi. W Lake Charles, 0.5 mi. N Maplewood, 2 (LSUMZ); Westlake, 2 (MSU); Hollywood, 3 (MSU). *Natchitoches Par.*: Flora, 1 (LSUMZ); Vowells Mill, 1 (LSUMZ); Provencal, 2 (LSUMZ); Kisatchie, 1 (LSUMZ); 1.6 mi. NW Flatwoods along La. Hwy. 119, 1 (USL); Natchitoches, 1 (NSU). *Vernon Par.*: Simpson, 2 (LSUMZ); 3 mi. S Leesville, La. Hwy. 171, 1 (MSU); 6 mi. E Leesville, 1 (LSUMZ).

The Pocket Gopher Family
Geomyidae

THIS FAMILY is exclusively North American. It contains eight genera, of which only one, *Geomys*, occurs in Louisiana. Pocket gophers are among the most highly specialized of all rodents. Their structural modifications equip them admirably for their predominantly subterranean existence. Although ratlike in general appearance, they possess massive shoulders and arms, as well as heavily clawed forefeet that enable them to dig their underground burrows. Spacious, fur-lined cheek pouches lie wholly outside the mouth. The huge incisors are likewise mostly outside the mouth cavity. As would be expected in a rodent living in tunnels beneath the surface of the ground, the hair is extremely short, the eyes are tiny and beadlike, and the ears are inconspicuous with almost no pinnae. Compensating for their apparently poor senses of sight and hearing, pocket gophers possess exceptionally sensitive tactile organs. The nose and the short tail, which is virtually naked except for a few hairs at the base, are particularly sensitive to touch.

Genus *Geomys* Rafinesque

Dental formula:

$$I\frac{0-0}{0-0},\ C\frac{1-1}{1-1},\ P\frac{1-1}{1-1},\ M\frac{3-3}{3-3} = 20$$

PLAINS POCKET GOPHER   Plate V
(*Geomys bursarius*)

Vernacular and other names. — The name most frequently heard applied to this species in Louisiana is "salamander." But zoologically this name is a gross misnomer. A true salamander belongs, of course, to the Amphibia, a group of vertebrates that includes the newts and mud puppies. In the eastern part of the Florida Parishes one hears the name gopher but never applied to this animal. There it is used for the large, dry land turtle or tortoise *Gopherus polyphemus.* Such unfortunate and complicating variations in the vernacular names of animals, particularly from one part of the country to another, are one of the main reasons arguing in favor of our binominal system of Latin names, whereby every kind of animal from a lowly, one-celled protozoan to the highest mammal has a unique scientific name consisting of two Latin or Latinized words, the first to designate the genus and the second to designate the species. The name *Geomys bursarius* is applied to only one species, the Plains Pocket Gopher, and to nothing else. It would mean the same thing to any zoologist whether in Siberia or Tombouctou. The generic name *Geomys* is derived from two Greek words and means earth mouse (*ge* or *geo,* earth and *mys,* mouse), an allusion to the animal's burrowing habits. The second word of the scientific name, *bursarius,* is of Latin derivation and means possessing a pouch (*bursa,* pouch and *arius,* connected with), referring to the fur-lined cheek pouches. The word Plains in the common name denotes the range of the species, which is predominantly in the central plains region of the United States; Pocket refers to the cheek pouches; and Gopher is from the French word *gaufre,* meaning honeycomb, and has to do with the network of tunnels that the animal digs in the earth.

Distribution.—This species occurs, chiefly in the Great Plains, from western Wyoming and southern and eastern South Dakota north through eastern North Dakota to extreme southern Manitoba, central northern Minne-

sota, and western Wisconsin south, in an irregular pattern, to northeastern Indiana, southeastern Missouri, eastern Arkansas, north-central and western Louisiana, central-southern Texas, and eastern New Mexico. (Map 28)

External appearance.—This species is a medium-sized gopher, 8 to 10 inches in total length. For a general description of its anatomical features, see the discussion of the family Geomyidae at the beginning of this account. The genus *Geomys* is easily distinguished by the two longitudinal grooves on each of the upper incisors, which, along with the two lower incisors, lie outside the mouth (Plate V). Individuals of this species vary greatly in color from a light brown to nearly black. All Louisiana specimens that I have examined show at least some irregular spotting of white on the face, chin, throat, forelegs, belly, or, rarely, the back.

Color phases.—True albinos are rare, but melanistic individuals are of rather frequent occurrence in Louisiana populations. For instance, 3 of 17 specimens taken from the fairly isolated population at Mer Rouge, in Morehouse Parish, are strongly melanistic.

Measurements and weights.—See the measurements for the various subspecies at the end of this account.

Cranial and other skeletal characters.—The two longitudinal grooves on the front surface of the upper incisors of *Geomys bursarius* are unique among the rodents of Louisiana. The lateral groove is deeper and more prominent than the medial groove. The robust skull is flat and broad, the zygomatic arches being widespread with their anterior borders almost at right angles to their lateral edges, as well as to the rostrum. The frontal bones are without postorbital processes, the sides of the massive, truncate rostrum are nearly parallel, and the ear openings into the skull are in the

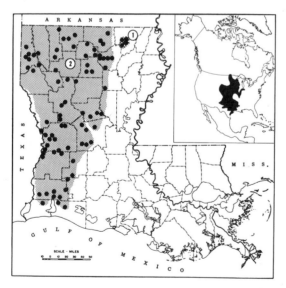

Map 28 Distribution of the Plains Pocket Gopher (*Geomys bursarius*). *Shaded areas,* known or presumed range within the state; •, localities from which museum specimens have been examined; *inset,* overall range of the species. 1. *G. b. breviceps;* 2. *G. b. dutcheri.*

form of elongate, tubular canals (Figure 87). According to Kennerly (1958), the os penis is simple with a bulbous base that tapers to a slightly knobbed distal end. It measures approximately 10 mm in length with a basal width of about 1.7 mm (Figure 72). The hip, or pelvic, girdle of a pocket gopher is extremely narrow, an important adaptation for turning in the narrow confines of the subterranean tunnel but a distinct disadvantage to the female when giving birth to her young. Indeed, in order for her to be able to do so, her pubic bones in the region of the symphysis begin to be resorbed at the time of her first pregnancy. By the time of parturition a large segment of the medial portion of the pubic bones has disappeared, thereby providing ample space for the passage of the embryos through the birth canal. The process of resorption of the pubic bones is said to be induced by ovarian hormones. Consequently, no comparable change takes place in the pelvic girdle of the male.

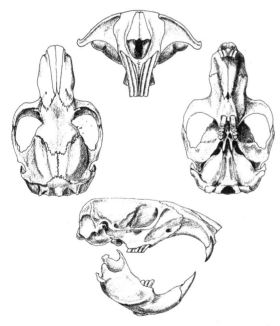

Figure 87 Skull of Plains Pocket Gopher *(Geomys bursarius dutcheri)*, ♀, Bossier City, Bossier Parish, LSUMZ 15487. Approx. natural size except uppermost figure, which is 2 ×.

Sexual characters and age criteria. — Pocket gophers exhibit marked sexual dimorphism in size, males being much the larger. Adult females never appear to attain the size of fully adult males. Immatures tend to have the hair denser, longer, and somewhat darker along the midline. Very young individuals are grayish.

Status and habits. — Except for a small isolated population a few miles west and southwest of Mer Rouge, in Morehouse Parish, the Plains Pocket Gopher is found nowhere in Louisiana east of the Ouachita and Little rivers. It is likewise absent from areas in the southern part of the state lying east of Allen and Jefferson Davis parishes, and even in these parishes it is confined to their western portions.

The critical factor governing the distribution of the species in the state is the availability of suitable soils. The animals can dig their subterranean tunnels only in sandy loam. Consequently, they are not found in river bottoms, where the soils are heavy and the groundwater level relatively high. Wide barriers of such unfavorable terrain often prevent their expansion into areas where soils meeting their requirements are found. For example, the soils of the northern parts of our Florida Parishes appear perfectly suitable for pocket gophers. But the species does not occur there, nor, as a matter of fact, anywhere in Mississippi nor in Alabama west of the Mobile and Tombigbee rivers. East of these rivers and their respective bottomlands pocket gophers of another species (*G. pinetis*) occur and are found all the way to the Savannah River (but not across it) and to the Atlantic Seaboard of Georgia and Florida.

Paradoxically, pocket gophers did succeed sometime in the past in crossing the bottomlands along the Ouachita River and Bayou Bartholomew to reach the isolated area of suitable soils in Morehouse Parish, but I have no explanation as to how they managed to do so.

In places where the species occurs in the

state it is generally common. The mounds of dirt that it brings to the surface from its underground excavations often dot the landscape. But, despite its prevalence in an area, few people ever have the opportunity to observe it because of its burrowing habits. It does not frequently leave the seclusion of its tunnels during the day except for brief moments when engaged in shoveling out dirt, and even then it is nervous and apprehensive, withdrawing quickly into its hole at the slightest sound or movement on the surface. When it finishes bringing up dirt to the mound, it tightly plugs the opening, thereby leaving no trace of the entrance. Much of its work is accomplished at night or in the periods of twilight in early morning and late evening. However, the frequency with which it is captured by owls, as revealed by studies of owl pellets, proves that it is sometimes abroad at night in search of food.

During most of the year a pocket gopher lives a solitary and highly territorial existence, with only one individual occupying a single system of tunnels. But with the approach of the breeding season, males will extend their tunnels in search of one occupied by a receptive female. While movement from one tunnel system to another is often accomplished by simply extending one labyrinth until it joins another, male pocket gophers come out on the surface exploring for openings to tunnel systems containing individuals of the opposite sex.

Although detailed studies are in progress on the breeding biology of pocket gophers in Louisiana, not much can now be said on the subject, pending the completion of these investigations. In eastern Colorado, Vaughan (1962), found that females of this species are in estrus in March and April; that pregnant females are found after 19 March, that most (72 percent) females taken in April have already bred, and that young are found from May to December. Dispersal of the young from the parental burrows begins in June, when some of these young are only one-third

grown. The infrequency of plural captures of young in a single burrow system suggests a high predispersal mortality among young. A strong intolerance exists between immature individuals and adults and between the immatures themselves. Many of the young show minor injuries that are perhaps sustained in intrafamilial fights. These conflicts may be responsible for the early dispersal of the young.

Little information is presently available on litter size in pocket gophers in Louisiana, but in a study of this species at nearby College Station, Texas, English (1932) found the number of young in a litter to average 2.36. According to Wood (1955), the newborn

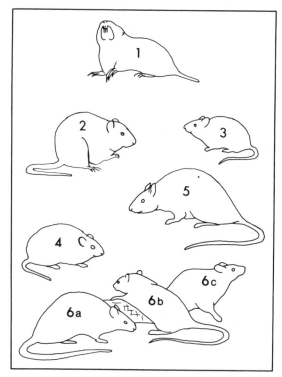

Plate V
1. Plains Pocket Gopher *(Geomys bursarius)*.
2. Eastern Wood Rat *(Neotoma floridana)*.
3. Marsh Rice Rat *(Oryzomys palustris)*.
4. Hispid Cotton Rat *(Sigmodon hispidus)*.
5. Norway Rat *(Rattus norvegicus)*.
6. Roof Rat *(Rattus rattus)*: **a,** *R. r. alexandrinus;* **b,** *R. r. rattus;* **c,** *R. r. frugivorus.*

young, which are less than two inches in length and which weigh between four and five grams, are naked except for the vibrissae and a few minute grayish hairs on the body. The skin is loose, wrinkled, and pink, and the eyes are closed. The cheek pouches are indicated by a line some four mm in length. By the time the young are as much as four inches in length and weigh nearly an ounce, the eyes are open and the cheek pouches are 10 mm in depth with a 12-mm opening. When about two months of age the young leave the parental burrow and wander off to establish burrows of their own.

The underground tunnels of our pocket gopher are four to eight inches below the surface, the usual depth being approximately six inches. A single system traced out by David L. Jones on 23 December 1967 near De Ridder, in Beauregard Parish, was, in the portion mapped, over 100 feet in total length. It contained 6 side chambers where excrement had been deposited and was associated with 11 fresh mounds of dirt on the surface. The various passageways converged into a central chamber that measured 11 inches long, 9 inches wide, and 8 inches deep. It was floored with a 1-inch layer of finely chewed plant material. (Figure 88)

The tunnels leading from the main horizontal passages to the surface are usually rather steeply sloped. In pushing or kicking the excavated earth out of its tunnel, a pocket gopher tends to pile more dirt laterally and to the rear, giving the mound a slight heart or fan shape, which provides a clue to the position and course of the descending tunnel. With this information anyone attempting to trap the animal knows where to position a shovel to remove a plug of earth to get down to the horizontal passageway. Special traps that operate in the narrow confines of the tunnel are usually required to catch a gopher. The proper procedure is to place two traps in the horizontal run, one on each side of the opening made by removing the plug of earth, since there is no way of knowing from

which direction the gopher will approach to repair the damage that it senses has occurred to its tunnel. Wiring the two traps together to a central stake is advisable, for it prevents the animal from pulling the trap in which it becomes ensnared deeper into its runway. (Figures 88 and 89)

Opening a tunnel to set traps perhaps creates drafts of air to which the gopher is averse and thereby brings about a fairly prompt response on the part of the animal. The trapper sometimes needs to wait only 25 to 30 minutes to make his catch. The late Vernon Bailey, one of our truly great mammalogists, once told me of a time when he was setting a gopher trap while a farmer was standing by looking on. Bailey said that just as he inserted the trap into one of the horizontal

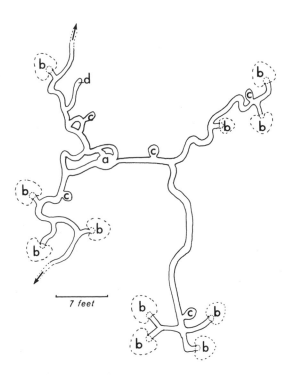

7 feet

Figure 88 Tunnel system of the Plains Pocket Gopher *(Geomys bursarius dutcheri)* excavated near De Ridder, Beauregard Parish, December 1967: (a) central chamber; (b) surface mounds; (c) excrement chambers; (d) blind tunnel ending near surface. (Based on D. L. Jones, manuscript)

Figure 89 Gopher traps set in a main horizontal runway.

runs the farmer asked, "How long does it take to catch one of them critters?" At that precise moment, Bailey felt the trap snap shut, so his reply to the question was, "You simply put it in and pull out the animal!" Obviously, the gopher was lying there in wait and immediately charged the foreign object being thrust into its passage. This incident must be a world's record for the time required for a catch to be made.

The pocket gopher feeds on a wide variety of roots, tubers, and other vegetable matter. It obtains most of these items by extending its tunnels, but its harvesting operations are sometimes carried out aboveground. Although it is alleged to be destructive to crops under some circumstances, I fail to find any justification for wholesale eradication programs that some agriculturists often advocate. In Louisiana, where the surface soil seldom becomes frozen for more than a few days at a time and where ample quantities of vegetable food are always available to the gopher, it does not need to stock its chambers with excessive quantities of reserve food to satisfy its appetite during extended periods of confinement, as do its northern relatives. Where indictments alleging damage are supported by evidence and proven to be true, I would certainly favor control measures, but not in uncultivated lands and scrubby areas where the little animal is doing no one any

harm and is instead performing a useful service to man. In excavating its burrow systems the animal loosens the subsoil and brings it to the surface, where it is exposed to weathering. This churning of the earth creates greater permeability and permits water and dissolved nutrients to percolate to greater depths.

Predators. — Because of its largely underground habits, our pocket gopher is not often captured by the usual predators of rodents except, as already noted, occasionally by owls. Weasels would be a menace, but they are rather rare animals in the areas of the state occupied by pocket gophers. Skunks doubtless capture a few, as do also, I am sure, various species of snakes such as the Common King Snake (*Lampropeltis getulus*), Prairie King Snake (*L. calligaster*), Racer (*Coluber constrictor*), Rat Snake (*Elaphe obsoleta*), Corn Snake (*E. guttata*), and the Pine Snake (*Pituophis melanoleucus*).

Parasites. — In eastern Texas, English (1932) found in 161 specimens examined that 119 were infested with the louse *Trichodectes geomydis*, 11 with the mite *Laelaps* sp., 23 with the roundworm *Protospirura ascaroidea*, and 8 with the flatworm *Hymenolepis* sp. In Kansas, Ubelaker and Downhower (1965) found the nematode *Capillaria hepatica* and two cestodes (*Aprostatandrya macrocephala* and *Paranoplocephala infrequens*), as well as two mites (*Hirstionyssus geomydis* and *Androlaelaps glasgowi*). Doran (1954a and b) lists a protozoan (*Eimeria geomydis*) and one additional cestode (*Ctenotaenia praecoquis*).

Subspecies

Davis (1940) recognized three races of the Plains Pocket Gopher in Louisiana, two of which he himself named: *G. b. breviceps* Baird, the isolated population in the vicinity of Mer Rouge, in Morehouse Parish; *G. b. dutcheri* Davis, in northwestern and central Louisiana, mainly east of the Red River, and in northeastern Texas, eastern Oklahoma, and southern and central-western Arkansas; and *G. b. pratincolus* Davis, of central-western and southwestern Louisiana and adjacent parts of eastern Texas. *G. b. breviceps* appears to be a fairly well-marked subspecies, and I have no doubts concerning its validity. But I can find little or no justification for recognizing more than one additional race in the state, and I therefore assign all other populations to the race *G. b. dutcheri,* a name that has page priority over *G. b. pratincolus.* The possibility exists that *dutcheri* should in turn be placed in the synonymy of *G. b. brazensis* Davis, but before this is done the matter would require careful reanalysis. According to Davis, *G. b. pratincolus* is distinguished by color and cranial characters, including "widely bowed zygomatic arches, short rostrum, and rich russet brown coloration, with nearly black dorsal stripe evident in most specimens." To be sure, some skulls of specimens from southwestern Louisiana have the zygomatic arches perceptibly more bowed than most specimens in population samples from northern Louisiana, but this may be only an age difference characteristic of old males. There appear to be absolutely no consistent color differences between the two series. Specimens from central-western and southwestern Louisiana, which Davis would assign to *pratincolus,* include samples that are indeed "rich russet brown," but so does also the series from Caddo, Bossier, Webster, Claiborne, Union, Bienville, Jackson, and Grant parishes, which Davis would refer to *dutcheri.* None of the alleged cranial differences other than the one pertaining to the degree to which the zygomatic arches are bowed, referred to above, appear to have any possible validity.

Geomys bursarius breviceps Baird

Geomys breviceps Baird, Proc. Acad. Nat. Sci. Philadelphia, 7, 1855:335; type locality, Prairie Mer Rouge, Morehouse Parish, Louisiana.

Geomys bursarius breviceps, Baker and Glass, Proc. Biol. Soc. Washington, 64, 1951:57.

Measurements and weights. — Skins of five males in the LSUMZ from Morehouse Parish averaged as follows: total length, 210 (177–225); tail, 54 (52–59); hind foot, 25 (23–29). Skulls of four males from the same source averaged as follows: greatest length, 41.2 (37.5–44.4); basilar length, 35.7 (31.5–38.3); mastoid breadth, 20.8 (19.0–22.4); zygomatic breadth, 25.4 (21.8–27.7); interorbital breadth, 6.7 (6.5–6.9); rostral breadth, 8.7 (8.0–9.5); length of nasals, 14.1 (11.8–16.1); alveolar length, 8.5 (7.5–9.3); length of diastema, 14.7 (12.3–15.7); width of foramen magnum, 5.8 (5.5–6.1); prepalatal length (for definition of this measurement, see Davis, 1940) 6.9 (5.5–7.1); width of auditory bulla, 5.8 (5.5–6.0); length of auditory bulla, 10.7 (9.9–10.9). Three males weighed 114, 174.4, and 212 grams. Skins of 11 females also from Morehouse Parish averaged as follows: total length, 203 (184–228); tail, 56 (53–64); hind foot, 24 (22–28). Skulls of 7 females averaged as follows: greatest length, 37.8 (34.0–41.2); basilar length, 32.7 (28.7–36.4); mastoid breadth, 19.8 (17.7–21.1); zygomatic breadth, 23.1 (19.8–25.4); interorbital breadth, 6.8 (6.6–7.0); rostral breadth, 8.4 (7.2–8.9); length of nasals, 12.3 (10.5–14.1); alveolar length, 7.9 (6.9–8.4); length of diastema, 13.1 (10.7–15.2); width of foramen magnum, 5.6 (5.4–5.9); prepalatal length, 6.4 (5.1–7.3); width of auditory bulla, 5.6 (4.9–5.9); length of auditory bulla, 10.3 (9.5–10.9). Weights of seven females averaged 101.7 (78.0–150.5) grams.

Specimens examined from Louisiana. — Total 114, as follows: *Morehouse Par.:* various localities in vicinity of Mer Rouge, 114 (19 LSUMZ; 7 USL; 45 USNM; 2 AMNH; 41 MCZ).

Geomys bursarius dutcheri Davis

Geomys breviceps dutcheri Davis, Bull. Texas Agri. Exp. Stat., 590, 1940:12; type locality, Fort Gibson, Muskogee County, Oklahoma.

Geomys bursarius dutcheri, Baker and Glass, Proc. Biol. Soc. Washington, 64, 1951:57.

Geomys bursarius pratincolus Davis, Bull. Texas Agri. Exp. Stat., 590, 1940:18; type locality, 2 miles east of Liberty, Liberty County, Texas.

212 ORDER RODENTIA

Measurements and weights. — Skins of 44 males from various localities in western and central Louisiana in the LSUMZ averaged as follows: total length, 210 (173–256); tail, 57 (46–69); hind foot, 25 (20–29). Skulls of 31 males from the same source averaged as follows: greatest length, 39.4 (34.4–44.9); basilar length, 33.4 (29.5–38.4); mastoid breadth, 19.7 (17.3–22.4); zygomatic breadth, 24.2 (20.9–29.7); interorbital breadth, 6.4 (6.1–6.8); rostral breadth, 8.4 (7.3–9.3); length of nasals, 13.7 (10.9–16.0); alveolar length, 8.2 (6.5–9.2); length of diastema, 13.8 (11.4–17.4); width of foramen magnum, 4.8 (4.6–5.6); prepalatal length, 6.6 (5.2–7.1); width of auditory bulla, 5.0 (4.6–5.9); length of auditory bulla, 10.1 (9.5–12.0). Weights of 6 males averaged 168.5 (103.5–201.0) grams. Skins of 41 females, also from western and central Louisiana and deposited in the LSUMZ, averaged as follows: total length, 239 (177–296); tail, 57 (45–67); hind foot, 24 (22–27). Skulls of 26 females from the same source averaged as follows: greatest length, 38.9 (36.0–41.0); basilar length, 31.5 (28.0–34.7); mastoid breadth, 19.3 (17.1–21.6); zygomatic breadth, 22.6 (19.8–25.0); interorbital breadth, 6.4 (5.8–7.0); rostral breadth, 8.0 (7.3–9.0); length of nasals, 12.5 (11.1–14.5); alveolar length, 7.8 (6.7–8.7); length of diastema, 13.0 (11.0–15.5); width of foramen magnum, 5.4 (4.9–5.9); prepalatal length, 5.6 (5.2–6.0); width of auditory bulla, 5.2 (4.7–5.5); length of auditory bulla, 10.9 (10.0–12.4). Weights of 17 females averaged 97.7 (74.0–143.0) grams.

Specimens examined from Louisiana. — Total 384, as follows: *Allen Par.:* 0.5 mi. NW Mittie on La. Hwy. 26, 1 (USL); 8 mi. SW Kinder, 1 (USL). *Beauregard Par.:* 2 mi. W De Ridder, 1 (LSUMZ); 1 mi. W De Ridder, 1 (LSUMZ); 5 mi. NE De Ridder, 5 (LSUMZ); Merryville, 2 (1 LSUMZ; 1 SLU); 1.5 mi. E De Ridder, 1 (LSUMZ); 7 mi. N Merryville, 1 (MSU); 8.9 mi. W Sugartown, 7 (USL). *Bienville Par.:* 5 mi. SW Ringgold, 2 (LSUMZ). *Bossier Par.:* Bossier City, 1 (LSUMZ); ca. 2 mi. E on levee near Red River, Bossier City, 1 (LSUMZ). *Caddo Par.:* Keithville, 4 (LSUMZ); 7 mi. WSW Shreveport, 5 (LSUMZ); 2 mi. N Blanchard, 1 (LSUMZ); 10 mi. S Shreveport on La. Hwy. 1, 4 (USL); Shreveport 16 (9 USNM; 7 LSUS); Oil City, 1 (LTU); 3 mi. S, 2.5 mi. W Blanchard, 2 (LSUMZ); LSU Shreveport, 4 (LSUS). *Calcasieu Par.:* various localities, 79 (18 LSUMZ; 48 MSU; 2 NLU; 9 USNM; 2 CAS). *Caldwell Par.:* 15 mi. SW Columbia, 1 (NLU). *Cameron Par.:* Cameron Farms, 12 mi. S Vinton, 2 (LSUMZ); State Hwy. 385, 1 mi. N Cameron–Calcasieu parish line, 4 (MSU); near edge of Sweet Lake, 17 mi. S, 5 mi. E Lake Charles, 1 (MSU); Gum Cove, Cameron Farms, 1 (USNM). *Claiborne Par.:* 1 mi. S Marsalis, 1 (LSUMZ); 1.5 mi. N Marsalis, 6 (LSUMZ); Homer, 1 (LSUMZ); 6 mi. S Summerfield, 2 (LSUMZ); Camp community, 3 (NLU). *De Soto Par.:* 16 mi. N Mansfield, 2 (LSUMZ). *Grant Par.:* Fishville, 2 (LSUMZ); Colfax, 1 (LSUMZ); Pollock, 3 (LSUMZ); 7 mi. SW Bentley on U.S. Hwy. 71, 2 (USL). *Jackson Par.:* 2 mi. E Jonesboro, 1 (LTU); Ansley, 1 (LTU); 0.7 mi. NE Weston on La. Hwy. 4, 7 (USL). *Jefferson Davis Par.:* 13 mi. N Iowa on La. Hwy. 383, 2 (USL). *Lincoln Par.:* Ruston, 18 (4 LSUMZ; 1 USNM; 13 LTU); 2 mi. N Tremont, 3 (LSUMZ); 8 mi. N Choudrant, 1 (LSUMZ); 2 mi. N Ruston, 1 (LSUMZ); 6 mi. S Bernice on U.S. Hwy. 167, 1 (USL); 1 mi. S Ruston, 29 (LTU); 1 mi. W Ruston, 2 (LTU). *Natchitoches Par.:* Provencal, 34 (30 LSUMZ; 4 USNM); Ashland, 1 (LSUMZ); Kisatchie, 4 (LSUMZ); Natchitoches, 7 (2 LSUMZ; 2 USL; 3 NSU); 2 mi. S Bellwood, 2 (USL). *Ouachita Par.:* Calhoun, 2 (1 LTU; 1 NLU); 15 mi. SW Monroe, 3 (NLU); 17 mi. W Monroe, 7 (NLU); 20 mi. W Monroe, 2 (NLU); 16 mi. W Monroe, 1 (NLU); West Monroe, 5 (NLU); 7 mi. W West Monroe, 6 (NLU); 6 mi. W West Monroe, 1 (NLU); 3 mi. W West Monroe, 5 (NLU); 8 mi. W Monroe, 2 (NLU); 2 mi. W Calhoun, 1 (NLU). *Rapides Par.:* 5 mi. WSW Lecompte, 2 (USL); Union Hill, 2 (USL); 0.5 mi. NE Gardner at Jct. La. Hwy. 28 and La. Hwy. 121, 1 (USL); Pineville, 2 (USNM); 3 mi. NW Glenmora, 1 (LSUMZ); 2 mi. NW Glenmora, 1 (LSUMZ); 1 mi. N Glenmora, 2 (LSUMZ); 2 mi. W Glenmora, 2 (LSUMZ); 3 mi. SW Boyce, 1 (LSUMZ). *Red River Par.:* 5 mi. NE Coushatta on La. Hwy. 155, 2 (USL). *Sabine Par.:* 13 mi. S Many, 4 (LSUMZ); 13 mi. SW Anacoco on La. Hwy. 392 (west bank of Bayou Toro), 2 (USL); 13.2 mi. W Anacoco at end Hwy. 392, 5 (TU). *Union Par.:* Marion, 3 (LSUMZ); 4.5 mi. NE Farmerville, 1 (LSUMZ); 6.9 mi. E Farmerville, 2 (USL). *Vernon Par.:* various localities, 23 (16 LSUMZ; 3 NLU; 4 USL). *Webster Par.:* Minden, 2 (LSUMZ); Claiborne Par. line on Haynesville–Shongaloo Hwy., 1 (LSUMZ); 0.5 mi. S Minden on U.S. Hwy. 80, 2 (USL); 5 mi. S Sibley, 1 (LTU). *Winn Par.:* 3 mi. SW Mill, 1 (USL).

The Pocket Mouse-
Kangaroo Rat Family
Heteromyidae

THIS FAMILY, also referred to as the heteromyid rodent group, comprises five genera and some 70 species. It is confined to North America and the extreme northwestern corner of South America. Four genera and 38 species occur in the area north of Mexico, mainly in prairies, dry plains, deserts, and other arid or semiarid habitats. The small population of the Hispid Pocket Mouse in western Louisiana represents the easternmost extension of the range of any member of the family. The family includes a number of diverse forms that reflect a progressive modification of the hind limbs for locomotion by leaping. The kangaroo rats, as the extreme example, are bipedal, using kangaroolike only their powerful hind legs in running, with the long tail serving as a balancing organ. All heteromyids possess fur-lined cheek pouches, and their skulls show many distinctive features, including greatly enlarged auditory bullae, weak or almost threadlike zygomatic arches, and nasals that extend well beyond the incisors.

Genus *Perognathus* Weid-Neuwied

Dental formula:

$$ I\frac{1-1}{1-1},\ C\frac{0-0}{0-0},\ P\frac{1-1}{1-1},\ M\frac{3-3}{3-3} = 20 $$

HISPID POCKET MOUSE   Plate VI
(*Perognathus hispidus*)

Vernacular and other names. — This species is so poorly known in Louisiana to the lay public that it has acquired no vernacular names. Indeed, I have never heard it called anything other than pocket mouse, even in parts of the country where it is common. The name refers to the fur-lined cheek pouches that all heteromyids possess. The generic name *Perognathus* comes from two Greek words (*pera*, pouch and *gnathus*, jaw) and again alludes to the presence of the cheek pouches. (Figure 90). The second part of the scientific name, *hispidus,* is the Latin word for spiny or rough and relates to the rather coarse fur of this particular group of heteromyid rodents in contrast to the soft, velvety hair of such members of the family as the kangaroo rats and kangaroo mice (*Dipodomys* and *Microdipodops*).

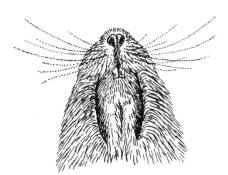

Figure 90 Facial features of the Hispid Pocket Mouse (*Perognathus hispidus*), showing particularly the openings to the cheek pouches.

213

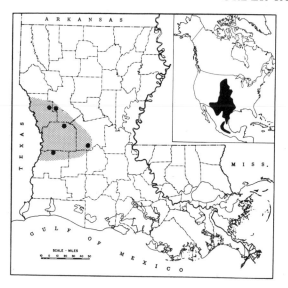

Map 29 Distribution of the Hispid Pocket Mouse *(Perognathus hispidus). Shaded area,* known or presumed range within the state; •, localities from which museum specimens have been examined; *inset,* overall range of the species.

Distribution. — The Hispid Pocket Mouse occurs from southern North Dakota south through the eastern part of the Great Plains to eastern Oklahoma, western Louisiana, southern Texas, southeastern Arizona, and central Mexico. In Louisiana a total of only seven specimens have been preserved from five localities in five parishes. (Map 29)

External appearance. — This mouse has the hairs of the upperparts light grayish basally. Some of these hairs are strongly tipped with ochraceous, while others are black tipped. Since the hairs are fairly stiff and at the same time by no means dense, some of the basal gray shows through and gives the overall color a somewhat grizzled ochraceous appearance. A line over the posterior corner of the eye and another that extends from the nose, along the lower edge of the face, the upper surface of the forelegs, and the sides of the body to the flanks and upperpart of the hind legs are a clear ochraceous. The underparts are white, sometimes with a very slight yellowish tinge. The tail is sharply bicolored,

blackish above and whitish below. The ears are dusky inside and buffy on the outside, except for an elliptical dusky spot on the inflexed edge.

Color phases. — None are known.

Measurements and weights. — Five adults (four males and one unsexed) from Louisiana in the LSUMZ averaged as follows: total length, 190.6 (184–210); tail, 92.8 (82–110); hind foot, 24.7 (23–27); ear, 9.5 (8–12); greatest length of skull, 28.9 (27.0–30.7); occipitonasal length, 28.4 (26.4–30.2); mastoid breadth, 13.9 (13.4–14.5); interparietal breadth, 8.0 (7.3–9.0); zygomatic breadth (based on the only 3 specimens in which the zygomata were unbroken), 14.5 (14.0–15.0); interorbital breadth, 6.7 (6.5–7.0); length of nasals, 10.9 (9.9–11.8); basilar length, 20.9 (20.0–22.0). A male from 5 mi. SE Tilden, McMullen County, Texas, weighed 31 grams, and a female from 8 mi. N Rockport, Aransas County, Texas, weighed 30 grams.

Cranial and other skeletal characters. — The skull of *Perognathus hispidus* is readily distinguished from that of any other Louisiana rodent of comparable size by the longitudinal groove on the front surface of each incisor. Only three others have grooved incisors, the much larger and more massive Plains Pocket Gopher (page 205) and the two diminutive harvest mice of the genus *Reithrodontomys* (page 230). The skull of *Perognathus hispidus* is also easily recognized by the weak zygomata, the much inflated auditory bullae, the greatly enlarged mastoids that form part of the dorsal surface of the skull, the reduced size of the occipital condyles, the extension of the nasals well beyond the incisors, and the greatly enlarged infraorbital foramina that lie on the lateral faces of the rostrum well ahead of the zygomatic arch with a perforation through the rostrum connecting the left and right foramina (Figure 91). The penis bone is slightly curved, is bulbous at the base, and tapers gradually from the base to near the distal end. A keel on the dorsal portion of the distal end, along with two processes that project ventrolaterally, give the bone, in end view, a trifid appearance. It measures about 14.8 mm (Burt, 1936). (Figure 72)

Sexual characters and age criteria. — Superficially the sexes are identical, and the immatures closely resemble the adults.

Status and habits. — This handsome mouse is apparently rare in Louisiana, but the possibility exists that intensive trapping in the western part of the state might show it to be more numerous than the few records of it would seem to indicate. Credit for its discovery in Louisiana goes to Professor Horace A. Hays, who first obtained it at Vowells Mill, in Natchitoches Parish, on 17 October 1939. He was at the time carrying on extensive trap-

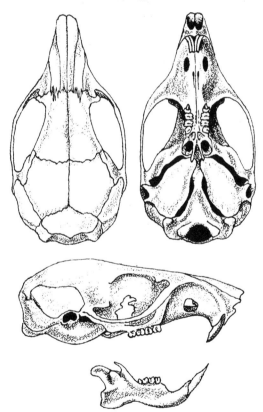

Figure 91 Skull of Hispid Pocket Mouse *(Perognathus h. hispidus)*, ♂, Vowells Mill, Natchitoches Parish, LSUMZ 1419. Approx. 2 ×.

ping operations for small mammals in western Louisiana, particularly in Natchitoches Parish, where he was then living. On 5 and 6 December 1939 he captured two more at the same locality, and on Christmas Day of that year he obtained another at Hutton, in Vernon Parish. In the following decade two additional specimens were obtained, one by H. E. Wallace near Fort Jesup, in Sabine Parish, and another by Charles Shaw near De Ridder, in Beauregard Parish. Finally, one was live-trapped by Brooke Meanley and Harold Derr on 22 November 1955 near Glenmora, in Rapides Parish. This last record was rather surprising to me because the habitat in the vicinity of Glenmora had not appeared to be suitable for the Hispid Pocket Mouse. However, this judgment was apparently in error, for Meanley informs me (pers. comm.) that he obtained four additional specimens in the grassy longleaf pine forest near Glenmora that he kept for a while in captivity and eventually released.

The collectors of the first six specimens mentioned have all informed me that they obtained their specimens in grassy fields, usually containing a heavy stand of broomsedge *(Andropogon* sp.). Pocket mice live in burrows in the ground that they dig themselves, piling up small mounds of excavated earth at the entrance. Generally, two or more entrances lead into the underground passages, but sometimes there is only one. In special subterranean chambers the mice store great quantities of seeds of grasses and weeds, transported from above ground in their cheek pouches. Little is known of their food habits in the state, but Wallace (pers. comm.) informs me that he counted 923 *Lespedeza sericea* seeds in the cheek pouches of the specimen that he caught near Fort Jesup. Blair (1937), studying the food habits of species in Oklahoma, found that 4 of 18 burrows examined contained the dried fecal pellets of the Eastern Cottontail *(Sylvilagus floridanus)*. He thought the pellets were being eaten by the pocket mice, since several showed signs of having been gnawed.

The number of young to a litter is said to vary from one to eight, and the number of litters a year to range from one to several.

Predators and diseases.—The normal life span in nature, if it is anything like that of other native mice, is probably less than a year because of predation. Hawks, owls, Coyotes, skunks, and snakes doubtless take a heavy toll. I am aware of no information dealing with the parasites and diseases of this species.

Subspecies

Perognathus hispidus hispidus Baird

Perognathus hispidus Baird, Mammals, in Repts. Expl. Surv. . . ., 8, 1857 [=1858]:421; type locality, Charco Escondido, Tamaulipas, México.
Perognathus hispidus hispidus, Lowery, Occas. Papers Mus. Zool., Louisiana State Univ., 13, 1943:242.

On the basis of geographical probability, specimens of this species from Louisiana could be referable either to the nominate race *hispidus* or to *P. h. spilotus* of extreme northern Texas, eastern Oklahoma, eastern Kansas, and southeastern Nebraska. The latter is said (Glass, 1947) to differ from *hispidus* as follows:

the total length averages over 200 mm instead of less; the length of the hind foot averages 25 mm instead of approximately 23; the occipitonasal length averages more than 30 mm instead of less; the interparietal is wider than the interorbital breadth instead of narrower; posteriorly the premaxillae and nasals extend equidistantly instead of the premaxillae extending further; and the ochraceous lateral line extends on to the forearm instead of the forearm being all white. In only two of the six characters do the Louisiana specimens unequivocally agree with *spilotus*. The interparietal is wider than the interorbital breadth and the ochraceous of the lateral line does encroach on the forearm. Perhaps if the Louisiana series were larger it would present a clearer picture, but I am inclined to continue to refer the present material to the race *hispidus,* as I did originally in 1943.

Specimens examined from Louisiana.—Total 7, as follows: *Beauregard Par.:* De Ridder, 1 (LSUMZ). *Natchitoches Par.:* Vowells Mill, 3 (LSUMZ). *Rapides Par.:* Glenmora, 1 (USNM). *Sabine Par.:* 2 mi. SE Fort Jesup, 1 (LSUMZ). *Vernon Par.:* Hutton, 1 (LSUMZ).

The Beaver Family
Castoridae

THE FAMILY Castoridae contains only one genus, *Castor.* The two species, the American Beaver *(C. canadensis)* and the European Beaver *(C. fiber),* are considered by some taxonomists to be conspecific, that is, to be members of only one species. The Old World

populations once occurred widely over much of Europe and Asia but are now reduced to a few small scattered colonies in France, Germany, Poland, Scandinavia, and Russia. In the New World, *C. canadensis* still occupies most of its overall original range from Alaska and northern Canada southward to the Rio Grande but in vastly reduced numbers and in far fewer places.

Beavers are among the largest of all rodents; they are certainly the largest in North America. They are massive and heavyset. Their most distinctive feature is, of course, the broad, flat tail. Numerous other anatomical features are worthy of note. The small eyes are equipped with nictitating membranes. The small ears are valvular, as are also the nostrils. The limbs are short, and each foot has five clawed digits, those of the hind feet being webbed and having the claws of the second and third toes, especially the second, split into double nails, presumably as an aid to grooming. The digestive and urogenital canals open into a common orifice, or cloaca.

Genus *Castor* Linnaeus

Dental formula:

$$I\frac{1-1}{1-1} \ C\frac{0-0}{0-0} \ P\frac{1-1}{1-1} \ M\frac{3-3}{3-3} = 20$$

AMERICAN BEAVER   Plate VIII
(*Castor canadensis*)

Vernacular and other names. — The generic name *Castor* is from the Greek word *kastor*, which comes over into the Latin, as well as the French, as *castor* and is the name by which the animal is called by French-speaking Louisianians. The second part of the scientific name, *canadensis*, is a Latinized word meaning of Canada. The name beaver is from the Anglo-Saxon word *beofar*. The adjective American serves to distinguish our species from the one of the Old World.

Distribution. — The range of the American Beaver originally extended from central-northern Alaska across most of Canada to Labrador and Newfoundland south to central California, northern Nevada, northern

Mexico, the Gulf Coastal Plain, and extreme northern Florida. It is now extirpated in many parts of its former range or greatly reduced in numbers. In some areas, such as Louisiana, management practices and protection have restored it in places where it once occurred. Its distribution in the state in the early 1930s was limited to the Amite and Comite rivers and their tributaries in East Baton Rouge, East Feliciana, St. Helena, Ascension, and Livingston parishes. Transplants of animals from this area that began in 1938 had resulted by 1958 in the establishment of 75 colonies in 21 parishes west of the Mississippi River, as well as in nearly all parts of the Florida Parishes (Noble, 1958). (Map 30)

External appearance. — The general form of the beaver is so well known that it requires little description (Plate VIII and Figure 92). In total length the adults range between 35 and 45

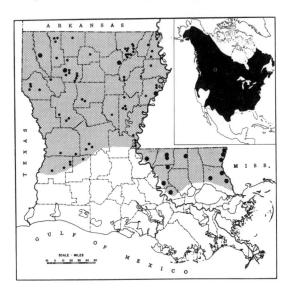

Map 30 Distribution of the American Beaver (*Castor canadensis*). *Shaded area*, present-day known or presumed distribution within the state; •, localities from which museum specimens have been examined; •, localities west of the Mississippi River to which beavers have been transplanted or to which transplants have spread (*fide*, Noble, 1958); *inset*, overall range of the species.

Figure 92   American Beaver *(Castor canadensis),* trapped near Big Sandy Creek, East Feliciana Parish, and released on Little Lake Lafourche, Morehouse Parish, July 1955. (Photograph by Joe L. Herring)

inches. The broad, flat tail measures 11 to 15 inches, 9 to 12 inches of which are densely scaled above and below. Most individuals weigh in the vicinity of 33 pounds, but even in Louisiana, where beavers tend to be smaller than northern representatives of the species, animals weighing 80 pounds are reported to occur (Dahlen, 1939; Thigpen, 1950; Smith, 1964). The pelage consists of a dense, grayish underfur that is overlaid dorsally with long, coarse, shiny guard hairs that give the coat a rich, glossy brown color. The underparts, lacking the guard hairs, are grayish tan. The feet and tail are black or brownish black. Paired anal scent glands, called castors, are present in both males and females. The substance secreted, referred to as castoreum, serves as the animal's "calling card." It has a not unpleasant odor and is sometimes used in the manufacture of perfumes. Other external features are included in the introduction to the family Castoridae.

Color phases.—Color variations or mutations of black, silvered, and white have been reported (Jones, 1923) to occur in the beaver, but I am aware of no such abnormalities among populations in Louisiana.

Measurements and weights.—Skins of two adult females in the LSUMZ from the Florida Parishes measured as follows: total length, 1,065 and 1,008; tail, 300 and 387; hind foot, 183 and 168. Four skulls of adult females from the same source averaged as follows: greatest length, 134.6 (132.2–137.3); mastoid breadth, 64.0 (61.3–67.0); zygomatic breadth, 102.0 (95.6–111.6); interorbital breadth, 25.3

(24.0–27.0); palatilar length, 75.3 (74.8–76.2); postpalatal length, 41.6 (40.5–43.5); alveolar length, 31.6 (31.0–32.5). The average weight of 82 beavers, of all ages, live-trapped in Louisiana in restocking operations, averaged 33.3 pounds, but, as noted above, extreme weights of 80 pounds are recorded.

Cranial and other skeletal characters. — The massive skull is easily distinguished by its size and by the strongly developed incisors, the hard enamel front surfaces of which are Orange-Rufous. The high-crowned cheek teeth possess flat, grinding surfaces with numerous folds of enamel (Figure 93). The penis bone is a bulbous, club-shaped structure proximally that tapers abruptly almost to the distal end (Figure 72). It somewhat resembles the shape of the head of a golf driving club. The head, or distal end, is slightly larger than the central portion of the bone. In adults that are

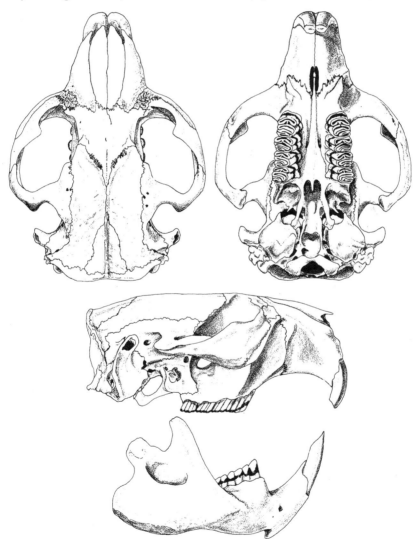

Figure 93 Skull of American Beaver *(Castor canadensis carolinensis),* ♀, 3 mi. N Fluker, St. Helena Parish, LSUMZ 3766. Approx. ½ ×.

four years old or older, the bone averages 34.4 mm in length with a dorsoventral height of 7.9 mm at the distal end (Friley, 1949b).

Sexual characters and age criteria.—So far as is known there are no color or cranial differences between the sexes. The young are the same color as the adults and are in other respects miniaturized editions of their parents. Bailey (1927) states that young weigh about one pound at birth, one and a half to two pounds when three weeks old, about four pounds at six weeks. He further states that one-year-old beavers generally weigh 25 to 30 pounds; two-year olds, 40 to 45 pounds; three-year olds about 50 pounds. However, these estimates were apparently based on samples that included northern beavers, which average appreciably larger than southern individuals. Consequently, some adjustments must be made if his age-weight classes are applied to Louisiana beavers. Friley (1949b) recognizes in Michigan beavers four distinct age classes: kits; yearlings; two- and three-year-olds; and four-year-olds and older. He demonstrates that a clear correlation exists between certain skull measurements and the size and shape of the baculum. For such correlations to be completely applicable to Louisiana animals, new averages would probably have to be determined on the basis of specimens taken at our latitude, since southern beavers average smaller than ones from Michigan.

Status and habits.—The early settlers in our state found the beaver widespread and fairly common, but the great demand for it in the fur trade led to its virtual extirpation long before the end of the nineteenth century. That the beaver once occurred over most of the state is evidenced by the great number of place-names and physiographic features that carry the animal's name. We find, for instance, the settlement by the name of Castor in Bienville Parish and one called Beaver in Evangeline Parish. There is a Beaver Lake in

Rapides Parish and a large number of Beaver creeks and Beaver or Castor bayous scattered here and there. As late as 1850 a few beaver were still to be found at Sicily Island, in central-eastern Louisiana, as revealed by the account of Kilpatrick in *De Bow's Review* (1852). Except for the colonies in the Florida Parishes, the species was nearly gone by the turn of the century, although a few may have persisted in Beauregard Parish until 1920 and in Morehouse Parish until 1925 (Noble, 1958). The only known places in the state that contained beavers as late as 1931 were those along the Amite and Comite rivers in certain parts of East Feliciana, St. Helena, East Baton Rouge, and Livingston parishes. There the animal multiplied, and, by the late 1930s, had increased to the point that complaints of damage in various forms began to accumulate. In response to these protests, biologists

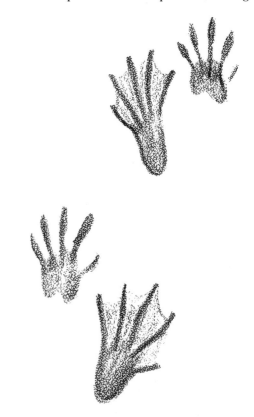

Tracks of American Beaver

with the Louisiana Wild Life and Fisheries Commission initiated a program of live-trapping beavers and then releasing them in other parts of the state. So successful were these operations that the species is now again fairly widespread and is, indeed, fairly common in some areas west of the Mississippi River, as well as in the Florida Parishes.

Perhaps no other animal contributed more to the development of this country and of Canada than did the beaver. Its hide once brought premium prices that prompted trappers to move farther and farther westward in search of new sources, and they were followed by settlers. Its fur was in such great demand that companies were formed to promote its capture. The Hudson's Bay Company is reported to have shipped more than 100,000 pelts a year for over two centuries, and during its best years to have marketed as many as 250,000 annually. Beaver pelts were the backbone of the fur trade even though other furbearers also were taken. Entire industries were dependent on the beaver as the source of their raw materials. Fine felt hats were made solely from its fur.

Seton (1929) estimates that the total original number of beavers in North America may have been as high as 400 million. It seems incredible that the species could have been brought so close to extinction. Yet when we recall that a prime blanket pelt (extra large) brought $65 or more, we see the basis of the strong motivation on the part of the trapper to obtain as many as possible. The situation today is quite different and herein lies our problem. With the decline in the market value of beaver pelts, the species began again in many places to increase. Prompted by aesthetic considerations, considerable effort was expended by many state game departments to reintroduce it in places where it had been extirpated. Thus, the biological pendulum started to swing in the other direction. Under protection beaver populations multiplied and with their increase came mounting complaints of alleged damage to crops and to

valuable forest trees. Not only does it cut down trees, but it girdles them in its feeding operations, and its dams impound water that drowns many more.

In 1939 the species was again so numerous in parts of the Florida Parishes that state biologists began to transplant many of them to other areas. Eventually, however, they concluded that live-trapping and removal, which is slow, expensive, and usually ineffective, could not keep up with complaints, and a trapping season was necessary to reduce the population. Consequently, the first trapping season of Louisiana beavers in 39 years was opened 1 January 1955 and covered 15 days. But according to O'Neil (pers. comm.) and records in the Fur Division of the Wild Life and Fisheries Commission, only in the neighborhood of 100 beavers were trapped in the first season. Most of the pelts brought a meager two dollars, the top price being no more than five dollars. The 1956 season was extended to cover the entire month of January, but fewer beavers were trapped than in 1955. The same held true for the 1957 and 1958 seasons. The present market value of beaver pelts simply does not justify the effort expended in trapping one of the animals.

Few, if any, other mammals of North America have elicited as much admiration on the part of the lay public as the beaver. It seems to personify the ultimate in diligence and hard work. The assiduous manner in which it goes about its business has given rise to such expressions as "busy as a beaver," and an enterprising person is often referred to as being an "eager beaver." The animal is indeed a prodigious worker. It fells trees of considerable diameter, cuts them into sections that it can manipulate, and drags or floats them to its dam. Hundreds of limbs and branches are similarly sectioned for placement in these structures, which are constructed with great engineering skill (Figures 94 and 95). When its dam is broken the beaver immediately proceeds to repair it, allowing no amount of discouragement to stand in its way. Noble (1958) tells of purposefully breaking the dams on Dunn Pond in East Feliciana Parish no less than seven times. Invariably they were repaired within a few days.

Because of the general flatness of the terrain in most of Louisiana, the waterways inhabited by beavers are usually slow-flowing, sluggish streams or bayous. Consequently, the dams that our beavers construct are often less massive than those built by beavers in other parts of the country. The smallest dam that Thigpen (1950) found in the Florida Parishes was 5 feet in length, 6 inches in height, and 14 inches wide at the base; the longest was 302 feet in length, 26 inches in height, and 45 inches in width at the base. Chabreck (1957), on the other hand, describes a dam at the Little Marsh colony, 2.1 miles west of Lacombe, in St. Tammany Parish, as being 1,391 feet long, 3 feet wide, and 1 foot in average height. On the upstream side of the dam a canal about 9 feet deep was formed when dirt was excavated by the beavers in the construction of the dam. The dam followed an irregular course across a marshy area, tying into natural features of the terrain such as logs, roots, cypress knees, and mounds of dirt wherever possible. When the dam was visited again, it had raised the water level on the upstream side by 10 inches.

Although the beaver in Louisiana often builds a conventional lodge in the open water that its dams impound, it perhaps more frequently places its den in one of the banks of the stream leading to the dam. These dens are sometimes rather difficult to locate because their entrance is below the water level. Where streams are fairly deep and maintain a steady year-around flow and where the banks on the streams are sufficiently high, the beaver's needs are adequately filled by constructing its den in one of the banks. Such a retreat provides adequate protection and shelter. The beaver in Louisiana does not need the massive lodges or dens that beavers in the northern part of the range use to store large required quantities of food to tide them over long periods of severe winter weather.

Three excellent studies have been made of the woody plants utilized by beavers in this state (Thigpen, 1950; Chabreck, 1957 and 1958; Noble, 1958). The last two of these involved systematic surveys of the kinds of trees available to the beaver and the degree to which each was utilized, thereby reflecting the beaver's preferences, both for food items or for building materials. Noble found not only that different sites offered different arrays of woody plants but that the beavers in different places seemed to show a preference for different species of plants. On the whole, however, loblolly pine, sweetgum, silverbell, sweetbay, and ironwood were among the woody plants most frequently utilized. Tupelogum, boxelder, yaupon, wax myrtle, chinaberry, witch-hazel, spruce pine, and various other woody plants are employed to one degree or another, depending on the location.

At one of the Sandy Creek colonies, in East Feliciana Parish, 18 species of woody plants were found by Noble (1958) in his study to be available to the beavers, but they made use of only 6. Sweetgum comprised 46.3 percent of the total plant material used, although it constituted only 16.0 percent of the woody plants in the stand near the colony. At another Sandy Creek colony, he found that silverbell comprised 52.5 percent of the total utilization. Sweetgum ranked second with 22 percent utilization, and ironwood was third with 10.5 percent. At these two colonies the

Figure 94   Beaver lodge on Lake Chaney, Webster Parish, June 1958. (Photograph by Robert E. Noble)

Figure 95 Yellow-poplar *(Liriodendron tulipifera)*, 19 inches in diameter, girdled by beaver on Beaver Creek, near Fluker, Tangipahoa Parish, April 1958. (Photograph by Robert E. Noble)

beavers had not built dams or pond lodges; so presumably most of the woody plants that had been barked, girdled, or felled were used for food.

The trees actually felled by a beaver are almost always in the one- to five-inch dbh (diameter breast high) class. Those girdled fell in higher dbh classes. Noble found that of all stems girdled at colonies in the Florida Parishes, 28.8 percent were in the 1–5 inch class, 48.6 percent were in the 6–10 inch class, 16.2 percent were in the 11–15 inch class, and 5.8 percent were in the 16–20 inch class. Thigpen (1950) found a 31-inch sweetgum girdled on the bank of the Amite River, in St. Helena Parish. The largest tree felled by a beaver ever recorded was a cottonwood in British Columbia that measured 5 feet and 7 inches in diameter above the cut (Hatt, 1944).

The beaver is not limited to feeding on the bark of woody plants. It sometimes consumes great quantities of oak mast, aquatic vegetation, and the stems of various weeds and other plants, including stinkweed, giant rag-weed, water hyacinth, cross vine, corn, and switch cane.

Although generally considered to be monogamous and to mate for life, the beaver is said to be occasionally polygamous, as are most rodents. A beaver family usually contains one adult male, one adult female, a one-year-old litter, and the succeeding litter. As yearlings young beavers are allowed to remain in the lodge or den with the new litter, but prior to the arrival of the next litter, they either leave on their own accord or are expelled (Grasse and Putnam, 1955). The beaver begins to breed in its second year, or when about 21 months old. The gestation period is approximately 120 days. The number of young in a litter ranges from one to eight but is usually two to four. The kits at birth have a coat of hair and the eyes are open. They nurse for approximately six weeks. Birth is said to occur in April or May, but little precise information is available on this subject pertaining to Louisiana populations.

The main sound that is identified with the adult beaver is the loud slap of the tail on the surface of the water. It serves as a danger signal to all other beavers within hearing distance. The sounds of its teeth gnawing on wood can be heard for several hundred feet on a quiet night and are often followed by a silence-shattering crash or loud splash as a tree plummets downward. Even though the adults have been heard to utter only a few groans or grunts while inside their dens, the kit, up to the age of six months, is quite vociferous. It has, according to Jackson (1961), "many cries, whines, and soft plaintive sounds, and other more vigorous tones of protest and anger."

The beaver is a marvelous creature, about which a great deal, both good and bad, can be said. Admittedly, it is sometimes inimical to man's best interest, but no considerations justify its wholesale elimination. Indeed, there are few people who could countenance its destruction except locally, under extreme circumstances. When a landowner can demonstrate to wildlife experts that unequivocal damage is being done by a beaver colony on his property, he should be allowed to dispose of the animals in any way he sees fit, provided, of course, that arrangements cannot be made to trap the offenders and to move them elsewhere. The aesthetic appeal of the animal to the general public is great, and this is attested to by the enthusiasm that has been expressed in some of our western and northern parishes when people have learned of successful reintroductions of the beaver to their areas. After all, in this day and age in which we are living, there must be some things worth preserving for reasons other than their dollar value.

Predators. — The beaver has few enemies. In the water an adult is capable of emerging victorious in an encounter with a dog. Perhaps it would fare poorly with an otter, but the latter is uncommon in most places where beavers occur. Fortunately, in Louisiana the beaver is not required to make long forays from its bailiwick in search of food to store for the winter and to get enough timber to construct elaborate dams and lodges. Consequently, it can usually reach the safety of its lodge or den when attacked by a Coyote, a Bobcat, or some other large predator. Panthers are now so rare in the state that they pose no problem. The kits are probably the most vulnerable members of the beaver clan. Alligators and large carnivores probably account for the loss of a few young.

Parasites and diseases. — Beavers seem to be remarkably free of both internal and external parasites. Perhaps the scarcity of endoparasites is attributable to the animal's habit of eating bark and sapwood. I find at least four trematodes (*Stichorchis subtriquetrus, Stephanoproraoides lawi, Renifer ellipticus,* and *Paraphistomum castori*) and seven nematodes (*Capillaria hepatica, Travassorius americanus, T. rufus, Castorstrongylus castoris, Filaria* sp., *Oxyuris* sp., and *Gongylonema* sp.) that have been reported from the beaver. Ectoparasites include two arachnids (the mite *Schizocarpus mingaudi* and the tick *Dermacentor albipictus*), a louse (*Trichodectes castoris*), and two parasitic beetles (*Platypsylla castoris* and *Leptinillas validus*). Beavers are also known to have been infected with rabies, tularemia, and a form of distemper. (Doran, 1955a; Fred Beckerdite, pers. comm.)

Subspecies

Castor canadensis carolinensis Rhoads

Castor canadensis carolinensis Rhoads, Trans. Amer. Philos. Soc., n.s., 19, 1898:420; type locality, Dan River, near Danbury, Stokes County, North Carolina.

The subspecies *carolinensis* occurs in the southeastern part of the United States north to southern Virginia, northern Ohio, Indiana, Illinois, and west to southeastern Iowa, eastern Missouri, eastern Arkansas, and Louisiana. I have made no attempt to study critically the presently available material from Louisiana, especially that from the reestablished populations in the western parts of the state. All but one of the colonies west of the Mississippi are believed to have been derived from *carolinensis* stock transplanted from the Amite–Comite river drainage of the Florida Parishes. Consequently, I feel justified in assigning all the material to *carolinensis* on the basis of geographical probability and in accord with standard taxonomic treatment.

Specimens examined from Louisiana. — Total 26, as follows: *Ascension Par.:* Gonzales, 1 (LSUMZ). *Bienville Par.:* 3 mi. N Kepler Lake, 1 (LTU). *Bossier Par.:* 1 mi. W Benton, 1 (LTU). *East Baton Rouge Par.:* Pride, 1 (LSUMZ); near Baton Rouge on Airline Hwy., 1 (LSUMZ). *East Feliciana Par.:* Clinton, 5 (LSUMZ); 5 mi. S Clinton, 2 (LSUMZ). *Ouachita Par.:* Russell Sage Game Mgt. Area, 3 (NLU). *St. Helena Par.:* 3 mi. N Fluker on Beaver Creek, 1 (LSUMZ). *St. Tammany Par.:* Honey Island swamp, NE Pearl River, 1 (USL); Bogue Falaya River, near Covington, 1 (NLU); Bogue Chitto Creek, 1 (LSUMZ). *Tangipahoa Par.:* locality unspecified, 1 (SLU). *Tensas Par.:* barrow pits, north St. Joseph on Miss. River, 1 (LTU). *Washington Par.:* locality unspecified, 1 (SLU); Pearl River, 1.5 mi. E Bogalusa, 1 (TU); 3 mi. E Angie, 1 (SLU); 1.5 mi. SE Varnado, 1 (SLU). *West Feliciana Par.:* Solitude, 1 (USL).

The New World Rat and Mouse Family
Cricetidae

THIS LARGE family of rats and mice is virtually worldwide in distribution. Cricetine rodents are absent only from Antarctica, the Austro-Malayan region, and certain islands, such as Iceland, Ireland, and a few islands in the Arctic. It comprises slightly more than 100 genera and some 730 species, of which 65 genera and approximately 574 species occur exclusively in the New World. Only 11 species are known from Louisiana, but these include several of our most common and best-known mammals, notably the immensely valuable muskrat.

The family is divided into two subfamilies, Cricetinae and Microtinae. In the first of these the three cheek teeth on each side of each jaw are highly variable. In some, two rows of enamel cusps are arranged longitudinally on the crowns, while in others the crowns are flat and the enamel is arranged in S-shaped loops. In the subfamily Microtinae, the crowns of the teeth possess sharp-angled folds of enamel in roughly triangular patterns (Figure 96). To this latter subdivision of the family belong the Common Muskrat *(Ondatra zibethicus)*, our Woodland Vole *(Microtus pinetorum)*, and the Prairie Vole *(Microtus ochrogaster ludovicianus)*.

Genus *Oryzomys* Baird

Dental formula:

$$\text{I}\frac{1-1}{1-1}\ \text{C}\frac{0-0}{0-0}\ \text{P}\frac{0-0}{0-0}\ \text{M}\frac{3-3}{3-3}=16$$

MARSH RICE RAT   Plate V
(*Oryzomys palustris*)

Vernacular and other names.—This species sometimes is called simply the rice rat, a name derived from its rice-eating habits. The adjective Marsh is required to distinguish it from numerous other species of the genus that occur farther south. The word Rat comes from the Anglo-Saxon word *raet*. The generic name is derived from two Greek words that translate as rice mouse (*oryza*, rice, *mys*, mouse), and also refers to the rice-eating habits of these rodents. The second part of the scientific name, *palustris*, is the Latin word for marshy, and alludes to the preferred habitat of the species.

Distribution.—*Oryzomys palustris* occurs from southeastern Pennsylvania and New Jersey along the Atlantic Seaboard and Gulf Coastal Plain to southern Florida and the northern Gulf Coast and west to southern Kentucky,

225

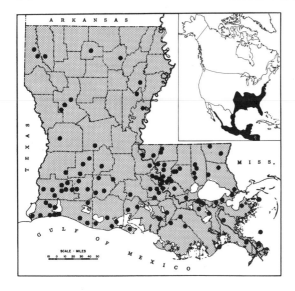

Map 31　Distribution of the Marsh Rice Rat (*Oryzomys palustris*). *Shaded area,* known or presumed range within the state; •, localities from which museum specimens have been examined; *inset,* overall range of the species.

southern Illinois, southeastern Missouri, southeastern Kansas, eastern Oklahoma, and southeastern Texas thence south to at least Nicaragua. (See Map 31)

External appearance.—This species is a small rat-sized rodent with a slender, scantily haired tail that is about equal in length to the head and body combined. The upperparts are dark grayish, heavily interspersed with black, especially on the head and the middle of the dorsum. The sides are washed with Ochraceous-Buff and the underparts are grayish white. The tail is bicolored, brownish above and white below, and the feet are white. There are four toes on the front feet, five on the rear. The eyes are moderately large and the ears are prominent.

Color phases.—No color phases are known.

Measurements and weights.—One hundred and twenty-two fully adult specimens in the LSUMZ from numerous localities in the state averaged as follows: total length, 237 (220–270); tail, 117 (100–137); hind foot, 28 (24–33); ear, 14 (10–18). Eighty-nine completely undamaged skulls from the same source averaged as follows: greatest length, 28.8 (26.2–32.8); cranial breadth, 12.8 (11.8–13.5); zygomatic breadth, 14.9 (13.5–16.7); interorbital breadth, 4.9 (4.4–5.5); length of nasals, 11.4 (9.5–13.7); length of diastema, 7.4 (6.4–8.6); palatilar length, 11.3 (8.9–13.6); postpalatal length, 10.2 (8.8–11.8). Twenty-six of these specimens showed an average weight of 51.6 (40.0–75.0) grams.

Cranial and other skeletal characters.—The skull of *Oryzomys palustris* can hardly be confused with the skull of any other Louisiana rodent except those of the genus *Peromyscus*. But a supraorbital ridge (on each side above the eye socket) that is prominent in *Oryzomys* is absent in *Peromyscus*. Also, the posterior border of the hard palate extends posteriorly beyond

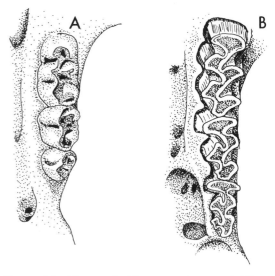

Figure 96　The molariform toothrow of (A) a cricetine rodent, *Oryzomys palustris texensis* (LSUMZ 16279), and (B) a microtine rodent, *Microtus pinetorum auricularis* (LSUMZ 1151), illustrating the two rows of cusps on the crowns of the former and the prismatic shape of the teeth of the latter. Approx. 10×.

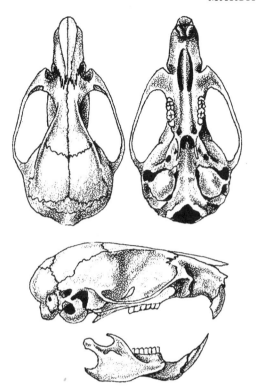

Figure 97 Skull of Marsh Rice Rat *(Oryzomys palustris texensis)*, ♀, 0.5 mi. S Plaquemine ferry, Iberville Parish, LSUMZ 15496. Approx. 2 ×.

the last molars, the interorbital constriction is wider, and the rostrum is wider and relatively shorter. The baculum is composed of an ossified base and shaft that is joined at its distal end by a three-pronged, cartilaginous process. Its length is approximately 5.1 mm and the diameter of the shaft is 0.87 mm at the base, tapering to 0.51 near the middle (Bailey, 1950). (See Figures 72 and 97.)

Sexual characters and age criteria. — The sexes are alike in size and color. Immatures are decidedly more brownish above and grayer, less whitish, below. Negus et al. (1961), in their superb detailed analysis of the rice rat population on Breton Island, near the mouth of the Mississippi River, recognized four age groups, as follows: juvenile, total length up to 210 mm and weight up to 32 grams; subadult, total length 205–220 mm, weight 30–50 grams; subadult–adult, total length 220–230 mm, weight 50–55 grams; adult, total length 230+ mm, weight 55 grams and over. Breeding can occur anytime after the postjuvenal molt.

Status and habits. — The Marsh Rice Rat apparently occurs widely over Louisiana, including the coastal marshes. It is abundant in the southern half of the state but decidedly uncommon in the northern portion, where it has been recorded at less than a dozen localities in only eight parishes. Systematic trapping would probably reveal its presence in parishes where it has not yet been taken, especially in the lowlands west of the Mississippi River (see Map 31). Its habitat is always rather wet, marshy places, such as grassy ditches, the edges of lakes and streams, and fields in which the soil is at least damp. It is seldom, if ever, found in dry fields or in well-drained woodlands far removed from water.

In the coastal areas of the state the species is abundant on the chenieres and along canal banks and even in the marshes themselves where the ground vegetation is not subject to high tides and flooding. At Baton Rouge it is extremely numerous on the flood plain west and south of Louisiana State University campus. There it inhabits ditches, the banks of drainage canals, the edges of sugarcane fields, and briar thickets, habitats that it shares in large measure with the Hispid Cotton Rat *(Sigmodon hispidus)* and two species of

Figure 98 Marsh Rice Rat *(Oryzomys palustris).* (Photograph by Richard K. LaVal)

harvest mice *(Reithrodontomys)*. In a broom-sedge *(Andropogon)* field on the eastern side of the campus, next to University Lake, where extensive collecting was carried on over a long period of years, we found the rice rat occupying that portion of the field lying adjacent to the lake with cotton rats occurring in the higher portions—evidence of the preference of the rice rat for wetter situations.

The nest is usually placed in a slight depression on the ground in a tangle of vegetation. It is comprised of finely shredded dry grasses and sedges, which are loosely woven together. Sometimes the nest is placed a few feet above the ground in a mass of vegetation, especially when the ground is inundated periodically. The food eaten by the rice rat consists mostly of seeds and the succulent parts of the various plants that are available. It consumes some insects and crustaceans and in captivity seems to relish small raw fish and bits of raw meat. In the salt marshes on the coast of Georgia, Sharp (1967) found that stomach analyses of rice rats taken in summer and fall indicated a predominance of animal food in their diets. Items consumed were chiefly insects and small crabs. In the same area Kale (1965) had already shown that rice rats prey on the eggs and young of the Long-billed Marsh Wren *(Telmatodytes palustris)*.

Rice rats are highly fecund and can produce as many as seven litters a year. However, as pointed out by Negus et al. (1961), the average life span of a female in the wild is probably much less than a year. So short an existence would not allow the average female much more than six months of reproductive activity during her life. Hence the average number of litters produced is probably no more than five or six. The gestation period is

Tracks of Marsh Rice Rat

25 days. Females are in estrus immediately following parturition and usually mate within 10 hours after giving birth to a litter. At birth the babies are blind, wrinkled, and covered with a sparse coating of fine, short hair on the legs and dorsal part of the body. The vibrissae are well-formed at birth and perhaps assist the young in finding a nipple. Svihla (1931) found that the weight of 13 newly born young averaged 3.14 (2.35–4.0) grams. The eyes open on the 6th day. Although Svihla states that weaning occurs on the 11th day, and Hamilton (1946b) noted that it takes place on the 11th to 13th day, Worth (1950b) found some individuals still nursing at the age of 20 days. Svihla further found that a 75-day old female gave birth to young, showing that she had reached sexual maturity when only 50 days of age.

The number of young in a litter of this species is usually thought to vary from two to five, but Negus et al. (1961), in their study of the population structure of rice rats on Breton Island, found that the number of young in a litter is density-dependent. Relying on counts of embryos in dead females that they examined, they found the mean number of embryos in 20 adult females collected from 1957 to 1959 to be 4.8. Ten of these counts were recorded from July to November 1957. The mean for this group was 3.7, with a range from 2 to 5. The remaining 10 counts were recorded during the period May 1958 to June 1959. The mean for this sample was 6.0, with a range of 4 to 7. The two groups with their contrasting number of embryos represented contrasting phases in the population growth of rice rats on the island. The low litter-size data of 1957 were obtained during a period of high density of rice rats. The other sample, however, was collected after the population had declined and had then begun to increase from a low population level. A further significant revelation of their study was the demonstration that following a low density level in the population in 1957, breeding and the production of litters continued unabated through the fall and winter of 1958 and into the spring of 1959. Little or no breeding takes place in this species between November and February.

Predators. — Rice rats are preyed upon heavily by all hawks and owls and by water snakes. Svihla (1931) considered the Water Moccasin *(Agkistrodon piscivorus)* and the Barn Owl *(Tyto alba)* two of the main enemies of *Oryzomys palustris* in the coastal marshes. Barn Owls on the Moore Ranch in Cameron Parish also feed extensively on rice rats, as evidenced by the large number of their skulls found in owl pellets that I have examined from that place. On Marsh Island, off the central Louisiana coast, rice rats are likewise heavily preyed upon by the Barn Owl. Analysis of 804 pellets revealed remnants of 1,008 vertebrate animals of which 984 (97.5 percent) were rice rats and 24 (2.5 percent) were small passerine birds (Jemison and Chabreck, 1962).

Parasites and diseases. — Morlan (1952) reports the presence on the rice rat of two lice *(Hoplopleura hirsuta* and *Polyplax spinulosa),* one flea *(Polygenis gwyni),* and eight mites and ticks *(Laelaps* sp., *Haemolaelaps glasgowi, Gigantolaelaps cricetidarum, Bdelonyssus bacoti, Bdelonyssus* sp., *Dermacentor variabilis, Trombicula splendens,* and *Listrophorus* sp.). Lumsden and Zischke (1961) found the trematode *Zonorchis komareki* in the bile duct of rice rats in the Bonnet Carré spillway at Norco. Doran (1955a) and Kinsella (1971) report three nematodes *(Mastophorus muris, Rictularia ondatrae,* and *Angiostrongylus schmidti)* in the species. Typhus has been reported in this species.

Subspecies

Oryzomys palustris texensis J. A. Allen

Oryzomys palustris texensis J. A. Allen, Bull. Amer. Mus. Nat. Hist., 6, 1894:177; type locality, Rockport, Aransas County, Texas.

The subspecies *texensis* occupies the western portion of the range of the species, in-

cluding all of Louisiana. It differs from the nominate race, *palustris,* in being on the average paler and having the skull usually narrower. Specimens from the southeastern corner of the state, including St. Tammany, Orleans, St. Bernard, and Plaquemines parishes, average appreciably darker than specimens from the remainder of the state. But I do not believe they merit any sort of nomenclatural recognition.

Specimens examined from Louisiana. — Total 791, as follows: *Acadia Par.:* 7 mi. SW Eunice, 2 (LSUMZ); 1 mi. NE Church Point, 3 (LSUMZ); 2 mi. N, 1 mi. W Crowley, 1 (LSUMZ); Mermentau, 4 (FM). *Allen Par.:* 3.5 mi. W Kinder, 1 (LSUMZ); 4.5 mi. NW Oberlin, W. Bay Game Mgt. Area, 1 (LSUMZ); just W Oakdale city limits, 3 (TU). *Ascension Par.:* 2 mi. E Gonzales, 1 (LSUMZ); 0.5 mi. S Gonzales, 1 (LSUMZ). *Assumption Par.:* 7 mi. SW Napoleonville, 1 (LSUMZ). *Beauregard Par.:* 1.5 mi. E Edith, 2 (LSUMZ). *Bossier Par.:* 1 mi. N Bossier, 1 (LTU). *Caddo Par.:* 2 mi. N Blanchard, 1 (LSUMZ); LSU Shreveport campus, 3 (LSUS). *Calcasieu Par.:* Iowa, 6 (3 LSUMZ; 3 USNM); 0.5 mi. W Chloe, 1 (LSUMZ); 2 mi. W Sulphur, 1 (LSUMZ); Sam Houston State Park, 1 (LSUMZ); 7.5 mi. N Toomey, 1 (LSUMZ); 1 mi. E Lake Charles, 5 (MSU); 5 mi. S, 2 mi. W Lake Charles, 3 (MSU); 10 mi. S Sulphur, 1 (MSU); Lake Charles, 6 (MSU); Moss Bluff, 1 (MSU); 7 mi. S Sulphur, 3 (NLU). *Cameron Par.:* Grand Cheniere, 4 (2 LSUMZ; 2 NSU); Cameron-Willow Island, 2 (LSUMZ); Lowry, 2 (LSUMZ); 1 mi. E Peveto Beach, 2 (LSUMZ); 7 mi. W Holly Beach, 3 (LSUMZ); Cameron, 4 (LSUMZ); 15 mi. W Holly Beach, 1 (LSUMZ); 12 mi. S Vinton, 7 (LSUMZ); 2 mi. N Cameron, 1 (LSUMZ); 3 mi. W Cameron, 1 (LSUMZ); 1 mi. W Cameron, 1 (LSUMZ); 8 mi. S Hackberry, 1 (LSUMZ); 4.3 mi. S, 7 mi. W Hackberry, 1 (LSUMZ); Rockefeller Refuge, 1 (SLU). *Catahoula Par.:* 0.25 mi. S Sicily Island, 2 (LSUMZ); 1 mi. N Sicily Island, 1 (LSUMZ). *Concordia Par.:* 1 mi. N Ferriday, 1 (LSUMZ). *East Baton Rouge Par.:* various localities, 228 (LSUMZ). *East Feliciana Par.:* 1 mi. W Jackson, 2 (LSUMZ); 4 mi. N Jackson, 1 (LSUMZ). *Evangeline Par.:* 0.5 mi. W Mamou, 1 (LSUMZ). *Franklin Par.:* 2 mi. E Gilbert, 1 (LTU). *Iberia Par.:* Marsh Island, 2 (1 LSUMZ; 1 CAS); Jeanerette, 2 (LSUMZ); 2 mi. NNE Avery Island, 3 (LSUMZ); Avery Island, 1 (CAS). *Iberville Par.:* North Island, Grand Lake, 1 (LSUMZ); 1.5 mi. S Indian Village, 1 (LSUMZ); 7 mi. S Plaquemine, 1 (LSUMZ); 6.15 mi. S, 3.7 mi. W Addis, 1 (LSUMZ); 0.5 mi. S Plaquemine ferry, 1 (LSUMZ); 11 mi. SSW LSU, 1 (LSUMZ). *Jefferson Par.:* Grand Isle, 5 (LSUMZ). *Jefferson Davis Par.:* 1.10 mi. N Fenton, 1 (LSUMZ); 1 mi. S Fenton, 6 (LSUMZ); Fenton, 1 (LSUMZ); Lacassine, 1 (LSUMZ); Jennings, 4 mi. W on N. Cutting Ave., 2 (LSUMZ) *Lafourche Par.:* 2 mi. S Thibodaux 4 (LSUMZ). *Lincoln Par.:* locality unspecified, 1 (LTU). *Livingston Par.:* 2.5 mi. N Denham Springs, 1 (LSUMZ); 8 mi. N Walker, 1 (LSUMZ); 3 mi. NNE Denham Springs, 4 (LSUMZ). *Natchitoches Par.:* Vowells Mill, 4 (LSUMZ);

Natchitoches, 219 (NSU); 2 mi. ESE Provencal, 5 (NSU). *Orleans Par.:* New Orleans, 4 (3 LSUMZ; 1 TU); 10 mi. E Gentilly, 1 (LSUMZ); Lake Catherine, 11 (3 TU; 2 USNM; 6 AMNH). *Ouachita Par.:* Monroe, 8 (2 LSUMZ; 6 NLU); West Monroe, 1 (NLU). *Plaquemines Par.:* Boothville, 1 (LSUMZ); Breton Island, 61 (57 TU; 4 CAS); Main Pass, Mississippi Delta, 2 (USNM); Burbridge, 18 (MCZ). *Pointe Coupee Par.:* 5 mi. E Batchelor, 1 (LSUMZ); 18 mi. NW Baton Rouge, 1 (LSUMZ). *Rapides Par.:* 1 mi. N Glenmora, 1 (LSUMZ). *St. Bernard Par.:* 10 mi. NE Shell Beach, 5 (LSUMZ); Brush Island, 6 (CAS). *St. Charles Par.:* 3 mi. WNW Paradis, 1 (LSUMZ); 3 mi. S Norco, 1 (LSUMZ); Couba Island, 1 (LSUMZ); Bonnet Carré Spillway, 5 (TU); Sarpy Wildlife Refuge, Norco, 3 (TU); 4.5 mi. SE Norco on Hwy. 61, 8 (TU). *St. James Par.:* 2 mi. N Gramercy, 4 (LSUMZ); 2 mi. SE Gramercy, 2 (LSUMZ). *St. Martin Par.:* 1 mi. W St. Martinville, 3 (NSU). *St. Mary Par.:* Morgan City, 6 (USNM). *St. Tammany Par.:* 10 mi. W Covington, 1 (LSUMZ); Slidell, 1 (LSUMZ); 10 mi. NE Covington, 1 (LSUMZ). *Tangipahoa Par.:* 2.5 mi. SW Hammond, 1 (LSUMZ); 2 mi. W Hammond, 3 (LSUMZ); 8 mi. N Hammond, 1 (LSUMZ); 2 mi. W Ponchatoula, 3 (SLU); 1 mi. N Ponchatoula, 1 (SLU); 4 mi. E Tickfaw, 2 (SLU); 2 mi. S Hammond, 1 (SLU); *Terrebonne Par.:* Gibson, 3 (1 LSUMZ; 2 MCZ); Houma, 1 (USNM). *Vermilion Par.:* 5 mi. W Redfish Point, 1 (LSUMZ); ca. 10.5 mi. NE Pecan Island, 5 (LSUMZ); 7.5 mi. W Pecan Island, 1 (LSUMZ); Cheniere au Tigre, 3 (CAS). *Vernon Par.:* Leesville, 1 (LSUMZ). *Washington Par.:* 10 mi. E Franklinton, 2 (SLU); 1 mi. W Warnerton, 4 (TU). *West Baton Rouge Par.:* 8 mi. W Miss. River Bridge, 1 (LSUMZ); 5 mi. W Port Allen, 1 (LSUMZ); 1 mi. S Port Allen, 4 (LSUMZ); 7 mi. W Port Allen, Hwy. 79, 2 (LSUMZ). *West Feliciana Par.:* 5 mi. NE St. Francisville, 1 (LSUMZ).

Genus *Reithrodontomys* Giglioli

Dental formula:

$$I\frac{1-1}{1-1} \ C\frac{0-0}{0-0} \ P\frac{0-0}{0-0} \ M\frac{3-3}{3-3} = 16$$

EASTERN HARVEST MOUSE   Plate VI
(Reithrodontomys humulis)

Vernacular and other names. — Since harvest mice are virtually unknown except to zoologists who study mammals, they have acquired no local or vernacular names. The generic name *Reithrodontomys* is derived from three Greek words that in combination mean groove-toothed mouse (*rheithron*, groove, *odous*, tooth, and *mys*, mouse), referring to the longitudinal grooves on the front surface of the upper incisors of this group of mice. The

second part of the scientific name of this species, *humulis*, involves what I believe to have been a spelling error on the part of Audubon and Bachman, who were responsible for the naming and the original description of this species. They applied the name *humulis*, although I think they intended *humilis*, the Latin word meaning small, since they refer to the animal as the "Little Harvest Mouse." Later, in their *Quadrupeds of North America* (1851–1854), they actually spelled the specific name "*humilis*." The word *humulis* denotes the plant called hops, but there is nothing to indicate that this meaning was intended by Audubon and Bachman. Despite the fact that their use of *humilis* in 1851 could be interpreted under the rules of the International Code of Zoological Nomenclature (Article 33a) as a "justified emendation," most subsequent workers have continued to preserve the original orthography, perhaps on the ground that Audubon and Bachman did not formally declare that they were emending the spelling. The subspecific name of one of the races inhabiting Louisiana, *merriami*, is a patronymic proposed by J. A. Allen to honor C. Hart Merriam, a famous mammalogist and the first head of the governmental organization now known as the Bureau of Sport Fisheries and Wildlife of the Department of the Interior.

Distribution. — The Eastern Harvest Mouse occurs in the southeastern part of the United States from Maryland, southern Ohio, Kentucky, and western Arkansas south to southern Florida, the northern Gulf Coast, and southeastern Texas. In Louisiana the species is rare or absent in the northern part of the state and uncommon and highly local in its distribution in the southern part. (Map 32)

External appearance. — This diminutive species is largely brownish above, though sometimes blackish down the middle of the back. The underparts are ash-colored. Specimens in fresh, unworn pelage show an indistinct lateral stripe that is light Pinkish Cinnamon. The small size and the longitudinal grooves on the front surface of the upper incisors immediately place individuals of this species in the genus *Reithrodontomys*. The Eastern Harvest Mouse can be distinguished from the Fulvous Harvest Mouse *(R. fulvescens)* by its diminutive size, duller coloration, and short tail (always less than 70 mm instead of considerably more).

Color phases. — No color phases are known.

Measurements and weights. — Twelve skins in the LSUMZ from various localities in the state averaged as follows: total length, 120 (115–132); tail, 56 (52–65); hind foot, 16 (11–18); ear, 10 (8–12). Fifteen skulls from the same source averaged as follows: greatest length, 19.1 (18.0–21.1); cranial breadth, 9.2 (8.7–10.3); zygomatic breadth (only 6 specimens), 9.3 (9.0–9.9); interorbital breadth, 3.1 (2.8–3.3); length of nasals, 6.7 (6.4–7.8); palatilar length, 7.4 (6.9–8.2); postpalatal

length, 6.5 (6.1–6.9). Five specimens averaged 7.8 (6.0–10.5) grams.

Cranial and other skeletal characters. — The grooved upper incisors are characteristic of all members of the genus *Reithrodontomys*. The skull of the present species is separable from that of *R. fulvescens* by its much smaller size. Furthermore, the last lower molar has the second primary enamel fold distinctly longer than the first primary fold, which extends less than halfway across the tooth. The so-called major fold is indistinct or absent. In *fulvescens* the first primary enamel fold is as long as or longer than the second primary fold and extends more than halfway across the tooth. The major fold is clearly visible. (Figure 101.) The penis bone is a slender rod with a slightly bulbous base. The shaft is somewhat curved dorsoventrally. It measures approximately 5.4 mm in length and the width of the bulbous base is only 0.4 mm (Blair, 1942). (Figures 72 and 99)

Sexual characters and age criteria. — Superficially the sexes are alike. Juveniles, in addition to being smaller, have the upperparts blackish gray with only a faint tinge of brown on the sides of the body. Their underparts are gray.

Status and habits. — The Eastern Harvest Mouse is decidedly uncommon in Louisiana. Three specimens have been taken in Caddo Parish, but the species has been recorded nowhere else in the northern half of the state. In southern Louisiana it has been captured in only eight parishes. Even East Baton Rouge Parish, where systematic trapping has been carried on for nearly 40 years, has yielded only 26 preserved specimens. Shadowen (1956), in an intensive study of the mammals residing in a 1.67-acre uncultivated field near the Louisiana State University campus between 25 January and 25 November 1955, captured only two males and five females of this species. At another study plot, consisting of 0.74 acres near Kleinpeter, approximately

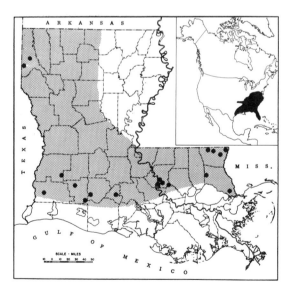

Map 32 Distribution of the Eastern Harvest Mouse *(Reithrodontomys humulis)*. *Shaded area,* known or presumed range within the state; •, localities from which museum specimens have been examined; *inset,* overall range of the species.

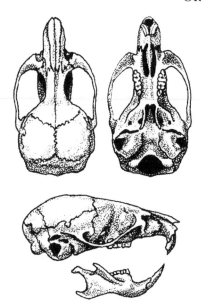

Figure 99 Skull of Eastern Harvest Mouse *(Reithrodontomys h. humulis)*, sex ?, 7.5 mi. S Varnado, Washington Parish, LSUMZ 10364. Approx. 2 ×.

10 miles below the University Campus, he found them somewhat more numerous for he captured 21 in a one-year period between November 1954 and November 1955. All these animals were marked and released, and many were subsequently recaptured as part of his efforts to learn something of the animal's home range, territoriality, and population density.

The Eastern Harvest Mouse lives in abandoned fields, usually those containing a heavy stand of broomsedge *(Andropogon)*. It also is found in weed-filled ditches, in briar thickets, and under tangles of honeysuckle. The nest is either on the ground or a foot or so above the ground in a clump of grass. Breeding probably occurs throughout the year with most births taking place between March and November. The gestation period is about 21 to 22 days, and litter size usually ranges from two to five young, the mean being approximately three. In view of the small size of the Eastern Harvest Mouse, the report of Dunaway (1962) of a female weighing 17 grams

that gave birth to a litter of 8 young with a combined weight of 7.77 grams is of considerable interest. Kaye (1959) found that newborn young weigh approximately 1.2 grams but that soon after their eighth week they have attained a body weight of 8 grams. According to Layne (1959), the eyes do not open until the seventh to tenth day and weaning takes place between the second and fourth weeks.

The food of this species consists almost wholly of weed seeds (Figure 100), grain, and

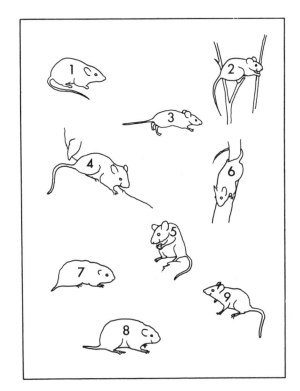

Plate VI
1. Hispid Pocket Mouse *(Perognathus hispidus)*.
2. Fulvous Harvest Mouse *(Reithrodontomys fulvescens)*.
3. Eastern Harvest Mouse *(Reithrodontomys humulis)*.
4. Cotton Mouse *(Peromyscus gossypinus)*.
5. White-footed Mouse *(Peromyscus leucopus)*.
6. Golden Mouse *(Ochrotomys nuttalli)*.
7. Woodland Vole *(Microtus pinetorum)*.
8. Prairie Vole *(Microtus ochrogaster)*.
9. House Mouse *(Mus musculus)*.

PRATT
1971

green vegetation, most of which is of no particular value to man. For this reason and because of its general scarcity in the state, it occupies no position of economic importance.

Predators. — Like all rats and mice that live in open fields, this species is doubtless preyed upon by hawks and owls, as well as other predators.

Parasites. — I am aware of only one external parasite that has been reported on the Eastern Harvest Mouse, the arthropod *Cheyletus eruditus*. The only internal parasite that I have found listed for this species is the trematode *Zonorchus komareki* (McKeever, 1971).

Subspecies

Reithrodontomys humulis humulis (Audubon and Bachman)

Mus humulis Audubon and Bachman, Proc. Acad. Nat. Sci., Philadelphia, 1, 1841:97; type locality, Charleston, Charleston County, South Carolina.
Mus humilis, Audubon and Bachman, Quadrupeds of North America, Vol. 2, 1851:103.
Reithrodontomys humulis, Osgood, Proc. Biol. Soc. Washington, 20, 1907:49.
Reithrodontomys humulis humulis, Crain and Packard, Jour. Mamm., 47, 1966:324.

I am following Hooper (1943) in his delineation of the races of this species. He considers specimens from east of the Mississippi River, in the Florida Parishes, referable to *humulis*, those west of the river, to *merriami*. The series from East Baton Rouge Parish is now large enough to show that it possesses the characters that Hooper ascribes to *humulis*. The limited material from the western part of the state seems to qualify as being representative of *merriami*. In *humulis* the dorsal coloration of the unworn adult winter pelage is a reddish brown without a pronounced blackish vertebral stripe, whereas in *merriami* the color of the upperparts is a mixture of black and Pinkish Buff or Light Pinkish Cinnamon, the black being predominant on the midback and usually forming a sharply defined middorsal stripe about one

Figure 100   Eastern Harvest Mouse *(Reithrodontomys humulis),* Owensville, Clermont County, Ohio, June 1957. (Photograph by Woodrow Goodpaster)

centimeter in width, extending from the forehead to the base of the tail.

Specimens examined from Louisiana. — Total 35, as follows: *East Baton Rouge Par.:* various localities, 25 (LSUMZ). *St. Tammany Par.:* 0.75 mi. S Goodbee, 1 (LSUMZ); 1 mi. N Slidell, 3 (TU). *Washington Par.:* 2 mi. N Angie, 3 (LSUMZ); 7.5 mi. W Varnado, 1 (LSUMZ); Warnerton, 1 (TU); Hackley, 1 (USNM)

Reithrodontomys humulis merriami J. A. Allen

Reithrodontomys merriami J. A. Allen, Bull. Amer. Mus. Nat. Hist., 7, 1895: 119; type locality, Austin Bayou, near Alvin, Brazoria County, Texas.
Reithrodontomys humulis merriami, A. H. Howell, N. Amer. Fauna, 36:21.

Specimens examined from Louisiana. — Total 14, as follows: *Acadia Par.:* Mermentau, 4 (FM). *Beauregard Par.:* 1.5 mi. E Edith, 1 (LSUMZ). *Caddo Par.:* 8 mi. W Shreveport, 1 (LSUMZ); 5 mi. NW Greenwood, 2 (LSUMZ). *Calcasieu Par.:* Edgerly, 1 (LSUMZ). *Jefferson Davis Par.:* Lake Arthur, 1 (LSUMZ); 1.1 mi. N Fenton, 2 mi. W Hwy. 165, 1 (LSUMZ). *Lafayette Par.:* Lafayette, 3 (USNM).

FULVOUS HARVEST MOUSE   Plate VI
(Reithrodontomys fulvescens)

Vernacular and other names.—Like the preceding species, the Fulvous Harvest Mouse has acquired no truly vernacular names. But, during the period when all subspecies were given English names, the form that occurs in Louisiana, *aurantius,* was called the Golden Harvest Mouse. The present-day practice is to drop vernacular names for subspecies and to maintain only one English name for a species as a whole. The derivation of *Reithrodontomys,* the first part of the scientific name, has already been given in the account of the preceding species. The second part, *fulvescens,* is a Latin word that means reddish yellow, tawny, or gold colored, and refers to the color of the sides of the body.

Distribution.—This species occurs from southeastern Kansas and southern Missouri south through the western three-fourths of Arkansas, southwestern Mississippi, all of Louisiana, most of Oklahoma, the eastern two-thirds and the southwestern portion of Texas, and southeastern Arizona, thence southward over most of Mexico to Nicaragua. In Louisiana it has been taken in all but a few parishes and it will undoubtedly, sooner or later, be found in those parishes. (Map 33)

External appearance.—The Fulvous Harvest Mouse is one of the most attractive of our small mammals. The color above is a golden brown, sometimes with a strong interspersion of black down the center of the dorsum. The distinctive feature, however, is the color of the sides of the face and body, which are a rich Vinaceous-Tawny or Ochraceous-Orange. The underparts are grayish white, strongly washed with Light Pinkish Cinnamon. The tail is longer than the head and body and in adults always measures over 80 mm. It is moderately bicolored, the top half being dark grayish, the lower half whitish. The long tail and the rich color of the sides make this species easily separable from the Eastern Harvest Mouse *(R. humulis).*

Color phases.—No color phases nor color aberrations are known.

Measurements and weights.—One hundred and thirty-three specimens in the LSUMZ from numerous localities in the state averaged as follows: total length, 158 (142–200); tail, 88 (80–100); hind foot, 19 (15–22); ear, 13 (9–16). One hundred and eight skulls from the same source averaged as follows: greatest length, 21.6 (19.6–23.1); cranial breadth, 10.1 (9.6–11.5); zygomatic breadth, 10.9 (9.8–11.7); interorbital breadth, 3.2 (3.0–3.7); length of nasals, 8.0 (6.2–9.0); palatilar length, 8.1 (6.6–9.2); postpalatal length, 7.6 (6.0–9.6). The weights of 49 specimens averaged 12.2 (8.5–17.8) grams. The average weight of 69 males trapped alive at Baton Rouge and then released was 9.6 (4.0–13.0) grams; that of 39 females, 9.6 (5.0–14.5) grams (Shadowen, 1956).

Cranial and other skeletal characters.—As in all members of the genus *Reithrodontomys,* the

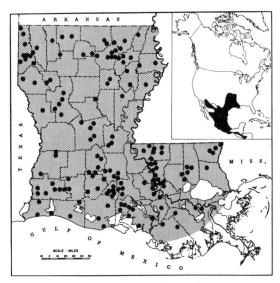

Map 33 Distribution of the Fulvous Harvest Mouse *(Reithrodontomys fulvescens). Shaded area,* known or presumed range within the state; •, localities from which museum specimens have been examined; *inset,* overall range of the species.

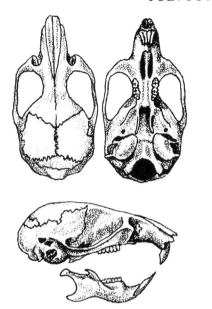

Figure 101 Skull of Fulvous Harvest Mouse (*Reithrodontomys fulvescens aurantius*), ♂ , jct. Perkins and Highland roads, East Baton Rouge Parish, LSUMZ 15228. Approx. 2 ×.

front surface of each upper incisor possesses a longitudinal groove. Only two other mice in Louisiana show this diagnostic feature, *R. humulis* and *Perognathus hispidus*. The latter is readily distinguished by other characters, such as the perforation in the rostrum between the infraorbital foramina, the extension of the nasals well beyond the incisors, and the enlarged auditory bullae. From *R. humulis* the skull of the present species is easily separated on the basis of its larger size, by its wider interpterygoid fossa, and by the pattern of the enamel folds on the last lower molar discussed, under the cranial characters of the Eastern Harvest Mouse. The penis bone of *fulvescens*, like that of *humulis*, is rod-shaped except for a slightly bulbous proximal end. The structure is approximately 8 mm in length with the bulbous base about 1 mm in diameter. (Figures 72 and 101)

Sexual characters and age criteria.—The sexes cannot be distinguished externally except by inspection of the sex organs and related structures. Immature individuals are some-what less ochraceous on the sides and darker above than subadults and adults. The difference between young and adults is much less well-marked than in *Peromyscus* and its related genera, such as *Ochrotomys*. Growth is rapid. Harvest mice, like all cricetine rodents, reach sexual maturity in two months or less, and an individual that survives for as long as a year can be said to have reached a "ripe old age."

Status and habits.—The Fulvous Harvest Mouse is one of the commonest of all the small mammals in the state. It occurs almost everywhere that it finds its suitable habitat of uncultivated fields, briar thickets on the borders of woodlands, dense tangles of low vegetation adjacent to fencerows, and similar situations. Dainty and otherwise attractive in appearance, it is always a source of delight to the mammalogist who finds one of them in his trapline. And this he is likely to do fairly often, for the species is widespread and sometimes abundant in open habitats where the cover of grasses or sedges is dense. In almost any field it is likely to be found with the Hispid Cotton Rat (*Sigmodon hispidus*) and the Least Shrew (*Cryptotis parva*), and, where the terrain is damp, with the Marsh Rice Rat (*Oryzomys palustris*). It is also common on the chenieres, canal banks, and other reasonably high ground in our coastal areas.

Nests of the Fulvous Harvest Mouse are usually built one to three feet above the ground. They are made of finely shredded blades of grasses and sedges, intricately woven into a compact mass about the size of a baseball. In most cases the nests house a pair of mice or else a female and her litter. The young at birth are pink, blind, hairless, and only slightly over one gram in weight (Svihla, R., 1930a).

In his excellent year-long study of rodent population dynamics in two uncultivated fields near the Louisiana State University Campus at Baton Rouge, Shadowen (1956) discovered many interesting facts about the Fulvous Harvest Mouse, including the following: the species was seldom active during the day; the peak density was 18.9 mice per acre; the size of the home range was larger in winter than in summer; the average home range of males was 12,384 square feet, that of females, 13,065 square feet; and the number of males was considerably higher than the number of females.

The food of this species consists almost entirely of weed seeds, but it occasionally eats green vegetable matter. Only rarely, if ever, does it do any significant damage to crops.

Predators.—Fulvous Harvest Mice are eaten by hawks and owls. Svihla (1930a) reported the remains of four in one pellet of a Barred Owl *(Strix varia),* and I have found them in the pellets of Barn Owls *(Tyto alba)* and in the stomachs of Red-tailed Hawks *(Buteo jamaicensis).* Svihla considered the Water Moccasin *(Agkistrodon piscivorus)* and the Racer *(Coluber constrictor)* to be possible major predators on Fulvous Harvest Mice in the coastal marshes, where I am sure the Marsh Hawk *(Circus cyaneus)* also takes many of them.

Parasites.—A perusal of the parasitological literature has revealed no information on the parasites of this species.

Subspecies

Reithrodontomys fulvescens aurantius J. A. Allen

Reithrodontomys mexicanus aurantius J. A. Allen, Bull. Amer. Mus. Nat. Hist., 7, 1895:137; type locality, Lafayette, Lafayette Parish, Louisiana.
Reithrodontomys fulvescens aurantius, A. H. Howell, N. Amer. Fauna, 36, 1914:48.

All populations of this species in Louisiana are referable to the subspecies *aurantius,* which occupies the easternmost part of the range of the species, occurring in Mississippi, Louisiana, Arkansas, southern Missouri, southern Kansas, eastern Oklahoma, and eastern Texas. It is the darkest and most richly colored of all the races of the species.

Specimens examined from Louisiana.—Total 709, representing all parishes except Concordia, East Carroll, Jackson, La Salle, Orleans, Point Coupee, St. Bernard, St. Helena, St. James, St. John the Baptist, Vernon, West Baton Rouge, and Winn (306 LSUMZ; 196 USL; 80 NLU; 33 LTU; 29 USNM; 17 SLU; 14 FM; 10 NSU; 7 LSUS; 6 ANSP; 6 MSU; 5 MCZ).

Genus *Peromyscus* Gloger

Dental formula:

$$I\frac{1-1}{1-1} \; C\frac{0-0}{0-0} \; P\frac{0-0}{0-0} \; M\frac{3-3}{3-3} = 16$$

WHITE-FOOTED MOUSE   Plate VI
(Peromyscus leucopus)

Vernacular and other names.—Although the mice of the genus *Peromyscus* are sometimes collectively called white-footed mice or deer mice, the former name is the one most often applied in a specific sense to the present animal. Some authors call it the Woodland White-footed Mouse to distinguish it from *Peromyscus maniculatus,* which they call the Prairie White-footed Mouse. I would prefer, as does Hall (1965), to reserve the name Deer Mouse for the latter species. There is no problem with respect to *P. gossypinus,* which has long been called the Cotton Mouse. The generic name *Peromyscus* is of questionable etymology. The last part of the word is derived from the Greek *myskos,* meaning little mouse. But the first part of the word is said by some authors to come from the Latin *pero,* meaning pointed, allegedly referring to the pointed shape of the skull produced by the long rostrum. Other authors say it is derived from the Greek *pera,* meaning pouched, possibly referring to the internal cheek pouches, or from the Greek *peros,* meaning maimed or deformed, which makes no sense at all. Unfortunately, the author of the genus, Gloger, supplied no clue as to his intentions. The second part of the scientific name, *leucopus,* is a combination of two Greek words, *leukon,* meaning white, and *pous,* meaning foot.

Distribution.—The White-footed Mouse occurs from Nova Scotia, southern Maine, and

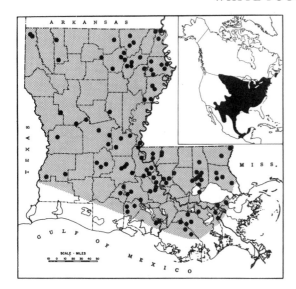

Map 34 Distribution of the White-footed Mouse *(Peromyscus leucopus). Shaded area,* known or presumed range within the state; •, localities from which museum specimens have been examined; *inset,* overall range of the species.

southern Ontario, across Michigan, Wisconsin, Minnesota, North Dakota, southern Saskatchewan, and southeastern Alberta south through the eastern United States and most of the Great Plains to central South Carolina, northern Georgia, central Alabama, most of Mississippi, southern Louisiana, eastern Texas, and southeastern Arizona, thence over most of north-central and eastern Mexico to Oaxaca and the western portion of the Yucatan Peninsula. In Louisiana it occurs more or less throughout the wooded regions of the state, often sympatrically with its close relative the Cotton Mouse. (Map 34)

*External appearance.—*All members of the genus *Peromyscus* can be immediately recognized by a combination of characters: their soft, lax pelage; huge, black eyes; large ears; bicolored tails; white feet; and white underparts that contrast sharply with their upperparts. The present species is very similar to the Cotton Mouse *(Peromyscus gossypinus)* but is smaller. The hind foot in adults is

usually 20 mm or less in length, while that of the adult Cotton Mouse is usually 21 mm or more. The two species are best distinguished by the size of the skull (see beyond).

*Color phases.—*No color phases nor color aberrations have been observed in specimens from Louisiana.

*Measurements and weights.—*Eighty-nine specimens from various localities throughout the state in the LSUMZ averaged as follows: total length, 157 (134–177); tail length, 69 (51–87); hind foot, 20 (17–21); ear, 15 (11–19). Seventy-eight skulls from the same source averaged as follows: greatest length, 25.8 (23.8–26.5); cranial breadth, 11.8 (11.2–13.0); zygomatic breadth, 13.5 (11.8–14.6); interorbital breadth, 4.1 (3.5–4.9); length of nasals, 9.1 (7.5–10.9); length of diastema, 6.6 (5.9–7.5); palatilar length, 10.1 (9.0–10.8); postpalatal length, 9.0 (8.2–10.2). Weights of eighteen individuals averaged 19.1 (17.0–23.0) grams.

*Cranial and other skeletal characters.—*The skull of *Peromyscus leucopus* is almost identical with that of *P. gossypinus* except for its slightly smaller size. Yet a comparison of the skulls of these two species provides the best way of telling them apart. In *P. leucopus* the total length of the skull of adults is close to 25.3 mm, but in *P. gossypinus* it is near 28.0 (Huckaby, 1967). Other skull measurements similarly show a smaller size in *P. leucopus*. The penis bone of this mouse is rod-shaped except for a slightly expanded base. Hooper (1958) gives its length as approximately 9 mm. (Figures 72 and 102)

*Sexual characters and age criteria.—*The sexes are indistinguishable except by inspection of the sex organs. The immatures, sometimes called "pups," are easily recognized as such by their gray dorsal coloration. Subadults acquire an admixture of brown and ochraceous, especially on the sides, but apparently

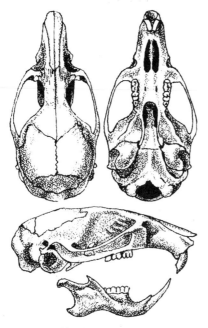

Figure 102 Skull of White-footed Mouse *(Peromyscus leucopus leucopus)*, ♂, 4 mi. SSE Gray, Plaquemines Parish, LSUMZ 16051. Approx. 2 ×.

only in old adults do the upperparts become uniformly rich brown.

Status and habits. — The White-footed Mouse is widely distributed and often abundant in the wooded areas of the state. In numerous places it occurs side by side with its close relative the Cotton Mouse *(P. gossypinus)*. The two species are structurally and behaviorly so similar that one cannot avoid wondering how they can occur in the same areas without a wholesale breakdown in the barriers that keep them from interbreeding. Several workers (Dice, 1937 and 1940; Bradshaw, 1965 and 1968; McCarley, 1964) have amply demonstrated that the two species will hybridize in the laboratory. There, however, each still shows a strong tendency, when given a choice, to mate with one of its own kind. Both Howell (1921) and McCarley (1954a and 1954b) believed that they were able to discern particular museum specimens that represent natural hybrids, but McCarley attributed

their overall rarity to an ecological isolating mechanism that serves under most circumstances to keep the two species apart. He contends that where the two come together in eastern Texas, *leucopus* is usually restricted to the uplands and *gossypinus* to the lowlands. In Louisiana, however, the reverse is often the case, and in many places the two species occur together. In the Tunica Hills, for example, I have found that *gossypinus* is by far the commonest form, but, in the bottomland on the opposite side of the Mississippi River, *leucopus* predominates. On the other hand, Huckaby (1967) reports taking 10 specimens of *leucopus* and 9 specimens of *gossypinus* in Tensas Parish in a large tract of homogeneous hardwood bottomland forest. At other bottomland sites in East Carroll, Madison, St. Landry, and West Baton Rouge parishes, populations of both species have been found together and specimens of both have been taken in the same trapline on the same night. If the two species were to hybridize, they would have ample opportunity to do so under these circumstances, in which ecological isolation such as McCarley found in eastern Texas appears to be lacking. A study of specimens from sympatric populations in Louisiana seldom poses any problem in assigning individual skins to one species or the other.

Even though the two species are extremely similar in appearance and obviously very closely related, there can be little or no doubt that they have arrived at a stage in their evolutionary development where they are, at

Tracks of White-footed Mouse

least for the most part, reproductively isolated. Consequently, where they come together, a minimum of interbreeding (and hence little gene flow) takes place between them. They are able, therefore, to maintain their morphological identities. Blair (1950) is of the opinion that spatial isolation in the Pleistocene was perhaps sufficient to allow the development of intrinsic isolating mechanisms between *leucopus* and *gossypinus*. Ethological, and possibly physiological, psychological, and, in some instances, ecological differences may now be operating in combination with the minor morphological differences between the two in reinforcing reproductive isolation.

The works of all the authorities cited above have contributed a great deal to a better understanding of what is taking place between these two sibling species, but much still remains to be learned. When we finally know more about the normal breeding behavior and interspecific interactions of these mice in the wild, we shall not only have gained a better insight into their relationship to each other, but we shall have arrived at a better understanding of the whole process of speciation.

The White-footed Mouse lives either in deep woods or in brushy areas in border situations. I have taken specimens of it most often by setting traps along rotten logs or at the base of stumps in the forest. It builds its nest under logs, in holes in trees, and sometimes in outbuildings. Occasionally, because it is at least semiarboreal (Figure 103), it will take over an old nest of a squirrel or bird. Wherever the mouse's nest may be located, on the ground or in a tree, it is lined with finely shredded, soft materials. It often builds more than one nest, and a new clean nest is generally constructed to hold the two to six naked young. Breeding occurs the year around, but most of the young are born in late fall and early winter. Pournelle (1952) has shown that at least part of the decrease in reproductive productivity in summer in

Figure 103 White-footed Mouse *(Peromyscus leucopus)*, Hamilton County, Ohio, June 1961. (Photograph by Woodrow Goodpaster)

P. gossypinus can be attributed to a decrease in the fertility of the male mice when the ambient temperature reaches or exceeds 89° F. I am confident the same thing would be true of *leucopus* and perhaps also of the Golden Mouse *(Ochrotomys nuttalli)*.

Young females begin to produce litters when only five or six weeks of age. The gestation period is about 21 days. A litter consists of one to seven, usually four, young. Although home ranges may overlap in part, most of the home range of a breeding female is not crossed by another breeding female. This fact suggests territorial behavior on the part of females while breeding (Burt, 1940).

Predators.—Few White-footed Mice live as long as a year, and probably few, if any, ever die of old age. From the time they first leave the nest, and even before, they are subject to predation by a host of enemies—owls, weasels, foxes, feral house cats, and snakes, to name a few.

Parasites and diseases.—Numerous external parasites are known to attack White-footed Mice, one of the most serious being the scab mite *(Sarcoptes scabiei),* which causes the ears and tail to become scabby and the hair to fall out. Burt (1940) was able to follow the case history of a female infested with the scab parasites. He first captured, marked, and released her on 3 June, at a time when she was nursing young. When he recaptured her on 18 June, she had begun to show the first signs of infection. Hair was falling from her forehead. On 30 June, she had lost the hair over most of her back and a scab was forming on her head. By 5 July her tail was swollen and she appeared to be in serious condition. On 28 July she was noticeably better, and by 12 August the hairs on her back had been mostly replaced, although her forehead was still bare and her tail slightly swollen. Less than two weeks later, on 25 August, she had recovered completely. In brief, the disease ran its course in approximately two months. Another important parasite is the botfly *(Cuterebra)* that lays its eggs on the skin, beneath which the larvae develop into what are popularly called "warbles." Few internal parasites have been recorded for this species. Doran (1954a and b, 1955a) lists two protozoans *(Chilomastix bettencourti* and *Giardia microti),* one cestode *(Choanotaenia peromysci),* and six nematodes *(Mastophorus muris, Rictularia onychomys, Syphacia peromysci, S. samorodini, Apiscularis americana,* and *Capillaria americana).*

Subspecies

Peromyscus leucopus leucopus (Rafinesque)

Musculus leucopus Rafinesque, Amer. Monthly Mag., 3, 1818:446; type locality, pine barrens of Kentucky.

Peromyscus leucopus, Thomas, Ann. Mag. Nat. Hist., ser. 6, 6, 1895:192.
Peromyscus leucopus brevicaudus Davis, Occas. Papers Mus. Zool., Louisiana State Univ., 2, 1939:1; Lowery, Occas. Papers Mus. Zool., Louisiana State Univ., 13, 1943:244.

Osgood (1909), in his analysis of the geographical variation in this species, recognized only the nominate subspecies, *P. leucopus leucopus,* in the southeastern United States and the Lower Mississippi Valley. He pointed out that, at least in the eastern part of the United States, the darkest specimens are from southern Louisiana, the lightest and richest colored are from the Upper Mississippi Valley and the northeastern states. The latter he referred to *P. l. noveboracensis.* The variation follows a fairly distinct north-south cline. Unfortunately, the type locality of *leucopus* is in Kentucky, near the middle of the cline. But this circumstance would not justify, in my opinion, naming the southern extremes. Indeed, I am not convinced that clinal variations either merit or require taxonomic recognition. Davis, in 1939, described the populations of this mouse in eastern Texas under the name *brevicaudus.* He contended that they were smaller than *leucopus* to the east, especially with regard to tail length and certain cranial measurements. Subsequently, in 1943, I referred material from western Louisiana to *brevicaudus.* But, after studying many additional specimens from the western part of the state and noting the great variation within population samples from elsewhere in Louisiana, I now believe that all White-footed Mice in the state are best treated as members of a single subspecies, *P. l. leucopus.* Moreover, McCarley (1959) contends that even in populations of White-footed Mice from eastern Texas he was unable to find any basis for recognizing *brevicaudus,* which he therefore synonymized with *P. l. leucopus.*

Specimens examined from Louisiana.—Total 342, as follows: *Avoyelles Par.:* 14 mi. ENE Marksville, Long Lake, 5 (LSUMZ); Evergreen, 2 (MSU); 2 mi. W Cottonport, 2 (USL); 1 mi. S Cottonport, 1 (USL); 3 mi. E Marksville, 7

(USL). *Beauregard Par.*: 1.5 mi. E Edith, 1 (LSUMZ); 6 mi. SE Fields, 1 (LSUMZ). *Caddo Par.*: 4 mi. NW Hosston, 2 (LSUMZ); 2.8 mi. W Hosston, 2 (LSUMZ); 2.5 mi. NW Hosston, 1 (LSUMZ). *Concordia Par.*: Point Breeze, 11 (LSUMZ). *East Baton Rouge Par.*: 10 mi. E LSU on Old Ham. Hwy. 2 (LSUMZ); Baton Rouge, 4 (LSUMZ); 3 mi. S LSU in Nelson's Woods, 1 (LSUMZ); 3 mi. E LSU Campus, 2 (LSUMZ); 8.4 mi. SE University, 2 (LSUMZ); 3.5 mi. E Baton Rouge, 2 (LSUMZ); near LSU Nursery, 1 (LSUMZ); 4 mi. SE BR Quail Hatchery, 1 (LSUMZ); 0.5 mi. W LSU, 1 (LSUMZ); 2 mi. SE LSU, 1 (LSUMZ). *East Carroll Par.*: 2.5 mi. W Gassoway on Macon Bayou, 1 (LSUMZ); 7 mi. E Oak Grove on La. Hwy. 11, 1 (LSUMZ). *East Feliciana Par.*: Jackson (Carr's Creek), 2 (LSUMZ); 3 mi. S Jackson, 2 (LSUMZ); 27 mi. N Baton Rouge, 1 (LSUMZ); 4 mi. N Jackson, 1 (LSUMZ). *Evangeline Par.*: 4 mi. WNW Ville Platte, 1 (LSUMZ); 8 mi. W Ville Platte, 1 (USL); 4 mi. W Easton, 4 (USL); 1.7 mi. W Easton, 1 (USL). *Franklin Par.*: 9 mi. S Delhi, 2 (NLU). *Iberia Par.*: Avery Island, 6 (4 TU; 1 CAS; 1 USNM); 5 mi. SE New Iberia, 1 (USL); 5 mi. W New Iberia, 1 (USL). *Iberville Par.*: 1.5 mi. S Indian Village, 1 (LSUMZ); ca. 6 mi. W, 1 mi. N Plaquemine, 2 (LSUMZ); 6.15 mi. S, 3.7 mi. W Addis, 1 (LSUMZ); 11 mi. SSW LSU, 2 (LSUMZ). *Jefferson Par.*: 4 mi. S Gretna, 2 (LSUMZ). *Lafayette Par.*: Lafayette, 2 (USNM); 6 mi. E Carencro, 1 (USL); 1 mi. S Lafayette, 16 (USL). *Lafourche Par.*: Raceland, 1 (LSUMZ); 9.4 mi. NE Thibodaux, 4 (LSUMZ); 4 mi. SE Gray, 5 (LSUMZ). *La Salle Par.*: 0.5 mi. W White Sulphur Springs, 3 (LSUMZ). *Lincoln Par.*: 2.5 mi. N Tremont, 1 (LSUMZ). *Livingston Par.*: 2 mi. S Watson, 1 (LSUMZ); Springfield, 2 (SLU). *Madison Par.*: Waverley, 2 (LSUMZ); 11 mi. SSW Tendal on Tensas River, 1 (LSUMZ); McGill Bend of Tensas River, 5 (LSUMZ); 7 mi. NE Tallulah on Miss. River, 1 (LSUMZ); McGill Bend of Tensas River about 2 mi. above line, 1 (LSUMZ); near Singer Tract, 1 (LSUMZ); Tallulah, 5 (USNM). *Morehouse Par.*: 4 mi. SE Bastrop, 1 (LSUMZ); Chemin a Haut St. Park, 1 (LSUMZ); 4 mi. N Collinston, 1 (LSUMZ); Mer Rouge, 5 (4 USNM; 1 USL). *Natchitoches Par.*: Vowells Mill, 1 (LSUMZ); Mora, 1 (LSUMZ). *Orleans Par.*: New Orleans, 1 (LSUMZ); 11 mi. E New Orleans on Hwy. 90, 1 (LSUMZ). *Ouachita Par.*: 1 mi. SE Monroe, 2 (NLU); Russell Sage Game Mgt. Area, 3 (NLU); Luna Fire Tower Area, 2 (NLU); 1 mi. N Monroe, 2 (NLU); Swartz, 2 (NLU); Monroe, 3 (NLU). *Plaquemines Par.*: 6 mi. SW Callender Naval Air Station, 2 (LSUMZ). *Pointe Coupee Par.*: 1 mi. W Erwinville, 8 (LSUMZ); 5 mi. E Livonia, 4 (LSUMZ); 2.7 mi. W Maringouin, 2 (LSUMZ); 18 mi. NW Baton Rouge, 1 (LSUMZ); 5 mi. E Batchelor, 1 (LSUMZ); 1 mi. S Old River locks on E. Bank of Miss. River, 3 (USL). *Rapides Par.*: 5 mi. E Lena Station, 1 (LSUMZ); 3 mi. SE Alexandria, 1 (USL); 3.6 mi. N Cheneyville, 1 (USL); 2 mi. N Avoyelles-Rapides Parish line off La. Hwy. 1, 2 (USL). *Richland Par.*: 3.2 mi. S, 1.6 mi. W Rayville, 2 (LSUMZ); East end of Clear Lake, 2 (NLU); 6 mi. S Rayville, Hwy. 47, 1 (NLU); 4 mi. N Holly Ridge, 1 (NLU); 5 mi. S Start, 1 (NLU). *St. Charles Par.*: 3 mi. WNW Paradis, 1 (LSUMZ). *St. John the Baptist Par.*: 4 mi. S Manchac, 2 (LSUMZ). *St. Landry Par.*: 0.5 mi. E Palmetto, 2 (LSUMZ); 7.25 mi. N, 2.75 mi. E Washington, 4 (LSUMZ); 9 mi. NNE Opelousas, 1 (LSUMZ); ca. 5 mi.

N, 2 mi. E Courtableu, 3 (LSUMZ); ca. 2 mi. NE Washington in Thistlethwaite, 1 (LSUMZ). *St. Martin Par.*: Lake Dauterive, 5 mi. E Loreauville, 5 (USL); 2 mi. S Parks, 9 (USL); 6 mi. NW St. Martinville, 1 (USL). *St. Mary Par.*: 1 mi. W Berwick, 1 (LSUMZ); Morgan City, 8 (USNM); 2.5 mi. E Morgan City, 4 (USL). *St. Tammany Par.*: Covington, 1 (SLU); 6 mi. N Covington, 1 (SLU); Abita Springs, 1 (SLU); Pearl River, 1 (SLU); 3 mi. NE Abita Springs, 1 (USL). *Tangipahoa Par.*: 12.6 mi. S Hammond, 1 (LSUMZ); 6 mi. NE Hammond, 1 (SLU); 1.5 mi. S Hammond, 2 (SLU); locality unspecified, 1 (SLU); 2 mi. E Ponchatoula, 3 (SLU); 1 mi. E Ponchatoula, 1 (SLU); 5.5 mi. NE Hammond, 1 (SLU); 7 mi. N Hammond, 1 (SLU); 4 mi. NW Husser, 5 (SLU); 8 mi. NNE Hammond, 8 (SLU); 3.5 mi. E Hammond, 4 (SLU); 3.25 mi. E Hammond, 1 (SLU); 3.75 mi. E Hammond, 6 (SLU); 3 mi. SW Hammond, 1 (SLU); 0.13 mi. N Hammond, 5 (SLU). *Tensas Par.*: 2.5 mi. NNW Waterproof, 1 (LSUMZ); 0.75 mi. NW Saranac on Clark Bayou, 3 (LSUMZ); 12 mi. W Waterproof, 1 (LSUMZ); 4 mi. S, 4 mi. S Newlight, 12 (LSUMZ). *Terrebonne Par.*: 0.75 mi. NE Houma, 1 (LSUMZ); 1 mi. NE Montegut, 2 (LSUMZ); 3.5 mi. SE Schriever, 2 (LSUMZ); Houma, 17 (USNM). *Union Par.*: 1 mi. S Haile, 1 (NLU). *Vernon Par.*: 5.5 mi. NNW Leesville, 1 (LSUMZ). *Washington Par.*: 2 mi. E Angie, 1 (SLU); 10 mi. E Franklinton, 2 (SLU). *Webster Par.*: 1.5 mi. SE Cotton Valley, 1 (LSUMZ); Lake Bistaneau St. Park, 2 (LSUMZ). *West Baton Rouge Par.*: 8 mi. W Miss. River Bridge, 1 (LSUMZ); 5 mi. W Brusly, 2 (LSUMZ); ca. 6.5 mi. W Brusly, Whiskey Bay, 1 (LSUMZ); 9 mi. W Miss. River on Hwy. 190, 1 (LSUMZ). *West Carroll Par.*: 2.5 mi. E Kilbourne, 2 (LSUMZ); locality unspecified, 1 (FM). *West Feliciana Par.*: 2 mi. N St. Francisville, 1 (LSUMZ); 8 mi. N St. Francisville, 1 (SLU); 3 mi. N St. Francisville, 1 (SLU).

COTTON MOUSE   Plate VI
(Peromyscus gossypinus)

Vernacular and other names.—The name Cotton Mouse has long been applied to this species. The second part of the scientific name, *gossypinus,* is from two Latin words (*gossyp,* a kind of plant, cotton, and *inus,* a suffix meaning belonging to) and implies a relationship to cotton or to cotton fields, where, interestingly enough, the Cotton Mouse is not regularly found. Perhaps Le Conte, who introduced the name, did actually capture one of these mice on the edge of a cotton field in his native state of Georgia and was thereby prompted to choose the name *gossypinus.* For a discussion of other names for this and related species, see the preceding account.

Distribution. — This mouse occurs primarily in the southeastern states from extreme southeastern Virginia, eastern North and South Carolina, southern Georgia, the western three-fourths of Tennessee, extreme southern Illinois, southeastern Missouri, most of Arkansas, and southeastern Oklahoma south to southern Florida, the northern Gulf Coast, and southeastern Texas. In Louisiana, it apparently occurs wherever there are large woods or forest. (Map 35)

External appearance. — The Cotton Mouse is almost an exact replica of the preceding species except for its slightly larger size. The large black eyes, large ears, white feet, the bicolored tail, and the rich, yellowish brown coloration of the upperparts of adults that contrast sharply with the white underparts all serve to distinguish it from any other mice that inhabit our state, except the White-footed Mouse. Its total length is approximately 170 mm (slightly over 4.5 inches) and that of the tail is about 77 mm (3 inches). The hind foot usually measures 22 mm or more.

Color phases. — Although no true color phases are known in this species, Vernon Bailey trapped on entirely melanistic individual at Houma on 8 May 1892 (USNM 33818).

Measurements and weights. — Two hundred and ninety-three skins from various localities throughout the state averaged as follows: total length, 171 (137–210); tail length, 75 (55–94); hind foot, 22 (20–26); ear, 17 (14-22). Two hundred and six skulls from the same source averaged as follows: greatest length, 28.3 (26.6–30.6); cranial breadth, 12.7 (11.7–13.6); zygomatic breadth, 14.2 (12.4–15.5); interorbital breadth, 4.5 (4.0–5.0); length of nasals, 11.0 (8.9–12.9); length of diastema, 7.5 (6.2–8.4); palatilar length, 11.0 (9.3–12.2); postpalatal length, 10.3 (8.7–11.6). Weights of one hundred and twenty-five individuals averaged 31.1 (25.0–45.0) grams.

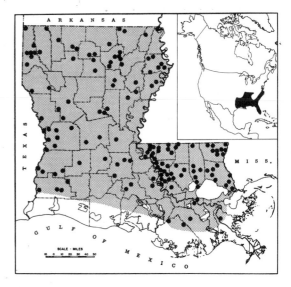

Map 35 Distribution of the Cotton Mouse (*Peromyscus gossypinus*). *Shaded area,* known or presumed range within the state; •, localities from which museum specimens have been examined; *inset,* overall range of the species.

Cranial and other skeletal characters. — The skull of *P. gossypinus* can hardly be confused with that of any other species in the state except *P. leucopus,* from which it differs by its larger size. In the present species the total length of the skull is usually over 28 mm, whereas in *leucopus* it is generally less than 25.4 mm. The length of the diastema, the width of the interorbital constriction, and the size of the auditory bullae all average significantly greater in *gossypinus* than in *leucopus.* The penis bone likewise is like that of *leucopus* except that, paradoxically, it seems to average slightly shorter than in *leucopus* (Hooper, 1958). (Figures 72 and 104)

Sexual characters and age criteria. — The sexes cannot be distinguished except by their genitalia. Like young White-footed Mice, the immatures are uniformly gray above. In the first postjuvenal molt they begin to acquire the first brown and ochraceous hairs on the upperparts, but only old adults possess a rich, uniform brown coloration above.

Status and habits. — The Cotton Mouse, at least in combination with its sibling the White-footed Mouse, is probably the commonest mammal in the state. The only rival to this claim would be the Hispid Cotton Rat *(Sigmodon hispidus)*, which is an inhabitant of uncultivated fields, grassy ditch banks, and brushy areas, whereas the two *Peromyscus* mice live in forest. I suspect that more of the latter kind of habitat is available in Louisiana and, consequently, our *Peromyscus* populations probably outnumber those of *Sigmodon hispidus*. In any event, the Cotton Mouse and the White-footed Mouse are both exceedingly abundant. For a discussion of the relationships of the two, the reader should refer to the previous account. Moreover, what has been said about the habits of the White-footed Mouse applies equally well to the present species. I am aware of no differences between them with respect to behavior although surely some dissimilarities await discovery.

Parasites. — Doran (1954b, 1955a and b) lists one nematode *(Protospirura numidica)*, one cestode *(Spirometra mansonoides)*, and four trematodes *(Eurytrema komareki, Scaphiostomum pancreaticum, Brachylaemus peromysci,* and *Alaria mustelae)* in this species.

Subspecies

Peromyscus gossypinus gossypinus (Le Conte)

Hesp[eromys]. *gossypinus* Le Conte, Proc. Acad. Nat. Sci., Philadelphia, 6, 1853:411; type locality, Le Conte Plantation, near Riceboro, Liberty County, Georgia.

Peromyscus gossypinus, Rhoads, Proc. Acad. Nat. Sci. Philadelphia, 48, 1896:189.

Peromyscus gossypinus nigriculus Bangs, Proc. Biol. Soc. Washington, 10, 1896:124; type locality, Burbridge, Plaquemines Parish, Louisiana.

Peromyscus gossypinus megacephalus, Osgood, N. Amer. Fauna, 28, 1909:138.

Peromyscus gossypinus megacephalus, Lowery, Occas. Papers Mus. Zool., Louisiana State Univ., 13, 1943:244; name applied to certain Louisiana specimens now treated as *P. g. gossypinus.*

Two subspecies of *Peromyscus gossypinus* have been reported to occur in Louisiana, *P. g. gossypinus* (Le Conte) and *P. g. megacephalus* (Rhoads). The former race has been alleged (Osgood, 1909; Lowery, 1943; Hall and Kelson, 1959) to occupy the area lying southeast of a line that might be drawn between the northeastern and southwestern corners of the state. The latter race has been said to range northwest of this line. The race *megacephalus* is supposed to differ from *gossypinus* in averaging larger and paler and in having a skull that is larger and more elongated, with the rostrum and nasals being especially longer. To test the hypothesis that mice from northwestern Louisiana are larger than mice from the southeastern corner of the state, I made statistical analyses of the measurements of 24 skins and 17 skulls of fully adult specimens from Caddo, Bossier, Claiborne, Morehouse, De Soto, Natchitoches, and Sabine parishes and of 198 skins and 149 skulls of adult mice from St. Landry, Pointe Coupee,

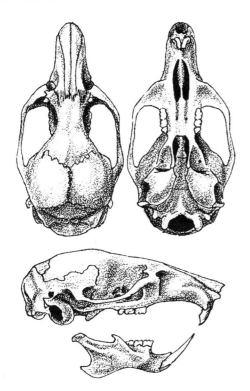

Figure 104 Skull of Cotton Mouse *(Peromyscus gossypinus gossypinus),* ♂, 0.5 mi. NNW Chipola, St. Helena Parish, LSUMZ 11863. Approx. 2 ×.

East Baton Rouge, East and West Feliciana, St. Helena, Tangipahoa, and Washington parishes. Neither the total length, length of the tail, nor any of the cranial measurements analyzed demonstrated any statistically significant differences between the two population samples. As a matter of fact, the means of these measurements in the two populations are identical or almost so. Likewise, I could find no indication of a color difference between the two populations. Consequently, I am satisfied that all specimens of the Cotton Mouse from the state are assignable to a single subspecies, namely, *gossypinus*. I do not mean necessarily to imply that *megacephalus* is not a valid race. An examination of much additional material, especially of topotypical specimens, would be required in order to resolve that question satisfactorily.

The sample mentioned above, from northwestern Louisiana, and the one from the southeastern corner of the state measured respectively as follows: total length, 171 (155–192) *vs* 171 (140–200); tail, 74 (65–84) *vs* 75 (55–92); hind foot, 22 (21–22) *vs* 22 (20–25); ear, 18 (15–20) *vs* 17 (14–22); greatest length of skull, 28.4 (26.7–30.5) *vs* 28.3 (26.6–30.6); cranial breadth, 12.7 (11.9–13.4) *vs* 12.7 (11.7–13.6); zygomatic breadth, 14.1 (13.2–14.9) *vs* 14.2 (12.4–15.2); interorbital breadth, 4.4 (4.2–4.7) *vs* 4.5 (4.0–4.9); length of nasals, 10.9 (10.1–12.5) *vs* 10.9 (8.9–12.9); length of diastema, 7.3 (6.8–7.7) *vs* 7.5 (6.2–8.4); palatilar length, 11.0 (10.4–11.4) *vs* 11.0 (9.6–12.0); postpalatal length, 10.2 (9.8–11.4) *vs* 10.2 (8.7–11.6). Fifteen specimens of the northwestern Louisiana sample averaged 30.1 (25.4–38.1) grams, while 74 specimens from the southeastern Louisiana sample averaged 32.0 (25.5–41.0) grams.

Specimens examined from Louisiana. — Total 763, as follows: *Allen Par.*: 4.5 mi. NW Oberlin, West Bay GMA, 5 (LSUMZ). *Ascension Par.*: Gonzales, 4 (LSUMZ). *Avoyelles Par.*: 2.5 mi. ENE Mansura, 5 (USL). *Beauregard Par.*: 1.5 mi. E Edith, 5 (LSUMZ); Brushy Creek at State Hwy. 143, 1 (LSUMZ); 8.5 mi. N Merryville on Bayou Ana-

coco, 2 (LSUMZ); 4 mi. E DeQuincy, 7 (LSUMZ); ca. 3.5 mi. SE Bancroft, 1 (LSUMZ). *Bienville Par.*: Bryceland, 1 (LSUMZ). *Bossier Par.*: 1.5 mi. W Bayou Bodcau on State Hwy. 562, 1 (LSUMZ); Fosters, 4 (USNM). *Caddo Par.*: 6 mi. N Shreveport on U.S. 71, 1 (LSUMZ); 10 mi. N Shreveport on U.S. 71, 1 (LSUMZ); 6 mi. SSW Shreveport, 2 (LSUMZ); 0.75 mi. E Zylks, 3 (LSUMZ); 0.5 mi. E Zylks, 1 (LSUMZ); ca. 1 mi. NW Spring Ridge, 2 (LSUMZ); 1 mi. NE Blanchard, 2 (LTU); 5 mi. W Blanchard, 1 (LTU). *Calcasieu Par.*: 7 mi. W Lake Charles, 1 mi. S Maplewood, 1 (LSUMZ); 6 mi. NNW Toomey, 1 (LSUMZ); 3 mi. E Moss Bluff, 1 (MSU); Lake Charles, 2 (USNM). *Caldwell Par.*: 1 mi. N Clarks, 1 (LSUMZ); 1 mi. E Clarks, 1 (LSUMZ); Clarks, 1 (USNM); Columbia, 13 (FM). *Catahoula Par.*: Harrisonburg, 2 (LSUMZ); Sicily Island, 1 (LSUMZ); 1 mi. N Sicily Island, 2 (LSUMZ). *Claiborne Par.*: 5.5 mi. N Arcadia, 1 (LSUMZ); 0.5 mi. S Marsalis, 2 (LSUMZ); locality unspecified, 1 (LTU). *De Soto Par.*: 3 mi. SSW Hunter, 3 (LSUMZ); 6.5 mi. N, 4.5 mi. E Pelican. 2 (LSUMZ). *East Baton Rouge Par.*: various localities, 249 (245 LSUMZ; 1 USNM; 3 USL). *East Carroll Par.*: 2.5 mi. W Gassoway on Macon Bayou, 1 (LSUMZ). *East Feliciana Par.*: various localities, 36 (LSUMZ). *Evangeline Par.*: 4 mi. WNW Ville Platte, 3 (LSUMZ). *Franklin Par.*: 2 mi. N Gilbert, 1 (NLU). *Grant Par.*: Fishville, 2 (1 LSUMZ; 1 TU); 2 mi. N Pollock, 1 (LSUMZ); 1 mi. E Fishville, 1 (TU); Stuart Lake, 2 mi. SW Fishville, 1 (USL). *Iberville Par.*: 7 mi. S LSU, 1 (LSUMZ); 6.2 mi. S, 3.7 mi. W Addis, 1 (LSUMZ); 11 mi. SSW LSU, 4 (LSUMZ); 2 mi. W Iberville, 1 (LSUMZ). *Jackson Par.*: 6 mi. SE Ruston, 1 (LTU); 10 mi. E Jonesboro, 1 (LTU). *Lafourche Par.*: 5 mi. NE Mathews, 1 (LSUMZ). *La Salle Par.*: 2 mi. SW White Sulphur Springs, 1 (LSUMZ), *Lincoln Par.*: Ruston, 13 (4 LSUMZ; 1 USNM; 8 LTU); 6 mi. E Ruston, 1 (LTU). *Livingston Par.*: Amite River, Magnolia crossing, 2 (LSUMZ); 1.5 mi. NE Hoo Shoo Too Rd., 2 (LSUMZ); Sandy Bayou and Amite River, 5 (LSUMZ); 4 mi. E Weiss, 1 (LSUMZ); 1 mi. E Magnolia, 1 (LSUMZ); 2.5 mi. S Livingston, 1 (TU). *Madison Par.*: Waverley, 6 (LSUMZ); 5 mi. SE Tallulah, 2 (LSUMZ); McGill bend of Tensas River, 2 (LSUMZ); 4 mi. E Tallulah, 2 (LSUMZ); 6 mi. SE Tallulah on U.S. Hwy. 80, 1 (LSUMZ); Bayou Macon, 9 mi. S Delhi, 5 (NLU); 8 mi. NW Tallulah, 5 (USNM). *Morehouse Par.*: Irvine Lake, 2 (LSUMZ); 1 mi. S Vaughn, 1 (LSUMZ). *Natchitoches Par.*: Kisatchie, 1 (LSUMZ); 1 mi. NE Natchitoches, 1 (LSUMZ); 1 mi. S Vowells Mill, 1 (LSUMZ); locality unspecified, 1 (LSUMZ); 2.5 mi. E Natchitoches, 1 (USL). *Ouachita Par.*: Monroe, 12 (9 LSUMZ; 3 NLU); 5 mi. E Monroe, 1 (LSUMZ); 8 mi. E, 1 mi. S Monroe, 1 (LSUMZ); Russell Sage GMA, 10 (NLU); Luna Fire Tower Area, 1 (NLU); 9 mi. SSW Monroe, 1 (NLU); Swartz, 2 (NLU); 1 mi. N Monroe, 1 (NLU); West Monroe, 4 (3 USNM; 1 LTU). *Plaquemines Par.*: Burbridge, 11 (6 USNM; 5 MCZ). *Rapides Par.*: 11 mi. W Forest Hill, 1 (LSUMZ); Lecompte, 4 (USNM). *Richland Par.*: 1 mi. E Bayou Lafourche, La. Hwy. 15, 1 (LSUMZ). *Sabine Par.*: Bayou Negreet on Sabine River, 9 (LSUMZ); 7.5 mi. SW Toro, 2 (TU). *St. Charles Par.*: 2 mi. W Norco, 1 (TU); Bonnet Carré Floodway, 1 (TU). *St. Helena Par.*: Amite River, 1 mi. W Chipola, 2 (LSUMZ); 5.5 mi. NNW Chipola, 9 (LSUMZ). *St. John the Baptist*

Par.: 4 mi. S Manchac, 2 (LSUMZ). *St. Landry Par.:* 0.25 mi. S Palmetto, 1 (LSUMZ); 6 mi. S Opelousas, 1 (USL); 1 mi. N Washington, 1 (USL); Melville, 1 (USL). *St. Martin Par.:* 5 mi. N St. Martinville, 1 (USL); island in Lost Lake, 5 mi. E Henderson, 3 (USL). *St. Tammany Par.:* 0.75 mi. S Goodbee, 1 (LSUMZ); 4 mi. NE Covington, 3 (LSUMZ); 4 mi. ESE Slidell, 1 (LSUMZ); 1 mi. E Fontainbleau St. Pk., 1 (LSUMZ); Fontainbleau St. Pk., 3 (LSUMZ); 1.5 mi. SW Pearl River, 1 (LSUMZ); 10 mi. NE Covington, 2 (LSUMZ); 3 mi. SW 90–190 junction on 90, 1 (LSUMZ); 5 mi. ENE Pearl River, 1 (LSUMZ); Covington, 1 (SLU); 8 mi. N Covington, 3 (SLU); 1.5 mi. SE Pearl River, 1 (SLU); 1 mi. E Pearl River, 2 (SLU); 5 mi. N Pearl River, 1 (USL). *Tangipahoa Par.:* Ponchatoula, 1 (LSUMZ); 4 mi. NE Kentwood, 7 (LSUMZ); 2.5 mi. ENE Kentwood, 2 (LSUMZ); 1.5 mi. S Robert, 3 (LSUMZ); 1 mi. W Robert, 1 (LSUMZ); 2.75 mi. SW Hammond, 1 (LSUMZ); 3 mi. E Ponchatoula, 1 (LSUMZ); 6.5 mi. N Hammond, La. Rt. 443, 6 (LSUMZ); 10 mi. S Hammond, 1 (LSUMZ); 12.3 mi. S Hammond, 1 (LSUMZ); 4 mi. N Hammond, 1 (SLU); locality unspecified, 1 (SLU);. *Tensas Par.:* 0.75 mi. N Saranac on Clark Bayou, 8 (LSUMZ); 4 mi. E, 4 mi. S Newlight, 10 (LSUMZ); Newelton, 1 (LTU). *Terrebonne Par.:* Houma, 7 (USNM); Gibson, 56 (MCZ). *Union Par.:* Bayou DeLoutre, Hwy. 2, 1 (NLU). *Vernon Par.:* Anacoco, 1 (LSUMZ); Leesville, 1 (LSUMZ); 0.75 mi. E Simpson, 1 (LSUMZ); 4 mi. W Pinewoods, 4 (LSUMZ); 4 mi. N Burr Ferry, 15 (LSUMZ); 8.5 mi. N Merryville on Bayou Anacoco, 2 (LSUMZ); 12 mi. WNW Anacoco, 1 (LSUMZ). *Washington Par.:* 1 mi. SW Varnado, 1 (LSUMZ); 12 mi. E Franklinton, 1 (LSUMZ); 4 mi. N Bogalusa, 8 (LSUMZ); 2 mi. S Bogalusa, 1 (LSUMZ); Bogalusa, 2 (LSUMZ); Angie, 1 (LSUMZ); 2 mi. N Angie, 1 (LSUMZ); 10 mi. E Franklinton, 3 (LSUMZ); 12 mi. WNW Bogalusa, 1 (LSUMZ); 10 mi. W Bogalusa, 10 (LSUMZ); 10 mi. WNW Bogalusa, 1 (LSUMZ); Sheridan, LSU Forestry Camp, 6 (LSUMZ); 2.5 mi. NE Varnado, 1 (SLU); 1 mi. W Warnerton, 6 (TU); near Varnado, Pearl River, 1 (TU); Warnerton, 1 (TU); Hackley, 5 (4 USNM; 1 FM). *Webster Par.:* Evergreen, 1 (LTU). *West Baton Rouge Par.:* 5 mi. W Brusly, 1 (LSUMZ). *West Carroll Par.:* 5 mi. W Oak Grove, 1 (LTU). *West Feliciana Par.:* Jackson, 1 (LSUMZ); Bains, 3 (LSUMZ); 10 mi. N St. Francisville, 1 (LSUMZ); 2 mi. SW St. Francisville, 1 (LSUMZ); 24.6 mi. N Baton Rouge, Route 67, 1 (LSUMZ); ca. 3.5 mi. N, 1 mi. E Weyanoke, 3 (LSUMZ); Lake Rosmound, 6 (LSUMZ); 5.6 mi. ENE St. Francisville, 1 (LSUMZ); 5.5 mi. ENE St. Francisville, 1 (LSUMZ); 5 mi. NE St. Francisville, 2 (LSUMZ); 8 mi. N St. Francisville, 2 (SLU); 1 mi. E Wakefield, 1 (SLU).

Genus *Ochrotomys* Osgood

Dental formula:

$$I\frac{1-1}{1-1}\ C\frac{0-0}{0-0}\ P\frac{0-0}{0-0}\ M\frac{3-3}{3-3}=16$$

GOLDEN MOUSE   Plate VI
(*Ochrotomys nuttalli*)

Vernacular and other names. — The Golden Mouse derives its sole vernacular name from the golden color of its upperparts. The generic name comes from the Greek word *ochra,* meaning yellow-ochre or golden, and *mys,* for mouse, hence golden mouse. The second part of the scientific name commemorates Thomas Nuttall (1786–1859), famous ornithologist, botanist, and overall naturalist, who was once Curator of the Harvard Botanical Gardens.

Distribution. — This species has almost the same distribution as the Cotton Mouse, *Peromyscus gossypinus,* though it occurs in the Allegheny Highlands, where the Cotton Mouse appears to be absent, and it extends only about halfway down the Florida Peninsula while *gossypinus* reaches Key Largo. In Louisiana it has a somewhat anomalous distribution in that it occurs in two disjunct areas, one in the Florida Parishes and the other in the north-central and northwestern parts of the state. (Map 36)

External appearance. — The Golden Mouse, as the name suggests, is an extremely attractive mouse. The upperparts are rich golden brown or orange-brown from the cheeks and top of the head along the dorsum, including the sides, to the rump. The rather large, brown ears are scantily furred inside and out with brown hairs and with a few black hairs around the upper rim. The underparts are white but are sometimes washed with Pale Pinkish Cinnamon. The feet are white and the tail is definitely but not sharply bicolored,

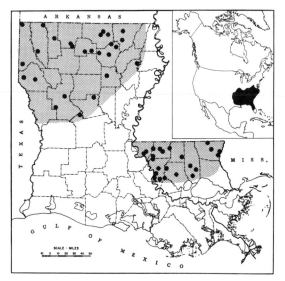

Map 36 Distribution of the Golden Mouse *(Ochrotomys nuttalli). Shaded area,* known or presumed range within the state; •, localities from which museum specimens have been examined; *inset,* overall range of the species.

brownish above and whitish below. The total length is six to seven inches, of which the tail makes up from three to three and one-half inches.

Measurements and weights. — Forty-five skins from various localities in Louisiana in the LSUMZ averaged as follows: total length, 176 (141–210); tail, 75 (65–89); hind foot, 18 (15–21); ear, 18 (14–20). Forty skulls from the same source averaged as follows: greatest length, 25.3 (24.1–26.5); cranial breadth, 12.0 (11.5–12.6); zygomatic breadth, 13.1 (12.2–14.0); interorbital breadth, 4.2 (4.0–4.6); length of nasals, 9.0 (8.0–11.0); length of diastema, 6.8 (6.0–7.6); length of maxillary toothrow, 3.8 (3.6–4.0); palatilar length, 9.3 (9.6–11.0); postpalatal length, 8.9 (8.2–10.0). Weights of 25 individuals averaged 17.8 (14.2–25.0) grams.

Cranial and other skeletal characters. — The Golden Mouse has long been regarded as being so different from other species of the genus *Peromyscus* that, as early as 1909, Os-

good erected a subgenus, *Ochrotomys,* just to accommodate it. More recently, though, especially after the penis bone (Figure 72) and phallus was found by Blair (1942) and Hooper (1958) to differ drastically in structure from those parts in other members of the genus, the subgenus *Ochrotomys* has been elevated by most workers to the rank of a full genus. It is monotypic, that is, it contains only the single species *O. nuttalli.* The skull (Figure 105) shows a number of features that set *Ochrotomys* apart from *Peromyscus*: the posterior palatine foramina (tiny paired openings on each side of the hard palate next to the second molars) are closer to the rear edge of the hard palate than they are to the anterior palatine foramina (large openings on the floor of the rostral part of the hard palate), and the front border of the infraorbital plate is perpendicular to the horizontal axis of the skull instead of bowed forward. The baculum has been described by Hooper (1958) as

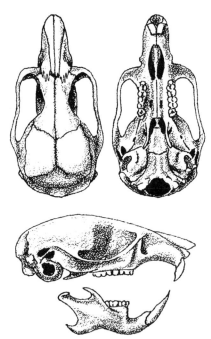

Figure 105 Skull of Golden Mouse *(Ochrotomys nuttalli lisae),*♂, 0.5 mi. W Bossier–Webster Parish line, 0.25 mi. S La.-Ark. line, Bossier Parish, LSUMZ 6708. Approx. 2 ×.

being ornate, with its base spade shaped, expanded laterally, flattened dorsoventrally, and concave ventrally. The distal part of the bone is club shaped, that is, blunt and comparatively broad. He gives the length of the bone as 3 mm. The glans is quite unlike that of other species (Plate XIV, in Hooper, 1958).

Sexual characters and age criteria. — Again, as in all cricetines, the sexes are indistinguishable except by the genitalia. Immature Golden Mice are quite unlike the adults, being dusky in color and with the golden color characteristic of adults confined to a modest indication on the head, shoulders, and sides of the body.

Status and habits. — If any mouse could be called beautiful, the Golden Mouse would certainly qualify. This petite and attractive denizen of our state is common in mixed stands of pines and hardwoods. Less frequently it is found in canebrakes and swampy woodlands, but wherever it occurs the habitat must contain good ground cover and an abundance of underbrush or masses of vines. Although most often encountered not far from a stand of pines, the Golden Mouse sometimes appears unexpectedly in places where it is not "supposed" to occur. For instance, I have trapped it south of the Louisiana State University campus at Baton Rouge in weed-filled ditches bordered by tangles of blackberry and honeysuckle. But such occurrences are exceptional.

The species is much more arboreal than its relatives and has been noted to climb 50 feet or more. It traverses limbs with great agility, using the tail to help maintain balance and also as a prehensile organ for grasping branches and twigs. It has even been observed hanging by its tail from a small branch in much the same fashion as does a prehensile-tailed monkey (Packard and Garner, 1964).

Nests are of two sorts, one for shelter and as a feeding station and the other mainly as a nursery. The former type is oval or globular in shape and is sometimes as much as 12 inches in outside dimensions although usually smaller. It is placed in the crown of a fallen tree, only a foot or so above the ground, or in a tangle of rattan and honeysuckle vines or in a crotch between two branches. Such nests serve primarily as feeding shelters, for the mice carry great quantities of seeds to them to be opened and the kernels eaten. Not infrequently they remodel old bird nests to convert them into such structures.

The nest used for rearing young, also oval or globular, is a woven structure of grasses, fibers, pine needles, and small leaves, carpeted with soft materials such as thistledown, rabbit hair, and feathers (Figure 106). It is placed a few inches to 15 feet above the ground in a bush or vine or on a horizontal branch. Occasionally a nest is in a hollow limb. But the home of the Golden Mouse is not always arboreal, for I have found its nest in rotten stumps and rarely under fallen logs. Nests generally contain from one to four adults, for the species is quite sociable in its relations with its own kind. Only when a female is ready to give birth to a litter are the other occupants required to go elsewhere. One investigator (Barbour, 1942) found a nest containing eight mice, all of which were captured and proved to be males, a veritable cricetine bachelor quarters.

The species breeds throughout the year. But, on the basis of immature specimens taken in Louisiana that I have seen in museum collections, I am inclined to believe that more young are produced in late fall and early winter than at any other time of the year. A similar conclusion was reached by McCarley (1954c) as a result of his studies of the Golden Mouse in eastern Texas. The gestation period is said to be 29 to 30 days and the litter size two or three.

The female is impregnated shortly after parturition, the next day according to one source. The young are naked and blind at birth, but on the 5th or 6th day dark brown hair appears over the back and hips. The eyes open between the 11th and 15th day, at which time the young are able to move about freely and with little effort. The mother broods the young in the nest. If she is compelled to leave the nest for any reason, she drags the babies along with them hanging to her teats. On the birth of the young, the parental male leaves the nest to take up residence elsewhere. (Goodpaster and Hoffmeister, 1954)

The average life-span of the Golden Mouse was found by McCarley (1958) to be only 6.5 months, but, of course, some individuals live much longer. The longevity record appears to be something over two and one-half years (Pearson, 1953). McCarley (1958) found the population density of this species in eastern Texas to be 2.2 mice per acre, but in a cutover loblolly-shortleaf pine forest near Ruston, in Lincoln Parish, Shadowen (1963) calculated the density per acre to be 2.79. The number of Golden Mice to be found in any given situation is directly correlated with the density of the ground cover and, particularly, the amount of underbrush and vine entanglements where the little animals can fulfill their propensity for a predominantly arboreal existence.

Predators. — No specific information is avail-

Figure 106 Golden Mouse *(Ochrotomys nuttalli)* and its nest. (Photograph courtesy of Woodrow Goodpaster and Donald F. Hoffmeister)

able that relates any particular predator to this species, although owls, hawks, and snakes doubtless take a heavy toll. The Golden Mouse has the peculiar habit, when routed out of one of its arboreal nests, of climbing to a height of 15 feet or more and then sitting on a limb peering at its harasser or even going about the business of nonchalantly grooming itself. I suspect such behavior makes it easy prey to a host of enemies. The mere fact that the average life-span of a Golden Mouse is less than seven months attests to heavy and early mortality within its ranks.

Parasites.—Very little seems to be known about the parasites of the Golden Mouse. Morlan (1952) reported two arachnids infesting the species in Georgia, *Bdelonyssus bacoti* and *Dermacentor variabilis.* I have found no reference to an internal parasite. My failure to do so certainly does not mean that the species is free of them, only that parasitologists have not yet investigated the matter.

Subspecies

Ochrotomys nuttalli lisae **Packard**

Ochrotomys nuttalli lisae Packard, Misc. Publ., Univ. Kansas Mus. Nat. Hist., 51, 1969:398; type locality, LaNana Creek bottoms, 1 mi. E Stephen F. Austin College Campus, Nacogdoches, Nacogdoches County, Texas.
Peromyscus nuttalli aureolus, Lowery, Occas. Papers Mus. Zool., Louisiana State Univ., 13, 1943:245; name applied to Louisiana populations now referred to *O. n. lisae.*

I have followed Packard (1969) in his conclusions concerning the pattern of geographical variation in this species. He assigns all Louisiana populations to the race *lisae,* which he delineates as occupying southern Illinois, southeastern Missouri, the western tip of Kentucky, extreme western Tennessee, eastern Arkansas, western Mississippi, all Louisiana, and eastern Texas.

Specimens examined from Louisiana.—Total 141, as follows: *Bossier Par.:* 0.5 mi. W Bossier–Webster line S Ark. line, 1 (LSUMZ); 1 mi. N Bossier, 1 (LTU). *Caddo Par.:* 0.5 mi. E Zylks, 2 (LSUMZ); 0.75 mi. E Zylks, 1 (LSUMZ); 2 mi. E Zylks, 1 (LSUMZ); ca. 1 mi. NW Spring Ridge, 1 (LSUMZ). *Caldwell Par.:* 6.75 mi. WNW Columbia, 1 (LSUMZ). *Claiborne Par.:* 0.5 mi. S Marsalis, 1 (LSUMZ); 5.5 mi. N Arcadia, 1 (LSUMZ). *De Soto Par.:* Mansfield, 1 (USNM); Longstreet, 1 (LSUS). *East Baton Rouge Par.:* various localities, 36 (35 LSUMZ; 1 MSU). *East Feliciana Par.:* Jackson, 1 (LSUMZ); 2.2 mi. S Clinton, 1 (LSUMZ); Highway 10 at Amite River, 2 (LSUMZ); 5 mi. SE Clinton, Idlewild exp. sta., 4 (LSUMZ); Clinton, 1 (TU); Lindsay, 2 (USNM). *Franklin Par.:* 9 mi. S Delhi, 5 (NLU). *Grant Par.:* 3 mi. W Georgetown, 1 (NLU). *Lincoln Par.:* 2 mi. N Tremont, 3 (LSUMZ); Ruston, 8 (1 USNM; 7 LTU); 1 mi. S Ruston, 4 (LTU). *Livingston Par.:* 2 mi. S Watson, 1 (LSUMZ); 2 mi. S Livingston, 1 (TU); 2.5 mi. S Livingston, 3 (TU). *Morehouse Par.:* 4 mi. SE Bastrop, 2 (LSUMZ); Chemin a Haut State Park, 2 (LSUMZ). *Natchitoches Par.:* 1 mi. S Vowells Mill, 1 (LSUMZ); 0.5 mi. W Natchitoches, 1 (NSU). *Ouachita Par.:* 1.5 mi. NW Sterlington, 1 (NLU); Monroe, 2 (NLU); Swartz, 1 (NLU); 0.5 mi. N Calhoun, 1 (NLU); Calhoun, 1 (NLU); West Monroe, 3 (NLU). *Rapides Par.:* 5 mi. E Lena Sta., 1 (LSUMZ). *St. Helena Par.:* 5 mi. NE Montpelier, 2 (LSUMZ). *Tangipahoa Par.:* 2 mi. E Ponchatoula, 1 (SLU); 1 mi. E Ponchatoula, 4 (SLU); 2 mi. S Ponchatoula, 2 (SLU); 8 mi. NNE Hammond, 1 (SLU); 1 mi. S Robert, 1 (SLU). *Union Par.:* 7 mi. NE Farmerville, 2 (LSUMZ); 3 mi. NE Farmerville, 2 (LSUMZ); 1 mi. S Haile, 1 (NLU). *Washington Par.:* Bogalusa, 1 (LSUMZ); 1.5 mi. SE Varnado, 1 (LSUMZ); 10 mi. NW Bogalusa, 9 (LSUMZ); Sheridan, LSU Forestry Camp, 6 (LSUMZ); Varnado, 1 (SLU); Hackley, 3 (FM). *Webster Par.:* Evergreen, 1 (LTU). *West Feliciana Par.:* near St. Francisville, 1 (LSUMZ); 5 mi. NE St. Francisville, 1 (LSUMZ).

Genus *Sigmodon* Say and Ord

Dental formula:

$$I\frac{1-1}{1-1} \ C\frac{0-0}{0-0} \ P\frac{0-0}{0-0} \ M\frac{3-3}{3-3} = 16$$

HISPID COTTON RAT   Plate V
(Sigmodon hispidus)

Vernacular and other names.—Among farmers and laymen this species commonly goes by the name of "field rat." Unfortunately they fail to discriminate between it and other native rats that occur in field situations. The adjective Hispid in the official vernacular name serves to distinguish the present species from several additional species of cotton rats that occur in the southwestern United States and in Mexico. The generic name, *Sigmodon,* is derived from two Greek words (*sigma,* the equivalent of the letter S, and *odous*, meaning tooth), and refers to the S-shaped enamel loops on the grinding surfaces of the cheek teeth. The second part of the scientific name, *hispidus*, is the Latin word for rough, and calls attention to the coarse texture of the animal's pelage. The word "rat" comes from the Anglo-Saxon word *raet*.

Distribution. — The Hispid Cotton Rat occurs across almost the entire breadth of the southern part of the United States from Virginia to southeastern California, thence southward over most of Mexico and Central America to Panama. In Louisiana it is virtually statewide in its occurrence. (Map 37)

External appearance. — Adults of this medium-sized, robust rat measure up to 12 inches in total length, of which the tail makes up less than 40 percent. The coarse pelage of its upperparts is blackish basally but with each hair tipped with yellow or light tan. Since

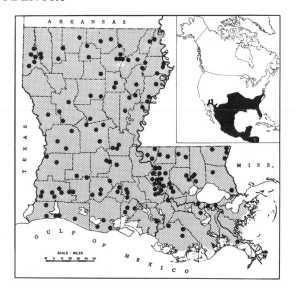

Map 37 Distribution of the Hispid Cotton Rat *(Sigmodon hispidus). Shaded area,* known or presumed range within the state; •, localities from which museum specimens have been examined; *inset,* overall range of the species.

Figure 107 Hispid Cotton Rat *(Sigmodon hispidus)*, Baton Rouge, East Baton Rouge Parish, November 1972. (Photograph by C. C. Lockwood)

some of the black shows through, the overall color takes on a strongly grizzled appearance. The sides are usually more uniformly tan, and the underparts are grayish white. The feet are grayish brown, as is also the scantily haired, somewhat scaly tail, which is indistinctly bicolored. The brown, thinly haired ears are not especially large, yet they extend well above the fur of the head and shoulders. (Figure 107)

Color phases.—Black, excessively brown, and albino individuals are known to occur in this species but are rare (Gardner, 1948; Sherman, 1951). The collections of the LSUMZ contain a perfect albino cotton rat from Mississippi. The specimen label states that the animal had pink eyes and that it was a female containing two embryos.

Measurements and weights.—Fifty skins and skulls of full adults in the LSUMZ from various localities in the state averaged as follows: total length, 267 (240–350); tail, 106 (90–138); hind foot, 31 (24–35); ear, 19 (14–24); greatest length of skull, 35.2 (33.1–40.6); cranial breadth, 15.1 (14.4–16.4); zygomatic breadth, 19.8 (18.4–22.3); interorbital breadth, 5.2 (4.6–5.9); length of nasals, 14.0 (11.9–15.9); length of diastema, 9.2 (8.7–11.2); palatilar length, 15.9 (14.1–18.8); postpalatal length, 12.1 (11.1–14.9). Weights of eight adults, all with total length in excess of 239 mm, averaged 170.0 (103.4–211.0) grams.

Cranial and other skeletal characters.—The skull of a cotton rat can be distinguished at a glance from that of any other Louisiana mammal by the characteristic S- or modified S-shaped pattern of the enamel folds surrounding the dentine on the grinding surface of the cheek teeth, especially of the second and third molars, and by the anterior projection of the upper outer corner of the zygomatic plate. The zygomata are strongly convergent anteriorly and the anteorbital

fossa in the maxillary plate is prominent. The penis bone is highly diagnostic. It is broadly spade shaped basally with a thin shaft that divides into a three-pronged distal end. It measures approximately 8.6 mm in total length and close to 3 mm in lateral width at the base. (Figures 72 and 108.) The glans is also unique among New World cricetine genera (Hooper, 1962).

Sexual characters and age criteria.—The sexes cannot be distinguished except by the sex organs. Immatures are miniature editions of the adults and are full grown by the time they are five months of age. The young are well developed at birth, weighing 6.5 to 8 grams. They are able to move about almost immediately and the eyes open within 24 hours (Svihla, 1929). They grow rapidly and appear on the runways when a week or less old and weighing only 10–20 grams (Odum, 1955).

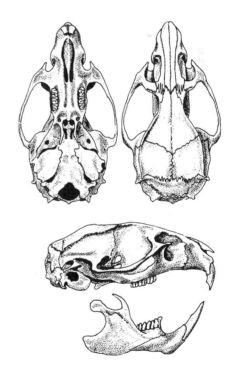

Figure 108 Skull of Hispid Cotton Rat (*Sigmodon hispidus hispidus*), ♂, University Campus, East Baton Rouge Parish, LSUMZ 2592. Approx. 1⅜ ×.

Status and habits.—The Hispid Cotton Rat is widely distributed and abundant throughout the state. It inhabits uncultivated fields, grassy ditches, thickets, and tangles of open vegetation (Figure 109). It also occurs widely in our coastal parishes but not often in the open marshes themselves. There it is more likely to be found on the high ground of chenieres and canal banks rather than in places where the terrain is wet. It seldom, if ever, inhabits hardwood forest, but it does occur in the bordering thickets. Pine forests with dense stands of broomsedge *(Andropogon)* are, however, occupied extensively.

One might argue that the Hispid Cotton Rat is the commonest native rodent in the state. The question is whether it is exceeded in numbers by the Cotton Mouse in combination with the White-footed Mouse, which together replace cotton rats ecologically. As I have indicated in the account of the Cotton Mouse, the state may have more woodland habitat than uncultivated fields and the kinds of places where cotton rats occur; hence the latter possibly ranks second insofar as total numbers are concerned. But, in any event, it is exceedingly numerous. And, for several reasons, it is one of our most important small mammals.

Cotton rats do considerable damage to farm products, sugarcane, truck crops, and garden plants. Moreover, they are probably one of the worst decimating factors, aside from hunting, operating on our Bobwhite populations. Since cotton rats live in the same places as this fine game bird, no one should be surprised that they interfere with its productivity and compete with it for food. Stoddard (1936) pointed to red, or thief, ants *(Solenopsis molesta),* which often enter the quail's eggs at pipping time, as one of the

Tracks of Hispid Cotton Rat

more important enemies of the Bobwhite, but he considered cotton rats high on his list of predators because of the great damage that they inflict by eating the eggs and chicks. The farmer, sportsman, and layman, as a result of long-standing and almost wholly unfounded prejudices, fail to realize that our birds of prey, hawk and owls, constitute the main force that keeps cotton rat and other small rodent populations in check.

The Hispid Cotton Rat, unlike most rodents, is active day and night. On many occasions, while setting a trapline, I have moved a few paces only to hear the snap of a trap that I had just set. On checking back I would find that I had already caught a cotton rat. The species is preyed upon heavily by hawks, as well as by owls and other predators. But the fact that hawks—the Red-tailed, the Marsh, and others—capture so many cotton rats is, I am sure, attributable in part to the diurnal activities of these rodents.

The average life-span of our cotton rat is probably less than six months. Consequently, in order to maintain its numbers, the species must, of necessity, be highly fecund. And so it is. Breeding occurs throughout the year, and litters are often born in rapid succession. Females come into estrus every 7 to 9 days and also immediately after giving birth to a litter. The gestation period lasts 27 days, and the number of young to a litter is 1 to 12, but usually 5 to 7. The testes descend into a shallow scrotum when the male is between 20 and 30 days of age, and the vagina opens when the female is between 30 and 40 days old. Breeding usually occurs when the young rat is no more than 2 months old and may take place at 40 days of age. The young are sometimes weaned on the fifth day, when they weigh 10–20 grams (Odum, 1955), but occasionally the young remain with the mother for 7 days or more (Goertz, 1965a). In any event, the ability of a young animal to fend for itself when no more than a week old must be of great survival value to the species as a whole. The main reason is that it permits the parents to get on with the business of producing another litter.

Shadowen (1956), in his excellent study of *Sigmodon hispidus* in uncultivated fields near Baton Rouge, found that the average home range was 4,924 square feet for males and 4,799 square feet for females. The length of the major axis averaged 137.3 feet for males and 145.8 feet for females. The home range appeared to increase in winter and decrease in summer. He contended that in one of his study areas the cotton rat population reached a peak density of 22.9 rats per acre in November, while in another plot a peak of 15.6 rats per acre was attained in June. On both plots the low point in the apparent population size occurred in late summer and early fall. His results may have been heavily influenced by his sampling techniques, which involved live trapping, marking, and releasing animals and their subsequent recapture. One might suspect, therefore, that some of the seasonal variation in the number of animals caught was due to seasonal variation in the amount of natural foods available, thereby affecting trapping success.

Most of the small runways that one often sees in grassy fields are the paths of cotton rats. These trails are used extensively by other species, such as rice rats, harvest mice, and shrews, if one judges by the frequency with which these other animals are caught in traps set in the runways. The nests consist of rather crudely constructed masses of grasses or of fibers stripped from stems of larger plants. They are placed in cup-shaped depressions in the ground or in clumps of broomsedge *(Andropogon)*. Sometimes they are located at the end of a shallow tunnel in the ground itself. I have also found nests under old boards and other pieces of discarded building materials.

Although cotton rats subsist mainly on vegetable matter, they also eat insects, crayfish, and other animal life. Often one finds a rat or mouse in a trap that has been partially devoured. The first thought is that a voracious shrew was the culprit, but I suspect that often a cotton rat is responsible. In captivity cotton rats seem to relish bits of meat.

Predators.—Comments have already been made concerning some of the birds that prey upon cotton rats, such as hawks and owls, which consume prodigious numbers of them. In addition, carnivores, including foxes, Coyotes, skunks, and weasels, rely on cotton rats for a large portion of their food intake. Years ago in Ruston when I was a student at what was then called Louisiana Polytechnic Institute, a Barn Owl had a nest in a hollow tree near the center of the campus. One night I watched the parent birds bring five cotton rats in rapid succession to one of the young owls. When the youngster took the fifth rat into its mouth there was no place for the morsel to go; so the little owl sat for a seemingly interminable period with the last rat's tail hanging out the corner of its mouth, waiting for its digestive processes to relieve

Figure 109 Hispid Cotton Rat *(Sigmodon hispidus)*, Brazos County, Texas. (Photograph by Richard K. LaVal)

the situation. More recently one of my own students described to me another incident of interest. He had been watching a cotton rat scurrying back and forth between a briar thicket and a fencerow. Having some traps in hand, he set one in the runway the rat was using in the hope of observing its reaction to the trap and the bait. He then concealed himself close by. Within 10 minutes the rat reappeared, circled the trap several times, sniffed the air, and then moved in to take the bait, whereupon it was caught by the hind foot. Before the observer could reach the trap, a Cooper's Hawk swooped in, grabbed the rat with the trap dangling, and flew off.

Parasites and diseases.—Parasitologists have paid more attention to the internal and external parasites of *Sigmodon hispidus* than to those of most other small mammals. I have found references to 2 protozoans, 2 lice, 11 fleas, 21 arachnids, 15 cestodes, 7 trematodes, and 18 nematodes that have been found in or on the Hispid Cotton Rat (Harkema, 1936 and 1946; Baylis, 1945; Harkema and Kartman, 1948; Morlan, 1952; Doran, 1954a and b, 1955a; Layne, 1968; Wenk, 1969; Kinsella, *in press*). A fungus disease is said to affect adults and to reach epidemic proportions in periods of wet weather coincident with high populations. Coccidiosis and tularemia have also been reported.

Subspecies

Sigmodon hispidus hispidus Say and Ord

S[igmodon]. *hispidus* Say and Ord, Jour. Acad. Nat. Sci. Philadelphia, 4, 1825:354; type locality, St. John's River, Florida.
Sigmodon hispidus texianus, Lowery, Occas. Papers Mus. Zool., Louisiana State Univ., 13, 1943:246; name applied to certain Louisiana specimens now treated as *S. h. hispidus.*

The nominate form of the species occurs in the Lower Mississippi Valley from central Missouri, western Tennessee, and Arkansas to Mississippi and Louisiana, thence eastward along the Gulf Coast to northern Florida, southeastern Georgia, and coastal South Carolina. In 1943, I referred specimens from Natchitoches, Sabine, and Vernon parishes, in extreme west-central Louisiana, to the race *texianus,* in full cognizance of their intermediacy toward *hispidus.* A reexamination of these specimens, along with the additional material obtained subsequently, has led me to the conclusion that all cotton rat populations in the state are best treated as *hispidus,* even though specimens from the west-central parishes show a tendency toward *texianus.*

Specimens examined from Louisiana.—Total 1,271, representing all parishes except Assumption, Bienville, De Soto, Lafourche, La Salle, Red River, St. Bernard, St. James, and St. John the Baptist (498 LSUMZ; 397 USL; 72 TU; 68 NLU; 61 LTU; 54 FM; 39 USNM; 34 NSU; 14 SLU; 11 MSU; 8 ANSP; 8 LSUS; 5 MCZ; 1 AMNH; 1 CAS).

Genus *Neotoma* Say and Ord

Dental formula:

$$I\frac{1-1}{1-1}\ C\frac{0-0}{0-0}\ P\frac{0-0}{0-0}\ M\frac{3-3}{3-3} = 16$$

EASTERN WOOD RAT   Plate V
(Neotoma floridana)

Vernacular and other names.—Wood rats are commonly called pack rats or trade rats because of their habit of picking up all kinds of things, particularly bright, metallic objects such as belt buckles, coins, nails, cartridges, and tin cans and carrying them off to their nest, about which more will be said later. The first part of the scientific name, *Neotoma,* is derived from two Greek words (*neos,* new, and *tomos,* cut) that mean "a new kind of mammal with cutting teeth," to distinguish it from another genus, *Mus,* in which it was originally placed. The second part of the name, *floridana,* is a Latinized word meaning of Florida, the state in which the species was first found. The adjective Eastern in the official common name is required to differentiate this species from other wood rats that occur in the western United States, Mexico, and Central America.

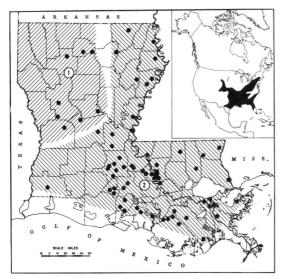

Map 38 Distribution of the Eastern Wood Rat (*Neotoma floridana*). *Shaded area,* known or presumed range within the state; •, localities from which museum specimens have been examined; *inset,* overall range of the species. 1. *N. f. illinoensis;* 2. *N. f. rubida.*

Distribution.—The Eastern Wood Rat occurs from western Connecticut, southeastern New York, over most of Pennsylvania, in southern Ohio and Indiana, northern and western Virginia, and the remainder of the southeastern United States (except parts of North and South Carolina and Georgia) south to central Florida and thence west to southern Missouri, Kansas, and eastern Colorado (with a finger-like projection through Nebraska into southern South Dakota), eastern Oklahoma, and eastern Texas. In Louisiana the distribution is probably statewide despite the fact that many parishes in certain parts of the state are still without definite locality records based on preserved specimens. (Map 38)

External appearance.—This species is easily identified by a number of salient characters. It usually approaches 18 inches in total length, of which the tail makes up slightly less than half. The pelage is dense and soft, the ears are exceptionally large, the eyes are large and bulging, and the tail is fairly well haired. The color above is blackish brown with an admixture of yellow on the sides and gray on the cheeks. The underparts are white and the tail is distinctly bicolored, blackish above and grayish white below. The white feet are comparatively small for a rat of its overall body size. The only species with which it can be confused is the Norway Rat *(Rattus norvegicus).* The latter differs in possessing much coarser pelage, smaller and less protruding eyes, smaller ears, a more elongated and pointed nose, a scantily haired or almost naked tail with rings or annulations, much larger feet, and underparts that are not pure white and do not sharply contrast with the upperparts. The female Norway Rat has six pairs of teats along the side of the belly instead of two pairs in an inguinal position.

Color phases.—No color phases or color aberrations are known in the Eastern Wood Rat.

Measurements and weights.—Fifty specimens in the LSUMZ from various localities in the state averaged as follows: total length, 382 (362–434); tail, 179 (157–210); hind foot, 39 (37–43); ear, 27 (20–30). Forty-six skulls from the same source averaged as follows: greatest length, 50.6 (47.9–55.6); cranial breadth, 19.2 (18.2–21.3); zygomatic breadth, 26.2 (23.9–27.6); interorbital breadth, 6.6 (6.1–7.4); length of nasals, 19.2 (18.6–22.5); length of diastema, 14.3 (13.0–15.6); palatilar length, 20.0 (18.8–23.4); postpalatal length, 19.4 (18.0–21.7). The average weight of 42 males obtained by Neal (1967) was 277.1 grams, that of 41 females was 247.9 grams. The largest male weighed 428 grams, the largest female, 364 grams.

Cranial and other skeletal characters.—The skull of *Neotoma floridana* can hardly be confused with that of any other mammal in Louisiana that possesses only 16 teeth, except the Norway Rat *(Rattus norvegicus)* and the Roof Rat *(Rattus rattus).* The two introduced house rats are easily separated from *Neotoma floridana* by the characters listed in Table 5 and by reference to Figures 110, 125, and 126. The penis

Table 5 Summary of Characters by Which Skulls of *Neotoma floridana* Can Be Distinguished from Skulls of *Rattus norvegicus* and *R. rattus*

Neotoma floridana	Rattus norvegicus and R. rattus
grinding surfaces of cheek teeth possess sharp-angled enamel folds surrounding the dentine, which in the middle loop of the first and second molars extend completely across each tooth	grinding surfaces of cheek teeth possess a series of cusps across the teeth, forming three longitudinal rows
hard palate does not extend beyond the last upper molars	hard palate extends well beyond the last upper molars
braincase smooth, without two parallel longitudinal ridges	two more or less parallel longitudinal temporal ridges on braincase
diastema only slightly longer than molariform toothrow	diastema nearly twice length of molariform toothrow
skull depressed between eye sockets	skull rather flat between eye sockets

bone is broadly truncate basally, tapering abruptly to a cylindrical shaft that is only slightly expanded distally (Figure 72). It measures approximately 7.4 mm in length (Hooper, 1960).

Sexual characters and age criteria. — The sexes are usually indistinguishable except by inspection of the genitalia. In adult males the testes descend into a shallow scrotum; adult

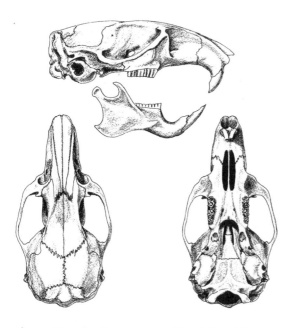

Figure 110 Skull of Eastern Wood Rat (*Neotoma floridana rubida*), ♂, 3 mi. S, 5 mi. E University, East Baton Rouge Parish, LSUMZ 15277. Approx. natural size.

females show a perforated vagina with the lips often red and swollen. In the female a prominent clitoris and its urethral papilla lie immediately ahead of the vagina and are often confused with a penis by a careless observer. Lactating females possess two pairs of conspicuous teats positioned in the groin region much as in the udder of a cow. Adults of both sexes possess scent glands in the center of the belly that exude a greasy yellow secretion that discolors the fur of the abdomen and leaves traces on the ground wherever the rat happens to wander. Young wood rats are much grayer than adults.

Status and habits. — In some parts of the state the Eastern Wood Rat is fairly common, but localities at which specimens have been taken are rather scattered, and whole groups of parishes still remain here and there where, for some reason, the species has not yet been recorded. I am confident that some of these apparent gaps reflect nothing more than insufficient collecting efforts. Although the species is quite numerous in hardwood bottomland forest, it is rare or absent in dry, wooded uplands where pine is one of the predominant trees. This factor could account for its absence in some areas. In the lowlands along the Mississippi River and in the central-southern part of the state it is common, even in the coastal marshes, where it lives on the chenieres and on tree islands bordering the banks of bayous. It frequently takes up residence in trappers' shacks, which are usually unoccupied by the owners during most of the year. When it invades one of these shacks it wrecks the contents, piling up quantities of shredded paper left by the trapper and assembling much debris to form its nests.

Wood rats are notorious for the large nests that they build, which are of two types. One is a sometimes massive pile of sticks on the ground at the base of a hollow tree or hollow log. Such nests are frequently built over old, abandoned armadillo holes. The largest ground structure observed by Neal (1967) in the course of his study of the ecology of wood rats in Louisiana was one that he found near Cheneyville, in Rapides Parish. It measured six feet in height and three feet in diameter at the base. Dens located on the ground are usually no more than three feet high and two feet in diameter at the base. The outer portions are constructed of materials obtainable nearby—twigs, sticks, stems of plants, and often the woody bases of palmetto leaves. (Figure 111)

The second type of nest is built above the ground, anchored to the limbs of a small sapling or sometimes placed in a tangle of vines. These arboreal dens are usually no higher up than 12 feet, but I have a record of one in Terrebonne Parish that was 25 feet from the ground in a large willow. The kind of nest that a wood rat builds varies from locality to locality and may be related to the degree of flooding to which the particular forest is subjected. Out of 271 nests studied by Neal (1967) at Cheneyville, Thistlethwaite, Shaw, and Plaquemine, 236 (85.6 percent) were on the ground, but of 43 nests he observed at Livonia and Gonzales, 32 (74.4 percent) were above ground level.

Most dens have more than one entrance leading to the nest proper, which lies in a cavity that measures between 5 and 13 inches in diameter. The nest is lined with finely shredded bark, newspaper (when available), leaves, and grass. The nest contains an alcove where the occupant sleeps and another where it piles a cache of palmetto seeds and other items of food. Although the feces are regularly deposited outside the den, a few may be piled in a single place within the nest chamber. The animals almost always urinate at a spot well removed from the den. The dens are seldom occupied by more than one adult, but the rat shares its retreat with a variety of other animals—turtles, snakes, mice, and numerous invertebrates, such as spiders and insects. Svihla and Svihla (1933) comment on the musty, sweet odor characteristic of wood rat nests.

Almost any kind of bright, shiny object can be found in wood rat nests. This propensity for carrying such items to its den is the basis for the rodent's name "pack rat." The fact that it often replaces a stolen object with something of its own gives rise to the name "trade rat." A camper is likely to find a pile of sticks, a bone, or some other item in place of a knife, a ring of keys, a watch, or other objects pilfered by one of these rodents. Woodsmen relate many tales, some of them purely fanciful, of feats performed by the "trade rat." One that I have heard tells of a wood rat stealing a half dollar but bringing back two quarters.

Wood rats are not nearly as prolific as most

Tracks of Eastern Wood Rat

rats and mice. Sexual maturity is attained much later, usually not until the animal is seven or eight months old. Only two or possibly three litters are produced annually, and the usual litter consists of two to four young but some contain as few as one or as many as six. Of nine pregnant females examined by Neal (1967), seven contained two embryos and two contained three embryos. One of the females was obtained on 26 June, three on 9 October, one on 19 December, and four on 22 December.

The gestation period does not appear to be known for certain but is probably somewhere between 30 and 39 days (Pearson, 1952). The young at birth are blind, deaf, and naked except for scattered hairs visible only under magnification. They weigh approximately one-half ounce. The upper and lower incisors have erupted through the gums and possess a peculiar modification that enables the young rat to attach itself to one of its mother's teats (Hamilton, 1953). The space between the teeth is a six-sided opening into which one of the elongated teats fits (Figure 112). The young attach themselves so firmly that the mother can dissociate herself only with difficulty and must resort either to pinching the baby on its neck or cheek with her teeth or else applying pressure with her foot and twisting it off (Schwartz and Schwartz, 1959). If forced to leave the nest hurriedly, she often scampers off dragging her offspring behind still hanging to her teats. When the young are two days old the ears unfold, but the external auditory meatus does not open until about the ninth day. By the fifteenth day, or slightly later, the eyes open. Weaning takes place after the twentieth day (Rainey, 1956).

The food of the Eastern Wood Rat consists almost entirely of vegetable matter, such as buds, seeds, nuts, roots, tubers, succulent herbs, grasses, and berries. In our bottomland swamps it relies heavily on oak mast. Snails and insects are also occasionally eaten. Seldom does it travel any great distance in search of food, for the home range is an area only about 100 feet in diameter.

The flesh of a wood rat is said to be delicious and even superior to that of a squirrel. In olden times doctors in rural areas often prescribed broth made from this rat just as chicken broth is recommended today for convalescents. Unlike the unsavory introduced house rats (*Rattus rattus* and *R. norvegicus*), which often frequent human garbage dumps, wood rats are always scrupulously clean and well-groomed and are almost wholly vegetarians. For this reason I never understood my wife's objections when a wood rat once took up residence in the utility room of our home. It seemed to me a nice animal to have

Figure 111   Eastern Wood Rat *(Neotoma floridana)* and its nest of sticks and leaves in a cavity at the base of a tree, Thistlethwaite Game Management Area, St. Landry Parish, winter 1972. (Photograph by J. G. Dickson)

Figure 112 Newborn Eastern Wood Rat *(Neotoma floridana)* showing the peculiar diamond-shaped space between the milk incisors, into which one of the mother's teats fits while the infant is suckling.

around. Needless to say, I lost the argument and had to get rid of our uninvited guest.

Predators.—Because of its almost exclusively nocturnal habits, the Eastern Wood Rat is probably not often captured by hawks, but I am confident that Great Horned and Barred Owls obtain many of them. Other predators, such as Coyotes, foxes, weasels, skunks, and even raccoons, are considered to be serious enemies of the Eastern Wood Rat. The Timber Rattlesnake *(Crotalus horridus)*, the Racer *(Coluber constrictor)*, and the Rat Snake *(Elaphe obsoleta)* likewise prey upon these rats. Only exceptionally large snakes are able to swallow an adult wood rat because of its size, but even a medium-sized snake, of course, has no difficulty with a young rat.

Parasites and diseases.—A cursory examination of the parasitological literature has revealed that the following ectoparasites have been reported from the Eastern Wood Rat: six ticks and mites, 10 chiggers, three fleas, two lice, and one cuterebra. The only internal parasites of which I have found mention are four nematodes *(Longistriata neotoma, Bohmiella wilsoni, Trichuris muris,* and *Nematodirus neotoma)* and two cestodes *(Andrya* sp. and *Hydatigera taeniaeformis).* Most members of the genus *Neotoma,* and the present species is no exception, are highly susceptible to infestation by cuterebra larvae, or warbles. The lar-

vae are most frequently found in the throat region, on the venter between the forelegs, and in the groin. Only occasionally is a cuterebra embedded under the skin of the dorsum or on the facial region. Bubonic plague is said to occur in wild wood rats.

Subspecies

Neotoma floridana rubida Bangs

Neotoma floridana rubida Bangs, Proc. Boston Soc. Nat. Hist., 28, 1898:185; type locality, Gibson, Terrebonne Parish, Louisiana.

Schwartz and Odum (1957) included all Louisiana within the range of *N. f. rubida.* Unfortunately, they did not examine the numerous specimens from Louisiana in the LSUMZ that were then available to them. All they looked at was a single specimen from Vernon Parish and eight from the central-southern portion of the state. Most of the eight were near-topotypes of *rubida,* and only one was from as far away as 90 miles from the type locality. Consequently, the fact that they found their meager material referable to *rubida* is not surprising. Specimens from the northwestern quarter of the state (Map 38) are clearly assignable to *illinoensis,* which differs from *rubida* in being much lighter colored on the dorsum. Although in 1943 I assigned specimens from northeastern Louisiana to *illinoensis,* I have since seen much additional material from that area and I now believe this population should be treated as *rubida.* Specimens from the Florida Parishes are not as dark as examples of *rubida* from the central-southern part of the state and may indicate a southward extension of the range of *illinoensis,* which apparently occurs over most of Mississippi. But I am referring this material provisionally to *rubida.*

Specimens examined from Louisiana.—Total 177, as follows: *Avoyelles Par.:* 2 mi. E Marksville, 1 (LSUMZ); 5 mi. S Cottonport, 3 (USL). *Calcasieu Par.:* Sulphur, 2 (NLU). *Concordia Par.:* 3 mi. N Shaw, 6 (LSUMZ); vicinity of Shaw, 1 (LSUMZ); Deer Park, 1 (LTU). *East Baton Rouge Par.:* various localities, 15 (LSUMZ). *East Carroll Par.:* 8 mi. S Lake Providence, 2 (NLU). *Franklin Par.:* 2 mi. S Lorelein, 1 (LSUMZ). *Iberia Par.:* 1 mi. E New Iberia, 1

(USL); Weeks Island, 11 (USL); Avery Island, 10 (2 TU; 8 CAS). *Iberville Par.:* 4 mi. E Iberville, 1 (LSUMZ). *Jefferson Par.:* H.P. Long Bridge, 1 (LSUMZ); 4 mi. S Gretna, 2 (LSUMZ). *Lafayette Par.:* Youngsville, 1 (USL); 3 mi. SE Lafayette courthouse, 1 (USL); 3.5 mi. E Lafayette courthouse, 1 (USL); 2 mi. SW Lafayette courthouse, 1 (USL); 5 mi. SW Lafayette, 8 (USL). *Lafourche Par.:* Golden Meadow, 1 (LSUMZ); 5 mi. NE Mathews, 2 (LSUMZ); Thibodaux, 1 (LSUMZ). *Madison Par.:* Tallulah, 4 (1 LSUMZ; 3 USNM). *Morehouse Par.:* Bonita, 1 (NLU). *Orleans Par.:* New Orleans, 1 (LSUMZ); 3 mi. E New Orleans, 2 (LSUMZ). *Ouachita Par.:* N of U.S. Hwy. 80 in Russell Sage Game Mgt. Area, 2 (NLU). *Plaquemines Par.:* 3 mi. SE Gretna, 1 (LSUMZ); Tulane Univ. Riverside campus, Belle Chasse, 3 (TU); Burbridge, 1 (MCZ). *Pointe Coupee Par.:* Lottie, 5 (LSUMZ); 8 mi. E Krotz Springs, 1 (LSUMZ). *Rapides Par.:* 10 mi. SE Alexandria, 1 (LSUMZ). *St. Bernard Par.:* Biloxi Marsh, 2 (TU). *St. Helena Par.:* 1 mi. S Greensburg, 1 (LSUMZ). *St. Landry Par.:* 0.5 mi. S Palmetto, 3 (LSUMZ); 9 mi. NNE Opelousas, Thistlethwaite Game Mgt. Area, 13 (LSUMZ); 7 mi. W Opelousas, 1 (USL); 5 mi. W Arnaudville, 1 (USL); Leonville, 1 (USL); 3 mi. NW Opelousas, 1 (USL); 2.5 mi. N jct. 190 on U.S. Hwy. 71, 1 (USL). *St.* (USL); 1 mi. SE Parks, 1 (USL); 1 mi. S Parks, 1 (USL). *St. Mary Par.:* Morgan City, 7 (USNM); Franklin, 2 (USNM); 3 mi. S Charenton, 1 (LSUMZ). *St Tammany Par.:* 4.6 mi. E Pearl River, 1 (LSUMZ). *Tangipahoa Par.:* Robert, 1 (LSUMZ). *Tensas Par.:* 0.75 mi. NW Saranac on Clark Bayou, 1 (LSUMZ); Newellton, 2 (LTU). *Terrebonne Par.:* 2.8 mi. E, 8 mi. N Donner, 1 (LSUMZ); 3.4 mi. E, 1.1 mi. N Donner, 1 (LSUMZ); 4.5 mi. E, 1.6 mi. N Donner, 1 (LSUMZ); 5 mi. W, 2.5 mi. S Schriever, 1 (LSUMZ); 3 mi. W, 1.45 mi. S Schriever, 1 (LSUMZ); Houma, 10 (USNM); Gibson, 3 (MCZ). *Washington Par.:* 8 mi. SW Franklinton, 1 (SLU); 10.5 mi. S, 2 mi. W Angie, 1 (SLU); locality unspecified, 1 (SLU). *West Baton Rouge Par.:* Erwinville, 1 (LSUMZ); 1 mi. S Port Allen, 2 (LSUMZ); 10 mi. W Miss. River Bridge on U.S. 190, 1 (LSUMZ). *West Feliciana Par.:* 3 mi. SE St. Francisville, 1 (LSUMZ).

Neotoma floridana illinoensis A. H. Howell

Neotoma floridana illinoensis A. H. Howell, Proc. Biol. Soc. Washington, 23, 1910:28; type locality, Wolf Lake, Union County, Illinois.

Specimens examined from Louisiana.—Total 22, as follows: *Bienville Par.:* Bryceland, 5 (LSUMZ). *Grant Par.:* Fishville, 1 (LSUMZ). *Lincoln Par.:* 2 mi. N Tremont, 1 (LSUMZ); Ruston, 2 (1 LTU; 1 USNM). *Natchitoches Par.:* Provencal, 3 (LSUMZ); Kisatchie, 5 (LSUMZ); 1 mi. N Kisatchie, 1 (NSU). *Rapides Par.:* 1 mi. N Boyce, 3 (LSUMZ). *Vernon Par.:* 0.5 mi. N Simpson, 1 (LSUMZ).

Genus *Microtus* Schrank

Dental formula:

$$I\frac{1-1}{1-1} \ C\frac{0-0}{0-0} \ P\frac{0-0}{0-0} \ M\frac{3-3}{3-3} = 16$$

PRAIRIE VOLE   Plate VI
(Microtus ochrogaster)

Vernacular and other names.—The Louisiana population of this species has also been called the Louisiana meadow mouse and the Louisiana vole. The generic name *Microtus* is derived from two Greek words (*mikros,* small, and *otos,* of ear) that together mean small-eared. The ears of this little animal and its close relative the Woodland Vole are indeed extremely small, barely extending above the surrounding fur. The second part of the scientific name, *ochrogaster,* is derived from two Greek words (*ochra,* yellow-ochre, and *gaster,* belly) meaning ochre-bellied. The subspecific name for the Louisiana form, *ludovicianus,* is

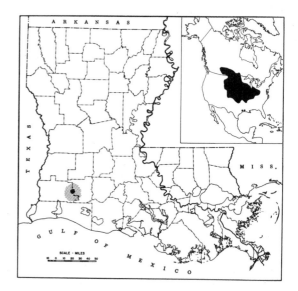

Map 39 Distribution of the Prairie Vole (*Microtus ochrogaster*). *Shaded area* and •, the presumed original distribution in the state and the one locality from which the species is known; *inset,* overall range of the species.

derived from Medieval Latin and means "of Louisiana." The word "vole" is of Scandinavian origin and is akin to the Norwegian word *voll*, meaning meadow or field, hence meadow mouse.

Distribution. — The Prairie Vole occurs from extreme western West Virginia across southwestern Ohio, southwestern Michigan, southern Wisconsin, southern and western Minnesota, southern Manitoba, southern Saskatchewan, and southeastern Alberta south through northern and western Kentucky, northwestern Tennessee, central Arkansas, and through the Great Plains from eastern Montana, the eastern two-thirds of Wyoming, eastern Colorado, and all of Kansas, to central Oklahoma, and with a disjunct population, at least formerly, in southwestern Louisiana and southeastern Texas. In the last-named area the species is known from only two localities, the prairie near the town of Iowa, in Calcasieu Parish, Louisiana, and Sour Lake, Hardin County, Texas. (Map 39)

External appearance. — The Prairie Vole is a small mouse that measures only five to six inches in total length, of which the tail makes up no more than 1.5 inches, or less than twice the length of the hind foot. As in other microtines, the body is thickset, the head is blunt, the legs are short, and the ears are extremely short and barely extending above the surrounding fur (Figure 113). The white, black,

Figure 113 Prairie Vole *(Microtus ochrogaster)*, Hamilton County, Ohio, August 1961. (Photograph by Woodrow Goodpaster)

and brown tips on some of the hairs give the basically dark gray coat a salt-and-pepper appearance. The venter is dark fulvous to dark buffy, and the tail is indistinctly bicolored, dusky above and buffy below.

Color phases. — No color phases are known, but both melanos and albinos occur occasionally, as well as various mutants representing dilution factors that produce "blonds."

Measurements. — The holotype and 21 additional adult paratypes of *M. o. ludovicianus* averaged as follows: total length, 143.6 (130–152); tail, 32.4 (29–38); hind foot, 18.4 (18–19). The skull of one of the adult paratypes (LSUMZ 2247, formerly USNM 96638) measured as follows: greatest length 26.9; zygomatic breadth, 14.9; mastoid breadth, 12.9; interorbital breadth, 4.0; length of nasals, 7.3; depth of braincase, 8.2; width of braincase, 12.3; length of maxillary toothrow, 6.9.

Cranial and other skeletal characters. — The skulls of Louisiana specimens are closely similar to those of *M. o. ochrogaster* from Kansas and Missouri. The only differences that appear to be constant are the slightly larger molars and the more rounded auditory bullae in *ludovicianus* (Figure 114). For other cranial characters of the genus see the account of *Microtus pinetorum* that follows the present account. The baculum of *M. ochrogaster* is shown in Figure 72.

Sexual characters and age criteria. — The sexes are superficially alike. The measurements of 12 adult males and 10 adult females from Louisiana show the females averaging slightly larger: total length, 146.2 as opposed to 141; tail, 33.1 as opposed to 31.8; hind foot, 18.8 as opposed to 18. However, the sample is much too small to inspire confidence. Mammae consist of three pairs, two inguinal and one pectoral. Two of the original 26 specimens from Louisiana were designated by

Bailey (1900) as immatures, and all their measurements are indeed small, but they agree closely with the remainder of the Louisiana series except that they are somewhat more dusky, less brownish, than the adults.

Status and habits. — The greatest unsolved mystery pertaining to our native mammals is the unexplained disappearance of the Louisiana population of the Prairie Vole. In 1899 the Biological Survey Collections, which are now incorporated into the collections of the United States National Museum, contained a vole skull that had been found in an owl pellet obtained at a place called Iowa Station (now known simply as Iowa), approximately 12 miles east of Lake Charles in Calcasieu Parish. No small microtine rodent was then known within 500 miles of southwestern Louisiana; so the matter merited immediate in-

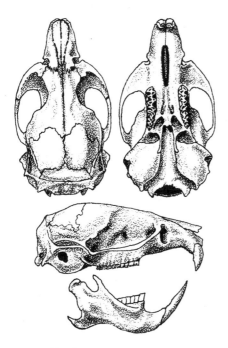

Figure 114 Skull of Prairie Vole *(Microtus ochrogaster ludovicianus),* ♂, Iowa Station, Calcasieu Parish, paratype, LSUMZ 2247 (formerly USNM 96638). Approx. 2 ×.

vestigation. Vernon Bailey was at the time engaged in the field work connected with his survey of the mammals of Texas. Consequently, in the spring of 1899, while en route to Texas, he stopped off at Iowa and by setting traps there on the nights of 6, 7, and 8 April, he obtained 26 specimens of the vole. Later, when he returned to Washington, he described the mouse as a new species, *Microtus ludovicianus*. In 1905, Bailey recorded a specimen captured by Ned Hollister at Sour Lake, Texas. But since then, no others have been seen or trapped.

My own first effort to obtain Prairie Voles at Iowa was in the summer of 1934. As far as I know, this may have been the first attempt made to trap a *Microtus* in Louisiana following Bailey's discovery in 1899. When I failed to catch one I was by no means discouraged, for I knew that all microtine rodents are extremely cyclic. The irruptions of lemmings in Scandinavia and in our own Arctic region at intervals of roughly four years are classical examples of this well-known cyclic population phenomenon. In some years the number of voles may be astonishingly high, but two years later the population has often declined so precipitously that only a few mice are to be found. With these facts in mind, I naturally presumed, in the summer of 1934, that I was merely in the trough between cyclic peaks, that, at a some later date, I would find the mouse without difficulty. But such was not the case. Since 1934 my students and I have set literally thousands of traps in the vicinity of Iowa, and at numerous other locations in southwestern Louisiana, all to no avail. Other mammalogists have also tried, notably Professor James Lane of McNeese State University in Lake Charles. He has been no more successful than others.

When traps are set in the vicinity of Iowa and in other attractive habitats in southwestern Louisiana, several species of small mammals are obtained in considerable numbers — Hispid Cotton Rats, Marsh Rice Rats, Fulvous Harvest Mice, Least Shrews, and

House Mice. One hundred and fifty mouse traps often yield a 10 to 15 percent catch, sometimes slightly more. The thought, therefore, occurred to me that it would be interesting to know what Bailey caught in 1899 besides *Microtus*. Reference to his field catalog, on file in the United States National Museum, shows that, in the three nights that he set traps at Iowa, he obtained one Fulvous Harvest Mouse and three Marsh Rice Rats in addition to the 26 voles. No clue is given as to the number of traps that he employed in his traplines, but knowing Bailey's thoroughness, as a field mammalogist, I suspect the number was considerable. The interesting fact is that he obtained no Hispid Cotton Rats or House Mice and only three Marsh Rice Rats and one Fulvous Harvest Mouse. Could it be that since 1899 the Hispid Cotton Rat and the other species have completely replaced the Prairie Vole?

Before Bailey's death I talked with him on several occasions in hopes of learning of some special technique that he might have employed. But, with a characteristic twinkle in his eyes, he would simply reply, "I got off the train at Iowa Station, walked out into the nearby prairie, and set my traps"!

Another possibility is that drastic changes may have taken place in the ecology of the prairie since 1899. To be sure some changes have taken place. The prairies of southwestern Louisiana have for more than a century been subjected to extensive grazing and to rice culture. Unquestionably, more land is now under cultivation and more acreage has been converted into pastures, yet despite this fact, even today many fields can be found with good stands of grass, and railroad rights-of-way provide in many places what appears to be a semblance of primeval prairie vegetation and excellent microtine habitat. Rice farmers often allow their fields to lie fallow for two years or more, during which time, in the absence of overgrazing, they develop a thick mat of grasses and sedges, interspersed with coffeebean and other shrubs.

A final hypothesis that would account for the disappearance of the Louisiana population of *Microtus,* if indeed it is extinct, is that it was wiped out by some epizootic disease that swept through its ranks. Although such a devastating epidemic could have occurred, we have no actual evidence that it did.

I find the idea that our Louisiana Prairie Vole is now extinct difficult to accept. But every time my associates and I set 150 mouse traps in some mousy-looking spot in southwestern Louisiana and fail once again the following morning to find one of these little mammals in the catch, I move closer to that conclusion.

Naturally, almost nothing is known about the life history and habits of the Prairie Vole in Louisiana, except by inference. The animal is quite similar to the Prairie Vole of the Upper Mississippi Valley and the central and northern Great Plains. Consequently, what is known about the species farther north (see Jameson, 1947; Martin, 1956; Schwartz and Schwartz, 1959) probably would apply in large measure to the population that once occurred here and hopefully still survives somewhere in the state. Bailey's notes on the species in 1905 in his *Biological Survey of Texas,* where he commented on the specimen obtained at Sour Lake, Texas, as well as on his capture of the original specimens in Louisiana, are worthy of quotation in full:

The prairie about Sour Lake is very similar to that just east of Lake Charles, La., where I found these little voles fairly numerous, living in the peculiar, flat mounds [physiographic features of the region that we now call pimple or Atta mounds] that are scattered over the low, damp prairie, and making their runways through the grass from one to another. Some of the mounds were perforated with a dozen or more of the little round holes, from each of which a smooth trail led away. A colony of a dozen or less of the voles, in some cases all adults, in others both adults and half-grown young, was usually occupying a mound. One female taken April 8 contained three well-developed embryos, and several others taken on the same date were giving milk. As usual in the females of this subgenus *(Pedomys)* the mammae were uniformly inguinal 2/2, pectoral 1/1. A few winter nests of grass were found on the surface of the ground where the standing grass had burned off, but the breeding nests apparently were all in the burrows below the surface.

Fire had recently run over most of the prairie and left the burrows exposed and the trails sharply defined over the blackened ground, but as the animals were caught as readily over the burned area as in the standing grass, the burrows are evidently a safe retreat in case of fire.

The stomachs of those caught contained only green vegetation, and along the runways grass and various small plants had been cut for food. As rice is the principal crop over these low prairies and as the ground is flooded while the rice is growing, this little vole is not likely to do serious damage.

No information is available on the predators, parasites, and diseases of this mouse during the time when it was common in the prairie region near Lake Charles and perhaps elsewhere in the southwestern part of the state. If, by chance, the species should ever be rediscovered anywhere in Louisiana, an intensive study of it would certainly be in order.

Subspecies

Microtus ochrogaster ludovicianus Bailey

Microtus ludovicianus Bailey, N. Amer. Fauna, 17, 1900:74; type locality, Iowa, Calcasieu Parish, Louisiana.
Microtus ludovicianus, Lowery, Occas. Papers Mus. Zool., Louisiana State Univ., 13, 1943:247.

I am now treating *ludovicianus* as a subspecies of *Microtus ochrogaster* rather than as a full species. When Bailey (1900) originally assigned specific rank to the Louisiana population, he stated: "There is no known and probably no actual intergradation or continuity of range between the two forms [*ludovicianus* and *ochrogaster*], and perhaps subspecific rank would show better the close relationship of *ludovicianus* to *austerus* [=*ochrogaster*]." Recent authors (including Hall and Cockrum, 1953; Hall and Kelson, 1959) have continued to consider *ludovicianus* as a nominal species, but, in the absence of any distinctive skull or color differences of more than subspecific quality, I can see no justification for doing so.

Specimens examined from Louisiana. — Total 26, as follows: *Calcasieu Par.:* Iowa Station, 26 (USNM, one now in LSUMZ).

WOODLAND VOLE   Plate VI
(Microtus pinetorum)

Vernacular and other names. — The Woodland Vole is often called the pine vole, pine mouse, or mole mouse. As explained under the previous species, the generic name *Microtus* is derived from two Greek words (*mikros*, small, and *otos*, ear) that together translate as "small of ear," a characteristic of all microtine rodents. The second part of the scientific name, *pinetorum,* is a Latin word that translates as "of pine woods," which is a misnomer as far as the animal's preferred habitat is concerned. The subspecies that occurs in Louisiana is *auricularis,* which is the Latin word meaning "pertaining to the ear or auricle of the ear." It was doubtless also designed to call attention to the small size of the pinnae.

Distribution. — The species occurs throughout the eastern one-third of the United States except for most of Maine, northern New Hampshire, northern Vermont, northern New York, the Upper Peninsula of Michigan, northern and central Wisconsin, all but the extreme southeastern corner of Minnesota, southern Georgia, and all of Florida. It ranges west to southwestern Iowa, extreme

Figure 115 Woodland Vole *(Microtus pinetorum),* Reelfoot Lake, Tennessee. (Photograph by Woodrow Goodpaster)

eastern Nebraska, eastern Kansas, eastern Oklahoma, and central Texas. In Louisiana it occurs in two disjunct parts of the state, the northern one-quarter and the Florida Parishes. (Map 40)

External appearance. — This small, thickset rodent, which is only four or five inches in total length, with a tail that barely exceeds an inch in length, is readily recognized by its blunt head, extremely short ears, tiny eyes, and reddish brown color above (Figure 115). The hair, both above and below, when parted is found to be gray except for the tips. But since the pelage is dense, the gray is concealed. The underparts are grayish on the throat but strongly fulvous on the remainder of the venter. The feet are grayish tan, and the tail is dark above and only slightly paler below. The only species with which it can possibly be confused is the Prairie Vole, which is grayish to blackish brown, not reddish brown.

Color phases. — Albinos occur rarely, as do also mutants representing various color dilution factors producing fawn-colored and dilute brown individuals (Owen and Shackleford, 1942). No true color phases affecting a sizable number of individuals in a population are known.

Measurements and weights. — Fifteen specimens in the LSUMZ from various localities in the state averaged as follows: total length, 116 (105–128); tail, 23 (18–27); hind foot, 16 (12–17); ear, 10 (9–13). Skulls from the same source (number of specimens in parentheses following the measurement in question): greatest length (7), 25.2 (24.6–26.9); cranial breadth (10), 12.9 (12.0–13.4); zygomatic breadth (9), 15.2 (14.9–15.5); interorbital breadth (13), 4.4 (4.1–4.6); length of nasals (12), 7.1 (6.7–7.7); length of diastema (13), 7.4 (7.0–7.8); palatilar length (12), 11.6 (11.1–12.3); postpalatal length (8), 10.5 (9.8–11.4). Two adult males from East Baton Rouge Parish weighed 25.2 and 26.0 grams.

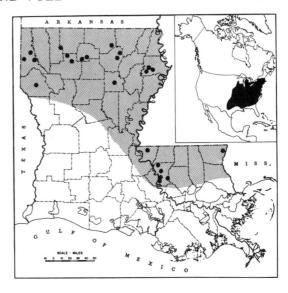

Map 40 Distribution of the Woodland Vole *(Microtus pinetorum). Shaded area,* known or presumed range within the state; •, localities from which museum specimens have been examined; *inset,* overall range of the species.

Cranial and other skeletal characters. — As in other microtine rodents, the cheek teeth are highly diagnostic, being clearly separable from those of cricetines by their sharp-angled enamel folds that give the toothrows zigzag lateral borders. The second upper molar has four discrete islands of dentine surrounded by enamel. By being more angular and in having the braincase decidedly flattened, the overall shape of the skull distinguishes this species from all other rodents in Louisiana that possess only 16 teeth. In these particulars *Microtus pinetorum* can be confused only with *M. ochrogaster ludovicianus.* In the latter the skull tends to be narrower in proportion to its length (width usually less than 56.6 percent of length instead of close to 65 percent of length as it is in *M. pinetorum*) and the interorbital width is narrower (4 mm or less instead of 4.5 mm or more). (Figure 116.) The baculum is broad at the base, narrowing abruptly to the distal end, which bears three small digitate processes (Figure 72). Hamilton (1946a) gives the following measurements:

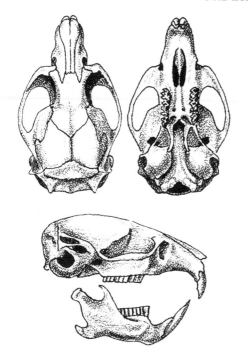

Figure 116   Skull of Woodland Vole *(Microtus pinetorum auricularis)*, ♂, 4 mi. S Lindsay, East Baton Rouge Parish, LSUMZ 2281. Approx. 2 ×.

overall length, 2.9 mm; length of stalk, 2.5 mm; width of base, 1.5 mm; length of median digital process, 0.27 mm.

Sexual characters and age criteria.—The sexes are superficially alike and can be consistently distinguished externally only by examination of the genitalia. Two pairs of mammae are located in a pectoral position. Paired scent glands are found on the hips and groins of both males and females. Immature Woodland Voles are grayish brown above and gray below. They begin to lose the juvenile pelage between their second and third week and acquire the full adult pelage between their seventh and tenth week (Benton, 1955).

Tracks of Woodland Vole

Status and habits.—The Woodland Vole appears to be common nowhere in Louisiana. Only 100 Louisiana specimens have so far been obtained, and these represent only twelve parishes, of which nine are in the northern part of the state and three in the Florida Parishes. Reference to Map 40 shows a wide expanse of parishes in the western two-thirds of the state from which locality records are completely lacking, and east of the Mississippi River in the Florida Parishes there are still more parishes in which the species has not been found than parishes in which it is known to occur. Although often called the pine mouse, it seems to prefer hardwood areas with a heavy layer of leaves and humus (Figure 117). Admittedly the species may occur in mixed pine-hardwood habitat, for I found it in such situations in Lincoln Parish, but I am not aware that anyone has ever taken it in a pure stand of pine. In the vicinity of Baton Rouge, it occasionally appears in a trapline set in a broomsedge field along the edge of a wood, but most often it is found inside a wooded area.

The little animal burrows underground passages just below the leaf mold and humus, or it utilizes the tunnels of other small mammals such as the mole. For the most part it is nocturnal or crepuscular, but occasionally it is

Figure 117   Woodland Vole *(Microtus pinetorum)*, Caddo Parish. (Photograph by Richard K. LaVal)

seen abroad in bright daylight. I once had in my yard a Woodland Vole whose tunnel opened at the base of a bird feeder just outside my living room window. During the noon hour I often watched the little creature emerge, nervously scurry out on the surface, grab a morsel of food such as a piece of cracked corn or a sunflower seed, and then rush back to its hole. In a few minutes it would appear again. The habitat in this particular situation is definitely deciduous upland forest, bordering on a bottomland swamp. The only pines present are a few ornamental plantings.

The nest is located in one of the burrows and is globular in shape, usually six to seven inches in diameter and four to six inches below the surface. It consists of shredded grass. The species appears to be highly tolerant of its own kind. Raynor (1960) reports finding three females and their respective litters in a single nest.

The species is sedentary in its habits, moving only short distances from its nest. Burt (1940) found the home range of 17 Woodland Voles in an oak-hickory woodland in southern Michigan to be 38 yards in average diameter, or about one-quarter of an acre in average area. The individual ranges varied from 15 yards in diameter to 93 yards. But Benton (1955), in a study of the species in New York, found that 13 individuals that were trapped three or more times, had an average home range only 21 yards in diameter and that no individual moved more than 50 yards.

Although the species appears to have a definite breeding season in the North from January to October (Benton, 1955), there is evidence that in Louisiana it produces young the year around. Most litters contain two to four young, but a record number of eight embryos has been reported (Roberts and Early, 1952). The gestation period is not known for certain, but it is probably somewhere near the 21 days determined for certain other species of the genus *Microtus*.

The Woodland Vole is of no great economic importance in our state because of its low numbers, but in the northern parts of its range, where a cover of snow often remains on the ground for long periods, it is said to do some damage by girdling orchard trees.

Predators. — The fossorial habits of the Woodland Vole doubtless protect it in some measure from predation, but in the northeastern United States, Pearson and Pearson (1947) and Latham (1951) found that it is eaten by six species of owls, four kinds of hawks, the Red Fox, Gray Fox, and Virginia Opossum. Parmalee (1954) and Packard (1961) report finding its skull in Barn Owl pellets in eastern Texas. Llewellyn and Uhler (1952) added the mink and raccoon to the predator list, and Barbour (1951) reported finding this vole in the stomach of a Rat Snake *(Elaphe obsoleta)*.

Parasites and diseases. — The acanthocephalan *Moniliformis clarki* is a serious internal parasite that occurs in the small intestines. Benton (1955) reports that one mouse had 22 of these worms packed in the intestine so tightly that adequate amounts of food would have not been able to pass. The worms reach a length of five inches, and when present in numbers can cause loss of weight and general debilitation. A nematode, *Trichuris* sp., has also been found in the small intestines. Doran (1954b) lists one cestode *(Catenotaenia pusilla)* as having been found in the species. Among the ectoparasites that have been recovered from the Woodland Vole are the louse *Hoplopleura acanthopa* and the mite *Haemolaelaps glasgowi*. Four other mites, including *Laelaps microti,* have been found on this mouse or in its nests, and the flea *Ctenophthalmus pseudagyrtes* has been detected on it (Benton, 1955). Very little is known of the diseases of this vole, but Hamilton (1938) describes an affliction that causes great patches of fur and epidermis to slough away, followed by encrustations about the eyes that produce blindness, by a general weakness of the hind quarters, and finally by death. Captive mice frequently succumb to unidentified ailments, but autopsy often shows an enlarged spleen, a necrotic appearance to the liver, or hard fecal material packed in the large intestines, indicating constipation (Benton, 1955).

Subspecies

Microtus pinetorum auricularis Bailey

Microtus pinetorum auricularis Bailey, Proc. Biol. Soc. Washington, 12, 1898:90; type locality, Washington, Adams County, Mississippi.
Pitymys pinetorum auricularis, Lowery, Occas. Papers Mus. Zool., Louisiana State Univ., 13, 1943:247.

All specimens from Louisiana are referable to *M. p. auricularis.*

Specimens examined from Louisiana.—Total 100, as follows: *Bienville Par.:* Bryceland, 2 (LSUMZ); 3 mi. W Gibsland, 3 (LSUMZ). *Caddo Par.:* 3 mi. N Blanchard, 1 (LSUMZ); 5 mi. NW Greenwood, 5 (LSUMZ); LSU Shreveport, 5 (LSUS). *Caldwell Par.:* Columbia, 2 (FM). *De Soto Par.:* Mansfield, 2 (USNM). *East Baton Rouge Par.:* University, 4 (LSUMZ); Baton Rouge, 3 (LSUMZ); 4 mi. S Lindsay, 1 (LSUMZ); 1 mi. S Baker, 1 (LSUMZ); 10 mi. E Baton Rouge, 1 (LSUMZ). *Lincoln Par.:* 1 mi. S Ruston, Tech Farm No. 1, 12 (LTU); Ruston, 4 (LTU); 7 mi. SW Ruston, 1 (LTU). *Madison Par.:* near Singer Tract, 1 (LSUMZ); 9 mi. S Delhi, 2 (NLU); 10 mi. SW Tallulah (Chicago Mills Game Mgt. Area), 1 (USL). *Ouachita Par.:* Monroe, 40 (NLU); Swartz, 5 (NLU). *Union Par.:* 6 mi. E Rocky Branch, 1 (NLU). *Washington Par.:* 6 mi. S Angie, 1 (SLU). *Webster Par.:* 10 mi. NE Minden, 1 (NLU). *West Feliciana Par.:* vic. Weyanoke, 1 (LSUMZ).

Genus *Ondatra* Link

Dental formula:

$$I\frac{1-1}{1-1}\ C\frac{0-0}{0-0}\ P\frac{0-0}{0-0}\ M\frac{3-3}{3-3} = 16$$

COMMON MUSKRAT   Plate VII
(Ondatra zibethicus)

Vernacular and other names.—The Common Muskrat is called *rat musqué* by our French-speaking inhabitants. In the Delacroix Island section of Louisiana, in St. Bernard and Plaquemines parishes, where Spanish was spoken for generations, the trappers call it *rata* or *rata almizcle,* the latter meaning musk-rat. The musk part of the name refers to the odor that both the male and the female emit.

If *Ondatra obscurus,* of Newfoundland, continues to be accorded full species rank and called the Newfoundland Muskrat, we must apply some adjective such as "Common" to muskrats elsewhere. To call *Ondatra zibethicus* simply "the muskrat" and *Ondatra obscurus* "the Newfoundland Muskrat" is analogous to the English calling their one species of wren "the Wren," as if there were no other wrens elsewhere in the world, and it appropriates a group name for a single species. The adjective "Common" is also needed if we continue to call *Neofiber alleni* of Florida the Round-tailed Muskrat. I would prefer, however, to use the name Florida Water Rat for the latter, and some authors have already done so.

The generic name *Ondatra* is a French Canadian word of Huron Indian origin and is masculine in gender and was so treated by the describer. Since adjectival specific names must agree with their genus in gender, the second part of the scientific name, the Latin adjective *zibethicus,* meaning musky-odored, must carry the masculine ending *us* rather than the *a* that the word *Ondatra* would seem to require. (Davis and Lowery, 1940)

The name of the subspecies in Louisiana is *rivalicius,* a Latin adjective meaning of, or belonging to, those who make use of the same brook or in this case bayou. The word is derived from *rivus,* meaning a brook, and the suffix *icius,* which when added to noun stems form adjectives meaning belonging to. The word was first used to refer to people who shared fishing rights but later other rights as well, hence the word "rival." Bangs, the author of the name *rivalicius,* failed to explain the etymology of the word or his reasons for using it, but he did repeatedly refer to the muskrat's abundance in the Louisiana bayou

Plate VII
Common Muskrat *(Ondatra zibethicus)* on edge of three-cornered sedge *(Scirpus olneyi)* marsh.

PRATT
1972

country and the fact that many of them are often found together in the bayous—hence sharing the same feeding areas.

Distribution.—The species is widespread in its occurrence in North America from northern Mexico to extreme northern Alaska and northern Canada. It is absent only in some areas bordering the Arctic Ocean and parts of Oregon, California, Nevada, Arizona, Texas, South Carolina, and Georgia, and from all of Florida. In Louisiana it is present throughout the southern part of the state and is especially numerous in the coastal marshes, but it occurs at least as far north as Avoyelles Parish. The recent appearance of the species in increasing numbers in northeastern Louisiana has probably resulted from a natural range extension from southeastern Arkansas. Its occurrence in Caddo Parish probably represents an introduction by man.

External appearance.—This rodent, when compared with the Nutria and American Beaver on the one hand and with the Eastern Wood Rat on the other, would be regarded as medium sized. Large individuals have a total length up to 22 inches, of which the tail comprises slightly less than half. The eyes are small and the ears are short, barely projecting above the surrounding fur. The lips close behind the incisor teeth, permitting the animal to gnaw underwater. The forefeet have four clawed toes and a thumb with a nail. The hind feet possess five clawed toes that are webbed at their bases. Fringes of stiff hairs on the edge of the web and along the sides of the toes and foot provide an additional adaptation for swimming. The scaly tail is vertically flattened. The pelage is made up of a dense, waterproof underfur overlaid with long, glossy guard hairs. The overall color is a rich, dark brown above, with the underparts, which lack guard hairs, grayish, shading to white on the throat. (Figure 118)

Color phases.—There are no color phases as such, but the Louisiana State University Museum of Zoology contains several examples of albinos and color mutants referred to as "blonds" that result from pigment dilution factors.

Figure 118 Common Muskrat *(Ondatra zibethicus).* (Photograph by Robert N. Dennie)

Measurements. — Fifty specimens in the LSUMZ from southern Louisiana average as follows: total length, 489 (440–526); tail, 215 (180–256); hind foot, 74 (63–87); ear, 19 (15–25). For body length and skull measurements of 158 males and 199 females, according to Gould and Kreeger (1948), see Table 6.

Cranial and other skeletal characters. — The skull of the Common Muskrat cannot be confused with that of any other rodent. The bones are thick and heavy, but the general features are typically microtine. The enamel folds of the surface of the upper cheek teeth are sharp-angled and surround four or more dentine islands. The postorbital process is prominent, and the interorbital constriction is conspicuously narrowed with a distinct ridge along the midline between the orbits. The distance between the inside of the incisors and the first cheek teeth is more than one-third the total basilar length. The zygomata are only slightly convergent anteriorly, and the sides of the rostum are essentially parallel except anteriorly. The baculum flares broadly at the base, which is roughly rectangular in shape, tapers abruptly to a thin shaft, and terminates in three fingerlike processes. Its overall length is approximately 9.2 mm and the basal width of the cylindrical shaft is approximately 5.9 mm (Hamilton, 1946a). (Figures 72 and 119)

Sexual characters and age criteria. — The sexes are superficially similar and can be easily confused, even when the genitalia are inspected. Gould and Kreeger (1948) have shown that males average only slightly larger than females (see Table 6). The clitoris of the female serves as a urinary papilla and is so prominent that it can be easily mistaken for a penis. It lies immediately ahead of the vagina. In virgin females the vagina is closed by a vaginal membrane, but in adults it can be opened by probing (Baumgartner and Bellrose, 1943). The space between the vagina and the anal opening is short and bare of hair, but the

space between the penis and anal opening is greater and always well furred (Figure 120).

Both males and females possess musk, or perineal, glands that are located on each side of the genitalia. In the male the ducts open within the foreskin of the penis, where the secretion mixes with the urine and is deposited on the feeding platforms and along trails. It is also expelled before and during the time of mating. The glands in the female do not enlarge to the extent that they do in the male.

Males lack teats whereas females possess

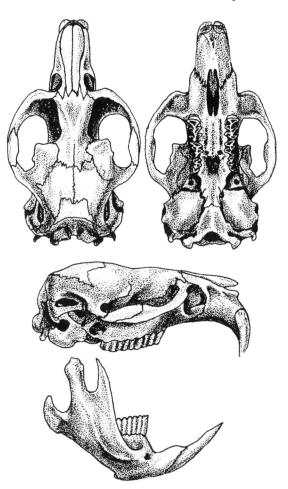

Figure 119 Skull of Common Muskrat *(Ondatra zibethicus rivalicius)*, ♂, University Campus, East Baton Rouge Parish, LSUMZ 2637. Approx. natural size.

Table 6   Body Length and Certain Skull Measurements of *Ondatra zibethicus* by Sex[1]

Character	Males		Females		Difference and Its Standard Error		
	No.	Mean	No.	Mean			
Body length	158	299.2	199	294.8	4.4	±	1.47
Total length of skull	158	66.33	199	64.82	1.51	±	.26
Condylo-alveolar length	158	39.21	199	38.27	.94	±	.17
Alveolar length	158	15.96	199	15.92	.04	±	.07
Condylo-postalveolar length	158	23.27	199	22.41	.86	±	.13
Length of diastema	158	22.88	199	22.24	.64	±	.11
Cranial breadth	158	22.47	199	22.09	.38	±	.06
Zygomatic breadth	155	41.43	193	40.44	.99	±	.18
Occipital height	158	16.90	197	16.51	.39	±	.08

[1]After Gould and Kreeger, 1948.

four, rarely five, pairs located along the sides of the venter between the front and hind legs. Two pairs are inguinal, two pairs pectoral. In dried muskrat skins, the nipple sites are readily identifiable on the leather side by thin, light-colored circular spots.

Schwartz and Schwartz (1959) point out a way of recognizing a muskrat less than 11 months of age. In adults the fluting on the labial side of the first upper molar does not reach the bony socket and the forward edge is slightly curved in profile. In animals under 11 months of age, the fluting reaches the bony socket and the forward edge of the tooth is nearly straight in profile (see also Gould and Kreeger, 1948). A method of recognizing immature muskrats that can be applied after the pelts have been prepared by the trapper for shipment to the fur market is by the bilaterally symmetrical pattern of dark and light areas on the flesh side of the skin. Pelts of adults display an irregular, spotted pattern of light and dark areas, indicating prime and unprime fur respectively. This method of aging is convenient because no carcasses are required. It is, however, not foolproof. Shanks (1948) compared the patterns of 455 pelts with the carcasses from which they came. He found 16 individuals with the juvenile type of pattern but with placental scars on the uterus that proved that they had already bred. Schofield (1955), on the other hand, after carefully testing all methods of aging muskrats, concluded that

the pattern of pelt-primeness was one of the most dependable criteria, and he endorsed its use. Extremely young muskrats are recognizable by the dark charcoal gray coloration of their dorsal fur, which lacks the rich brown guard hairs characteristic of adults.

Status and habits.—The Common Muskrat is indeed common in Louisiana, especially in the coastal marshes in the southern portion of the state. Any species of mammal that yields on the average nearly one million pelts annually to the Louisiana fur trade must certainly be classified as truly abundant. Our muskrat has earned this distinction (Table 2).

The species is by no means confined to the

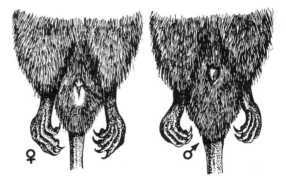

Figure 120   Genitalia of the Common Muskrat. Note in the female, on the left, the prominent clitoris, which can be easily mistaken for a penis, and the area largely bare of hair between the vagina and the anal opening. In the male, on the right, the area between the penis and the anal opening is not bare of hair.

coastal marshes, for it ranges inland where rice is grown and is found along bayous and in lakes throughout most of southern Louisiana. The extent of its occurrence north of the marshes is shown in Map 41. In the freshwater lakes around Baton Rouge it is common but is seldom seen by the layman. Its tunnels honeycomb the edges of these lakes, but because it does not build houses, as it does in the marshes, it escapes general notice.

A specimen record from Black Bayou Lake in northern Caddo Parish could represent a natural population, but the strong possibility exists that its presence there resulted from introductions by man. The situation in northeastern Louisiana is, however, quite different. Although entirely unknown in this section of the state until the middle part of the 1960s, the muskrat has in the last eight years become increasingly numerous there. I am informed by Varney Robinson, fur dealer at Oak Grove, and by his brother Fenton Robinson, that no less than 600 hides of the Common Muskrat were obtained in the 1972–1973 season in East and West Carroll, Morehouse, Richland, Madison, Tensas, and Franklin parishes, and that a few were taken along the Ouachita River in Caldwell Parish. According to the Robinsons, the rat first made its appearance in northeastern Louisiana near Gassoway, in extreme northern East Carroll Parish, in about 1965, when a few hides found their way into the hands of fur buyers. Since then the species has spread southward in ever-increasing numbers. The consensus is that these rats came into Louisi-

ana from the rice fields of southeastern Arkansas and that the increase of rice culture in recent years in northeastern Louisiana has favored the buildup of their populations in the aforementioned parishes.

The ecology of the Louisiana muskrat has been extensively studied by Ted O'Neil. His book *The Muskrat in the Louisiana Coastal Marshes,* published in 1949 by the state agency now known as the Louisiana Wild Life and Fisheries Commission, is a monumental contribution to the subject and should be consulted by anyone interested in any aspect of the biology of the species. The work is also a classic treatise on general marsh ecology. Louisiana has over 4 million acres of marsh, but, according to O'Neil, 80 percent of the muskrats are produced on approximately 1 million acres of brackish marsh where three-cornered grass *(Scirpus olneyi)* is the dominant or subdominant vegetation.

O'Neil subdivides the coastal marshes into three distinct types, Delta, Subdelta, and Prairie. No sharp lines of demarcation between the three types of marsh are readily apparent. Sometimes one type grades into another almost imperceptibly, requiring a person with extensive marsh experience and expertise in marsh ecology to recognize the indicators that appear when these marsh types merge from one into the other. Map 1 (page 46) shows the general distribution of the three marsh types, but because of local edaphic factors each type often contains patches of one or both of the other types. Space does not permit a discussion here of

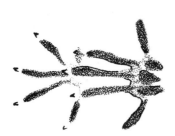

Tracks of Common Muskrat

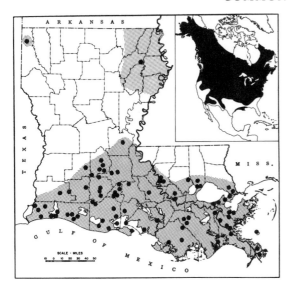

Map 41 Distribution of the Common Muskrat *(Ondatra zibethicus). Shaded area,* known or presumed range within the state; •, localities from which museum specimens have been examined; *inset,* overall range of the species.

the characteristics of each of the marsh types, but the subject is thoroughly covered in the work by O'Neil. Suffice it to say that the heaviest concentration of muskrats occurs in the Subdelta marshes. In the 20 years for which records are available in the period of 1924 through 1946, a total of 14,933,806 muskrats were trapped on the holdings of the Delacroix Corporation, 129,000 acres of brackish marsh in which three-cornered grass *(Scirpus olneyi)* is the dominant plant.

O'Neil's treatment (1949) of the basic features of the life history of the Louisiana muskrat is so interesting and informative that I am taking the liberty of quoting him extensively as follows:

House Construction.

The near-sea-level type habitat of the Louisiana muskrat *(Ondatra zibethicus rivalicius)* has made house-building a necessity with this subspecies, though where burrows are possible the coastal muskrats revert almost entirely to burrowing.

Muskrats that have been live-trapped in the marsh area and removed to upland ponds have immediately abandoned the practice of building grass houses and resorted to the digging of ground burrows. Only occasionally in large, flat, marsh-type ponds can the house building tendency of the muskrats be observed, and these houses are generally constructed around stumps, logs, or shrubs.

The average 'rat house ranges in size from three to six feet in diameter at the base, and from two to four feet in height. The material used is that nearest at hand. The most unsubstantial houses studied were those of the Delta marshes, constructed of alligator grass and cattail. The most substantial houses are constructed of the brackish marsh species, three-cornered grass and wiregrass ... [Figure 121]. Many of the houses in the deeper fresh marshes are constructed of poor material such as bull tongue or delta potato. A mixture of three-cornered grass and wiregrass makes the best building material.

The essential purpose of a muskrat house is to afford the animal a comparatively dry nest with an even temperature; it is located above normal high tides. Nests are not built at random in luxuriant and undisturbed marsh as they generally appear to be to the casual observer. However, in over-populated areas and in marshes where muskrats have been harassed by storm tides or drought, nests may be constructed overnight, camped in for several days or weeks, then vacated.

A few scattered pairs of muskrats are generally directly responsible for a colony or colonies that "spring-up" in large unpopulated areas. Muskrats cannot exactly be classed as monogamous, but they do have definite family characteristics. A pair of young adults, the female bearing young for the first time, will often travel several hundred yards from an over-populated area to construct a house in an entirely new unpopulated area. The new house, quite a makeshift affair, is nothing more than a heap of grass thrown up on top of the standing vegetation with a short trench cut in the marsh floor leading to the center of the house ... [Figure 122]. An opening is bored in the straw heap from underneath and a nest rounded-out, usually 10 to 15 inches above the water level. Muskrats are cooperative and diligent workers and construct a livable house within a couple of hours. The house, plunge hole, and deep underground runs, especially one or two main escape runs, are fairly complete before the first young are born. Houses containing newly-born muskrats or females ready to give birth to young can be pointed out in the marsh for they are almost always freshly-worked and plastered with muddy peat that has been obtained from the ooze around the base of the house. The female leaves her young warmly packed in a ball of finely-shredded grass while she is out feeding or working on the house. Observations by the author indicate that, within five to ten days after giving birth to a litter, the female is ready to copulate again.

During the time that the first litter is suckling, the male lives in a separate nest located in the opposite side of the house. At this period of family activity, the house has taken on a "lived-in" appearance. Normally, about the time the first litter reaches kit stage, with guard-hair beginning to appear, the second litter is born in a new nest that has been constructed in the same house. These new nests are constructed until the original house contains three or four—enough to take care of as many litters—and from then on the same nests are used over and over again. The adult pair and the kits all work at the perpetual job of maintaining the house. The marsh around the base of the

house becomes increasingly cut-up and serves as a moat, preventing marsh fires from destroying it.

The investigator has no record of young muskrats being burned in a nest. However, this might easily happen during extreme drought and undoubtedly does happen when fires occur during major droughts. In cases where a house is afire, the female always insulates the nest with a wet mud pack, if it contains young.

Normally, when the first litter reaches sexual maturity and the second litter reaches kit stage, the first litter is evidently forcibly evicted by the original pair, to mate and repeat this same process.

In a good three-cornered grass marsh that is sparsely populated, the young pairs will build their houses within 20 to 30 feet of the original house. In some instances a pair will just build an apartment connected with the original house. This is what the trappers call a "double house," and the

Figure 121 Muskrat house on Rockefeller Refuge. (Photograph by Joe L. Herring)

entire group of 10 to 15 or more muskrats live and work together. However, the usual number living together in one house is a pair of old adults and two to four young.

These colonies continue to expand until they meet with other expanding colonies, tunneling and destroying the food supply until the muskrats denude the entire area of practically all vegetation and cut the marsh floor to a depth of 18 to 20 inches. This is what is termed a peak population and the slide toward the trough in the population curve is inevitable.

Floating-turf House: The floating marsh . . . or "flotant," as it is called in Louisiana, fluctuates with the water levels and the buoyancy of the turf tends to maintain a constant minus three to a minus ten inch water level. The space between the surface of the water level and the surface of the marsh floor is where the nests are constructed, and there is very little true surface activity in a floating marsh of this type. Practically all of the feeding runs and traveling lanes are concealed in the turf and only during near-freezing weather does the 'rat show a great deal of surface movement. Muskrats in the floating marsh have semi-burrowing tendencies; their nests are constructed at the end of a tunnel just at the surface of the marsh floor and are similar to those constructed in the grass houses. Where the water level is only minus two to minus three inches, the 'rats are forced to build small, poorly-constructed grass houses . . . .

Bayou Bank and Levee Burrows: Bayou banks and levees that extend 10 to 15 inches above normal high tide levels for the marshes are perforated with muskrat runs and burrows. Muskrats construct nests in the levees and bayou banks in the same manner in which they build nests in the floating marsh. In the areas where a great many banks and levees are inhabited by muskrats, censusing is more difficult as the number of obvious grass houses in the marsh is not a true indicator of the number of muskrats inhabiting the area . . . .

Rice Field Burrows: The Louisiana muskrat has never been found inhabiting streams subject to frequent floodings. However, the coastal prairie or rice belt is inhabited by some fair sized colonies of 'rats. Rice field trappers catch from 300 to 500 'rats per season in the sections where the habitat is better for 'rats. The impounded areas that are used for reservoirs maintain constant water levels, making bank-burrowing possible. During the rice growing season most of the reservoir and canal bank inhabitants migrate to the flooded rice fields. Many construct houses from rice straw, produce a litter, and return to the burrows when threshing season commences.

The Common Muskrat reproduces the year around, reaching the highest degree of sexual activity in November and March and the lowest in July and August. In the period 1942 to 1945, O'Neil examined 645 females during the trapping season and found that 16 percent contained embryos, showing that the production of young during the winter months is substantial. He gives the average number of young per litter, presumably at different localities, as 3.6 to 3.8, and he estimates that the adult female generally produces five to six litters each year but is capable of having seven or eight litters. The gestation period is stated to be 26 to 28 days. Muskrats in the kit stage, which is reached between the sixth and eighth weeks, have been found with large embryos, but they seldom carry more than one or two young.

Although sometimes abroad in the middle of the day, muskrats usually remain in their beds from daylight to the middle of the afternoon, when they commence to forage for food. In their selection of food items they are largely opportunists, taking what is most readily available. Again I rely on the authority of O'Neil (1949) for the following information on the foods eaten by muskrats in their various habitats.

In the Delta marshes approximately 70

Figure 122 Interior of muskrat house with top layer of vegetation removed, revealing two holes through which the occupants gained entrance by way of underwater passages. The bright area in the center marks the location of the living quarters. (Photograph by Erol Barkemeyer)

percent of the diet is made up of cattails *(Typha);* 15 percent, alligator grass *(Alternanthera philoxeroides);* and 10 percent, roseau cane *(Phragmites communis)* shoots, delta duck potato *(Sagittaria platyphylla)* and dog-tooth grass *(Panicum repens).* The remaining 5 percent consists of crabs, crayfish, mussels, and small fish.

In the brackish Subdelta marshes about 80 percent of the muskrat's diet consists of three-cornered grass; 10 percent, wiregrass *(Spartina patens);* and 5 percent, hogcane *(Spartina cynosuroides).* The remaining five percent is made up of cattails, roseau cane shoots, giant bulrush *(Scirpus californicus),* coco grass *(Scirpus robustus),* oystergrass *(Spartina alterniflora),* spike rush tubers *(Eleocharis* sp.*),* delta duck potatoes, pickerelweed *(Pontederia cordata),* and black rush *(Juncus roemerianus).* Small amounts of animal life, mainly crustaceans and mollusks, are also eaten.

Figure 123 Muskrat run in a well-cared-for marsh on the State Wildlife Refuge, Vermilion Parish. (Photograph from Louisiana Wild Life and Fisheries Commission files)

In the fresh Subdelta marshes, foods eaten are as follows: paille fine *(Panicum hemitomon),* 70 percent; cattails, 10 percent; wapato *(Sagittaria latifolia)* and roseau cane shoots, 10 percent; *Sagittaria* sp., giant bulrush, and softstem bulrush *(Scirpus validus),* 5 percent; animal matter, 5 percent.

In the Prairie marshes, the plants eaten by the muskrat include: three-cornered grass, 70 percent; coco grass and wiregrass, 10 percent; and hogcane, 10 percent. The remaining 10 percent is made up of cattails, bulrushes, lake grass *(Paspalum distichum),* salt grass *(Distichlis spicata),* dwarf spike rush *(Eleocharis parvula),* Gulf Coast spike rush *(E. cellulosa),* black rush, pickerelweed, *Sagittaria* sp., roseau cane shoots, oystergrass, and alligator weed.

Muskrats living in freshwater lakes, flooded rice fields, and bayou banks north of the coastal marshes feed on whatever is available, including *Sagittaria,* pickerelweed, spike rush, rice cutgrass *(Leersia hexandra),* alligator weed, paille fine, clover *(Trifolium* spp.*),* and even field grasses. According to O'Neil, animal life makes up a larger percentage of the food of these rats than of any of the marsh muskrats. Remains of freshwater mussels, clams, crayfish, and even fish are found on the feeding platforms.

Muskrat populations fluctuate considerably from year to year, as is clearly evident from records on file in the Fur Division of the Wild Life and Fisheries Commission. Table 2 and Figure 3 show the annual catches from 1913 to 1973.

In poor trapping years, muskrats make up 50 percent or less of the income derived from the furs produced, but in good years (before the ascendancy of the Nutria to top position) they have accounted for as much as 80 to 90 percent of the total value. The latter, in the last decade, has amounted to between 2 and 6.7 million dollars. The all-time record year was the season of 1945–1946, when fur sales yielded 15.6 million dollars. Peak years generally reflect wet years when the vegetation was lush and the marshes were in optimum condition. Low production reflects drought, hurricanes, floods, pollution, and other adverse factors that profoundly influence the productivity of the marshes, the least of which is by no means the vegetational eat-outs that follow periods of maximum muskrat abundance.

Ideal marsh conditions for muskrats are often the product of a delicate balance between climax wiregrass *(Spartina patens)* and three-cornered grass *(Scirpus olneyi),* the latter being the choice food of muskrats. When natural or man-made changes, such as burning or saltwater high tides, kill off the wiregrass, three-cornered grass takes over and blankets an area. When such an event occurs, the muskrat population multiplies rapidly, soon reaches immense proportions, and then proceeds to eat virtually every blade of grass over wide expanses of the marsh. As food starts to diminish, the population pendulum begins to swing in the other direction, and within a matter of months the number of muskrats has again reached the bottom of a trough. (Figures 123 and 124)

In the last two decades still another factor has entered the picture of marsh ecology and has unquestionably exerted an influence on muskrat productivity. I refer, of course, to the introduction of the Nutria in 1937 and its subsequent rapid increase in numbers. In the trapping season of 1943–1944, the Nutria first appeared in fur-catch records with 436

Figure 124 Muskrat eat-out on McIlhenny property, Vermilion Parish, 1944. (Photograph by Erol Barkemeyer)

being reported by trappers. The pelts sold for an average price of 50 cents. But two years later the number increased to 8,784, and in ten years, during the 1953–1954 season, the species yielded 160,654 pelts. In the 1969–1970 season the catch was up tenfold to 1,604,175 with a total value of $3,826,680. Although the Louisiana muskrat had long reigned as the state's number one fur animal, it relinquished this position to the Nutria in the 1961–1962 season, when trappers caught 912,890 Nutria but only 632,558 muskrats. The ascendancy of the Nutria involves the interplay of an array of intricate environmental factors and is by no means a simple case of interspecific competition in which the muskrat emerged second best. Since the early 1950s a general deterioration has occurred in many of our marshes and has often been followed by a succession in the vegetation that has culminated in conditions much more favorable to the Nutria. Alligator weed, for example, provides adequate Nutria habitat but is almost worthless in maintaining muskrats. For detailed accounts of changes in the marsh, see O'Neil (1968) and Palmisano (1967 and 1970).

Predators. — Muskrats, especially immatures, are subject to severe predation by a host of enemies. O'Neil (1949) gives a list of animals that prey upon muskrats in order of their importance, as follows: mink, raccoons, owls, alligators, ants, Marsh Hawks, Cottonmouth Moccasins, bullfrogs, garfish, bowfins, snapping turtles, black bass, crabs, hogs, Domestic Cats, and dogs.

Parasites and diseases. — The parasites of the Louisiana muskrat have been studied in some detail by Penn (1942). He found the mite *Tetragonyssus spiniger* abundant in the nests and reported that trappers complain that their clothing frequently becomes infested with the vermin even though the mites seem to die soon after leaving the fur of the host. Several larvae of the flesh-fly *Sarcophaga* sp.

were found in the fur of two muskrats from Willow Bayou in Cameron Parish. Penn recorded two protozoans (*Giardia ondatrae* and *Trichomonas* sp.), four nematodes (*Capillaria hepatica*, *Rictularia* sp. [cf. *R. ondatrae* Chandler, 1941, from muskrats in southeastern Texas], *Physaloptera* sp., and *Longistriata adunca*), one cestode (*Hymenolepis evaginata*), three trematodes (*Nudacotyle novicia*, *Echinochasmus schwartzi*, and *Paramonostomum crotali*). The last-named parasite was found in 9.3 percent of 1,032 muskrats examined (Penn and Martin, 1941). Additional internal parasites that have been found in muskrats in other parts of the range of the species include 4 protozoans, 34 trematodes, 7 cestodes, and 16 nematodes (Doran, 1954a and b, 1955a and b). Many diseases are known to affect muskrats — coccidiosis, tularemia, and leukemia, as well as a fungus infection that is often fatal to the young.

Subspecies

Ondatra zibethicus rivalicius (Bangs)

Fiber zibethicus rivalicius Bangs, Proc. Boston Soc. Nat. Hist., 26, 1895:541; type locality, Burbridge, Plaquemines Parish, Louisiana.
Ondatra rivalicia, Miller, Bull. U.S. Nat. Mus., 79, 1912:232.
Ondatra z[ibethicus]. rivalicius, Davis and Lowery, Jour. Mamm., 21, 1940:212.

Muskrats of Louisiana average smaller than northern populations and are referable to the race *rivalicius*, which was once, without justification, accorded the rank of a full species. Although all Louisiana and southeastern Texas specimens, as representative of a single population, are assignable to *rivalicius*, there is, nevertheless, geographical variation from east to west along the central-northern Gulf Coast. Specimens from southeastern Texas and southwestern Louisiana are the darkest of all, while those from near the type locality in southeastern Louisiana are somewhat paler. Specimens from Baton Rouge are much more heavily suffused with brown than are coastal rats, but they are much darker than specimens from the north-central and

Table 7 Body Length and Certain Skull Dimensions of *Ondatra zibethicus* by Region[1]

Character	Eastern Louisiana			Midwestern Louisiana			Western Louisiana		
	No.	Mean	SE	No.	Mean	SE	No.	Mean	SE
Body length	89	283.5 ±	1.3	46	293.7 ±	1.5	222	302.7 ±	.76
Total length of skull	89	62.83 ±	.21	46	64.07 ±	.27	222	66.85 ±	.13
Condylo-alveolar length	89	37.20 ±	.14	46	37.80 ±	.18	222	39.57 ±	.09
Alveolar length	89	15.60 ±	.07	46	15.71 ±	.15	222	16.11 ±	.05
Condylo-postalveolar length	89	21.60 ±	.12	46	22.12 ±	.14	222	23.39 ±	.07
Length of diastema	89	21.65 ±	.10	46	21.89 ±	.12	222	23.00 ±	.06
Cranial breadth	89	21.57 ±	.07	46	21.81 ±	.09	222	22.63 ±	.05
Zygomatic breadth	87	39.41 ±	.17	43	40.06 ±	.21	218	41.63 ±	.10
Interorbital breadth	89	6.14 ±	.05	46	6.29 ±	.07	222	6.35 ±	.04
Occipital height	89	16.20 ±	.07	46	16.31 ±	.11	222	16.95 ±	.05

[1]After Gould and Kreeger, 1948.

northeastern states and adjacent parts of Canada (Davis and Lowery, 1940). When a sufficient number of specimens from northeastern Louisiana become available for study, they may show that this population is referable to *O. z. zibethicus*. Gould and Kreeger (1948) have shown that muskrats from southeastern Louisiana marshes are significantly smaller in body length and in most skull measurements than are those from western Louisiana (Table 7), but this clinal trend would hardly merit nomenclatorial recognition.

Specimens examined from Louisiana. — Total 354, as follows: *Acadia Par.:* 1 mi. W Eunice, 1 (USL); 5 mi. S Crowley, 1 (USL); 2 mi. N Crowley, 1 (USL); 5 mi. N Crowley off La. Hwy. 13, 4 (USL). *Allen Par.:* 5 mi. E Oberlin, 4 (LSUMZ). *Ascension Par.:* Sorrento, 1 (LSUMZ). *Avoyelles Par.:* Coco Lake, Moreauville, 1 (LTU). *Caddo Par.:* Black Bayou Lake, 2 mi. E Vivian, 1 (NSU). *Calcasieu Par.:* 4 mi. S Lake Charles, 1 (MSU); Iowa, 2 (USNM). *Cameron Par.:* Grand Cheniere, 1 (LSUMZ); 3 mi. W Cameron, 1 (LSUMZ); Little Cheniere Rd., 1 (MSU); 0.5 mi. from east end of Cheniere Pardue, 1 (MSU); Johnsons Bayou, 1 (USL); 17 mi. E Cameron, 5 (USL); Calcasieu Lake, 6 (USNM); locality unspecified, 2 (USNM); Leedyburg, 1 (USNM); Sabine Nat. Wildlife Refuge, 6 (USNM); Gum Cove, 1 (USNM); Bowman Camp [near mouth of Black Bayou], 4 (USNM). *East Baton Rouge Par.:* various localities, 69 (LSUMZ). *Evangeline Par.:* 8 mi. NW Ville Platte, 1 (LSUMZ); Basile, 1 (USL); 4 mi. W Ville Platte, 1 (USL); 1.5 mi. N Chataignier, 1 (USL); 4 mi. SE Mamou, 1 (USL). *Iberia Par.:* Marsh Island, east end of island, 2 (LSUMZ); Avery Island, 6 (CAS). *Iberville Par.:* 20 mi. SW Port Allen, 1 (LSUMZ). *Jefferson Par.:* Grand Isle, 1

(LSUMZ); 4.5 mi. NW Barataria, 1 (LSUMZ). *Jefferson Davis Par.:* 5 mi. N Jennings, 1 (LSUMZ). *Lafayette Par.:* 13 mi. N Lafayette, 1 (USL); 8.4 mi. SW Lafayette, 1 (USL); 3 mi. SW Lafayette, 1 (USL); 10 mi. SW Lafayette courthouse, 2 (USL); 11 mi. SW Lafayette courthouse, 1 (USL); 9 mi. SW Lafayette courthouse, 2 (USL); 2 mi. NE Lafayette, 1 (USL). *Lafourche Par.:* 1 mi. SW Raceland, 1 (LSUMZ); 5 mi. S, 4 mi. W Raceland, 1 (LSUMZ); 2.5 mi. W Raceland on Hwy. 90, 1 (USL). *Livingston Par.:* Springfield, 1 (SLU). *Orleans Par.:* New Orleans, 4 (2 LSUMZ; 2 TU); Lake Catherine, 4 (1 TU; 3 USNM); Industrial Canal at U.S. Hwy. 90, 1 (TU); Chef Menteur, 1 (USNM). *Plaquemines Par.:* Davant, 7 (LSUMZ); 0.5 mi. NW Port Sulphur, 1 (LSUMZ); Happy Jack, 1 (SLU); 2 mi. NW Pointe-a-la-Hache, 1 (SLU); 4 mi. S Port Sulphur, 1 (SLU); Bohemia, 1 (SLU); 0.5 mi. N Port Sulphur, 1 (SLU); Tulane Univ. Riverside campus, Belle Chasse, 1 (TU); Belair, 20 (USNM); Breton Island, 1 (USNM); Johnson's Pass, 15 mi. above mouth of Miss. River, 10 (USNM); Octave Pass, Miss. Delta Camp Bayou, 2 (USNM); Burbridge, 7 (MCZ). *Richland Par.:* 1 mi. E Delhi, 1 (NLU). *St. Bernard Par.:* Reggio, 1 (LSUMZ); Shell Beach, 1 (LSUMZ); 0.8 mi. W Delacroix, 1 (LSUMZ). *St. James Par.:* 10 mi. NW Vacherie, 1 (LSUMZ). *St. John the Baptist Par:* Akers, 1 (SLU). *St. Landry Par.:* 6 mi. S Opelousas courthouse, 6 (USL); 7 mi. W Opelousas, 1 (USL). *St. Martin Par.:* Lake Farin, approx. 4 mi. N Henderson, 1 (USL). *St. Mary Par.:* 0.5 mi. from tip of Cypremort Pt., 1 (USL); Cypremort Pt., 29 (USL); 6 mi. S Ellerslie on La. Hwy. 317, 9 (USL); Morgan City, 37 (USNM). *St. Tammany Par.:* Salt Bayou, 2 (LSUMZ); Slidell, 1 (USNM). *Tangipahoa Par.:* 3 mi. E Ponchatoula, 1 (LSUMZ); 1.5 mi. E Ponchatoula, 1 (SLU); Manchac, 1 (SLU); Ponchatoula, 1 (SLU). *Terrebonne Par.:* Grand Caillou, 2 (LSUMZ); 30 mi. SW Houma, 8 (LSUMZ); Houma, 4 (USNM). *Vermilion Par.:* Abbeville, 2 (LSUMZ); 23 mi. S Abbeville, 1 (LSUMZ); 3 mi. S Boston, 12 (LSUMZ); 2 mi. SE Gueydan, 1 (LSUMZ); 8 mi. S Intracoastal City, 1 (USL); 2 mi. N Gueydan, 1 (USL); Pecan Island, 1 (USL); 20 mi. SW Abbeville, 5 (USNM); Cheniere au Tigre, 9 (ANSP).

The Old World Rat and Mouse Family
Muridae

ALTHOUGH THE members of this family occur virtually throughout the Old World, none of its species are found in the Western Hemisphere except as a result of man's activities. Almost wherever man has gone in historic time he has taken with him the House Mouse (*Mus musculus*) and the Norway and Roof Rats (*Rattus norvegicus* and *R. rattus*). Consequently, the family is now nearly worldwide in distribution. It comprises, according to Walker (*in* Walker et al., 1968), 100 genera. The number of species is problematical because so many of the some 570 named forms in the genus *Rattus* are of uncertain systematic rank. But exclusive of the genus *Rattus* the family contains approximately 290 species.

The introduced murids of Louisiana consist solely of the House Mouse and the Norway and Roof Rats. They differ from the Louisiana cricetine rodents in having the nose more pointed, the eyes smaller, and the tail naked with prominent annulations. The crowns of the cheek teeth possess three longitudinal rows of cusps (not always evident in worn teeth) that render them easily separable from those of cricetines possessing cusps, for in the latter animals the cusps are arranged in only two longitudinal rows.

Members of this family are of great economic importance. They damage crops, destroy or despoil great quantities of foods and stored grains, and harbor diseases to which man is susceptible. On a worldwide basis the damage caused by Norway and Roof Rats alone amounts to billions of dollars, and they are responsible for inestimable human misery.

Genus *Rattus* Fischer

Dental formula:

$$\text{I}\frac{1-1}{1-1} \ \text{C}\frac{0-0}{0-0} \ \text{P}\frac{0-0}{0-0} \ \text{M}\frac{3-3}{3-3}=16$$

ROOF RAT
(*Rattus rattus*)

Plate V

Vernacular and other names. — This species, depending mainly on the subspecies being referred to, has also been called the black rat, the Alexandrine rat, and the white-bellied rat. The nominate subspecies, *Rattus rattus rattus*, is all black and was, therefore, appropriately called the Black Rat back in the days when English names were given to subspecies. But that practice has now been wisely discontinued. Unfortunately, some recent authors have applied the name Black Rat to the species as a whole. I object to this name on the ground that it is too closely associated by previous usage with one of the subspecies. Moreover, it is grossly inappropriate, since most examples of the species are not black but are instead gray, often with pure white bellies. The name Alexandrine rat has been applied to the subspecies *Rattus rattus alexandrinus*, which was described on the basis of specimens from Alexandria, Egypt. The appelation white-bellied rat has been used for the white-bellied subspecies, *R. r. frugivorus*. The word *frugivorus* means fruit-eating. Although in Louisiana the subspecies has no particular predilection in its diet for fruit, in some places, such as in Florida, where much fruit is grown, it does do great damage and is even called the fruit rat. Of the various vernacular names that have been used for the species as a whole, I believe Roof Rat is unequivocally preferable to all others, and I recommend its adoption as the official name. It is based on the animal's habit of living in the upperparts of buildings and is therefore eminently appropriate. It also has the advantage of being a name that has been extensively used for the species in previous literature and it is not associated with any particular subspecies. When speaking of this species and the Norway Rat together the general term house rat is sometimes applied. The generic and specific names, *Rattus rattus,* are the Latinized form of the Old English word *raet.*

Distribution.—The Roof Rat is almost worldwide in distribution. It occurs in urban areas of the eastern and southern United States, in seaport cities and towns of western North America, and over most of Mexico and Central America. It has even found its way to South America. The species has been recorded from numerous localities in Louisiana. (Map 42)

External appearance—This species is a medium-sized rat with a total length up to 18 inches, of which the tail comprises more than half. The subspecies differ markedly in color. *R. r. rattus* is wholly black; *R. r. alexandrinus* is grayish brown above and gray below; *R. r. frugivorus* is similar to *alexandrinus* except that it is white or yellowish white below. The species can be confused only with the Norway Rat *(R. norvegicus),* but can be easily distinguished externally by its tail, which is always longer than the head and body. In *R. norvegicus* the tail is shorter than the head and body, and, in addition, the animal is somewhat more robust than *R. rattus.*

Color phases.—No color phases as such are known, yet I am confident that albinos and

other mutants occur occasionally. I do not know how one would distinguish a melanistic example of either *R. r. frugivorus* or *R. r. alexandrinus* from a normal *R. r. rattus,* which is black. I am fully aware of the suggestion by Caslick (1956) that the three so-called subspecies may be nothing more than color phases. Along with most mammalogists, I find the evidence for this conclusion unconvincing.

Measurements and weights.—Twenty-five specimens in the LSUMZ from various localities in the state averaged as follows: total length, 401 (371–436); tail, 216 (190–250); hind foot, 37 (31–40); ear, 21 (17–27); greatest length of skull, 42.1 (39.9–46.9); cranial breadth, 17.1 (16.1–18.9); zygomatic breadth, 19.7

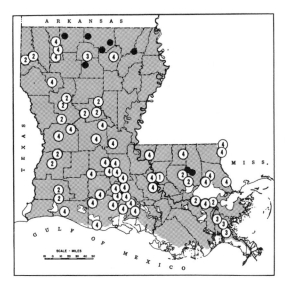

Map 42   Distribution of the introduced Roof Rat *(Rattus rattus)* in Louisiana. *Shaded area,* known or presumed range within the state. Numbered localities represent places from which museum specimens have been examined and identified to subspecies as follows: 1. *R. r. rattus, R. r. alexandrinus,* and *R. r. frugivorus;* 2. *R. r. alexandrinus* and *R. r. frugivorus* only; 3. *R. r. alexandrinus* only; 4. *R. r. frugivorus* only. Additional specimens unidentified to subspecies have been examined from most of the above numbered localities. •, additional localities from which specimens have been examined but not identified to subspecies.

(18.3–22.3); interorbital breadth, 6.0 (5.4–6.9); length of nasals, 15.2 (13.3–17.7); length of diastema, 11.2 (10.1–12.8); length of maxillary toothrow, 6.5 (5.5–7.0); palatilar length, 18.8 (17.1–21.6); postpalatal length, 14.6 (13.7–15.2). Two adults weighed 202 and 206 grams, respectively.

Cranial and other skeletal characters.—As noted in the introduction to the family Muridae, the crowns of the upper cheek teeth are characterized by having the cusps arranged in three longitudinal rows. The skulls of this species and the Eastern Wood Rat are similar in overall size but are easily distinguished by the characters listed in Table 5 on page 256. The skull of *R. rattus* differs from that of *R. norvegicus* in having the temporal ridges on each side of the parietals bowed outward instead of more or less parallel. Also, the length of the parietals measured along the temporal ridges is decidedly less than the greatest distance between the ridges, whereas in *norvegicus* the length of the parietals is approximately equal to the greatest distance between the ridges. Finally, in *rattus* the diastema (the gap between the posterior border of the incisors and the first cheek tooth) is decidedly less than twice the length of the cheek toothrow. In *norvegicus* the diastema is nearly twice the length of the cheek toothrow. The baculum is spatulate on the proximal end and possesses a slight constriction in the shaft near the distal end. (Figures 72 and 125)

Sexual characters and age criteria.—The sexes are superficially nearly identical, but in adult males the testes descend into a shallow scrotum and in the adult females the vagina is perforate. The testes, unlike those of cricetines, do not regress between periods of breeding activity. The mammae usually number five on each side, but occasionally six pairs are present. Paired scent glands that give off an unpleasant, musky odor are located just inside the anus. Immature individ-

uals of *Rattus rattus* are grayer and much duller in color than adults. They also lack long guard hairs.

Status and habits.—The Roof Rat is probably abundant in most towns and cities throughout the state despite the fact that specimen records are lacking from many localities. One reason for the paucity of specimens is that mammalogists dislike the task of skinning and preparing house rats, which are often dirty, diseased, and infested with fleas. Although the species is generally found in close association with man and his garbage, it sometimes occurs fairly far from human habitations. I once found a pair occupying a nest of grasses and sedges three feet above ground level in a coffeebean plant in the marsh ap-

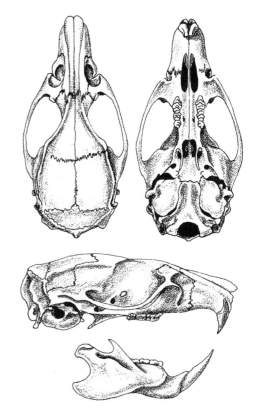

Figure 125   Skull of Roof Rat *(Rattus rattus rattus)*, ♀, 1 mi. S University, East Baton Rouge Parish, LSUMZ 2716. Approx. 1½×.

proximately one mile north of the town of Creole, in Cameron Parish. When it comes in contact with the Norway Rat in buildings it usually occupies the attics while the Norway Rat takes over the lower levels. In general the Norway Rat supplants the Roof Rat in the more favorable situations, such as the vicinity of wharves, warehouses, sewers, refuse dumps, and garbage cans containing discarded food.

The species is said to have arrived in Latin America with the first Spanish conquistadores in the early 1500s and to have come ashore at Jamestown with the English colonists in 1607. The Norway Rat apparently did not reach our shores until much later, but because of its more aggressive habits it drove the Roof Rat from many localities. The all-black subspecies *R. r. rattus* seems to have been particularly unsuccessful in competing with the Norway Rat and is now exceedingly rare. I had never seen an example of the subspecies *rattus* from Louisiana until 31 December 1943, when L. S. Bridges, a long-time pest control operator in Baton Rouge, brought me specimens that he had captured that day on the premises of the Esso Refinery on the north side of the city. They were the first that he had ever found in his many years of rat extermination in the area. Since the Port of Baton Rouge and the associated industrial complex are served by a steady flow of ships from foreign ports, the specimens obtained by Bridges may have been part of a fairly recent introduction. In subsequent years a few additional specimens of *R. r. rattus* have been obtained in the state, but it appears to be scarce. The overwhelming majority of specimens of the Roof Rat from Louisiana are referable to *R. r. frugivorus,* the white-bellied race, but a few typical examples of *R. r. alexandrinus,* the gray-bellied form, are known from scattered localities. (Map 42)

The two species of commensal house rats are similar in their general habits, which are discussed more fully under the next species, the Norway Rat. The Roof Rat is, never-

theless, different in some respects. It is a much better climber than the Norway Rat, a fact that perhaps explains its tendency to reside in attics and other high places. I have often seen rats of this species crawling along power and telephone lines across city streets, and I have watched them climbing trees, vines, and downspouts of gutters to gain access through the eaves of houses to their attic retreats. When not competitively excluded from ground level habitats by Norway Rats, they often move back and forth from their second-story haunts to underground nests in tunnels beneath buildings. More frequently, however, they build their nests in the walls.

Breeding takes place throughout the year, but most of it occurs in early spring and again in early summer. Roof Rats are said to have fewer litters and fewer young to a litter than the Norway Rat. Predators are essentially the same in the two species and are treated under the Norway Rat.

Parasites. — Doran (1954a and b, 1955a) lists 6 protozoans, 11 cestodes, and 5 nematodes as occurring in this species. Morlan (1952) found the Roof Rat in Georgia to be the host of 3 lice, 2 flies, 11 fleas, and 31 mites and ticks.

Subspecies

Rattus rattus rattus Linnaeus

[*Mus*] *rattus* Linnaeus, Syst. nat., ed. 10, 1, 1758:61: type locality, Uppsala, Sweden.
Rattus rattus, Hollister, Proc. Biol. Soc. Washington, 29, 1916:126.

Specimens examined from Louisiana. — Total 10, as follows: *East Baton Rouge Par.:* various localities, 10 (LSUMZ).

Rattus rattus alexandrinus (E. Geoffroy Saint-Hilaire)

Mus alexandrinus E. Geoffroy Saint-Hilaire, Catalogue des mammifères du Muséum National d'Histoire Naturelle, Paris, 1803:192; type locality, Alexandria, Egypt.
R[*attus*]. *rattus alexandrinus,* Hinton, Jour. Bombay Nat. Hist. Soc., 26, 1918:63.

Specimens examined from Louisiana.—Total 19, as follows: *Beauregard Par.:* Singer, 1 (MSU). *Caddo Par.:* Shreveport, 1 (LSUMZ). *Calcasieu Par.:* Lake Charles, 1 (MSU). *East Baton Rouge Par.:* LSU, 2 (LSUMZ). *Grant Par.:* 0.25 mi. S Pollock, 1 (USL). *Lincoln Par.:* Ruston, 1 (LSUMZ). *Livingston Par.:* 1 mi. N Holden, 1 (LSUMZ). *Natchitoches Par.:* Kisatchie, 8 (LSUMZ). *Orleans Par.:* New Orleans, 1 (LSUMZ). *Plaquemines Par.:* 2 mi. W Pointe-a-la-Hache, 1 (LSUMZ). *St. Charles Par.:* 1 mi. SE Norco, 1 (TU).

Rattus rattus frugivorus (Rafinesque)

Musculus frugivorus Rafinesque, Précis des decouverts et travaux somiologiques . . ., 1814:13; type locality, Sicily.

R[*attus*]. *rattus frugivorus,* Hinton, Jour. Bombay Nat. Hist. Soc., 26, 1918:65.

Specimens examined from Louisiana.—Total 111, as follows: *Acadia Par.:* 8 mi. S Crowley, 1 (USL). *Avoyelles Par.:* 2 mi. W Cottonport, 1 (USL). *Beauregard Par.:* De Ridder, 2 (MSU). *Caddo Par.:* 3 mi. S, 2 mi. W Blanchard, 1 (LSUMZ). *Calcasieu Par.:* Lake Charles, 6 (1 LSUMZ; 5 MSU); 11 mi. S Lake Charles, 1 (MSU). *Cameron Par.:* Creole, 1 (LSUMZ). *East Baton Rouge Par.:* various localities, 40 (LSUMZ). *Evangeline Par.:* Basile, 1 (USL); Ville Platte, 1 (USL). *Grant Par.:* Colfax, 1 (LSUMZ). *Iberia Par.:* USDA Livestock Expt. Farm, Willow Woods, 2 (LSUMZ); Avery Island, 2 (TU); New Iberia, 1 (USL). *Iberville Par.:* 0.5 mi. N Plaquemine, 1 (LSUMZ). *Jefferson Par.:* 2 mi. W Metairie, 1 (LSUMZ); 2–3 mi. N Kenner, 1 (LSUMZ); Metairie, 2 (SLU); Kenner, 1 (SLU). *Lafayette Par.:* Lafayette, 2 (USL); 2 mi. S Lafayette, 1 (USL); 6 mi. S Lafayette, 1 (USL). *Livingston Par.:* 3 mi. SE Springfield, 1 (LSUMZ). *Natchitoches Par.:* Provencal, 1 (LSUMZ); Kisatchie, 7 (LSUMZ). *Orleans Par.:* New Orleans, 4 (2 LSUMZ; 1 TU; 1 USL). *Ouachita Par.:* Monroe, 1 (LSUMZ). *Rapides Par.:* 3 mi. SE Libuse, 1 (LSUMZ); 4 mi. W Lecompte, 1 (USL). *Red River Par.:* 5 mi. N Coushatta, 2 (LSUMZ). *St. Charles Par.:* Norco, 1 (TU). *St. Helena Par.:* 0.5 mi. E Greensburg, 1 (LSUMZ). *St. Landry Par.:* Opelousas, 1 (LSUMZ); 7 mi. W Opelousas, 1 (USL); 6 mi. S parish courthouse, 1 (USL); 10 mi. N Opelousas, 1 (USL); Arnaudville, 1 (USL). *St. Martin Par.:* 0.5 mi. E Breaux Bridge, 1 (USL); St. Martinville, 1 (USL); 0.5 mi. S Parks, 1 (USL). *St. Mary Par.:* 1 mi. S on La. Hwy. 318 from junct. La. Hwy. 318 and U.S. Hwy. 90, 1 (USL). *St. Tammany Par.:* 8 mi. S Covington, 1 (USL). *Vermilion Par.:* Abbeville, 1 (USL); Pecan Island, 1 (USL); Gueydan, 1 (USL). *Vernon Par.:* 0.5 mi. N Simpson, 2 (LSUMZ); Leesville, 1 (MSU). *Washington Par.:* Bogalusa, 1 (LSUMZ). *Webster Par.:* Sibley, 1 (USL). *West Baton Rouge Par.:* 2.5 mi. S Port Allen, 1 (LSUMZ). *West Feliciana Par.:* 5.6 mi. ENE St. Francisville, 1 (LSUMZ).

Additional specimens of *Rattus rattus* from Louisiana not identified to subspecies.—Total 58, as follows: *Caddo Par.:* Shreveport, 1 (LSUS). *Claiborne Par.:* 7 mi. NW Homer, 1 (LTU). *East Baton Rouge Par.:* Baton Rouge, 1 (USNM). *Grant Par.:* Georgetown, 1 (LTU). *Jackson Par.:* 1 mi. E Ansley, 1 (LTU). *Lincoln Par.:* Ruston, 7 (LTU); 1

mi. S Ruston, 11 (LTU); 3 mi. W Ruston, 1 (LTU). *Natchitoches Par.:* Kisatchie, 1 (USNM); Natchitoches, 3 (NSU). *Plaquemines Par.:* Port Sulphur, 3 (SLU); 2 mi. W Pointe-a-la-Hache, 1 (SLU); Burbridge, 4 (MCZ). *Richland Par.:* 5 mi. E Oak Ridge, 1 (NLU). *St. Tammany Par.:* Pearl River, 1 (SLU); 3 mi. E Covington, 1 (SLU); Covington, 1 (SLU); 3 mi. NE Covington, 3 (SLU). *Tangipahoa Par.:* Hammond, 4 (SLU); 1 mi. S Hammond, 2 (SLU); 2 mi. W Natalbany, 1 (SLU); 1 mi. W Hammond, 1 (SLU); locality unspecified, 1 (SLU). *Union Par.:* 1 mi. N Bernice, 1 (LTU); 12 mi. E Farmerville, 1 (LTU); 10 mi. E Farmerville, 1 (LTU); 12 mi. S Farmerville, 1 (LTU). *Washington Par.:* 3 mi. E Angie, 1 (SLU). *Webster Par.:* 1 mi. E Minden, 1 (LTU); 10 mi. N Minden, 1 (LTU).

NORWAY RAT   Plate V
(*Rattus norvegicus*)

Vernacular and other names.—The Norway Rat has also been called brown rat, wharf rat, house rat, and sewer rat. The name Norway Rat and the specific name *norvegicus,* the Latin word for the country Norway, are both misnomers, for the species is of Asiatic origin. After spreading into western Europe, it was probably first brought to America from England. Though sometimes alleged to have come from Norway, the material on which the original description was based was from England. As previously explained, the word "rat" is derived from the Medieval English word *raet,* and the generic name, *Rattus,* is the Latinized rendition of the same word.

Distribution.—The species is a native of the Old World, originally Asia, but through the agencies of man has now spread over Europe, Africa, Australia, and most of the Western Hemisphere; hence it is virtually cosmopolitan in distribution. In Louisiana it doubtless occurs in all cities and towns and is often found in rice and cane fields well removed from human habitations. (Map 43)

External appearance.—The Norway Rat is so well known that it hardly requires description. The total length ranges up to 18 inches, of which the tail is always slightly less than half. The color is brownish above and grayish below, and the hair is coarse. The large ears are naked, and the tail is sparsely haired, scaly, and prominently annulated. The only other rats with which it can be easily confused are *Rattus r. alexandrinus* and *R. r. frugivorus,* but the latter are more grayish above and the underparts of *frugivorus* are pure white or yellowish white, not gray. The main distinction is, however, the length of the tail relative to the length of the head and body. In the Norway Rat the tail is always less than half the total length; in the Roof Rat it is always greater.

Color phases. — The white rat that is utilized for experimental purposes in virtually every biological laboratory in the world is an albino derivative of the Norway Rat. Other mutations in the species found in biological laboratories range from white, through spotted black and white, to nearly all-black individuals. Black Norway Rats have been reported in wild populations but not in Louisiana.

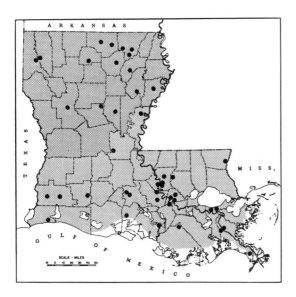

Map 43 Distribution of the introduced Norway Rat *(Rattus norvegicus)* in Louisiana. *Shaded area,* known or presumed range within the state; •, localities from which museum specimens have been examined. Probably occurs wherever there is human habitation.

Measurements and weights. — Twenty-five skins in the LSUMZ from various localities in the state averaged as follows: total length, 401 (360–457); tail, 189 (160–220); hind foot, 40 (35–46); ear, 20 (15–25). Twenty skulls from the same source averaged as follows: greatest length, 49.3 (43.9–51.5); cranial breadth, 18.4 (17.2–20.0); zygomatic breadth, 23.0 (20.4–27.7); interorbital breadth, 6.5 (6.1–7.0); length of nasals, 17.7 (15.8–19.6); length of diastema, 13.7 (12.1–15.2); length of maxillary toothrow, 6.6 (6.4–7.4); palatilar length, 22.5 (20.5–24.6); postpalatal length, 16.5 (15.7–18.5). Weights of three individuals averaged 269 (256–272) grams.

Cranial characters. — The skull of *Rattus norvegicus* is similar to that of *Rattus rattus* but differs in certain important particulars, which are discussed in the previous account. (See page 282 and Figures 125 and 126.)

Sexual characters and age criteria. — The information on sexual and age characters given for the Roof Rat applies equally well to the Norway Rat and need not be repeated here. One additional point seems to be worthy of mention. Davis (1948) found that the size at which Norway Rats reach sexual maturity varies from one locality to another. The testes of 50 percent of the males of a Maryland farm population descended into the scrotum when the rats were about 18 weeks old and weighed approximately 136 grams. But males of a Baltimore city population reached sexual maturity at 119 grams. In the female farm rats, 50 percent of the population were capable of breeding, as evidenced by having perforated vaginae, when they reached 88 grams. The corresponding weight for city females was 105 grams.

Status and habits. — Norway Rats were introduced into Europe from Asia sometime around 1727, when hordes of them are alleged to have crossed the Volga River. At about the same time, they arrived in England on ships from Asia. They increased rapidly and soon overran all western Europe, whence they were carried by ships to ports throughout the world. The first to appear in what is now the United States arrived around the time of the American Revolution. The Norway Rat is now found in every state of the Union, including Alaska and Hawaii, and in all the countries of the Western Hemisphere. The species is now probably resident in every town and city in the state, as well as in rural communities. It is also sometimes abundant in farm buildings and even in cultivated fields well removed from human dwellings.

The species is extremely unattractive and obnoxious. It possesses only one redeeming

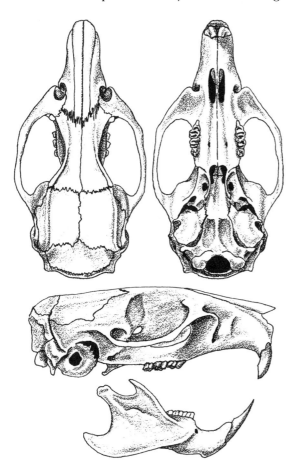

Figure 126 Skull of Norway Rat *(Rattus norvegicus)*, ♀, 0.5 mi. E Schriever, Terrebonne Parish, LSUMZ 13735. Approx. 1½ ×.

feature, which will be mentioned later. It causes great damage to man's property, destroys immense quantities of his food resources, spreads several potentially fatal diseases, and is an all-around nuisance. Its numbers sometimes become so great in our towns and cities that expensive programs of control are required even to hold the species in check. Such efforts are usually at best only a temporary source of relief. The manager of a large supermarket in Baton Rouge once reported that rats caused $1,500 damage each month to his stock. The figure is not surprising in view of the fact that 159 house rats were killed at this store in one night. No one has attempted to estimate the number of commensal house rats in any of our cities, such as New Orleans, Baton Rouge, or Shreveport, but the figure is unquestionably high. The number of rats in the city of Baltimore in 1944 was said to be somewhere in the neighborhood of 400,000, but diligent efforts on the part of the community to clean up rat havens resulted in a marked decline in the number of rats to an estimated 60,000 by 1949. This reduction was accomplished primarily by renovating substandard dwellings and by wholesale improvements of sanitary conditions (Davis and Fales, 1950).

Although poisoning campaigns are often necessary, the only permanent method of eradicating rats is by permanently eliminating deposits of trash and garbage and by otherwise altering the environment to provide less harborage for the rats. Davis (1948) has shown that only about 5 percent or fewer of all Norway Rats live for more than one year. Whether the number of rats in a given population increases or decreases depends on the number of young produced minus the number of rats that die. The regulatory factors are shown by Davis (1951c) to be environmental conditions, predation, and competition. An excerpt from his 1951 paper follows:

1. *Environment*—Rats require a variety of foods and harborage conditions. Studies of the rats' preferences as to garbage, for example, indicate

that Norway rats must have a substantial amount of grain (usually available in the form of bread) in their diet. Rats will starve in the midst of plenty of raw or cooked vegetables. Other studies give us some information about the requirements for nesting places, shelter, defecatoria, and refuges. In general, the number of rats is directly proportional to the amount of the factor that is available. Thus, if 10 gm. of food will support 1 rat, then 100 gm. will support 10 rats and 1,000 gm. will support 100 rats. This implies that all other factors are in excess of the requirements of the population. But if, in the presence of 1,000 gm. of food, the number of nesting sites will only support 10 rats, then the food will be in excess and nesting sites will limit the population. It is fundamental to recognize that at a particular place some factor is limiting the number of rats. The practical problem of finding the factor is as yet unsolved and requires much more study of the biology of rats.

Another point requires emphasis. It has been stated that the number of rats is directly proportional to the characteristics of the environment. This means that the effect of a change in a factor does not depend upon the density of the rat population. For example, removal of 10 per cent of the food should reduce the rat population by 10 per cent whether there are 100 or 1,000 rats. Because of this relation, environmental conditions are said to be density-independent.

2. *Predation* — For present purposes, the word predation may be used to include all causes of death whether it be by cats, bacteria, viruses, traps, or poisons, because the result of these agencies is dead rats. In contrast to the effects of environmental factors, predation depends upon the density of the population. At high populations a high proportion may be removed by the cats or traps. At low populations, a low proportion may be removed. This is no more than a statement of the law of diminishing returns, but it has a very important consequence. As the population decreases, the proportional effect of predators also decreases. This phenomenon is readily apparent to epidemiologists who are aware of the difficulty of transmission of a pathogen in sparse populations.

The effect of predators therefore is said to be density-dependent.

3. *Competition* — As the population increases toward the capacity of the environment the individuals compete for the available food, space, and mates. The results of competition become more intense as the population increases, and hence are density-dependent. In nature, rats have adjusted themselves so well to predation that it is rarely sufficiently intense to hold the population at a level below the conditions of intense competition.

Some of the data concerning the results of competition may be presented here. Davis . . . [1951d] found that reduction of the rat population by trapping half of the rats in a city block was followed by a doubling of the pregnancy rate in a couple of months. Furthermore, Davis . . . [1951a] also concluded that the survivors rapidly gained weight. The reduction of the population by trapping reduced the competition for food and mates so that there were better conditions for the survivors. Actually, the removal merely made room for more rats.

Calhoun . . . [1949] observed some of the behavioristic results of competition. The rats in a colony develop a definite social ranking. The highest ranking members win the most fights and have first choice of food and mates. As the population increases relative to its food supply, the higher ranking members still get adequate food, but the low members begin to starve. Low ranking females have poor reproductive success and progeny from low ranking females have little chance to grow normally.

Competition, like predation is density-dependent in its effects. These two factors are true controlling factors because, in a given environment, as the population increases, the proportional effect increases, thereby causing a greater control on the population.

. . . Predation, then, is rarely sufficiently intense to do much good. Indeed, it usually merely reduces the population to the level of maximum reproduction and causes the maximum rate of conversion of garbage into rat flesh.

The intensity of competition depends upon the

Tracks of Norway Rat

density of population. As the population increases, relative to the capacity of the environment, the competition increases and, unlike predation, cannot be diverted to other species. This level of competition is inherent in the psychology of the rats. Fortunately, means to increase competition are readily available in sanitation procedures. Reduction of the food and harborage increases competition and thereby decreases the rat population. The important point is that, in contrast to predatory methods, sanitation is feasible on a scale that will reduce the rat population. The biological principle governing predation and competition is the same but the practicabilities are greatly different. Given our present conditions as regards housing, human failings, wages, etc., it is generally feasible to increase competition but rarely feasible to increase predation.

The home range of the Norway Rat is seldom more than 100 feet in its greatest dimension (Davis et al., 1948). It will cross alleys and enter adjacent buildings but will not often venture across a city street let alone move from one city block to another. This fact has an important bearing on its population structure. Overpopulation does not as a rule precipitate mass movements in search of relief from competition. The rats simply stay where they are and most of them succumb to starvation, disease, or death in combat with their own kind.

The average number of litters per year is 5, and the number of young to a litter is generally somewhere from 7 to 11, but as few as 2 and as many as 22 are on record. It is mathematically possible, should all offspring survive, for one pair to produce a total of 359,709,482 descendants in 3 years and, at the end of only 10 years, a grand total of 48,319,698,843,030,344,720 descendants (Cockrum, 1962). Fortunately, not all offspring do survive.

The young at birth are blind and naked. The eyes open between 14 and 17 days later, and weaning takes place at about three weeks of age. Most Norway Rats begin to breed in their third or fourth month. Males are capable of breeding the year around, but the greatest reproductive activity occurs from March to May and in October and November. Gestation lasts 21 to 26 days. The longer periods occur when the female is nursing another litter and implantation is thereby delayed.

The one redeeming feature of the Norway Rat is the role that its white mutant, the laboratory rat, has played indirectly in alleviating human suffering and in contributing to the solution of innumerable biological problems. The white rat and its other genetic variations are employed in nearly all biological laboratories throughout the world in both medical and basic research. Great discoveries in diverse fields, including immunology, pathology, epidemiology, genetics, and physiology, have been made as a result of experiments utilizing the various laboratory strains of *Rattus norvegicus*. In this way the Norway Rat has made some amends for the devastation, havoc, and suffering that it has inflicted on mankind.

Predators. — Cats and other carnivores manage to capture a few of these rats, and occasionally an owl or hawk, under special situations, will succeed in doing so, but the worst enemy of a Norway Rat is another Norway Rat. The species is extremely pugnacious, and the considerable intraspecific fighting leads to heavy mortality among the young and adults alike.

Parasites and diseases. — Norway and Roof Rats are subject to several diseases that result in many deaths. Indeed, mortality from intraspecific conflicts, death from starvation, and deaths from diseases are collectively the major decimating factors operating against the two species of commensal house rats. Both species are infested with many external and internal parasites. A cursory perusal of parasitological literature reveals that either the Norway Rat or the Roof Rat, and in most cases both, have been found to serve as host to the following parasites, which are too numerous to itemize species by species here: 4 lice, 4 beetles, 2 flies, 22 fleas, 49 mites and ticks, 21 nematodes, 13 cestodes, 9 trematodes, 2 acanthocephalans, and 20 protozoans. Some of the last-named organisms, along with certain bacteria and spirochetes, are highly pathogenic. (Harkema, 1936; Yeh and Davis, 1950; Worth, 1950a; Davis, 1951a and b; Morlan, 1952; Schiller, 1952; Doran, 1954a and b; 1955a and b.) Among the diseases spread by house rats, some through the agency of the parasites that they harbor, are murine typhus, bubonic plague, leptospirosis, and rat-bite fever. As I have previously noted, rat-borne diseases have taken more human lives than all the wars and revolutions in which man has ever fought in his recorded history. On more than one occasion plague has reduced the population of western Europe by one-fourth or more, and a succession of serious outbreaks in the United States such as those that recurred at San Francisco (1902–1941), Galveston (1920–1922), and New Orleans (1912–1926). The New Orleans epidemic is discussed in detail by Link (1955).

Rattus norvegicus (Berkenhout)

Mus norvegicus Berkenhout, Outlines of the natural history of Great Britain and Ireland, 1, 1769:5; type locality, England.
Rattus norvegicus, Hollister, Proc. Biol. Soc. Washington, 29, 1916:126.

The Norway Rat is monotypic, that is, it has no recognized subspecies.

Specimens examined from Louisiana.—Total 106, as follows: *Ascension Par.*: Burnside, 1 (LSUMZ); 2 mi. E La. Hwy. 30 on La. Hwy. 73, 2 (LSUMZ); Donaldsonville Fairgrounds, 2 (LSUMZ); 0.75 mi. E Gonzales, 2 (SLU). *Avoyelles Par.*: 1 mi. N Bunkie, 2 (LSUMZ). *Bossier Par.*: 1 mi. N Bossier City, 1 (LSUMZ); Bossier city dump, 1 (LTU). *Caddo Par.*: Shreveport, 1 (LTU). *Calcasieu Par.*: Lake Charles, 1 (LSUMZ); 5 mi. W Sulphur, 3 (MSU). *Cameron Par.*: 12 mi. W Cameron, 1 (LSUMZ). *Concordia Par.*: Ferriday, 1 (LTU). *East Baton Rouge Par.*: Baton Rouge, 15 (LSUMZ); University, 22 (LSUMZ); 2 mi. S Baker, 1 (LSUMZ); 1 mi. SE LSU, 1 (LSUMZ). *Franklin Par.*: Wisner, 1 (NLU). *Grant Par.*: Georgetown, 1 (LTU). *Iberia Par.*: Avery Island, 1 (CAS). *Iberville Par.*: 0.5 mi. N Plaquemine, 1 (LSUMZ); 7 mi. S Plaquemine, 1 (LSUMZ). *Jefferson Par.*: Metairie, 1 (SLU). *Jefferson Davis Par.*: Jennings, 1 (LSUMZ). *La Salle Par.*: Jena, 1 (LTU). *Lincoln Par.*: Ruston, 6 (LTU); 1 mi. SW Ruston, 1 (LTU). *Livingston Par.*: 4 mi. N Denham Springs, 1

(LSUMZ). *Morehouse Par.*: 6 mi. N Oak Ridge, 1 (LSUMZ); 4 mi. W Oak Ridge, 1 (NLU); Perryville, 1 (NLU). *Natchitoches Par.*: Natchitoches, 4 (NSU). *Orleans Par.*: New Orleans, 7 (6 LSUMZ; 1 TU). *Ouachita Par.*: 1 mi. N Monroe, 1 (NLU); Monroe, 4 (NLU); West Monroe, 1 (USNM). *Plaquemines Par.*: Phoenix, 1 (LSUMZ); 2 mi. N Myrtle Grove, 1 (SLU); Tidewater (4 mi. SW Venice), 1 (SLU). *Richland Par.*: Archibald, 1 (NLU). *St. Martin Par.*: 10 mi. E Lafayette, Hwy. 90, 1 (LSUMZ); 4 mi. SE Breaux Bridge, 1 (LTU). *St. Mary Par.*: Morgan City, 1 (USNM). *Tensas Par.*: 14 mi. NW St. Joseph, 1 (LTU). *Terrebonne Par.*: 0.5 mi. E Schriever, 1 (LSUMZ). *Union Par.*: 12 mi. E Farmerville, 1 (LTU); 6 mi. N Farmerville, 1 (LTU). *Washington Par.*: 1.5 mi. N, 0.3 mi. W Bogalusa, 1 (LSUMZ). *West Baton Rouge Par.*: ca. 1 mi. S Port Allen, 1 (LSUMZ).

Genus *Mus* Linnaeus

Dental formula:

$$\text{I}\frac{1-1}{1-1} \ \text{C}\frac{0-0}{0-0} \ \text{P}\frac{0-0}{0-0} \ \text{M}\frac{3-3}{3-3} = 16$$

HOUSE MOUSE   Plate VI
(*Mus musculus*)

Vernacular and other names.—The name House Mouse is self-explanatory other than to say that the word "mouse" comes from the Anglo-Saxon word *mus,* which descended through Latin from the ancient Sanskrit word *musha,* meaning thief. The generic name is a direct application of the Latin word *mus,* for mouse. The specific name *musculus* is the Latin word for little mouse and refers to the small size of the animal.

Distribution.—The House Mouse originated in central Asia but is now almost worldwide in distribution, having first invaded Europe and then followed man in his colonization of the remainder of the world. In Louisiana there is probably not a town or farm from which it is absent. (Map 44)

External appearance.—This small mammal is so well known that it would hardly require detailed description except for the fact that it can be confused with certain native mice. It measures five to eight inches in its total

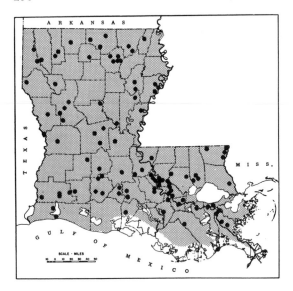

Map 44 Distribution of the introduced House Mouse *(Mus musculus). Shaded area,* known or presumed range within the state; •, localities from which museum specimens have been examined. Probably occurs wherever there is human habitation.

length, of which the tail comprises approximately one-half. The snout is elongated, the eyes are small and only slightly protruding, the ears are large and naked, and the tail is nearly naked and scaly. The color is variable, brown or grayish brown above, grading to buff, gray, or white on the belly. The body fur is short, not long and lax as it is in cricetine mice, such as the members of the genera *Peromyscus, Reithrodontomys,* and *Microtus.*

Color phases. — No color phases in the normal sense are known, but in laboratory strains of the House Mouse albinos and numerous other variations, including waltzing and hairless mutants, are frequent. Albinos and some of the mutant variations have been found in wild populations in Louisiana.

Measurements and weights. — Twenty-five selected specimens in the LSUMZ from various localities in the state averaged as follows: total length, 163 (151–186); tail, 78 (69–92); hind foot, 17 (12–20); ear, 13 (10–18); greatest length of skull, 22.0 (20.4–22.9); cranial breadth, 10.2 (9.2–11.9); zygomatic breadth, 11.2 (10.3–13.0); interorbital breadth, 3.6 (3.4–4.2); length of nasals, 7.9 (7.1–9.0); length of diastema, 6.0 (5.2–6.7); length of maxillary toothrow, 3.4 (3.1–3.7); palatilar length, 9.5 (9.0–10.2); postpalatal length, 7.5 (6.1–9.6). Weights of 25 specimens averaged 26.7 (19.1–28.7) grams.

Cranial and other skeletal characters. — The skull approximates in size that of the harvest mice *(Reithrodontomys),* but it lacks the grooves on the face of the incisors, and the crowns of the cheek teeth have three rows of longitudinal cusps instead of two. The first cheek tooth is larger than the second and third molars combined, and the incisors possess a terminal notch that is conspicuous in lateral view (Figure 127). The hind feet each have five prominent toes with nails, the forefeet only four toes and a vestigial thumb. The baculum is spoon shaped at the base with a slight subterminal constriction in the shaft (Figure 72).

Sexual characters and age criteria. — The sexes are superficially indistinguishable. In color the young are like the adults except that the pelage is somewhat longer and more lax but not as much as in cricetines.

Status and habits. — The House Mouse is sometimes exceedingly abundant, around both human dwellings and other urban quarters, as well as on farms and in field situations. There are few of us who have escaped having House Mice invade our homes at one time or another. We hear them scurrying about in our attics or diligently gnawing away at the woodwork in a wall or closet, and we find their

Tracks of House Mouse

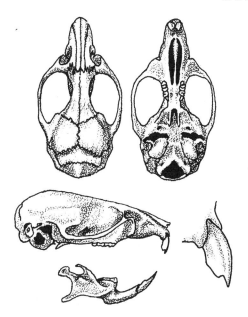

Figure 127 Skull of House Mouse *(Mus musculus)*, ♂, 2 mi. E jct. hwys. La. 30 and 73, LSUMZ 15559. Approx. 2 ×, except for enlargement of incisor, which is 6 × and shows the terminal notch characteristic of this species.

telltale little black droppings on shelves and in other places. Ordinarily a few carefully set mouse traps serve to catch the intruders. But in such places as restaurants, supermarkets, warehouses, and farm buildings their elimination is often difficult if not virtually impossible.

Population densities of House Mice fluctuate rather drastically over a period of years. I have already mentioned on page 172 the phenomenal concentration of House Mice that developed in Kern County, California, in 1926 (Hall, 1927). Although no population explosions of such magnitude are known to have occurred in Louisiana, or as a matter of fact anywhere else, heavy concentrations are nevertheless on record. For example, one of my students, Paul Matthews *(in litt.)*, found by applying both the Hayne (1949) and the Zippin (1958) methods of estimating total population that the computed number of House Mice in a field on the north side of the LSU campus in the fall of 1967 was nearly 500

mice per acre. Another student, Jay H. Huner *(in litt.)*, obtained 902 House Mice in five nights in January 1969 in a 13.4-acre tract of unharvested sorghum one mile south of Port Allen. Using the Hayne (1949) method of estimating total population from trapping success, he found that the density in his study area was approximately 97 mice per acre. Applying the Zippin (1958) method to Huner's raw data, I obtained close agreement, 96.5 mice per acre, with Huner's original calculated density of 97.

The House Mouse is a prolific breeder. It produces young throughout the year, particularly when residing in warm buildings and other sheltered situations. It reaches sexual maturity at an early age of 5 to 7 weeks, after which it comes into heat every 4 to 6 days. The gestation period is usually between 19 and 21 days in length, but when the female is lactating it may be 2 or 3 days longer. Although 13 to 14 litters a year are achievable, the usual number is 5 to 10. The number of young to a litter may be as low as 3 and as high as 12, but is most commonly 5 or 6. The young are blind and naked at birth but are fully furred by the time they are 10 days old. The eyes open on the 14th day and weaning takes place at about 3 weeks of age.

Because of its smaller size, the species is not as serious a pest as its large relatives, the Roof and Norway Rats. But it is still responsible for great economic loss. While the mouse does not consume a great deal of food, it despoils immense quantities with its urine and feces and by the dirty tracks it leaves wherever it goes. In a home or in a department store, it damages clothing, and it often gnaws on furniture and upholstery. In Louisiana alone the destruction that it causes probably amounts to well over a million dollars a year. But worse still is the fact that it spreads disease, notably murine typhus, as well as such infections as rickettsial pox. The latter is carried by the mite *Allodermanyssus sanguineus*, which the mouse harbors.

The albino strain of the House Mouse is

sold extensively in pet shops, and it makes, I suppose, an interesting pet for small children. But the species plays its most truly important role as a laboratory animal. To the fields of genetics and medical research and testing it has contributed immeasurably. Large colonies are easily and economically maintained under laboratory conditions and, because the species reproduces itself rapidly, it is an ideal experimental animal.

When elimination or control of House Mice that are present in small numbers is required, trapping is the best method to employ. Ordinary mouse traps, when set at right angles to a wall or at a mouse's entrance into a room, usually result in captures. But, if more drastic measures are required, such as the use of toxic poisons and fumigation, the services of a pest control operator should be obtained. Not only are poisons dangerous to use where small children and pets are present, but some poisons can be administered only by licensed operators under the provisions of rigorous state laws. There is, however, one additional measure that the householder can resort to other than using mouse traps and that is to find where the mice are gaining entrance to the house. Sealing or otherwise cutting off these avenues often easily eliminates the mice. U.S. Fish and Wildlife Leaflet 349, entitled *Control of House Mice,* can be obtained by writing to the U.S. Fish and Wildlife Service, Washington, D.C. 20240. A booklet entitled *Control of Domestic Rats and Mice,* by B. F. Bjornson and C. V. Wright, issued as Public Health Service Publication No. 563, can be obtained for 50 cents by writing to the U.S. Government Printing Office, Washington, D.C. 20402. A limited number of copies of a similar publication with the same title can be obtained by writing to the Department of Health, Education, and Welfare, Center for Disease Control, Atlanta, Georgia 30333. Information is also available by addressing an inquiry to the U.S. Fish and Wildlife Service, Bureau of Sports Fisheries and Wildlife, Agricultural Center, Louisiana State University, Baton Rouge, Louisiana 70803.

Predators. — House Mice that live in residential and business districts of our towns and cities are preyed upon by Domestic Cats and house rats and occasionally by owls. In rural and field situations they are subjected to heavy predation on the part of numerous carnivores and all hawks and owls.

Parasites. — The House Mouse is the host for many internal and external parasites, and it is subject to many diseases. The list of parasites and diseases is too long to justify summarization here, especially since the information is available in papers by Heston and by Dingle (1941); see also Doran (1954a and b, 1955a).

Subspecies

Mus musculus domesticus Rutty

Mus domesticus Rutty, Essay Nat. Hist. County Dublin, 1, 1772:281; type locality, Dublin, Ireland.
Mus musculus domesticus, Schwarz and Schwarz, Jour. Mamm., 24, 1943:65.

Mus musculus brevirostris Waterhouse

Mus brevirostris Waterhouse, Proc. Zool. Soc. London, 1837:19; type locality, Maldonado, Uruguay.
Mus musculus brevirostris, Schwarz and Schwarz, Jour. Mamm., 24, 1943:64.

I have made no attempt to identify subspecifically the Louisiana specimens of *Mus musculus* now available in museum collections (over 250 in the LSUMZ alone). But in 1943 Ernst Schwarz examined 52 specimens from Louisiana that were then in the collections of the Louisiana State University Museum of Zoology and made the following statement *(in litt.)* with regard to the series: "The mice in Louisiana are a mixed group made up of *domesticus* primarily, but also with a definite percentage of *brevirostris.* Hybrids are not uncommon, although the majority of the material at hand can be referred to one or the other of the two subspecies. Ecological segregation tends to restrict *brevirostris* to the open and *domesticus* to houses and their neighborhood. However, this segregation is not complete." He further remarked to me that the *domesticus* type of House Mouse is to be expected in Louisiana because of the early French settlement of this general area. He considered that *domesticus* was originally introduced from the population native to northern Spain, France (except the Mediterranean littoral), the British Isles, and western Germany. He regarded *brevirostris* as the type of House Mouse that occurs in the area of the Iberian peninsula, the Mediterranean coast of France, and the Italian peninsula. According to Schwarz, it was extensively introduced into Central and South America, as well as the southern part of the United States.

Specimens examined from Louisiana.— Total 945, representing all parishes except Bienville, Caldwell, East Carroll, East Feliciana, Jackson, Madison, Richland, St. Bernard, St. Helena, St. John the Baptist, and Winn (395 USL; 294 LSUMZ; 69 MCZ; 66 NSU; 47 NLU; 29 LTU; 19 SLU; 10 LSUS; 8 TU; 4 MSU; 3 USNM; 1 CAS).

The Coypu and Hutia Family
Capromyidae

THE ONLY representative of the suborder Hystricomorpha and the family Capromyidae that occurs in Louisiana is the introduced and now abundant Nutria or coypu from South America. Characteristics of the suborder Hystricomorpha have already been discussed on page 171. The family Capromyidae contains eight genera, seven of which consist of species that are called hutias; the eighth is *Myocastor,* which contains only one species, the Nutria. The hutias are predominantly West Indian and are believed to have been transported to and between the islands by natives in pre-Columbian times for use as food. Three of the seven genera of hutias, consisting of four species, are extinct. Their morphology is known solely on the basis of their skeletal remains from caves, but one was apparently still extant at the time of the arrival of the Spanish in the New World.

The living hutias are adapted to a terrestrial life, while the Nutria is modified for an aquatic existence. Nutrias are excellent swimmers and spend much of their time in the water. The toes of the hind feet are webbed, and the young ride on the back of the mother as she swims.

Some authorities place the Nutria in a separate family, Myocastoridae, but it is here treated as constituting a monotypic subfamily, Myocastorinae, in the family Capromyidae.

Genus *Myocastor* Kerr

Dental formula:

$$I\frac{1-1}{1-1}\ C\frac{0-0}{0-0}\ P\frac{1-1}{1-1}\ M\frac{3-3}{3-3}=20$$

NUTRIA   Plate VIII
(Myocastor coypus)

Vernacular and other names.— The word "nutria" is a Spanish cognate of the Latin word *lutra,* and these names have been applied to various aquatic mammals, including the present species and the Nearctic River Otter. The generic name is derived from two Greek words (*mys,* for mouse, and *kastor,* for beaver) that translate as mouse beaver. The specific name *coypus* is the Latinized form of coypú, a name in the language of the Araucanian Indians of south-central Chile and adjacent parts of Argentina for an aquatic mammal that was possibly this species. It was taken over by the Spaniards and applied as a vernacular to the present species, and the word has been used in some American literature for the Nutria. Molina, who named and first described the species, was a Chilean naturalist who was doubtless familiar with the Araucanian Indian etymology of the word "coypú." In Louisiana the name "swamp beaver" is sometimes heard applied to this species, but more often it is called the Nutria.

Distribution.—The species occurs normally in temperate South America, whence it has been introduced widely in the United States (Map 45) and western Europe. Its present distribution in Louisiana is nearly statewide (Map 46).

External appearance.—The Nutria is a large robust rat, almost equal in size to a beaver. It measures up to 40 inches in total length, of which the tail always makes up somewhat less than half. The pelage consists of a dense, soft, slate-colored underfur that is overlaid with long, glossy, dark brown or yellowish brown guard hairs. The sides of the face and the body are often a rich golden yellow, and the chin and the tip of the muzzle are white. The short ears, seldom more than an inch in length, are almost buried in the surrounding fur. The tail is cylindrical (compressed laterally in the muskrat) and is scantily haired

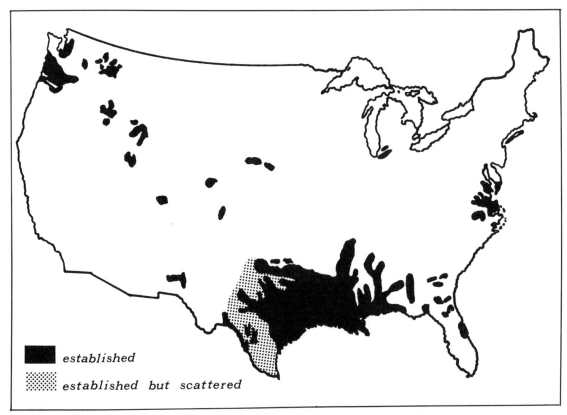

Map 45   Nutria distribution in the United States in 1966. (After Evans, 1970)

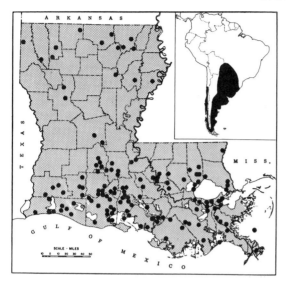

Map 46 Distribution of the introduced Nutria *(Myocastor coypus). Shaded area,* known or presumed range within the state; •, localities from which museum specimens have been examined; *inset,* overall range of the species in South America.

except at the base. The forelegs and feet are blackish and possess four long toes and a vestigial thumb, whereas the hind feet have five toes, all of them connected by webs except the pollex, which is free.

Color phases.—Both albinos and melanos are of rather frequent occurrence, but no true color phases are known in this species.

Measurements and weights.—Eight skins in the LSUMZ from various localities in southern Louisiana averaged as follows: total length, 940 (837–1010); tail, 344 (300–450); hind foot, 131 (100–150); ear, 27 (21–30). Eleven skulls from the same source averaged as follows: greatest length, 116.3 (109.6–125.6); mastoid breadth, 50.1 (46.4–56.3); zygomatic breadth, 71.9 (68.4–79.6); interorbital breadth, 31.4 (28.4–34.9); length of nasals, 43.4 (40.2–47.4); length of diastema, 32.9 (30.6–36.2); length of maxillary toothrow, 29.1 (27.6–30.8); palatilar length, 60.3 (56.8–66.4); postpalatal length, 32.9

(30.0–36.0). Data derived from Table 7 in Kays (1956) indicate that 599 adults weighing between 8 and 20 pounds averaged approximately 12 pounds.

Cranial characters.—The skull of the Nutria is so distinctive that it can be recognized at a glance by certain diagnostic features. It is massive with greatly elongated paroccipital processes that curve slightly forward at the distal end. The infraorbital foramina are immense, large enough to admit the end of one's little finger. The huge incisors are reddish orange on the anterior surface. The palate and the cheek teeth are strongly convergent anteriorly, the latter being markedly hypsodont (high-crowned) and decreasing in size from back to front. The lower edge of the mandible and that of the angular process are strongly flared laterally. (Figure 128)

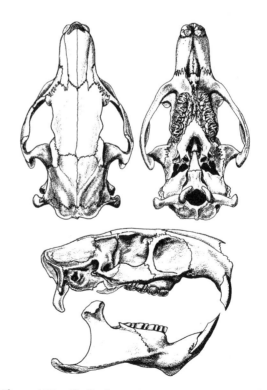

Figure 128 Skull of Nutria *(Myocastor coypus),* ♂, 1 mi. E Red Cross, Pointe Coupee Parish, LSUMZ 6456. Approx ½ ×.

Sexual characters and age criteria.—The sexes of the Nutria are distinguishable only by the genitalia and by the presence or absence of teats and mammary glands. The latter are located so far laterad that they are conspicuous when the female is viewed from above, giving rise to the often repeated erroneous statement that in the Nutria the mammae are located on the back. Immatures are like the adults except for their smaller size and duller coloration.

Status and habits.—The Nutria is now abundant in the southern part of Louisiana and it has been introduced into lakes and streams in almost all parts of the state. The first Nutrias are said to have been released in the Louisiana marshes in the early 1930s near New Orleans and to have all been recovered by trappers shortly thereafter. Then in 1938 the late E. A. McIlhenny imported 20 Nutria from Argentina and placed them in a specially constructed enclosure at Avery Island, Iberia Parish. Here they reproduced prolifically, and many escaped or perhaps were released intentionally. By 1943 the species was widespread in the southern part of the state. During the trapping season of 1943–1944, Nutrias first appeared on the Louisiana fur market but in limited numbers. Only two years later the number turned in by trappers was up to 8,784. In the 1969–1970 season no less than 1,604,175 Nutria were trapped. The skins, plus the 9,500,000 pounds of Nutria meat that was salvaged, returned an income of $4,586,680 to the state's trapping industry. In the 1972–1973 season 1,611,623 Nutria were trapped and brought $6,737,775, which, combined with the $800,000 derived from the meat, yielded an income of $7,537,774. In slightly more than 30 years since its introduction, the Nutria had risen not only to a place of eminence in our state's fur trade but to the number one position, for it had now outstripped the muskrat in value.

During the period of its ascent to its present position of importance as a furbearer, the Nutria was the subject of considerable controversy. Some people extolled its virtues from the standpoint both of its value as a fur animal and of the alleged role it plays in the control of aquatic weeds that often clog our streams and lakes. On the other side of the ledger it was condemned for doing great damage to dikes and levees and to crops, particularly of rice and sugarcane, and for having caused the decline in muskrat populations.

At first the Nutria (Figure 129) was not looked upon with any great esteem either by the trapper or the fur dealer, for its pelt brought less than a lucrative return for the work required in obtaining, preparing, and marketing it. The trapper is able to skin a muskrat where it is caught on the trapline in

Plate VIII
1. American Beaver *(Castor canadensis).*
2. Nutria *(Myocastor coypus).*
3. Nearctic River Otter *(Lutra canadensis).*

the marsh, and the operation requires only a few minutes (the record in competition is 20 seconds). But a Nutria must generally be brought back to the trapper's camp, where it can be placed on a skinning table and tackled with a long-bladed sharp knife. Marshmen have now improved their skinning techniques and have learned better how to stretch and dry their pelts. These factors, combined with the improved market value of the Nutria, have now brought the species into favor with most trappers.

As agents in the control of aquatic plants, Nutria have been vastly overrated. More often than not they eat vegetation that man does not want controlled, passing up water hyacinths, alligator weed, coontail, bladder-wort, and other objectionable plants that they are supposed to destroy. In 1957 Hurricane Audrey pushed thousands of Nutria inland out of the coastal marshes into the rice and cane fields, where they proceeded to wreak havoc upon these crops. Nutria normally eat about 2.5 to 3.5 pounds of food a day, but in sugarcane fields they often gnaw or cut far more stalks than they can possibly utilize. Their burrows sometimes result in serious damage to levees in areas being flooded for rice culture, as well as to protection levees around cane fields.

Since muskrats and Nutria live mainly in different kinds of marshes, the former being primarily an animal of certain types of salt or brackish marsh, the latter of freshwater situ-

Figure 129   Nutria *(Myocastor coypus)*, 5 mi. SE Sorrento, Ascension Parish. (Photograph by Robert H. Potts, Jr.)

ations, direct competition between them is of serious concern only in a few places where both happen to be present together in large numbers. During years of peak muskrat production, about 85 percent of the rats live in approximately one-quarter of the state's available marshlands, the part covered with stands of three-cornered grass *(Scirpus olneyi)*, the muskrat's preferred food item. Beginning in the 1940s this type of marsh has been gradually changing or even disappearing, as a result of drought, hurricanes, industrialization, or damage by the rats when they become too numerous for the available food supply, thereby causing extensive eat-outs.

From the all-time record high of more than eight million pelts in the 1945–1946 season, muskrats began a drastic decline that reached a low of only slightly more than 200,000 pelts in 1964–1965. The harvest of Nutria simultaneously increased from 436 in 1943–1944 to over a million and a half in 1964–1965. But federal and state biologists who studied the decline of the muskrat in this period could find only circumstantial evidence that it was related in any way to the concurrent increase of the Nutria. Other drastic vacillations in muskrat populations are known to have occurred fairly often over the years prior to the establishment of the Nutria. For example, the hurricane of 1915 and the severe drought in 1924–1925 each caused great damage to the marshlands, and each was followed by a crash in the muskrat population.

I do not mean to imply that competition does not exist where Nutria and muskrats now occur in the same marsh. Two similar animals living side by side in a marsh compete for things other than food. They compete for living space or perhaps simply for high spots to which they can retreat for survival during high water. But when a catastrophe, such as the flooding of the marsh with salt water during a hurricane, destroys a large segment of the available food, severe competition for that which remains is bound to ensue.

Although the muskrat and the Nutria often appear to share a marsh in complete harmony, with the muskrat population showing a general upward trend, the removal of the Nutria often makes the muskrat population skyrocket (Evans, 1970). So it would appear that, even though Nutria seldom injure muskrats physically, their harassing tactics and competition for food and living space is sufficient to prevent muskrats from increasing at their full potential.

Whatever may be the ultimate impact of Nutria on muskrats, the former are here to stay. Presently they occupy a place of eminence in the economy of our state. The interest of the state will be best served if both they and muskrats continue to thrive, even in places where they occur side by side. For further information on the role of the Nutria in the Louisiana fur industry, see the chapter dealing with that subject (page 21 *et seq.*).

The Nutria, as is usual in the case of ro-

Tracks of Nutria

dents, is quite prolific. The number of young in a litter ranges from one to nine with the average being four and a half. The babies at birth are fully furred and the eyes are open. Evans (1970) states that miscarriages are common, estimating that sometimes only 60 percent of the embryos survive to parturition. The gestation period is about 130 days in duration. Within a day or two following the birth of a litter the female is again in estrus. Otherwise she comes into heat every 24 to 26 days and remains in that state for 1 to 4 days.

Sexual maturity in the Nutria is reached before the animal is full grown. When the food supply is ample and habitat conditions are good, Nutria sometimes begin breeding as early as the fourth month, but more often not until they are close to eight months old. Evans (1970) found that young males born in early summer may breed within four to six months, while those born in early winter do not commence reproductive activity until their seventh or eighth month. Some females wean their young at about five weeks but others nurse up to seven weeks.

Although adept at moving about on land, Nutria are more at home in the water, where they swim with agility, usually with the head and back out of the water and with the tail trailing on the surface. In the coastal marshes they are often seen moving about leisurely in the daytime, but their period of greatest feeding activity is at night except when hunger sends them on forays in search of food at any hour of the day. Nutria are strict vegetarians, consuming their food both on land and while floating in the water, where they shove aquatic plants to their mouths with their forepaws. They often amass quantities of readily available vegetation into a sort of platform, which they use as a place of feeding and resting, or as a toilet for their meticulous grooming operations. Not infrequently a muskrat house is preempted for these purposes, as well as for nesting quarters. Nutria can dig their own burrows in canal banks or levees, but frequently they take over old burrows of armadillos, muskrats, and other Nutria. These tunnels are either simple, short underground passages with one entrance or else rather complicated long tunnels, extending 50 feet or more and with multilevel entrances and compartments or ledges, where crude nests made of vegetation are formed.

Predators. — The alligator is the Nutria's chief predator, and it takes a heavy toll. The young, however, are captured by a variety of denizens of the swamps and marshes, including turtles, gars, large snakes, and birds of prey.

Parasites and diseases. — In the southern part of the state 80 to 90 percent of the Nutria are infected with the nematode *Strongyloides myopotami,* which fur farmers claim restricts reproduction and causes mass mortality to young and old. In addition, Babero and Lee (1961) and Lee (1962) found in Louisiana Nutria three trematodes (*Heterobilharzia americana, Echinostoma revolutum,* and *Psilostomum* sp.), one cestode (*Anoplocephala* sp.), one acanthocephalan (*Neoechinorhynchus* sp.), and four nematodes (*Trichostrongylus sigmodontis, Longistriata maldonadoi, Strongyloides myopotami,* and *Trichuris myocastoris*). Fur trappers in southern Louisiana occasionally pick up a rash called "marsh itch" or "nutria itch" from handling Nutria. It is caused by the roundworm *Strongyloides myopotami* and produces a severe rash on the hands, arms, legs, and face, or wherever the immature, larval forms of the organism penetrate the skin. (Burk and Junge, 1960; Lee, 1962; Little, 1965). Attention by a doctor is recommended when the symptoms first appear.

Subspecies

Myocastor coypus bonariensis
(E. Geoffroy Saint-Hilaire)

Myopotamus bonariensis E. Geoffroy Saint-Hilaire, Ann. Mus. Hist. Nat., Paris, 6, 1805:82; type locality, Argentina.
Myocastor coypus bonariensis, Thomas, Ann. Mag. Nat. Hist., ser. 8, 20, 1917:100.

The original Nutrias that were introduced at Avery Island came from Argentina and were doubtless of this subspecies.

Specimens examined from Louisiana. — Total 259, as follows: *Acadia Par.:* 6 mi. S Crowley, 1 (LSUMZ); 5 mi. S Crowley, 1 (USL); 4.5 mi. S Crowley, 1 (USL); 6 mi. S Rayne, 1 (USL); 1 mi. E Rayne, 1 (USL). *Ascension Par.:* Bayou Manchac, 1 (LSUMZ); 0.75 mi. E Gonzales, 1 (SLU). *Assumption Par.:* Bayou l'Ourse swamp, 1 (LSUMZ). *Bossier Par.:* 1 mi. W Caplis, Red River, 1 (LSUMZ). *Caddo Par.:* Oil City, Caddo Lake, 1 (LTU). *Calcasieu Par.:* 2 mi. S Holmwood on La. Hwy. 14, 1 (MSU); 1.5 mi. N Intracoastal Canal on Vinton Drainage Ditch, 2 (MSU); English Bayou, 1 (MSU); 4 mi. W Lake Charles, 1 (NLU). *Caldwell Par.:* 5 mi. NE Columbia, 1 (NLU). *Cameron Par.:* 4.5 mi. N Hackberry, 1 (LSUMZ); Lacassine Refuge, 12 (1 LSUMZ; 4 USL; 7 USNM); Grand Cheniere, 1 (LSUMZ); 3 mi. W Cameron, 1 (LSUMZ); near Creole in marsh, 1 (USL); Cameron, 1 (USL); Johnsons Bayou, 2 (MSU); Little Cheniere Rd., 1 (MSU); 3 mi. S Gibbstown ferry, 5 mi. N Creole, 1 (MSU); 1 mi. E, 6 mi. N Creole, 1 (MSU); near Cheniere Pardue, 1 (MSU); 6 mi. S Creole ferry, 1 (MSU). *Claiborne Par.:* Corney Lake, 2 (LTU). *East Baton Rouge Par.:* 6 mi. S University, 1 (LSUMZ); 7 mi. S University, 1 (LSUMZ); 5 mi. S University Sta., 2 (LSUMZ); 1 mi. NE University, 1 (LSUMZ); 2.75 mi. S, 0.25 mi. W University, 1 (LSUMZ); 1 mi. S LSU Union, 1 (LSUMZ); Baton Rouge LSU campus, 1 (LSUMZ); Baton Rouge airport, 2 (USL). *Evangeline Par.:* 8 mi. NW Ville Platte in Miller's Lake, 1 (LSUMZ); 4 mi. W Ville Platte, 3 (USL); Ville Platte, 1 (USL); 1.5 mi. N Chataignier, 1 (USL); 1 mi. S Mamou, 1 (USL); 0.5 mi. S Mamou, 1 (USL); 4 mi. SE Mamou, 1 (USL). *Franklin Par.:* 2 mi. N Gilbert, 1 (NLU). *Iberia Par.:* 2 mi. S Weeks Island, 1 (LSUMZ); Weeks Island, 1 (LSUMZ); 3 mi. SE New Iberia, 1 (USL); Jeanerette airport, 1 (USL); Spanish Lake, 1 (USL); 8 mi. W New Iberia, 1 (USL); 5 mi. N New Iberia, 1 (USL). *Iberville Par.:* 4 mi. WNW Jack Miller, 1 (LSUMZ); 5 mi. S Ramah, 2 (LSUMZ). *Jefferson Par.:* Metairie, 2 (1 LSUMZ; 1 SLU); 4.5 mi. NW Barataria, 1 (LSUMZ); Grand Terre Research Station, 1 (LSUMZ); w. side Miss. River bridge tollgate, 1 (USL). *Jefferson Davis Par.:* 5 mi. S Elton, 1 (LSUMZ); Jennings, 1 (LSUMZ); 2.5 mi. N Roanoke, 1 (MSU); Lake Arthur, 1 (LTU). *Lafayette Par.:* 2 mi. E Lafayette courthouse, 7 (USL); Lafayette, 1 (USL); 4.3 mi. S Lafayette, 1 (USL); 2 mi. S Lafayette courthouse, 1 (USL); 5 mi. SW Lafayette courthouse, 1 (USL); 3 mi. SW Lafayette, 1 (USL); 9 mi. SW Lafayette courthouse, 1 (USL). *Lafourche Par.:* 3 mi. S Golden Meadow, 1 (LSUMZ); 1.5 mi. W Lockport, 1 (USL); 6 mi. SW Raceland, 1 (USL); 0.5 mi. NW Lockport, 1 (USL); 3 mi. E Larose, 1 (USL). *Livingston Par.:* 0.5 mi. N Killian, 1 (SLU); 1 mi. S Denham Springs, 1 (LSUMZ). *Morehouse Par.:* 7 mi. W Oak Ridge, 1 (NLU); Horseshoe Lake, 2 (NLU); Wham Break, 1 (NLU). *Natchitoches Par.:* Black Lake, 1 (NLU); Natchitoches, 2

(NSU). *Orleans Par.:* New Orleans, 2 (LSUMZ); 3 mi. W Chef Menteur Pass along L & N Railroad, 1 (USL); 6 mi. SW Chef Menteur Pass along Intracoastal Waterway, 2 (USL). *Ouachita Par.:* Wham Break, 2 (NLU). *Plaquemines Par.:* 1 mi. SE Carlisle, 2 (LSUMZ); 2 mi. S Wills Point, 1 (LSUMZ); Pass a Loutre Game Mgt. Hqs., 1 (LSUMZ); middle of South Breton Island, 2 (USL); Bohemia, 2 (SLU); Port Sulphur, 1 (SLU); Tulane Univ. Riverside campus, Belle Chasse, 2 (TU). *Pointe Coupee Par.:* 1 mi. E Red Cross, 1 (LSUMZ); Morganza, 1 (LSUMZ). *Rapides Par.:* 3 mi. off Lecompte-Forest Hill Hwy. on Bayou Boeuf Rd., 1 (USL); 3 mi. S Cheneyville, 1 (LTU). *Richland Par.:* 5 mi. E Oak Ridge, 1 (NLU). *St. Bernard Par.:* marsh east Chalmette near Lake Borgne, 1 (USL). *St. Charles Par.:* 2 mi. NW New Orleans Int. Airport, 1 (USL); Boutte, 1 (SLU); 0.5 mi. S Norco, 1 (SLU); Sarpy's Swamp, 6 mi. W of U.S. Hwy. 50 on U.S. Hwy. 61, 2 (TU). *St. John the Baptist Par.:* 6.5 mi. S Manchac, 1 (LSUMZ); Akers, 1 (SLU). *St. Landry Par.:* 0.25 mi. NE Port Barre, 1 (LSUMZ); 2 mi. W Leonville on Bayou Teche, 1 (USL); 2.5 mi. W Opelousas, 1 (USL); 6 mi. S Opelousas courthouse, 1 (USL); 10 mi. E Opelousas courthouse, 1 (USL); 1 mi. E Eunice, 1 (USL); 3 mi. NW Washington, 1 (USL); 8 mi. W Opelousas, 1 (USL). *St. Martin Par.:* Grand Lake area, 1 (LSUMZ); 5 mi. S St. Martinville, 1 (USL); Lake Martin, 4 (USL); 3.5 mi. NE Morgan City on Lake Palourde levee, 1 (USL); 1 mi. S St. Martinville, 1 (USL); 1 mi. S Parks, 1 (USL). *St. Mary Par.:* Bellisk, 1 (LSUMZ); Berwick, 1 (LSUMZ); s. shore Cote Blanche Island, 8 (USL); Cypremort Pt., 7 (USL); 6 mi. S Ellerslie, 2 (USL); Southern Bayou on Avoca Island, 8 mi. S ferry, 3 (USL); Belle Isle, 7 (USNM). *St. Tammany Par.:* 0.3 mi. N, 0.1 mi. W Pearl River, 1 (LSUMZ); 0.5 mi. S, 0.4 mi. E Pearl River, 1 (LSUMZ); North Shore, 1 (SLU); 4 mi. W Madisonville, 1 (SLU); 1 mi. N Slidell on Hwy. 11, 1 (TU). *Tangipahoa Par.:* 8 mi. E Ponchatoula, 1 (SLU); 1 mi. N Manchac, 1 (SLU); 4 mi. N Robert, 1 (SLU); 2 mi. W Natalbany, 1 (SLU); 5 mi. E Ponchatoula, 1 (SLU); Manchac, 1 (SLU). *Tensas Par.:* Newellton, 1 (LTU). *Terrebonne Par.:* 3 mi. SE Houma near Bayou du Large, 1 (LSUMZ); Chacahoula, 1 (LSUMZ); 30 mi. SW Houma, 1 (LSUMZ). *Union Par.:* Lake D'Arbonne, 3 mi. SE Farmerville, 1 (LTU); Finch Lake, north Farmerville near Haile, 2 (LTU); 3 mi. E Farmerville, 1 (LTU). *Vermilion Par.:* Henry, 1 (LSUMZ); McIlhenny estate, 1 (LSUMZ); 5 mi. W Red Fish Point, 1 (LSUMZ); approx. 3 mi. S Belle Isle, 1 (LSUMZ); 3 mi. N Intracoastal City, 27 (LSUMZ); 2 mi. SE Gueydan, 1 (LSUMZ); 5 mi. N Intracoastal City, 1 (LSUMZ); 5 mi. S Abbeville, 1 (USL); Bayou Chene near Intracoastal City, 1 (USL); 2 mi. N Gueydan, 1 (USL); 3 mi. N Leroy, 1 (USL); Esther, 2 (USL); 1.5 mi. SW Intracoastal City, 1 (USL); Red Fish Point, 1 (USL); 3 mi. NW Indian Bayou, 1 (USL); 4 mi. S Erath, 1 (USL); 3 mi. S Maurice, 1 (USL); Intracoastal City, 1 (USL); Pecan Island, 5 (USL); 5 mi. S Kaplan, 1 (USL). *Washington Par.:* Varnado, 1 (LSUMZ). *Webster Par.:* 10 mi. SW Sibley, Lake Bistineau, 1 (LTU); Minden, 1 (LTU). *West Baton Rouge Par.:* 4 mi. S Port Allen, 1 (LSUMZ).

Whales and Dolphins

Order Cetacea

THE MEMBERS of the order Cetacea are wholly aquatic and as a group occur throughout the oceans and seas of the world. Occasionally some of them ascend rivers and enter large inland lakes. The order is comprised of three suborders: the extinct Archaeceti, which had teeth that were differentiated into incisors, canines, and molars; the Odontoceti, an extant group of living toothed whales, porpoises, and dolphins, in which the teeth, variable in number, are usually undifferentiated and the skull is nonsymmetrical and greatly telescoped; and the Mysticeti, a living group of whales, toothless as adults, that have in the mouth plates of baleen (modified mucous membranes) through which small organisms are strained for food. According to my tally, living cetaceans comprise eight families, 36 genera, and some 90 species, but among cetologists these figures vary because of divergent taxonomic treatments.

Although these three taxa are regarded as suborders in the single order Cetacea by most mammalogists, serious questions have been raised concerning their monophyletic origin. No known forms bridge the gaps between them, and no fossils belonging to any one of them would appear to have been capable of

having given rise to either of the other. The similarities between them could be interpreted as retention of primitive characters or as the result of convergent adaptations to the pelagic life that all three share (Rice, 1967). However, despite the fact that the differences between them are greater than those separating other universally recognized orders of mammals, such as the Artiodactyla and Perissodactyla, I believe that they are best retained as suborders of the order Cetacea. Rice (1967), in treating them as three separate orders, admits that our present knowledge neither proves a monophyletic origin for the Cetacea nor provides evidence incompatible with such an origin.

Cetaceans as a group are distinguished by a number of features that clearly set them apart from all other mammals. Particularly significant are the modifications that have taken place in their skeleton, notably in the skull, which has undergone considerable telescoping and rearrangements of its elements (Figure 130). In most forms the skull is greatly elongated anteriorly into a pronounced rostrum that, along with the tapering posterior end of the body, accentuates the animal's fusiform shape. Other features of

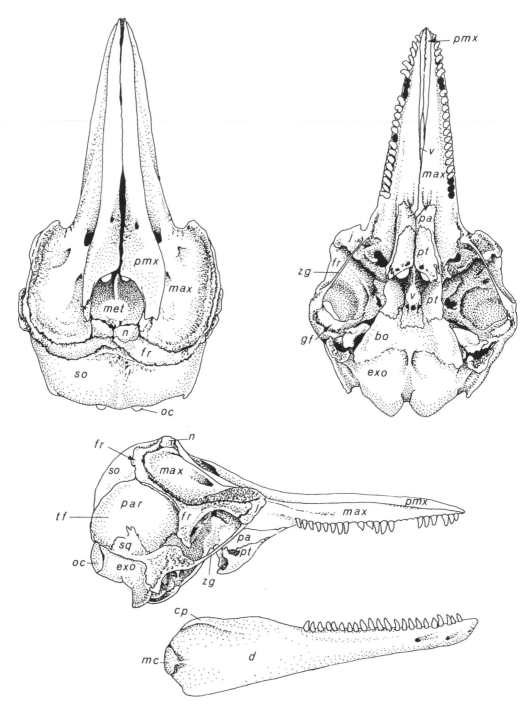

Figure 130 Drawing of skull of Atlantic Bottle-nosed Dolphin *(Tursiops truncatus)*, LSUMZ 17034, identifying some of the important bones: *bo,* basioccipital; *cp,* coronoid process; *d,* dentary; *exo,* exoccipital; *fr,* frontal; *gf,* glenoid fossa; *j,* jugal; *max,* maxillary; *mc,* mandibular condyle; *met,* mesethmoid; *n,* nasal; *oc,* occipital condyle; *pa,* palatine; *par,* parietal; *pmx,* premaxillary; *pt,* pterygoid; *so,* supraoccipital; *sq,* squamosal; *tf,* temporal fossa; *v,* vomer; *zg,* zygomatic arch.

the cranium include elaborate modifications of the nasal passages that permit more efficient respiratory processes. These and other structural developments have been accompanied by drastic physiological adaptations, such as those that allow cetaceans in quest of food to descend to great depths in the inky darkness of the ocean where the pressure on their bodies is 100 tons to the square foot—and to stay down for over an hour. The end product of the evolutionary history of cetaceans has been a kind of mammal that is indeed unique. It is unexcelled in its adaptations for a wholly aquatic existence.

To be sure, the cetaceans share, in a superficial way, certain morphological features with the order Sirenia, which includes the Dugong, Sea Cow, and the manatees. The similarities are believed, however, to represent evolutionary convergence resulting from independent adaptations to an aquatic environment. The tail in both groups is dorsoventrally flattened, and the tail of the Dugong possesses lateral flukes like those of the whales and dolphins, as did also the tail of the recently extinct Steller's Sea Cow, of the Bering Straits. In both orders, the pectoral appendages are modified as flippers, but in the cetaceans the joints beyond the shoulder are immobile, and the limb serves only as a steering, balancing, and caressing organ, although in captivity they can be trained to use the limbs together to grasp large objects such as a basketball. In the Sirenia the elements of the forelimb are less rigidly fused and indeed are used in directing food toward the mouth and even in cleaning the mouth (Figure 222); the limb also provides the main means of propulsion. The cetaceans possess a fibrous layer of tissue beneath the skin filled with fat and oil, the blubber that protects them against the icy waters in which they often live, but sirenians have only a modest amount of fat beneath the surface of the skin and it is without free oil.

The fact that all wholly aquatic mammals, as well as the seals and their allies, share certain structural features is not surprising, because they in turn have characters in common with lower forms of aquatic life such as fishes. But the presence of gill slits and gills, the vertical position of the tail (Figure 131), the skin that is rough or covered with scales, the presence of two pairs of ventral fins instead of one, and the fact that several rows of teeth are found in each jaw all serve to set fishes apart from mammals. Both fishes and most cetaceans possess a dorsal fin, but in the latter the dorsal fin contains no bony elements and hence is not connected to the vertebral column except by muscles and connective tissue.

In the toothed whales the nasal passages open to the exterior through a single blowhole on the top of the head, in baleen whales, through a double blowhole. The mammae consist of a single pair with the teats located in deep depressions, one on each side of the vulva. The sacral vertebrae are similar to the adjacent caudal and lumbar vertebrae, but the caudal vertebrae possess chevron processes on their ventral surface. The cervical vertebrae are flattened and frequently fused. The pelvic girdle, which is reduced to two small bones embedded in tissue and not connected with the spinal column, provides only for the attachment of muscles associated with the external sex organs.

One of the truly remarkable anatomical features of cetaceans is the structure known as the melon, located on the flat surface of the forehead anterior to the blowhole. It gives the head its bulbous crown, greatly exaggerated is some species (see especially Figure 168 depicting *Globicephala macrorhyncha*, the Short-finned Pilot Whale). The melon is a huge mass of waxy, oily, gristly stuff called the spermaceti organ, a name derived from the Medieval Latin words *sperma ceti*, meaning whale sperm, and based on the erroneous belief that the substance was the coagulated semen of the whale. The white crystalline waxy solid that separates from the sperm oil consists principally of cetyl palmitate and other esters of fatty acids. It has been used in

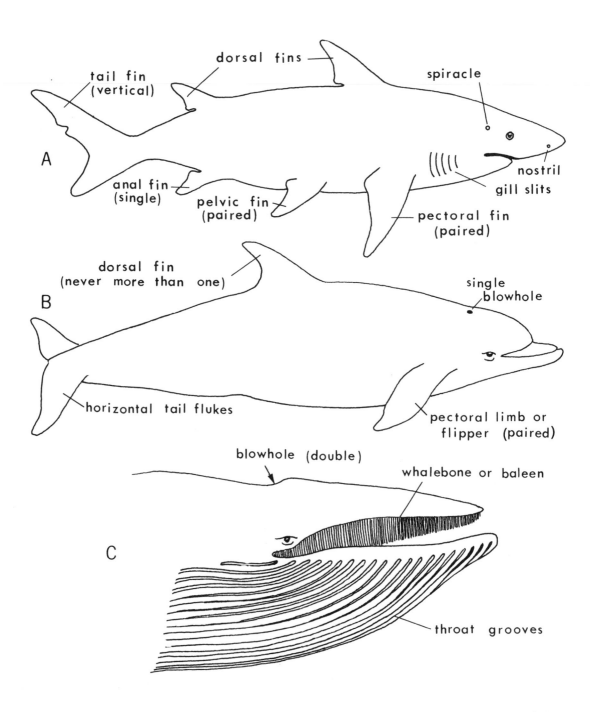

Figure 131 Some major differences between fishes and cetaceans: (A) a shark; (B) a toothed whale or dolphin; (C) head of a rorqual, or baleen whale.

the manufacture of candles, ointments, and fine lubricants. F. G. Wood, formerly of Marine Studios, Marineland, Florida, and now Senior Scientist with the Naval Undersea Research and Development Center of the Department of the Navy in San Diego, California, has suggested that the melon may act as an acoustical lens for concentrating the sounds received by the cetacean's sonar system (Norris, 1964:329; Schevill, 1964:309).

The brains of cetaceans are relatively large, and, according to Lilly (1958), that of the Atlantic Bottle-nosed Dolphin weighs up to 1,700 grams, while that of man averages 1,450 grams. Dolphins are accredited by some with being exceptionally intelligent. Dr. Lilly (1961) is convinced of the possibility that man may be able in time to establish some form of two-way communication with dolphins, but other investigators are incredulous to say the least.

Whales and dolphins frequently become stranded on beaches, sometimes in large numbers. Attempts by man to redirect floundering animals back toward the open sea seldom meet with success. No one has yet discovered why these compulsive strandings occur. Some authors have gone so far as to suggest that an "ancestral memory" is at work, that because the progenitors of cetaceans were land mammals, the modern forms sometimes become obsessed with a desire to get back to land. But, as attractive as such an explanation might appear to be, it is hardly plausible. Excellent accounts of mass strandings are available, including those of Alpers (1961), which are especially illuminating.

Except in the case of the Atlantic Bottlenosed Dolphin, which occurs commonly in our coastal waters, most of the records of cetaceans along our shores are of animals stranded on our beaches. Representatives of two suborders, 4 families, 8 genera, and 11 species are now known from Louisiana's shores or its offshore waters, but several additional species have been recorded stranded on the Gulf Coast, both to the east and west of

Louisiana, and will doubtless sooner or later be found in Louisiana's offshore waters or stranded on one of our beaches. Anyone finding a whale or dolphin that he is reasonably sure is not the common Atlantic Bottlenosed Dolphin should immediately contact one of the biologists at the nearest state college or university or call the Museum of Zoology at Louisiana State University, Baton Rouge (Area 504, 388-2855). Most of the cetaceans are poorly represented in museum collections everywhere.

Two examples will serve to illustrate why the preservation of skeletal parts and the subsequent identification of stranded whales and dolphins is a matter of great scientific importance. A dolphin obtained on Padre Island on the lower Texas coast in 1969 (James et al., 1970) turned out to be one of the few specimens in the museums of the world of the Pygmy Killer Whale *(Feresa attenuata)*. This species was named in 1875 on the basis of two skulls of unknown origin in the British Museum (Natural History). More than 75 years elapsed before the next specimen turned up, on the coast of Japan. This small whale has been recorded again on a few occasions: from Japanese waters, the Pacific Ocean off Hawaii, and Costa Rica, the South Atlantic and Indian oceans, and the North Atlantic off Senegal. The Padre Island specimen was the first from the western Atlantic. Similarly, a dolphin that I found on the beach in Cameron Parish in 1970 proved to be the second record of *Stenella coeruleoalba* for the Gulf of Mexico. These two instances clearly demonstrate the value of determining the identity of all stranded cetaceans.

Anyone finding a stranded whale or dolphin should attempt first to answer certain basic questions about the animal and to make key measurements that will provide museum personnel with clues as to what kind of cetacean is involved. The museum staff can then decide whether an attempt to salvage the specimen is justified. Working with decomposing whale flesh is not a pleasant undertaking, but, if the animal involved is a rarity, no dedicated mammalogist would allow unpleasant odors to deter him from doing what has to be done. Immediate action is required, because cetaceans spoil rapidly, and therefore important details, such as the animal's true color, can be lost within a matter of hours after its death.

Further on in this account are a series of questions and suggested measurements to be used as a guide in taking down preliminary data on stranded or captured cetaceans. They are based mainly on two sources. One is a set of instructions issued by the Committee on Marine Mammals of the American Society of Mammalogists (Kenneth S. Norris, ed., 1961). The other source is a booklet issued by the British Museum (Natural History) entitled *Guide for the Identification and Reporting of Stranded Whales, Dolphins, and Porpoises on the British Coast* by F. C. Fraser (1966).

In the British Isles the program of asking people to report stranded cetaceans has proven to be immensely successful; and, as a result, much new and valuable information has been assembled on this interesting and poorly known group of mammals. Some species originally thought to be rare or absent in British waters were found to be actually common, and at least two new species were added to the British list.

I would hope that readers of this account will help spread the word that the Louisiana State University Museum of Zoology is extremely desirous of being advised of any cetaceans stranded on our shores. The examination of carcasses, in which decomposition is not too far advanced, permits analysis of the stomach contents and inspection of the other internal organs to find out what foods are being eaten by the species, and, in the case of females, to determine whether embryos are present and thereby to shed light on the breeding biology of the species involved. The preservation of the skulls and skeletal parts of a stranded cetacean provides a basis for a positive identification and, at the same time, a permanent record of its occurrence at a particular place. In brief, anyone reporting a stranded whale or dolphin and thereby causing it to be preserved in a museum is indirectly contributing to the sum of knowledge about a group of animals concerning which surprisingly little is known.

Here are the questions that should be asked and, if possible, answered on finding a dead or stranded whale or dolphin:

1. Is the tail horizontal to the axis of the body as in Figure 131? If the tail is vertical and the animal possesses gill slits, containing gills, two pairs of both dorsal and ventral fins, several rows of teeth, and a rough or scaly skin, it is a fish, probably a shark, and should *not* be reported. In a toothed whale or porpoise, gill slits are absent, only one row of teeth is present in each side of each jaw, or the teeth may be buried in the gums or absent, the skin is smooth, and there is only one pair of flippers on the side of the body.
2. Is a blowhole present on the top of the head? Is it single or double?
3. Does the mouth contain whalebone or baleen? If so, what is the color of the whalebone plates and what is the color of the hairy fringes?
4. Is the skin of the throat marked by numerous deep grooves, as in Figure 131?
5. Does the mouth contain teeth? If so, state:
   (a) whether they occur in both jaws or in the lower jaw only;
   (b) the number of teeth, *on one side,* in the upper jaw;
   (c) the number of teeth, *on one side,* in the lower jaw;
   (d) if the teeth are few, what is their position on the jaw?
   (e) the diameter of the largest tooth;
   (f) whether the teeth are circular or spade shaped in cross section.
6. What is the shape of the head? For in-

stance, is a beak absent, or is a beak present? Is the forehead much swollen and bulbous as in the Short-finned Pilot Whale illustrated in Figure 168?

7. What is the color of the skin? Are white markings present and, if so, what are their positions? Are the underparts different in color, or shade of color, from the upperparts?

8. What is the condition of the specimen? Is it badly decomposed or is it relatively fresh?

9. If possible take photographs or make a rough sketch of the side view of the animal, showing the relative positions of the eyes, the dorsal fin, and the pectoral flippers, along with the shape of the tail flukes, with careful attention to the presence or absence of a median notch.

10. Is its location such that its head, flippers, or complete skeleton could be secured for the museum, if wanted?

11. How long can you delay its disposal by any sanitation crews while waiting for a representative of the museum to reach the scene?

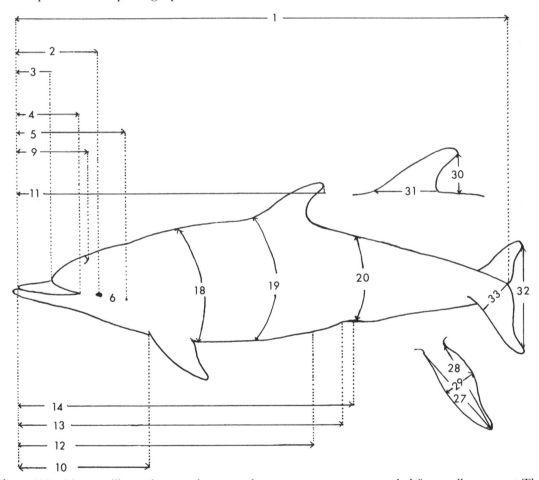

Figure 132   Diagram illustrating certain external measurements recommended for small cetaceans. The numbers refer to the measurements listed on page 308. All measurements, except girth, should be taken in *straight lines* parallel to the axis of the body, and not over the curvature of the body. When direct line measurements are employed, as is often done in the case of the large baleen and sperm whales, be sure to specify that this was done.

CETACEAN DATA RECORD
(Particularly important items are preceded by an asterisk)

*Species, if known_____

*Number of individuals_____

*Sex, if determined_____

*Date of stranding, capture, or discovery

    (state which)_____

*Time of stranding or capture_____

*Observer_____

*Locality_____

_____

_____

*Condition of specimen(s)_____

_____

*Remarks (circumstances surrounding

    stranding or capture)_____

_____

_____

*Number of teeth (or baleen plates):

    Right upper_____ Left upper_____

    Right lower_____ Left lower_____

*Number of throat grooves_____

*Color (indicate pattern by diagram)_____

_____

_____

Measurements (See Figure 132)

Body:

  *1. Total length (tip of upper jaw to deepest part of notch between flukes or middle of posterior margin of tail, if no notch is present) _____

  *2. Distance from tip of upper jaw to center of eye _____

    3. Distance from tip of upper jaw to melon (fatty protuberance on forehead of some cetaceans) _____

    4. Length of gape (tip of upper jaw to angle of gape) _____

    5. Distance from tip of upper jaw to external auditory meatus (ear opening, which is often minute) _____

    6. Distance from center of eye to external auditory meatus _____

    7. Distance from center of eye to angle of gape _____

    8. Distance from center of eye to center of blowhole _____

  *9. Distance from tip of upper jaw to blowhole or to midpoint between blowholes when double _____

*10. Distance from tip of upper jaw to anterior insertion of flipper _____

*11. Distance from tip of upper jaw to tip of dorsal fin _____

 12. Distance from tip of upper jaw to umbilicus _____

 13. Distance from tip of upper jaw to midpoint of genital aperture _____

*14. Distance from tip of upper jaw to center of anus _____

 15. Length of projection of lower jaw beyond upper (if reverse is true, so state) _____

 16. Length from tip of upper jaw to posterior extremity of throat creases _____

 17. Length of throat creases if present: maximum _____
                             minimum _____

 18. Girth, on transverse plane intersecting base of flipper _____

*19. Maximum girth (describe where measured in distance from tip of upper jaw) _____

 20. Girth, on transverse plane intersecting anus _____

Apertures:

 21. Dimensions of eye (length × height) _____

 22. Length of mammary slits (left and right) _____

 23. Length of genital slit _____

 24. Length of anal opening _____

 25. Dimensions of blowhole(s) _____

 26. Diameter of external auditory meatus _____

Appendages:

*27. Length of flipper from anterior insertion to tip _____

*28. Length of flipper from axilla to tip _____

*29. Width of flipper (maximum) _____

*30. Height of dorsal fin _____

 31. Length of base of dorsal fin _____

*32. Width of tail (fluke tip to fluke tip) _____

*33. Distance from nearest point on anterior border of flukes to notch _____

*34. Depth of notch between flukes (if none, so state) _____

The Beaked Whale Family
Ziphiidae

THESE MEDIUM-SIZED cetaceans, consisting of five genera and some 14 species, are distinguished by a combination of anatomical features: their total length ranges from 15 to 40 feet; the nasal aperture is median and shaped like a half moon; the occipital crest is high and prominent; one pair of longitudinal grooves on the throat are divergent posteriorly, nearly meeting anteriorly; the tail lacks a median notch on its posterior edge between the flukes; two to seven cervical vertebrae are fused; numerous vestigial teeth are embedded in the gums of both the upper and lower jaws; functional teeth, which are never more than four nor less than two, are confined to the lower jaw in all North American species (sometimes difficult to see in an animal in the flesh, especially in females and juveniles). The teeth may be either conical and located at the anterior end of the mandibular symphysis or spade-shaped and located beside the symphysis or posterior to it.

Genus *Ziphius* G. Cuvier

Dentition. — Two conical teeth are present, one at the tip of each ramus of the mandibular symphysis.

GOOSE-BEAKED WHALE
(*Ziphius cavirostris*)

Figure 133

Vernacular and other names. — The name Goose-beaked Whale is based on the shape of the forehead and rostrum in this species. The generic name *Ziphius* is the Latinized version of the Greek word *xiphios*, meaning swordfish and is of uncertain intent in the present connection unless the name was designed to call attention to the long, narrow rostrum. The specific name *cavirostris* comes from two Latin words (*cavus*, meaning hollow or excavated and *rostrum*, meaning snout or beak) and probably alludes to the great depressions on each side of the skull at the base of the maxillae. The species is sometimes called Cuvier's Beaked Whale because it was first named and described by Baron Georges Cuvier, famous French comparative anatomist and founder of the science of paleontology.

Distribution. — The species is present in most of the seas of both the Northern and Southern hemispheres. In the North Atlantic it occurs from Rhode Island south to the Gulf of Mexico, and in the North Pacific from Alaska south to California. Its known occurrence in Louisiana waters is limited to a single specimen stranded on one of the Chandeleur Islands in May 1969 (see beyond).

External appearance. — The Goose-beaked Whale usually measures between 18 and 26 feet in total length. The basic color is dark grayish black except for the face and upper back, which are often cream colored. Some individuals are said to have only the rostrum and top of the lower jaw cream colored, the remainder of the body being light grayish brown with blotches of dark gray or even white below. The dorsal fin is located at approximately two-thirds the distance between the tip of the snout and the midpoint between the tail flukes. Its forward edge is slightly convex, the hind edge concave.

Measurements and weights.—The one Louisiana specimen (LSUMZ 15609) measured as follows: total length, 18 feet; tip of snout to base of dorsal fin, 12 feet, to eye, 28 inches, to blowhole, 26 inches, to base of pectoral fin, 52 inches; width of tail flukes, 22 inches. A few of the skull measurements of this specimen are as follows: overall length, 925 mm; greatest depth, 404 mm; greatest width, 470 mm; length of ramus of lower jaw, 791 mm. A female, slightly more than 21 feet in length, weighed 5,905 pounds (Walker et al., 1968).

Cranial and other skeletal characters.—The skull of *Ziphius cavirostris* is strikingly different from that of other cetaceans. The following are particularly diagnostic features: the relatively narrow rostrum; the steeply elevated cranial crest; the fact that the nasal bones largely obscure the external nares when viewed from above; the massive size posterior to the premaxillae, which form a distinct cup around the external nares; and the single pair of functional teeth, which lie anterior to the mandibular symphysis and which in females usually do not extend above the gums. In addition, the mesial laminae of the ptery-goids are greatly enlarged, the lower jaw protrudes beyond the tip of the maxilla, and three or four cervical vertebrae are fused. (Figure 134)

Status and habits.—The only Goose-beaked Whale recorded in our state's coastal waters is one found on 15 May 1969 stranded on the outer beach five miles south of the north end of the Chandeleur Island chain (Gunter and Christmas, in press). Through the sterling efforts of J. Y. Christmas of the Gulf Coast Research Laboratory, at Ocean Springs, Mississippi, the head was removed and later cleaned. It was finally deposited in the collections of the Louisiana State University Museum of Zoology. I am aware of only two other records of the occurrence of this species on the northern Gulf Coast, a 12-foot 8-inch female that stranded near Fort Walton, on 10 December 1964 (Caldwell et al., 1965), and an adult that beached during Hurricane Carla in September 1961 on Galveston Island (Schmidly and Reeves, in press). There are several instances of its occurrence on the western coast of the Florida Peninsula (Moore, 1953).

The habits of the Goose-beaked Whale are poorly known. The members of the species are said by some authors to travel in groups of 30 to 40 individuals, and to feed together in fairly close association, but other writers assert that the species is usually solitary, thus

Figure 133   Goose-beaked Whale *(Ziphius cavirostris).*

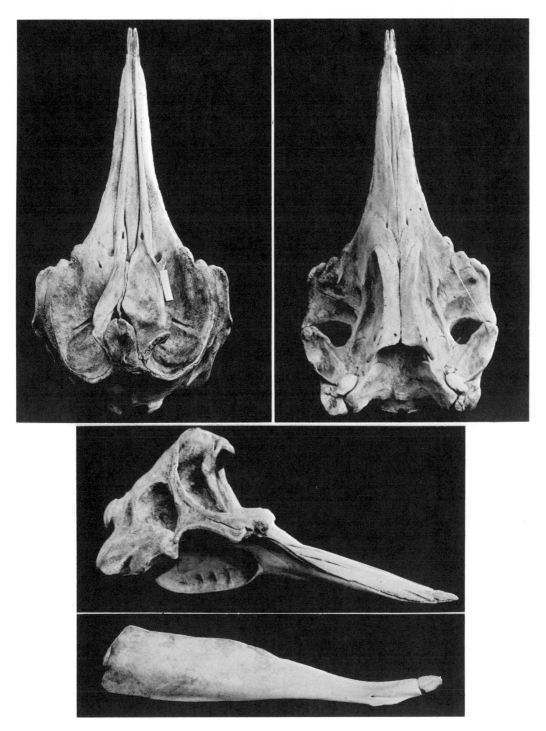

Figure 134 Skull of Goose-beaked Whale *(Ziphius cavirostris)*, LSUMZ 15609, 5 miles south of north end of Chandeleur Island. Total length of skull 925 mm or 36.4 inches. (Photograph by J. Harvey Roberts)

highlighting how uncertain is our information regarding even the most general features of the animal's life history. It is alleged to remain underwater for periods exceeding 30 minutes. Squids apparently are its preferred food. A female measuring slightly over 21 feet in length was found to have the remains of 1,304 squids in its stomach. The young are born after a gestation period of approximately one year and measure about one-third the length of the mother. (Walker et al., 1968)

Ziphius cavirostris G. Cuvier

Ziphius cavirostris G. Cuvier, Recherches sur les ossemens fossiles . . . , ed. 2, 5, 1823:352; type locality, Fos, delta of Rhone River, France.

The species is monotypic, that is, no subspecies are recognized.

Specimens examined from Louisiana.—Total, 1, as follows: *St. Bernard Par.:* outer beach, 5 mi. S of the north end of the main island of the Chandeleur Island chain (LSUMZ).

[Genus *Mesoplodon* Gervais]

Dentition.—Two laterally flattened, spade-shaped teeth located in the lower jaw, one on each side in alveoli adjacent to the mandibular symphysis.

[ANTILLEAN BEAKED WHALE]
(Mesoplodon europaeus)

This exceedingly rare beaked whale is known from the Gulf of Mexico on the basis of two records. One was stranded on the Gulf Coast of Florida at Boca Grande, near latitude 26°42′ N, in April, 1959, and was, at the time, the twelfth known example of the species. Another specimen from Padre Island, latitude 27°15′ N, Texas, found in September 1946 and originally reported as *Mesoplodon densirostris* (Blainville) by Gunter (1955), has subsequently proven to be *M. europaeus.* As such, it represented the thirteenth specimen

to be identified and the third adult male of the species known to science (Moore, 1960). The taxonomic characters of this species, particularly those features that serve to distinguish *M. europaeus* from True's Beaked Whale, *M. mirus,* are thoroughly analyzed by Moore and Wood (1957) and Moore (1960, 1966, and 1968). *M. europaeus* has a total length of approximately 15 feet. It possesses two teeth, which are located one on each side of the posterior end of the mandibular symphysis, with the posterior lip of the alveolus

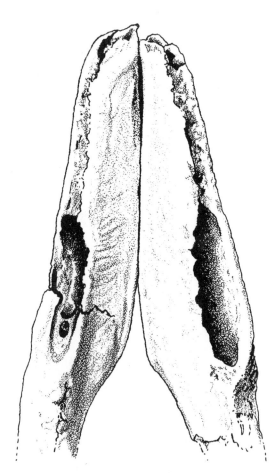

Figure 135   Anterior end of the mandible of Antillean Beaked Whale *(Mesoplodon europaeus)*, showing the relative positions of the two alveoli to the symphysis. (Drawing is based on photograph in Moore, 1960, of Boca Grande, Florida, specimen, AMNH 182649.)

nearly always well behind the posterior margin of the symphysis. The teeth are broadly flattened basally but taper abruptly above the gum line to an apex. The dimensions of the teeth are approximately 96 × 51 × 15 mm. (Figures 135 and 136.) In *M. mirus* the posterior lip of the alveolus is anterior to the posterior margin of the mandibular sym-

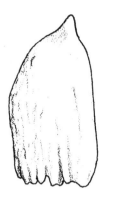

Figure 136 Left tooth of Antillean Beaked Whale *(Mesoplodon europaeus)*. *Left,* lateral view with anterior edge of tooth on right side; *right,* posterior view. (Drawing is based on photographs in Moore, 1960, of Padre Island, Texas, specimen, AMNH 150037.)

physis by a distance ranging from 140 to 167 mm.

The genus *Mesoplodon* is easily distinguished from *Ziphius* by the presence in the latter of two conical teeth located at the end of the mandibular symphysis instead of having two laterally flattened and spade-shaped teeth that lie, as in *Mesoplodon,* beside the mandibular symphysis.

Any beaked whale found stranded on the coast of Louisiana, especially any that possess only two laterally flattened teeth, one on each side of the lower jaw, should be reported immediately to the Louisiana State University Museum of Zoology in order that the skull and other skeletal parts may be salvaged.

Mesoplodon europaeus (Gervais)

Dioplodon europaeus Gervais, Zoologie et paléontologie Francaises . . . , ed. 1, 1848–1852 [=1852]:4 [a *nomen nudum* according to recent authors].

Dioplodon europaeus Gervais, Histoire naturelle des Mammifères (Paris), 2, 1855:320; type locality, English Channel.

Dioplodon gervaisi Deslongchamps, Bull. Soc. Linn. Normandie, 10, 1866:176.

Mesoplodon europaeus, Flower, Proc. Zool. Soc. London, 1877[=1878]:684.

The Sperm Whale Family
Physeteridae

THIS FAMILY, as defined by some authorities, contains only one genus and a single species, *Physeter catodon.* The modern trend among mammalogists is, however, not to place the small sperm whales of the genus *Kogia* in a separate family Kogiidae, as was long done,

but to treat them as a subfamily Kogiinae in the family Physeteridae.

The Giant Sperm Whale is the largest of the toothed whales. Males attain 65 feet in total length, females only 40 feet or slightly more. The Giant Sperm Whale is distinguished from all other cetaceans by the rectangular shape of its head when viewed in profile, by the absence of a dorsal fin, and by the relatively small size of the pectoral appendages. Other structural details are given in the account of the species.

The small sperm whales of the subfamily Kogiinae differ from the one species, *Physeter catodon*, of the subfamily Physeterinae, as follows: the total length is 6.9 to 10.9 feet instead of 38 to 65 feet; the body shape is porpoiselike instead of anteriorly exaggerated (the conical-shaped head of *Kogia* is among the shortest of all cetaceans instead of one-third the total length and thus one of the longest, as it is in *Physeter*); the blowhole is on top of the head above the eyes instead of far forward at end of the snout; the rostrum is the shortest of all living cetaceans; the teeth are thin, curved, and sharp pointed instead of massive and blunt. In addition, there are numerous postcranial differences, including the presence of a well-formed and falcate dorsal fin instead of a dorsal fin that is rudimentary or absent.

Genus *Physeter* Linnaeus

Dentition.—Twenty to 25 conical teeth, 5 to 8 inches long, are located in each side of the lower jaw. Numerous vestigial teeth are embedded in the gums of the upper jaw.

GIANT SPERM WHALE   Figure 137
(Physeter catodon)

Vernacular and other names.—This species is also called the cachalot. The generic name *Physeter* is the Greek word meaning blower, and refers to the whale's habit of making a vapor spout when it exhales air from its lungs on surfacing. The specific name *catodon* is a combination of two Greek words (*kata*, lower, and *odon*, tooth), no doubt suggested by the long row of teeth in the lower jaw. The adjectival noun Sperm in the vernacular name pertains to the spermaceti or sperm oil obtained from the melon on the animal's forehead. The adjective Giant serves to distinguish this huge species from other kinds of sperm whales, such as the Dwarf and Pygmy Sperm Whales of the genus *Kogia*.

Distribution.—The Giant Sperm Whale occurs throughout the oceans of both the Eastern and Western hemispheres, ranging from the Arctic to the Antarctic, but occurring most commonly in the temperate and tropical latitudes of the Atlantic and Pacific oceans. The species once occurred in numbers in the Gulf of Mexico, including Louisiana's offshore waters, but is now rare anywhere in the Gulf. (Maps 47, 48, and 49)

External appearance.—This large whale, the male of which sometimes attains a length of 65 feet or slightly more, is bluish black except

Figure 137 Miniature reproduction of Giant Sperm Whale *(Physeter catodon)* in Los Angeles County Museum of Natural History. (Photograph courtesy LACMNH)

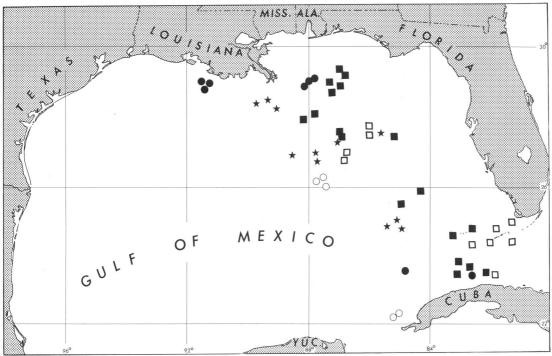

Map 47   Occurrences of the Giant Sperm Whale *(Physeter catodon)* in the northern half of the Gulf of Mexico between 1761 and 1920, based on logbook records of whaling ships. Each plot point represents the approximate location of a whaling vessel on a day when one or more Giant Sperm Whales were taken: ○ , March; ● , April; ★, May; ■ , June; □ , July. (After Townsend, 1935)

for occasional small areas of white on the lower jaw and venter. The head is rectangular in profile and accounts for approximately 35 percent of the animal's total length. The dorsal fin is replaced by a hump and by a series of longitudinal ridges on the posterior part of the back. The lower jaw is small, narrow, and decidedly shorter than the snout. The pectoral flippers are exceedingly small. This combination of characters clearly separates the Giant Sperm Whales from all other cetaceans.

Color phases.—No color phases as such are known, but albinos are on record. Moby-Dick, of Herman Melville's story, was an albino Giant Sperm Whale.

Measurements and weights.—A Giant Sperm Whale stranded at Sabine Pass in 1910 mea-sured in feet as follows: total length from tip of snout to posterior extremity of tail flukes, 63.4; tip of snout to base of pectoral appendage, 24.5; tip of snout to angle of gape, 17.1; tip of lower jaw to gape, 10.8; dorsoventral diameter of flat end of snout, 10.3; circumference in front of pectoral appendages, 37; width across tail flukes, 16.6 (Newman, 1910). A Giant Sperm Whale nearly 60 feet in length weighed 58.7 tons; males commonly weigh 35 to 50 tons (Walker et al., 1968).

Cranial characters.—The cranium is massive, disproportionately so in comparison with the postcranial skeleton. Each ramus of the lower jaw possesses 20 to 25 large conical teeth that fit into depressions in each side of the upper jaw, many of which contain vestigial teeth buried in the gums. The rostrum is broad and flat, providing a shelf for the huge

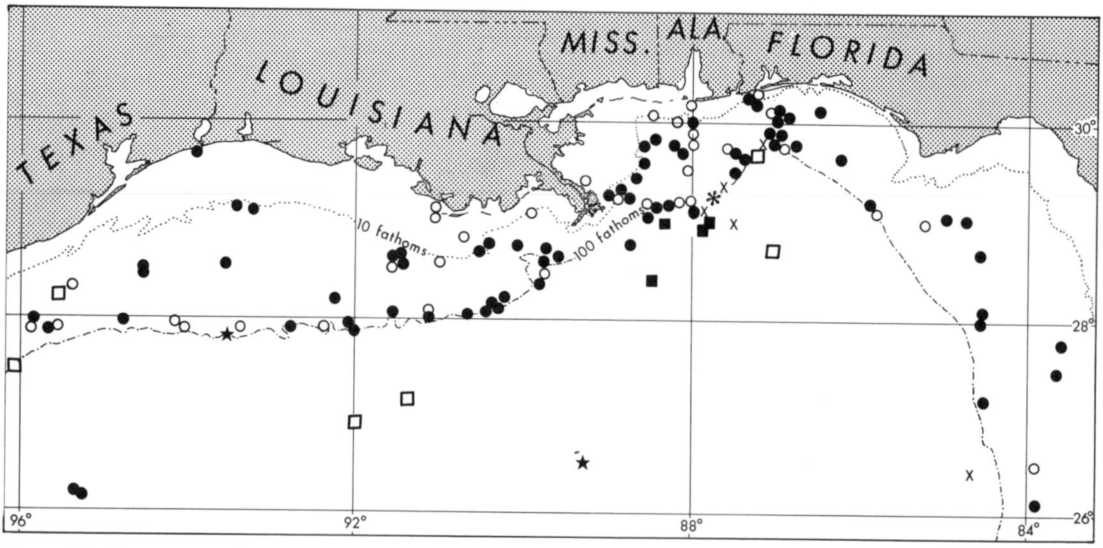

Map 48 Cetaceans observed in the northern Gulf of Mexico from aboard the U.S. National Marine Fisheries Service research vessels, the M/V *Oregon* and M/V *Oregon II*, 1950–1970; ●, Spotted Dolphin *(Stenella plagiodon);* ○, Atlantic Bottle-nosed Dolphin *(Tursiops truncatus);* ★, False Killer Whale *(Pseudorca crassidens);* □, Common Dolphin *(Delphinus delphis);* ■, Giant Sperm Whale *(Physeter catodon);* ✻, Fin-backed Whale *(Balaenoptera physalus);* ×, unidentified whales. Not plotted were 97 unidentified dolphins, only a few of which were appreciably beyond the 100 fathom line. (Data courtesy of Richard B. Roe, Exploratory Fishing and Gear Research Base, Bureau of Commercial Fisheries Service, Pascagoula, Mississippi.)

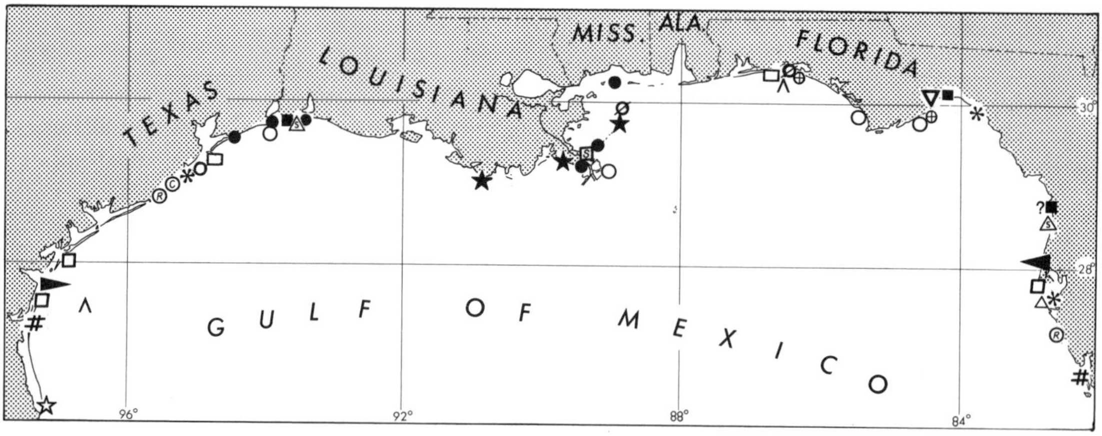

Map 49 Locations at which certain cetaceans referred to in the text have stranded or have been observed on the northern Gulf Coast: ∅, Goose-beaked Whale *(Ziphius cavirostris);* #, Antillean Beaked Whale *(Mesoplodon europaeus);* ✻, Rough-toothed Dolphin *(Steno brenadensis);* O, Giant Sperm Whale *(Physeter catodon);* □, Pygmy Sperm Whale *(Kogia breviceps);* ▭, Dwarf Sperm Whale *(Kogia simus);* Ⓢ△, Gray's Dolphin *(Stenella coeruleoalba);* ▶, Bridled Dolphin *(Stenella frontalis);* ⊕, Long-beaked Dolphin *(Stenella longirostris);* ☆, Pygmy Killer Whale *(Feresa attenuata);* ∧, Common Killer Whale *(Orcinus orca);* ◀, Risso's Dolphin *(Grampus griseus);* ●, Fin-backed Whale *(Balaenoptera physalus);* ⓢ, Sei Whale *(B. borealis);* ▽, Bryde's Whale *(B. edeni);* ©, Blue Whale *(B. musculus);* ■, Little Piked Whale *(B. acutorostrata);* ★, unidentified rorquals *(Balaenoptera* sp. ?); △, Humpback Whale *(Megaptera novaeangliae);* ®, Black Right Whale *(Balaena glacialis).*

spermaceti organ, or melon. The single nasal aperture is located to the left of the median line, with the result that the spout is directed obliquely forward. (Figures 138 and 139)

Sexual characters and age criteria.—Extreme sexual dimorphism is exhibited by this species, in which the female seldom attains a size much greater than half that of the adult male, although exceptions are known (Nishiwaki, 1972, cites a female that was nearly 56 feet in length). The penis, which is retractable, measures six feet in length (Figure 140). In the female the urethral and vaginal orifices are

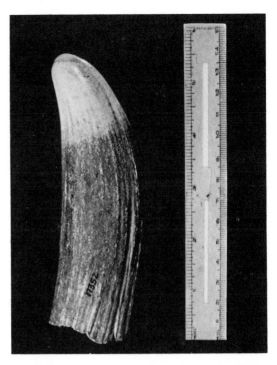

Figure 139   Tooth of Giant Sperm Whale *(Physeter catodon)*, LSUMZ 17352, locality unknown. (Photograph by Joe L. Herring)

separate. The teats, which are two in number, are located in deep, longitudinal slits on each side of the vagina. Sexual maturity in the female is reached when she attains a length of approximately 28 feet, that is, in about her eighth year. Most females follow a five-year reproductive cycle. They mate in May of year I, give birth after a gestation period of 16 months in September of year II, nurse the calf for two years, or until September of year IV, rest for eight months, and are reimpregnated in May of year V. Some females, though, are on a two-year cycle and are reimpregnated in May of year III while they are still nursing (Scheffer, 1969). Usually the newborn calf is 14 feet long and tips the scale at one ton.

Status and habits.—Giant Sperm Whales were once quite numerous in the Gulf of Mexico, enough so to justify full-scale whaling opera-

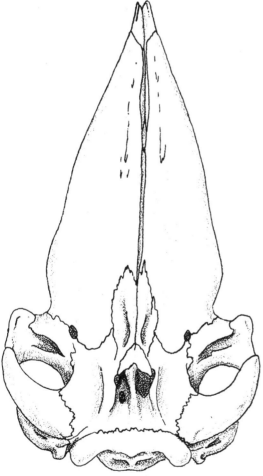

Figure 138   Skull of Giant Sperm Whale *(Physeter catodon)*. Approx. ¹⁄₂₅ ×. (After Hall and Kelson, 1959)

tions. Townsend (1935), in summarizing 160 years of whaling, sifted through logbooks kept by captains of American whaling ships. He drafted charts of the oceans of the world on which he plotted the latitude and longitude and indicated the time of the year and the kinds of whales killed. Somewhat surprising were the Giant Sperm Whale maps, which show great clusters of dots in the Gulf of Mexico, many of them directly off the coast of Louisiana, mainly in the period from April through July but with some catches recorded in March (Map 47).

Although no longer common in the Gulf of Mexico and thus no longer subject to commercial exploitation there, the species has been observed on several occasions in recent years off the mouth of the Mississippi River by fishermen and by personnel of the National Marine Fisheries Service exploratory research vessels, the M/V *Oregon* and its successor M/V *Oregon II*. The positions at which Giant Sperm Whales were observed from aboard these vessels in the period 1950–1969 are shown in Map 48. In addition, Mr. Lester J. Plaisance of Golden Meadow, while deep-sea fishing 20 to 25 miles off South Pass of the mouth of the Mississippi River in July 1959, obtained a short color film strip of several Giant Sperm Whales. The film has been inspected by Drs. F. C. Fraser, N. A. Macintosh, R. Gambell, and S. G. Brown, in the British Museum (Natural History), and the identification of the whales confirmed.

To my knowledge only two strandings have occurred on the central northern Gulf Coast that are a matter of record. One was an adult male that foundered approximately eight miles off the mouth of Sabine Pass on 8 March 1910 (Newman, 1910). Captain Cott Plummer, member of a long line of distinguished mariners of the Sabine area, had himself rowed in a small boat alongside the 63-foot 5-inch monster, whereupon he climbed aboard its back and succeeded in attaching a heavy hawser around its tail. The dying whale was then towed to Port Arthur, where it proceeded to attract legions of curious spectators from throughout East Texas and southern Louisiana (Figure 140). Gate receipts showed that over 32,000 people viewed the whale in one day. On 20 March, after advanced decomposition had rendered the carcass an odiferous mass of flesh that could be smelled for miles around, it was towed back into the Gulf to a point off the Houston ship channel, where it was skinned and the hide mounted on a wooden frame. Captain Plummer placed the "exhibit" on a barge with the idea of "taking it around-the-world," but it wound up somewhere on the Mississippi River, where the barge sank, whale and all.

Although the whale was actually brought ashore on the Texas side of the Louisiana-Texas state line, it was, for all practical purposes, in waters adjacent to Louisiana when it was first found floundering in shallow water off Sabine Pass. As a matter of fact, whether the middle of the Sabine River or the west bank of that river is the boundary between Texas and Louisiana is still the subject of litigation between the two states.

The other stranding took place on 6 November 1960 at the mouth of Thomasin Bayou near the outlet of the North Pass of Pass a Loutre in the Mississippi River delta (latitude 29°13′ N, longitude 89°03′ W), where it was inspected by Curtis Roberts of Venice, Louisiana. Mr. Roberts has described the animal to me in detail, particularly the disproportionately small lower jaw and its teeth, and there is no doubt regarding its identity as a Giant Sperm Whale. It was said to be about 40 feet in length.

Two particularly memorable books have been written on the Giant Sperm Whale. One is Herman Melville's *Moby-Dick*, perhaps the greatest epic of the sea ever composed. The other is V. B. Scheffer's *Year of the Whale*. The first of these classics reflects the author's superb knowledge of all the then known facts pertaining to whales and whaling. The imaginative portion of the book tells of Captain

Ahab's long and crazed quest for a ferocious and monstrous beast, an albino Giant Sperm Whale, that maimed him on an earlier voyage and against which he swore vengeance. But the book is more than fiction. Page after page is chock-full of whaling lore and facts about whales. The second book, the one by Scheffer, is a modern account by a scientist of repute that incorporates most of what is known today about the Giant Sperm Whale. Like *Moby-Dick,* it has a central imaginary figure, a baby whale named Little Calf that the author carries through the first year of its life. This part is fictional, but it is based on unadulterated facts. Interwoven into the story, in the most delightful and masterful prose, is a sensitive, revealing commentary on the life history and behavior of the species. The book is nature writing at its best.

The Giant Sperm Whale is a deep diver. It has been known to have become entangled in transoceanic telephone cables at depths of nearly three-quarters of a mile. Its dives not only take it down through zones of green and purple twilight to utter blackness below but may last anywhere from twenty minutes to over an hour. When it surfaces it may blow 20 to 70 times before making another descent. The spout is produced by the condensation of the moisture in the warm, stale air that is being exhaled, combined, according to some authorities, with a mucous foam from the sinuses.

The food of the Giant Sperm Whale consists of a variety of squids, octopuses, sharks, skates, and bony fishes. The species is highly gregarious and often travels in aggregations numbering 100 or more. As many as 1,000

Figure 140 Male 65-foot-5-inch Giant Sperm Whale *(Physeter catodon)* that was found floundering eight miles off the mouth of Sabine Pass on 8 March 1910. It was towed to Port Arthur, where it is here being prepared for public display. (Photograph courtesy of Trost Studio, Port Arthur)

individuals have been recorded in congregations of years gone by. The sperm whale family is a loose social group of about 30 individuals, which are part of the great school comprised of many families. The basic units, called harems by whale hunters, include young males and young females, pregnant cows, cows nursing calves, and an old bull, the harem master, who usually stations himself on the perimeter of the herd.

A gummy substance called ambergris forms in the large intestine of the Giant Sperm Whale and no other cetacean, and was once the subject of speculation as to its origin when it was found floating in pieces of various sizes on the surface of the sea or washed ashore on a beach. When first expelled from the digestive tract it possesses the consistency of thick grease. After exposure to the air it hardens and acquires its sweet, earthy smell. The word ambergris refers to the variegated orange-yellow and grayish color of the substance. Though it was once believed to have medicinal qualities, its modern usage has been in connection with the manufacture of perfumes. Of special interest here is the fact that the word ambergris is said to have given rise to the name of an island, Ambergris Cay, just south of the Gulf of Mexico, off extreme northeastern British Honduras, because of the quantities of the substance gathered along its shores, thereby reflecting the Giant Sperm Whale's former abundance in Gulf and Caribbean waters.

Physeter catodon Linnaeus

[*Physeter*] *Catodon* Linnaeus, Syst. nat., ed., 10, 1, 1758:76; type locality, Kairston, Orkney Islands, Scotland (by restriction, Thomas, Proc. Zool. Soc. London, 1911:157).

Specimens examined from Louisiana. — No specimens or parts thereof preserved, but see pages 317–318.

[Genus *Kogia* Gray]

Dentition. — Eight to 16 sharp, curved teeth are present in each side of the lower jaw.

[PYGMY SPERM WHALE]   Figure 141
(*Kogia breviceps*)

In a recent revision of the genus *Kogia,* Handley (1966) has demonstrated beyond any reasonable doubt that two well-defined species, *breviceps* and *simus,* are recognizable instead of only one, as was previously generally believed. Of the two, *breviceps* is decidedly the larger, and the two species differ in several other important particulars. According to Handley, some of the more distinctive differences between adult *breviceps* and adult *simus* are, respectively: total length in feet from snout to notch in flukes, 8.6–10.9 versus 6.9–8.6; weight, 700–898 pounds versus 299–378 pounds; dorsal fin, low and posterior to center of back versus high and near center of back; condylobasal length, 391–469 mm versus 262–302 mm; mandibular teeth, 12–16 (rarely 10 or 11) pairs versus 8–11

Figure 141   Pygmy Sperm Whale *(Kogia breviceps).*

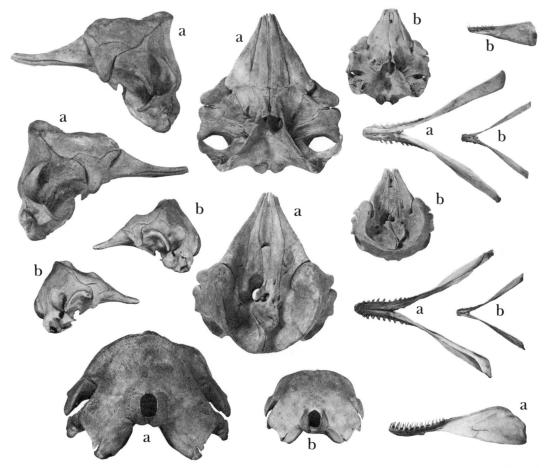

Figure 142   Skulls and mandibles (8 views) of the two species of *Kogia*: (a) *K. breviceps*, USNM 22559, adult male (condylobasal length, 443 mm), Dam Neck Mills, Virginia; (b) *K. simus*, USNM 21627, adult male (condylobasal length, 279 mm), Kitty Hawk, North Carolina. (Photographs courtesy of Charles O. Handley, Jr., from Handley, 1966)

(rarely 13) pairs; mandibular symphysis, long (86–120 mm) and ventrally keeled versus short (37–46 mm) and without ventral keel (Figure 142). The bracket-shaped mark on each side of the head, between the eyes and the flippers, is apparently present in both species. (Figure 143)

Specimens of the genus *Kogia* found on the northern Gulf Coast and identified by Handley (pers. comm.) as *breviceps* are a female stranded on Mustang Island, Texas, on 27 December 1948 (Gunter et al., 1955), a male also stranded on Mustang Island, 15.8 miles south of the south jetty at Port Aransas,

Texas, on 28 January 1965 (Raun et al., 1970), and a female that gave birth to an offspring while stranded at St. Petersburg Beach, Florida, on 26 November 1949 (Moore, 1953). A specimen found on Padre Island, Texas, on 7 February 1954 (Gunter et al., 1955) was questionably of this species.

Kogia breviceps (Blainville)

Physeter breviceps Blainville, Ann. Anat. Phys., 2, 1838:337, pl. 10; type locality, Cape of Good Hope, South Africa.

Kogia breviceps, Gray, Zoology of the voyage of H.M.S. Erebus and Terror . . . , 1 (Mammalia), 1846:22.

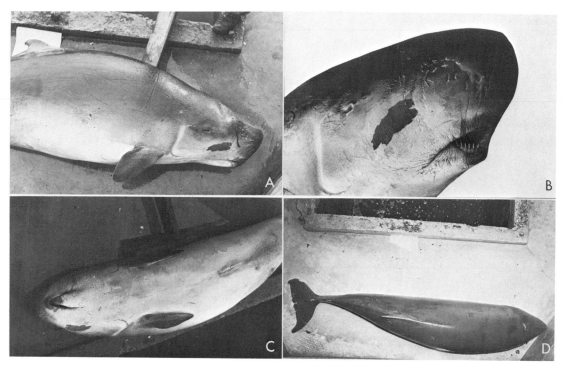

Figure 143   Immature Pygmy Sperm Whale *(Kogia breviceps)* stranded near Imperial Beach, California: (A) lateral view, showing bracket-shaped mark behind the eye and parasitic copepod; (B) close-up oblique view of mouth showing teeth; (C) ventral view of mouth showing small lower jaw; (D) dorsal view showing position of dorsal fin posterior to center of the body. (Photographs courtesy of Carl L. Hubbs and Scripps Institution of Oceanography)

Figure 144   Dwarf Sperm Whale *(Kogia simus)*. (Portrait by H. Douglas Pratt is based on photographs by David K. Caldwell of animals that stranded on the Atlantic Coast of Florida in November 1970 and February 1973.)

[DWARF SPERM WHALE]   Figure 144
(Kogia simus)

Only two specimens from the northern Gulf Coast are considered by Handley (pers. comm.) definitely referable to *simus*. One was a seven-foot individual that stranded on Galveston Island, Texas, on 6 December 1957 (Caldwell et al., 1960). The other was a specimen found on the beach 3 miles east of Destin, Okaloosa County, Florida, on 1 April 1958, and reported by Caldwell et al., 1960, as *breviceps* but which is now considered by Caldwell (pers. comm.) as an example of *simus*.

Diagnostic differences between this species and *K. breviceps* are given in the preceding account and illustrated in Figure 142.

Kogia simus (Owen)

Physeter (Euphysetes) simus Owen, Trans. Zool. Soc. London, 6, 1866:30, pls. 10–14; type locality, Waltair, Madras.

Kogia simus, Beddard, Mammalia, 1902:367.—Yamada, Sci. Rep. Whales Res. Inst., 9, 1954:37, figs. 3, 5, 8–12 and plate.—Handley, in Whales, dolphins, and porpoises, 1966:68.

Cogia [sic] *simus,* Ogawa, Botany and Zoology [in Japanese], 4, 1936: 2017–2024 [not examined].

The Porpoise and Dolphin Family
Delphinidae

SOME MAMMALOGISTS divide among several families the genera that are here included in the family Delphinidae. Fraser and Purvis (1960b) erected the family Stenidae to accommodate *Steno* and *Sousa,* but later Fraser expressed the opinion that further studies might require that even *Sousa* be accorded familial rank. Rice (1967) tentatively referred *Sotalia* to Stenidae, and he placed *Neophocaena, Phocoena,* and *Phocoenoides* in a separate family, Phocoenidae. Because of obvious uncertainties and the general lack of agreement on the familial classification of this group of cetaceans, I am employing the traditional arrangement of Hall and Kelson (1959) and other authors, wherein all these genera are placed in the single family Delphinidae. Of the six genera mentioned above, none has yet been found in Louisiana waters; so the matter is of consequence in the present connection only in showing the scope of the family Delphinidae as it is here employed. Moreover, the genus *Steno* is almost certain to appear sooner or later in Louisiana's coastal waters, for *S. bredanensis* has been recorded on the coasts of both Florida and Texas.

The family Delphinidae, in its broad sense, contains in North American waters 13 genera and 30 species. Of these, only 5 genera and 6 species have been recorded along our coast or in our offshore waters. The family is characterized by a number of anatomical characters, including the following: size small to medium; tail flukes with a medial notch; pectoral appendage sickle-shaped; dorsal fin usually well developed; mandibular symphysis less than 20 percent as long as the rami; numerous functional teeth in upper and lower jaws; maxilla not crested; and two cervical vertebrae fused.

The common name dolphin is usually applied to any of the small, toothed whales of the family Delphinidae having a beaklike snout and a slender body form, whereas the name porpoise generally refers to groups having a blunt snout and a stocky body. But these considerations are not always heeded. The trim, long-snouted members of the genus *Stenella*, two species of which are definitely known to occur in our offshore waters, are called porpoises by some authors (Hall and Kelson, 1959; Hall, 1965; and others), when strictly speaking they should be called dolphins, as I have done herein. Unfortunately, the name dolphin causes some confusion because it is also applied to an important group of fishes, members of the genus *Coryphaena*. One species, *C. hippurus*, is common in Louisiana waters.

[Genus *Steno* Gray]

Dentition.—Twenty to 27 teeth are located in each toothrow. They are spatulate and curved forward basally. Their sides are rough, with the texture of sandpaper. The maximum width of the teeth is approximately 10 mm, but their diameter at gum line is only 6 to 7 mm.

[ROUGH-TOOTHED DOLPHIN]　　Figure 145
(Steno bredanensis)

The Rough-toothed Dolphin has been recorded only three times in the Gulf of Mexico. The United States National Museum contains a skeleton from Tampa obtained on 22 March 1902 (Miller, 1924); on the night of 29 May 1961, a group of 16 stranded on the upper east coast of the Gulf of Mexico be-

Figure 145　Rough-toothed Dolphin *(Steno bredanensis)*.

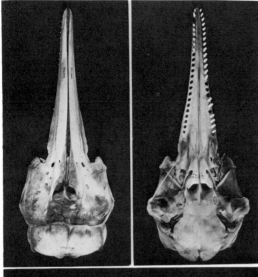

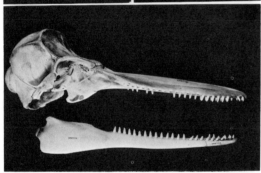

Figure 146 Skull of Rough-toothed Dolphin *(Steno bredanensis)*, USNM 395770. Total length of skull 533 mm (21 in.). (Photograph by Joe L. Herring)

tween Fenholoway River and Cow Creek, approximately one mile NNW of Rock Island, Taylor County, Florida (Layne, 1965); and a single individual stranded on Galveston Island in late June, 1969 (Schmidly and Reeves, in press).

This species is purplish black above with yellowish white spots on the sides. The beak and ventral surfaces are white, tinged with purple and rose color. The snout is not set off from the head by a groove or angle of demarcation surrounding the base, as it is in *Tur-*

siops, but is long, narrow, and laterally compressed. The mandibular symphysis is one-fourth the length of the rami. The skull resembles that of *Tursiops truncatus*, but the rostrum is decidedly longer with its distal one-third prominently attenuated (Figure 146). The sandpaperlike lateral surfaces of the teeth are highly diagnostic. The total length of the animal is approximately eight feet.

Steno bredanensis (Lesson)

Delphinus bredanensis Lesson, Histoire naturelle . . . des mammifères et des oiseaux découverts depuis 1788, cétacés, 1828:206; type locality, coast of France.
Steno bredanensis, Miller and Kellog, Bull. U.S. Nat. Mus., 205, 1955:657. — Ellerman and Morrison-Scott, Checklist of Palearctic and Indian mammals, 1758 to 1946, 1955:734.

Genus *Stenella* Gray

Dentition. — Thirty-four to 56 teeth are present on each side of both the upper and lower jaws.

SPOTTED DOLPHIN   Figure 147
(Stenella plagiodon)

Vernacular and other names. — The name Spotted Dolphin is applied to this species because of the fine flecking of white spots sometimes found on its dorsum and the black spots on its venter. It has also been called the long-snouted dolphin because of its long snout. The name spotted porpoise that is used by some authors is an inconsistent application of the term porpoise in place of dolphin and the name should be suppressed (see page 324). The word "dolphin" is akin to the Greek *delphin,* a name used by the Greeks in referring to small cetaceans. The word "porpoise" is derived from the Medieval Latin *porcopiscis,* meaning pig fish, through the Medieval French *porpois* and the Medieval English *porpoys.* The generic name *Stenella* comes from *Steno,* the generic name of another dolphin, plus the diminutive suffix -*ella,* meaning little, and hence means little *Steno.* The specific name *plagiodon* is a combination of two Latinized Greek words (*plagio,* meaning slanting, and *odon,* meaning tooth) and is possibly an allusion to the fact that the tips of the teeth are slightly incurved.

Distribution. — The species is found in the Atlantic Ocean from North Carolina southward to the Gulf of Mexico and the mainland shores of the Caribbean Sea thence to waters off South America. It is common in the Gulf of Mexico, in offshore waters. (Maps 48 and 50)

External appearance. — The Spotted Dolphin is dark purplish gray above, paler below. As the name implies, it is flecked with small spots of white on the dorsum. The venter is densely spotted or blotched with gray. The area immediately posterior to the blowhole is often only sparsely and finely spotted. The lower jaws (from tip to the corner of the mouth) and the tips and oral edges of the upper jaws are white, blotched with gray. In some individuals the light areas have a pinkish appearance, which Caldwell (1960) once thought to be associated with breeding condition, but he now tells me he is not so sure.

Measurements and weights. — The total length of this species sometimes reaches seven feet. Measurements of a single skull from Holly Beach, Cameron Parish, in the collections of McNeese State University at Lake Charles,

Figure 147   Spotted Dolphin *(Stenella plagiodon).* (Courtesy of Marineland of Florida)

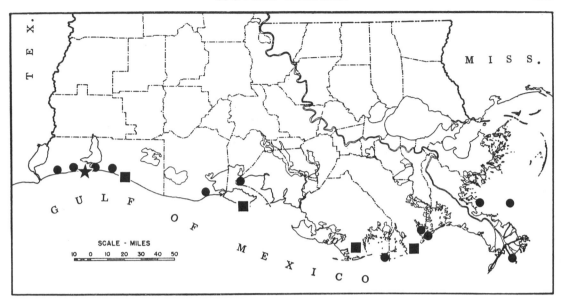

Map 50 Locations on the coast of Louisiana from which specimens have been examined of: ●, Atlantic Bottle-nosed Dolphin *(Tursiops truncatus);* ★, Spotted Dolphin *(Stenella plagiodon)* and Gray's Dolphin *(S. coeruleoalba);* and ■, Short-finned Pilot Whale *(Globicephala macrorhyncha).*

are as follows: length of rostrum, 250; width of rostrum at base, 94; width of rostrum 60 mm anterior to level of anteorbital notches, 65; width of rostrum at the middle, 54; width of premaxilla at same point, 29; tip of rostrum to nares, 295; tip of rostrum to median posterior border of pterygoids, 315; preorbital width, 171; postorbital width, 187; orbital width, 171; zygomatic breadth at posterior points of attachment, 185; greatest width of premaxilla, 75; width of braincase across parietals, 145; greatest width of temporal fossa, 68; greatest height of temporal fossa, 36; greatest height of skull, 105; internal cranial length, 133; length of upper right toothrow, 214+; length of upper left toothrow, 208+. Kellogg (1940) states that adults weigh approximately 280 pounds.

Cranial and other skeletal characters.—The skull of this species is easily separable from that of *S. coeruleoalba* on the basis of its somewhat larger size in most dimensions and its fewer teeth (34–37 instead of 44–50). It has fewer vertebrae (68–69 instead of approximately

76). The teeth of *S. plagiodon* measure approximately five mm in diameter, while those of *S. coeruleoalba* are about three mm in diameter. The temporal fossa of *S. plagiodon* is considerably larger than that of *S. coeruleoalba.* The pectoral appendages of *S. plagiodon* are said to be larger than those of *S. coeruleoalba,* but, David K. Caldwell informs me that two specimens of each that he measured showed the appendages of *plagiodon* to be only slightly if any larger. (Figure 148)

Sexual characters and age criteria.—The sexes are quite similar in appearance. The young differ from the adults in being light gray above and whitish below. They are completely devoid of any spotting (Caldwell and Caldwell, 1966).

Status and habits.—The Spotted Dolphin is apparently common in Louisiana's offshore waters. It is definitely known to prefer blue water. I have received frequent reports from deep-sea fishermen operating off the mouth of the Mississippi River and off Grand Isle, describing dolphins they have seen that seem unmistakably to be this species. Observations of the Spotted Dolphin made aboard the commercial fisheries research vessels, the M/V *Oregon* and the M/V *Oregon II*, in the northern part of the Gulf of Mexico, including the observations of Caldwell (1960), are plotted in Map 48. It is often found in pods of a dozen or more that break water in unison in graceful arcs.

Very little information is available on the biology of the species. At Marineland of Florida, near St. Augustine, Spotted Dolphins have been maintained for periods of time, but not nearly so successfully as have bottle-nosed dolphins. The Spotted Dolphin has been said to make far fewer sounds than does

the bottle-nose, which is, of course, renowned for its repertoire of clicks, creaks, blats, whistles, and other vocalizations. On the contrary, Wood (1954), Caldwell and Caldwell (1971b), Caldwell et al. (1973), and others have demonstrated a variety of low intensity, pulsed sounds and narrow-band whistles emitted by this species. Some of the differences between the Spotted and Atlantic Bottle-nosed Dolphins in the sounds they produce may have ecological significance, for the former species lives in offshore waters and the latter resides most of the time in protected inshore waters.

The small calf often swims close alongside its mother between a pectoral flipper and a caudal fluke, the two rising and sinking in unison. Often a suckerfish, *Remilegia,* which is sometimes as much as two feet in length, attaches itself to the Spotted Dolphin (Mahnken and Gilmore, 1960) and other cetaceans as well (Rice and Caldwell, 1961). Suckerfish adhering to a foreflipper or to the side of the tail doubtless gave rise to the erroneous belief that the dolphin calf grasps a flipper or a caudal fluke of the mother and is towed along.

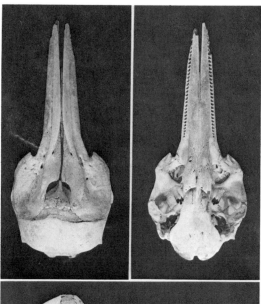
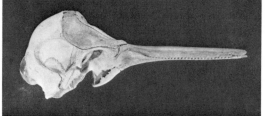

Figure 148 Skull of Spotted Dolphin *(Stenella plagiodon),* MSU 2139, Johnsons Bayou, Cameron Parish, spring 1963. Total length of skull 425 mm (16.75 in.). (Photograph by Joe L. Herring)

Stenella plagiodon (Cope)

Delphinus [*(Delphinorhynchus)*] *pernettensis* Blainville, *in* Desmarest, Nouv. Dict. Hist. Nat., 9, 1817:154.

Delphinus Pernettyi, Desmarest, Mammalogie, 1822:543 [proper emendation of spelling of specific name].

Delphinus plagiodon Cope, Proc. Acad. Nat. Sci. Philadelphia, 18, 1866:296; type locality, unspecified.

Stenella plagiodon, Kellogg, Nat. Geog., 77 (1), 1940:83.

The species is monotypic. Hershkovitz (1966) believes that the name *Delphinus pernettensis* Blainville, 1817, along with its valid emendation to *D. pernettyi,* Desmarest, 1822, applies to this dolphin. If he is correct, *Delphinus pernettyi* Blainville would have priority over *Delphinus plagiodon* Cope, 1866. Were it not for considerations mentioned beyond, the name of the species would become *Stenella pernettyi* (Blainville). However, as pointed out by Caldwell and Caldwell (1966), no useful purpose is served at this late date by

Figure 149 Gray's Dolphin *(Stenella coeruleoalba).*

reviving *pernettyi.* The drawing of the dolphin by the collector Pernetty on which Blainville based the name *pernettensis* (=*pernettyi*) is much too crude to be identified with certainty as being the same animal as *plagiodon.* Nor can the name be assigned with confidence to any other known delphinid. Moreover, there is no need to try to do so, for under Article 23b of the International Code of Zoological Nomenclature for 1964, the name *pernettyi* is a *nomen oblitum* inasmuch as more than 50 years had elapsed since it was employed, even by Philippi in 1893. The only other major authors to consider the name *pernettyi,* aside from Hershkovitz (1966), were True (1885) and Fraser (1950). True definitely considered unidentifiable the dolphin to which this name had been applied. Fraser also discussed the shortcomings of Pernetty's drawing, and he has informed me (pers. comm.) that he too regards the animal depicted in the drawing unidentifiable. If neither of these two expert cetologists could find any basis for considering the name *pernettyi* applicable to the animal we now call *plagiodon,* I would think that the name *pernettyi* should be permanently suppressed. Furthermore, under Article 23b(ii) of the Rules, as a *nomen oblitum,* it is not to be used unless the Commission so directs.

Specimens examined from Louisiana.—Total 1, as follows: *Cameron Par.:* Holly Beach (cranium only, MSU).

GRAY'S DOLPHIN

Figure 149

(Stenella coeruleoalba)

Vernacular and other names.—The scientific name *Stenella coeruleoalba* (Meyen) replaces *Stenella styx* (Gray) for reasons of nomenclatural priority. The latter has long gone by the name of Gray's Dolphin or Gray's porpoise and, for this reason, I have retained the name Gray's Dolphin as the vernacular name for this species, despite the fact that I have seen the names Meyen's dolphin, blue dolphin, blue-white dolphin, and euphrosyne dolphin applied to it. The specific name *coeruleoalba* comes from two Latin adjectives (*coerulea,* dark-colored or dark blue, and *alba,* white), and refers to the animal's coloration.

Distribution.—In the Atlantic, Gray's Dolphin occurs from waters off southern Greenland to the mouth of Río de la Plata and from the North Sea to waters off southern Africa. In the Pacific it is found from the Bering Sea to California waters and from Siberian waters to those off New Zealand. I know of only two records from the Gulf of Mexico, the ones reported here (see beyond and Map 49).

External appearance.—This small dolphin is bluish black above, white below. A narrow dark band runs from the eye along the side to the vent, with a short stripe branching off above the base of the pectoral appendage. A double bar of pigment extends from the flipper toward the eye, but the lower bar does not quite reach the eye. Dark pigment encircles the eye, and the region of the mouth is pigmented. The darkly pigmented flippers insert on white portions of the body. (Fraser, 1966; Fraser and Noble, 1970)

Measurements. — I am indebted to Dr. F. C. Fraser for the external dimensions of two British specimens, a male and a female, that measured, respectively, as follows: total length, 1,968.5 (6.46 feet), 1,663.7 (5.46 feet); tip of upper jaw to center of eye, 349.2 and 292.1; tip of upper jaw to angle of gape, 292.1 and 203.2; tip of upper jaw to blowhole, 349.2 and 279.4; tip of upper jaw to anterior insertion of flipper, 742.9 and 660.4; length of flipper, 222.2 and 215.9; width of flipper, 85.7 and 76.2; height of dorsal fin, 146.0 and 152.4; length of base of dorsal fin, 247.6 and 228.6; width of tail (fluke tip to fluke tip), 361.9 and 342.9. The single skull from Louisiana has the following dimensions: condylobasal length, 380; length of rostrum from level of the anteorbital notches, 220; width of rostrum at base, 92; width of rostrum 60 mm anterior to anteorbital notches, 62.5; width of rostrum at middle, 54; width of premaxillae at middle of rostrum, 26; tip of rostrum to anterior edge of nares, 265; tip of rostrum to median posterior edge of pterygoids, 277; preorbital width, 165; postorbital width, 182; orbital width, 165; greatest length of temporal fossa, 52; greatest height of temporal fossa, 36; width of blowhole, 44; zygomatic breadth, 180; greatest width of premaxillae, 74; width of braincase across parietals, 142; length of temporal fossa, 57; height of temporal fossa, 40; greatest height, 100; internal length of cranium, 111; number of teeth, upper right, *ca.* 45; number of teeth, upper left, *ca.* 45; length of upper toothrows, 195; posterior end of upper toothrows to tip of premaxilla, 195; number of teeth lower right toothrow, *ca.* 45; number of teeth lower left toothrow, 47; length of lower right tooth-

row, 208; length of mandible, 338; coronoid height, 59.9; length of symphysis, *ca.* 32.

Cranial characters. — See previous account and Figure 150.

Status and habits. — A carcass that my wife and I found on the edge of the surf four miles west of Holly Beach on 26 April 1970 furnishes the second record of this species for the Gulf of Mexico. It was immediately recognized as something unusual, if for no other reason than its great number of small, prominently recurved teeth (approximately 45 in each toothrow, both above and below). The animal had been dead for at least several days and hence showed none of its original

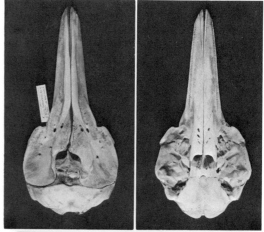

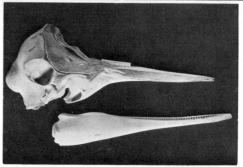

Figure 150 Skull of Gray's Dolphin *(Stenella coeruleoalba),* LSUMZ 15607, 4 miles W Holly Beach, Cameron Parish, 26 April 1970. Total length of skull, 380 mm (15 in.). (Photograph by Joe L. Herring)

color, being almost black. We were able to retrieve the skull and most of the other skeletal parts. Since we did not have with us a tape or ruler, we were not able to obtain precise measurements, but we estimated the total length to be six feet. Later I took the skull and part of the skeleton to the British Museum (Natural History) in London, where F. C. Fraser, world authority on cetaceans, kindly examined it with me and positively identified it as the present species. The only other record of this species for the Gulf is that of Caldwell and Caldwell (1969) of an individual found on the coast about 50 miles north of Tampa Bay on 19 October 1968, following the passage of Hurrican Gladys. It was reported under the name *Stenella styx*.

Stenella coeruleoalba (Meyen)

Delphinus coeruleo-albus Meyen, Nova Acta Acad. Caesareae Nat. Curios., 16 (2), 1833:609, 610, pl. 43, fig. 2; type locality, Río de la Plata, Argentina.

Stenella coeruleo-albus [sic], Tomilin, Mammals of USSR and adjacent lands, 9 (Cetacea), 1957:554, figs. 96–98.

Stenella caeruleoalba [sic], Scheffer and Rice, A list of the marine mammals of the world. U.S. Fish and Wildl. Serv., Special Sci. Rept.—Fisheries, 431, 1963:6.

Delphinus styx Gray, in The Zoology of the voyage of H.M.S. Erebus and Terror..., 1 (Mammalia), 1846:39, pl. 21.

Stenella caeruleoalba [sic], Rice and Scheffer, A list of the marine mammals of the world. U.S. Fish and Wildl. Serv., Special Sci. Rept.—Fisheries, 579, 1968:8.

The species is generally regarded as being monotypic, but Scheffer and Rice (1963) recognize *S. c. euphrosyne* (Gray) as a valid race. *Stenella coeruleoalba* is the *Stenella styx* (Gray) of most authors, including Miller and Kellogg (1955) and Hall and Kelson (1959).

Specimens examined from Louisiana.—Total 1, as follows: *Cameron Par.*: 4 mi. W Holly Beach (skull and partial body skeleton, LSUMZ).

[BRIDLED DOLPHIN]
(*Stenella frontalis*)

The only record of the Bridled, or Cuvier's, Dolphin from the Gulf of Mexico is that of three individuals found stranded on Padre Island, in the vicinity of Yarbrough Pass,

Texas, on 12 September 1971 (Schmidly et al., 1972).

The nomenclature and taxonomy of the species of the genus *Stenella* are still confused, but cetologists are indebted to F. C. Fraser, as in so many other instances, for his elucidation of the characters of *frontalis* (Fraser, 1950). It is from Fraser's comprehensive "Atlantide" paper that the following brief notes are extracted. The color of the dorsum of *frontalis* is black, with many irregular small gray flecks. The black of the back extends forward as a tapered area with the apex on the basal two-thirds of the snout. The snout anterior to the black area and the margins of the upper jaws are whitish pink but are otherwise without pigment. The forward ends of the lower jaws are white but possess dark spots posteriorly. Dark areas surround each eye and run forward to the base of the snout. A dark band, with a subsidiary dark line above, extends from the insertion of each flipper to the angle of the mouth. The pectoral flippers are dark on both surfaces and insert on the gray of the sides. The boundary between the black of the upperparts and the gray of the sides extends tailwards from the base of the snout, curving regularly at first upward well above the eye, then downward so that in the region between the flipper insertion and the dorsal fin it is at about one-third the body depth from the ventral margin. It curves upwards again but is not clearly defined towards the caudal peduncle. Ventrally, apart from the whiteness of the chin, the general color is dark gray with abundant darker spots and fewer scattered white spots. The dental formula of Fraser's "Atlantide" specimen of *frontalis*, showing the number of teeth visible in the animal in the flesh, plus the additional teeth found in the cleaned skull was:

$$\frac{38+3-38+4}{38+1-38+1}$$

S. frontalis bears some resemblance to *S. plagiodon*, mainly in that it exhibits some degree of spotting. The latter species, though,

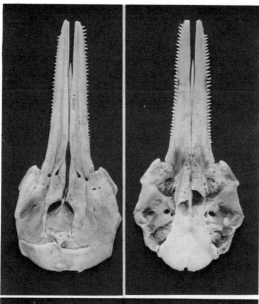

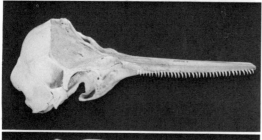

Figure 151 Skull of Bridled, or Cuvier's, Dolphin (*Stenella frontalis*), TCWC 25575, Yarbrough Pass, Padre Island, Texas, 12 September 1971. Total length of skull, 384 mm (15.12 in.). (Photograph by Joe L. Herring)

differs in having only the tip of the chin white in adults, in showing a dark band from the angle of the mouth to the flipper, and in possessing a white instead of gray belly. The skull of one of the Padre Island specimens is depicted in Figure 151.

Stenella frontalis (G. Cuvier)

D[elphinus]. *frontalis* G. Cuvier, Le regne animal . . . , 1, 1829:288; type locality, off Cape Verde Islands, North Atlantic Ocean.
Stenella frontalis, Fraser, Atlantide Rept. No. 1, 1950:61.

[LONG-BEAKED DOLPHIN]   Figure 152
(Stenella longirostris)

The stranding of 36 Long-beaked Dolphins on Dog Island, Franklin County, Florida, on 23 September 1961 (Layne, 1965), was apparently the first record of the occurrence of this species in the Gulf of Mexico. The site where the stranding took place is located about four miles off the mainland at Carabelle, between St. George Island on the west and Alligator Point on the east.

As is often the case, there was little indication of the cause of the stranding. One of the animals, however, had a small hole about the size of a .38 caliber bullet on the side of the tail stock, which suggested the possibility that the dolphins had been shot at either before or after the stranding. By the time of the first examination, the specimens had undergone some postmortem change in coloration as a result of drying, exposure to the sun, or discoloration from blood, which should be considered in evaluating the descriptions given by Layne. The color of the specimens was generally dark gray to blackish on the back and upper sides. One of the females showed the most distinctive color pattern, and its details were most evident on the side upon which the carcass had lain. This individual had a series of small, dark, relatively discrete pigment spots extending horizontally from the eye to a point about halfway to the origin of the dorsal fin and a rather broad band of dark gray passing obliquely from the eye to the base of the flipper. This band was bordered anteriorly by a rather conspicuous patch of light gray. The eye was ringed with black, which extended forward to meet the similar mark from the opposite side of the forehead. Counts of the visible teeth in a female (now FSU 284) and a larger male (now UF 7861) were made in the flesh, and the numbers obtained were, respectively:

$$\frac{48-48}{52-50} \text{ and } \frac{45-44}{49-49}$$

Figure 152 Facsimile reproduction in fiberglass of Long-beaked Dolphin *(Stenella longirostris)* in the Louisiana State University Museum. The animal (now preserved as a complete skeleton, LSUMZ 17017) stranded just east of Fort Walton, Okaloosa County, Florida, on 23 March 1972. (Photograph by Joe L. Herring)

The alveoli counted in the cleaned skull of the male were:

$$\frac{56-56}{49-49}$$

The only other record of the species in the Gulf of Mexico is that of an adult male that stranded on the beach just east of Fort Walton Beach, Florida, on 23 March 1972. It was alive when found and an attempt was made to move it to the tanks of the Gulfarium in Fort Walton Beach, but it died en route. The specimen was turned over to the LSUMZ where a reproduction of it was made in fiber glass and the complete skeleton was preserved. Before postmortem changes in the coloration had taken place the animal was described as being grayish black above and whitish below. The all-black pectoral appendages inserted at the line of demarcation between the dark upperparts and the light-colored venter. No blackish lines extended, as is often the case in some species of *Stenella*, from the eyes toward the points of insertion of the pectoral fins. There were, though, shallow but elongated blotches of grayish white extending from the belly over the posterior dorsal borders of the pectoral fins and

another on the sides beneath the dorsal fin. The lips were blackish and a narrow black line, barely discernible against the grayish black ground color, extended from the anterior corner of each eye around the base of the rostrum for a distance of approximately four inches (in other words, the two lines on each side did not meet in the center of the forehead). The body and cranial measurements of this specimen are given in Tables 8 and 9.

Table 8 Selected Absolute and Proportional Body Measurements of Adult Male *Stenella longirostris* (LSUMZ 17017), Stranded near Fort Walton, Okaloosa County, Florida, 23 March 1972.

	Mm	Percent of TL
Total length	1865	100.0
Distance from tip of upper jaw to center of eye	300	16.0
Distance from tip of upper jaw to angle of gape	266	14.2
Distance from tip of upper jaw to external auditory meatus	343	18.3
Distance from center of eye to external auditory meatus	45	2.4
Distance from center of eye to center of blowhole	127	6.8

Table 8 Continued	Mm	Percent of TL
Distance from tip of upper jaw to blowhole	296	15.8
Distance from tip of upper jaw to anterior insertion of flipper	437	23.4
Distance from tip of upper jaw to posterior tip of dorsal fin	1060	56.8
Distance from tip of upper jaw to umbilicus	883	47.3
Distance from tip of upper jaw to genital aperture	1170	62.7
Distance from tip of upper jaw to center of anus	1328	71.2
Girth, on transverse plane intersecting base of flipper	796	42.6
Length of eye	18.3	0.9
Height of eye	8.0	0.4
Width of blowhole	20.1	1.0
Length of flipper from anterior insertion to tip	285	15.1
Length of flipper from axilla to tip	220	11.7
Maximum width of flipper	96	5.1
Height of dorsal fin	181	9.7
Length of base of dorsal fin	326	17.4
Width of tail (fluke tip to fluke tip)	427	22.8
Depth of notch between flukes	18.0	0.9

Table 9   Skull Measurements of Adult Male *Stenella longirostris* (LSUMZ 17017), Collected Near Fort Walton Beach, Okaloosa County, Florida, on 23 March 1972.

	Mm	Percent of Condylobasal Length	Percent of Parietal Width
Condylobasal length	399	100.0	297.3
Length of rostrum from level of anteorbital notches	250	62.7	186.3
Width of rostrum at base	75.5	18.9	56.3
Width of rostrum 60 mm anterior to level of anteorbital notches	53.7	13.5	40.0
Width of rostrum at middle	45.0	11.2	33.5

Table 9 Continued	Mm	Percent of Condylobasal Length	Percent of Parietal Width
Width of premaxillae at same point	18.2	4.5	13.6
Tip of rostrum to anterior edge of nares	276	69.2	205.7
Tip of rostrum to median posterior border of pterygoids	298	74.7	222.1
Preorbital width	143.4	36.0	106.9
Postorbital width	155.3	38.9	115.7
Orbital width	141.0	35.3	105.1
Width of blowhole	40.0	10.0	29.8
Zygomatic breadth at posterior points of attachment	141.7	35.5	105.6
Greatest width of premaxillae	65.4	16.4	48.7
Width of braincase across parietals	134.2	33.6	100.0
Greatest length of temporal fossa	44.9	11.3	33.5
Height of temporal fossa	39.2	9.8	29.2
Greatest height of skull	125	31.3	93.1
Internal length of cranium	88.3	22.1	65.8
Number of teeth, upper right	58	–	–
Number of teeth, upper left	59	–	–
Length of upper right toothrow	224	56.1	166.9
Length of upper left toothrow	225	56.3	167.7
Posterior end of upper right toothrow to tip of maxilla (l. and r.)	226	56.6	168.4
Number of teeth, lower right	53	–	–
Number of teeth, lower left	54	–	–
Length of lower right toothrow	212	53.1	158.0
Length of lower left toothrow	211	52.8	157.2
Posterior end of lower toothrow to tip of mandible (l. and r.)	215	53.8	160.2
Length of mandible	343	85.9	255.6
Coronoid height	55.4	13.9	41.3
Greatest length of symphysis	45.2±	11.3	33.7

The extremely long rostrum, which is greater than twice the length of the cranial portion of the skull, is highly diagnostic in this species (Figure 153). For further details concerning the skull, other skeletal features, and pigmentation of *Stenella longirostris*, see Van Bree (1971) and Perrin (1972). I am indebted to William F. Perrin, David K. Caldwell, and F. C. Fraser for assistance in the identification of the Fort Walton Beach specimen.

Stenella longirostris (Gray)

Delphinus longirostris Gray, Spicilegia zoologia . . . , 1, 1828:1; type locality, unknown.
Stenella longirostris, Iredale and Troughton, Mem. Australian Mus., 6, 1934:66.

Genus *Delphinus* Linnaeus

Dentition. — The teeth are small, acute, and numerous, 47 to 65 in each toothrow.

COMMON DOLPHIN   Figure 154
(*Delphinus delphis*)

Vernacular and other names. — The Atlantic Ocean population of this species is called the Atlantic Dolphin when *D. d. bairdii* of the Pacific Ocean is treated as a full species under the name of Pacific Dolphin. It is frequently called the saddleback dolphin. Both the generic and specific names stem from the Greek word *delphis*, meaning a dolphin.

Distribution. — *Delphinus delphis* occurs in the western Atlantic from off Newfoundland to the Gulf of Mexico and Caribbean Sea thence southward to waters off southern Argentina. It is found in the eastern Atlantic from Iceland, Norway, and the Baltic Sea south to the Mediterranean and Black seas thence to waters off southern Africa; also in the Indian Ocean and the South Pacific off Australia, New Zealand, Peru, and Chile.

External appearance. — The Common Dolphin measures up to eight feet but is more often about six feet in total length. The dark color of the back does not extend far down on the sides. The black pectoral flippers originate from the light portion of the sides, but a dark streak extends forward from their base to the lower jaws. The eyes are surrounded by black circles (on a grayish background) from which black lines run forward to the base of the snout. Alternating dark and light streaks of yellowish pigment are prominent on the

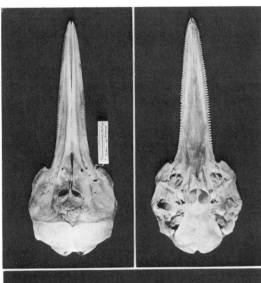
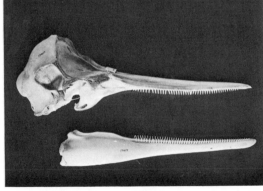

Figure 153 Skull of Long-beaked Dolphin (*Stenella longirostris*), LSUMZ 17017, stranded near Fort Walton, Okaloosa County, Florida, 23 March 1972. Total length of skull, 399 mm (15.8 in.). (Photograph by Joe L. Herring)

sides, the uppermost dark streak joining the dark portion of the back below the dorsal fin. The tail peduncle and the flukes are black.

Measurements. — A female specimen from New York, now in the American Museum of Natural History, measured in inches as follows: total length from tip of snout to notch in tail flukes, 84; tip of snout to base of pectoral appendage, 19.5; tip of snout to base of dorsal fin, 36; tip of snout to vent, 60.5; circumference at front of pectoral appendages, 36.5; circumference in front of dorsal fin, 45; circumference at vent, 27; tip of snout to corner of mouth, 11.5; outside curve of dorsal fin, 16; outside curve of pectoral appendage, 14. The skull measurements of four specimens from New York (Stoner, 1938) averaged as follows: total length, 428 (412–440); length of rostrum from anteorbital notches, 256 (247–266); breadth of rostrum at base of anteorbital notches, 89 (82–93); breadth of rostrum at middle, 64 (49–95); length of toothrow, 214 (197–226); distance fom last tooth to anteorbital notch, 47 (41–50); distance from tip of rostrum to anterior margin of nares, 308 (297–320); dis-

tance from tip of rostrum to posterior edge of pterygoids, 315 (305–325); breadth of skull between orbits, 167 (161–175); breadth of skull between posterior margins of temporal fossae, 142 (139–144); length of temporal fossae, 72 (69–75); depth of temporal fossae, 52 (49–58); length of mandible, 366 (355–380); length of mandibular symphysis, 44 (43–46); length of mandibular toothrow, 223 (216–231). The number of teeth in the four skulls were, respectively:

$$\frac{48-46}{48-51} \quad \frac{45-45}{45-46} \quad \frac{?-49}{47-47} \text{ and } \frac{49-49}{50-48}$$

Cranial and other skeletal characters. — In each toothrow this species usually has 48 to 50 teeth, which are sharply acute and only about one-tenth of an inch in diameter. The slender rostrum is considerably longer than the cranial portion of the skull and the palate has two deep lateral grooves on the median side of the toothrows. The number of vertebrae is 73 to 76. (Figure 155)

Status and habits. — To my knowledge no Common Dolphin has ever been taken in the Gulf

Figure 154 Common Dolphin *(Delphinus delphis).*

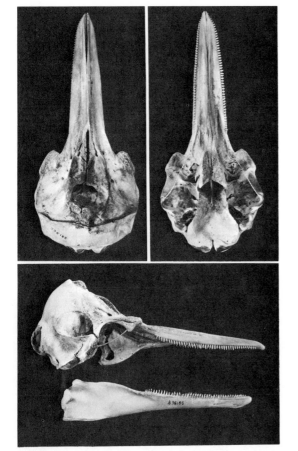

Figure 155 Skull of Common Dolphin *(Delphinus delphis)*, USNM 336185, off Costa Rica, 08°04′ N, 84°17′ W. Total length of skull, 448 mm (17.6 in.). (Photograph by Joe L. Herring)

of Mexico or found stranded anywhere on its shores or coastal islands, but at least seven sightings of the species have been made by competent observers. It was recorded by personnel of the M/V *Oregon*, the Bureau of Commercial Fisheries research vessel, as follows: 27 November 1950, at latitude 27°27′ N, longitude 96°04′ W, 36 nautical miles off Mustang Island, Texas; 29 November 1950, at latitude 28°14′ N, longitude 95°34′ W, 31 nautical miles off Matagorda Peninsula, Texas; 7 August 1952, at latitude 28°43′ N, longitude 87°00′ W, 98 nautical

miles south of Santa Rosa Island, Florida; 19 August 1952, at latitude 29°35′ N, longitude 87°18′ W, 45 nautical miles south of entrance to Pensacola Bay, Florida; 19 August 1952, at latitude 29°05′ N, longitude 87°55′ W, 63 nautical miles east of the mouth of Pass a Loutre of the Mississippi delta and 69 nautical miles south of the entrance to Mobile Bay, Alabama. Twelve individuals were also recorded by David K. Caldwell (1955) on 13 June 1954 at latitude 27°00′ N, longitude 92°00′ W, 132 nautical miles SSW Isles Dernieres, off Terrebonne Parish, Louisiana. On the same day, Caldwell observed "several more" at latitude 27°10′ N, longitude 91°22′ W, 117 nautical miles SSW Isles Dernieres. (Map 48)

The Common Dolphin is distinctively colored and should be looked for by deep-sea fishermen in our offshore waters. It is a rapid swimmer and has been clocked at 18 knots. Although generally regarded as a pelagic species, it has been known to ascend rivers for considerable distances. In October 1936 two were stranded, one 73 miles, the other 145 miles, up the Hudson River (Stoner, 1938).

Delphinus delphis Linnaeus

[*Delphinus*] *Delphis* Linnaeus, Syst. nat., ed. 10, 1758:77; type locality, European seas.

This species is polytypic, if one treats the northern Pacific population, *bairdii*, and the Black Sea population, *ponticus*, as races of *Delphinus delphis* rather than as full species. When specimens from the Gulf of Mexico finally become available they will doubtless prove referable to *D. d. delphis*.

Specimens examined from Louisiana. – None.

Genus *Tursiops* Gervais

Dentition.—Twenty-one to 25 large, smooth teeth are present in each toothrow.

ATLANTIC BOTTLE-NOSED DOLPHIN
(*Tursiops truncatus*)   Figure 156

Vernacular and other names.—This species is also called the common bottlenose dolphin, the bottlenosed dolphin, and the bottlenose. The specific name *truncatus* is the New Latin word meaning cut off, or truncated, and refers to the relatively short rostrum, or beak, in comparison to that of some delphinids.

The adjective Atlantic is required to differentiate between this species and the one from the Pacific Ocean, *T. gillii.* The word "bottlenosed" alludes to the shape of the snout, which resembles an old-fashioned gin bottle.

Distribution.—Atlantic Bottle-nosed Dolphins occur in coastal waters of the western Atlantic from Massachusetts to Florida, the Gulf of Mexico, the Lesser Antilles, and Venezuela. In the eastern Atlantic it is found from waters off Norway and in the Baltic Sea to the Mediterranean and Black seas, thence southward to waters off Senegal and the Congo. The overall range is not shown but see Maps 48 and 50.

Figure 156   Atlantic Bottle-nosed Dolphin *(Tursiops truncatus).* (Courtesy of Marineland of Florida)

External appearance.—This dolphin is purplish gray to clear gray dorsally, whitish ventrally except for the underside posterior to the vent, which is dark like the back; the sides are light gray. The head tapers abruptly into a snout of a modest three-inch length with a groove, or crease, surrounding its base. Although the total length may reach 12 feet, most individuals seen in Gulf waters are less than 9 feet from the tip of the snout to the notch in the tail flukes. The dorsal fin, located in the middle of the back, is moderately high with the tip directed posteriorly and rather acutely pointed.

Color phases.—No true color phases are known for this species, but an albino has been recorded in waters off South Carolina (Essapian, 1962), and Professor H. D. Hoese of the University of Southwestern Louisiana informs me (pers. comm.) that he found a nearly white bottle-nose stranded at Holly Beach, in Cameron Parish, Louisiana, on 1 February 1969. The specimen was too badly decomposed to show if the eyes were pink.

Measurements.—I am indebted to David K. Caldwell (pers. comm.) for the following external measurements of five Florida specimens comprising two males (one from Steinhatchee, the other from Port St. Joe) and three females (two from near Tampa, the third from Steinhatchee). The measurements in centimeters of these specimens are, respectively, as follows: tip of upper jaw to deepest part of notch between flukes, 254.0, 183.3, 247.4, 209.5, 226.0; tip of upper jaw to center of eye, 29.0, 26.7, 28.6, 28.8, 26.0; tip of upper jaw to angle of gape, 24.2, 24.3, 25.3, 25.2, 22.0; tip of upper jaw to external auditory meatus, 34.1, 34.5, 37.4, 32.3, 33.0; center of eye to auditory meatus, 5.4, 5.9, 8.3, 3.6, 7.1; center of eye to angle of gape (direct), 6.4, 5.4, 5.8, 5.5, 6.3; tip of upper jaw to blowhole along midline, 26.6, 28.1, 29.2, 29.6, 27.5; tip of upper jaw to anterior insertion of flipper, 49.4, 45.1, 50.9, 48.1, 43.5;

tip of upper jaw to tip of dorsal fin, 147.5, 145.6, 145.4, 123.5, 141.0; tip of upper jaw to midpoint of umbilicus, 110.5, 87.3, 113.0, 105.5, 97.0; tip of upper jaw to midpoint of genital aperture, 154.0, 117.5, 164.0, 145.0, 153.0; tip of upper jaw to center of anus, 181.3, 130.7, 176.0, 148.0, 158.5; projection of lower jaw beyond upper, 0.7, 0.9, 0.7, 1.1, 2.1; thickness of blubber, middorsal at anterior insertion of dorsal fin, 1.3, 1.2, 1.7, 0.7, 1.4; thickness of blubber, midlateral at midlength, 1.4, 1.6, 1.0, 0.3, 0.9; thickness of blubber, midventral at midlength, 1.6, 1.7, 1.4, 0.7, 1.1; girth, on a transverse plane intersecting axilla, 124.0, 86.0, 97.0 (very emaciated), 96.0 (somewhat emaciated), 119.0 (bloated); maximum girth, 142.0, 92.4, 106.5 (very emaciated), 110.0 (somewhat emaciated), 139.0 (bloated); girth on a transverse plane intersecting anus, 87.0, 52.0, 62.4 (very emaciated), 53.4 (somewhat emaciated), 77.0; length of mammary slits, ——, ——, L3.4 R2.9, L1.2 R1.1, L1.8 R1.7; length of genital slit, 15.7, 13.2, 16.4, 13.2, 18.0; width of blowhole, 3.5, 2.9, 3.2, 3.4, 2.9; length of blowhole, 1.8, 0.9, 2.1, 1.9, 1.7; length of flipper from anterior insertion to tip, 42.0, 35.6, 41.7, 35.5, 38.7; maximum width of flipper, 16.4, 13.1, 15.9, 12.9, 15.0; height of dorsal fin, 22.4, 18.9, 22.1, 16.6, 22.3; length of base of dorsal fin, 39.1, 30.0, 34.4, 31.6, 39.2; width of flukes, tip to tip, 59.5, 49.8, 53.0, 40.9, ——; distance from nearest point on right anterior border of flukes to notch, 16.8, 13.6, 16.1, 13.8, ——.

Five fully adult but unsexed skulls in the LSUMZ collections from the coast of Louisiana or adjacent waters measured in millimeters as follows: condylobasal length, 435, 432, 438, 435, 438; length of rostrum from level of anteorbital notches, 245, 248, 245, 240, 250; width of rostrum at base, 114.8, 118.0, 116.0, 124.0, 118.6; width of rostrum 60 mm anterior to level of anteorbital notches, 94.2, 85.1, 89.7, 94.3, 82.0; width of rostrum at middle, 70.0, 68.5, 68.0, 78.0, 62.5; width of premaxillae at the same point, 38.4, 35.0,

46.6, 49.0, 32.8; tip of snout to anterior edge of blowhole, 304, 300, 297, 290, 302; tip of snout to median posterior border of pterygoids, 300, 300, 298, – –, 305; preorbital width, 197, 199, 200, 207, 183; postorbital width, 230, 230, 228, 235, 222; orbital width, 208, 205, 205, 223, 200; width of blowhole, 54.4, 53.0, 55.0, 54.3, 57.4; zygomatic breadth at posterior points of attachment, 207, 205, 210, 223, 208; greatest width of premaxillae, 84.0, 83.7, 85.5, 92.0, 92.0; width of braincase across parietals, 172.2, 168.5, 181.0, 166.7, 185.0; length of temporal fossa, 106.5, 107.9, 100.0, 118.2, 105.5; height of temporal fossa, 70.2, 73.9, 77.2, 83.7, 70.2; greatest height, 175, 176, 175, 183, 175; internal length of cranium, 145, 142, 158, 149, 148; number of teeth upper right, 22, 24, 24, 23, 20+; number of teeth upper left, 19, 24 + 1, 25, 22, 20 + 1; length of upper right toothrow, 199, 205, 205, 203, 207; length of upper left toothrow, 203, 206, 201, 206, 200; posterior end of upper right toothrow to tip of premaxilla, 216, 214, 212, 210, 217; posterior end of upper left toothrow to tip of maxilla, 219, 212, 215, 207, 215; number of teeth lower right, 23, 22 + 1, 21, 20, 20 + 2; number of teeth lower left, 19, 22, 24, 22, 20 + 2; length of lower right toothrow, 215, 210, 195, 190, 212; length of lower left toothrow, 198, 208, 200, 198, 213; posterior end of lower right toothrow to tip of mandible, 227, 223, 210, 212, 216; posterior end of lower left toothrow to tip of mandible, 217, 219, 210, 213, 217; length of mandible, 383, 382, 375, 387, 383; coronoid height, 80.5, 86.0, 84.0, 87.3, 87.3; length of symphysis, 54.8, 58.5, 49.3, 48.5, 45.0.

Cranial and other skeletal characters. — The skull of *Tursiops truncatus* possesses a number of distinctive features: the pterygoid bones are in median contact; the postorbital process of the frontals is triangular; the posterior end of the vomer is narrow and rectangular; the posterior margins of the orbital plates of the maxillae are rounded; the mandibular sym-

physis is less than one-fifth the length of the ramus; and the mandibular condyle is less than one-half as long in its greatest dimension as the height of the mandible at the coronoid process. The teeth are approximately 8 mm in diameter (Figure 157). The number of vertebrae is 61 to 64.

Sexual characters and age criteria. — As in all cetaceans, the sexes can be distinguished by the number and arrangement of the abdominal openings. In the male a recess into which the penis retracts lies just ahead of the anal aperture. In the female a single opening that contains the urinary, vaginal, and anal orifices, is flanked laterally by two thin, longitudinal slits in each of which a teat is located. Newborn dolphins of this species measure

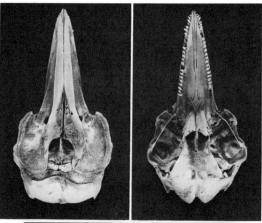

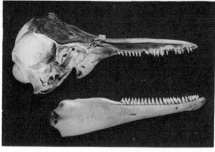

Figure 157 Skull of Atlantic Bottle-nosed Dolphin *(Tursiops truncatus)*, LSUMZ 17034, East Jetty, Cameron Parish, 19 May 1972. Total length of skull, 457 mm (18 in.). (Photograph by Joe L. Herring)

approximately three feet seven inches in length. The age attained by *Tursiops* in the wild is believed to be at least between 25 and 35 years. As the animals grow old their teeth often exhibit excessive wear and may be flattened almost to the gum line.

Status and habits. — The Atlantic Bottle-nosed Dolphin occurs commonly in the waters along the coast of Louisiana. It is, as a matter of fact, the only cetacean that is normally seen in our state's inshore waters. The species is found in greatest numbers in the vicinity of passes connecting the larger bays with the Gulf, but it is likewise present in back bays, where the salinity of the water is often quite low. I also have reports that the bottle-nosed dolphin has appeared as far up the Mississippi River as New Orleans. Joseph Billiot, of the Louisiana Wild Life and Fisheries Commission, informs me that he observed it in 1956 at the Canal Street Ferry in downtown New Orleans. This was one of the years in which a wedge of salt water extended up the bottom of the Mississippi River channel as far as New Orleans and seriously affected the chloride content of the city's water supply. I have talked with several river pilots, who have the responsibility of bringing ships up the river from Pilottown, and I have been told that dolphins are frequently seen in the lower reaches of the river and sometimes extend more than 100 miles upstream. In 1924 Captain G. J. Spence (pers. comm.) saw them at Meraux, just below Chalmette, and Captain Harry Post (pers. comm.) states that he has seen them as far up the river as Point a la Hache. Both of these pilots have spent many years on the river between New Orleans and Pilottown, where the river pilots take over from the bar pilots who bring ships to and from the open Gulf.

The bottle-nose is frequently seen just beyond the surf in the open Gulf itself. Occasionally it wanders much farther offshore but is seldom found beyond the 100-fathom line. In the few places where a highway leads to one of our Gulf beaches, dolphins, or porpoises as they are more often called, surfacing just beyond the breakers, are a sight familiar to nearly all Louisianians. Unfortunately, their numbers now appear to be reduced in comparison with their former abundance. The decline is attributable to a variety of factors, two of which are wanton slaughter with rifles by deep-sea commercial and sports fishermen and by deaths resulting from the explosion of seismographic charges in our offshore waters. The species deserves complete protection by law, a status that it enjoys under the Marine Mammal Protection Act of 1972.

Most of what is known about the biology of this species has been derived from studies of it in captivity. It is, perhaps, the most adaptable of all cetaceans to life in an aquarium. Not only is it capable of adjusting to artificial conditions, but it can also be trained to perform rather remarkable acrobatic feats that stimulate great spectator interest. Consequently, oceanaria are steadily increasing in number in many parts of the country, and, almost invariably, trained bottle-nosed dolphins are the main attractions of these establishments. I was nevertheless surprised not long ago in Paris to find that a bottle-nosed dolphin was one of the stars in the feature act of the fabulous night show at the *Moulin Rouge*.

But the saltwater aquaria, with their performing dolphins, are more than places of public entertainment. Many of them, such as Marineland Studios, of Marineland, Florida, have pioneered in dolphin research by making their facilities and their personnel available to scientists wishing to study their captive cetaceans. Whereas the intimate details of the life history of an animal that lives in the sea would ordinarily be extremely difficult to come by under natural conditions, much has been learned, especially about the bottle-nosed dolphin, as a consequence of investigations made at these commercial oceanaria. Not only have insights been gained with respect to the general habits of dolphins and to their physical attributes, behavior, social organization, the sounds they produce, and their system of intraspecific communication, but many details have been acquired about their reproductive processes. Their courtship antics and matings have been observed, their pregnancies have been followed on a day-to-day basis, and the actual births of their young have been photographed. So, despite the fact that bottle-nosed dolphins are animals of the sea, with all that such an existence entails when it comes to learning facts relating to their biology, perhaps more is now known about them than is known about some of the small mammals that live in the forest near our suburban homes. That such is true is amply demonstrated by the recent (1972) appearance of David K. and Melba C. Caldwell's delightful and highly informative book *The World of the Bottle-nosed Dolphin,* in which they summarize the abundance of present-day knowledge of the species.

The height of the bottle-nose's sexual activity occurs in March and April (McBride and Kritzler, 1951). The males fight viciously during this period and acquire many battle scars (Gunter, 1954). The first step in the formation of the brief pair bond takes place when the male begins to exhibit a preference for the swimming company of a particular female and remains with her for prolonged periods of times. Posturing on the part of the male is an initial feature of precopulatory behavior, which also includes stroking, rubbing, nuzzling, mouthing, jaw clapping, and yelping. The male often positions himself in front of the female with his body in an S-shaped curve, with head up and flukes down (Figure 158). Then the two swim in close proximity, often with their bodies in contact. Intromission occurs when the female rolls over on her side, presenting her ventral surface to the male. The process lasts for 10 seconds or less, but may be repeated several times at intervals of one to eight minutes for a half hour or more. The copulatory activities

occur at night or in the early hours of morning, at least in the confines of aquaria, but their timing may be influenced by the distractions that feeding periods and spectators cause during the day. Mating behavior in the wild is unknown. (Tavolga and Essapian, 1957)

During the second half of a pregnancy, the female begins to withdraw from the other animals and to seek the companionship of her mother or occasionally other pregnant females. She becomes slower and more deliberate in her movements. By the end of the fourth month her abdomen is noticeably more rounded, and, at seven months following conception, her mammary glands exhibit perceptible growth and development. But all dolphins are extremely streamlined and the females do not show the distortions in body shape exhibited by most pregnant mammals. In the eleventh and twelfth months body flexion is often observed, and in the last few weeks before parturition a change in respiratory habits takes place. The female rises frequently to the surface, hovers there, and respires several times in succession. Feeding continues normally, but the number of defecations per day increases, presumably because of the pressure of the enlarging uterus and its fetus on the other internal organs. During the last month the movements of the fetus inside the mother are easily seen.

The gestation period in captivity is at least 12 months (McBride and Kritzler, 1951). The long gestation period has the same functional value as it does in the large herbivores. The infant dolphin is extremely precocial and is

Figure 158 Posturing position of the Atlantic Bottle-nosed Dolphin *(Tursiops truncatus)* in courtship. (After Tavolga and Essapian, 1957)

thus immediately able at birth to rise to the surface of the water to breathe and to follow its mother.

In the wild the bottle-nose probably bears one offspring a year, beginning probably with its sixth year, but in captivity it generally produces one every two years. Births occur at our latitude from February through May (Gunter, 1942; McBride and Kritzler, 1951; Moore, 1953). The fetus usually emerges flukes first, and the whole birth process seldom requires much more than 20 minutes, sometimes much less. When the process lasts an hour or more a stillbirth often results. At the moment the fetus is extruded, the female whirls suddenly around to face the infant, ready to help if aid is needed. The action also serves to snap the umbilical cord. The newborn dolphin weighs approximately 30 pounds and is approximately three and a half feet in length. Its eyes are open and it has no teeth, but 11 to 12 blond vibrissae are present above the lips. The infant swims expertly from the moment of birth and immediately moves to the surface for its first breath of air. Should it show any hesitancy in doing so, it is lifted to the surface by the mother, who exhibits great solicitude for her offspring. Numerous instances of mother dolphins supporting stillborn or otherwise dead babies at the surface are on record (McBride, 1940; Hubbs, 1953).

Within moments of its birth and after obtaining the first draft of air into its lungs, the baby dolphin seeks out one of its mother's two teats, and it usually succeeds provided she has passed the placenta. Then, as soon as it takes hold of a nipple, the mother, by muscular contractions, squirts milk from the mammary sinuses into the infant's mouth. Weaning, at least in captivity, does not occur until a year and a half later (McBride and Kritzler, 1951).

During the first month of its life the infant stays close to its mother. Its typical rest position is under one of the mother's pectoral fins with its head touching her underparts. Soon, though, the youngster begins to develop confidence and commences to slip away from its mother, especially when she drops asleep, to engage in playful frolics, all to the consternation of the mother when she awakens to find the youngster missing. She immediately begins to emit a series of whistled calls that continue until the baby has returned to her protective custody.

In all the captive groups studied by McBride and Hebb (1948), the bottle-nose males were found to exhibit a well-developed and fairly stable hierarchy of intraspecific dominance that was correlated with size. Jaw clapping was one of the main forms of intimidation employed. But, despite an established order of dominance within a group of males, fighting among them occurred frequently. The females did not appear to compete in, or to form part of, the hierarchy. Tavolga (1966) carried her studies of social interactions in a captive colony much further, and she was able to designate a particular adult male as the alpha individual, a group of females as beta individuals, juvenile males as gamma individuals, and infants as delta, or omega, individuals in the overall social system. Except during the breeding season, the adult male usually swam alone and was normally peaceful, aggressive only when provoked. Among the females, one individual was the center of activities and the leader of the other females. Another one of the females was quite passive and a recluse. The juvenile males spent most of their time together and, because of their submissive position, were usually rejected as sexual partners by the mature females. Their copulatory attempts were almost always abortive. As a result, many of the juvenile males engaged in homosexual activities. Animals of the youngest group were the most submissive and during the first months stayed close to their mothers, after which they swam and played with the other animals of their age group.

The food of the bottle-nose on the northern Gulf Coast consists of the common and easily available fishes. According to Gunter (1942), the mullet *(Mugil cephalus)* is the principal component of its diet, but it also eats the puffer, sheepshead, needle-gar, black drum, spotted trout, sand trout, flounder, spot, and croaker, and it is also known to consume shrimp. In a later paper Gunter (1951) provided evidence that the bottle-nose eats quantities of shrimp. He quoted from a personal communication that he had received from E. A. McIlhenny under date of 8 February 1944. The words of this well-known naturalist merit, I believe, repetition here with slight paraphrase:

For many years it has been the custom of this large plantation to use Porpoise or Dolphin skins for the making of the leather harness used on the mules in plowing the fields. The reason for using porpoise skin was that it contains so much oil that it remains soft and does not harden with the excessive rains we have all during the summer. Each spring, for many years, I would go out into the passes connecting Vermilion Bay with the Gulf and harpoon porpoises to obtain their skins. On March 12, 1938, I harpooned three large porpoises in Southwest Pass between Marsh Island and the mainland connecting Vermilion Bay and the Gulf. The pass is from 80 to 100 ft. deep and about three miles long. These porpoises were skinned within one-half hour after being caught, and upon opening them I found from one to two gallons of extra-large shrimp in their stomachs. These shrimp were so fresh that after washing in several waters, they were cooked and used for food. Not one of all the shrimp taken from these porpoises was mangled. They were evidently swallowed whole, and probably swallowed only a few minutes before the porpoises were caught. On a number of other occasions I have found shrimp in the stomachs of porpoises.

The observations by McIlhenny explain why bottle-nosed dolphins so often become entrapped in the nets of Louisiana shrimp fishermen and cause much resented damage. Perhaps the dolphins become ensnared while pursuing the shrimp and not necessarily the fish the shrimpers previously blamed.

Reference has already been made to the large repertoire of sounds made by the dolphins and to the fact that these sounds are used in a system of intraspecific communication. The great Greek scholar Aristotle (384–322 B.C.), although demonstrating in his writings a considerable knowledge of the anatomy and behavior of dolphins, had overlooked the two tiny pinholes behind and slightly below the eyes that mark the entrance to the canal leading to the cochlea of the dolphin's middle ear, and hence probably wondered how they could hear even though he gave them credit for being able to do so. And Pliny the Elder (A.D. 23–79), four hundred years later was still baffled, for he wrote, according to one translation of his words, that dolphins "are obviously able to hear" but that "precisely how they do it is a riddle." The matter is still something of a riddle for some experts believe that the external ear openings are nonfunctional and that hearing is through the jaws.

One of the commonest sounds made by the bottle-nose is a series of rapidly reiterated whistles that emanate from the region of the blowhole, as do all underwater sounds made by dolphins. The call is uttered almost continuously while a group is swimming together and when playing. The baby dolphins start to whistle as soon as they are born, and each quickly learns to recognize the whistles of its own mother. At feeding times dolphins make a variety of mewing and rasping sounds, and on other occasions a sound resembling that of a rusty hinge. All seem to have special meanings or functions and are used to assist the dolphins in locating each other, in avoiding obstructions, in finding their food, and in expressing their feelings. The use of the sounds, some of which are ultrasonic (Lilly, 1966), in a system of echolocation had long been suspected by dolphin trainers and by students of dolphin behavior. But it remained for William E. Schevill and Barbara Lawrence (1956), working at the Woods Hole Oceanographic Institute with a *Tursiops* obtained from Marineland of Florida, and for W. N. Kellogg (1958 and 1961), working at the Marine Laboratory of Florida State University, again with animals furnished by Marine Studios, to demonstrate conclusively on the basis of carefully controlled experiments that echolocation is involved. Since then further elaborate experiments have been performed by these and other researchers, showing that the dolphin's sonar not only enables it to locate objects but to discriminate between objects of different sizes and shapes.

To do all these things, the dolphin requires a hearing apparatus of great sensitivity, and one that operates in a medium of water as well as air. With such an apparatus the dolphin is superbly equipped. The hearing capacity of dolphins that was to Pliny the Elder "a riddle" to which he could supply no answer, is now a matter reasonably well understood by experts, thanks to the brilliant anatomical research performed at the British Museum (Natural History) by F. C. Fraser and P. E. Purvis (1960 a and b) and to work done by P. Vigoureux (1960), F. W. Reysenbach de Haan (1960, 1966), Norris et al. (1961), and others.

From the time of the early Greeks the claim has often been made that dolphins sometime save humans from drowning or from attacks by sharks. Unfortunately, none of the cases cited provide incontrovertible evidence by present standards. We now know that dolphins will support members of their own kind at the surface to permit them to breathe when they are in distress. Dolphins do this for their youngsters, and they do it for their adult companions when the need arises. Siebenaler and Caldwell (1956), writing of bottle-nosed dolphins, describe such an instance of cooperation that they witnessed at the Gulfarium, an oceanarium located near Fort Walton Beach, Florida. A number of Atlantic Bottle-nosed Dolphins had been captured and were being transferred from the deck of the collecting vessel into a holding tank. Three adults had already been unloaded, and a fourth, also an adult, was being lifted from the deck by the tail. As it was swung over a fence into the tank, it struggled and struck its head, just behind the eye, on a post, and was stunned. The moment it was released it sank in seven feet of water. Two of the three dolphins already in the pen immediately assumed a position (Figure 159) on each side of the injured animal, placing their heads beneath its pectoral appendages to buoy it to the surface in an apparent effort to allow it to breathe while it remained partially stunned. The third dolphin stood by. Although the original two had to desert the injured dolphin to blow, they continued to return to it until it had sufficiently recovered from the injury to swim about unassisted.

Numerous similar instances of cooperation among dolphins are on record, but to say that they will do the same for humans is another matter. In 1943 on the coast of Florida occurred an incident that has been widely publicized. The wife of an attorney was bathing alone on a private beach when she was caught by the undertow, and before she could get back to the beach, the current pulled her down and her situation became desperate. "With that," she afterwards wrote, "someone gave me a tremendous shove." She landed on the shore and when she turned to express her thanks to her rescuer, no one was there. But a dolphin was leaping a few yards away. A man, who had seen the last part of the event, came rushing to her side exclaiming that she had appeared to be a "dead body" when he first saw her in the water, and that a dolphin had shoved her ashore. The question

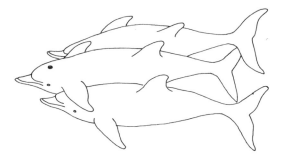

Figure 159 Two Atlantic Bottle-nosed Dolphins in assisting position to support ailing dolphin in the middle. (After Siebenaler and Caldwell, 1956)

is not so much whether the dolphin had done just that, but whether it might not have also done the same thing to a log or a mattress floating in the surf.

The swimming speed of the bottle-nose has been grossly exaggerated. Aristotle, who was not often wrong in what he had to say about dolphins, stated that they are the "fleet-est of all animals, marine or terrestrial." Certainly when land animals are considered, this statement is patently untrue, for cheetahs can achieve a speed up to 70 m.p.h. Although some accounts have claimed for the dolphin such speeds as 60 knots, a more reasonable figure is probably no more than 20 knots. Indeed, 10.7 knots (12.2 m.p.h.) is the most that Moore (1953) was able to verify on the part of a frightened and hard-pressed bottle-nose.

In captivity the Atlantic Bottle-nosed Dolphin is extremely playful and some of its self-devised games show a relatively high degree of intelligence. Fish may be chased, caught, let go, and chased again. Brown and Norris (1956) give an account of two dolphins trying to get a moray eel (Scorpaena guttata) out of its hiding place in the rocks, where they could not quite reach it. One of the dolphins left the scene, caught a scorpion fish, so called because of the sharp spines along its back with poison glands at their base. The dolphin then proceeded to poke the fish, spines forward, into the eel's retreat. The eel shot out into the open, and the dolphins played with it before finally letting it go.

The bottle-nose, aside from being playful and remarkably intelligent, also tends to imitate what it sees other dolphins doing (Caldwell et al., 1965; Brown et al., 1966; and others). It is therefore easily trained to perform fairly complicated tasks, and to do so without reward other than the apparent satisfaction of having achieved success. The performances of Flipper, who was a bottle-nosed dolphin (I understand there were several individuals involved), on hundreds of thousands of television screens, bear witness to

this uncanny ability. In view of the bottle-nose's attributes one should not be surprised that it ranks as the number one cetacean among the species displayed in public aquaria, nor that it has been designated for special study by the United States Navy as a possible collaborator in the performance of duties of vital interest to the armed services.

Predators.—Perhaps the only natural enemy of bottle-nosed dolphins are sharks, particularly some of the larger and more rapacious species. But against these adversaries the dolphin appears to be more than a match. In a confrontation it does not bite, as does the shark, but instead usually kills or seriously injures its foe by ramming it in the region of the gills. Not always, though, does it emerge unscathed, for Caldwell et al. (1965) describe the capture of a bottle-nose near Destin, Florida, that showed on its side the unmistakable but nearly healed tooth marks of a large shark (see also Wood et al., 1970). The chief predator of bottle-nosed dolphins is, however, modern man. The shooting of them with rifles by so-called sportsmen, notably fishermen, is absolutely inexcusable. Seldom is the dolphin killed outright by a rifle bullet. It simply dies slowly from secondary infection or other complications and eventually washes up on a beach. No argument against this kind of slaughter, such as that the dolphin is beneficial to man and therefore deserves his protection, is required. I am objecting on the ground that the wanton killing of any kind of wildlife is to be condemned. Dolphins are not only extremely interesting animals because of their many adaptations to their particular way of life, but they are an integral part of the world in which we live. The irresponsible, unsportsmanlike act of shooting at them simply because they represent a moving target, is contemptible. Consequently, bottle-nosed dolphins should be specifically protected by state law and the molesting of them in any fashion should be punishable by a severe penalty. Fortunately, they are now protected under federal legislation by the Marine Mammal Protection Act of 1972.

Parasites.—Three papers by Stephen G. Zam, David K. Caldwell, and Melba C. Caldwell summarize their findings concerning the internal and external parasites of *Tursiops truncatus,* as well as other odontocete cetaceans, in Gulf and other waters (see Caldwell et al., 1968; Zam et al., 1971; Caldwell et al., 1971; and the included bibliographies). They report finding in or on *Tursiops truncatus* seven species of trematodes *(Campula oblonga, C. delphini, C. palliata, C.* sp., *Nasitrema delphini, Pholeter gastrophilus, Braunina cordiformis),* one cestode *(Diphyllobothrium* sp.), five nematodes *(Anisakis typica, A. tursiopis, A. marina, Halocercus lagenorhynchi, Gnathostoma* sp.), one unidentified holotrich ciliate, two barnacles *(Conchoderma auritum* and *Xenobalanus globicipitus),* and five unidentified arthopods of the family Cyamidae.

Subspecies

Tursiops truncatus truncatus (Montagu)

Delphinus nesarnack Lacépède, Hist. nat. cétacées, 1804:xliii, 307, pl. 15, fig. 2; type locality, North Atlantic.

Tursiops nesarnack, Hershkovitz, Fieldiana, Zool., 39, 1961:550.

Delphinus truncatus Montagu, Mem. Wernerian Nat. Hist. Soc., 3, 1821:75, pl. 3; type locality, Duncannon Pool, near Stoke Gabriel, approximately 5 mi. up River Dart, Devonshire, England.

Tursiops truncatus, Gray, List Mammals British Museum, 1843:105.

Although numerous subspecies of *Tursiops truncatus* have been described, Hershkovitz (1966) recognizes only two, *truncatus* and *aduncus.* The latter occurs, insofar as its range in waters off the Western Hemisphere is concerned, from Baja California to Chile, and off Brazil and Argentina. Populations in the Gulf of Mexico are, on the basis of geographical probability, referable to the nominate race, *truncatus.*

Hershkovitz (1961) has shown that *Delphinus nesernack* Lacépède (1804) has priority over *Delphinus truncatus* Montagu (1821), but he also points out (1966) that the former name has had virtually no currency since its introduction. Contrariwise, the name *Tursiops truncatus* (Montagu) was used extensively during the last century and almost exclusively ever since for the Atlantic Bottle-nosed Dolphin. It has been employed widely as the technical name for this species of dolphin in both technical and popular literature, and for this reason, Hershkovitz contends that, in the interest of nomenclatural stability, the name *Tursiops truncatus* (Montagu) should be treated as a *nomen conservandum.* I heartily concur.

Specimens examined from Louisiana.—Total 25, as follows: *Cameron Par.:* 4 mi. W Holly Beach, 2 (LSUMZ); 1 mi. W Holly Beach, 1 (LSUMZ); 0.25 mi. E East Jetty, S Cameron, 1 (LSUMZ); 2 mi. S Cameron, 1 (LSUMZ); 3 mi. W Holly Beach, 1 (LSUMZ); Cameron, East Jetty, 2 (LSUMZ); 11 mi. W Holly Beach, 1 (MSU); 4 mi. SW Creole, 1 (MSU); Holly Beach, 3 (2 MSU; 1 USL); Rutherford Beach, 2 (1 LSUMZ; 1 USL); 1.5 mi. W Holly Beach, 1 (USL). *Jefferson Par.:* Grand Isle, 1 (LSUMZ); Bayou St. Honore, Caminada Bay, 1 (LSUMZ). *Plaquemines Par.:* Port Eads, 1 (LSUMZ); Telegraph Point, Breton Sound, 1 (LSUMZ); Breton Island, 1 (USL). *St. Mary Par.:* Cypremort Point, 1 (USL). *Terrebonne Par.:* Timbalier Isle, 1 (LSUMZ). *Vermilion Par.:* 1 mi. E Cheniere au Tigre, 1 (USL); 3 mi. E Cheniere au Tigre, 1 (USL).

[Genus *Feresa* Gray]

Dentition.—Nine to 11 large, conical teeth are present in each maxillary toothrow, 11 to 13 in each mandibular toothrow. The maximum diameter of the teeth is approximately 8 mm.

[PYGMY KILLER WHALE]   Figure 160
(Feresa attenuata)

This species is one of the rarest of all mammals on the basis of the number of specimens of it that have found their way into the museums of the world. Although it was long known solely on the basis of two skulls of unknown origin in the British Museum (Natural History), records of it are now scattered across the Pacific Ocean, and it has been found in the Indian Ocean and in the South Atlantic off West Africa. The first record of the species in the Gulf of Mexico, and indeed anywhere in Western Atlantic waters, was that of an individual found stranded on Padre Island (ca. latitude 26°04′ N, longitude 97°09′ W) on 21 January 1969 (James et al., 1970). Almost at the same time, specifically in the spring of 1969, David K. and Melba C. Caldwell acquired a skull from a specimen taken off St. Vincent in the Lesser Antilles. Their report of it (Caldwell and Caldwell, 1971a) furnished the second record for western Atlantic waters and the first for the Caribbean and West Indian region. (Yamada, 1954; Cadenat, 1958; Fraser, 1960; Nishiwaki et al., 1965; Pryor et al., 1965; Nishiwaki, 1966; Perrin and Hubbs, 1969; Best, 1970; Caldwell et al., 1971)

The Pygmy Killer Whale, which has also been called the slender blackfish, is dark gray, almost black, except for margins of white about the lips and a white area in the anal region. A pale gray or sometimes white area also lies between the flippers. The head is bluntly and evenly rounded and the snout overhangs the tip of the lower jaw. A ventral groove running along the middle of the abdominal surface extends from the anus toward and beyond the navel (Nishiwaki, 1966). The total length of males is between seven and nine feet; females average slightly smaller. The rostrum is somewhat less than one-half the total length of the skull. The skull is depicted in Figure 161.

Feresa attenuata Gray

Feresa attenuata Gray, Jour. Mus. Godeffroy (Hamburg), 8, 1875:184; type locality, unknown.
Feresa occulta Jones and Packard, Proc. Biol. Soc. Washington, 69, 1956:167.—Hall and Kelson, Mammals N. Amer., 1959:830.

[Genus *Orcinus* Fitzinger]

Dentition.—From 10 to 13 teeth are present in each toothrow; the roots of the teeth are bluntly rounded, the crowns inwardly recurved.

[COMMON KILLER WHALE]   Figure 162
(*Orcinus orca*)

The Common Killer Whale has been recorded several times in the southeastern part of the Gulf of Mexico, in the vicinity of the Keys and off the southern tip of the peninsula, but evidence of its occurrence in the northern part of the Gulf is limited to one specimen and a sight record. A carcass was found on the beach 6.5 miles east of East Pass, near Destin, Okaloosa County, Florida, on 27 May 1956 (Caldwell et al., 1956), and Gunter (1954) mentions the sighting of one 35 miles southeast of Port Aransas, Texas, in the summer of 1951.

Figure 160   Pygmy Killer Whale (*Feresa attenuata*).

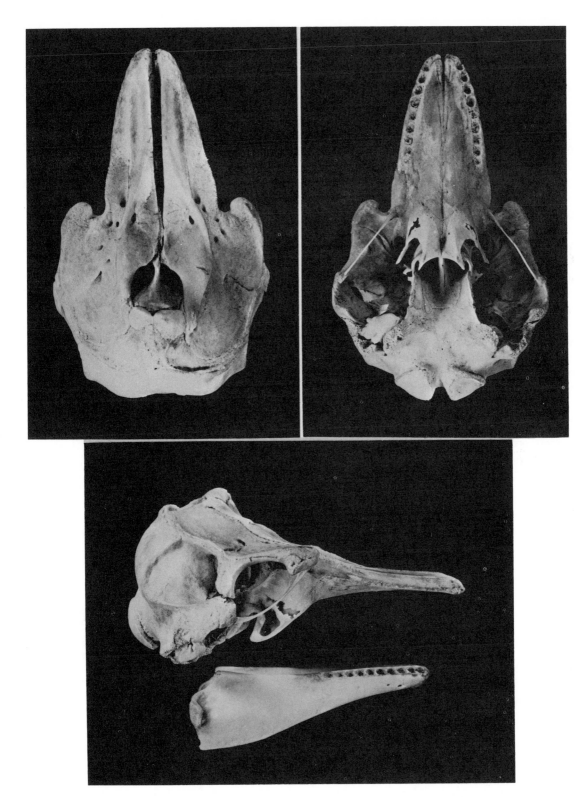

Figure 161 Skull of Pygmy Killer Whale *(Feresa attenuata)*, Pan American Univ., Padre Island, Texas, 21 January 1969. Total length of skull, 351 mm (13.8 in.). (Photograph by Joe L. Herring)

The Common Killer Whale is the largest of the delphinids, with males sometimes reaching 30 feet in total length. It is easily identified by its distinctive color pattern and by its high, sharply erect dorsal fin, which is straight on its posterior border instead of falcate to one degree or another as in other delphinids. The color is basically black dorsally, except for a prominent white patch behind the eye and another below the dorsal fin. The chin, throat, and belly are white, but posteriorly the white of the belly extends up onto the black of the sides as conspicuous lobes. The undersides of the tail flukes are white, but the pectoral appendages are black on both surfaces. The latter structures are large, ovate, and rounded on their distal ends. The skull is likewise distinctive. The rostrum is broad basally, being wider than one-half its length at a point opposite the base of the anteorbital notches. The pterygoid bones are not in contact. The skull is depicted in Figure 163.

Orcinus orca (Linnaeus)

[*Delphinus*] *Orca* Linnaeus, Syst. nat., ed. 10, 1, 1758:77; type locality, European seas.
Orcinus orca, Palmer, Proc. Biol. Soc. Washington, 13, 1899:23.

Genus *Pseudorca* Reinhardt

Dentition.—From 8 to 11 large teeth, circular in cross-section, are located in each toothrow. At gum line the teeth measure up to one inch in diameter.

FALSE KILLER WHALE   Figure 164
(*Pseudorca crassidens*)

Vernacular and other names.—This species is also called the lesser killer whale and the false killer. The generic name *Pseudorca* is based on the Greek word *pseudos,* meaning false, and the Latin word *orca* for a kind of whale. The specific name *crassidens* is a combination of two Latin words (*crassus,* thick, and *dens,* tooth) and means thick tooth, doubtless referring to the large diameter of these structures.

Distribution.—The False Killer Whale appears to be fairly widespread in the warm and temperate seas of the world. It is known largely from mass strandings. Observations of living animals suggest that it is mainly a deep-sea inhabitant. There are only three records for the Gulf of Mexico (see beyond, under *Status and habits*).

Figure 162   Common Killer Whale *(Orcinus orca).*

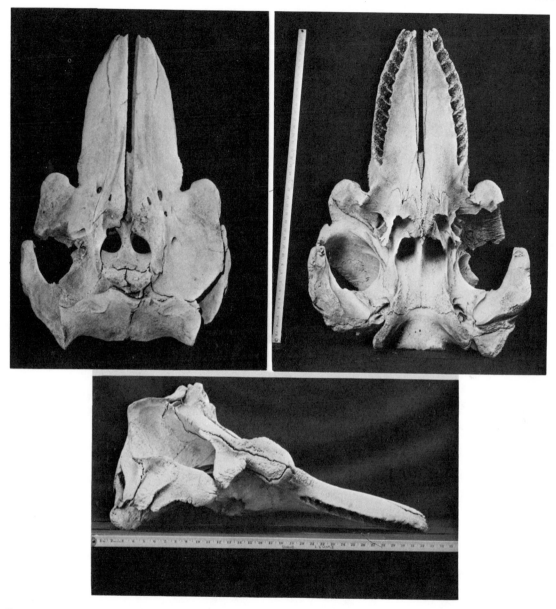

Figure 163 Skull of Common Killer Whale *(Orcinus orca),* TU 4171, Florida. Total length of skull, 985 mm (38.8 in.). (Photograph by Joe L. Herring)

External appearance.—Males attain a length of 18 feet or slightly more, whereas females measure up to 16 feet. The species is easily recognized by a combination of salient features: the color is a uniform black except for varying amounts of white on the lips, chin, and belly; the head is evenly rounded with no trace of a beak such as that possessed by many delphinids, and it lacks the bulbous swelling possessed by *Globicephala macrorhyncha;* the length of the snout prominently exceeds the length of the lower jaw; the dorsal fin is located about midway of the body length and its tip points backward; the pectoral appendages taper to a point and measure about one-tenth of total body length (in the somewhat similar Short-finned Pilot Whale the appendages are about one-fifth the total body length).

Measurements. – The single whole specimen known from the Gulf of Mexico is a female that measured (in part) as follows (Bullis and Moore, 1956): length from snout to tail notch, 4,210; snout to posterior margin of blowhole, 437; snout to eye, 401; snout to angle of gape, 391; snout to dorsal fin insertion, 1,618; snout to anus, 2,622; center of eye to ear aperture, 131; length of gape, 344; length of dorsal fin to tip, 489; vertical height of dorsal fin, 298; length of flipper, 524; width of flipper at base, 209; breadth of blowhole, 55; head depth at blowhole, 360; body depth at point of insertion of dorsal fin, 684.

Cranial and other skeletal characters. – The skull may be characterized as follows: short with a broad rostrum; breadth of rostral portion of

Figure 164   False Killer Whale *(Pseudorca crassidens)* with Atlantic Bottle-nosed Dolphins *(Tursiops truncatus)* in the background. This false killer, named "Lucie," was one of the more than 150 that stranded near Fort Pierce, Florida, on 11 January 1970. (Photograph courtesy of Nat Fain and Marineland of Florida)

premaxillae equal to, or greater than, proximal portion; rostrum truncate distally; pterygoid bones in contact; palatines elongated laterally (Figure 165). Approximately 50 vertebrae are present. (Hall and Kelson, 1959)

Status and habits. — Three published records of the occurrence of the False Killer Whale in the Gulf of Mexico and its environs are available. One is based on a skeleton of an animal found stranded a few miles east of Havana (Aguayo, 1954). Another is a specimen captured by Harvey J. Bullis, Jr., and the crew of the U.S. Fish and Wildlife Service's commercial fisheries exploratory vessel the M/V *Oregon* on 30 April 1955 at latitude 26°30′ N, longitude 89°15′ W (Bullis and Moore, 1956). This position is approximately 150 nautical miles south of the mouth of the Mississippi River. At the time of the animal's capture another individual was observed nearby. Finally, Schmidly and Reeves (in press) report the harpooning of a False Killer Whale 20 to 30 miles beyond Flower Gardens Banks, 120 miles SSE of Galveston, almost due south of Cameron, Louisiana, in 1961. The skeleton of the specimen is now in the Houston Museum of Natural Science.

Statements in the literature about the biology and habits of this species are not always in perfect agreement. For instance, the species has been said by some authorities always to travel in groups of several hundred individuals, a notion that may have stemmed from records of mass strandings. Although the species was first described in 1846 on the basis of a subfossil skeleton found in the Lincolnshire Fens, near Stamford, England, it was not seen in the flesh in Britain until 1927, when a school of about 150 stranded in Dornoch Firth, Sutherlandshire, in northern Scotland (Fraser, 1966). Other recorded cases of large strandings involve 75 to 300 individuals each (see, for example, Caldwell et al., 1970). But the two False Killer Whales observed in the Gulf of Mexico by the crew of the M/V *Oregon*, noted above, bear witness to the fact that the species is not always found in large pods.

One usually authoritative source dogmatically states that the False Killer Whale feeds solely on large pelagic fishes, but Bullis found only the remains of squids in his Gulf specimen. Judging from the size of the fragments that he recovered from the stomach, he estimated the weight of a whole squid of the kind eaten would have been 15 to 20 pounds.

Breeding is said to occur at no particular time of the year, for the dissection of pregnant females stranded in the same month has revealed fetuses varying greatly in the stage of embryonic development. Further study, though, may reveal that the present meager information on reproductive seasons is altogether fallacious. Unfortunately, one cannot study the breeding biology of an oceanic cetacean with the ease that one studies a mammal on land, or even with the facility that captive cetaceans, most notably Atlantic Bottle-nosed Dolphins, have been investigated in numerous marine studios.

Pseudorca crassidens (Owen)

Phoecaena crassidens Owen, A history of British fossil mammals and birds, 1846:516; type locality, Stamford, England (subfossil).
Pseudorca crassidens, Reinhardt, Overs. Danske Vidensk. Selsk. Forh., 1862:103–152, figs. 1–4.

The species is monotypic. Should an analysis of *Pseudorca crassidens* populations on a worldwide basis fail to reveal any significant geographical variation in the species, I would be indeed surprised. But so far as I know such a study has not yet been done, and no subspecies have been described.

Specimens examined from Louisiana. — None.

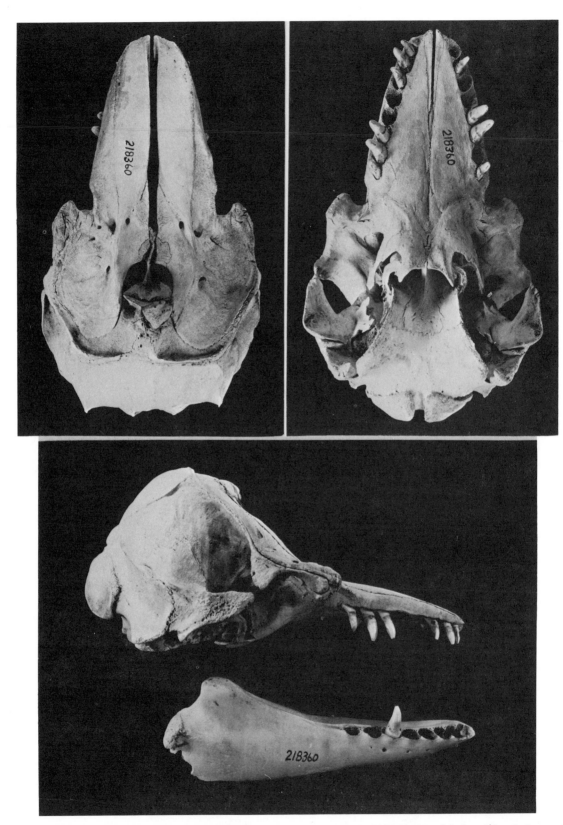

Figure 165 Skull of False Killer Whale *(Pseudorca crassidens)*, USNM 218360, 20 mi. below Miami, shore of Biscay [=Biscayne] Bay, Florida. Total length of skull, 575 mm (22.6 in.). (Photograph by Joe L. Herring)

[Genus *Grampus* Gray]

Dentition.—Two to nine teeth in each of the lower toothrows, usually none in the upper jaw but exceptions do occur (Nishiwaki, 1972).

[RISSO'S DOLPHIN]   Figure 166
(*Grampus griseus*)

Risso's Dolphin occurs in most of the oceans of the world but in the western North Atlantic seldom ranges farther south than the waters off New Jersey. I know of only one record for the Gulf of Mexico. An individual was found stranded on a sandbar at the mouth of the Anclote River near Tarpon Springs, Pinellas County, on 12 June 1966 (Paul, 1968).

The species is also sometimes called the grampus (from Latin roots by way of various cognates including the Old French *gras peis,* which means fat fish). An elaboration of this name, gray grampus, refers to the general grayish coloration, though the underparts are paler and the appendages and flukes are black. The body is often covered with light-colored scratchlike marks, which are most concentrated on the head and dorsal surfaces. The forehead is prominently blunt and bulging with no snout extending beyond the melon, and the dorsal fin is high and strongly recurved. The skull has several particularly diagnostic features. The rostrum is greatly expanded anterior to the anteorbital notches and tapers to an obtuse termination; each mandibular ramus possesses two to nine teeth clustered in the region of the symphysis; and teeth in the upper jaw are either lacking or when rarely present are exceedingly small (Figure 167). The total length of the animal is most commonly between 9 and 13 feet. For detailed external and cranial measurements, see True (1889), Orr (1966), and Paul (1968).

Grampus griseus G. Cuvier

Delphinus griseus G. Cuvier, Ann. Mus. Nat. Hist., Paris, 19, 1812:13–14; type locality, Brest, France.
Grampus griseus, Hamilton, Jardine's Naturalist's Library, Mammalia, 6, 1837:233.—Ellerman and Morrison-Scott, Checklist of Palearctic and Indian mammals, 1758 to 1946, Brit. Mus. (Nat. Hist.), 1951: 739–741.—Schevill, Jour. Mamm., 35, 1954:123–124.

The genus and the species are presently

Figure 166   Risso's Dolphin *(Grampus griseus).* (Portrait by H. Douglas Pratt, based on drawings in Fowler, 1874, and Richard, 1936)

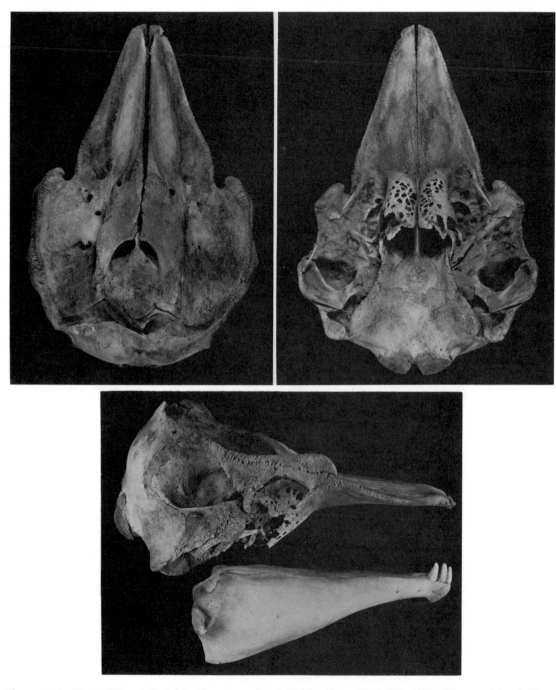

Figure 167 Skull of Risso's Dolphin *(Grampus griseus)*, Caldwell no. SV-2-GG, off St. Vincent Island, West Indies, summer 1970. Total length of skull, 466 mm (18.3 in.). (Photograph by Joe L. Herring)

both treated as being monotypic, although some authorities think the eastern Pacific population is deserving of at least subspecific status under the name *G. g. stearnsii* Dall. The taxonomic relationships of the genus *Grampus* are uncertain, but I am including it here provisionally between *Pseudorca* and *Globicephala*.

Genus *Globicephala* Lesson

Dentition. — From 7 to 11 conical teeth are located in each toothrow and are confined to the anterior portions of the jaws.

SHORT-FINNED PILOT WHALE
(Globicephala macrorhyncha)

Figure 168

Vernacular and other names. — This species is also called the short-finned blackfish. The name pilot whale is based on the habit of the genus of traveling in groups led by large bulls. The adjective Short-finned refers to the inferior length of the pectoral appendages in comparison to those of *Globicephala melaena*, of the North Atlantic. The members of the genus *Globicephala* have long been called blackfish. This name is partially appropriate because of the animal's basically all-black color, but the "fish" part of the word is, of course, a misnomer. The generic name comes from two words (the Latin *globus*, globe or ball, and *cephalus*, the Latinized Greek word for head), meaning globe head, and was inspired by the globose shape of the forehead. They are also sometimes called the pothead. In whaling literature pilot whales are often called by the Scottish name caa'ing or ca'ing whales, which has been said to refer to the grunting call that they are sometimes heard to emit. Fraser (1949), however, denies this explanation of the name. He says that it has to do with the practice on the part of whalers in driving and herding the animals.

Distribution. — The Short-finned Pilot Whale occurs in Atlantic waters. On the western side it extends from the latitude of New Jersey thence southward to waters surrounding the West Indies and to the Gulf of Mexico.

External appearance. — The total length is generally no more than 15 feet but sometimes reaches 20 feet. It is identifiable by the following characters in combination: the bulbous, globular shape of the head and the lack of a beak; by the position of the blunt, posteriorly curved dorsal fin, which is located well anterior to the middle of the body; by the slender, pointed pectoral fins; and by the presence of

Figure 168   Short-finned Pilot Whale *(Globicephala macrorhyncha).*

seven to nine teeth in each toothrow. All the stranded pilot whales that I have examined on Louisiana beaches have been solid black, probably because of postmortem color changes that occur rapidly following death in cetaceans. Consequently, I do not know whether the gray kedge-shaped mark on the throat and belly, described by Sergeant (1962) and others, is present. (Figures 168 and 169)

Measurements and weights.—An adult that was found stranded on the coast of Cameron Parish, in southwestern Louisiana, measured in feet as follows (Paul Duhon, pers. comm.):

total length, from tip of snout to notch in the tail flukes, 13.3; tip of snout to base of dorsal fin, 3.5; snout to anterior corner of eye, 0.8; height of dorsal fin, 13.4; length of pectoral appendage measured from base along the curve to the tip, 2.5; straightline depth of pectoral appendage, 19.4; width across tail flukes, 3.3. The total length of 12 pilot whales stranded at the mouth of Bayou Lafourche (Lowery, 1943) averaged 10.5 (5.3–18.4). The measurements of two skulls, one from the mouth of the Mermentau River in Cameron Parish, the other from the mouth of Bayou Lafourche (both LSUMZ), are, respectively, as follows: greatest length, 580, ——; con-

Figure 169 Eastern Pacific Pilot Whale *(Globicephala scammoni),* a species closely resembling our own Short-finned Pilot Whale *(G. macrorhyncha).* (Photograph courtesy of Marineland of Florida)

dylobasal length, 565,———; length of rostrum from anteorbital notches, 292, 290; basal width of rostrum, 240, 248; width of rostrum 60 mm anterior to anteorbital notches, 224, 240; width of rostrum at middle, 173, 193; width of premaxillae at same point, 173, 195; tip of snout to blowhole, 386, 395; tip of snout to median posterior edge of pterygoids, 367, 370; preorbital width, 390, 403; post-orbital width, 415, 445; orbital width, 380, 400; width of blowhole, 96, 103; zygomatic breadth, 415, 439; greatest width of pre-maxillae, 180, 200; width of braincase across parietals, 230, 260; length of upper right toothrow, 140, 134; tip of premaxillae to posterior edge of right toothrow, 172, 165; mandibular length, 475, 479; coronoid height, 130, 133; length of symphysis, 57, 57. The numbers of teeth are as follows: max-illary, 8, 8 on each side; lower right, 8, 7; lower left, 9, 8. According to Walker et al. (1968), a male 13.4 feet in length was esti-mated to weigh 1,480 pounds.

Cranial characters.—The skull is large and massive in proportion to the size of the body. The breadth of the rostrum basally is great, being over 80 percent of its length. The pre-maxillae are much expanded laterally, cov-ering the lateral and anterior edges of the maxillae when viewed from above. In *G. me-laena* of the northern Atlantic, the pre-maxillae do not project over the edges of the maxillae. (See also Paradiso, 1958, and Ser-geant, 1962.) An average-sized tooth mea-sures approximately 18 mm in greatest diameter and 47 mm in total length, only about 18 mm of which extend above the gum line. (Figure 170)

Sexual characters and age criteria.—The sexes can best be separated by examination of the primary sex organs. Adult males greatly ex-ceed adult females in length and bulk, but a young male is often no larger than many of the females in the gam to which it is attached. Males, according to Walker et al. (1968),

reach sexual maturity at about 13 years of age, when about 13 feet in length. Females first breed when about six years old and no more than nine feet long. The gestation pe-riod is approximately one year and the calves at birth are eight to nine feet long.

Status and habits.—The Short-finned Pilot Whale is apparently fairly common in the northern part of the Gulf of Mexico, judging by the comparative frequency with which it is found dead or dying on beaches from Florida to Texas. Moore (1953) and Layne (1965), on the basis of known strandings, both consid-ered it the commonest cetacean in Gulf wa-ters adjacent to Florida, and I have numerous records of strandings on our own coast, sug-gesting that the species is regular in its occur-rence in Louisiana's offshore waters. I must admit, though, that I have never had the good fortune to see one alive. This whale's habit of feeding in deep water probably ac-counts for its failure to appear close to shore. Opposite most of the Louisiana coastline, the continental shelf slopes gradually away from the shore. Only near the mouths of the Mis-sissippi River does a precipitious drop-off in the shelf take place close to shore. In the course of the evolution of the delta of the Mississippi, land has built up seaward and, at the end of South Pass, now extends to within seven miles of the 100-fathom line. On the other hand, off southwestern Louisiana the 100-fathom contour line lies approximately 150 miles from the shore. No one, therefore, should be surprised that sightings of pilot whales are rather frequently made by fish-ermen near the mouth of South Pass of the Mississippi River but are seldom made else-where along our coast.

My only especially noteworthy experience with the species was on 19 August 1939, when I was taken by personnel of the Gulf Oil Company to inspect 49 whales lying on a beach several hundred yards west of Pass Fourchon of the delta of Bayou Lafourche. The stranding occurred on or about 5 Au-

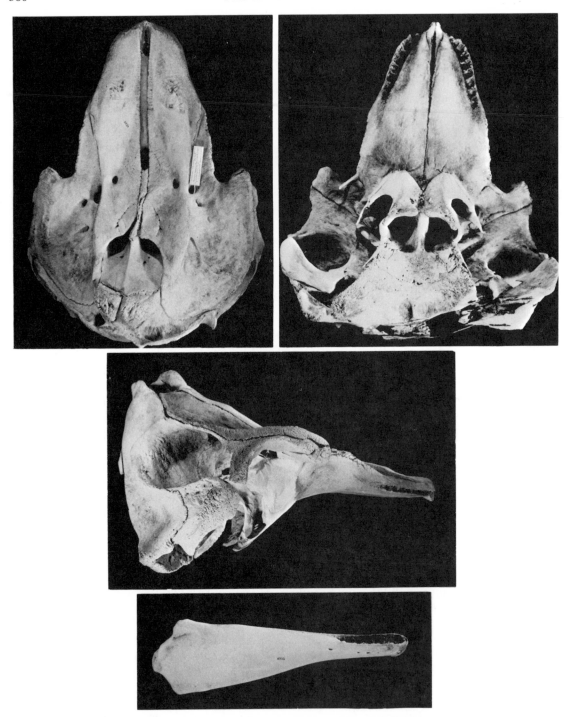

Figure 170 Skull of Short-finned Pilot Whale *(Globicephala macrorhyncha)*, LSUMZ 15912, stranded 5 miles east of the mouth of Mermentau River, Cameron Parish, 11 December 1970. Total length of skull, 550 mm (21.5 in.). (Photograph by Joe L. Herring)

gust, at the time of the passage of a severe hurricane that struck the eastern Louisiana coast. Since I did not examine the carcasses until two weeks after the catastrophe, the whales were by then partially buried in the sand and in an advanced stage of decomposition. I was, however, able to identify them as Short-finned Pilot Whales and to obtain the skull of one of them for the Louisiana State University Museum. The bodies were strewed along a two-hundred-yard stretch of beach, and all lay immediately behind the natural levee of the shore, as if the whales had been driven close to land by the high tides and winds associated with the hurricane and left stranded when the waters receded. But this conclusion, which implies involuntary beaching of the whales, is based solely on conjecture; they may, as so often appears to be the case, have been following a panicked leader.

Pilot whales are highly gregarious and are often seen in schools of more than a hundred individuals. One such aggregation was photographed near the mouth of the Mississippi River by Lester Plaisance, long-time deep-sea fisherman residing in Golden Meadow. Mr. Plaisance informs me that he believes he has most often seen this species in midsummer and that other offshore fishermen with whom he has spoken have the same impression. He is, however, suspicious that this consensus regarding the predominately summer occurrence of these whales may be biased by the fact that deep-sea fishing is carried on mainly from May to September, and the fishing rodeos, in which so many of the fishermen participate, occur in midsummer. At any rate, Layne (1965) found that a preponderance of strandings on the west coast of Florida also occurred in spring and summer.

These animals appear to have a natural tendency to follow a leader, hence the name pilot whale. This behavior has been commented upon by numerous authors, one of whom (Gunter, 1954) goes so far as to state that the leader makes a bellowing, grunting sound that the others follow, even when it takes them into shallow waters where the entire gam may become stranded and die.

The food of pilot whales consists of various kinds of squids and fishes. They are said to breed in winter and to have a gestation period of approximately one year. According to Walker et al. (1968), females mature at about six years of age, but males take about seven years longer. The species is polygamous.

The North Atlantic species of the genus, *G. melaena,* is today the principal support of the Newfoundland whaling industry, with several thousand being killed annually. The average yield per animal is said to be approximately 40 gallons of blubber oil and about two gallons of head and jaw oil. The mean for the southern species, *G. macrorhyncha,* would probably be considerably less because of its smaller size. Fortunately, though, I know of no present-day exploitation of *macrorhyncha,* at least in waters of the United States.

Globicephala macrorhyncha Gray

Globicephalus macrorhynchus Gray, *in* Zoology of the voyage of H.M.S. Erebus and Terror . . . , 1 (Mammalia), 1846:33; type locality, "South Seas."
Globicephala macrorhyncha, Fraser, Atlantide Rept., 1, 1950:49.
Globicephala ventricosa, Lowery, Occas. Papers Mus. Zool., Louisiana State Univ., 13, 1943:257.

This species is monotypic. Hershkovitz (1966) goes so far as to synonymize *macrorhyncha* under the larger, distinctively marked species of the North Atlantic, *G. melaena,* despite the fact that Sergeant (1962) and others have demonstrated ample justification for recognizing two well-defined species.

Specimens examined from Louisiana. — Total 4, as follows: *Cameron Par.:* 5 mi. E mouth Mermentau River, 1 (LSUMZ). *Iberia Par.:* Marsh Island, near Oyster Bayou, 1 (LSUMZ). *Lafourche Par.:* mouth of Bayou Lafourche, 1 (LSUMZ). *Terrebonne Par.:* marsh below Houma, 1 (LSUMZ).

The Fin Whale Family
Balaenopteridae

THIS FAMILY is one of three that make up the suborder Mysticeti. The name means moustached whales, and is based on the racks of brushy baleen in their mouths. In these cetaceans the embryonic teeth are replaced in postfetal life by baleen, or whalebone, which is modified mucous membrane. It consists of a series of thin plates, one behind the other, suspended from each side of the palate into the mouth cavity. The outer edges of the plates are smooth, the inner edges frayed into long, brushlike fibers. Small forms of aquatic life, such as crustacea and mollusks, known collectively as zooplankton, are taken into the enormous mouth, the cavity of which can be greatly expanded by means of the pleats and furrows in the chin and throat. Then by a piston action of the huge tongue the water is squeezed through the baleen plates, and out the sides of the mouth, leaving the food entrapped by the plates, which serve as sieves or strainers. Gray Whales of the monotypic family Eschrichtidae, now confined to the Pacific Ocean, possess 130 to 180 yellowish white plates on each side of the palate, and the two rows do not meet in front as they do in the whales of the other two families. The members of the family Balaenidae, which includes the Black Right Whale and the bowhead, or Greenland Right Whale, are distinguished by their stocky body and huge head, by the absence of throat furrows, by the fusion into one compact unit of all seven cervical vertebrae, by the shape of the flippers, and by

certain features of the baleen plates, which may be up to 12 feet in length and number more than 350 on each side of the upper jaw.

The present family, Balaenopteridae, contains two genera and six species of whales: the Fin-backed Whale *(Balaenoptera physalus)*, Bryde's Whale *(B. edeni)*, Sei Whale *(B. borealis)*, Little Piked, or Minke, Whale *(B. acutorostrata)*, Blue Whale *(B. musculus)*, and Humpback Whale *(Megaptera novaeangliae)*. Some authorities consider that the Blue Whale also merits placement in a separate genus, *Sibbaldus*.

The baleen plates of the fin whales range in length from one to three feet, and their color is white, yellow, or grayish black. Ten to a hundred deep longitudinal furrows or pleats are present on the throat and chest. These whales are often called rorquals. One member of the family, the Blue Whale, is the largest animal ever to have lived on the face of the earth. It has been known to attain a length of 100 feet and a weight of 176 tons.

Although baleen whales are decidedly uncommon in the Gulf of Mexico, one is occasionally seen offshore or found stranded on our coast. Unfortunately, of the several carcasses or skeletal remains that have been found on the coast of Louisiana, only three can be identified to species with absolute certainty. In the case of the others no detailed descriptions and measurements were recorded, no skeletal parts were preserved, and no photographs were made showing diagnos-

FIN-BACKED WHALE363

tic features that would enable a cetologist to determine the species involved. The meager skeletal material and the few positive sight records that are available from Louisiana all pertain to three species of the family Balaenopteridae, but other members of the family have been recorded from the northern Gulf Coast, as detailed beyond, and on 30 January 1972 a Black Right Whale (*Balaena glacialis*) of the family Balaenidae stranded on the Texas Coast near Freeport (see account on page 378).

Genus *Balaenoptera* Lacépède

Dentition.—In postfetal life the teeth are replaced entirely by a series of baleen plates, the number, length, and color of which vary according to species.

FIN-BACKED WHALE   Plate IX
(Balaenoptera physalus)

Vernacular and other names.—This species is also called the finback, fin, or finner whale, and the common or razorback rorqual. The word rorqual is said to mean "whales having folds or pleats" on the throat and anterior underparts. The word razorback refers to the fact that the posterior middorsal region is distinctly and acutely ridged. The word fin in the other vernacular names refers to the presence of a dorsal fin, which is characteristic of the members of the family Balaenopteridae but which is absent in all but one of the species of the family Balaenidae. The generic name *Balaenoptera* is derived from two Latin words (*balaena*, a whale, and *ptera*, a fin), meaning fin whale. The specific name *physalus* is the New Latin word for a rorqual whale, and was derived from the Greek word *physalos*, which meant a kind of whale.

Distribution.—The Fin-backed Whale occurs in the Atlantic, Pacific, and Indian oceans from equatorial waters to pack ice in both the Northern and Southern hemispheres. In the Gulf of Mexico the species is apparently present in only limited numbers.

External appearance.—The finback reaches a length of up to 80 feet but generally no more than 65. It is gray or blackish gray above, white below. Especially diagnostic is the asymmetry of the color of the head region, which immediately separates it from other rorquals. In the forepart of the body a shifting of the pigment to the left side has taken place, so that the right lower jaw is without color, while the left is pigmented. This asymmetry in color extends to the remainder of the head and to the shoulders, and even the inside of the mouth is affected, the right jaw being pigmented while the left is without pigment. Even the tongue shows more pigment on its right side, as do also the baleen plates, for the blades on the anterior one-third of the right side are white, the remainder on that side and all on the left side being dull blue-gray or purplish, with only streaks of white and yellow. The frayed edges of the baleen, both of the white and pigmented blades, are a uniform yellowish white. The dorsal fin is located closer to the tail than to the center of the body. The pectoral appendages are small, being only about one-ninth the total length of the body. (Fraser, 1949)

Cranial characters.—The rostrum, when viewed from above, is triangular instead of largely parallel sided, a feature that serves to separate this species from the Blue Whale (*Balaenoptera musculus*) but not from the other members of the genus. The number of baleen plates varies from 350 to 400, the longer ones being approximately 36 inches exclusive of the bristles. (Figure 172)

Sexual characters and age criteria.—The total length of females is usually several feet less than that of males of comparable age. The gestation period is approximately 360 days, and at birth the young whale is slightly more than 20 feet in length. Weaning occurs in about six months, by which time the offspring has doubled its length. Sexual maturity is attained when the animal is beginning its third year of life and has attained a length of about 58 feet.

Status and habits. — The Fin-backed Whale, or common rorqual as it is often called, definitely occurs in waters off Louisiana, but positive records of it there and elsewhere in the Gulf are few. The Louisiana State University Museum of Zoology has a single blade of baleen from a whale that stranded in 1928 or thereabouts on one of the barrier islands in the Chandeleur Chain off the coast of Mississippi and southeastern Louisiana. It was identified by the late Remington Kellogg, noted cetologist and former Curator of Mammals of the United States National Museum, as belonging to *Balaenoptera physalus.* Through the help of several present-day residents of the Mississippi Gulf Coast, notably Arthur V. Smith of Pascagoula, I have been able to learn some of the details about this whale. It was actually discovered on Breton Island by Elmer Gautier and a companion. Later it was towed to Biloxi with the idea of exhibiting it for a fee. The salvagers also thought they could recover a quantity of ambergris at great financial profit to themselves, an impossibility from the outset, for ambergris is found only in the large intestine of the Giant Sperm Whale. But the whole venture collapsed when the whale began to decompose and authorities ordered it towed back into the Gulf.

Other unequivocal records of the Fin-backed Whale in Louisiana waters are represented by photographs of stranded individuals and by skeletal material in the LSUMZ. One specimen is the left ramus of a lower jaw of a Fin-backed Whale that was found in the marsh west of Venice, in Plaquemines Parish, in 1968 by Robert Rodrigue and given to Randolph A. Bazet of Houma, who later turned it over to the LSUMZ.

A photograph (Figure 173) is available of a finback that stranded near Sabine Pass in early December 1924. When discovered, the animal was still alive. While an attempt was being made to tow the whale into Sabine Pass and to Port Arthur, the hawser broke and the whale foundered on the jagged rocks of the jetty and died. It was then towed ashore and exhibited for two weeks. According to a news account, two other whales, both dead when found, stranded at the mouth of Sabine Pass at the same time, and one allegedly measured 85 feet in length.

Another whale was found on one of the islands in the Isles Dernieres group, off Terrebonne Parish, in 1915. I originally reported (Lowery, 1943) the date of this stranding as 1916 on the basis of information

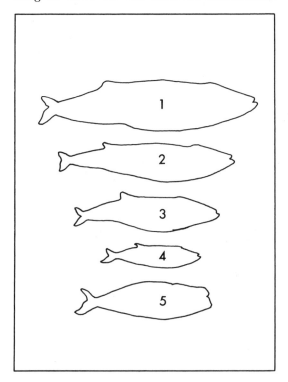

Plate IX
Baleen whales:
1. Blue Whale (*Balaenoptera musculus*).
2. Fin-backed Whale (*Balaenoptera physalus*).
3. Sei Whale (*Balaenoptera borealis*).
4. Little Piked Whale (*Balaenoptera acutorostrata*).
5. Black Right Whale (*Balaena glacialis*).

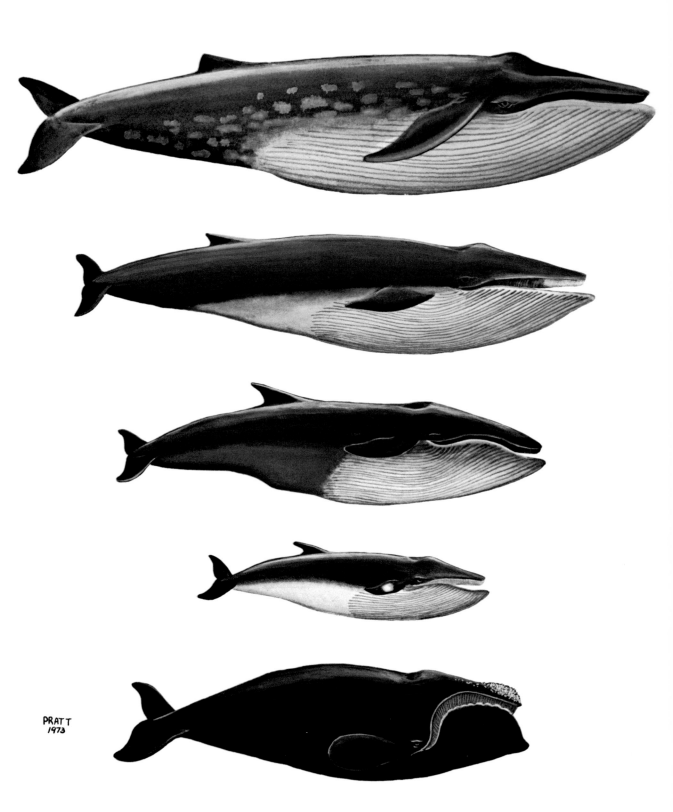

PRATT
1973

supplied by the Louisiana State Museum in New Orleans, but Randolph A. Bazet now informs me that the year was definitely 1915. It was then that he first saw a jawbone and several vertebrae of this whale displayed in Houma. Later the jawbone was transferred to Audubon Park in New Orleans, where I photographed it in 1942. The pictures were sent to Remington Kellogg, who expressed the opinion that the jaw was probably that of a finback. Unfortunately, the jaw can no longer be found at Audubon Park and was evidently discarded, but the photographs of it are on permanent file in the Louisiana State University Museum of Zoology.

Still another whale, reported to have been 64 feet in total length, was found dead on Pelican Island, off Plaquemines Parish, latitude 29°15′ N, longitude 89°35′ W, on 13 February 1917. A photograph of the whale appeared in the *Times-Picayune* three days later, but it had been taken from such an

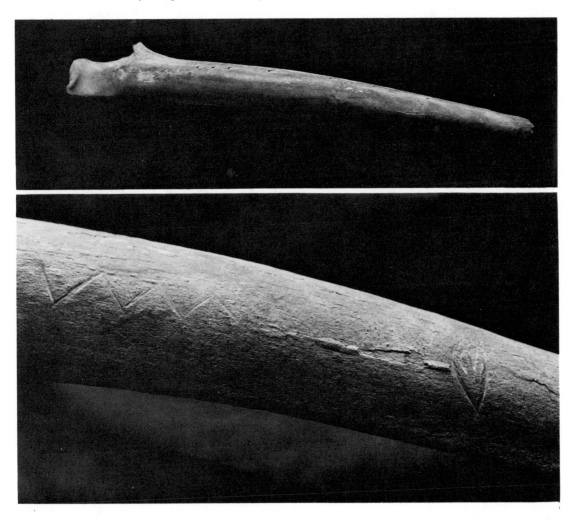

Figure 171 Ramus of lower jaw of a yet unidentified whale of the genus *Balaenoptera* found by W. M. Gosselin on 1 June 1973, partially buried in the sand of the Gulf shore, 6 miles west of Holly Beach. Note the mysterious hieroglyphics carved in the bone, apparently the work of an Indian sometime in the distant past. (Photograph by James R. Jeansonne)

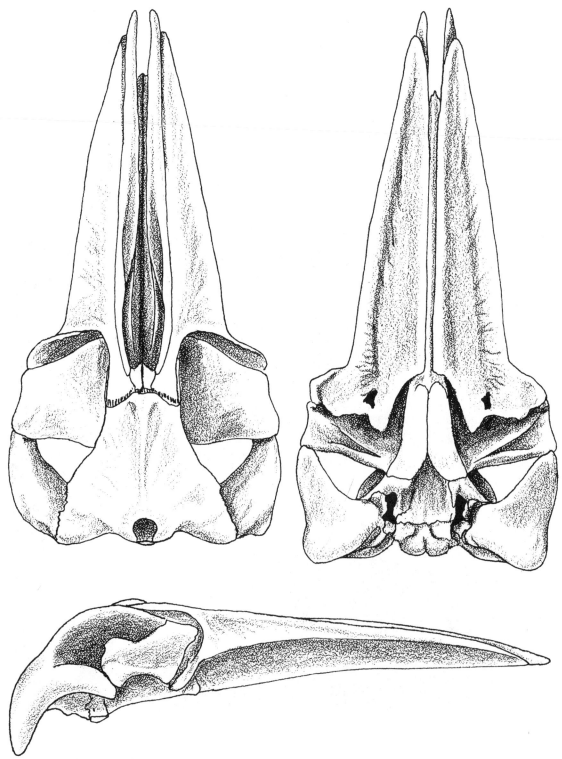

Figure 172 Skull of Fin-backed Whale *(Balaenoptera physalus)*. Approx. ¹⁄₂₀ ×. (Drawings based on photographs of USNM 16039, in True, 1904)

Figure 173 Fin-backed Whale *(Balaenoptera physalus)* that stranded near the mouth of Sabine Pass in early December 1924. Note rows of baleen plates in roof of the mouth and the pleated throat and anterior underparts. The specimen was reported to have measured slightly over 58 feet in length and to have weighed approximately 25 tons.

angle that not a single identifying feature can be detected. The whale was, though, almost certainly one of the species of the genus *Balaenoptera*.

The only other records of Fin-backed Whales in the Gulf of Mexico of which I am aware are what appeared to have been a dead newborn calf found on the beach 22 miles east of Galveston on 21 February 1951 (Breuer, 1951; Gunter, 1954) and a 45-foot subadult male that stranded on the Mississippi Sound side of Ship Island on 7 April 1967 (Gunter and Christmas, 1973, in press).

The rorqual sometimes travels in gams consisting of 100 or more individuals, but occasionally only 2 or 3 are seen together. It swims with great agility and is capable of achieving speeds up to 29 m.p.h. when pressed. Nothing is known of the food habits of this species in the Gulf of Mexico, but judging from its diet elsewhere it presumably eats pelagic organisms, consisting of various species of shrimp, copepods, mollusks, and small, swarming fish. Two methods of taking food are employed: gulping a large mouthful of zooplankton or fishes and swallowing, or skimming near the surface until the mouth is full, then squeezing the water out through the baleen plates and swallowing.

Subspecies

Balaenoptera physalus physalus (Linnaeus)

[*Balaena*] *Physalus* Linnaeus, Syst. nat., ed. 10, 1, 1758:75; type locality, Spitzbergen seas.
Balaenoptera physalus physalus, Tomilin, Compte. Rend. Acad. Sci., U.S.S.R., 54, 1946:468.

Some authors consider the Fin-backed Whales of the Southern Hemisphere to be subspecifically distinct from those of the Northern Hemisphere.

Specimens examined from Louisiana. — A single blade of baleen from a specimen found in 1928 or thereabouts on Breton Island, one of the islands in the Chandeleur chain off the coast of Mississippi and southeastern Louisiana; and a ramus of a lower jaw found in the marsh west of Venice, in Plaquemines Parish, in 1968. This material is deposited in the LSUMZ.

SEI WHALE   Plate IX
(Balaenoptera borealis)

Vernacular and other names. — The name Sei Whale comes from the old Norse word *seihval* (*seje*, meaning coalfish, and *hval*, whale) and is said to be based on the habits of this species in following the coalfish in search of food. The species is also sometimes called Rodolphi's rorqual and the pollack whale. The specific name, *borealis*, means northern, in reference to what was once believed to be its greater abundance in the cold waters of high latitudes, which is not necessarily the case.

Distribution. — This species occurs widely throughout the oceans of the world and may be found in relatively warm waters. Records of its known occurrence in the Gulf of Mexico are limited to two strandings, one on the coast of Veracruz, the other from Louisiana as detailed beyond.

External appearance. — This large whale has the body powerfully built, much more so than the Little Piked Whale (*Balaenoptera acutorostrata*), and the Fin-backed Whale (*B. physalus*). The dorsal fin possesses a deeply concave hind margin, is longer and more conspicuous than that of the Fin-backed, and is located somewhat closer to the center point of the body. The 32 to 60 throat grooves terminate about halfway between the tip of the flipper, when it is laid alongside the body, and the navel, whereas in the Fin-backed Whale they extend to, or almost to, the navel. The color is bluish black above and whitish below, except on the chin, which is gray. The sides are grayish. The undersurface of the flukes and flippers are never white, being usually heavily pigmented with gray. The peduncle is strongly compressed, forming a thin dorsal and ventral ridge, and it joins the flukes abruptly. Viewed laterally the rostrum is slightly curved, recalling the arcuate form of the head of the Black Right Whale (*Balaena glacialis*) but to a much lesser degree. The baleen, or whalebone, is mostly black, with the fringes grayish white.

Measurements.—The length of the Sei Whale may extend up to 60 feet but is usually no more than 50 feet. The body is deepest in the region opposite the pectoral appendages. Selected measurements of the one Louisiana specimen, referred to later in this account, were: total length, 37 feet 3 inches; length of left flipper, 4 feet 7 inches; breadth of flukes, 7 feet. The ventral grooves numbered 48. (Negus and Chipman, 1956)

Cranial characters.—The skull of the Sei Whale is decidedly more massive than that of the Little Piked, or Minke, Whale *(Balaenoptera acutorostrata),* but it resembles closely that of the Fin-backed Whale *(B. physalus).* Much additional study seems to be required to ascertain the range of variation in the skulls of the Sei and Fin-backed Whales. Selected measurements of the skull shown in Figure 174 were as follows: total length, 2,119 plus; greatest breadth (squamosal), 1,335; length of squamosal, 530; breadth of supraoccipital, 940; length of supraoccipital to anterior border of foramen magnum, 675 plus; least width of supraoccipital, 270; width of occipital condyles, 250.

Status.—The only published records prior to this account of the Sei Whale in the waters of the Gulf of Mexico are two skeletons in the National Museum of Mexico labeled as having been taken at Campeche in the southern extremity of the Gulf of Mexico sometime prior to 1915 (Miller, 1928) and that of an animal found stranded in the vicinity of Tecolutla, Veracruz, in 1969 (Villa R., 1969). There is, however, an earlier specimen extant, a male that was found dead several miles northeast of Boothville, near the mouth of Fort Bayou on Breton Sound, in Plaquemines Parish, Louisiana, in April 1956. It was first examined by Dr. Norman C. Negus and Robert K. Chipman, both then of Tulane University. Before any of the skeletal parts could be salvaged, Hurricane Flossie intervened, and the carcass was torn to bits and scattered over the marsh. Fortunately, the skull and other

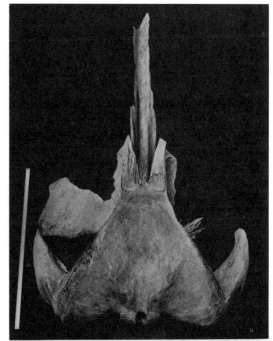

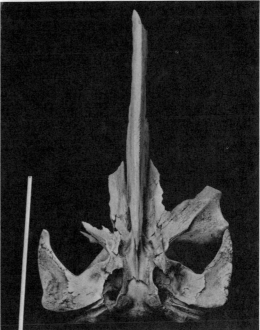

Figure 174 Skull of Sei Whale *(Balaenoptera borealis),* Tulane 4172, from an animal that stranded several miles northeast of Boothville, near mouth of Fort Bayou where it empties into Breton Sound, Plaquemines Parish, April 1956. Approx. ¹/₂₇ ×. (Photograph by Joe L. Herring)

elements were retrieved and now repose in the Tulane University collection. Although first thought to be a Little Piked Whale (*Balaenoptera acutorostrata*) and so reported by Negus and Chipman (1956) and by Waldo (1957), my study of the specimen has shown it to be an example of the Sei Whale (*Balaenoptera borealis*). The suggestion that it might be an example of Bryde's Whale (*B. edeni*) does not appear supported by any evidence, a conclusion in which I am joined by Charles O. Handley, Jr., and James Mead. One especially diagnostic character mentioned by Negus and Chipman (1956) is the fact that the ventral grooves extended only halfway between the end of the flipper and the navel, whereas in *edeni* they extend to or beyond the navel (Omura, 1959).

Balaenoptera borealis Lesson

Balaenoptera borealis Lesson, Compléments des oeuvres de Buffon ou histoire naturelle des animaux rares, 1 (Cétacés), 1828:342; type locality, coast of Holstein near near Grömitz, Schleswig-Holstein, Germany.

Specimens examined from Louisiana.—Total 1, as follows: *Plaquemines Par.:* near mouth of Fort Bayou on Breton Sound (TU).

[BRYDE'S WHALE]
(Balaenoptera edeni)

Vernacular and other names.—The vernacular name Bryde's Whale was first used for this species by the zoologist Olsen when he described *Balaenoptera brydei* in 1912 and was applied by him in honor of Johan Bryde, founder of the South African whaling industry. Although the correct scientific name is *B. edeni* Anderson (1878), the name Bryde's Whale has been retained as the vernacular designation. I am uncertain as to the etymology of the specific name *edeni*.

Distribution.—Bryde's Whale occurs widely throughout most of the oceans of the world, but, because of the relatively short time that it has been recognized as a valid species and the fact that it is often confused with the Sei

Whale (*B. borealis*), precise locality records are sparse in some areas where it is surely present. It has been recorded from Grenada in the West Indies, but I know of only one positive record of its occurrence in Gulf waters, as detailed in this account.

External appearance.—This large whale has the body remarkably elongated and not nearly as powerful in build as that of the Sei Whale or Little Piked Whale. The dorsal fin is smaller than that of the Fin-backed Whale, and the rear margin is concave. The color of the upperparts is blue-black, that of the underparts bluish gray on the throat and white or yellowish white posteriorly, often with a gray band across the belly in front of the navel. The flippers are dark bluish gray on both surfaces. The blades of baleen are comparatively broad and have a concave inner margin. The longest blades are about 19 inches in length and their color anteriorly, extending back some 27 inches, is generally white, frequently with gray stripes; the remainder of the baleen is grayish black. A pair of lateral longitudinal ridges on the dorsal surface of the snout are said to be highly diagnostic of this species. (Olsen, 1913)

Measurements.—According to Fraser (1949), Bryde's Whale averages 42 feet in total length, the maximum length recorded being slightly over 49 feet.

Status.—The only record of this species in the Gulf of Mexico is of a 38-foot individual that stranded alive near Panacea, Wakulla County, Florida, on 2 April 1965 (Rice, 1965). Two days later the Coast Guard towed the animal to sea, and it reportedly swam away. A photograph of the animal, showing the pair of lateral longitudinal ridges on the dorsal surface of the snout, was examined by Dr. Hideo Omura of the Whales Research Institute, Tokyo, and he concurred with Rice in the identification.

Another specimen (LSUMZ 17027), a complete cranium, obtained by the late Edouard Morgan in June 1954 on Chandeleur Island, Louisiana, is considered by F. A. Fraser of the British Museum (Natural History) to be referable to *edeni*. But other experts who have studied photographs of the skull (Figure 175) are hesitant to express a definite opinion.

Balaenoptera edeni Anderson

Balaenoptera edeni Anderson, Anat. zool. res. . . . London, 1878:551 et seq.; type locality, Gulf of Martaban, Burma.
Balaenoptera brydei Olsen, Tidens Tegn., Nov. 12, 1912 [Norwegian newspaper, not seen]; Proc. Zool. Soc. London, 1913:1073–1090.

[BLUE WHALE]   Plate IX
(Balaenoptera musculus)

Baughman (1946) identified a whale that washed ashore on the Texas coast between Freeport and San Luis Pass, in Brazoria County, on 17 August 1940, as an example of this species. He reached this conclusion on the basis of his critical study of excellent photographs that appeared in Houston newspapers following the stranding. He considered as especially significant the reported length of 70 feet; the proportional length of the mouth to the total length being 1 to 4.5; the proportional length of the distance from the tip of the jaw to the pectoral axilla being 35 percent of the total length; the shape of the head, which he said matched closely the description and photographs in True (1904); and the shape and height of the dorsal fin.

Although there can be little doubt that Baughman's identification was correct, the record is nevertheless not as completely satisfactory as it would have been had the carcass been examined by a competent cetologist at the time of the stranding (Baughman did not see the animal himself) or especially if the skull had been salvaged, permitting its reexamination by any subsequent workers who might have reason to question the identification.

The Blue Whale, which also goes by the

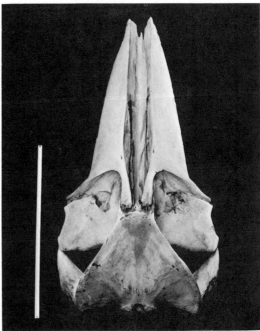
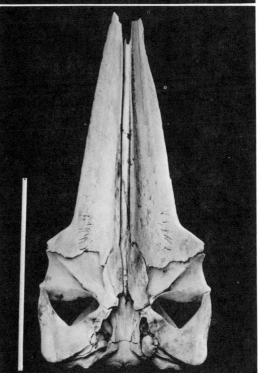

Figure 175 Skull of a still unidentified species of the genus *Balaenoptera,* LSUMZ 17027, that stranded on Chandeleur Island, Louisiana, in June 1954. F. A. Fraser of the British Museum (Natural History) has suggested that the skull may be referable to Bryde's Whale *(B. edeni)*. Approx. $\frac{1}{22}$ ×. (Photograph by Joe L. Herring)

names of sulphur-bottomed and Sibbald's whale, is not only the largest of all living animals but probably also the largest animal to have ever inhabited the earth. A length of 70 feet is about average, but some individuals measuring over 100 feet are on record and 80-foot Blue Whales are not uncommon. The species differs structurally from other rorquals to such a marked degree that it is often placed in the separate genus *Sibbaldus*.

The color of the upperparts, as the vernacular name implies, is a dark bluish gray. The lower surface resembles the back except that it is often lighter in tint and sometimes is yellowish, hence the name sulphur-bottomed whale. The yellow color is adventitious, being produced by layers of diatomes, a form of unicellular, colonial algae that occasionally cover the whale's abdomen. Both above and below the animal is often freckled and mottled with whitish spots. The pectoral appendages are dark above and white below and about one-seventh the body length. The snout is broad with the jaws almost parallel to within a short distance of the tip. The dorsal fin is relatively low and only slightly concave posteriorly and is located closer to the tail flukes than to the center of the body. Like *B. physalus,* but unlike *B. borealis* and *B. acutorostrata,* the throat grooves extend to or beyond the navel. The baleen plates are entirely black and the longest blades are usually less than 30 inches in length. Particularly diagnostic is the great length of the lower jaw, which makes up over one-fourth of the total length. The distance from the tip of the snout to the axilla is also both relatively and absolutely greater than it is in other rorquals, being 35 percent or more of the total length. The skull is quite distinct from the skull of other rorquals. When viewed from above the rostrum is parallel-sided posteriorly, curving inward at the tip. The occipital shield touches the nasals and the parietals are in contact with the maxillae.

Under present international agreements regulating the commercial exploitation of cetaceans, the Blue Whale may be making a slight comeback. It was indeed at one time perilously close to extinction. Notwithstanding the long-held belief that the species is generally confined to icy waters of high latitudes in both hemispheres, it does occasionally move into warmer oceans, especially during the breeding season.

Balaenoptera musculus (Linnaeus)

[*Balaena*] *Musculus* Linnaeus, Syst. nat., ed. 10, 1, 1758:76; type locality, the Firth of Forth, Scotland.
Balaenoptera musculus, True, Whalebone whales of the western North Atlantic, Smithson. Contrib. Knowledge, 33, 1904:149.

LITTLE PIKED WHALE   Plate IX
(Balaenoptera acutorostrata)

Vernacular and other names.—This species also goes by the names Minke whale and lesser rorqual. It is indeed the smallest of all the rorquals. The word piked doubtless refers to the shape of the dorsal fin, which resembles a pike or halberd. The name Minke whale is said to have been derived from the name of a Norwegian whaler by that name who had the reputation of always harpooning small whales (Ash, 1962).

Distribution.—This rorqual is found in most of the oceans and seas of the world, mainly in Arctic, Antarctic, and Temperate zone waters. In the North Atlantic it ranges from Baffin Bay and waters off Greenland south to the Antilles, and in the North Pacific from the Bering Sea to waters off Baja California. Only two instances of its occurrence in the northern part of the Gulf of Mexico are known, although Layne (1965) reported what may have been this species stranded off Pine Island, near the town of Bayport, Hernando County, Florida, approximately 40 miles north of Tampa Bay, on 14 November 1962.

External appearance.—The Little Piked Whale is distinguished by its relatively small size (seldom exceeding 30 feet in total length) and by the color of its baleen, which is entirely whitish or yellowish white. A highly dis-

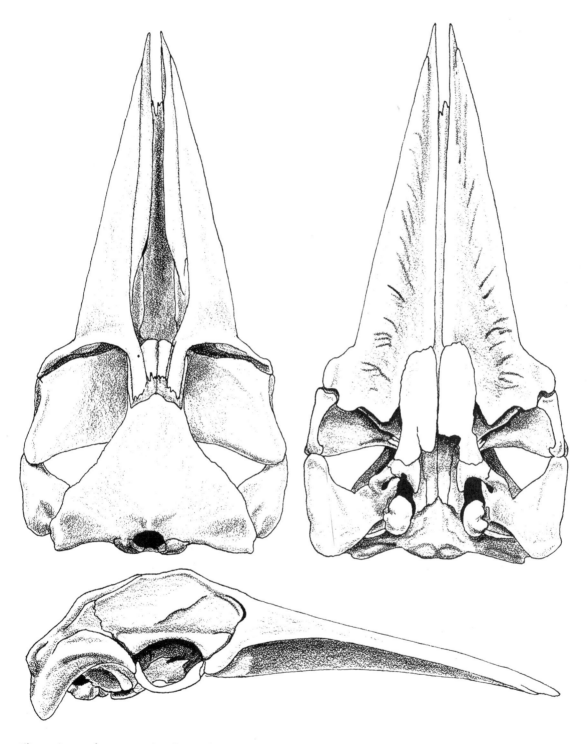

Figure 176 Skull of Little Piked Whale *(Balaenoptera acutorostrata)*. Approx. ⅛ ×. (Drawings based on photographs of USNM 20931, in True, 1904)

tinctive color feature is the patch of white on the dorsal surface of the flipper, which contrasts sharply with the otherwise blue-gray of the upperparts. The underparts, including the undersurface of the flippers, are white. The snout is shorter and more triangular when compared with that of other members of the genus. The flippers measure approximately one-eighth of the body length, and the dorsal fin is prominent and falcate. The throat grooves number about 50 and extend to only approximately halfway between the ends of the flippers and the navel. More than 300 blades of baleen are present on each side of the upper jaw. Each blade, excluding the bristles, is slightly less than a foot in length. As noted above, the color of the baleen is highly diagnostic.

Cranial characters.—The skull of *B. acutorostrata* is easily separable from that of other members of the genus by the straight, rather than outwardly bowed, lateral borders of the rostrum when viewed from above (Figure 176).

Status.—I know of only two records of the Little Piked Whale anywhere on the northern Gulf Coast. A skeleton was found in 1940 embedded in the mud of Spring Creek, Wakulla County, Florida, south of Tallahassee (Moore, 1953), and an individual whale 20 to 30 feet in length stranded or washed ashore five miles west of Holly Beach, in Cameron Parish, on or about 1 January 1966. Unfortunately, I did not learn about the animal until after it had been burned and the remains buried by a sanitation crew. A poor photograph, made by a local resident, showed very little in the way of diagnostic features, but everyone with whom I talked who had seen the animal reported the color of the baleen to be yellowish in color. I was, however, recently elated to find that one of the rami of the lower jaw had been salvaged by a science teacher in the Hackberry High School, Mr. Robert F. Henry, who generously turned the specimen over to me for study and for de-

posit in the LSUMZ. Mr. Henry also independently asserted that the color of the baleen was entirely yellowish. A critical examination of the jawbone leaves little or no doubt that the animal was indeed a Little Piked Whale.

A specimen reported by Negus and Chipman (1956) to be of this species that was found near the mouth of the Mississippi River in April 1956 now proves to be an example of the Sei Whale *(Balaenoptera borealis)*. For further details concerning this specimen, see the account of the Sei Whale (page 369).

Balaenoptera acutorostrata Lacépède

Balaenoptera acuto-rostrata Lacépède, Histoire naturelle des cetacées . . ., 1804:xxxvii; type locality, European seas = Cherbourg, France.

The species *Balaenoptera acutorostrata* is presently regarded as monotypic, but some of the species now included in its synonymy may eventually prove to be valid races.

Specimens examined from Louisiana.—Total 1 (ramus of lower jaw only): *Cameron Par.:* 4 mi. W Holly Beach (LSUMZ).

[Genus *Megaptera* Gray]

Dentition.—No teeth present in postfetal life; approximately 300 nearly black baleen plates.

[HUMPBACK WHALE]   Figure 177
(Megaptera novaeangliae)

The only record of this species in the Gulf of Mexico is that of an individual observed, studied, and photographed at close range on 8 April 1962 off Egmont Key at the mouth of Tampa Bay (Layne, 1965).

The Humpback Whale is easily identified by a number of salient features. Measuring up to 50 feet in total length, it is black above and on the sides and is generally white on the venter; the pectoral appendages are white below and on the forward half or more of the dorsal surface; the flippers are exceedingly

Figure 177 Humpback Whale *(Megaptera novaeangliae)*.

long (approximately one-third the length of the body) with their anterior edge prominently serrated; knoblike protuberances are located on the anterodorsal part of the snout and on the lower jaw; the dorsal fin is small, falcate, and situated two-thirds the length of the body posterior to the snout. The skull is depicted in Figure 178.

Megaptera novaeangliae (Borowski)

Balaena Novae Angliae Borowski, Gemeinnüzzige Naturgeschichte des Thierreichs ..., 2 (1), 1781:21; type locality, coast of New England.
Megaptera novaeangliae, Kellogg, Proc. Biol. Soc. Washington, 45, 1932:148.

[The Right Whale Family]

[Balaenidae]

THE MEMBERS of this family, which include the Greenland, Black, and Pygmy Right Whales, are readily distinguished from the rorquals of the family Balaenopteridae by their enormous heads, by the extremely long baleen plates, by the curved, arcuate shape of the mouth (Plate IX), and by the absence of throat furrows. The baleen plates in the genus *Balaena* are grayish black or black and measure up to 12 feet in length. When viewed head on the lower jaws are U-shaped. The Greenland Right Whale, or bowhead, *Balaena mysticetus,* once occurred in great numbers in Arctic waters, but by the early 1900s it had been almost exterminated by the whaling industry. The species comprising this genus are now protected by international agreement and seem to be increasing in numbers. The only member of the family known from the Gulf of Mexico, and at that on the basis of only two definite records, is the Black Right Whale *(Balaena glacialis)*. This species, often considered mainly an inhabitant of cold waters, is, however, sometimes found in warmer seas than is the bowhead, as evidenced by its occurrence, although infrequently, in the Gulf of Mexico.

Because the members of this family were

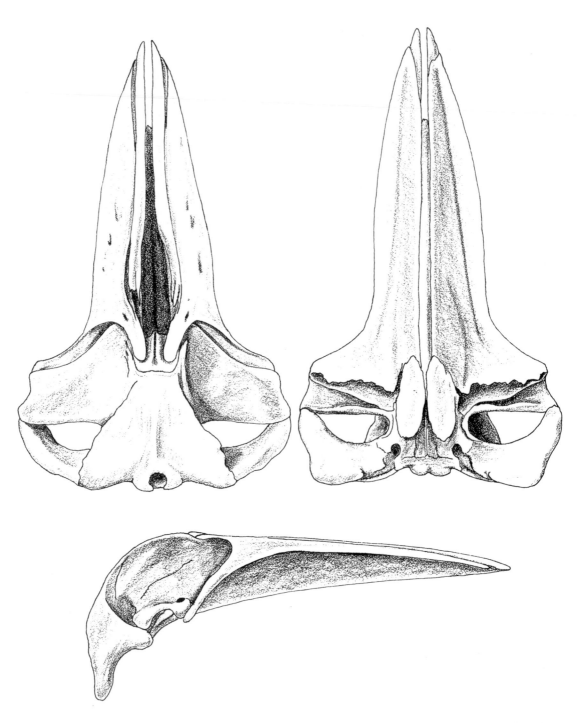

Figure 178 Skull of Humpback Whale *(Megaptera novaeangliae)*. Approx. ¹/₂₄ ×. (Drawings based on photographs of USNM 21492 and an unnumbered skull from Cape Cod, in True, 1904)

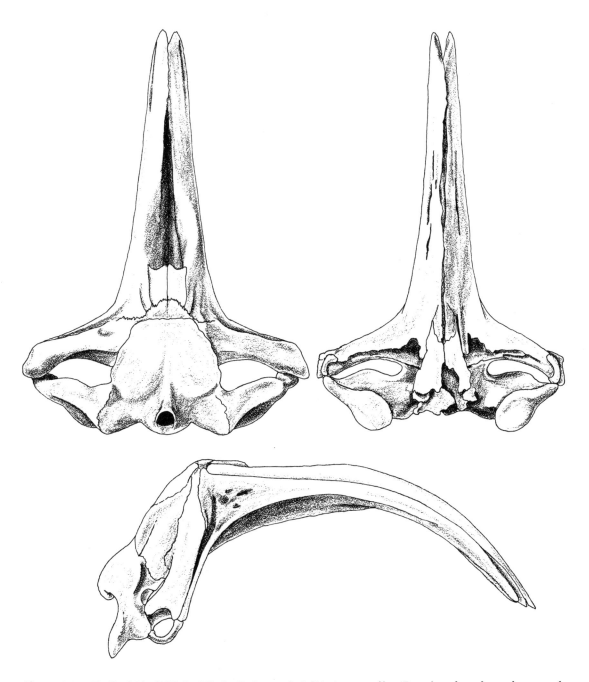

Figure 179 Skull of Black Right Whale *(Balaena glacialis)*. Approx. ¹⁄₂₉. (Drawings based on photographs of USNM 23077, in True, 1904)

notoriously slow swimmers, were usually abundant, yielded copious amounts of oil, and had a highly buoyant carcass not inclined to sink when harpooned, they were considered the right whale to pursue, hence their vernacular name "right whales."

[Genus *Balaena* Linnaeus]

Dentition. — No teeth present except in the embryo; replaced in postfetal life by 250 or more baleen plates on each side of upper jaw that are grayish black or black.

[BLACK RIGHT WHALE]   Plate IX
(Balaena glacialis)

The Black Right Whale has only twice been recorded in the Gulf of Mexico. Moore and Clark (1963) report the occurrence of two of these huge cetaceans off New Pass, near Sarasota, Florida, on 10 March 1963. More recently, on 30 January 1972, one washed ashore approximately five miles northeast of the jetties on Surfside Beach near Freeport, Brazoria County, Texas (Schmidly et al., 1972).

Like other members of the family Balaenidae, this whale has the jaws highly arched when viewed from the side. The head is huge and makes up at least one-fourth of the body length. The animal possesses neither dorsal fins nor throat pleats, and the top of the head has a rough, horny projection called the "bonnet." The largest blades of baleen are sometime six feet or more in length, and their color, including the fringes, is black. The pectoral appendages are large and rounded and much broader than those of the rorquals, or fin whales. The skull is depicted in Figure 179.

Balaena glacialis **Borowski**

Balaena glacialis Borowski, Gemeinnüzzige Naturgeschichte des Thierreichs . . . , 2 (1), 1781: 18; type locality, North Cape, Norway.

FLESH-EATING MAMMALS
Order Carnivora

THE ORDINAL name Carnivora comes from two Latin words (*carnis*, flesh, and *vorare*, to devour) and means flesh eating. One might infer that the members of this group consume only red meat in their diets. Such, however, is not the case. Bears, for instance, include in their fare great quantities of berries, at least during certain times of the year, and their passion for honey is notorious. Skunks eat insects; raccoons relish crayfish and crabs; aardwolves of Africa consume termites, carrion beetles, and maggots; binturongs of southeast Asia feed mainly on fruit; and hyenas are great scavengers. Yet despite these dietary departures, carnivores are indeed basically flesh eaters and as such they are well adapted. Not only do many of them possess the physical prowess and other attributes needed to capture prey, but their teeth are specialized for seizing, holding, and tearing apart the flesh of the animals they subdue. The incisors are long and sharp and the so-called carnassials (the fourth upper premolar over the first lower molar) are unusually well developed (especially so in the cats, least so in the bears) and aid in shearing flesh.

The members of the order are terrestrial except for the Polar Bear (*Thalarctos maritimus*), the Sea Otter (*Enhydra lutris*), and to some extent the river otters (*Lutra*). The smallest living carnivore is the Least Weasel (*Mustela rixosa*), which is only eight or nine inches long with a weight of less than a quarter of a pound. The largest is one of the Alaskan races of the Brown Bear (*Ursus arctos*), with a length of eight to nine feet and a weight of more than 1,700 pounds. Carnivores have at least four clawed toes on each foot, and in the cats, except for the cheetah, the claws can be fully retracted into sheaths. The dogs and the cats walk on their toes, that is they are digitigrade, while most other carnivores walk on the soles of their feet and are classed as plantigrade.

Carnivores are worldwide in distribution except for Antartica and some oceanic islands. The order comprises seven families of which five are represented in Louisiana. The only families that are absent from our state are not found anywhere in the entire Western Hemisphere except as introductions. They are Hyaenidae, the African and southwest Asian hyenas, and Viverridae, the Old World civets, genets, mongooses, and their allies. But of the world's 101 genera and some 247 species of carnivores Louisiana has a decidedly unimpressive number—9 genera and 14 species. North America down through Panama has 27 genera and 61 species.

The Doglike Carnivore Family
Canidae

THIS FAMILY of some 14 genera and 35 species is worldwide in distribution except for New Zealand, New Guinea, Celebes, Polynesia, the Moluccas, Taiwan, Madagascar, the West Indies, most oceanic islands, and Antartica. The Dingo *(Canis dingo)* of Australia is probably only a breed of dog that was introduced to that continent by aboriginal immigrants. It had certainly been there a long time when Europeans first arrived.

The Domestic Dog is often referred to as *Canis familiaris,* as if it were a distinct species. Instead it is probably a derivation of the wolf or of several wolflike canids with which man has come in contact over the last ten thousand years or so of his history. Charles Darwin, in attempting to trace the evolution of the dog, finally gave up in despair, saying that its ancestry was concealed "in the mist of antiquity."

Admittedly much time has elapsed since primitive man can first be pictured cuddling an orphaned puppy wolf and finding it responsive to his care and comforting to himself. The union was a happy one, for the ancestral dog possessed many qualities that equipped it for successful domestication. Above all it had a desire to please its master, and it seemed to enjoy doing the things asked of it, such as participation in the hunt. And for man, it served all important functions as scavenger of camp leftovers and as a sentinel to warn his owner of intruders.

In the time that has gone by since the first domestication was achieved, there has been ample opportunity for genetic change, both by design and by accident, and thus the multiplicity of breeds is no more cause for wonder than what man has accomplished by selective breeding of the domestic pigeon and the chicken, which are descendants of, respectively, the Rock Dove *(Columba livia)* and the Red Junglefowl *(Gallus gallus).* The diversity of breeds has prompted some students of the subject to try to implicate jackals in the dog's ancestry, but the present consensus of experts is that only the wolf and wolflike canids could have provided the matrix from which all dogs are made.

Genus *Canis* Linnaeus

Dental formula:

$$I\frac{3-3}{3-3} \ C\frac{1-1}{1-1} \ P\frac{4-4}{4-4} \ M\frac{2-2}{3-3} = 42$$

COYOTE   Plate X
(Canis latrans)

Vernacular and other names. — The common name Coyote comes from the Aztec Indian word *coyotl* and should be pronounced "ky-ó-tee" as it is in the Spanish version rather than "ky-oát." The generic name *Canis* is the Latin word for dog, and the specific name *latrans* is

380

a Latin participle meaning barking, which refers to some of the sounds made by the animal.

Distribution.—The Coyote ranges widely in North America from extreme northern Alaska across western and southern Canada to southern Quebec, thence southward through the western two-thirds of the United States to southern Mexico (except the Yucatan Peninsula), Guatemala, Honduras, El Salvador, Nicaragua, and northwestern Costa Rica. In Louisiana it occurs in the northern and central parts of the state as far south and east as East Baton Rouge Parish. (Map 51)

External appearance.—This doglike canid, which closely resembles a small German Shepherd, requires little description other than to tell how it differs from the true wolves, especially the Red Wolf *(Canis rufus)*. Small specimens of the Red Wolf are difficult to distinguish from large examples of the Coyote, and hybrids occur frequently be-

tween the Coyote and the Red Wolf, as well as between each of these species and the Domestic Dog. The nose pad of the Coyote tends to be smaller than that of the Red Wolf, being one inch or less in diameter instead of 1.25 inches or more. The heel pad is also smaller in the Coyote. The color of the Coyote is variable. Typically the upperparts are light gray to pale yellow with the guard hairs broadly tipped with black, particularly along the center of the dorsum. The muzzle, back of the head, base of the ears, and nape are strongly tinged with yellow or buff. The color of the muzzle of the Red Wolf tends to be light, with the white of the lips extending well up on the sides of the muzzle. In contrast, the area of white around the lips of the Coyote is narrow and sharply demarcated. Wolves usually have light areas around the eyes but with a tan spot present over each eye, thereby adding to the almond or slanted appearance (Riley and McBride, 1972). The upperside of the tail is like the back, but on the underside it is whitish near the base, then pale yellowish; the tip is black. The throat, belly, and inside of the legs are whitish to pale yellow, but the outside of the hind legs is somewhat reddish. The overall color of the entire upperparts tends to be more rufescent in the Red Wolf than in the Coyote. Color, though, is of little value in distinguishing the Coyote from the Red Wolf, or from certain breeds of the Domestic Dog. This difficulty is further complicated by the occurrence of hybrids. The tail of the Coyote is usually held low when the animal is running, whereas in the Red Wolf it is nearly always held horizontal. As pointed out by Riley and McBride (1972), of all the external characteristics, the long legs of the Red Wolf are one of the most striking distinctions between it and the Coyote. People who know the two species frequently comment on the "legginess" of the wolf. The ears are also longer in the Red Wolf. In 14 specimens from Liberty, Chambers, and Galveston counties, Texas, the ears averaged 5.01 (4.5 – 5.5) inches, while in 25

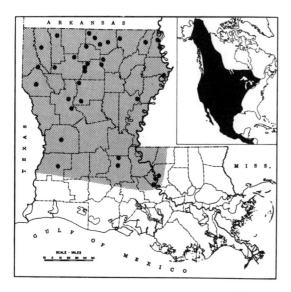

Map 51 Distribution of the Coyote *(Canis latrans)*. *Shaded area,* known or presumed range within the state; •, localities from which museum specimens have been examined; *inset,* overall range of the species.

specimens of the Coyote the ears averaged only 4.18 (3.5–4.5) inches. (Figure 180)

Color phases.—Coat color in the Coyote is highly variable, with some individuals being excessively blackish. Although a black color phase is common in the Red Wolf, it is rare in the Coyote. Albinos also appear occasionally in litters of Coyotes.

Measurements and weights.—Three males and two females in the LSUMZ measured as follows: total length, 1,194 (1,000–1,238); tail, 355 (300–440); hind foot, 199 (175–210); ear, 108 (97–120). Weights generally range between 18 and 30 pounds, but much heavier individuals are on record (Young, *in* Young

and Jackson, 1951). I am indebted to Ronald M. Nowak (pers. comm.) for the following skull measurements of 26 Louisiana Coyote specimens, including those in the LSUMZ: greatest length, 204.0 (190.0–222.0); condylobasal length, 191.7 (184.0–207.0); zygomatic breadth, 103.6 (94.0–110.0). Thirty-one skulls of known Coyotes (30 from San Luis Potosí and one from Eddy County, New Mexico) in the LSUMZ measured by Nowak averaged, respectively, as follows: 190.9 (171.0–207.0); 178.3 (162.0–189.0); 95.7 (86.0–106.0).

Cranial and other skeletal characters.—The skull of the Coyote cannot always be distinguished from that of the Red Wolf or the Domestic

Figure 180   Coyote *(Canis latrans)*, ♂, Colorado. (Photograph by Robert W. Wright)

Dog, but in the Coyote the anteroposterior diameter of the canines is usually less than 11 mm, whereas in the case of wolves and most dogs it is more than 11 mm. According to Jackson (*in* Young and Jackson, 1951), when the lower jaw is in place the canines of the Coyote often extend well below a line drawn through the anterior mental foramina of the mandibles. In the Red Wolf and Domestic Dog the tips of the canines usually fall above this line. Paradiso and Nowak (1971) question the validity of this character. In general the Coyote possesses a longer and narrower rostrum, but some dog skulls are not distinguishable by this means. In a Coyote skull, the length of the upper toothrow is three times or more the width of the palate (the distance between the inner borders of the alveoli of the first premolars). (Figures 181 and 183)

Sexual characters and age criteria.—Males are easily identified by the presence of the penis and scrotum. A penis bone, or baculum, is present. Females normally possess five pairs of teats. Males average slightly larger than females. In the first years of the Coyote's life the upper and lower incisors have lateral lobes on each side of the teeth. By the time the animal is five or six years old the tips of these teeth have usually worn down to the level of the lateral lobes, and the canines display a similar degree of wear. Both sexes have a scent gland on top of the tail about two inches from the base. It measures approximately one inch long and slightly more than a quarter of an inch in width. The odor produced by the secretion of this gland provides a means of intraspecific recognition. The strongly scented urine performs the same function.

Status and habits.—Coyotes, or Coyotelike animals, appeared in Louisiana sometime after 1942, mainly after 1950. They are now, however, widespread throughout the wooded uplands of the northern and western parts of the state. The Louisiana Wild Life and Fish-

eries Commission's *Biennial Report for 1966–1967* (1968), showed that 269 Coyotes were trapped in the period 1 December 1965 to 30 November 1966; and 306 were obtained in the period 1 December 1966 to 15 November 1967. The presence of wolves in the state in the same periods was not mentioned.

Paradiso (1968) and Paradiso and Nowak (1971) point out that before the drastic disruption of the environment by man, three species of canids, the Gray Wolf *(Canis lupus),* the Red Wolf *(C. rufus),* and the Coyote *(C. latrans)* maintained themselves in North America as discrete, viable species and showed no evidence of interspecific hybridization. But sometime prior to 1930, probably before the turn of the century, the concurrent decline in the Red Wolf population and extensive modifications of the habitat led to a breakdown in the isolating mechanisms between the Red Wolf and the Coyote and resulted in the formation of a hybrid swarm on the Edwards Plateau of central Texas. By 1940 this hybrid population had spread eastward to eastern Texas, eastern Oklahoma, and western and southern Arkansas, filling in the void created by the virtual extirpation of the Red Wolf. In the next decade it entered Louisiana and promptly occupied most of the state west of the Mississippi River. The eastward movement is still in progress, as evidenced by the appearance recently of these Coyotelike animals in the Florida Parishes.

Coyotes can mate with Domestic Dogs and produce fertile progeny, which are often called "coy-dogs." Mengel (1971), who raised two generations of these creatures in captivity, doubts that hybrid swarms of coy-dogs are ever likely to develop in the wild. The breeding seasons of the male and female Coy-

Tracks of Coyote

ote are synchronized; both come into heat in late winter. Dogs as a group have no annual reproductive cycle. Females come into breeding condition twice a year on the average but without clear group correlation with any certain times of the year. To cope with this situation, males are constantly in breeding condition. Thus any time a male dog chances upon a female Coyote in heat, he is ready to impregnate her. In contrast, the male Coyote that meets a female dog in rut seldom happens to be capable of copulating. The result is

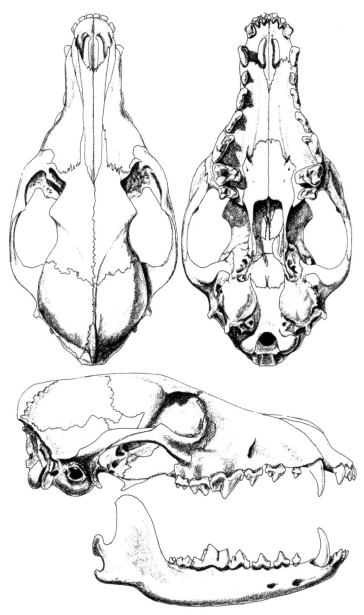

Figure 181 Skull of Coyote *(Canis latrans frustror)*, ♂, 4 mi. NNE Foules, Catahoula Parish, LSUMZ 16153. Approx. ½×.

that coy-dogs born in the wild usually have a dog as their male parent. They are at a marked disadvantage, because Father Dog has no interest in his pups whereas Father Coyote is a model parent, who shares the work of feeding the young. Male coy-dogs apparently share the male dog's disinterest in his offspring as well as the male Coyote's restricted season of rut. For some unfathomable reason, however, the period of heat comes earlier in the season in the coy-dog than in the Coyote, so much earlier that animals of coy-dog parentage are born at a time of year unfavorable for survival in the wild, particularly in northern latitudes. These factors combine to minimize the flow of dog traits into the Coyote gene pool.

The arrival of the Coyote in Louisiana was in the opinion of most people by no means a welcome addition to our mammal fauna. It is detested by farmers and sportsmen because of its sometimes justifiably alleged depredations on livestock, poultry, and game birds. But so often the villain is not the Coyote. Packs of wild dogs are frequently the culprits, especially with regard to livestock, and they are much more destructive. Moreover, when damage is being done by the Coyote, it may be the work of a small percentage of the population, the "bad guys" as we would say in modern parlance. Probably no more than 20 percent of the Coyote's total food intake consists of items that are of any real importance to man. The examination of 770 Coyote stomachs in Missouri (Korschgen, 1957) showed the following items and their percentages by volume: rabbits, 53.7; rats and mice, 8.7; other wild mammals, 7.5; livestock, 8.9; poultry, 11.3; wild birds, 0.5; known carrion, 5.8; insects, 0.8; plants, 2.0; and miscellaneous, 0.8. Unfortunately, one cannot always tell if the food in a stomach represents an animal killed by the Coyote or merely one eaten as carrion. Moreover, I suspect that Korschgen's percentages may have been based on a biased sample. To have obtained 770 stomachs, he surely must have utilized many animals that had been killed because of suspected depredations, thereby perhaps accounting in some measure for the high percentage of poultry in the sample. Sportsmen would doubtless point to the 53.7 percentage of rabbits as justification for their position with regard to the Coyote, but game management experts would be quick to point out that rabbit populations in Missouri are still optimal despite the number taken by the Coyote. In any event, much of the destruction caused by the Coyote is compensated for by the good it does in eating rats and mice, in killing old, sick, and injured animals that are not fit to survive, and in acting as a scavenger. Hall (1946) and others have shown conclusively that predator control in most cases is biologically unsound, that it often results in disastrous consequences by upsetting the balance of nature.

The vocalizations of the Coyote consist of barks, yelps, and yaps in a series that often starts out as a solo performance but becomes a chorus as other members of the pack join in the refrain. The calls increase in volume and usually end in a somewhat mournful squall. A sudden clap of thunder or any other loud noise will frequently set off one of these vocal renditions. I for one would rue the day when such sounds are no longer part of the world in which we live. At present we can still at least occasionally pay a visit to places where Coyotes roam. Of course, one nowadays seldom goes long without hearing the sound of a Coyote for it is a standard sound effect in nearly all TV westerns.

For the most part the Coyote is an animal of fairly open, brushy, country or idle farm lands bordered by timber. Its dens are in gulleys, under roots or overhanging banks, and in places studded with thickets and dense cover. Where the soil is loose and sandy it often digs a hole in the ground or usurps a hole formerly occupied by a fox. The breeding season is in late winter and early spring. The gestation period lasts for 58 to 63 days. The young usually number from five to seven and are blind and helpless at birth. Litters containing 19 pups are on record (Young, *in* Young and Jackson, 1951). The eyes open sometime within two weeks after birth and by the end of the third week the pups venture forth out of the den for the first time. Weaning occurs at about eight weeks of age. Both parents care for the young. At first the male brings food to the female and she tears it into small bits before portioning it out to the pups. Later both parents bring food to the den in their stomachs, disgorging it at the entrance in a partially digested state. When the young are 8 to 12 weeks old, the task of teaching the young to hunt for themselves begins.

Predators. — The Coyote has only one serious enemy — man with his rifle, traps, and poison bait. Sometimes overzealous predator control agencies launch poison bait campaigns against Coyotes, but such a practice should never be allowed. It is nonselective and results in the deaths of many unintended victims, including pets, furbearers, and livestock not to speak of untold numbers of smaller animals that cause no harm.

Parasites and diseases. — Coyotes are frequently heavily infested with fleas. In severe cases they are sometimes forced to seek out new dens to obtain relief. Eads (1948), in his study of the ectoparasites of Coyotes in eastern Texas, examined 90 young and adult individuals and found three fleas (*Pulex irritans, Echidnophaga gallinacea,* and *Orchopeas sexdentatus*), seven species of ticks (*Amblyomma americanum, A. inornatum, A. maculatum, Dermacentor variabilis, Ixodes cookei, I. scapularis,* and *Octobius megnini*), one species of mallophaga (*Trichodectes canis*) and one biting louse (*Heterodoxus spiniger*). Other external parasites reported in the literature include three fleas (*Oropsylla arctomys, Arctopsylla setosa,* and *Cediopsylla simplex*). The mange mite (*Sarcoptes scabiei*) commonly infects the Coyote, at times so severly that it causes the almost complete loss of hair. Coyotes are also the host for an

array of protozoans, flukes, tapeworms, roundworms, and thorny-headed worms. The following parasites have been reported, mainly by Erickson (1944): four species of trematodes (*Alaria mustelae, A. oregonensis, Opisthorchis pseudofelineus,* and *Nanophyetus salmincola*), nine species of cestodes (*Mesocestoides kirbyi, Multiceps multiceps, M. packii, Taenia hydatigena, T. krabbei, T. laruei, T. pisiformis, T. rileyi,* and *Hydatigera laticollis*), 10 species of nematodes (*Ancylostoma caninum, Dioctophyme renale, Oslerus osleri, Mastophorus muris, Molineus patens, Passalurus nonannulatus, Physaloptera rara, Protospirura numidica, Rictularia splendida,* and *Toxascaris leonina*), and one species of acanthocephalan (*Oncicola canis*).

Subspecies

Canis latrans Say

Canis latrans Say, *in* Long, Account of an exped.... to the Rocky Mts...., 1, 1823:168; type locality, Engineer Cantonment, about 12 mi. SE of the present town of Blair, Washington County, Nebraska, on the West bank of the Missouri River.

Although no less than 19 subspecies of *Canis latrans* are recognized by Young and Jackson (1951) and by Hall and Kelson (1959), there is little justification for attempting to assign Louisiana specimens to one of the subspecies, since the population in the state is believed to be the product of a hybrid swarm originating initially in central Texas between *C. latrans texensis* Bailey and *C. rufus rufus* Audubon and Bachman that moved eastward, incorporating *C. l. frustror* Woodhouse of eastern Texas and finally extending into Louisiana. Skulls of Louisiana specimens of the Coyote are intermediate in size between pre-1900 *Canis latrans* and early specimens of *Canis rufus,* thereby possibly pointing to their hybrid origin (Paradiso, 1968 and Paradiso and Nowak, 1971). The fact that *frustror* has been described as "the largest of the Coyotes" could, perhaps, be attributable to introgression of genes of *Canis rufus gregoryi,* which once occurred commonly in eastern Texas and is still present there, albeit in marginal numbers. No less an authority than Vernon Bailey (1905) regarded *frustror* as more nearly related to *Canis rufus* than to *Canis latrans,* but I know of no modern systematist who subscribes to this conclusion.

Specimens examined from Louisiana. — Total 66, as follows: *Beauregard Par.:* 6 mi. E Singer, 1 (MSU). *Bienville Par.:* 3 mi. N Arcadia, 1 (LTU); Arcadia, 1 (LTU); Toulon Creek, 3 mi. NW Ringgold, 1 (LTU). *Bossier Par.:* 8 mi. E Benton, Dean's Point, 1 (LSUMZ); locality unspecified, 1 (LTU). *Caddo Par.:* Wallace Lake, 1 (LTU). *Catahoula Par.:* 4 mi. NNE Foules, 1 (LSUMZ). *De Soto Par.:* locality unspecified, 7 (1 LSUMZ; 6 LTU). *East Baton Rouge Par.:* 5 mi. N Baton Rouge, 1 (LSUMZ). *East Carroll Par.:* Lake Providence, 1 (LTU); 4 mi. N Transylvania on Miss. River, 1 (LTU). *Jackson Par.:* Quitman, 1 (LTU); 1 mi. NE Clay, 1 (LTU); Punkin Center, 5 mi. N Jonesboro, 1 (LTU). *Lincoln Par.:* Hico, 1 (LTU). *Madison Par.:* Diamond Point, 2 (NLU). *Morehouse Par.:* 7 mi. W Beekman, 1 (NLU). *Natchitoches Par.:* west bank of Red River, S Natchitoches, 1 (LSUMZ); locality unspecified, 9 (LTU); Tauzin Island [mouth of Bayou Pierre], 1 (NSU). *Ouachita Par.:* Calhoun, 1 (LTU); 6 mi. SW Calhoun, 1 (LTU). *St. Landry Par.:* 1 mi. N Big Cane, 1 (USL); 9 mi. NE Opelousas, 1 (USL); Thistlethwaite Game Mgt. Area, 3 (LSUMZ). *Union Par.:* 7 mi. E Farmerville, 1 (LTU); between Spearsville and Farmerville, 1 (LTU); 3 mi. S Farmerville, 1 (LTU); 5 mi. S Marion, 1 (LTU); 2 mi. N Farmerville, 1 (LTU); E Lillie, 3 mi. N La. 558 from jct. La. 15, 1 (USL). *Vernon Par.:* locality unspecified, 1 (MSU). *Webster Par.:* locality unspecified, 6 (1 LSUMZ; 5 LTU); Indian Creek, 1 (LTU). *West Baton Rouge Par.:* Port Allen, 3 mi. W Miss. River Bridge, 1 (LSUMZ); ca. 2 mi. W Addis, 1 (LSUMZ). *West Carroll Par.:* locality unspecified, 1 (NLU). *Winn Par.:* 2 mi. S Brewster Mill, 5 (LTU); 1 mi. N Wheeling, 1 (NSU).

RED WOLF
(Canis rufus)

Plate X

Vernacular and other names. — The name Red Wolf refers to the overall rufescence that tinges the coat of the animal. This feature is also responsible for the second part of the scientific name, *rufus,* which is the Latin word for reddish. The word "Red" in the name is, however, misleading, for the normal color phase of the species is basically gray. Because of the prevalence of a black color phase, the name black wolf has often been applied to the species, especially in the northeastern part of the state. The subspecific name of the population that occurs in Louisiana, *gregoryi,* is in honor of Tappan Gregory, an accomplished,

pioneer wildlife photographer, who obtained superb flash pictures of wolves in northeastern Louisiana in the 1930s (Figure 182) and who authored in 1935 the paper entitled "The Black Wolf of the Tensas."

Distribution. — This species originally occupied a rather restricted range in the Mississippi Valley from central-western Illinois and north-central Indiana south through southern Missouri, Arkansas, eastern Oklahoma, and doubtless western Kentucky and western Tennessee to the Gulf Coast of Louisiana and Mississippi, thence west to central Texas and east through Alabama to the Atlantic Coast in Georgia and Florida. It is now virtually extirpated in all its former range. Meager numbers are still definitely present in parts of southwestern Louisiana and extreme southeastern Texas, and a few possibly exist in the Ozark Mountain region of Arkansas and extreme eastern Oklahoma. (Map 52)

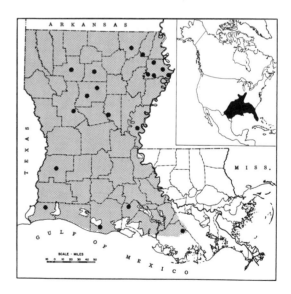

Map 52 Distribution of the Red Wolf *(Canis rufus). Shaded area,* former range within the past two decades; •, localities from which museum specimens have been examined; *inset,* overall former range of the species, which is now extirpated almost everywhere.

External appearance. — The Red Wolf closely resembles the Coyote but is decidedly larger and more robust. The rhinarium, or nose pad, and the feet are larger, and the Tawny or rufescent element in the pelage is more predominant. The normal color phase is a varying mixture of Cinnamon-Buff and Tawny interspersed with gray and black. The middorsal area is heavily overlaid with black, as is the dorsal portion of the tail, which is tipped with black. The muzzle, ears, and the nape are distinctly Tawny. The cheeks are grayish and the underparts vary from whitish to Pinkish Buff.

Color phases. — A black color phase is of fairly frequent occurrence. The color is solid black except that small median white patches some-

times are present on the chin, throat, and pectoral region; the inguinal area and inside of the legs are often extensively suffused with white hairs. (See Figure 182)

Measurements and weights. — According to E. A. Goldman (*in* Young and Goldman, 1944) the external measurements of two adult males from Barren, Missouri, were, respectively, as follows: total length, 1,650, 1,473; tail vertebrae, 393, 362; hind foot, 254, 235. The weight of an adult male from the same locality was 67 pounds. Two adult females from Barren measured, respectively, as follows: total length, 1,448 and 1,442; tail vertebrae, 355 and 343; hind foot, 216 and 222. The weight of a female from the same locality was 50 pounds. Riley and McBride (1972) state

Figure 182   A melanistic Red Wolf *(Canis rufus),* photographed by the late Tappan Gregory in Tensas Parish in November 1934. (Courtesy of William Beecher and the Chicago Academy of Sciences)

that recently collected Red Wolves from the Texas Gulf Coast prairies fall within the 40-to 60-pound range, with individuals weighing over 60 pounds being rare; the heaviest weighed 76 pounds. Of 14 adults captured in Chambers County, Texas, between 1968 and 1970, weights of males ranged from 46 to 62 pounds (average, 52.25 pounds) and those of females ranged from 45 to 54 pounds (average, 46.7 pounds). The skull measurements of the type of *C. r. gregoryi,* an adult male from Mack's Bayou, 18 miles southwest of Tallulah, Louisiana, and another male from 15 miles northwest of Tallulah, again according to Goldman's data, were, respectively, as follows: greatest length, 250.0, 249.5; condylobasal length, 228.0, 230.7; zygomatic breadth, 126.0, 124.7; width of rostrum, 34.7, 35.9; interorbital breadth, 40.6, 42.6; postorbital constriction, 32.4, 36.7; length of mandible, 173, 176; height of coronoid process, 68.6, 67.8; crown length of maxillary toothrow, 103.8, 104; crown length (outer side), of upper carnassial, 23.5, 24.9; crown width, 12, 12; anteroposterior diameter of first upper molar, 15.9, 16.1; transverse diameter of first upper molar, 20.6, 22.6; crown length of lower carnassial, 26.5, 26.8. For additional skull measurements of *C. r. gregoryi,* see Young and Goldman (1944).

Cranial and other skeletal characters.—For a discussion of the distinctions between the skull of *C. rufus* and that of *C. latrans,* see the account of the latter, page 383. Also see Figures 181 and 183.

Status and habits.—Excellent historical summaries of the former occurrence of the Red Wolf in Louisiana have already been compiled by Nowak (1967, 1970, 1972), who cites numerous accounts by early residents and travelers to show that the species was once not only widely spread over the state but also common. Perhaps, according to Nowak, the initial recorded observation of a wolf in Louisiana was by the explorer Iberville, who

stated that just after entering the mouth of the Mississippi River in 1699 he saw a "Mexican Wolf." Antoine Le Page du Pratz (1758), Louisiana's first full-fledged naturalist, who lived in New Orleans and Natchez from 1718 to 1734, also noted the presence of wolves in Louisiana. The following excerpt from Nowak (1967) describes other early accounts of wolves in the state:

According to Skipwith (1892:42, 46, 51), the wolf was one of the animals with which the pioneers of East Feliciana Parish, southeastern Louisiana, had to contend in the 1795–1802 period.

During the early settlement, 1818–1819, of Claiborne Parish in the northwestern part of the state, "black wolves" were said to be among those predators which were "plentiful, also troublesome to stock" (*Historic Claiborne '65,* 1965:29, quoting the *Webster Tribune,* March 1879). One pioneer of the region commented: "Wolves, too, abounded. It was common to see them, of moonlight nights, traveling around the house or cow pen.... They were fearful on young pigs and calves" (Harris and Hulse, 1886:46). An immigrant to the Athens area of Claiborne Parish in 1847 reported: "Wolves were numerous and very troublesome. They were fond of pigs, and many a night have I heard a pig squeal as Mr. Wolf was flying off with him to the thicket. They disappeared in 1852...." (*Ibid.*:64).

At the June 1, 1846 meeting of the Police Jury of Sabine Parish, west central Louisiana, a resolution was adopted providing for the payment by the Police Jury of two dollars for every wolf killed in the parish. However, at the October, 1848 meeting, this bounty law was repealed (Belisle, 1912:109, 111).

In 1871 when staying with a settler in the backwoods of Calcasieu Parish, southwestern Louisiana, Samuel Lockett... [1873:54] heard noises outside the house, saw injured hogs later, and was told by the settler that wolves had attacked a drove of hogs near the house. The author's phrasing suggests, however, that such an incident was unusual, and he makes no other references to the wolf in his detailed survey of the state.

Describing Morehouse Parish in 1885, C. T. Dunn (p. 57) states that wolves were "still to be found in sparsely settled sections and secluded swamps." That same year, wolves were included in a listing of the "liberal share of game" then present in Bienville Parish (Howell and Richardson, 1885: 15).

That the wolf could still occasionally increase to rather high numbers, even in long-settled regions, is suggested by an article in the October 1, 1896 *Shreveport Times* (p. 6). It was claimed that in the Spring Ridge section of Caddo Parish, just southwest of Shreveport, wolves were "so numerous that those who have sheep, hogs and cows, with young calves, in the woods are justly alarmed, as the wolves have already destroyed a lot of sheep and pigs, and have even killed some grown cows.... It is thought by even the most conservative that there are at least 50 wolves in this immediate neighborhood.... Up to four years ago a wolf was an unheard of thing in this part of the woods."

The vast Atchafalaya River swamp region of southern Louisiana was reported by an old sportsman's magazine in 1898 (W.H.R., p. 11) to have "a sprinkling of the great timber wolf...." The wolf was also said to be present in 1907 in the vicinity of the town of Bogalusa in Washington Parish, southeastern Louisiana (Williamson and Goodman, 1939, 1: 421).

The species probably held out most successfully in the northeastern quarter of the state. Here the densely wooded swamps and nearly impenetrable canebrakes along the Mississippi, Tensas, Ouachita, and related watercourses, provided a wilderness refuge for the wolf and a number of other large, rare species, well into the present century.

In his October, 1907 expedition into the Tensas River bottoms, Theodore Roosevelt (1908: 59–60) found wolves to be common and heard stories of their depredations on sheep and young hogs. He reported also that "the hunters all bore them a grudge, because if a hound got lost in a region where wolves were at all plentiful they were almost sure to find and kill him before he got home. They were fond of preying upon dogs, and at times would boldly kill the hounds right ahead of the hunters." The President states he heard no stories of humans ever being molested, but did hear of a case in which one wolf rushed at a man who had already shot another.

In the vicinity of Rayville, Richland Parish, wolves were reported to be "exceedingly bad" in 1911 (F. G. G., 1911: 534), and depredations were reported also in the Tiger Bayou section of Catahoula Parish in the years 1916–1917 (Jenkins, 1933: 125).

As late as the 1930s, the Red Wolf was still present in some numbers in various localities in Louisiana, especially in the northeastern part of the state. In 1933–1935, when I was studying Ivory-billed Woodpeckers in Madison Parish, my companions and I often found wolf tracks and other sign, and on several

occasions while camping in the heart of what was then a vast, virgin hardwood bottomland forest in Madison Parish, we would hear wolves at night howling not far away. But that magnificent stand of trees, from which at that time not even a matchstick had been removed, is today a thing of the past. First the large trees were felled, then the smaller ones, and now only soy beans grow over much of the area that formerly provided sanctuary for countless numbers of wild creatures. The Ivorybills are gone, and so are the wolves, the panthers, the Swallow-tailed Kites, and most of the other spectacular denizens of that original forest primeval. What a pity! The thought of what was allowed to occur gives one cause for grief. My grandsons will probably never have the thrill, as have I, of standing in a secluded forest under a giant tree, sporting a dbh greater than the expanse of one's arms, with an Ivory-bill pecking away overhead, and catching one of the chips before it hit the ground. Imagine holding a piece of wood personally autographed by an Ivorybill! The youngsters of today might sometime hear the howl of a wolf, but it will likely be in a city zoo after the shriek of a nearby police siren has stimulated one of these sad creatures into one of its vocalizations. But so much for that ...

The present status of the Red Wolf is dismal to say the least. Somewhere in the Tensas or the Atchafalaya basins there may be a few remaining, but I am not too optimistic. Fortunately, the evidence is good that small numbers are still present in extreme southwestern Louisiana in Cameron and Vermilion parishes. Those on the Moore-Odom Ranch, in northwestern Cameron Parish, are given protection, but any wolf outside the bounds of this property had better be on its guard if it is to survive. The wolf is rigidly protected now that its name is on the list of "rare and endangered species." The killing of any species on this roster, if discovered, will result in severe penalties—and in a federal court.

The howl of the wolf has been described

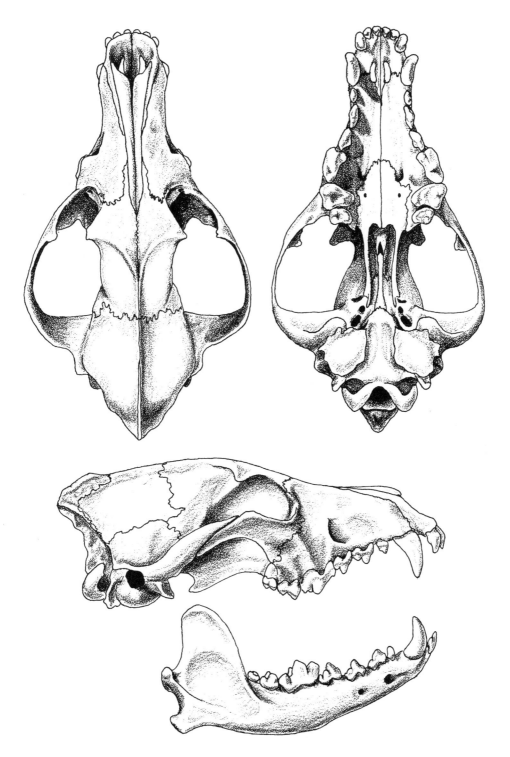

Figure 183 Skull of Red Wolf *(Canis rufus gregoryi)*, ♂, Houma, Terrebonne Parish, LSUMZ 7006. Approx. ½ ×.

vividly in the words of an unnamed writer for the National Livestock Historical Association (1905) as follows: "The cry of the lobo is entirely unlike that made by any other living creature; it is a prolonged, deep, wailing howl, and perhaps the most dismal sound ever heard by human ear. Anyone who has heard it when the land was covered with a blanket of snow and elusively lighted by shimmering moonlight, never will forget this strange, trembling wolf cry that came to him through the cold and stillness of the dead of a winter night on the range."

Young (*in* Young and Goldman, 1944) has this to say about the sounds made by wolves, and I consider the passage worth quoting here:

> ...the wolf utters five distinct vocal sounds, each with a different significance. The first in our classification is a high, though soft, and plaintive sound similar to the whine of a puppy dog, and is used mostly at or near the opening of the wolf den, particularly when the young whelps are out playing around. It seems to indicate solicitude for the offspring, and is made mostly by the adult female. As the pitch of this whine varies so much, it seems ventriloquial.
>
> The second is a loud, throaty howl, seemingly a call of loneliness. It is best heard in zoological parks where either the male or female wolf is confined alone; and when a sudden loud noise such as the sound of a whistle, a clap of thunder, or the clang of a fire-bell causes the animal to give voice. This long, lonesome-sounding howl is also uttered during the breeding season (December, January, and February).
>
> The third sound is a loud, deep, guttural, though not harsh, howl, apparently the call of the chase, that generally is given by an adult male and is answered by other wolves on the hunt for food. It might be termed a call for assembling a group of wolves of the same vicinity. This guttural howl may at times be followed by a loud bark or two similar to that of a Newfoundland dog.
>
> The fourth call is that of the chase, made after starting prey, as a deer or moose. This utterance is not quite so throaty or guttural as the third, and consists of several short, deep, barking sounds rapidly made, very similar to that of a pack of hounds on a very hot scent.
>
> The fifth call is that of the kill. A bulldog of the heavier breed makes a similar sound when in hard

combat with a rival. It is a deep snarl produced by exhaust of air through the wolf's partially opened mouth as it hangs on with teeth sunken into the flesh of its victim.

The howl or lonesome call of the wolf is generally given early in the evening, or again in the morning shortly before break of day. During even the coldest winter nights, wolf howling may be quite common.

The breeding biology of the Red Wolf is not significantly different from that of the Coyote. It reaches sexual maturity between its second and third year, and once that mating takes place the pair bond appears to be for life. Copulation occurs in late winter and the whelps are born after a gestation period of 60 to 63 days, at least on the basis of what is known about the Gray or Timber Wolf *(Canis*

Plate X
1. Coyote *(Canis latrans)*.
2. Gray Fox *(Urocyon cinereoargenteus)*.
3. Red Fox *(Vulpes fulva)*.
4. Red Wolf *(Canis rufus):* **a,** melanistic phase; **b,** normal phase.

lupus). The female has ten teats in two rows, one on each side of the abdomen. The pups, which average seven in number, are covered with soft hair at birth, but the eyes are closed for five to nine days.

Little documented information is available on the food habits of wolves in Louisiana, although one can be quite certain that they preyed heavily on deer before the arrival of the white man, and that the residual population does so even now. Livestock has also suffered some depredation at the hands of these cunning creatures, which often operate in packs.

By far the best work on the life history and habits of wolves is the book by Stanley P. Young and Edward A. Goldman entitled *The Wolves of North America* that was published in 1944 by The American Wildlife Institute. Anyone desiring a comprehensive coverage of the subject should consult that work.

Parasites and diseases. — Information on the internal parasites of the Red Wolf seems to be lacking, and that pertaining to external parasites is meager. Randolph and Eads (1946) and Eads (1948) report the occurrence of three fleas *(Echidnophaga gallinacea, Pulex irritans,* and *Ctenocephalides felis)* and five ticks and lice *(Dermacentor variabilis, Amblyomma americanum, Heterodoxus spiniger, Ixodes cookei,* and *Rhopalopsyllus coxi)* in the Red Wolf in Texas. According to Paradiso and Nowak (1972), most pups acquire hookworms and are so weakened by them that they cannot keep up with their parents. They often die indirectly as a result of hookworms, and the adults have a shortened life span because of hookworms and heartworms and are often anemic with low-level infections of one sort or another.

Subspecies

Canis rufus gregoryi Goldman

Canis rufus gregoryi Goldman, Jour. Mamm., 18, 1937:44; type locality, Mack's Bayou, 3 mi. E Tensas River, 18 mi. SW Tallulah, Madison Parish, Louisiana.

The subspecies *gregoryi* formerly occupied the central part of the range of the species in the Mississippi Valley from Illinois and Indiana south through western Kentucky and Tennessee, southern Missouri, and most of Arkansas to eastern Texas, Louisiana, Mississippi, and the Gulf Coast. Compared with *rufus* to the west in Texas and eastern Oklahoma and with *floridanus* to the east in Alabama, Georgia, and Florida, *gregoryi* is somewhat intermediate in characters, as well as in geographical position. Goldman, in describing the race, designated as the type specimen a wolf obtained in 1905 in Madison Parish, 18 miles southwest of Tallulah. Also available to him were 229 additional skins and skulls, or skulls only, from Morehouse, West Carroll, Winn, Jackson, La Salle, Beauregard, and Iberia parishes, as well as from numerous localities in other parts of the defined range.

The use of the name *Canis rufus* instead of *Canis niger* is based on Nowak (1967) and Paradiso (1968), who pointed out that the name *Canis niger,* employed since 1942 (Harper, 1942) for the Red Wolf, dates from Bartram's *Travels . . . of* 1791. In 1957, however, the International Commission on Zoological Nomenclature, in Opinion 447, placed Bartram's *"Travels"* on its list of invalid works, thereby rejecting all Bartram names and making *niger* unavailable either as a specific or subspecific name. The next oldest name for the Red Wolf is *Canis rufus* Audubon and Bachman (1851).

Specimens examined from Louisiana. — Total 31, as follows: *Beauregard Par.:* locality unspecified, 2 (USNM). *Bienville Par.:* locality unspecified, 1 (LSUMZ) [may be *rufus* × *latrans* hybrid, *fide* Nowak]. *Cameron Par.:* below Gum Cove, 1 (LSUMZ). *Concordia Par.:* 20 mi. SW Vidalia, 1 (USNM). *Iberia Par.:* 12 mi. N Avery Island, 2 (USNM). *Jackson Par.:* locality unspecified, 1 (USNM). *La Salle Par.:* Little River, 1 (LSUMZ). *Madison Par.:* Tallulah, 5 (2 LSUMZ; 3 USNM); 5 mi. S Vicksburg bridge, 1 (LSUMZ); Indian Lake, 23 mi. SW Tallulah, 1 (USNM); Macks Bayou, 3 mi. E Tensas River, 18 mi. SW Tallulah, 1 (type of *gregoryi,* USNM); locality unspecified, 1 (USNM); 15 mi. NW Tallulah, 2 (USNM). *Morehouse Par.:* Mer Rouge, 2 (1 USNM; 1 MCZ). *Natchitoches Par.:*

west bank of Red River, S Natchitoches, 1 (LSUMZ). *Terrebonne Par.:* Houma, 1 (LSUMZ). *Vermilion Par.:* west end of Pecan Island, 1 (LSUMZ). *West Carroll Par.:* 10 mi. SW Floyd, 1 (USNM). *Winn Par.:* Sikes, 2 (USNM); locality unspecified, 2 (USNM); near Winnfield, 1 (USNM).

Genus *Vulpes* Bowdich

Dental formula:

$$I\frac{3-3}{3-3}\ C\frac{1-1}{1-1}\ P\frac{4-4}{4-4}\ M\frac{2-2}{3-3}=42$$

RED FOX Plate X
(Vulpes fulva)

Vernacular and other names.—The word Red of the common name refers to the animal's general color. The word Fox is the Anglo-Saxon name for *Vulpes vulpes* and its close relatives. The generic name *Vulpes* is the Latin word for fox, and *fulva* is the Latin word for reddish yellow. The female of the species is often called the vixen. Among our French-speaking inhabitants foxes go by the name *renard*.

Distribution.—The range of the Red Fox extends over most of North America from the Arctic Ocean to the southern United States. The species is absent only in the extreme southeastern United States, much of the Great Plains, coastal western Canada, coastal Oregon and California, the Great Basin, Arizona, and southern Texas. In Louisiana it occurs statewide except for the extreme southern parishes. (Map 53)

External appearance.—Typical examples of this small doglike canid are easily recognized by their predominantly reddish yellow color. The tail is reddish yellow except for the terminal four or five inches, which are black, tipped with white. The nose pad is black, as are also the ears, legs, and feet. The cheeks, throat, and belly are white. The ears are well furred and rather long and are held erect with the inner surfaces directed forward. The pupil of the eye is vertically elliptical, while that of the Coyote and Red Wolf are circular. The muzzle is elongated and pointed.

Color phases.—Several color phases occur in the species. The cross fox gets its name from the black cross formed by a line down the middle of the back and another dark line across the shoulders. The so-called silver fox is a black color phase in which some of the hairs of the dorsum are tipped with white, producing a frosted appearance. When the white-tipped hairs are lacking an animal of this phase is called a black fox. Many of these mutant color variants are extremely valuable in the fur trade and consequently they are subject to selective breeding on fox farms. Albinos are exceedingly rare and the few on record either died at birth or soon thereafter.

Measurements and weights.—Red Foxes range in total length between three and four feet, of which the tail makes up 12 to 17 inches. Seven specimens in the LSUMZ from various

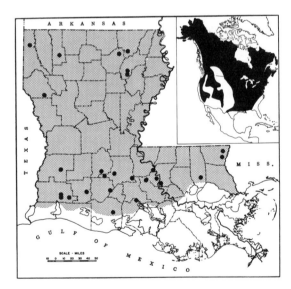

Map 53 Distribution of the Red Fox (*Vulpes fulva*). *Shaded area,* known or presumed range within the state; •, localities from which museum specimens have been examined; *inset,* overall range of the species.

localities in Louisiana, Mississippi, and southern Arkansas averaged as follows: total length, 975 (853–1,140); tail, 310 (300–432); hind foot, 151 (135–173); ear, 85 (76–95). Eight skulls from the same sources averaged as follows: greatest length, 140.0 (132.5–154.5); condylobasal length, 125.8 (119.2–140.5); cranial breadth, 46.8 (45.7–47.8); zygomatic breadth, 70.8 (66.2–75.5); interorbital breadth, 23.4 (19.5–24.8); breadth of rostrum, 20.1 (19.0–21.0); palatal breadth, 32.5 (29.5–35.0); palatilar length, 68.6 (65.2–75.0); maxillary toothrow, 61.7 (58.6–64.5). Adults weigh between 12 and 19 pounds, with males being somewhat heavier than females.

Cranial and other skeletal characters. — The skull of the Red Fox is distinctly canid in appearance and is readily distinguished from the skull of its close relative, the Gray Fox, by the following characters: the depression above the postorbital process is shallow instead of deep as in the Gray Fox; the uppersurface of the cranium possesses two distinct ridges that begin at the postorbital processes and converge toward the rear to lie three-eighths of an inch or less apart at the anterior edges of the parietals, whereas in the Gray Fox the ridges are much more pronounced and are U-shaped, lying considerably more than three-eighths of an inch apart at the anterior edges of the parietals; the lower edge of the mandible lacks a notch toward the rear that is conspicuous on the mandible of the Gray Fox; and, finally, the upper incisors are distinctly lobed on each side (evenly rounded in the Gray Fox). A penis bone is present and is slightly over two inches in length. It is keeled and the tip is sometimes slightly bifurcate. (Figures 184 and 185)

Sexual characters and age criteria. — Except for the external sex organs and the slightly larger size and weight of males, the sexes are alike. The voice of the dog fox is a short yelp, terminating in a throaty, gargling sound; the call of the vixen is more of a yapping scream. Females have eight mammae, two pectoral, four abdominal, and two inguinal. According to Sullivan and Haugen (1956), age determination of foxes is possible by means of X ray of the forefeet.

Status and habits. — The Red Fox is by no means abundant, but it is nevertheless widely distributed over most of the state. It is considerably more numerous now than it was 20 or 30 years ago, perhaps because of repeated introduction by fox-hunting clubs. As a matter of fact, one cannot avoid the suspicion

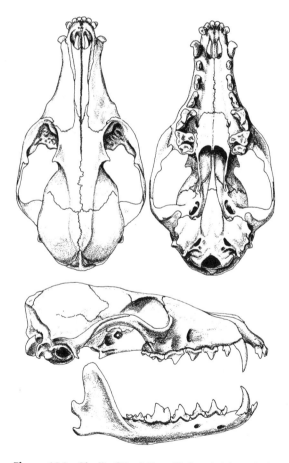

Figure 184   Skull of Red Fox (*Vulpes f. fulva*), ♀, 7 mi. NW Baton Rouge, West Baton Rouge Parish, LSUMZ 16152. Approx. ½ ×.

that the presence of the Red Fox in Louisiana is mainly, if not entirely, a consequence of introductions. Whatever may be the case, the species is certainly now well established, not only in the northern part of the state, but in southern Louisiana as well.

The favorite haunts of the Red Fox are mixed oak-pine wooded uplands interspersed with farms and pastures, but it also occurs in the vicinity of cane fields in our bottomlands. South of Baton Rouge I have often seen it skulking in ditches adjacent to freshly mowed hayfields. The ditches probably contain concentrations of rats and mice.

Although principally nocturnal, the species is often abroad in daylight and still more often at twilight. The normal gait is a brisk walk, but it often breaks into a running trot

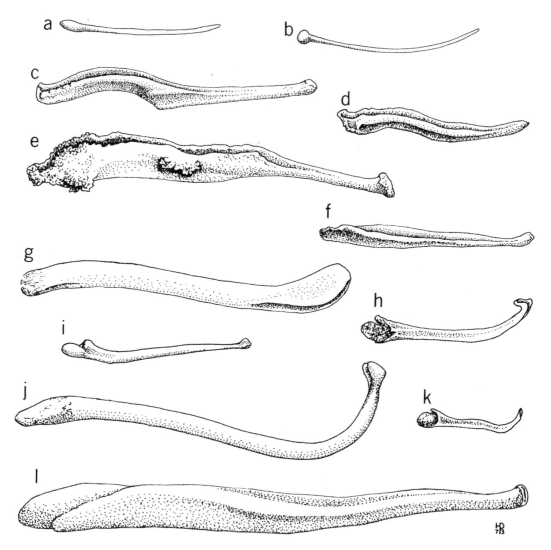

Figure 185 Penis bones of certain carnivores: (a) Striped Skunk; (b) Spotted Skunk; (c) Coyote; (d) Gray Fox; (e) Red Wolf; (f) Red Fox; (g) Nearctic River Otter; (h) North American Mink; (i) Ringtail; (j) Northern Raccoon; (k) Long-tailed Weasel; (l) American Black Bear. Scale: a and b, 2 ×; all others, 1 ×. (a, b, f, and l, after Burt, 1960; others based on specimens in the LSUMZ.)

Tracks of Red Fox

or into a gallop when pursued. Speeds of 32 miles an hour have been recorded, but often the animal moves considerably slower. Even though averse to water, the Red Fox swims quite well and in the fashion of a dog, with its head slightly elevated and with its tail trailing directly behind. Its reputation as an alert, sly, and cunning creature is proverbial and is the basis for numerous expressions whereby these qualities are attributed to humans.

Red Foxes, usually the vixens, excavate their dens in banks or gullies, or they take over the dens of another animal, such as those of armadillos, enlarging and extending the cavities to suit their own needs. The tunnels almost invariably have more than one entrance and are occasionally occupied by more than one pair of foxes. The burrow is "ordinarily 15 or 20 feet long, and sometimes reaches a length of 40 feet or more, with the den at least 3 feet beneath the surface" (Jackson, 1961).

Breeding takes place in late fall or early winter and the 2 to 10 young are born after a gestation period of approximately 53 days. A nest found in an excavation in a bale of hay in a hayloft on the Callicott Farm near Walls, in West Baton Rouge Parish, on 25 January 1972, contained seven kits not more than two days old. That night six of them were moved by one of the adults, presumably the vixen. The seventh was recovered the following day in a weakened condition, but it was subsequently raised as a pet by Mrs. Anne Penn of Baton Rouge.

The young at birth are covered with dense, fine, grayish hair, and the eyes are closed until the ninth day. By the end of the third week, the pups begin to play outside the den under the watchful protection of one or both of the parents. Weaning takes place when the young foxes are about two months of age, and they leave the den when six months old (Jackson, 1961).

On the basis of an analysis of 886 stomachs of the Red Fox in Missouri, Schwartz and Schwartz (1959) reported that 68.2 percent of the animal's diet by volume consisted of small mammals, mainly rabbits, rats, and mice. Young livestock and poultry amounted to 19 percent, while wild birds, known carrion, insects, plant materials, and miscellaneous trace items made up the remaining 12.8 percent. The loss of livestock and poultry is probably not as serious as the 19 percent would suggest, because some of this amount was doubtless carrion.

There are two schools of thought concerning the value of the Red Fox. It is held in high esteem by fox hunters, who favor it above the Gray Fox because of its craftiness, its greater ability to elude or extend the hounds, and its lesser tendency to tree. Others admire the species because of its known beneficial habit of eating great quantities of rats and mice, thereby helping to keep these destructive rodents in check. And still others look upon it as a valuable furbearer. On the other side of the ledger, the Red Fox is in disfavor with farmers because it often makes raids on poultry and small livestock, such as lambs and young pigs. Perhaps the worst mark against it is its susceptibility to rabies, a disease which sometimes reaches epidemic proportions.

Predators. — Man and his dogs are virtually the only predators of the Red Fox, though some kits may be lost to Great Horned Owls and other predators on their first excursions outside the den.

Parasites and diseases. — Numerous internal parasites are recorded from the Red Fox, including 15 species of trematodes, 9 cestodes, and 28 nematodes (Stiles and Baker,

1935; Erickson, 1944). External parasites include a louse *(Felicola vulpis),* an ear mite *(Otodectes cynotis),* the mange mite *(Sarcoptes scabiei),* two ticks *(Ixodes hexagonus* and *I. marxi),* and six species of fleas *(Ctenocephalides canis, Hoplopsyllus affinis, Pulex irritans, Cediopsylla simplex, Oropsylla arctomys,* and *Megabothris wagneri).* Red Foxes are also subject to diseases such as coccidiosis, distemper, and, as already mentioned, rabies.

Subspecies

Vulpes fulva fulva (Desmarest)

Canis fulvus Desmarest, Mammalogie, 1, 1820:203; type locality, Virginia.
Vulpes fulvus, DeKay, The zoology of New York, Mammalia, 1842:44.

No less than 12 subspecies of *Vulpes fulva* have been recognized in the United States and Canada. The nominate race, *fulva,* occupies most of the eastern United States west to Minnesota, Iowa, central Kansas, and central Oklahoma and south to central western and southeastern Texas, southern Louisiana, southern Mississippi, southern Alabama, and the Piedmont of Georgia and North and South Carolina. Some authorities consider *V. fulva* and *V. vulpes* of Europe conspecific.

Specimens examined from Louisiana.—Total 32, as follows: *Acadia Par.:* 1.5 mi. NW Duson, 1 (USL). *Beauregard Par.:* 0.5 mi. S Longville, 2 (MSU). *Caddo Par.:* 3 mi. W Gilliam, 1 (LTU). *Calcasieu Par.:* 7.8 mi. S Lake Charles, 1 (MSU); 5 mi. S Lake Charles, 1 (MSU); 1 mi. S Holmwood, 1 (MSU). *De Soto Par.:* 15 mi. SE Mansfield, 1 (LSUMZ). *East Baton Rouge Par.:* 3 mi. S University, 3 (LSUMZ); 4 mi. S University, 1 (LSUMZ). *Evangeline Par.:* 4 mi. SE Mamou, 3 (USL). *Iberia Par.:* Coteau Community, 1 (USL). *Iberville Par.:* 2 mi. NW Bayou Grosse Tete bridge, 1 (LSUMZ). *Jefferson Davis Par.:* 5 mi. E Welsh, 1 (MSU). *Morehouse Par.:* 5 mi. W Oak Ridge, 1 (NLU). *Ouachita Par.:* 7 mi. N Monroe, 1 (NLU). *Richland Par.:* 2.5 mi. S Mangham, 1 (NLU); 6 mi. S Mangham, 1 (NLU). *St. Landry Par.:* 4 mi. S Krotz Springs, 1 mi. w. River Rd., 1 (LSUMZ); 4 mi. NE Eunice, 1 (USL); 5 mi. W Opelousas, 1 (USL). *Tangipahoa Par.:* 7.5 mi. E Ponchatoula, 1 (SLU). *Vermilion Par.:* 6 mi. S Abbeville, 1 (USL). *Washington Par.:* 4 mi. W Bogalusa, 1 (LSUMZ); 1 mi. SE Varnado, 1 (SLU). *Webster Par.:* 2 mi. S Minden, 1 (LTU). *West Baton Rouge Par.:* 7 mi. NW LSU, 1 (LSUMZ); Arbroth, 1 (LSUMZ).

Genus *Urocyon* Baird

Dental formula:

$$\text{I}\frac{1-1}{1-1}\ \text{C}\frac{3-3}{3-3}\ \text{P}\frac{4-4}{4-4}\ \text{M}\frac{2-2}{3-3}=42$$

GRAY FOX   Plate X
(Urocyon cinereoargenteus)

Vernacular and other names.—The generic name *Urocyon* is derived from two Greek words *(oura,* tail, and *kyon,* dog) and means tailed dog. The specific name is a combination of two Latin words that translates as silvery gray *(cinereus,* ash-colored or gray and *argenteus,* silvery). So, part of the technical name and part of the vernacular name refer to the basic color of the animal.

Distribution.—The species occurs widely in every continental state of the Union except Alaska, Washington, Idaho, Montana, and Wyoming. It appears to be absent in eastern Oregon, northern Nevada, northwestern Utah, extreme eastern Colorado, and the western parts of North Dakota, South Dakota, Nebraska, Kansas, and Oklahoma, as well as in northern Maine. The species ranges southward over Mexico and Central America (except for northern Honduras, eastern Nicaragua, and extreme eastern Costa Rica) to northwestern South America. (Map 54)

External appearance.—The Gray Fox is easily distinguished from the Red Fox by its basically gray coloration, somewhat smaller size, and black-tipped tail. The anterior end of the muzzle, the borders of the lower jaw and chin, the center of the dorsum, and the upper portion of the tail, including its tip, are black, strongly interspersed with gray. The back of the ears, the sides of the neck, a band across the lower throat, the legs, and the border between the gray of the flanks and the white of the belly are rich fulvous, tawny, or sometimes deep ferruginous or reddish brown.

The center of the abdomen, the borders of the upper jaw, and the throat are white.

Color phases.—Coat color in the Gray Fox is quite stable, with mutations exceedingly rare. Jones (1923) reports the occurrence of black variants.

Measurements and weights.—Gray Foxes range in total length from 30 to 44 inches, of which the tail accounts for 11 to 15 inches. Adults weigh approximately 14 pounds, though a 19-pound specimen is on record. Sixteen adults in the LSUMZ from various localities in Louisiana averaged as follows: total length, 887 (780–1,088); tail, 322 (275–380); hind foot, 129 (120–143); ear, 69 (62–75). Sixteen skulls in the LSUMZ from various localities in Louisiana averaged as follows: greatest length, 117.9 (111.7–124.6); cranial breadth, 44.2 (42.8–45.2); zygomatic breadth, 65.1 (61.8–68.8); interorbital breadth, 23.7 (21.7–25.1); length of nasals, 41.2 (36.7–44.8); palatilar length, 54.7

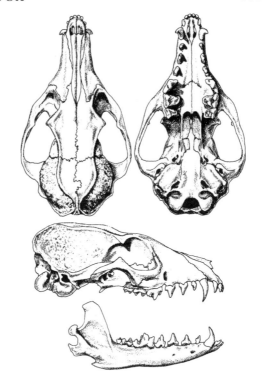

Figure 186   Skull of Gray Fox (*Urocyon cinereoargenteus floridanus*), ♀, 4 mi. N Port Barre, St. Landry Parish, LSUMZ 11271. Approx. ½×.

(51.0–58.0); postpalatal length, 51.4 (48.2–54.6).

Cranial and other skeletal characters.—See discussions of the differences between the skull of the Gray Fox and that of the Red Fox under the latter species. In the Gray Fox the penis bone is typically more deeply keeled and its anterior portion is narrower than in the Red Fox; penis bones of adults measure 54 to 60 millimeters in length (Petrides, 1950). (Figures 185 and 186)

Sexual characters and age criteria.—Like the Red Fox, the sexes of the Gray Fox are indistinguishable except by examination of the genitalia. Kits of the Gray Fox are blackish at an early stage but soon begin to show the pattern of the adults. Petrides (1950) found that the length of the penis bone of adults

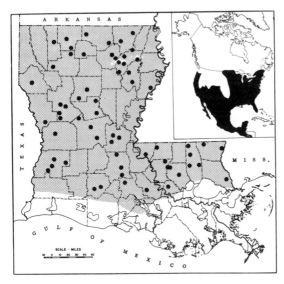

Map 54   Distribution of the Gray Fox (*Urocyon cinereoargenteus*). *Shaded area,* known or presumed range within the state; •, localities from which museum specimens have been examined; *inset,* overall range of the species.

Figure 187 Gray Fox *(Urocyon cinereoargenteus),* Sciote County, Ohio, August 1962. (Photograph by Woodrow Goodpaster)

averaged 57 ± 2.6 millimeters, with a weight of 528 ± 100 milligrams, while that of the juveniles averaged 51 ± 1.7 millimeters, with a weight of 280 ± 62 milligrams. In contrast, the penis bone of adult Red Foxes averaged 54.5 ± 1.5 millimeters, with a weight of 492 ± 103 milligrams, that of the juvenile, 55.4 ± 4.8 millimeters, with a weight of 342 ± 70 milligrams.

Status and habits. — The Gray Fox is fairly common throughout the uplands and other favorable habitats in the state. It prefers mixed pine-oak woodlands bordering upon pastures and fields with weed patches (Figure 187). Because of its usually crepuscular and nocturnal habits it is not often seen. Occasionally it is found abroad in daylight, but ordinarily it spends its days in a dense thicket, in a hollow tree or log, or in an underground den. The Gray Fox regularly climbs trees, shinning up the trunk to a limb, and then jumping from branch to branch as it ascends higher. This it does to escape enemies, such as a pack of hounds, or to reach a bird's nest.

The gait of the Gray Fox is said to be the basis of the proverbial "fox trot." Although I am of the proper vintage to be familiar with this kind of dance, I am not clear as to its similarity to the movements of the fox. Occasionally the Gray Fox will break into a run and in a short spurt attain a speed of close to 30 miles per hour, but ordinarily it travels much slower. Even at reduced speeds it cannot continue for extended periods and, when under pursuit by dogs, will often seek safety in a hole or climb a tree.

Mating in the Gray Fox in Louisiana is believed to occur in late winter although no precise information on the species in the state is available. The two to seven, usually three to five, finely furred young are born after a gestation period of 53 to 63 days. The eyes are closed at birth but open when the kit is 9 to 12 days old. For the first five or six weeks the female cares for the young, but after that time the male brings food to the family. The pups begin to hunt for themselves by the time they are three months old but often remain together as a family unit until the following autumn. They begin to breed the next year. The parents, as a rule, maintain their pair bond throughout the year.

The food of the species is primarily flesh but also includes considerable plant material such as berries, fruits, corn, and acorns. The high protein element of its diet consists of rats, mice, rabbits, and adult and larval insects. Although it is occasionally guilty of taking the Bobwhite and destroying its eggs, I do not believe the Gray Fox can be considered a serious factor in lowering Bobwhite population levels. Likewise, it is known to destroy some poultry, but in the vicinity of the farmyard it is just as likely to kill destructive feral house rats and house mice.

The primarily southern sport of fox hunt-

Tracks of Gray Fox

ing is a subject about which much has been written. Audubon and Bachman (1851–1854) had this to say:

Fox-hunting, as generally practised in our Southern States, is regarded as a healthful manly exercise, as well as an exhilarating sport, which in many instances would be likely to preserve young men from habits of idleness and dissipation. The *music* of the hounds, whilst you breathe the fresh sweet morning air, seated on a high-mettled steed, your friends and neighbours at hand with light hearts and joyous expectations, awaiting the first break from cover, is, if you delight in nature and the recreation we are speaking of, most enliven-ing. . . .

We will now . . . try to make you familiar with the mode of hunting the Gray Fox generally adopted in Carolina and Louisiana. The hounds are taken to some spot where the animal is likely to be found, and are kept as much as possible out of the "drives" frequented by deer. Thickets on the edges of old plantations, briar patches, and deserted fields covered with broom-grass, are places in which the Fox is most likely to lie down to rest. The trail he has left behind him during his nocturnal rambles is struck, the hounds are encouraged by the voices of their masters, and follow it as fast as the devious course it leads them will permit. Now they scent the Fox along the field, probably when in search of partridges, meadow-larks, rabbits, or field-mice; presently they trace his footsteps to a large log, from whence he has jumped on to a worm-fence, and after walking a little way on it, has leaped a ditch and skulked toward the borders of a marsh. Through all his crooked ways the sagacious hounds follow his path, until he is sud-denly aroused, perchance from a sweet, dreamy vision of fat hens, geese, or turkeys, and with a general cry the whole pack, led on by the staunch-est and best dogs, open-mouthed and eager, join in the chase. The startled Fox makes two or three rapid doublings, and then suddenly flies to a cover perhaps a quarter of a mile off, and sometimes thus puts the hounds off the scent for a few minutes, as when cool and at first starting, his scent is not so strong as that of the red fox; after the chase has continued for a quarter of an hour or so, however, and the animal is somewhat heated, his track is followed with greater ease and quickness and the scene becomes animating and exciting. Where the woods are free from underbrush, which is often the case in Carolina, the grass and bushes being burnt almost annually, many of the sportsmen keep up with the dogs, and the Fox is very frequently in sight and is dashed after at the

horses' greatest speed. He now resorts to some of the manoeuvres for which he is famous; he plunges into a thicket, doubles, runs into the wa-ter, if any be at hand, leaps on to a log, or perhaps gets upon a worm-fence and runs along the top of it for a hundred yards, leaping from it with a desperate bound and continuing his flight in-stantly, with the hope of escape from the relentless pack. At length he becomes fatigued, he is once more concealed in a thicket where he doubles hurriedly; uncertain in what direction to retreat, he hears, and perhaps sees, the dogs almost upon him, and as a last resort climbs a small tree. The hounds and hunters are almost instantly at the foot of it, and whilst the former are barking fi-ercely at the terrified animal, the latter determine to give him another chance for his life. The dogs are taken off to a little distance, and the Fox is then forced to leap to the ground by reaching with a long pole, or throwing a billet of wood at him. He is allowed a quarter of an hour before the hounds are permitted to pursue him, but he is now less able to escape than before; he has become stiff and chill, is soon overtaken, and falls an easy prey, turning however upon his pursuers with a growl of despair, and snapping at his foes until he bites the dust and the chase is ended.

Annual fur records maintained by the Lou-isiana Wild Life and Fisheries Commission do not differentiate between the two species of foxes, but, even so, the fluctuations in the annual catch and the price that fox pelts brought to the trapper are revealing (Tables 2 and 10). Although the annual take of any furbearer is influenced by the anticipated market value of the pelts, I am at a loss to explain the large catches of the late 1930s and the 1940s, when the price to the trapper was low. In the more northern parts of the United States, populations of the Gray Fox are said to fluctuate, reaching a peak density about once in a ten-year period.

Predators. — The Gray Fox has few enemies other than man and his dogs. Bobcats, Coy-otes, and Great Horned Owls probably get some of the pups when they first venture forth into the open. Only rarely is a fox killed along one of our highways. The fur is now of such low value that only a few are trapped for their pelts.

Table 10 Comparative Trapping Records of Foxes in Louisiana, 1939–1969, with the
Average Price of Pelts

	Avg. Catch Per Year	Avg. Price
1939–40 through 1948–49	3,819 (1,727–8,534)	$ .58 (.50–$ 1.25)
1949–50 through 1958–59	347 (25–1,250)	.29 (.15– .45)
1959–60 through 1968–69	230 (95–324)	1.35 (.50– 3.00)
1969–70 through 1972–73	813 (242–1,899)	5.25 (3.00– 10.00)

Parasites.—According mainly to Stiles and Baker (1935) and Erickson (1944), the following external and internal parasites have been found associated with the Gray Fox: one louse *(Felicola quadraticeps),* three ticks and mites *(Ixodes cookei, I. ricinus, I. scapularis, Otodectes cynotis),* six fleas *(Oropsylla simplex, O. arctomys, Ctenocephalides felis, C. canis, Echidnophaga gallinacea, Pulex irritans),* one trematode or fluke *(Alaria arisaemoides),* six cestodes or tapeworms *(Diphyllobothrium latum, Mesocestoides litteratus, M. variabilis, Multiceps serialis, Taenia pisiformis, T. serrata),* eight nematodes or roundworms *(Ancylostoma caninum, Crenosoma vulpis, Capillaria aerophila, Physaloptera praeputialis, Spirocerca lupi, Toxascaris leonina, T. canis, Uncinaria* sp.), and one acanthocephalan or thorny-headed worm *(Pachysentis canicola).* It is subject to tularemia and rabies.

Subspecies

Urocyon cinereoargenteus floridanus Rhoads

Urocyon cinereoargenteus floridanus Rhoads, Proc. Acad. Nat. Sci. Philadelphia, 47, 1895:42; type locality, Tarpon Springs, Hillsboro County, Florida.

Some 15 subspecies of the Gray Fox are recognized. The race *floridanus* occupies the extreme southeastern United States from southern South Carolina, southern Georgia, and Florida across southern Alabama and Mississippi, and all of Louisiana to eastern Texas. It is characterized by its smaller size, its relatively shorter hind foot, tail, and ears, and its somewhat harsher pelage.

Specimens examined from Louisiana.—Total 76, as follows: *Acadia Par.:* 3 mi. SE Egan, 1 (USL); 2 mi. N Crowley, 1 (USL). *Ascension Par.:* 0.5 mi. W jct. Hwy. 427 and Hwy. 67, Gonzales, 1 (LSUMZ). *Avoyelles Par.:* 5 mi. W Cottonport, 1 (USL); 3 mi. E Marksville, 1 (USL); 5 mi. N Marksville, Bayou Spring, 1 (USL). *Beauregard Par.:* 10 mi. S De Ridder, 1 (LSUMZ); 3 mi. SSE Singer on Hickory Creek, 1 (MSU); 3.5 mi. SE Singer, 1 (MSU); Dry Creek, 1 (MSU). *Bienville Par.:* 2 mi. W Ringgold, 1 (LTU); 20 mi. W Ruston on I-20, 1 (LTU). *Claiborne Par.:* Homer, 1 (LSUMZ); Hurricane Community, 1 (NLU). *De Soto Par.:* locality unspecified, 1 (LTU). *East Baton Rouge Par.:* 2.5 mi. NW Pride, 1 (LSUMZ); locality unspecified, 1 (SLU). *East Feliciana Par.:* 6 mi. N Jackson, 1 (LSUMZ). *Franklin Par.:* 2 mi. N Gilbert, 1 (NLU); 4 mi. S Baskin on Hwy. 15, 1 (NLU). *Grant Par.:* Selma, 1 (LSUMZ); locality unspecified, 1 (MSU); 3.5 mi. E Pollock, 1 (USL). *Iberville Par.:* 0.5 mi. E Bayou Sorrel, 2 (USL). *Lincoln Par.:* Ruston, 1 (LTU). *Livingston Par.:* 3.5 mi. SE Livingston, 1 (LSUMZ); 1 mi. W Watson, 1 (LSUMZ); 3.5 mi. E Watson, 1 (LSUMZ). *Morehouse Par.:* Boeuf River Swamp, 1 (NLU); 4 mi. S Bastrop, 1 (NLU). *Natchitoches Par.:* Provencal, 1 (LSUMZ); Bellwood, 1 (LSUMZ); 1 mi. N Kisatchie, 1 (USL); 6 mi. S Natchitoches, 1 (NSU). *Ouachita Par.:* Monroe, 7 (2 LSUMZ; 5 NLU); 6 mi. SE Monroe, 2 (NLU). *Rapides Par.:* 5 mi. W Alexandria, 1 (LSUMZ); 2.7 mi. N Lecompte, 1 (LSUMZ); 1 mi. S Lena, 1 (USL). *Richland Par.:* 6 mi. S Delhi, 1 (LTU); between Girard and Start on U.S. Hwy. 80, 1 (NLU); 13 mi. SW Mangham, 1 (NLU); 5 mi. NW Rayville, 1 (NLU); 3 mi. W Archibald, 1 (NLU); Clear Lake, 5 mi. S Start, 1 (NLU); 5 mi. E Oak Ridge, 1 (NLU). *Sabine Par.:* 13 mi. S Many, Hwy. 171, 1 (LSUMZ). *St. Landry Par.:* 4 mi. N Port Barre, 1 (LSUMZ); 3 mi. N Arnaudville on east side of Bayou Teche, 1 (USL); Big Cane, 1 (USL). *St. Martin Par.:* Cecelia, 1 (USL). *St. Tammany Par.:* 2 mi. N Pearl River, 1 (SLU). *Tangipahoa Par.:* 10 mi. S Kentwood, 1 (LSUMZ); 6 mi. W Folsom, 1 (SLU); Kentwood, 1 (SLU); Independence, 2 (TU). *Tensas Par.:* St. Joseph, 1 (LTU). *Union Par.:* 1 mi. E Spencer, 1 (NLU); Corney Creek, 5 mi. N Bernice, 1 (LTU). *Vernon Par.:* 3 mi. S Leesville, Hwy. 171, 1 (MSU). *Washington Par.:* 1 mi. N Varnado, 1 (LSUMZ); 0.5 mi. N Warnerton, 1 (LSUMZ); 1 mi. W Warnerton, 1 (TU). *Webster Par.:* Evergreen, 1 (LTU). *West Feliciana Par.:* St. Francisville, 1 (LSUMZ). *Winn Par.:* Winnfield, 1 (LSUMZ); S Hwy. 471 between Atlanta and Verda, 1 (MSU).

The Bear Family
Ursidae

THIS FAMILY comprises seven genera and approximately nine species: the Spectacled Bear *(Tremarctos ornatus)* of the Andes; the Asiatic Black Bear *(Selenarctos thibetanus);* the three or more Brown or Grizzly Bears *(Ursus)* of Holoarctic distribution; the American Black Bear *(Euarctos americanus);* the Polar Bear *(Thalarctos maritimus)* of circumpolar distribution; the Malayan Sun Bear *(Helarctos malayanus)* of southeastern Asia; and the Sloth Bear *(Melursus ursinus)* of India, Ceylon, and the foothills of the Himalayas. Bears are large, powerfully built mammals with short limbs and a short tail. The eyes are small, the ears short, and the pelage dense and shaggy. They walk on the soles of their feet and sometimes stand erect and move forward on their hind limbs.

Bears of the temperate and cold regions become exceedingly fat in late autumn and then retreat to a den, where they sleep for more or less extended periods. Physiologists prefer not to call these periods of sleep hibernation, because the body temperature is not depressed, most of the body processes remain fairly active, and the bears can usually be aroused. Indeed, in warm regions, such as in Louisiana, bears sometimes leave their dens for short periods during favorable spells of weather in midwinter. In mammals that enter into true hibernation the body temperature drops close to the ambient temperature, the respiratory rate and other physiological processes slow down drastically, and the animal can be awakened only with extreme difficulty.

Genus *Euarctos* Gray

Dental formula:

$$\text{I}\frac{1-1}{1-1}\ \ \text{C}\frac{3-3}{3-3}\ \ \text{P}\frac{4-4}{4-4}\ \ \text{M}\frac{2-2}{3-3} = 42$$

AMERICAN BLACK BEAR   Plate XIII
(Euarctos americanus)

Vernacular and other names. — The American Black Bear derives its common name in part from its color. The word bear comes from the Medieval English word *bere.* The Louisiana form of the species was for a long time accorded full species rank and was called the Louisiana Black Bear, *Euarctos luteolus.* The generic name *Euarctos* is a combination of two Greek words *(eu,* true or typical, and *arktos,* bear) that means typical bear. The specific name *americanus* is the Latinized form of the word American *(americ* plus the Latin suffix *anus,* which, when added to a noun stem, forms an adjective meaning belonging to). It was given to the species by the zoologist who first described the American Black Bear and who evidently wished to emphasize that it occurred only on this continent. The subspecific name *luteolus* is the stem of the Latin word *luteus,* plus the diminutive ending *olus* and means yellowish. Its intent is unfathomable, for only the muzzle has even a hint of yellowness. The word American also provides a way of distinguishing the name of this black bear from that of the so-called Asiatic Black Bear, *Selenarctos thibetanus.*

403

Distribution. — The species was formerly widespread in North America from northern Alaska and northern Canada, including Newfoundland, south to central northern Mexico. It was apparently absent only in certain arid regions of the southwestern part of the United States and northwestern Mexico. In many areas, the species has now been extirpated or its numbers have been vastly reduced. In Louisiana only a few remain, but the distributional picture is confused as a result of introductions from Minnesota in the 1960s. (Map 55)

External appearance. — The general appearance of the American Black Bear is so well known that little description of it is required. It is a huge, bulky animal with long, dense, and glossy black hair. The tail is exceedingly short and inconspicuous, but it is well haired.

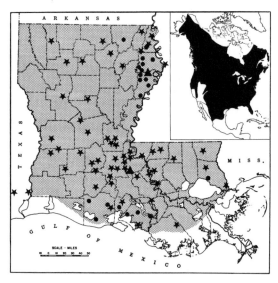

Map 55 Distribution of the American Black Bear (*Euarctos americanus*). *Shaded area,* known or presumed range in the state; •, localities from which museum specimens have been examined; ▲, localities where 161 bears imported from Minnesota have been released since 1964; ★, localities from which bears have been reliably reported since 1964, probably mainly on the basis of released animals or the progeny thereof; *inset,* overall range of the species.

The facial profile is rather blunt, the eyes small, and the nose pad broad with large nostrils. The muzzle is yellowish brown, and a white patch is sometimes present on the lower throat and chest. There are five toes on both front and hind feet with short, curved claws.

Color phases. — Brownish or cinnamon color phases are common in the Rocky Mountains and other western parts of the range, and additional variants occur in northwestern Canada and Alaska — chocolate brown, bluish, and creamy white. A single litter may contain both normal-colored young and either brown or cinnamon phases. I am not aware of the occurrence of any of these color variations in bears in Louisiana. Albinism has been reported in the species in Canada. According to Standley (1921), an old female that was pure white with pink eyes gave birth to four cubs, "one white, with red eyes, and red nails, like herself; one red [reddish brown?] and two black."

Measurements and weights. — An adult or subadult male from Mandeville, St. Tammany Parish, LSUMZ 9045, measured as follows: total length, 1,573 (61 in.); tail, 94 (3.6 in.); hind foot, 225 (8.8 in.); ear, 142 (5.5 in.). It weighed 225 pounds. E. A. McIlhenny told me of a bear that he obtained at Avery Island on 12 November 1916 that weighed 677 pounds. An adult female from one mile SW Gueydan, Vermilion Parish, LSUMZ 9819, measured as follows: total length, 1,913 (74.7 in.); tail, 87 (3.4 in.); hind foot, 230 (9 in.); ear, 135 (5.3 in.). It weighed 230 pounds. Skulls of two adult males from Louisiana in the LSUMZ measured as follows: total length, 320, 286; basilar length, 291, 262; interorbital breadth, 75.5, 64.9; zygomatic breadth, 171.0, 162.7; length of nasals, 82.8, 75.4; palatilar length, 150.6, 145.4; postpalatal length, 134, 116; mastoid breadth, 148.3, 122.9. The skulls of one subadult and one adult female from the same source measured, respectively, as follows: total length,

267 and 290; basilar length, 256 and 293; interorbital breadth, 59.4 and 62.8; zygomatic breadth, 156.4 and 173.9; length of nasals, 66.3 and 77.5; palatilar length, 133 and 143; postpalatal length, 107.3 and 132.6; mastoid breadth, 117.8 and 127.8.

Cranial and other skeletal characters. — The skull, as would be expected in such a large carnivore, is massive. The canine teeth are long and pointed and the horizontal length of the last upper molar is more than one and one-half times its width and its length is considerably greater than that of the upper carnassial (P^4). The crowns of the molariform teeth are blunt and flat and thus serve to crush rather than to shear. Old individuals seldom have a full complement of 42 teeth since the anterior premolars of both the upper and lower jaws are often lost. The palatal plate extends posteriorly well beyond the last molars, but in canids and certain other carnivores such is not the case. A penis bone is present, as is also an os clitoris (Jellison, 1945). (Figures 185 and 188)

Sexual characters and age criteria. — In color the sexes are alike. Males average somewhat larger than their mates, but a broad overlap in mensural characters is evident. Fifteen years is probably the normal life span of bears in the wild, but in captivity they have been known to live almost twice that long. Advanced age is clearly indicated in the skulls of old animals by the wear on the teeth and by the actual loss of some of the premolars.

Status and habits. — Little information is available concerning the status of the bear in Louisiana at the time of early colonization, but it was probably then fairly numerous, serving as food both for the Indians and the white settlers. Le Page du Pratz (1758) tells of hunting bears in the canebrakes along the Mississippi River as early as 1750.

In writing of a hunt in Tensas Parish (Figure 189), in which he had participated while President of the United States, Theodore Roosevelt (1908) informs us that a few years earlier black bears had been extraordinarily plentiful in the swamps and canebrakes on both sides of the lower Mississippi. By the time of his visit in October 1907, their numbers were already greatly diminished, perhaps because of the work of the old southern planters, who for a century past had "followed the bear with horse and hound and horn in Louisiana, Mississippi, and Arkansas." The old ex-slave who was in charge of the dogs on the 1907 hunt had killed or assisted in the killing of more than 3,000 bears! Roosevelt's account (1908) of the adventure is so evocative of the not-so-distant

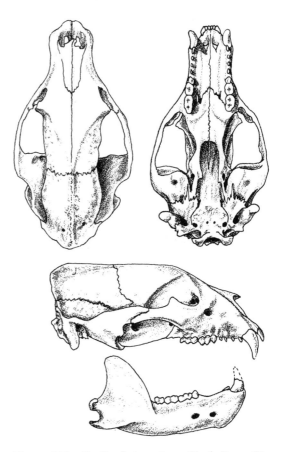

Figure 188 Skull of American Black Bear (*Euarctos americanus luteolus*), ♂, Avery Island, Iberia Parish, LSUMZ 2734. Approx. ³⁄₁₆×.

Louisiana past that the temptation to quote it at length is irresistible.

...when the Father of the Waters breaks his boundaries he turns the country for a breadth of eighty miles into one broad river, the plantations throughout all this vast extent being from five to twenty feet under water.... Conditions are still in some respects like those of the pioneer days. The magnificent forest growth which covers the land is of little value because of the difficulty in getting the trees to market.... In consequence, the larger trees are often killed by girdling.... At dusk, with the sunset glimmering in the west, or in the brilliant moonlight when the moon is full, the cottonfields have a strange spectral look, with the dead trees raising aloft their naked branches....

Beyond the end of cultivation towers the great forest. Wherever the water stands in pools, and by the edges of the lakes and bayous, the giant cypress loom aloft, rivalled in size by some of the red gums and white oaks. In stature, in towering majesty, they are unsurpassed by any trees of our eastern forests.... The canebrakes stretch along the slight rises of ground, often extending for miles.... They choke out other growths, the feathery, graceful canes standing in ranks, tall, slender, serried, each but a few inches from his brother, and springing to a height of fifteen or twenty feet. They look like bamboos; they are well-nigh impenetrable to a man on horseback; even on foot they make difficult walking unless free use is made of the heavy bush-knife. It is impossible to see through them for more than fifteen or twenty paces, and often for not half that distance. Bears make their lairs in them, and they are the refuge for hunted things....

The most notable birds and those which most interested me were the great ivory-billed woodpeckers. Of these I saw three, all of them in groves of giant cypress; their brilliant white bills contrasted finely with the black of their general plumage. They were noisy but wary, and they seemed to me to set off the wildness of the swamp as much as any of the beasts of the chase....

Bears vary greatly in their habits in different localities.... Around Avery Island, John McIlhenny's plantation, the bears only appear from June to November; there they never kill hogs, but feed at first on corn and then on sugar-

Figure 189 Photograph made on President Theodore Roosevelt's bear hunt in Louisiana in 1907. Arrows point, reading left to right, to John A. McIlhenny, President Roosevelt, and Governor John M. Parker. (Photograph in the LSU Archives)

cane, doing immense damage in the fields, quite as much as hogs would do. But when we were on the Tensas we visited a family of settlers who lived right in the midst of the forest ten miles from any neighbors; and although bears were plentiful around them they never molested their corn-fields. . . .

It is only in exceptional cases . . . that these black bears, even when wounded and at bay, are dangerous to men, in spite of their formidable strength. Each of the hunters with whom I was camped had been charged by one or two among the scores or hundreds of bears he had slain, but no one of them had ever been injured, although they knew other men who had been injured. Their immunity was due to their own skill and coolness; for when the dogs were around the bear the hunter invariably ran close in so as to kill the bear at once and save the pack. Each of the Metcalfs had on one occasion killed a large bear with a knife, when the hounds had seized it and the man dared not fire for fear of shooting one of them. . . .

A large bear is not afraid of dogs, and an old he, or a she with cubs, is always on the lookout for a chance to catch and kill any dog that comes near enough. . . . If a man is not near by, a big bear that has become tired will treat the pack with whimsical indifference. The Metcalfs had recounted to me how they had once seen a bear, which had been chased quite a time, evidently make up its mind that it needed a rest and could afford to take it without much regard for the hounds. The bear accordingly selected a small opening and lay flat on its back with its nose and all its four legs extended. The dogs surrounded it in frantic excitement, barking and baying, and gradually coming in a ring very close up. The bear was watching, however, and suddenly sat up with a jerk, frightening the dogs nearly into fits. Half of them turned back-somersaults in their panic, and all promptly gave the bear ample room. . . .

We waited long hours on likely stands. We rode around the canebrakes through the swampy jungle, or threaded our way across them on trails cut by the heavy wood-knives of my companions; but we found nothing. Until the trails were cut the canebrakes were impenetrable to a horse and were difficult enough to a man on foot. On going through them it seemed as if we must be in the tropics; the silence, the stillness, the heat, and the obscurity, all combining to give a certain eeriness to the task, as we chopped our winding way slowly through the dense mass of close-growing, feather-fronded stalks. Each of the hunters prided himself on his skill with the horn, which was an essential adjunct of the hunt, used both to summon and

control the hounds, and for signalling among the hunters themselves. The tones of many of the horns were full and musical; and it was pleasant to hear them as they wailed to one another, backwards and forwards, across the great stretches of lonely swamp and forest.

. . . Clive Metcalf and I separated from the others and rode off at a lively pace between two of the canebrakes. After an hour or two's wait we heard, very far off, the notes of one of the loudest-mouthed hounds, and instantly rode toward it, until we could make out the babel of the pack. Some hard galloping brought us opposite the point toward which they were heading. . . . The tough woods-horses kept their feet like cats as they leaped logs, plunged through bushes, and dodged in and out among the tree trunks; and we had all we could do to prevent the vines from lifting us out of the saddle, while the thorns tore our hands and faces. . . . we rode ahead, and now in a few minutes were rewarded by hearing the leading dogs come to bay in the thickest of the cover . . . . we threw ourselves off the horses and plunged into the cane. . . . Clive Metcalf, a finished bear-hunter, was speedily able to determine what the bear's probable course would be, and we stole through the cane until we came to a spot near which he thought the quarry would pass. . . . Peering through the thick-growing stalks I suddenly made out the dim outline of the bear coming straight toward us. . . .

I fired for behind the shoulder . . . and . . . at the crack of the rifle the bear stumbled and fell forward, the bullet having passed through both lungs. . . .

This was a big she-bear, very lean, and weighing two hundred and two pounds. . . .

John McIlhenny had killed a she-bear about the size of this on his plantation at Avery's Island the previous June. Several bear had been raiding his corn-fields, and one evening he determined to try to waylay them. After dinner he left the ladies of his party on the gallery of his house while he rode down in a hollow and concealed himself on the lower side of the corn-field. Before he had waited ten minutes a she-bear and her cub came into the field. The she rose on her hind legs, tearing down an armful of ears of corn which she seemingly gave to the cub, and then rose for another armful. McIlhenny shot her; tried in vain to catch the cub; and rejoined the party on the veranda, having been absent but one hour.

Black bears seem always to have been mainly animals of the heavily wooded areas in the central part of the state from the Arkansas border south along the Mississippi River to the Atchafalaya Basin and the swamps of the central southern part of the state. The few native bears that remain today are largely confined to this vast expanse of hardwood bottomland terrain. A female was captured at Cote Blanche, in St. Mary Parish, on 29 April 1961, and another, also a female, was killed one mile southwest of Gueydan, in Vermilion Parish, on 30 October 1964. The latter specimen is now part of the Canebrake Habitat Exhibit in the LSU Museum.

Occasionally, in fairly recent years, a bear has appeared in the Florida Parishes. As late as July 1953 one wandered into a populated subdivision on the eastern edge of Baton Rouge, possibly out of the Homochitto Swamp of southern Mississippi. And in 1963 one was shot within the limits of Mandeville, in St. Tammany Parish. Perhaps it had come out of the swamps along the Pearl River.

In the summers of 1964 through 1967 agents of the Louisiana Wild Life and Fisheries Commission trapped 161 black bears in Cook County, Minnesota, transported them to Louisiana, and released them as part of a restocking program. One of the two release sites was on the Lottie Wildlife Protection Association Hunting Club property between Lottie and Krotz Springs, in Pointe Coupee Parish, where 130 of the bears were liberated. The remaining 31 were released on the Tensas River northwest of Somerset near the Tensas-Madison parish line. All the animals were tagged in order that they could later be identified as part of the restocking experiment. The known mortality of 30 individuals in subsequent years has resulted from 12 being struck on highways, nine being illegally shot, five dying from overdoses of tranquilizers, one being killed by a train, and three succumbing to unknown causes. Verified reports indicate that the bears dispersed widely from the two release sites into numerous parishes in Louisiana, as well as into adjacent states. No less than seven turned up in Mississippi (Canton, McComb, Meadville, Vicksburg, Mason, Yazoo City, and Liberty), three in eastern and southeastern Texas (north of Houston, at Vidor, and near Fort Hood), and one in southern Arkansas (near El Dorado).

In the period from September 1970 to October 1971, David Taylor, as part of his graduate program in the School of Forestry and Wildlife Management at Louisiana State University (Taylor, 1971), captured six bears in Pointe Coupee and St. Landry parishes for use in a radiotelemetry study of their movements. Four of these bears, a male and three females, were part of the original group of Minnesota transplants, one was a cub of one of the captured females, and the sixth was an untagged adult male that was in all probability an offspring of one of the original releases. After the bears were captured, they were tranquilized, weighed, color-marked with plastic ear streamers, fitted with radio transmitters, and released. Their subsequent movements were followed and recorded. Monitoring was achieved by using a handheld antenna and receiver on foot or in a vehicle patroling roads and levees or in a low-flying airplane. The bears were located by compass triangulation based on two or more "fixes."

The results achieved by Taylor were dra-

Tracks of American Black Bear

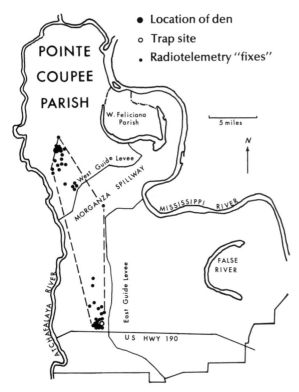

- Location of den
- Trap site
- Radiotelemetry "fixes"

POINTE
COUPEE
PARISH

W. Feliciana
Parish

5 miles

N

West Guide Levee

MORGANZA SPILLWAY

MISSISSIPPI RIVER

ATCHAFALAYA RIVER

East Guide Levee

FALSE
RIVER

U S HWY 190

Figure 190 Map of minimum home range (39,040 acres) of an American Black Bear *(Euarctos americanus)* in Pointe Coupee Parish, based on radiotelemetry "fixes." (Adapted from Taylor, 1971)

matic. He found that the minimum home range of the five bears that he was able to monitor sufficiently varied from 4,365 to 39,040 acres, with an average of 13,795 acres. Figure 190 shows the recorded locations of the individual with the largest minimum home range. The minimum home range for two adult males averaged 27,440 acres, while that of two adult females averaged only 4,866 acres. The known distances traveled from the release site varied greatly, but one bear moved 20 miles in 73.5 hours. Three of the bears being monitored by Taylor were known to have "hibernated," that is, to have gone into a state of torpor, which lasted from 74 to 124 days. The extreme dates for this condition was 17 November and 20 March. There was considerable evidence that some of the

bears moved in and out of their dens during the "hibernating" period. One of the females under surveillance gave birth to cubs while denned.

Weight changes in the bears varied considerably. One adult male gained 60 pounds in 31 days between 3 October and 4 November for an average of 1.94 pounds per day. During the same period an adult female gained 1.04 pounds per day, while her cub's weight increased 15 pounds in 15 days. At the other extreme another adult male lost 35 pounds in the period of the last two weeks of June and the first three weeks of July, for an average daily loss of .88 pounds per day.

Bears den in a variety of situations, including road culverts, hollow logs, and tree cavities. They are excellent climbers and often have their dens in cavities well above ground level (Figure 191). Taylor found one of his six-year-old female bears in a hollow tupelogum *(Nyssa aquatica)* that stood in the water at the edge of a slough. The entrance hole was 30 feet above ground and the dbh measured five feet. The bear was found to be sitting upright in the bottom of the cavity, which was at ground level. This bear later moved to a new den in a large baldcypress *(Taxodium distichum)* a mile away. This tree was well over 100 feet in height and had a 7-foot dbh. The cavity entrance was calculated to be 96 feet above the ground! It was estimated to be 24 inches in diameter. How deep the cavity extended could not be ascertained. Another bear, a male, was found denned in a hollow green ash *(Fraxinus pennsylvanica)*. The tree had a 5-foot dbh and was a hollow shell with the opening 8 feet from the ground. The bear was curled in the bottom of the cavity some 6 inches above ground level. A second of Taylor's female bears denned in a hollow baldcypress that was 60 feet tall with a dbh of 7 feet. The lower edge of the opening was 30 feet above ground and was 14 inches wide and 14 feet long.

The sexes separate after mating. Because of the delayed implantation of the blastocyst,

Figure 191   The American Black Bear *(Euarctos americanus)* is an excellent climber. Lottie Wildlife Protection Association Hunting Club property, Pointe Coupee Parish, 1968. (Photograph by Joe L. Herring)

the period of gestation is 100 to 210 days in duration. Parturition often occurs while the mother is in "hibernation." The number of cubs varies from one to five but is usually two, less frequently three (Matson, 1952). They weigh less than a pound at birth. The youngsters remain with the mother through the following fall, sometimes longer. The female first gives birth at three years of age, usually to a single offspring. She breeds in alternate years thereafter.

Bears are omnivorous. Although occasionally eating meat or animal matter in one form or another, such as mice and squirrels, they are perhaps as much as 95 percent vegetarians. Oak mast, especially that of live oak (*Quercus virginiana*), field corn, muscadines, and blackberries are consumed in large quantities, as is honey when available. Depredations on livestock are negligible, but bears often do serious damage on farmlands to corn crops and beehives. In other parts of their range, black bears are said to consume carrion, but J. D. Newsom (pers. comm.) informs me that he and his students could find no evidence of their doing so in the study area in Pointe Coupee Parish, even though the bears were purposely provided with ample opportunity. In the vicinity of human dwellings they often become a nuisance by getting into garbage cans.

Predators. — Black bears have virtually no enemies except man and his dogs. In days gone by an occasional cub may have been lost to wolves and panthers, but these large predators are now too rare to merit consideration. A she-bear is highly solicitous of her young and their safety and is not likely to allow any molestation of them.

Parasites. — Apparently little attention has been given to the parasites of black bears.

The occurrence of the tick *Dermacentor venustus* and the flea *Chaetopsylla setosa,* as well as the trematode *Nanophyetus salmincola* and the cestode *Diphyllobothrium latum,* have been reported. The nematode *Trichinella spiralis* has been found prevalent in black bears; and therefore bear meat, like pork, should be well cooked before eating to avoid the danger of trichinosis. (Stiles and Baker, 1935)

Subspecies

Euarctos americanus luteolus Griffith

Ursus luteolus Griffith, Class Quadrimembra, order Carnivora, general and particular description of carnivorous animals . . . , 1821:236; type locality, Louisiana.

Ursus americanus Pallas, . . . Spicilegia zoologica . . . , fasc. 14, 1780:5.

Euarctos americanus, Gray, Proc. Zool. Soc. London [for] 1864 [= 1865]:692.

Euarctos americanus luteolus, Miller and Kellogg, Bull. U.S. Nat. Mus., 205, 1955:693.

The original black bears of Louisiana were referable to the subspecies *luteolus,* but the stock introduced from Minnesota is unquestionably assignable to *americanus,* which differs in being slightly larger and in having minor cranial distinctions. The Minnesota bears or their progeny would doubtlessly interbreed freely with any of our native bears still remaining in the state wherever they come into contact.

Specimens examined from Louisiana. — Total 44, as follows: *East Carroll Par.:* Monticello, 1 (NLU). *Iberia Par.:* Avery Island, 11 (1 LSUMZ; 10 USNM). *Madison Par.:* ca. 8 mi. S Waverly, 1 (LSUMZ); Indian Lake, 1 (USNM); Tallulah, 5 (USNM); 21 mi. SW Tallulah, 1 (USNM); 20 mi. SW Tallulah, 1 (USNM); 14 mi. W Tallulah, 1 (USNM). *Morehouse Par.:* Mer Rouge, 4 (1 MCZ; 3 USNM). *St. Mary Par.:* Cote Blanche, 1 (LSUMZ); Franklin, 1 (USNM); Morgan City, 1 (USNM). *St. Tammany Par.:* Mandeville, 1 (LSUMZ). *Tensas Par.:* Newellton, 1 (USNM); Newlight, 2 (USNM); 8 mi. NW Newlight, 1 (USNM); 25 mi. NW St. Joseph, 1 (USNM); Tensas Bayou, 4 (USNM). *Vermilion Par.:* 1 mi. SW Gueydan, 1 (LSUMZ); Abbeville, 3 (USNM); 25 mi. SW Abbeville, 1 (USNM).

The Raccoon Family
Procyonidae

THE FAMILY Procyonidae includes the raccoons and their allies, the Ringtail, or cacomistle, coatis, Kinkajou, olingos, and, according to the traditional view, the pandas of southeastern Asia. The only representatives of the family in Louisiana are the Northern Raccoon and the Ringtail, the latter species being exceedingly rare within our borders. Procyonids possess five toes, with nonretractile or at best only semiretractile (*Bassariscus*) claws, on both the front and hind feet, and they walk on the sole of the foot with the heel touching the ground. Although properly classified as carnivores, most members of the family are omnivorous in their feeding habits. The molariform teeth lack the shearing quality of those of cats, the crowns being low with rounded cusps.

Genus *Bassariscus* Coues

Dental formula:

$$I\frac{3-3}{3-3} \ C\frac{1-1}{1-1} \ P\frac{4-4}{4-4} \ M\frac{2-2}{2-2} = 40$$

RINGTAIL
Plate XII
(*Bassariscus astutus*)

Vernacular and other names. — The name Ringtail is highly appropriate, for it calls attention to one of the most striking features of the species. The name ring-tailed cat is also applied but is a misnomer, for the animal is definitely not a cat. Another frequently heard appellation is cacomistle, which comes from the Nahuatl word *tlacomiztle* (*tlaco,* half, and *miztli,* mountain lion). The generic name *Bassariscus* is derived from two Greek elements (*bassaris,* fox, and the diminutive suffix, *iskos*) that in combination mean little fox. The specific name *astutus* is a Latin word meaning cunning.

Distribution. — The species occurs from southwestern Oregon, through most of California, southern Nevada, most of Utah, western Colorado, and southern Kansas south over the southwestern United States, rarely eastward as far as Louisiana, thence to southern Mexico. Its occurrence in Louisiana is now based on several definite records. (Map 56)

External appearance. — This small somewhat catlike mammal is pale yellowish above except for black-tipped guard hairs on the face, nape, and shoulders and down the center of the back. An area of white or pale yellowish white nearly surrounds a narrow black eyering. The bushy tail is nearly encircled by eight or nine alternating black and white rings (the black rings do not quite meet on the white undersurface). The tip of the tail is black. The feet and underparts are white, sometimes slightly tinged with pale buff. The nails are semiretractile and the soles of the feet are hairy, whereas in *Procyon* they are naked. In color the sexes are alike.

Measurements and weights. — Total length in this species ranges from about 27 to 31 inches, of which the tail makes up more than half. The shoulder height is approximately 6 inches. The weight is usually between 2 and 2.5 pounds. The measurements that follow are of specimens in the LSUMZ from the Mexican state of San Luis Potosí. Two males measured: total length, 758, 792; tail, 383, 384; hind foot, 77, 76; ear, 49, 47. Three females measured as follows: total length, 687, 680, 732; tail, 341, 340, 352; hind foot, 66, 61, 71; ear, 46, 46, 47. Skulls of three males measured: greatest length, 81.0, 83.0, 81.7; cranial breadth, 37.3, 35.9, 34.6; zygomatic breadth, 47.7, 55.4, 49.8; interorbital breadth, 14.6, 16.0, 16.6; palatilar length, 35.1, 35.9, 35.5; postpalatal length, – –, 38.8, 38.0. Four females measured: greatest length, 76.2, 73.5, 78.0, 79.1; cranial breadth, 35.2, 34.9, 34.0, 34.9; zygomatic breadth, 43.1, 41.7, 48.9, 48.6; interorbital breadth, 14.5, 14.1, 15.1, 14.4; palatilar length, 32.3, 31.5, 33.2, 35.2; postpalatal length, 34.4, 31.8, 35.1, 34.9.

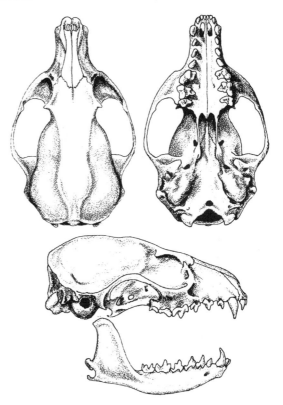

Figure 192 Skull of Ringtail *(Bassariscus astutus flavus)*, ♂, Kerr County, Texas, TAMU 3689. Approx. ⅔ ×.

Cranial and other skeletal characters. — The skull is elongate, the rostrum short and narrow. The postorbital process of the frontal is well developed, relatively more so than in *Procyon*. The palatal shelf extends only slightly posterior to the last molariform tooth, whereas in *Procyon* it extends well beyond. The second upper molar is decidedly smaller than the first and is almost triangular in shape (subquadrate in *Procyon*). The penis bone is approximately two inches in length, virtually straight, basally keeled, and slightly spatulate terminally. (Figures 185, 192, and 194)

Status and habits. — The Ringtail has long been reported to occur near the Sabine River, along the Louisiana–Texas border, on the basis of hides received by fur dealers from

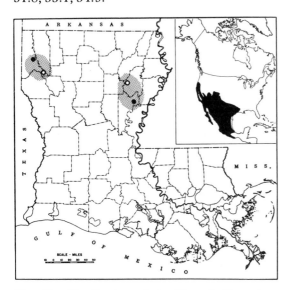

Map 56 Distribution of the Ringtail *(Bassariscus astutus)*. •, localities from which museum specimens have been examined; ○, localities from which the species has been reliably reported; *inset,* overall range of the species.

that area (Arthur, 1928), but such evidence is not acceptable, for the pelts could have originated in eastern Texas. There are, though, other records of the species from the state. One was captured in 1954 near Caspiana, in extreme southeastern Caddo Parish, and in March of the same year John Blanchard photographed a Ringtail (Figure 193) found 14 miles west of Winnsboro, in the Liddieville community. In neither instance was the animal preserved. On 4 December 1959 an actual specimen was obtained three miles south of Sicily Island, in Catahoula Parish. The label on the specimen, which was obtained by R. D. Cupit, states that it was "caught in a trap set in a tree overhanging [Buck Bayou] lake." And, on 18 November 1963, a Ringtail was trapped on the edge of Shreveport, in Caddo Parish, by Donald Waldron, a high school student at the time. I see no reason why these

records should be questioned as natural occurrences of the species in our state. Obviously, though, it is quite rare.

The Ringtail is an extremely attractive small carnivore that is said to make a delightful household pet, especially if the animal is a female captured when young. In the wild it is mainly nocturnal. It climbs well and its movements are quick and agile. Its diet includes rats, mice, birds, a wide variety of insects, and some plant material. The young are born in May and June with the number per litter varying from one to five (usually three or four). The babies are said to weigh about 28 grams at birth, and in the early stages of their development they are cared for by the female only. After the youngsters are three weeks old both parents sometimes bring food to them. Although weaning does not take place until the fourth month, the young begin to forage with the parents when only eight weeks old (Walker, *in* Walker et al., 1968).

Figure 193 Ringtail *(Bassariscus astutus)*, 14 miles northwest of Winnsboro, in the Liddieville community, March 1954. (Photograph by John Blanchard)

Subspecies

Bassariscus astutus flavus Rhoads

Bassariscus astutus flavus Rhoads, Proc. Acad. Nat. Sci. Philadelphia, 45, 1894:417; type locality, Texas, exact locality unknown.

With only two specimens available from the state, subspecific allocation is almost solely on the basis of geographical probability. But these two specimens are quite similar to the Texas material with which they have been compared.

Specimens examined from Louisiana.—Total 2, as follows: *Caddo Par.:* Shreveport, 1 (LSUMZ). *Catahoula Par.:* 3 mi. S Sicily Island, Buck Bayou Lake, 1 (LSUMZ).

Genus *Procyon* Storr

Dental Formula:

$$I\frac{3-3}{3-3} \ C\frac{1-1}{1-1} \ P\frac{4-4}{4-4} \ M\frac{2-2}{2-2} = 40$$

NORTHERN RACCOON   Plate XI
(*Procyon lotor*)

Vernacular and other names.—Because seven nominal species of raccoons are currently recognized, the one occupying continental North America requires an adjectival modifier in its name to distinguish it from the insular forms of the Tres Marias Islands, Bahamas, and Lesser Antilles, as well as from the so-called Crab-eating Raccoon of Panama and South America. The word Northern, in my opinion, best serves this purpose. Early French settlers called the raccoon *chat sauvage,* and Cat Island, off the Mississippi and southeastern Louisiana coast, is said to have gained its name from these strange "cats" that Iberville and his men found inhabiting the place. The word raccoon comes from an Indian name, variously spelled *arocoun, arakun, arrathkune,* and *aroughcun.* The Choctaw Indians of the Gulf Coast called it *shoui,* as some French-speaking trappers in coastal Louisiana do to this day. The Biloxi Indians termed it *atuki* (Arthur, 1928). In Louisiana people generally call the species simply "coon." The generic name *Procyon* is of Greek derivation (*pro,* before, and *kyon,* dog) and reflects the fact that raccoons are considered close to the primitive stock that also gave rise to the canid-ursid lines. The specific name *lotor* is the New Latin word meaning washer and refers to the misconception that raccoons always wash their food before eating it.

Distribution.—The species occurs across southern Canada and most of the United States, except for portions of the Rocky Mountains and arid regions of the southwest, thence south over Mexico and Central America to Panama. In Louisiana its presence is statewide. (Map 57)

External appearance.—Raccoons are rather stocky, short-legged grayish to blackish animals, often with a strong suffusion of yellow, that are about the size of a large beagle. The heavily furred tail is ringed alternately with five or six blackish and yellowish rings, the tip being black. A prominent black mask extends across the face from the jowls through the eyes and is bordered above and below by white. The white area above the mask is, however, interrupted by a black line down the center of the muzzle between the eyes. The whitish-rimmed ears, which are prominent and held erect, are blackish on the basal part of their posterior surface. The eyes are medium in size and black. The lower sides and belly are much grayer than the center of the dorsum, which is always strongly grizzled with black. The guard hairs of the nape are sometimes rich yellowish brown basally, pro-

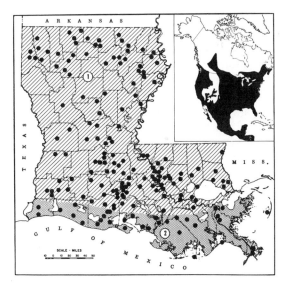

Map 57 Distribution of the Northern Raccoon (*Procyon lotor*). *Shaded areas,* known or presumed range within the state; •, localities from which museum specimens have been examined; *inset,* overall range of the species. 1. *P. l. varius;* 2. *P. l. megalodous.*

ducing an indistinct patch of this color that becomes more evident when the hairs are erected.

Color phases.—Considerable individual variation is evident in the color of the upperparts, some examples being exceedingly black, others being quite gray or yellowish. Albinos occur occasionally, and I have seen a specimen that I would definitely consider an example of xanthochroism (LSUMZ 2144). In regard to the latter condition, though, one must exercise caution, for the pelage of raccoons living in salt marshes of coarse grasses and sedges, often becomes excessively worn and extremely yellowish prior to the annual molt.

Measurements and weights.—Skins of 23 males in the LSUMZ from various localities in Louisiana, averaged as follows: total length, 739 (607–1,032); tail, 252 (200–300); hind foot, 104 (82–128); ear, 56 (42–66). Eleven females from the same source averaged as follows: total length, 706 (610–820); tail, 233 (210–283); hind foot, 96 (75–112); ear, 51 (45–62). The skulls of 45 males averaged as follows: greatest length, 109.6 (93.6–116.0); cranial breadth, 47.8 (43.9–53.5); zygomatic breadth, 66.8 (57.8–81.7); interorbital breadth, 22.1 (19.3–26.7); length of nasals, 33.3 (30.4–38.5); palatilar length, 63.1 (57.7–70.9); postpalatal length, 34.1 (30.2–39.8). The skulls of 28 females averaged as follows: greatest length, 106.9 (101.2–112.0); cranial breadth, 46.5 (43.8–49.4); zygomatic breadth, 63.3 (55.8–68.2); interorbital breadth, 21.3 (18.6–23.5); length of nasals, 32.4 (28.7–35.2); palatilar length, 60.9 (56.8–64.3); postpalatal length, 32.8 (28.7–36.0). Few weights of Louisiana specimens are available, the largest being a 21-pound male, which is probably close to the maximum.

Cranial and other skeletal characters.—No other carnivore of Louisiana, except the ex-ceedingly rare Ringtail *(Bassariscus astutus),* is characterized by possessing 40 teeth. Canids and bears have 42, mustelids 34 or 36, and felids 30. Consequently, this feature by itself serves to distinguish this species and the Ringtail. The skull of the latter is, however, much smaller than that of any raccoon except very young individuals. By the time the permanent teeth begin to erupt in August, immatures of *Procyon lotor* exceed in size of skull even the oldest examples of *Bassariscus astutus.* The skull of a raccoon is rather massive for the size of the animal, and the hard palate extends posteriorly well beyond the molariform toothrow. The canines, at the rim of the alveoli, are oval in cross section. The carnassials and molars are large and tuberculate with rounded cusps and are about as long as they are broad. The first upper premolar is single rooted, the second and third two rooted, and the fourth three rooted. The mandible is heavy with an evenly rounded inferior edge, the symphysis is short, and the coronoid process rises high and curves posteriorly over the condyle. The penis bone is well developed and; in fully adult animals, measures 3.3 to 4.8 inches in length. It has a pronounced L-shaped terminal bend that ends in two lobes separated by a median groove. The overall shape of the bone is slightly sigmoid. (Figures 185 and 194)

Sexual characters and age criteria.—Adult males average somewhat larger than females, but the sexes are otherwise alike except for the primary and secondary sex organs. Males possess a penis bone and females have three pairs of teats, one pectoral, one abdominal, and one inguinal. Supernumary teats are not unusual. The young grow rapidly and are soon indistinguishable from adults except for

Plate XI
Northern Raccoon *(Procyon lotor)* in a magnolia *(Magnolia grandiflora)* with an anole, or chameleon *(Anolis carolinensis),* camouflaged against a green leaf in foreground.

PRATT
1972

exceedingly mature individuals. Juveniles begin to erupt their permanent teeth in August. Females that have bred and borne young have well-developed, blackish, and wrinkled teats, and the uterus shows placental scars where embryos have been attached. When a female is lactating her teats become flaccid. Other sex and age criteria are reviewed by Stuewer (1943) and Petrides (1956).

Status and habits. — The Northern Raccoon is abundant and widespread in Louisiana, occurring in every parish. It ranks as one of the state's main furbearers. In the 57 seasons from 1913 to 1973 for which records are available (Table 2), the annual catch of raccoons ranged from a low of 50,790 pelts in the trapping season of 1967–1968 to a high of 821,733 in 1916–1917 and averaged 137,764. In the 44 seasons between 1928 and 1973 when prices paid to the trapper are on

record, the amount per pelt averaged $1.68, ranging between a low of 27 cents in the 1949–1950 season to a high of $7.25 in the 1928–1929 season.

As I write this account, eight raccoons are feasting at a feeding station only 10 feet from my living room window, and in the yard of my next door neighbors, Harvey and Eleanor Roberts, no fewer than 22 are assembled in a similar nocturnal gathering for a handout. Our feeding every night of more than two dozen coons, plus several possums, is not unlike maintaining a pack of as many dogs. A sack of cracked corn and a 25-pound bag of dog biscuits are consumed every few days! But it is worth every cent that it costs, for it has provided us and our friends with great enjoyment and entertainment, and it has permitted us to establish a delightfully close relationship with one of the state's most interesting woodland inhabitants. No two individuals are exactly alike either in color or personality. Nearly every one has been given a name, inspired by some feature of its color or else some unique behaviorism displayed — "Easter" stands on her haunches and reminds us of an Easter Island monolithic statue, "Outfielder" is especially adept at catching dog biscuits in her paws, "Bandit" invariably holds his right paw above his head when begging, "Dolichocephalic," shortened to "Dolly," possesses an unusually long head, "Tiny Tip" has an abbreviated black tip to his tail, "Boobtube Tavvy" climbs onto the ledge beneath the living room window where she watches the TV screen inside with obvious curiosity, and so on down the roster. (Figure 195)

Ordinarily raccoons are not so abundant in residential areas, but our situation is exceptional in that we are partially surrounded by a heavily wooded swamp that provides retreats that are both ample and secluded, capable of supporting a large population of raccoons. Several well-defined coon trails lead to our yards from the woods, and in late afternoon the animals begin to arrive from nearly every direction.

Raccoons produce only one litter a year.

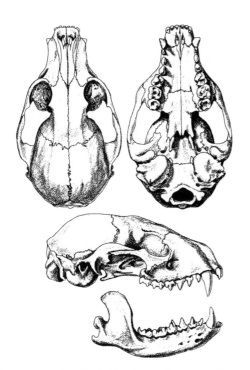

Figure 194 Skull of Northern Raccoon *(Procyon lotor varius>megalodous)*, ♂, 3 mi. WNW Paradis, St. Charles Parish, LSUMZ 16147. Approx. ½×.

Figure 195 Raccoons partaking of a nightly handout on a feeder outside the author's living room window in Baton Rouge. (Photograph by Joe L. Herring)

even at some distance away, causes an immediate response—a frozen posture with ears erect and facing toward the sound. But they give the youngsters little or no help. For instance, one of our feeders is a rough cypress slab 5 feet long and 2 feet wide that rests on legs 12 to 14 inches in height. We once watched a female on her first visit to the feeder with her four babies. One was smaller and obviously weaker than the others and seemed incapable of climbing on top of the feeder. The female simply watched its struggles but made no effort to assist it. The number of young in a litter, at least by the time the mother first brings them to the feeders, is usually three or four, but we have seen one group of five, and litters numbering seven are on record. The babies at birth weigh between two and three ounces and are blind. Their eyes do not open until they are nearly two weeks old, sometimes several days older. Almost from birth they show the characteristic black facial mask that is typical of the species.

As previously noted, raccoons are omnivorous. They relish crayfish, and in our coastal marshes they devour crabs, snails, and clams, as well as small fishes. In agricultural areas they sometimes raid cornfields and do some damage. Frogs, earthworms, and a wide variety of insects, such as grasshoppers, crickets, and beetles, constitute a significant portion of their diet, but they also consume much vegetable food that includes acorns, berries, watermelons, canteloupes, tomatoes, and tender shoots and buds of many trees and other plants. The only food item that we have

Mating apparently occurs in December and January, occasionally later in females less than a year old. The young are generally born in late April or early May after a gestation period of 63 days. We have seldom observed females bringing their young to our feeders before late June or early July. The mothers are highly solicitous of their offspring, at least to the extent of maintaining over them an alert vigil. The bark of a dog,

Tracks of Northern Raccoon

placed on our feeders that the coons completely ignore is raw onions, but they eat leftover cooked foods containing onions as seasoning. To a coon the *piece de resistance* is beef suet, yet dog biscuits and such items as chicken carcasses seem to rank high.

Reference to the animal's diet leads to a consideration of the commonly heard statement that raccoons always wash their food before eating it. The legend is totally without foundation. Raccoons frequently drink water when they are eating, provided it is available nearby. But they seldom take food to the water, even when close at hand. Raccoons rarely pick up food directly from the ground with their lips. Instead they lift it with their forepaws and transfer it to the mouth. When sitting in water and excavating crayfish and other forms of aquatic life from holes and from beneath partially submerged roots and logs, they will sometimes dip what they catch in the water a few times, but I would not interpret these actions as a washing procedure. Although a large concrete pond with circulating water lies adjacent to the main raccoon feeding station in our yard, I have never seen a raccoon take an item of food the short distance of less than two feet to wash it before eating it.

Raccoons are comical creatures. I once watched one that was unaware of my presence as it moved slowly along the edge of a small bayou exploring here and there for food. When it came to a hollow tree on the edge of the stream, into a basal cavity of which water extended, it stuck one of its paws inside and seemed to be feeling for whatever it might find. Suddenly it jerked out its paw as if it had been bitten and then sat back on its haunches with its paw held in front of its face as if inspecting it for possible injury.

The main feeding platform outside our living room window will accommodate only eight fairly large coons at one time, but others are usually struggling to find a place to climb on, often amid much snarling and other aggressive behavior. Once a coon finds a spot on the log it will frequently turn its back to its closest competitor thereby exposing its nape and hindquarters and at the same time avoiding an eye-to-eye confrontation, which would normally elicit violent aggressive responses. And it will often crowd the adjacent coon by backing into it. One would think that such shenanigans would invite attack, but I have never seen a coon making a nuisance of itself in this manner bitten in the process. This behavior is reminiscent of actions ascribed to the dog by Konrad Lorenz (1963), who points out that a dog will cease to attack a canine adversary once that its opponent has assumed a submissive posture by turning its back and exposing its nape.

Although raccoons are mainly active in late afternoon and at night, I often find them coming to our feeders in broad daylight when the food on the platforms has been put there for the birds. They like to swim and in the hot summer months are frequently found bathing in the pond in the middle of the day. One individual, now no longer with us, and who was dubbed "Spray Belly," had a habit of climbing atop a continuously spouting fountain in the middle of the pond and allowing the water to strike her underparts. Raccoons are excellent swimmers. I have heard it said that most dogs can whip most coons on dry land, but in the water most coons can give a good account of themselves against the best of dogs. Raccoons are also excellent climbers (Figure 196). Their dens are often in hollows in trees 30 or 40 feet above the ground. They descend a tree either by backing down or by coming down head first. When the weather is excessively warm, I often see coons sprawled out on a limb with their feet drooped over the sides.

Coon hunting has long been a favorite pastime in the South, as well as elsewhere. It provides exciting night sport and is done with or without dogs. The eyes of a treed coon glow bright yellow in the beam of a headlight. For some reason I have never eaten raccoon, but I am told that it is delicious when prop-

Figure 196   Northern Raccoon *(Procyon lotor)*. (Photograph by Robert N. Dennie)

erly prepared, which is true, I suppose, of any kind of food. The scent glands under the legs and along the spine near the rump should be carefully removed, as should most of the fat, and the meat parboiled before final roasting. Raccoons can be legally hunted and trapped only during a season regulated by the Louisiana Wild Life and Fisheries Commission.

The species is occasionally somewhat detrimental to man's interest. It is known to damage corn, especially sweet corn, and it is also fond of watermelons and tomatoes. The last time I tried to grow tomatoes in my yard, some unidentified culprit was taking bites out of them just about the time they began to ripen. Our coon population was suspect, but I was never able to pin the blame with certainty. But, "So what?" I say to myself, "the worms would have probably gotten the tomatoes anyway." I shall keep my coons and buy my tomatoes at a grocery store.

Predators. — Raccoons have few enemies, although some unlucky individuals, especially babies, are doubtless killed by Coyotes and other large predators, including dogs. In olden days when panthers and Red Wolves roamed our forest, the life of the coon was probably fraught with great danger from these large carnivores. But now one of the important decimating factors operating against the raccoon is the automobile on our highways. To be sure, trapping takes the heaviest toll of all, but it represents the utilization of a harvestable crop from which economic benefits to man accrue.

Parasites and diseases. — The literature dealing with the internal parasites of raccoons is much too voluminous to itemize in detail here. I find references to the occurrence in *Procyon lotor* of no less than 4 protozoans, 25 flukes, 4 tapeworms, 14 roundworms, and 5 thorny-headed worms (Stiles and Baker, 1935; Chandler, 1942b; Clark and Herman, 1959; Jordan and Hays, 1959; Layne et al.,

1960; Malek et al., 1960; and numerous others). External parasites include five mites (*Dermacentor variabilis, Ixodes cookei, I. diversifossus, I. marxi, I. texanus*), one sucking louse (*Polyplax spinulosa*), five lice (*Trichodectes crassus, T. malis, T. procyonis, T. vulpis,* and *Suricatoecus octomaculatus*), and nine fleas (*Chaetopsylla lotoris, C. coloris, Orchopeas wickhami, O. howardii, Odontopsyllus multispinosus, Ctenocephalides felis, C. canis, C. serraticeps,* and *Cediopsylla simplex*).

Subspecies

Procyon lotor megalodous Lowery

Procyon lotor megalodous Lowery, Occas. Papers Mus. Zool., Louisiana State Univ., 13, 1943:225; type locality, Marsh Island, Iberia Parish, Louisiana.

This subspecies of the raccoon occurs throughout the extensive marshes of southern Louisiana. It is distinguished from adjacent populations of *P. l. varius* Nelson and Goldman and *P. l. fuscipes* Mearns by its much more yellowish (less grayish) overall color, by its more pronounced black middorsal line, and by the conspicuously larger size of its molariform teeth.

Specimens examined from Louisiana. — Total 220, as follows: *Cameron Par.:* locality unspecified, 19 (LSUMZ); 25 mi. S Lake Charles, 1 (LSUMZ); Rockefeller Refuge, 2 (1 LSUMZ; 1 MSU); Johnsons Bayou, 3 (MSU); Lacassine Refuge, 9 (USNM); Grand Cheniere, 4 (USNM); Sabine National Wildlife Refuge, 12 (USNM). *Iberia Par.:* Marsh Island, 9 (2 LSUMZ; 7 USNM). *Jefferson Par.:* Grand Terre, 4 (LSUMZ). *Lafourche Par.:* 2.5 mi. W Raceland, 1 (USL); 1 mi. E Larose, 1 (USL). *Plaquemines Par.:* mouth of Miss. River, 1 (LSUMZ); 2 mi. NE Belair, 10 (LSUMZ); near Belair, 8 (LSUMZ); coastal marsh below Pilottown, 2 (LSUMZ); Delta Refuge, 4 (3 LSUMZ; 1 USL); Belair, 1 (LSUMZ); 2 mi. N Pilottown, 1 (LSUMZ); Pass-a-Loutre, 1 (LSUMZ); Tulane Univ. Riverside Campus, Belle Chasse, 3 (TU); Burbridge 3 (MCZ); Pilottown, 1 (USNM). *St. Bernard Par.:* Toca, 1 (LSUMZ); Chandeleur Islands opp. North Island, 1 (USL). *St. Mary Par.:* Belle Isle, 68 (10 USL; 58 USNM); south side of Cote Blanche Island, 1 (USL); marsh adjacent to Cote Blanche Island, 2 (USL); 6 mi. S Ellerslie, 3 (USL); Berwick Bay, 1 (USL); 0.5 mi. W Morgan City, Berwick Bridge, 1 (USL); Southern Bayou on Avoca Island, 8 mi. S ferry, 4 (USL); Morgan City, 5 (USNM); 3.8 mi. N, 0.5 mi. E Cypremort Pt., 1 (LSUMZ). *Terrebonne Par.:* Timbalier Island, 1 (LSUMZ); Grand Caillou, 1 (LSUMZ); 3 mi. NE Schriever, 1 (USL); Gibson, 2 (MCZ). *Vermilion*

Par.: 6 mi. N, 5 mi. E Pecan Island, 1 (LSUMZ); Esther, 1 (USL); 2 mi. W Esther, 3 (USL); 3 mi. N Intracoastal City, 1 (USL); Pecan Island, 8 (1 USL; 7 USNM); 1.5 mi. NW Esther, 1 (USL); 22 mi. SW Abbeville, 1 (USNM); 23 mi. SW Abbeville, 1 (USNM); 24 mi. SW Abbeville, 1 (USNM); State Wildlife Refuge, 9 (USNM).

Procyon lotor varius Nelson and Goldman

Procyon lotor varius Nelson and Goldman, Jour. Mamm., 11, 1930: 456; type locality, Castleberry, Conecuh County, Alabama.

Although Hall and Kelson (1959) show the range of *P. l. fuscipes* Mearns extending across northern Louisiana, I am inclined to refer all populations of the raccoons in Louisiana north of the coastal marshes to *P. l. varius*.

Specimens examined from Louisiana.—Total 215, as follows: *Acadia Par.:* 3 mi. N Crowley, 1 (USL); 6 mi. S Basile, 1 (USL). *Ascension Par.:* 3 mi. S Donaldsonville, 1 (LSUMZ). *Assumption Par.:* 7 mi. SW Napoleonville, 1 (LSUMZ). *Avoyelles Par.:* 1.5 mi. on Belledeaux-Hessmer Hwy. 1 (LSUMZ); 5 mi. W Cottonport, Lake Pearl, 1 (USL). *Beauregard Par.:* 4 mi. W Sugartown, 1 (LSUMZ); De Ridder, 1 (LSUMZ). *Bienville Par.:* 2 mi. NW Saline, 1 (NSU); Kepler Creek Swamp north Castor Refuge, 1 (USL); Ringgold, 7 (1 USL; 6 USNM); Arcadia, 1 (USNM); Castor, 2 (USNM). *Bossier Par.:* 6 mi. E Princeton, 1 (LSUMZ); 10 mi. SE Bossier City, 1 (LSUMZ); Benton, Bodcau Hunting Reservation, 1 (LSUMZ); 8 mi. N Benton, 1 (LTU). *Caddo Par.:* 1.5 mi. N Cross Lake, 1 (NLU); 7.75 mi. NW Shreveport, 1 (LTU); Shreveport, 1 (LSUMZ). *Calcasieu Par.:* Lake Charles, 2 (1 LSUMZ; 1 MSU); 4 mi. S Sulphur, 1 (LSUMZ); 1.5 mi. E Sulphur, 4 (1 LSUMZ; 3 USNM); 2 mi. N Gillis, 2 (LSUMZ); 5 mi. W Sulphur, 1 (MSU); Choupique, 1 (MSU); Iowa, 1 (USNM). *Catahoula Par.:* 0.5 mi. W, 2 mi. S Utility, 1 (LSUMZ); 4 mi. NNE Foules, 1 (LSUMZ); Catahoula Swamp, 2 (USNM). *Claiborne Par.:* 10 mi. E Homer, 1 (LSUMZ); 4 mi. SE Lisbon, 1 (LSUMZ); Middlefork Bottom, 1 (NLU); Junction City, 1 (USNM). *Concordia Par.:* Buckney Bayou, 10 mi. NE Ferriday, 1 (NSU). *East Baton Rouge Par.:* various localities, 16 (LSUMZ). *East Carroll Par.:* 5 mi. W Roosevelt, 1 (LTU). *East Feliciana Par.:* 0.5 mi. N Reiley, 1 (LSUMZ). *Evangeline Par.:* 4 mi. N Ville Platte, 1 (LSUMZ); Ville Platte, 1 (USL); 8 mi. N Ville Platte, 1 (USL); 4 mi. W Ville Platte, 1 (USL); 1.5 mi. N Chataignier, 1 (USL); 2 mi. S Easton, 1 (USL); 4 mi. N Basile, 1 (USL). *Franklin Par.:* 2 mi. N Gilbert, 1 (NLU); 4 mi. S Delhi, 1 (NLU); 9 mi. S Delhi, 1 (NLU). *Grant Par.:* Georgetown, 1 (LTU); 1.5 mi. NE Pollock, 1 (USL). *Iberia Par.:* Jeanerette, 1 (USL); 6 mi. W New Iberia, 1 (USL). *Iberville Par.:* St. Gabriel, 1 (LSUMZ); 2 mi. E Iberville, 2 (LSUMZ); 3 mi. E Iberville, 1 (LSUMZ); Bayou Goula, 1 (LSUMZ); 1 mi. S Grosse Tete, 2 (LSUMZ); 7.5 mi. SW Ramah, 1 (LSUMZ); Plaquemine, 1 (LSUMZ). *Jefferson Davis Par.:* 0.3 mi. N, 1 mi. W Fenton, 1 (LSUMZ); 3 mi. W Lake Arthur, Morgan

Plantation, 1 (LSUMZ). *Lafayette Par.:* 10 mi. S Lafayette, 1 (LSUMZ); Lafayette, 3 (2 USL; 1 USNM); 4.3 mi. S Lafayette, 1 (USL); 3 mi. W and 0.5 mi. S Lafayette, 1 (USL). *La Salle Par.:* Hwy. 165, 1 mi. SW Olla, 1 (NLU). *Lincoln Par.:* 1 mi. W Vienna, 1 (LTU). *Livingston Par.:* 4.5 mi. NW Port Vincent on Amite River, 1 (LSUMZ); 3 mi. S Denham Springs, 1 (LSUMZ); 1 mi. S Holden, 1 (LSUMZ); Watson, 1 (LSUMZ); 2 mi. W Denham Springs, 1 (LSUMZ); Holden, 1 (SLU); 3 mi. E Frost, 1 (SLU). *Madison Par.:* Tallulah area, 11 (1 NLU; 10 USNM); 10 mi. NW Tallulah, 1 (USNM); 28 mi. SW Tallulah, 1 (USNM). *Morehouse Par.:* 5 mi. N Oak Ridge, 1 (USNM); 2 mi. S Oak Ridge, 1 (NLU); Bastrop, 1 (NLU); 7 mi. W Beekman, 1 (NLU); 3 mi. N Bastrop, 1 (NLU). *Natchitoches Par.:* Provencal, 3 (LSUMZ); 3 mi. NE Provencal, 1 (NSU). *Ouachita Par.:* Monroe, 2 (1 LTU; 1 NLU); Russell Sage Game Mgt. Area, 1 (NLU); 12 mi. SW Monroe, 1 (NLU). *Pointe Coupee Par.:* 2 mi. NW Lottie, 1 (LSUMZ). *Rapides Par.:* Alexandria, 1 (LSUMZ); 10 mi. NW Alexandria, 1 (LSUMZ); 15 mi. S Alexandria, 1 (LSUMZ); 10 mi. N Glenmora, 1 (USL); 4 mi. E Cheneyville, 1 (LTU). *Richland Par.:* 4 mi. NW Rayville, 1 (NLU); 5 mi. S Start, 1 (NLU). *St. Charles Par.:* Bonnet Carré Spillway, 1 (LSUMZ); 3 mi. WNW Paradis, 1 (LSUMZ); 3 mi. E Bonnet Carré Spillway, 1 (USL). *St. James Par.:* 2 mi. SE Gramercy, 1 (LSUMZ); 6 mi. NE Gramercy, 1 (LSUMZ). *St. John the Baptist Par.:* 0.5 mi. ESE Laplace, 1 (LSUMZ); 7 mi. E Akers, 1 (SLU); Manchac, 2 (SLU). *St. Landry Par.:* Morrow, 1 (LSUMZ); 8 mi. E Lebeau, 1 (LSUMZ); 10 mi. S Krotz Springs, 1 (LSUMZ); 7 mi. N Port Barre, Foy Lake, 1 (LSUMZ); 2 mi. E Leonville, 1 (USL); near Beggs, 4 mi. N Washington, 1 (USL); Opelousas, 4 (USL); 4 mi. E Chataignier, 1 (USL); 3 mi. NW Washington, 1 (USL). *St. Martin Par.:* Cade, 2 (USL); 3 mi. NW Cade, 1 (USL); 3.25 mi. E Breaux Bridge, 1 (USL); Lake Martin, 1 (USL); Cypress Island, 1 (USL); St. Martinville, 1 (USL); Nina community, 1 (USL); 3 mi. SE Parks, 1 (USL); 8 mi. SW Cecilia, 1 (USL); 4 mi. W Cecilia, 1 (USL). *St. Tammany Par.:* 4 mi. W Slidell, 1 (LSUMZ); 1.5 mi. E, 1.3 mi. S Pearl River, 1 (LSUMZ); 1.1 mi. E, 1.2 mi. S Pearl River, 1 (LSUMZ); 4 mi. NW Pearl River, 1 (SLU); Hwy. 21, 0.7 mi. S Waldheim, 1 (TU); Mandeville, 1 (TU). *Tangipahoa Par.:* 5 mi. N Robert, 1 (LSUMZ); 3 mi. W Ponchatoula, 1 (SLU). *Tensas Par.:* 20 mi. S, 3 mi. W Tallulah, 1 (LSUMZ); Lake Ridge, 25 mi. NW St. Joseph, 1 (USNM). *Union Par.:* Hwy. 15, West Sterlington, 1 (NLU); 20 mi. N Ruston, 1 (USL); near Farmerville, 1 (LTU); D'Arbonne Bayou and Lake, 1 (LTU). *Vermilion Par.:* 4 mi. NW Indian Bayou, 1 (USL); 4 mi. S Abbeville, 1 (USL); Abbeville, 7 (USNM); Gueydan, 3 (USNM); Kaplan, 2 (USNM). *Vernon Par.:* 3 mi. S Simpson, 1 (LSUMZ); 3 mi. S Leesville, 1 (MSU); 0.5 mi. E Newllano, 1 (USL). *Washington Par.:* 2 mi. E Varnado, 2 (LSUMZ); 2 mi. SE Varnado, 1 (LSUMZ); 2 mi. W, 1 mi. S Varnado, 1 (LSUMZ); Varnado, 1 (SLU). *Webster Par.:* 2 mi. S Minden, 1 (LTU). *West Baton Rouge Par.:* 5 mi. W Port Allen, 1 (LSUMZ); 3 mi. SW Port Allen, 1 (LSUMZ). *West Carroll Par.:* 7 mi. E Epps, 1 (LTU). *West Feliciana Par.:* St. Francisville, 1 (LSUMZ); 5 mi. W St. Francisville, 1 (LSUMZ). *Winn Par.:* Atlanta, 1 (USL).

The Mustelid Family
Mustelidae

THE FAMILY Mustelidae includes 25 genera and nearly 70 species. It is nearly worldwide in distribution, occurring everywhere with the exception of Australia, Madagascar, Antarctica, and most oceanic islands. Some members, such as the weasels, are strictly terrestrial; others, like the American Marten *(Martes americana)*, are arboreal; and still others, like mink and the river otters, are semiaquatic. The Sea Otter *(Enhydra lutris)* is almost wholly aquatic, seldom leaving the water. Included in the family is the Wolverine *(Gulo gulo)*, which is generally regarded as one of the most ferocious and powerful of all mammals pound for pound. Most mustelids possess powerful anal scent glands, in which regard the skunks are notorious.

Genus *Mustela* Linnaeus

Dental formula:

$$\text{I}\frac{3-3}{3-3} \quad \text{C}\frac{1-1}{1-1} \quad \text{P}\frac{3-3}{3-3} \quad \text{M}\frac{1-1}{2-2} = 34$$

LONG-TAILED WEASEL　　　　　Plate XII
(Mustela frenata)

Vernacular and other names. — The name weasel is derived from the Anglo-Saxon word *wesle.* The modifier long-tailed is based on the fact that the tail of this species is relatively and actually longer than that of other species in the genus. According to Arthur (1928), it was known to the Biloxi Indians as *iskixpa,* a diminutive form of the name they gave to the mink. The early French settlers called it *fouine,* meaning sneak, a noun applied to the animal because of its skulking, stealthy habits. The generic name *Mustela* is the Latin word for weasel. The specific name *frenata* comes from the Latin word *frenum,* which means bridle. It refers to the black mask possessed by weasels in the more southern portions of the range of the species from Mexico to South America.

Distribution. — The Long-tailed Weasel occurs from southern Canada to Peru. It is absent only in a small area in the southwestern United States and northwestern Mexico. In Louisiana it appears to be highly local in its occurrence. (Map 58)

External appearance. — This small, long-bodied mustelid with its short legs, distinctive color, and long tail can hardly be confused with any other mammal, except weasels of some other species, none of which occur within several hundred miles of our state. The upperparts are uniformly brown, the underparts yellowish or yellowish white except for the white chin. The terminal one-fourth or one-third of the well-furred tail is black. The body hair is moderately fine, rather short, and not dense.

Color phases. — Long-tailed Weasels in the northern part of the range of the species generally turn white in winter except for the

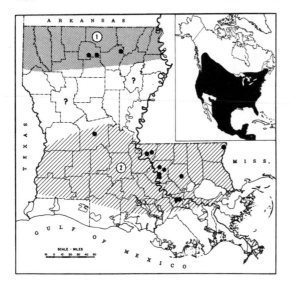

Map 58 Distribution of the Long-tailed Weasel (*Mustela frenata*). *Shaded areas,* known or presumed range within the state; •, localities from which museum specimens have been examined; *inset,* overall range of the species. 1. *M. f. primulina;* 2. *M. f. arthuri.*

black-tipped tail, but this change into a white winter pelage never occurs in Louisiana. As far as I am aware no color phases in the usual sense are known in this species.

Measurements and weights.—Four males in the LSUMZ from various localities in the state averaged as follows: total length, 452 (410–495); tail, 143 (131–153); hind foot, 47 (45–50); ear, 23 (20–27). Skulls of seven males from the same source averaged as follows: greatest length, 51.3 (48.3–53.6); cranial breadth, 21.6 (20.2–22.1); zygomatic breadth, 27.8 (26.3–28.9); breadth across the postorbital processes, 14.3 (12.9–16.2); interorbital breadth, 11.2 (9.8–11.8); palatilar length, 20.3 (18.4–22.0); postpalatal length, 25.6 (24.7–26.7); height of coronoid, 14.0 (13.3–15.5). Males weigh up to nine ounces, females usually half that amount.

Cranial and other skeletal characters.—The skull of a weasel is readily identified by a combination of characters, including its small size,

the greatly inflated auditory bullae, which are much longer than they are wide, the extension of the hard palate well beyond the end of the toothrow, and the transversely dumbbell-shaped upper molars. The middle lower incisor on each side is larger than those adjoining it, and it is especially enlarged on its posterior edge. The skulls of a weasel and a mink are similar, but the former is decidedly smaller and the auditory bullae are relatively and absolutely much longer. The penis bone is an inch or slightly more in length, expanded and flattened laterally at the base, curved terminally, slightly skewed to one side at the tip, and markedly grooved dorsally on its terminal half. (Figures 185 and 197)

Sexual characters and age criteria.—The sexes are indistinguishable except by their primary

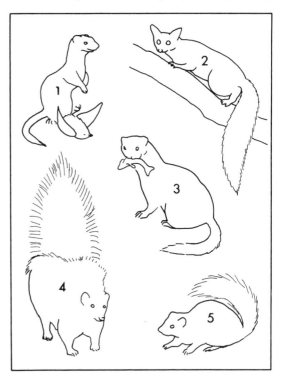

Plate XII
1. Long-tailed Weasel (*Mustela frenata*).
2. Ringtail (*Bassariscus astutus*).
3. North American Mink (*Mustela vison*).
4. Striped Skunk (*Mephitis mephitis*).
5. Spotted Skunk (*Spilogale putorius*).

PRATT
1972

and secondary sex organs. According to Hamilton (1933), permanent dentition has been attained by the time the young weasel is 75 days of age. In adults the penis bone is greatly enlarged basally, although laterally flattened; in juveniles not only is the bone short but the base is not much larger than the main part of the shaft. The four pairs of teats in the female are postabdominal and inguinal in location.

Status and habits.—The Long-tailed Weasel is extremely local in its distribution in the state and must be considered quite rare. As is evident from Map 58, specimens are available from only 14 localities, but I have reports of its occurrence in several additional places in the Florida Parishes, and Arthur (1928) mentions its presence at Laurel Hill, West

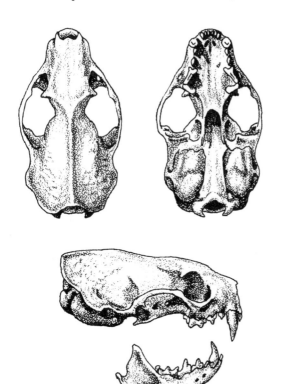

Figure 197 Skull of Long-tailed Weasel *(Mustela frenata arthuri),* sex ?, 10 mi. N Baton Rouge, East Baton Rouge Parish, LSUMZ 2221. Approx. natural size.

Feliciana Parish; Greensburg, St. Helena Parish; Braithwaite, Plaquemines Parish; Geismar, Assumption Parish; and French Settlement, Livingston Parish. The type specimen of the subspecies *M. f. arthuri* Hall was taken at Remy, and others at Convent, in St. James Parish. I strongly suspect that the species may be somewhat more numerous than the few records of it would seem to indicate, despite the fact that I have actually seen only two weasels in Louisiana in the nearly 50 years during which I have pursued an intense interest in the mammals of the state. One morning shortly after sunrise I saw a weasel loping across the edge of the Louisiana State University campus in Baton Rouge, and years ago, when I resided just south of the campus, I had a weasel living under my house. How long it had been there before I first detected its presence and how long it remained I have no way of knowing, for our boarder was so stealthy and elusive that I seldom saw it. Elsewhere in the range of the species I have on several occasions excited the curiosity of a weasel and enticed it to come within a few feet of me by the squeaking notes that I make in imitation of a nestling in distress for the purpose of calling up small songbirds.

The little animal is extremely courageous and does not hesitate to attack ferociously species much larger than itself. Its movements are swift and nimble and in a flash it leaps on the back of its adversary or intended prey and sinks its sharp teeth into the skull or severs the jugular vein. The main food of a weasel consists of rats and mice, but occasionally it captures small birds that habitually feed on the ground. Although it is probably seldom successful in overtaking a full-grown Cottontail or Swamp Rabbit, it does prey heavily on young ones. Its quest for small mammals sometimes brings it to the farmyard, where it not only catches Norway and Roof Rats but also commits depredations on poultry. It is alleged to kill more than it needs and only to eat the brains and lap the blood of its victims, but evidence of such wasteful selectivity is not altogether conclusive. It definitely does not "suck the blood" of animals that it kills.

The senses of smell, sight, and hearing are all well developed, and each is brought into play as the animal pursues its prey with unrelenting persistence. Its intended victim does not have many places it can go where the weasel cannot follow. Although not an expert climber, as is the marten, it will scramble up trees if required. Similarly it is not a good swimmer, but when the occasion demands, it will take to the water for short distances. In this latter regard it differs markedly from its close relative the mink, which is as much at home in the water as it is on land. When running, a weasel progresses in a lope that one writer said reminded him of the movements of a measuring worm because of the height to which the back was arched.

Information on the breeding biology of the Long-tailed Weasel in Louisiana is totally lacking. Elsewhere the species has been found to breed in the summer, and the single litter of up to nine (generally five to seven) young are born the following spring, usually in April. The length of the gestation period varies from 205 to 337 days. But this extended period is the result of the fertilized eggs' remaining in a state of prolonged quiescence in the uterus before implantation takes place. Once the embryo begins its development it comes to full term in 27 days or less (Wright, 1942, 1947, 1948 a and b). The den of a weasel is usually in a rotten log, hollow stump, or a hole in the ground, sometimes an old mole run that has been enlarged. The nest chamber is floored with grasses and mouse or shrew fur, and not infrequently contains scats over which fresh layers of grass are placed, giving the nest a stratified appearance (Polderboer et al., 1941).

Predators. — Because of its agility and its pugnacious disposition the Long-tailed Weasel is not often molested by predators. In other parts of its range it has been known to have been eaten by Great Horned Owls, Barred Owls, rattlesnakes, black snakes, and occa-sionally foxes. Dogs and Coyotes are said to catch a few.

Parasites. — External parasites that have been found on this weasel somewhere within its range include 12 fleas (*Orchopeas sexdentatus, Nosopsyllus fasciatus, Ceratophyllus wagneri, C. vison, C. labis, Hytrichopsylla dippiei, Nearctopsylla brookei, Neopsylla inopina, Oropsylla poelantis, O. rupentris, Dactylopsylla bluei, Foxella ignota*), four lice (*Trichodectes retusus, T. kingi, Neotrichodectes mephitidis, N. minutus*), and three ticks and mites (*Dermacentor variabilis, Ixodes hexagonus, I. scapularis, I. cookei*). Known internal parasites include one tramatode (*Alaria taxideae*), one cestode (*Taenia tenuicollis*), and eight nematodes (*Filaroides martis, Molineus mustelae, M. chabaudi, M. patens, Physaloptera maxillaris, Skrjabingylus nasicola, Capillaria mustelorum, Filaria perforans*), and no acanthocephalans (Stiles and Baker, 1935; Erickson, 1946; Morlan, 1952; Voge, 1955; Jackson, 1961; Schmidt, 1965).

Subspecies

Mustela frenata arthuri Hall

Mustela noveboracensis arthuri Hall, Proc. Biol. Soc. Washington, 40, 1927:193; type locality, Remy, St. James Parish, Louisiana.

This subspecies occupies all the southern half of Louisiana and adjacent parts of southeastern Texas and southern Mississippi. It differs from *M. f. primulina* Jackson in having "narrower bullae, which are much less inflated on their anteromedial faces, a less marked postorbital constriction, a braincase which is narrower across the mastoid region and broader anteriorly, and a skull, which, in longitudinal axis, has the dorsal outline markedly more convex" (Hall, 1951).

Specimens examined from Louisiana. — Total 13, as follows: *East Baton Rouge Par.:* Baton Rouge, 1 (LSUMZ); University, 3 (LSUMZ); 10 mi. N Baton Rouge, 1 (LSUMZ); 6 mi. ENE Baton Rouge, 1 (LSUMZ). *Livingston Par.:* Springville, 1 (MVZ). *Rapides Par.:* 4 mi. W Woodworth, 1 (LSUMZ). *St. James Par.:* Convent, 1 (USNM); Remy, 1 (MVZ). *Washington Par.:* 4.25 mi. S Angie, 1 (SLU). *West Feliciana Par.:* Bains, 1 (LSUMZ); 6 mi. NE St. Francisville, 1 (LSUMZ).

Mustela frenata primulina Jackson

Mustela primulina Jackson, Proc. Biol. Soc. Washington,
   26, 1913:123; type locality, 5 mi. NE Avilla, Jasper
   County, Missouri.

E. Raymond Hall has kindly examined the
specimen from Swartz, in Ouachita Parish,
and found that it is assignable to *primulina*
even though it shows evidence of intergradation with *arthuri*. I consider the specimens
from Ruston and Choudrant, in Lincoln
Parish, likewise referable to *primulina*.

Specimens examined from Louisiana. — Total 3, as follows:
Lincoln Par.: Choudrant, 1 (LSUMZ); west of Ruston, 1
(LTU). *Ouachita Par.:* Swartz, 1 (LSUMZ).

NORTH AMERICAN MINK   Plate XII
(*Mustela vison*)

Vernacular and other names. — The word
"mink" is possibly traceable to the Medieval
English *mynk* or the Swedish *menk*. The adjectival modifier North American in the vernacular name is required to differentiate the
species of the New World from the mink of
the Old World, *Mustela lutreola*. The name
Sea Mink is applied to the now extinct form
that until about 1860 inhabited New Brunswick and coastal Maine and which is currently
accorded full species rank under the name
Mustela macrodon. The generic name *Mustela*
is the Latin word for weasel. The fact that
both the weasels and minks are treated as
members of the same genus shows that they
are regarded as being closely related. The
specific name *vison* is of somewhat doubtful
origin. It probably comes from the Icelandic
or Swedish word *vison,* which means a kind of
weasel. The subspecific name for the population of mink that occurs in Louisiana, *vulgivaga,* is a Latin adjective with several
meanings, including roving, changeable, and
variable. Unfortunately, the author of the
name did not give his reasons for using it.
According to Arthur (1928), it was called
iskixpa by the Biloxi Indians and *toni* by the
Choctaws. The name *belette,* applied to it by
the early French colonists, is still heard today
in our Acadian country.

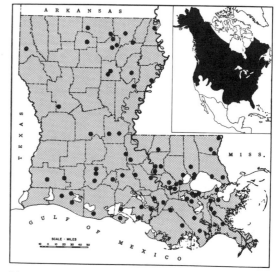

Map 59 Distribution of the North American
Mink (*Mustela vison*). *Shaded area,* known or presumed range within the state; •, localities from
which museum specimens have been examined;
inset, overall range of the species.

Distribution. — The species is found widely
across most of Canada, thence south over all
the United States except for Arizona and
parts of California, Nevada, Utah, New Mexico, and western Texas. Its distribution in
Louisiana is statewide. (Map 59)

External appearance. — This moderate-sized
mustelid has a total length that ranges from
19 to 23 inches, of which the tail comprises
approximately one-third. It is considerably
longer and heavier bodied than a weasel, and
its color is decidedly darker, being a fairly
uniform dark brown, above and below, except for white blotches on the chin and throat
and occasionally a few white blotches or
streaks on the chest, abdomen, and anal region. The brown of the upperparts varies
from burnt umber to chocolate, that of the
legs and underparts being somewhat lighter
in color. The tail is dark brown basally but
becomes progressively blacker toward the tip.
It is much more bushy than that of a weasel,
and the entire pelage is denser and more
glossy. The thick underfur and the abundant,
long, somewhat oily guard hairs render the

animal's pelage water-resistant and permit it to lead a semiaquatic existence. The summer pelage is thinner and somewhat paler than the prime coat into which the animal molts in late summer and early fall. The feet are densely furred except for the pads on the tips of the toes and on the soles of the feet, and the toes are semiwebbed. A pair of anal glands, typical of all mustelids, secretes a substance with a strong fetid odor, particularly during the breeding season and also when the animal is excited. I, for one, find the smell more obnoxious than that of the skunks.

Color phases. — No true color phases are known among wild mink, but I have seen so-called "blond" pelts in which the coat color is a light tan. This color mutation is probably attributable to a dilution factor. Albinos occur infrequently. Fur-farm breeding has shown that coat color is potentially highly variable. Even such novelties as ambergold, sapphire, and gunmetal have been produced.

Measurements and weights. — Twenty-nine males (10 in the MCZ from Burbridge, Plaquemines Parish, *fide* Bangs, 1895; 19 in the LSUMZ from various localities in Louisiana) averaged as follows: total length, 568 (504–680); tail, 184 (167–200); hind foot, 68 (60–79); ear (19 specimens), 23 (19–27). Two males weighed 2.2 and 2.6 pounds. Respective averages of five females in the LSUMZ from various localities in Louisiana were as follows: 517 (488–580); 172 (152–185); 52 (50–57); 23 (21–25). A single female weighed 1.7 pounds. The skulls of 19 males in the LSUMZ from various localities in Louisiana averaged as follows: greatest length, 65.9 (60.4–70.6); cranial breadth, 28.0 (26.6–29.2); zygomatic breadth, 37.4 (33.5–42.3); breadth across postorbital processes, 17.9 (15.7–20.8); interorbital breadth, 15.1 (12.5–18.7); palatilar length, 28.1 (25.0–31.4); postpalatal length, 30.2 (25.6–33.1); height of coronoid process, 18.6 (16.4–19.8). The skulls of six females from

the same source averaged as follows: greatest length, 57.6 (54.7–60.1); cranial breadth, 26.1 (25.4–27.4); zygomatic breadth, 33.0 (30.3–33.9); breadth across postorbital processes, 15.6 (14.7–16.2); interorbital breadth, 12.8 (11.8–13.2); palatilar length, 25.0 (23.4–25.9); postpalatal length, 27.1 (26.7–28.1); height of coronoid, 15.8 (14.2–16.6).

Cranial and other skeletal characters. — The skull of *Mustela vison* is identifiable by the following combination of characters: the presence of 34 teeth, the dumbbell shape of the last upper molar, the long, greatly inflated auditory bullae, and the extension of the palate well beyond the last molar. It resembles the skull of the Long-tailed Weasel but is much larger; the breadth of the postorbital constriction is relatively much greater and the braincase is relatively much wider and more expanded. The penis bone likewise resembles that of the Long-tailed Weasel but is appreciably longer (1.75 inches), decidedly hooked at the end, deeply furrowed on the terminal half of its dorsal surface, and slightly furrowed laterally on its basal portion. (Figures 185 and 198)

Sexual characters and age criteria. — The sexes are alike except for the primary and secondary sex organs and the slightly larger size of the males. The bacula of males only one year old or younger are appreciably shorter, being less than 1.75 inches in length and lacking the greatly enlarged basal portion with a dorsal process that is found in fully mature individuals (Petrides, 1950). Pelts of adult males can be recognized by the presence of a distinct penis scar and those of adult females by the prominent nipple sites.

Status and habits. — The North American Mink, one of Louisiana's principal furbearers, is fairly abundant and widespread over the state and doubtless occurs in every parish. It is particularly numerous in the tupelogum swamps along the Mississippi River and in the

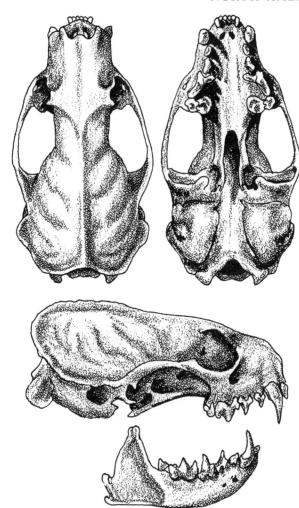

Figure 198 Skull of North American Mink *(Mustela vison vulgivaga)*, ♂, Lake Verret, Assumption Parish, LSUMZ 3427. Approx. natural size.

Atchafalaya Basin, as well as in the freshwater to brackish marshes of coastal Louisiana. St. James Parish alone produces from 10 to 14 thousand mink annually. The abundance of the species in bottomland swamps has been attributed to the prevalence of crayfish, one of its main items of food in these areas.

In the 57 years for which records are available, beginning with the season of 1913–1914 and continuing through the season of 1972–1973, no fewer than 4,592,221 mink were obtained by Louisiana trappers. The average annual catch was 80,565, ranging from a low of 18,805 in the 1963–1964 season to a high of 184,552 in the 1950–1951 season. Likewise the average annual price that trappers received for their mink pelts fluctuated widely, ranging from a low of $1.50 in the 1932–1933 season to a high of $15.00 in the 1945–1946, the 1947–1948, and the 1954–1955 seasons. The average price over the 44 years for which records are available was $6.58

Mink are never found far from water (Figure 199). They are especially numerous in our coastal marshes, and they live along wooded streams and on the edges of lakes. Their dens are located under fallen logs, in hollow stumps, and in old abandoned muskrat houses and Nutria burrows. The nest is lined with grasses, feathers, fur, and any other soft materials that are obtainable. Only one litter a year is produced, usually in early spring. The gestation period, as is apparently the case with all mustelids, is sometimes greatly prolonged and may last as much as 75 days because the fertilized eggs can remain free in the uterus in a dormant state for an extended period before finally becoming implanted and resuming their development leading to full-term fetuses. This latter process requires 30 to 32 days (Enders, 1952). The number of young in a litter is usually only 3 or 4. Large litters, such as one of 17 that has been reported to occur, are probably more than one litter combined. At birth the kits are less than four inches in length, blind, and scantily covered with fine, whitish hair. When the kits are about five weeks old the eyes open, and weaning takes place soon

Tracks of North American Mink

thereafter. The male mink is said to mate with more than one female but to stay with the last one and to assist with the raising of the young.

In their feeding habits mink are intermediate between terrestrial weasels and aquatic otters. On the one hand, they eat rats, mice, rabbits, and occasionally birds, and on the other, they consume fish, crabs, crayfish, frogs, and almost any kind of aquatic life that is available to them. Perhaps their favorite food item is crayfish—crawdads or crawfish as they are commonly called. Under certain circumstances an individual mink may enter a farmyard to prey on poultry or to kill domestic ducks on a farm pond, but these depredations are not nearly as serious as those for which weasels are sometimes responsible.

Predators.—Mink have few enemies but occasionally, especially when they are young, they fall victim to a Bobcat or a Great Horned Owl. In the coastal marshes, though, the Alligator is a formidable enemy and is said by trappers to take a heavy toll.

Parasites.—Mink harbor many internal parasites, including no less than 23 trematodes, two cestodes, 24 nematodes, and two acanthocephalans. External parasites include

Figure 199 North American Mink *(Mustela vison),* Hamilton County, Ohio, October 1963. (Photograph by Karl H. Maslowski)

five mites (*Ixodes cruciaris, I. hexagonus, I. spinosa, I. ricinus, I. kingii*), nine fleas (*Ceratophyllus oculatus, C. vison, C. acomantis, Hystrichopsylla dippiei, Nearctopsylla hyrtaci, Oropsylla arctomys, Orchopeas howardii, Nosopsyllus fasciatus, N. genalis*), and two lice (*Lipearus dissimilis, Trichodectes retusus*). (Stiles and Baker, 1935; Erickson, 1946; Gorham and Griffith, 1952; Schacher and Faust, 1956; Clark and Herman, 1959; Sogandares-Bernal, 1961; Lumsden and Zischke, 1961; Miller and Harkema, 1964; and others.)

Subspecies

Mustela vison vulgivaga Bangs

Putorius (Lutreola) vulgivaga Bangs, Proc. Boston Soc. Nat. Hist., 26, 1895:539; type locality, Burbridge, Plaquemines Parish, Louisiana.

Mustela vison vulgivaga, Miller, Bull. U.S. Nat. Mus., 79, 1912:102.

The currently recognized pattern of geographical variation in mink populations, particularly in the southeastern part of the United States, is, I believe, in need of considerable further study and possible reevaluation. To accomplish this objective, one would first need to assemble as large a sample as possible of skins and skulls from the entire geographical area under consideration. The specimens would then have to be segregated according to sex, season, and age class. Finally, the resulting measurements would require statistical analysis. Until such a study has been made, I am recognizing the race *vulgivaga* on a provisional basis only.

Specimens examined from Louisiana.—Total 153, as follows: *Acadia Par.:* 2 mi. W Crowley, 1 (USL); Cartville, 4 (MCZ); Pointe aux Loups Springs, 6 (MCZ). *Ascension Par.:* Blind River, 2 (LSUMZ); Sorrento, 1 (LSUMZ); 2 mi. SSW Sorrento, 1 (SLU). *Assumption Par.:* Lake Verret, 1 (LSUMZ); Labadieville, 1 (LSUMZ). *Avoyelles Par.:* near Coco Lake, Moreauville, 1 (LTU); Lake Pearl, 5 mi. W Cottonport, 1 (USL). *Caddo Par.:* Interstate 20, 13.5 mi. W Red River, 1 (LSUMZ). *Calcasieu Par.:* Carlyss, 1 (MSU); locality unspecified, 12 (USNM); Iowa, 7 (USNM). *Caldwell Par.:* Castor Creek, Grayson, 1 (LTU); Columbia, 1 (FM). *Cameron Par.:* near Lacassine Refuge, S Intracoastal Canal, 1 (LSUMZ); Intracoastal Canal near crossing of La. Hwy. 27, 1 (MSU); 1 mi. E, 7 mi. N

Creole, 1 (MSU); 13 mi. E Creole, 1 (MSU). *Catahoula Par.:* 15 mi. NE Harrisonburg, 1 (LSUMZ). *East Baton Rouge Par.:* 5 mi. E Baton Rouge, 1 (LSUMZ); 3 mi. S University, 2 (LSUMZ); 3 mi. SE Baton Rouge, 1 (LSUMZ); 1 mi. S University, 1 (LSUMZ); Baton Rouge, 1 (LSUMZ); Elbow Bayou, 3 mi. S Baton Rouge, 1 (LSUMZ); ca. 0.6 mi. S, 3.5 mi. E University, 1 (LSUMZ). *East Carroll Par.:* Lake Providence, 1 (LTU). *East Feliciana Par.:* 9 mi. SSW Clinton, Comite River, 1 (LSUMZ). *Franklin Par.:* locality unspecified, 1 (NLU); 2 mi. N Gilbert, 1 (NLU). *Iberia Par.:* Marsh Island, 1 (LSUMZ). *Iberville Par.:* 2 mi. S Plaquemine, 1 (LSUMZ); 2 mi. W Atchafalaya Spillway [7.5 mi. SW Ramah], 1 (LSUMZ). *Jefferson Par.:* 3 mi. W Kenner, U.S. Hwy. 61, 1 (TU). *Lafayette Par.:* Lafayette, 1 (USL). *Lafourche Par.:* 12 mi. SSE Golden Meadow, 1 (LSUMZ); 3 mi. E Larose, 1 (USL). *Lincoln Par.:* 7 mi. N Ruston, 1 (LTU). *Livingston Par.:* 2 mi. E Clio, 1 (SLU). *Morehouse Par.:* Mer Rouge, 4 (USNM). *Natchitoches Par.:* locality unspecified, 1 (LSUMZ); 3 mi. N Kisatchie, 1 (NSU). *Orleans Par.:* east Chef Menteur Pass between Lake Pontchartrain and Lake Borgne, 1 (LSUMZ). *Ouachita Par.:* 8 mi. N Monroe, 1 (NLU); Swartz, 1 (NLU); Monroe, 1 (NLU). *Plaquemines Par.:* Belair, 9 (6 LSUMZ; 3 USNM); 1 mi. NW Port Sulphur, 1 (LSUMZ); Tulane Univ. Riverside Campus, Belle Chasse, 3 (TU); Burbridge, 11 (MCZ). *Pointe Coupee Par.:* 3 mi. W Lottie, 1 (LSUMZ). *Rapides Par.:* Cocodine Lake Swamp, 5 mi. SW Lecompte, 1 (NSU). *Richland Par.:* 5 mi. E Oak Ridge, 1 (NLU). *St. Bernard Par.:* Toca Village, 1 (LSUMZ); locality unspecified, 3 (LSUMZ). *St. Charles Par.:* 3 mi. WNW Paradis, 1 (LSUMZ); Paradis, 1 (SLU); Rt. 61, 4 mi. W New Orleans Int. Airport, 1 (TU); 3 mi. E Norco, 4 (TU); Good Hope, 1 (TU); 5 mi. W Paradis, 1 (LSUMZ). *St. James Par.:* 10 mi. NW Vacherie, 1 (LSUMZ); 11.6 mi. SE Sorrento, 1 (TU); locality unspecified, 1 (USNM). *St. John the Baptist Par.:* 7 mi. E Akers, 1 (SLU); w shore Lake Maurepas at Blind River, 2 (LSUMZ). *St. Landry Par.:* Melville, 1 (USL). *St. Martin Par.:* Nina community, 1 (USL). *St. Mary Par.:* Franklin, 1 (LSUMZ); 1.2 mi. N, 2.5 mi. E Cypremort Pt., 1 (LSUMZ); 6 mi. S Ellerslie, 1 (USL); Cypremort Point, 1 (USL); Berwick Bay, 1 (USL); Morgan City, 7 (USNM). *St. Tammany Par.:* 3 mi. SW Lacombe, 1 (LSUMZ); 1.1 mi. E, 1.2 mi. S Pearl River, 1 (LSUMZ). *Tangipahoa Par.:* 2 mi. E Manchac, 1 (SLU); Ponchatoula, 1 (TU). *Tensas Par.:* Lake Bruin, St. Joseph, 1 (LTU). *Terrebonne Par.:* 1 mi. NE Montegut, 2 (LSUMZ); 0.5 mi. E Schriever, 1 (LSUMZ). *Union Par.:* Corney Bayou, Bernice, 1 (LTU); 16 mi. N Monroe, 1 (LTU). *Vermilion Par.:* Pecan Island, 1 (USL). *Washington Par.:* 2 mi. NW Augie, 1 (LSUMZ); 8 mi. SW Bogalusa, 1 (LSUMZ); 16 mi. S Angie, 1 (SLU).

Genus *Spilogale* Gray

Dental formula:

$$I\frac{3-3}{3-3} \quad C\frac{1-1}{1-1} \quad P\frac{3-3}{3-3} \quad M\frac{1-1}{2-2} = 34$$

SPOTTED SKUNK   Plate XII
(*Spilogale putorius*)

Vernacular and other names. — In Louisiana and elsewhere the Spotted Skunk is frequently called the civet cat, which is an unfortunate misnomer. The animal is neither a cat nor one of the true civets, which belong to an Old World group of carnivores. Both the Spotted and the Striped Skunks are also often called polecats, probably because early settlers in America confused them with the similarly foul-smelling mustelids that are called polecats in the Old World. The word "skunk" is an Algonquian Indian word, and spotted refers to the numerous white streaks and spots on an otherwise black coat. To avoid supplying a modifier to distinguish the present species from *Spilogale pygmaea,* long known as the Pygmy Spotted Skunk, I recommend that the name of the latter, a rare form occurring in western and southwestern Mexico, be changed to Pygmy Skunk. The Pygmy Skunk is currently known from less than a dozen museum specimens, one of which is in the Louisiana State University Museum of Zoology. The generic name *Spilogale* comes from two Greek words (*spilos,* spot, and *gale,* weasel) that together mean spotted weasel, thereby calling attention to the close relationship of skunks and weasels. The second part of the scientific name, *putorius,* is from the Latin word *putor,* meaning fetid odor. Its appropriateness is evident to anyone familiar with skunks.

Distribution. — *Spilogale putorius* in a broad sense (that is, including the nominal species *S. gracilis* and *S. angustifrons*) has a patchy distribution with records as far north as south-

western British Columbia, northern Idaho, South Dakota, northern Minnesota, Kentucky, West Virginia, and south-central Pennsylvania and as far south as Costa Rica. It has been found in the majority of the intervening states and countries but not in most of the Atlantic coastal plain nor in western Sonora, Sinaloa, Nayarit, Veracruz, and lowland Central America. Spotted Skunks in Louisiana are local in their distribution and are presently known only in the southern part of the state. (Map 60)

External appearance. — This small skunk, which is not much larger than a muskrat, is easily identified by the shape and pattern of the white markings on its otherwise glossy jet black pelage. The most conspicuous of these, in spite of the name of the species, are four pairs of long, more or less symmetrically arranged streaks. White marks short enough to approximate spots are usually confined to one on the top of the head, one on the chin, a pair on the back, a pair on the rump, and a

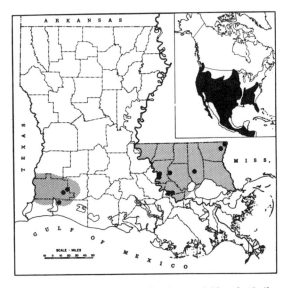

Map 60 Distribution of the Spotted Skunk (*Spilogale putorius*). *Shaded areas,* known or presumed range within the state; •, localities from which museum specimens have been examined; *inset,* overall range of the species.

pair on the basal portion of the sides of the tail, which is tipped with white. (Figure 200)

Color phases. — No color phases or color mutations are known in this species.

Measurements and weights. — All measurements are of specimens in the LSUMZ from various localities in southern Louisiana. Four males averaged as follows: total length, 478 (440–511); tail, 166 (148–186); hind foot, 44 (39–48); ear, 18 (17–22). Skulls of males measured as follows (number of specimens in parentheses following designation of measurement): basilar length (4), 49.8 (46.9–50.9); condylobasal length (4), 56.0 (52.6–57.3); occipitonasal length (3), 53.0 (52.5–53.7); zygomatic breadth (3), 34.4 (33.1–36.0); mastoid breadth (5), 29.9 (28.6–30.8); interorbital breadth (5), 15.4 (14.3–16.0); postorbital breadth (4), 15.2 (14.0–16.3); palatilar length (4), 19.0 (18.6–19.8); postpalatal length (4), 29.3 (28.0–30.6); cranial height (4), 18.6 (17.5–19.7); maxillary toothrow (5), 17.7 (17.4–18.3). Two females measured as follows: total length, 390, 480; tail, 128, 185; hind foot, 43, 47; ear, 21, 25; basilar length, 47.0, 48.0; condylobasal length, 52.6, 53.0; occipitonasal length, 49.4, 49.4; zygomatic breadth, 31.5, 34.0; mastoid breadth, 28.5, 29.2; interorbital breadth, 14.5, 14.9; postorbital breadth, 13.4, 14.7; palatilar length, 18.9, 19.1; postpalatal length, 28.0, 29.0; cranial height, 17.6, 18.2; maxillary toothrow, 16.6, 17.3. No weights are available on any of the Louisiana specimens, but in Missouri the species is said to weigh up to two and three-quarters pounds.

Cranial and other skeletal characters. — The skulls of Spotted and Striped Skunks differ from the skulls of other mustelids in Louisiana by the following combination of characters: the presence of 34 teeth; the more or less rectangular shape of the upper molars; and the termination of the hard palate opposite the posterior borders of the molariform toothrow. The skull of *Spilogale* differs from that of *Mephitis* as follows: size much smaller (basilar length 51 mm or less instead of 52 mm or more); frontal and parietal regions of equal height instead of arched above the orbits; postorbital processes prominent instead of barely indicated; mastoid region and auditory bullae much more inflated; and inferior border of mandible nearly straight instead of possessing a decided step beneath the coronoid. The penis bone is weak, only slightly curved, without pronounced terminal enlargements, and less than three-fourths of an inch in length. (Figures 185 and 201)

Sexual characters and age criteria. — The sexes are alike except for the primary and secondary sex organs. Females average smaller and possess five pairs of teats, two pectoral, two abdominal, and one inguinal. The young at birth show the color pattern of the adults.

Status and habits. — The Spotted Skunk, although rare or absent in the Atchafalaya Basin, is fairly numerous in southeastern and southwestern Louisiana. For some reason it is not known to occur anywhere in the northern half of the state. Its absence there is inexplicable since it occurs in numerous localities in adjacent counties of Mississippi (Cook, 1945) and of Arkansas (Sealander, 1956). Although I am aware that Spotted Skunks can be fairly common in an area and still escape detection, this fact would hardly account for the complete absence of records in our northern parishes. But even in southern Louisiana, where both species occur, Striped Skunks appear to be decidedly more numerous than the present species. One factor that contributes to this impression is the frequency with which Striped Skunks are killed along our highways. Contrariwise, I have never seen a dead Spotted Skunk on a road. H. Douglas Pratt, whose drawings and paintings illustrate this book, found one on the side of Interstate Highway 55, two miles west of Indepen-

PRATT
1972

dence, in Tangipahoa Parish, and a couple of others have been reported to me, including one by Ted O'Neil from near Holmwood, in Calcasieu Parish. Just why Striped Skunks suffer heavy highway mortality and Spotted Skunks are virtually immune is a mystery, but some behavioral difference is probably involved. Perhaps Striped Skunks panic and bolt from the shoulder of the road into a car's pathway or maybe they simply freeze when caught in the beam of the headlights. Spotted Skunks, on the other hand, may habitually back away from the blinding glare of headlights or conceivably they tend to avoid foraging along highways.

Records of fur sales in the state, which date back to 1913, do not always differentiate between Striped and Spotted Skunks. Indeed, in only six seasons in the period covered is a distinction made. In the 1928–1929 season the reported take was 7,279 Spotted Skunks and 31,661 Striped Skunks. In the years 1939–1943 and in 1966–1967, when again a breakdown was made between the two species, the annual take of Spotted Skunks paradoxically averaged only 205, the low being 23 in 1966–1967. In these years the annual catch of Striped Skunks averaged 11,945, the low being 260, again in 1966–1967. I am consequently at a loss to explain the high catch of 7,279 Spotted Skunks in the 1928–1929 season. In the period from the 1925–1926 season through the one of 1969–1970, the price paid for Striped and Spotted Skunks has varied considerably. The highest price ever paid for a skunk was $1.25 in the 1928–1929 season, but the average has been only 40 cents. The figure dropped as low as 8 cents in the 1932–1933 season. In the 44 years covered by the records, the average annual take was 13,234, with a high of 122,679 in 1931–1932 and a low of 6 in 1970–1971. If, however, we separate the first 22 seasons for which appropriate information is available, 1928–1951, the average annual take was 27,878 compared to an average of 378 over the last 22 years.

These figures suggest that a marked diminution in the skunk population has taken place in the last two decades. I suspect, though, that the decline in the catch may be nothing more than a consequence of trappers having turned their attention to more lucrative pursuits, such as trapping for Nutria. Another consideration, and one that makes the annual fur records unreliable as a basis for discerning actual short-term fluctuations in population density in a given species, is the fact that fur dealers sometimes hold back great quantities of furs, waiting for a demand on the part of the industry and a concurrent rise in the market value of the pelts. Such was apparently the case with skunks in the 1970–1971 season, when records of the Fur

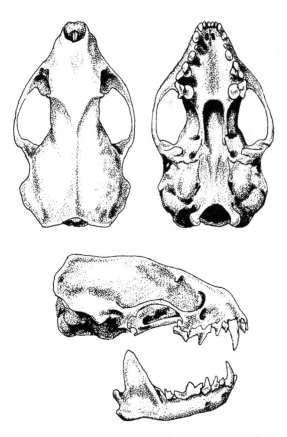

Figure 201 Skull of Spotted Skunk (*Spilogale p. putorius*), ♀, 2 mi. SE Burnside, Ascension Parish, LSUMZ 3408. Approx. natural size.

Figure 200 Spotted Skunk (*Spilogale putorius*). (Portrait by H. Douglas Pratt)

Division of the Wild Life and Fisheries Commission show the payment of the severance tax on only *six* skunks! Quite obviously fur dealers that year held back their skunk pelts awaiting a more favorable market.

The secretion of the anal glands of a Spotted Skunk is strongly scented and pungent and is said by some to be irritating to the eyes, although I have not personally found it so. The glands are located one on each side of the anus to which they are each connected by a duct.

Members of the American Society of Mammalogists who attended the society's annual convention when it was held at Louisiana State University in Baton Rouge in 1939 will recall the role played by a captive Spotted Skunk on the end of a leash as part of the banquet entertainment. This particular skunk was one of two obtained on the edge of the campus by a mammalogy student two weeks prior to the meeting. We brought the animals into the laboratory, etherized them, and tied off their scent glands to render them innocuous. The operation was really quite simple. It involved making two longitudinal incisions in the skin parallel to and about half an inch lateral to the anus, lifting the skin on each side of the slits to expose the two ducts leading from the scent glands to the anus, ligaturing the ducts with pieces of surgical catgut, and finally suturing the incisions in the skin and applying a disinfectant. Within a week both skunks were completely recovered but unable to use their scent glands, hence

ready for the part that one of them was to play in the banquet skit.

Another method of deodorizing a skunk is the surgical removal of the two glands in their entirety, but this is something of a major operation. And Vernon Bailey (1937) describes a technique, which he recommends for use with young animals only one or two months old. It involves going into the ducts through their anal orifices with a small hook and drawing out the lining to the point where it can be cut.

When a Spotted Skunk is confronted by an enemy it arches the tail over the back, conspicuously displaying the white tip. This aggressive action is often sufficient to warn and deter its molester, but if not, the little animal will sometimes do a handstand with its rump high in the air. According to Johnson (1921), the animal emits a fine spray of scented vapor from its vent when in this position, but Walker (1930) claims to have observed the handstand behavior "hundreds of times" on the part of Spotted Skunks in Oregon "without having once noted any attempt on the part of the animals to throw scent from that position." Walker goes on to say that a Spotted Skunk assumes only one stance "during the actual or threatened act of discharging the scent, a horseshoe shaped position with both head and rear facing the object of its attention." Nearly all writers on the subject agree that the handstand behavior is mostly an aggressive, bluff reaction.

Although the Striped Skunk is often seen

Tracks of Spotted Skunk

abroad in daytime, I have never observed a Spotted Skunk except at twilight, and then only rarely, for the species is mostly nocturnal. The little animal makes its home in a variety of situations, including hollow logs, old armadillo holes, and shallow depressions scratched under the roots of trees, as well as under old buildings, sheds, piles of lumber, and under similarly protected places.

Little information is available on the reproductive habits of the species in the state, but it is believed to breed in the winter and to give birth to its young in early spring. The length of the gestation period is not known for certain but is probably close to 60 days. The litters usually contain four to five young, but sometimes as few as two or as many as six. The babies are naked at birth, but the hair has already begun to form in the skin and therefore shows the black and white pattern that will soon be evident. By the 21st day the fur is dense and the youngsters are quite similar in markings to the mother, but the eyes do not open until they are 32 days old. They can emit musk when 46 days of age, and weaning occurs at about 54 days. By the time they are slightly more than three months old, they have attained full size. (Crabb, 1944 and 1948)

The food of Spotted Skunks includes a wide variety of plants and animals. Insects, such as grasshoppers, crickets, and ground beetles, make up a large part of their diet, but they also eat rats, mice, salamanders, frogs, and crayfish. Fruits and corn are favorite items, when available, and they occasionally capture small birds and eat bird eggs. They eat poultry regularly only as carrion, but chicken eggs are relished. When eating a chicken egg, Spotted Skunks bite into the end or the side and lap the contents. Sometimes, though, they experience difficulty with large eggs, yet ingeniously solve the problem by straddling the egg, holding it with the forelegs, then passing it beneath their abdomen and giving it a quick kick to the rear with one of the hind limbs, thus propelling it with

force against anything it might strike. The process is repeated until the egg is finally cracked (Van Gelder, 1953).

Because of its overall beneficial feeding habits, the Spotted Skunk must be regarded as a highly desirable species. The odoriferous product of its scent glands certainly gives man no excuse for killing it. I find the odor not entirely offensive. Only when skinning one of the animals and inadvertently cutting one of the ducts leading from a scent gland do I consider the smell completely repugnant. Should the scent spill onto the animal's pelt or contaminate the preparator's hands, it can be easily removed by strong soap and water to which household ammonia has been added.

Predators.—The main enemy of Spotted Skunks is the dog, but sometimes they are able to escape by climbing a fence post or a small sapling. Should the skunk succeed in discharging its musk into the dog's eyes, the substance causes great irritation and pain, allowing the skunk to make its getaway. Great Horned Owls occasionally capture one, as probably do Barred Owls also. The ornithological collections at Louisiana State University contain several Great Horned Owl specimens that still retain the odor of skunk, providing incontrovertible evidence of the owls' diet prior to their demise for the sake of science. Foxes, Coyotes, and Bobcats have been known to prey on skunks, and, unfortunately, man wantonly kills them for no good reason.

Parasites and diseases.—The following external and internal parasites have been reported to infest the Spotted Skunk: two biting lice (*Trichodectes osborni, Neotrichodectes mephitidis*); two fleas (*Echinophaga gallinacea, Ctenocephalides felis*), one tick (*Ixodes cookei*); one trematode (*Alaria taxideae*), two cestodes (*Mesocestoides latus, Oochoristica oklahomensis*); four nematodes (*Ascaris columbaris, Crenosoma mephitidis, Physaloptera maxillaris, Skrjabingylus chitwoodorum),* and two acanthocephalans *(Centrorhynchus wardae, Echinopardalis macrurae).* In the southwestern United States particularly the species is highly susceptible to rabies and is sometimes called the hydrophobia cat. But, fortunately, I know of no positive case of rabies in the Spotted Skunk in Louisiana. The comparatively few instances of rabid skunks from the state have all been from northern Louisiana, where only the Striped Skunk is known to occur. It is said to contract tularemia in the western states. (Stiles and Baker, 1935; Erickson, 1946; Holloway, 1956; *fide* U.S. Public Health Service, New Orleans.)

Subspecies

Spilogale putorius putorius (Linnaeus)

Viverra Putorius Linnaeus, Syst. nat., ed 10, 1, 1758:44; type locality, South Carolina (Howell, 1906).
Spilogale putorius, Coues, Bull. U.S. Geol. and Geog. Sur. Territories, ser. 2, 1, 1875:12.
Spilogale putorius putorius, Crain and Packard, Jour. Mamm., 47, 1966:324–325.
Spilogale indianola Merriam, N. Amer. Fauna, 4, 1890:10.
Spilogale indianola, Lowery, Occas. Papers Mus. Zool., Louisiana State Univ., 13, 1943:233.

I am following Van Gelder (1959) in his excellent revision of the genus *Spilogale,* wherein he synonymized *S. indianola* Merriam under *S. putorius* (Linnaeus). He referred the Louisiana material that was then available to him to the race *putorius,* but he considered these specimens, as well as those from southeastern Texas and southwestern Mississippi, intermediate between *S. p. putorius* and *S. p. interrupta* (Rafinesque). The latter ranges south in the central Great Plains, entirely west of the Mississippi River, to Arkansas and northeastern Texas, whereas the former occupies the area in which the species occurs east of the Mississippi River except for the Florida peninsula, inhabited by *S. p. ambarvalis.* Additional material from southern Louisiana bears out Van Gelder's original conclusion that our Spotted Skunks are close to *putorius.*

Specimens examined from Louisiana.—Total 19, as follows: *Ascension Par.:* 7 mi. SW Sorrento, 1 (LSUMZ); 2 mi. SE Burnside, 1 (LSUMZ);. *Calcasieu Par.:* 7.5 mi. SE Lake Charles, 1 (TU); Iowa, 4 (USNM). *Cameron Par.:* 16 mi. S, 1.5 mi. W Lake Charles, 1 (MSU). *East Baton Rouge Par.:* Baton Rouge, 2 (LSUMZ); University, 3 (LSUMZ). *Livingston Par.:* 16 mi. NE University, 1 (LSUMZ). *Tangipahoa Par.:* 1 mi. S Natalbany, 1 (SLU); 1 mi. W Hammond, 1 (SLU). *Washington Par.:* 2 mi. N Angie, 1 (LSUMZ); 3 mi. S Varnado, 1 (SLU); 0.75 mi. S Varnado, 1 (SLU).

Genus *Mephitis* G. Cuvier

Dental formula:

$$I\frac{3-3}{3-3} \ C\frac{1-1}{1-1} \ P\frac{3-3}{3-3} \ M\frac{1-1}{2-2} = 34$$

STRIPED SKUNK   Plate XII
(Mephitis mephitis)

Vernacular and other names.—The origin of the name skunk has already been explained under the previous account. The other name most frequently heard applied to this species in Louisiana is polecat, which comes from the Middle English *polcat* (probably from the Middle French *poul,* cock, plus the English word *cat*) and possibly alludes to the habit of sometimes molesting poultry that is attributed to the European polecat, with which our skunks were allied by early settlers. The official designation Striped Skunk calls attention, of course, to the two prominent longitudinal stripes down the back of most individuals. Other names include striped pussycat, wood-pussy, smell-cat, and two-striped skunk. The early French settlers of Louisiana called it *bete puante,* or "stinking beast" (Le Page du Pratz, 1758). The species name *Mephitis mephitis,* a case of absolute tautonymy, is a repetition of the Latin word meaning foul odor, referring to the effects produced by the secretion of the animal's anal glands. The name *mesomelas* for one of the subspecies that occur in Louisiana is a combination of two Greek words (*mesos,* median, and *melas,* black), possibly referring to the black area between the two longitudinal white stripes. The Latin adjective *elongata,* applied to the other subspecies, means elongated and refers to the allegedly longer tail in this population.

Distribution.—The species occurs widely throughout southern Canada and the United States south to northern Mexico but is locally absent in some places, such as the desert areas of the southwestern United States. In Louisiana it is widespread over most of the state but is paradoxically absent in the extreme southeastern section. (Map 61)

External appearance.—This striking carnivore is about the size of a Domestic Cat. The somewhat triangular-shaped head tapers to the bulbous nose pad. The small eyes are black and beady, the ears short and rounded. The legs are short, somewhat longer in the rear than in front, and the tail is fairly long and bushy. The head, body, and tail are largely glossy black except for a narrow white stripe on the forehead and a white stripe beginning on the pate and extending to the shoulders where it often divides into two longitudinal stripes of variable length and width. The amount of white in the tail differs from specimen to specimen, the tip being sometimes entirely white. Occasionally there is a small patch of white on the chest. No two Striped Skunks appear to be exactly alike in the extent of white in the pelage. Occasional individuals have the white reduced to a small spot on the head. These are referred to in the fur trade as "stars" and bring a premium price. Other individuals from the same population have the upperparts, including the tail, predominantly white. The five toes on the forefeet are provided with long, curved claws; the five toes on the hind feet have somewhat shorter and straighter claws. The animal is plantigrade. (Figure 202.) The striking feature, as with all skunks, is the presence of two well-developed anal glands that produce a yellowish, oily, phosphorescent substance that is highly malodorous. Chemically it is known as butylmercaptan ($C_4 H_9 SH$).

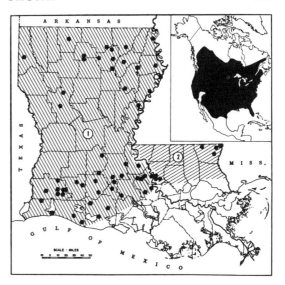

Map 61 Distribution of the Striped Skunk (*Mephitis mephitis*). *Shaded areas,* known or presumed range within the state; •, localities from which museum specimens have been examined; *inset,* overall range of the species. 1. *M. m. mesomelas;* 2. *M. m. elongata.*

Color phases.—Although the Striped Skunk is highly variable, particularly with regard to the length and breadth of the two longitudinal white stripes, which may sometimes be totally lacking, no true color phases are known. As one might expect, albinos crop up occasionally.

Measurements and weights.—Six males and six females in the LSUMZ from various localities in the state averaged as follows: total length, 592 (566–695); tail, 264 (203–430); hind foot,

Figure 202 Striped Skunk (*Mephitis mephitis*), Balmorhea, Reeves County, Texas. (Photograph by Richard K. LaVal)

61 (57–65); ear, 25 (18–30). Seventeen skulls from the same source averaged as follows: basilar length, 56.4 (52.5–60.6); condylobasal length, 60.8 (58.9–69.7); occipitonasal length, 61.3 (57.9–65.6); zygomatic breadth, 40.2 (31.3–45.7); mastoid breadth, 34.2 (32.0–38.3); interorbital breadth, 19.1 (17.5–21.0); postorbital breadth, 18.3 (17.0–21.1); palatilar length, 24.2 (22.7–26.9); postpalatal length, 32.3 (29.2–36.3); cranial height, 23.0 (21.7–23.9); length of maxillary toothrow, 21.3 (19.7–22.7). In nine of the above measurements the males averaged slightly larger than the females, but in four (total length, tail, condylobasal length, and length of toothrow) the females averaged larger; in two measurements the averages were identical. Unfortunately, the sample is too small for the differences to be considered significant. Weights of 242 Striped Skunks captured in the Atchafalaya Basin in late 1959 and early 1960 were reported by Adams et al. (1964). The 137 males ranged from 511 to 2,258 grams, the average being 1,305 grams (2.87 pounds). Eighteen (13.1 percent) of the males weighed less than 1,000 grams, 89 (65.0 percent) from 1,001 to 1,500 grams, 26 (19.0 percent) from 1,501 to 2,000 grams, and 4 (2.9 percent) over 2,001 grams. The 105 females ranged in weight from 512 to 1,489 grams, the average being 1,033 grams (2.27 pounds). Forty-six (43.8 percent) of the females weighed less than 1,000 grams, 59 (56.2 percent) weighed over 1,000 grams.

Cranial and other skeletal characters. — The distinctions between the skull of *Mephitis mephitis* and that of *Spilogale putorius* are discussed under the latter species and need not be repeated here (see page 433 and Figure 203). The penis bone of the Striped Skunk is slightly curved and ranges from 21 to 23

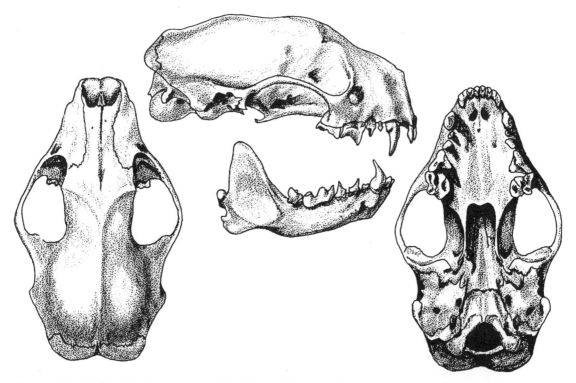

Figure 203 Skull of Striped Skunk *(Mephitis mephitis mesomelas),* ♂ 10 mi. S Kinder, Jefferson Davis Parish, LSUMZ 3406. Approx. natural size.

millimeters in length. The basal and terminal ends are slightly enlarged (see Figure 185). The skull of one specimen in the LSUMZ from West Baton Rouge Parish possesses only two pairs of upper premolars, instead of the normal three pairs, thereby reducing the total number of teeth to 32 as in the Hog-nosed Skunk *(Conepatus mesoleucus)*, which is known to occur in eastern Texas to within 35 miles of the Louisiana line (Davis, 1966) but which has not yet been found within our borders. The abnormal skull, along with its accompanying skin, is otherwise typical of *Mephitis mephitis*.

Sexual characters and age criteria.—The sex ratio of 702 Striped Skunks captured alive in the Atchafalaya Basin from February 1959 through December 1962 was 93 females per 100 males (Adams et al., 1964). The sexes are alike except for their primary and secondary sex organs. The usual number of mammae in females is 12, but Verts (1967) found numerous instances of supernumerary teats in Striped Skunks in Illinois. Of 23 females examined, 12 (52.1 percent) possessed other than the "normal" six pairs (6+6); five had 11 (5+6); one had 12 (5+7); four had 13 (6+7); one had 14 (7+7); and one had 15 (7+8). Verts also discussed various methods that have been employed to age Striped Skunks, most of which he found not completely satisfactory. He did find that he could determine ages within narrow limits by using a combination of several criteria. The most important of these were determination of the whelping season in the areas from which skunks were collected, the degree of closure of the epiphyses, and the dry weights of the lenses of the eyes. In Illinois most females give birth to their litters in May. A sample of skunks collected in any month contains animals in age groups approximately one year apart. For example, a skunk collected in October would be either 5 months old, 17 months old, 29 months old, or older in steps of 12 months. Verts found that ossification of the epiphyses was completed at about 8 or 9 months of age. Finally, he cited evidence from his own studies showing that the dry weights of eye lenses of Striped Skunks appeared to increase with age. He also concluded that his calculated mortality rates of Striped Skunks were for several reasons unreliable, but he expressed the opinion that probably less than 35 percent of the Striped Skunks born attain one year of age.

Status and habits.—The Striped Skunk is virtually statewide in its occurrence except that it is unrecorded in extreme southeastern Louisiana in Assumption, Terrebonne, Lafourche, St. James, St. Charles, St. John the Baptist, Jefferson, Orleans, Plaquemines, and St. Bernard parishes. Wherever the species is found its presence is made abundantly evident by its telltale scent, and it is also frequently seen dead on the shoulders of our highways as a result of having been struck by passing automobiles. It is not a species that can easily go undetected. Consequently, I was incredulous when I first became aware of its scarcity, if not complete absence, in southeastern Louisiana. Randolph Bazet, of Houma, who is extremely knowledgeable concerning the animal life of his section of the state, informs me that he has never seen a skunk in over 50 years' residence in Terrebonne Parish. Similarly, the county agents of Lafourche, Jefferson, Plaquemines, and St. Bernard parishes report the paucity or non-existence of skunks in their areas. In parts of the state where it is present the species is generally quite common and is certainly one of the state's better known small carnivores. In the upper and middle Atchafalaya Basin the species is extremely numerous. George E. Sanford (pers. comm.) reports capturing over 30 alive in one night along the Atchafalaya levee south of Krotz Springs while engaged in the Louisiana State University leptospirosis investigations of the 1960s. Altogether some 1,500 specimens were taken alive, mainly in the Atchafalaya Basin, in the course of this study.

As already discussed on page 435 under the account of the previous species, the state's trapping records do not always separate the annual take of Striped Skunks from that of Spotted Skunks, but we know that the former make up the bulk of the skunks marketed in any given year. In the years 1939–1943 and in 1966–1967, when a breakdown was made between the two species, the annual catch of Striped Skunks averaged 11,945, while that of the small, spotted species averaged only 205. In 1928–1929, though, 31,661 Striped Skunks were handled by fur dealers, along with a record number of 7,279 Spotted Skunks. Another peak year was 1931–1932, when a combined total of 122,679 Louisiana skunks were placed on the market by fur dealers. The drastically lower take in recent years probably reflects the reduced demand for skunk pelts rather than a diminution in the skunk population. But the matter deserves careful study by wildlife experts, for although skunks are of secondary importance as fur animals, they are nevertheless highly beneficial to the farmer as consumers of rats, mice, and noxious insects. Moreover, except for occasionally befouling the air with the product of their scent glands, the little animals are delightful members of our fauna. They make charming pets when they have been operated upon (see previous account for techniques by which skunks can be rendered inoffensive). The wanton killing of skunks simply because they possess scent glands that they use in their defense is reprehensible beyond all words.

Striped Skunks are not quick to use their scent; they usually do so only after much provocation. When faced by an antagonist, they arch their backs, elevate the tail, erect the hairs thereon, and sometimes stamp their forefeet on the ground. The sound created by the foot stamping can often be heard several feet away when performed on hard ground. Another warning behavior is that of standing and walking on the forefeet. Although decidedly more characteristic of Spotted Skunks, the action is occasionally observed in the present species (Seton, 1929). Verts (1967), in his exhaustive study of the Striped Skunks in Illinois, observed handstands only on the part of young animals raised in captivity.

The odor of a skunk's spray has been described as a "mixture of strong ammonia, essence of garlic, burning sulphur, a volume of sewer gas, a vitriol spray, a dash of perfume, all mixed together and intensified a thousand times." These words are from the pen of no lesser an authority than Ernest Thompson Seton, but I regard his diagnosis something of an exaggeration, according to my own sense of olfactory perception. I must admit that a full blast of the scent is potent, but I do not object to it when it is dilute. A whiff of a skunk as one drives along a monot-

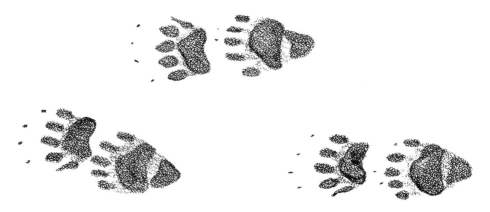

Tracks of Striped Skunk

onous, modern highway, particularly at night, has a certain fragrance that I find "interesting" to say the least. Perhaps I consider it so because of the thoughts that it conjures up in my mind of the docile little animal from which it comes, peacefully going about its business with almost utter disdain for even the giants of the world in which it lives. One squirt from its scent gun is sufficient to send the most ferocious dog yelping in agony from burning eyes and nostrils and retching with nausea. Consequently, most of its would-be enemies have learned to avoid it. Some dogs, though, never seem to profit by experience, for they come home time and again with their tails between their legs, reeking with the smell of skunk.

The scent is expelled from two tiny nipples, located just inside the anus, marking the outlets of the two ducts leading from the scent-producing muscle-encapsulated glands, lying adjacent to the anus. No urine at all is involved. The musk is discharged either as an atomized spray of almost invisible droplets or as a short stream of rain-sized drops propelled toward the adversary while the animal is turning slightly so that the jet of musk covers an arc of 30 to 45 degrees. By this means skunks greatly increase the probability of hitting what they are aiming at. Just before scenting occurs, the body is turned in a *U* position with the head and tail facing the intruder. The musk usually travels 8 to 10 feet, rarely up to 14. The smell, on the other hand, can be detected for much greater distances, as much as a mile and a half downwind. A common misconception is that skunks cannot emit musk when held suspended by the tail. My advice is don't try it! If, however, the skunk is grasped with one hand behind the head and the other at the base of the tail, and the tail is held depressed to the ground, the little animal is unable to discharge its scent, at least in any appreciable quantity (Figure 204).

Striped Skunks appear to be unusually silent animals, but they nevertheless make low churrings, growls, shrill screeches, birdlike twitters, and dovelike cooings (Seton, 1929).

The home of the Striped Skunk is located in a variety of situations. Customarily it is in an old, discarded den of a fox, armadillo, or some other mammal, or it may be in a refuse dump, in a stump, or beneath an old building. Little is known about the breeding season of Striped Skunks in Louisiana, but elsewhere the females come into heat and mate in late winter. The period of gestation is variable, sometimes extending to 75 days, but ordinarily it is 62 to 66 days in duration. A litter usually consists of 6 or 7 kits but may comprise as few as 2 or as many as 10. The youngsters are born with the eyes and ears closed. The skin is pinkish, only thinly covered with short, fine hair that nevertheless clearly shows the animal's black and white color pattern. The eyes and ears open before the end of the first month, and shortly there-

Figure 204 Striped Skunk *(Mephitis mephitis)*, cannot easily "spray" when the tail is depressed. (Photograph by Joe L. Herring)

after the little skunks are able to discharge a small amount of the scent fluid. Weaning is complete at about 8 to 10 weeks of age, but usually after the female has already taken the babies hunting for the first time. The offspring stay with the mother until fall, when family ties begin to dissolve.

The Striped Skunk is highly beneficial in its feeding habits, consuming as it does great quantities of insects, along with many rats and mice. It also eats frogs, salamanders, and crayfish, and, in fall and winter, small amounts of plant material are ingested. It has a fondness for bees and may occasionally raid a farmer's hives. But the Striped Skunk is no climber; so if the hives are elevated three or four feet above ground they are probably safe. Occasionally an individual skunk becomes perverted and kills poultry, but not when the birds are properly housed at night. Eggs, though, are relished, and any that a skunk finds on the ground are greedily eaten. But such transgressions are rather infrequent and are more than compensated for by the skunk's importance otherwise as an economic asset.

Predators and other decimating factors. — Even though Great Horned Owls, Coyotes, Bobcats, foxes, and dogs are among the animals known to prey on the Striped Skunk, one encounter is generally all that is required to deter further molestation on the part of most of these potential enemies. Insecticides, including chlorinated hydrocarbons, are suspected of being highly toxic to skunks and to have been responsible for the diminution in their numbers in some areas. But this suspicion requires substantiation by field studies on the part of competent biologists. Striped Skunks suffer heavy mortality by being struck by cars along our highways.

Parasites and diseases. — The external parasites of the Striped Skunk include 10 fleas *(Orchopeas howardii, Oropsylla arctomys, Chaetopsylla lotoris, Opisocrostis bruneri, Megabothris asio, Ctenophthalmus pseudagyrtes, Nosopsyllus fasciatus,*

Ctenocephalides felis, Cediopsylla simplex, Pulex irritans), one louse *(Trichodectes mephitidis),* and 5 ticks and mites *(Ixodes cookei, I. hexagonus, I. kingi, Amblyomma americanum, Dermacentor variabilis).* Internal parasites known from the Striped Skunk are summarized by Babero (1960) and Verts (1967). They include 3 protozoans, 9 trematodes, 10 cestodes, 25 nematodes, and 6 acanthocephalans. Leptospirosis is highly prevalent in the Striped Skunk population of Louisiana. Roth et al. (1963) report isolating *Leptospira* in 57.4 percent of 650 specimens of skunks examined from the state. For this reason, despite the attractiveness of the little animals as pets, bacteriologists at Louisiana State University recommend against keeping them in captivity even though many other kinds of mammals, including livestock, also act as reservoirs for the several serotypes of *Leptospira* involved in the epizootiology of leptospirosis. In many parts of the range of the Striped Skunk rabies is highly prevalent. In Louisiana, fortunately, no rabid skunks were found until 1960. Since that time 33 additional cases have been confirmed, all from parishes in the northern part of the state (*fide* U.S. Public Health Service, New Orleans).

Subspecies

Mephitis mephitis mesomelas Lichtenstein

Mephitis mesomelas Lichtenstein, Darstellung neuer oder wenig bekannter Säugethiere . . . , 1832:pl. 45, fig. 2; type locality, Louisiana.

Mephitis mephitis mesomelas, Hall, Carnegie Inst. Washington, Publ. 473, 1936:66.

Mephitis mephitica scrutator Bangs, Proc. Biol. Soc. Washington, 10, 1896:141; type locality, Cartville, Acadia Parish, Louisiana.

This small race of the Striped Skunk occupies the area from southeastern Kansas and southwestern Missouri south through eastern Oklahoma, eastern Texas, western Arkansas, and Louisiana west of the Mississippi River.

Specimens examined from Louisiana. — Total 89, as follows: *Acadia Par.:* 5 mi. S Crowley, 1 (USL); Cartville, 3 (1 USNM; 2 MCZ); Pointe aux Loups Springs, 4 (MCZ). *Caddo Par.:* 10 mi. W Shreveport, 1 (LTU). *Calcasieu Par.:*

Lake Charles, 1 (LSUMZ); Lake Charles near McNeese State College, 1 (LSUMZ); 2.5 mi. N Gillis, 1 (LSUMZ); 7 mi. & Lake Charles, 1 (LSUMZ); 5 mi. S Lake Charles, 1 (MSU); 7.5 mi. S McNeese College on Hwy. 14, 1 (MSU); 7.8 mi. S Lake Charles, 5 (MSU); 3 mi. SW Sulphur, 1 (MSU); Holmwood, 1 (MSU); 15 mi. E Lake Charles on I-10, 1 (USL); locality unspecified, 12 (USNM). *Cameron Par.:* Cameron, 1 (LSUMZ); 3.5 mi. S Intracoastal Canal at Gum Cove Road, 1 (MSU); 2 mi. E Grand Chenier on Hwy. 82, 2 (MSU); Rockefeller Wildlife Refuge, 2 (MSU); Leedyburg, 1 (USNM). *Claiborne Par.:* locality unspecified, 1 (LTU). *Concordia Par.:* 4 mi. NE Ferriday, 1 (LTU). *East Carroll Par.:* Lake Providence, 1 (LTU). *Evangeline Par.:* Chicot State Park, 1 (USL). *Franklin Par.:* Extension, 1 (LSUMZ); 9 mi. S Delhi, 1 (NLU). *Iberville Par.:* 3 mi. S Plaquemine, 1 (LSUMZ); 1.5 mi. N Bayou Sorrel, 1 (LSUMZ); 1.5 mi. E Ramah, 1 (LSUMZ); 0.5 mi. S Grosse Tete, 1 (LSUMZ). *Jefferson Davis Par.:* 10 mi. S Kinder, 1 (LSUMZ); Intersection state roads 26 and 102, 1 (MSU). *Lafayette Par.:* 13 mi. N Lafayette, 1 (USL); 3 mi. W, 0.5 mi. S Lafayette, 1 (USL). *Lincoln Par.:* 6 mi. S Ruston on Hwy. 167, 1 (LTU); Ruston, 1 (LTU). *Madison Par.:* Tallulah, 1 (LSUMZ); 16 mi. S Delta Point, 1 (LSUMZ). *Morehouse Par.:* 3 mi. N Oak Ridge, 1 (LSUMZ). *Natchitoches Par.:* Provencal, 2 (LSUMZ); Campti, 1 (LSUMZ). *Ouachita Par.:* Monroe, 3 (1 LSUMZ; 2 NLU); 15 mi. S Monroe, 1 (NLU); Sterlington, 2 (NLU); 4 mi. N Monroe, Hwy. 165, 1 (NLU); Calhoun, 1 (NLU). *Pointe Coupee Par.:* 18 mi. NW Baton Rouge, 1 (LSUMZ). *Richland Par.:* Rayville, 1 (NLU). *St. Landry Par.:* Melville, 1 (LSUMZ); 1 mi. N Sunset, 1 (USL); 5 mi. S Hwy. 90 on Contableau-Henderson Levee, 1 (USL); 1 mi. E Eunice, 1 (USL). *St. Martin Par.:* Cade, 1 (USL). *Tensas Par.:* 5 mi. S St. Joseph, 1 (LTU). *Union Par.:* 1 mi. S, 1 mi. W Sterlington, 1 (NLU). *Vermilion Par.:* Perry, 1 (LSUMZ); 5 mi. S Gueydan, 1 (USL); 20 mi. SW Abbeville, 1 (USNM); 21 mi. SW Abbeville, 2 (USNM); 22 mi. SW Abbeville, 1 (USNM). *West Baton Rouge Par.:* 4 mi. E Rosedale, 1 (LSUMZ); ca. 1 mi. S Port Allen, 1 (LSUMZ).

Mephitis mephitis elongata Bangs

Mephitis mephitica elongata Bangs, Proc. Boston Soc. Nat. Hist., 26, 1895:531; type locality, Micco, Brevard County, Florida.
Mephitis mephitis elongata, Howell, N. Amer. Fauna, 45, 1921:39.

This subspecies of the Striped Skunk, which is characterized especially by being long-tailed, occurs in the extreme southeastern United States from Virginia south through North Carolina, extreme eastern Tennessee, South Carolina, southeastern Georgia, and Florida west along the Gulf Coast through southern Alabama to the Florida Parishes of Louisiana. Although the sample is much too small, the tails of three females from the Florida Parishes do average

considerably longer than nine males and females from west of the Mississippi River: 330 (260–430) as opposed to 241 (203–291) mm.

Specimens examined from Louisiana. — Total 9, as follows: *East Baton Rouge Par.:* Baton Rouge, 2 (LSUMZ); 6 mi. SE University, 1 (LSUMZ); 3 mi. S LSU on Highland Rd., 1 (LSUMZ); 4 mi. S University, 1 (LSUMZ). *Washington Par.:* 12 mi. E Franklinton, 1 (LSUMZ); 2.5 mi. N Angie, 1 (LSUMZ); 3 mi. NW Angie, 1 (SLU); 1 mi. W Warnerton, 1 (TU).

Genus *Lutra* Brünnich[1]

Dental formula:

$$\text{I}\frac{3-3}{3-3}\ \text{C}\frac{1-1}{1-1}\ \text{P}\frac{4-4}{3-3}\ \text{M}\frac{1-1}{2-2}=36$$

NEARCTIC RIVER OTTER   Plate VIII
(Lutra canadensis)

Vernacular and other names. — The word otter is of Old World origin. In Old English the animal was called *otor,* in Middle English, *oter.* The German *otter,* the Danish *odder,* and the Swedish *utter* are cognates. The French name for it, *loutre,* was applied by Iberville, Bienville, and other early French explorers and colonists to some of the bayous and other waterways they found in the southern part of the state. Bayou la Loutre and Pass a Loutre were noted for the number of otters seen in them, and for this reason were so named. The substantive modifier River serves to differentiate these freshwater otters from the Sea Otter *(Enhydra lutris).* The word Nearctic is required to distinguish *Lutra canadensis* from other river otters, such as the Southern River Otter *(L. annectens)* of Middle and South America, the Prince of Wales River Otter *(L. mira)* of southern Alaska, and the river otters of the Old World. The generic name *Lutra* is the Latin word for otter, and the specific name *canadensis* is the word Canada with the Latin suffix *ensis,* meaning belonging to, and

[1.] In a recent revision of the genus *Lutra,* van Zyll de Jong (1972) concluded that New World otters are generically distinct from *Lutra* of the Old World and for the former he resurrected the name *Lontra* Gray.

is a reference to the country from which the species was first described.

Distribution. — The Nearctic River Otter is distributed throughout most of Canada and the continental United States. It is absent only from portions of extreme northern Alaska and Canada and from arid regions of the southwestern United States. In Louisiana the species was probably once found statewide in favorable situation, but now it is decidedly more local in its occurrence because of drastic alterations in much of its original habitat. (Map 62)

External appearance. — This long-bodied, short-legged, semiaquatic mustelid cannot be confused with any other mammal in the state. Included among its distinctive features are its sleek brown pelage with dense, oily underfur overlaid by glossy guard hairs; its cylindrical body with a thick neck, flattened head, and long, heavy tail, which is flat on the bottom, thick at the base, and tapered toward the tip;

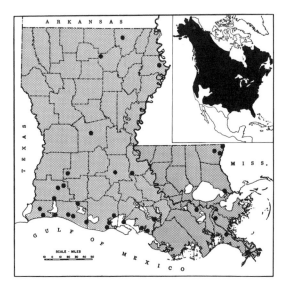

Map 62 Distribution of the Nearctic River Otter (*Lutra canadensis*). *Shaded area,* known or presumed range within the state; •, localities from which museum specimens have been examined; *inset,* overall range of the species.

and its five fully webbed toes on each foot. The nose pad is prominent, the whiskers long, the eyes and ears small, and, except for the pads on the toes and soles, the undersurfaces of the feet are well furred. The underparts are light brown or grayish, hence decidedly paler than the upperparts. The muzzle and throat are silvery gray.

Color phases. — Some individuals are decidedly darker than others but probably do not represent color phases in the true sense. Albinos are rare.

Measurements and weights. — Three males in the LSUMZ from St. Charles, Plaquemines, and Cameron parishes averaged as follows: total length, 1,129 (1,118–1,150); tail, 444 (420–470); hind foot, 129 (115–140); ear, 24 (23–25). Four skulls of males from the same source averaged as follows: greatest length, 111.8 (108.9–115.1); mastoid breadth, 68.6 (65.3–74.2); zygomatic breadth, 72.3 (68.9–78.1); postorbital breadth, 21.7 (20.7–22.4); interorbital breadth, 25.1 (22.9–26.3); palatilar length, 52.0 (48.8–53.3); postpalatal length, 48.9 (46.7–51.4); length of maxillary toothrow, 39.0 (37.9–40.0); cranial height, 39.6 (37.4–41.5). Three females in the LSUMZ from Washington, Plaquemines, and Cameron parishes averaged as follows: total length, 978 (900–1,113); tail, 358 (317–400); hind foot, 112 (101–126); ear, 23 (22–23); greatest length of skull, 108.6 (106.9–110.4); mastoid breadth, 65.2 (63.8–66.7); zygomatic breadth, 67.6 (63.7–72.8); postorbital breadth, 19.6 (17.5–22.5); interorbital breadth, 23.0 (22.5–23.5); palatilar length, 48.3 (45.8–50.3); postpalatal length, 47.6 (46.3–49.0); maxillary toothrow, 36.4 (33.5–38.7); cranial height, 38.0 (37.7–38.2).

Cranial and other skeletal characters. — The skull of the Nearctic River Otter is 4.2 to 4.5 inches in length and is broad and flat. The rostrum is broader than it is long, the auditory bullae are small and depressed, and the nasals are noticeably constricted behind the postorbital processes. The middle incisor on each side of the lower jaw is larger than the other two and is inserted in its alveolus slightly behind them. This species is the only carnivore in Louisiana with a total of 36 teeth and the only one with 5 molariform teeth on each side of both the upper and lower jaws. The last upper premolar and the upper molar are large and more or less rectangular. The penis bone has the shape of a thinly compressed S but occasionally is J-shaped. The proximal end is rough, and the distal end is laterally flattened with a deep ventral urethral groove. The central part of the shaft is smooth and triangular in cross section with the apex on the dorsal side. Three penis bones from adult Louisiana specimens measured 87, 88, and 91 mm in length. (Figures 185 and 205)

Sexual characters and age criteria.—The sexes are alike except for the primary and secondary sex organs and the slightly larger size of the males. Friley (1949a) distinguished four age classes on the basis of penis bone morphology: older adults, younger adults, immatures, and very young. He found that in his "older adults" the length of the penis bone in millimeters averaged 96.1 (87.9–106.4) with a standard deviation of ±4.31. In the "immatures" the length averaged 79.7 (68.5–90.0) with a standard deviation of 0.82. The penis bones of his "younger adults" were intermediate in size. In bacular length, three specimens from Louisiana that I regard as adults are definitely smaller than Friley's "older adults," but I am not surprised because otters

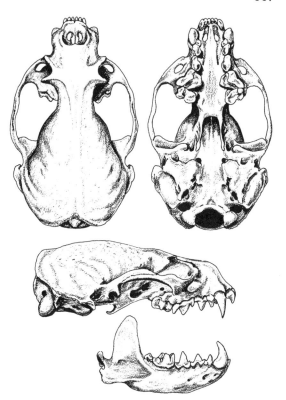

Figure 205 Skull of Nearctic River Otter *(Lutra canadensis lataxina)*, ♂, Belair, Plaquemines Parish, LSUMZ 1582. Approx. ½×.

from Louisiana are definitely smaller in overall body size than those of Michigan, where Friley's material was taken. One would therefore expect the skeletal elements of Louisiana otters to average smaller.

Status and habits.—Otters are still fairly numerous in certain parts of the state, notably in some of our coastal marshes and in the Atchafalaya Basin, but elsewhere, because so much of their original habitat along streams and rivers has been destroyed, their numbers are probably only a fraction of what they were in days gone by. Even so the species ranks as one of our major furbearers, not because of the number of pelts obtained but because of the high price that a pelt brings on the fur market. In the 31 trapping seasons

Tracks of Nearctic River Otter

since 1940–1941, the number of otter pelts reported by fur dealers has averaged 4,023, ranging from a low of 1,728 in the first year of this period to a high of 8,484 in the 1962–1963 season. The amount paid the trapper for an otter pelt has varied considerably from year to year, but since the 1940–1941 season the price has averaged $14.74, the low being $5.00 in the first year of this period, the all-time high being $42.00 in the 1972–1973 season. This record price exceeded by $17.00 the previous high of $25.00 of the 1970–1971 season. Consequently, we can now anticipate an upsurge in the number of otters taken as more and more trappers turn their attention to this lucrative market. One otter would be worth as much as 10 muskrats or Nutrias to the trapper fortunate enough to obtain it.

Otters are never found far from water, and it is there that they spend most of their time. They possess many adaptations for an aquatic existence (Figures 206 and 207). The toes are fully webbed, the pelage is water-resistant, and the nostrils and ears can be closed when the animal is submerged. The position of the eyes, near the top of the head, permits the animal to see above the surface while it swims with most of the body beneath the surface of the water. River otters feed almost entirely on frogs, turtles, snakes, fish, and aquatic invertebrates, notably crayfish and crabs. Birds and small land mammals, such as rats and mice, are occasionally eaten. Studies have shown that the fish they consume are largely nongame species.

Otters are among the most playful of all mammals and take particular delight in sliding down mud banks into the water. Slides in the bank of a stream are sometimes used so extensively over such prolonged periods of time that they become deep chutelike troughs. Generally the "game" is played by a group of otters that toboggan headfirst down the slide one after the other and then race to get around to the top of the course to repeat the process again.

Otters make delightful pets when raised from a young age. The main problem is that they often cache some of the fish they are fed, which in time begin to smell. They are quite kittenlike in their behavior, playfully rolling and tumbling for what seems to be the sheer fun of it. The body is highly muscular, especially in the shoulder and neck region, and the senses are extremely acute, particularly those of scent and touch. When swimming they move ahead rapidly almost without a ripple, and they dive with little splashing, in sharp contrast to a muskrat or beaver.

Little is known about the reproductive biology of otters in Louisiana, but they are said to breed in late fall. The species is generally amiable and sociable for most of the year, but when the mating season arrives the males habitually engage in fierce combats. Delayed implantation results in the gestation period being extended to as much as 270 days. One female is known to have finally given birth to her young after 380 days. The blastocysts remain loose and unattached in the uterus before finally implanting and resuming embryonic development. Litters normally consist of one to three kits, but a litter of six is on record. The den is almost invariably dug in the bank of a canal or stream with the entrance beneath the surface of the water but with the grass-lined nest chamber above high-water level. Otters in the coastal marshes, so I am told, sometimes use old muskrat houses and, now that the Nutria has become plentiful in our coastal parishes, frequently utilize old Nutria burrows. The youngsters remain

Figure 206 Nearctic River Otter (*Lutra canadensis*). (Photograph by Lloyd Poissenot)

with the mother until they are nearly full grown.

The species is nowhere ever as numerous as the mink, possibly because of its lower breeding potential and the fact that it requires a larger home range. Food is certainly not a limiting factor, nor does trapping, at least in the southern part of the state, appear to be adversely affecting its numbers. I believe, on the other hand, that a strong case could be made for stopping all trapping for otter in northern Louisiana. However, before any such action is taken an intensive study of its present status on a statewide basis should be conducted.

Predators and other decimating factors. — Otters have few enemies other than man. Even the kits, because of the manner in which they are carefully guarded by the mother, are probably only rarely taken by Great Horned Owls and other birds of prey or by larger carnivores. In the coastal marshes I have no doubt that alligators capture a few of the youngsters, but I know of no evidence to this effect. Otters are occasionally drowned in fishermen's hoop nets. St. Amant (1959) states that trappers estimate that they take about 40 percent of the state's otter population each season. I doubt that this is true, but if it is, I would think that it applies only to the situation in our coastal marshes.

Parasites. — The only parasites that I have found listed as occurring in or on this species are three trematodes (*Euparyphium melis, Baskirovitrema incrassum, Enhydridiplostomum alaroides*), one cestode (*Ligula intestinalis*), three nematodes (*Dracunculus insignis, Strongyloides lutrae, Physaloptera* sp.), and one tick (*Ixodes hexagonus*). (Erickson, 1946; Greer, 1955; Layne et al., 1960; Sawyer, 1961; Little, 1966.)

Subspecies

Lutra canadensis lataxina (G. Cuvier)

Lutra lataxina G. Cuvier, Dictionnaire des sciences natu-

relles..., 27, 1823:242; type locality, South Carolina.
Lontra canadensis lataxina, van Zyll de Jong, Life Sci. Contri., R. Ontario Mus., 80, 1972:80, 82, 85.
Lutra canadensis texensis Goldman, Proc. Biol. Soc. Washington, 48, 1935:184.—Lowery, Occas. Papers Mus. Zool., Louisiana State Univ., 13, 1943:233.

In his recent taxonomic and systematic revision of the Nearctic and Neotropical river otters, van Zyll de Jong (1972) reduced some 19 subspecies of *canadensis* to a total of only seven that he considered worthy of recognition. He assigned all populations of eastern Canada and adjacent parts of the United States to the nominate race *L. c. canadensis,* and he included under the race *L. c. lataxina* the remainder of the populations of the United States east of the Rocky Mountains.

Specimens examined from Louisiana. — Total 49, as follows: *Allen Par.:* Clear Creek, 2 mi. N Reeves, 1 (MSU). *Cal-*

Figure 207 Nearctic River Otter (*Lutra canadensis*), 5 mi. SE Sorrento, Ascension Parish. (Photograph by Robert H. Potts, Jr.)

casieu Par.: Calcasieu River, 1 mi. S Moss Bluff, 1 (MSU); mouth of Choupique Bayou, 1 (MSU); 1 mi. W Sam Houston St. Park, 1 (MSU). *Cameron Par.:* Cameron, 1 (LSUMZ); Grand Cheniere, 1 (LSUMZ); 17 mi. E, 2 mi. S Cameron, Hog Bayou, 1 (LSUMZ); 5 mi. W Holly Beach, 1 mi. N Hwy. 82, 1 (MSU); Sabine Refuge, 4 (USNM); locality unspecified, 1 (USNM); Vermilion Par. line on Hwy. 82, 1 (LSUMZ); Lacassine Bayou, 1 (USNM). *Madison Par.:* Tallulah, 4 (USNM). *Morehouse Par.:* south end of Bussy Break, north Bastrop, 1 (NLU). *Ouachita Par.:* Pleasant Valley, Eureka Area, 1 (NLU). *Plaquemines Par.:* Belair, 2 (LSUMZ); below Venice on Pass a Loutre, 1 (LSUMZ); 1 mi. S Port Sulphur, 1 (SLU). *Pointe Coupee Par.:* 4.5 mi. SE Krotz Springs, 1 (LSUMZ). *Rapides Par.:* 6.5 mi. SSW Alexandria, 1 (NSU). *St. Charles Par.:* Luling, 1 (LSUMZ). *St. John the Baptist Par.:* locality unspecified, 1 (SLU). *St. Landry Par.:* Opelousas, 1 (USL). *St. Mary Par.:* Belle Isle, 2 (USL); 6 mi. S Ellerslie, 1 (USL); Cypremort Point, 4 (USL); Morgan City, 1 (USNM). *St. Tammany Par.:* 2 mi. S Slidell, U.S. Hwy. 11, 1 (TU). *Tangipahoa Par.:* locality unspecified, 1 (SLU). *Vermilion Par.:* Belle Isle, 2 (USNM); Cheniere au Tigre, 2 (USNM); 5 mi. S Boston, 1 (LTU). *Washington Par.:* 4.5 mi. E Angie (Thomas Creek), 1 (LSUMZ); 3.25 mi. SE Varnado, 2 (SLU); 1 mi. NE Sheridan, 1 (SLU).

The Cat Family
Felidae

THE CATS are nearly worldwide in distribution, for they are absent only in Antarctica, Australia, Madagascar, and the West Indies and on most oceanic islands. Some authorities include all cats, except the cheetahs *(Acinonyx),* in the genus *Felis.* Others recognize six or more genera to accommodate the some 36 species that occur in one part of the world or another. Only 2 species are definitely known from Louisiana, the Cougar *(Felis concolor)* and the Bobcat *(Lynx rufus),* but some evidence points to the possibility that the Jaguar *(Felis onca)* and the Ocelot *(Felis pardalis)* may have occasionally found their way into the state.

The family contains animals of many sizes, ranging from the small Domestic Cat *(Felis catus)* to the huge and spectacular Lion *(Felis leo)* and Tiger *(Felis tigris).* But, despite this diversity, all are easily recognizable as cats. All walk on their toes, which number five on the forefeet, four on the hind feet; the toes possess sharp, strongly curved, completely retractile claws except that in the cheetahs the claws are semiretractile, blunt, and only slightly curved; the eyes are large and the pupils are capable of being retracted to vertical slits except again in the cheetahs, in which the pupils are always round; the skull is highly arched in the frontal region; the rostrum is short; and the face is flattened. The feet, excluding the pads, are well furred and assist the animals in the silent stalking of their prey. The baculum is vestigial or absent. The incisors are unspecialized, the canines long and sharp, and the carnassials well developed for shearing food. The tongue is covered with sharp-pointed, recurved papillae suited for helping to hold its prey. All cats are excellent climbers and when pursued often ascend trees. Most of them stealthily stalk their prey or lie in wait for it and kill the victim if it is large by a sudden rush, culminating in a leap on its back and the sinking of their teeth into its neck. Cats possess acute senses of hearing, smell, and sight. They prey on any animal they can overpower and this often includes animals much larger than themselves. As evidence of muscular power, a Tiger is reported to have moved a buffalo more than 45 feet, though 13 men could not drag it.

Genus *Felis* Linnaeus

Dental formula:

$$\mathrm{I}\frac{3-3}{3-3}\ \mathrm{C}\frac{1-1}{1-1}\ \mathrm{P}\frac{3-3}{2-2}\ \mathrm{M}\frac{1-1}{1-1}=30$$

[JAGUAR]   Figure 208
(Felis onca)

Vernacular and other names. — This species is often called the American tiger. Other names include spotted cat and, in Mexico and other parts of Latin America, *el tigre*. The word *jaguar*, or *yaguar*, was taken over by the Spaniards and Portuguese from *jaguara* and *yaguara*, two names that South American Indians applied to this large feline. The generic name *Felis* is the Latin word for cat. The specific name of the Jaguar, *onca*, was introduced into zoological nomenclature by Linnaeus, but its etymology is uncertain. The word may have been derived from the Middle English *unce* or *once* or the Old French *once*, all of which have been employed as names for various wild cats. The English word "ounce" has been used for a leopard, particularly the Snow Leopard (*Uncia uncia*). In Brazil the Portuguese names *onça verdadeira* and *onça pintada* designate the Jaguar, whereas the names *onça parda* and *onça vermelha* denote the Cougar.

Distribution. — The Jaguar occurs regularly from the tropical lowlands of western, eastern, and southern Mexico south through the lowlands of Central America and over a large part of South America. The basis of its provisional inclusion on the Louisiana list is discussed beyond. (Map 63)

External appearance. — This large, powerful cat is easily distinguished by its massive size and its beautiful golden-buff, black-spotted coat. Many of the spots are in the form of blotches of irregular shape or rosettes, each of the latter with a small black spot in the center.

Color phases. — Melanistic individuals occur occasionally in this species and are probably indicative of a low-frequency color phase.

PRATT
1972

Figure 208   Jaguar *(Felis onca)*.

Albinos are known to appear in South American populations, and indeed one of these mutants provided the basis for Fitzinger's description in 1869 of *Panthera onca alba,* a name that is invalid for several reasons.

Measurements and weights. — Adult males range in total length from 1,625 to 2,240 mm (64–88 in.), of which the tail makes up approximately 35 percent. Their weights range from 140 to 250 pounds. Females measure 1,400 to 1,850 mm (55 to 73 in.) and weigh 100 to 180 pounds (Leopold, 1959). The skull of a Jaguar from British Honduras (LSUMZ) of unrecorded sex (but almost certainly a male) measures as follows: total length, 225+; zygomatic breadth, 182; width of postorbital processes, 73.3; interorbital breadth, 47.1; width of rostrum at canines, 78.2; greatest width of nasals, 44.0; width of nasals at junction with frontals, 23.2; alveolar length of maxillary toothrow, 80.3; crown length of upper carnassial, 29.0; width of interpterygoid fossa, 24.2; anteroposterior diameter of upper canine, 23.4; height of coronoid, 80.2.

Cranial characters. — The skull of *Felis onca* is large and massive, much more so than that of *Felis concolor,* from which it is readily distinguished by its greater size and decidedly angular shape. Although sexual dimorphism in size is considerable in all the large cats, the skull of a small female Jaguar is still larger than that of a large male Cougar, at least among North American populations. (Figure 209)

Sexual characters and age criteria. — Males are noticeably larger than their mates. Jaguar kittens are heavily spotted, as are also baby Cougars, but in the Cougar the spots are lost in adolescence, whereas in the Jaguar they are retained throughout life.

Status. — Several reports of the occurrence of the Jaguar in Texas are available, some as

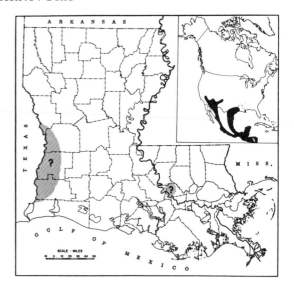

Map 63 Distribution of the Jaguar *(Felis onca).* *Shaded areas,* possible places where the species may have once occurred in the state; *inset,* overall range of the species.

close to Louisiana as the vicinity of Beaumont, in Orange Couty, and another south of Jasper, in Jasper County, both localities less than 30 miles from the Louisiana–Texas state line (Bailey, 1905; Taylor, 1947). The only evidence of the presence of the species in Louisiana is provided by two old newspaper accounts, recently uncovered by Ronald M. Nowak (1973) and discussed by him in some detail. The stories appeared in the newspaper the *Donaldsonville Chief* in 1886. They contain a description of a huge cat killed on Bay Ridge near New River, in Ascension Parish, on 1 June of that year. The first article (12 June 1886, p. 2) tells how the animal was suspected of killing livestock and was hunted down and killed. Before it was seen it was thought to be a panther. The account goes on to say: ". . . the animal measured eight feet in length, from tip of the nose to end of tail, two and a half feet in height, and weighed over 200 pounds." It was described as having had "a very broad breast, indicating great muscular power," and resembling "the jaguar, or American tiger, more than a panther."

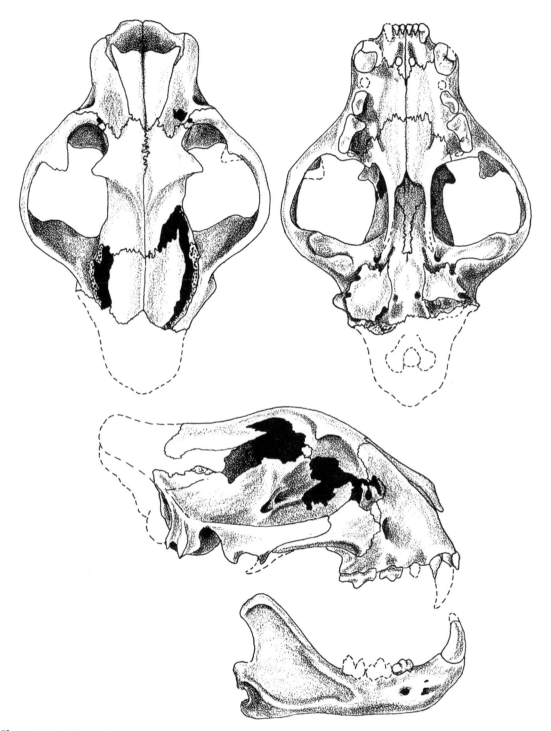

Figure 209 Skull of Jaguar *(Felis onca veraecrucis)*, sex ?, Regalia, Stann Creek District, British Honduras, LSUMZ 8652. Approx. ⅜ ×.

The second article (3 July 1886, p. 2) reiterated that the animal was "at first supposed to be a panther, but subsequently ascertained to be an American tiger." Two witnesses are quoted, one asserting that the animal "was discovered to be an American tiger, measuring 7 feet, 10 inches in length, 3 feet high, and weighed 250 pounds." As Nowak (1973) points out, the various accounts of the people involved differ in enough details to suggest that the cat may not have been accurately measured and weighed. But the alleged weights, both being over 200 pounds, would point definitely to its being a Jaguar and not a Cougar. Instances of Cougars weighing over 200 pounds are on record, but they are highly exceptional, and only rarely does one exceed 175 pounds.

The one somewhat disturbing fact is that in neither of the accounts in the *Donaldsonville Chief* is anything said about the animal's having spots. The Cougar, or panther, was, however, well known to rural Louisianians during the period in question, and one had actually been shot several years earlier in the same general locality. The animal killed on 1 June 1886 was said by the persons involved to be a different species; the names "American tiger" and "jaguar" applied to the animal in the newspaper articles are used synonymously, as they are today by English-speaking inhabitants throughout the range of *Felis onca*.

Considerable evidence points to the presence of the Jaguar in the eastern United States during the Pleistocene (Simpson, 1941b; Kurten, 1965; Oesch, 1969), and some authors (Harlan, 1825; Baird, 1859; Simpson, 1941a; and others) were convinced of its occurrence east of Texas in historic times but were unable to cite specific records. The mere fact that the big cats have actually been killed in Texas less than 30 miles from our borders, makes their former existence in Louisiana a virtual certainty, for the species is notoriously wide-ranging in its quest for food. The Jaguar killed at Jasper, Texas, was obtained a few years prior to 1902 and hence

was subsequent to the one described from near Donaldsonville, Louisiana, in 1886.

In the absence of an unequivocal record of the occurrence of the Jaguar in Louisiana, I have included the species on a provisional basis and, to indicate this status, I have placed the species heading in brackets.

Subspecies

Felis onca veraecrucis Nelson and Goldman

Felis onca veraecrucis Nelson and Goldman, Jour. Mamm., 14, 1933:236; type locality, San Andrés Tuxtla, Veracruz, México.

Since Nelson and Goldman referred Texas material to *veraecrucis*, any Jaguars that formerly ranged into Louisiana from Texas were probably also assignable to that subspecies. I admit, though, that the matter is somewhat academic.

Some authors place the Jaguar in the genus *Panthera* Oken, along with the Lion, Tiger, and Common Leopard, but this generic name is now generally regarded as being invalid. Other authors (for example, Stains, *in* Anderson and Jones, 1967) use the generic name *Leo* Brehm for most of the so-called great cats: Jaguar, *Leo onca;* Tiger, *Leo tigris;* Common Leopard, *Leo pardus;* Snow Leopard, *Leo uncia;* Lion, *Leo leo.* I have elected to retain these cats and the Cougar in the genus *Felis*.

Specimens examined from Louisiana. — None.

[OCELOT]   Figure 210
(Felis pardalis)

Vernacular and other names. — The Ocelot is often called the tiger-cat. In Latin America it goes by the name *ocelote* or *tigrillo*.

Distribution. — Although the species doubtless once ranged across southern Texas as far east as Louisiana, it is now essentially confined to the tropical lowlands of eastern, western, and southern Mexico, including the Yucatan Peninsula, thence southward through Central

America and over a considerable part of the northern half of South America. For the evidence in support of its possible former occurrence in Louisiana, see the discussion under *Status*. (Map 64)

External appearance. — The Ocelot is considerably larger than a large Domestic Cat. The upperparts have numerous dark, elongated, black-rimmed markings that contrast with the grayish to cinnamon ground color and run obliquely down the sides. The head has several black spots and there are two black stripes on the neck; the ground color is paler on the sides than on the dorsum; the underparts and inner surfaces of the legs are whitish, with numerous black spots; the tail is blotched and barred with black.

Measurements and weights. — The total length of the animal ranges from 850 to 1,215 mm (33.5 to 49 in.); tail, 300 to 435 (12 to 17 in.). Weights range from 10 to 25 pounds. Females are only slightly smaller than males. The skulls of four adult males in the LSUMZ ranged as follows: greatest length, 130.7 to 142.3; zygomatic breadth, 83.9–94.7; length of maxillary toothrow, 39.5–43.5; least interorbital breadth, 22.8–25.2; least postorbital breadth, 29.4–33.8. The skull of a single female measured, respectively, as follows: 127.1, 81.6, 40.1, 21.8, 29.5.

Cranial characters. — The skull of *F. pardalis* is similar to that of *Lynx rufus* but larger. The braincase is decidedly more inflated, the postorbital constriction is broader, the nasal processes of the premaxillae are broader and more abruptly attenuated, the notching of the suborbital edge of the palate is deeper and more prominent, and the premolars are 3–3/2–2, as is the case in all members of the genus *Felis,* instead of 2–2/2–2, the formula in the genus *Lynx.* (Figure 211)

Status. — In all probability the Ocelot once ranged as far east as Louisiana, but unequivocal records of its occurrence in the state are lacking. Harlan, writing in 1825, gave the range of the species as "Mexico, and the south-western parts of the United States, particularly Louisiana . . . ." Audubon and Bach-

Figure 210   Ocelot *(Felis pardalis).*

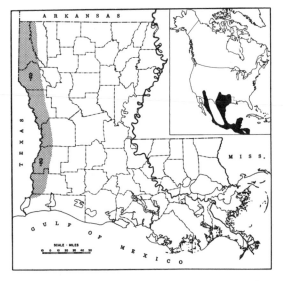

Map 64 Distribution of the Ocelot *(Felis pardalis).*
Shaded area, section of the state into which the
species may have once roamed; *inset,* overall range
of the species.

man (1851–1854) stated: "We have heard of
an occasional specimen of this cat having
been obtained in the southern parts of Louisiana." In 1855 Pucheran, in describing *Felis
albescens,* a name that later became the valid

designation of the subspecies of Ocelot occurring in Texas and northeastern Mexico,
based his diagnosis on a male specimen, obtained by the well-known naturalist Trudeau,
"originaire de l'état d'Arkansas, dans la Louisiane." Brass (1911) proposed the name *Felis
ludoviciana* (the translation of which is "Louisiana cat") as a substitute name for *Felis limitis*
Mearns. Both names are now relegated to the
synonymy of *F. p. albescens.* Brass stated that
the species occurs "In Louisiana, Arkansas
und Texas...." Seton (1929), in his map
delineating the range of the Ocelot, shows a
locality dot in southwestern Louisiana without specifying his basis for doing so, but he
does say that "in primitive days" the species
had been taken in Louisiana. Bailey (1905)
cited several reports of the occurrence and
killing of the Ocelot in southeastern Texas, as
far east as Beaumont and Jasper, both within
30 miles of the Louisiana–Texas state line. As
late as 1943, Goldman stated that the range
of *F. p. albescens* formerly included parts of
Arkansas and western Louisiana, and, in
1959, Hall and Kelson plotted the range of
the species into western Louisiana and southwestern Arkansas. These references to the

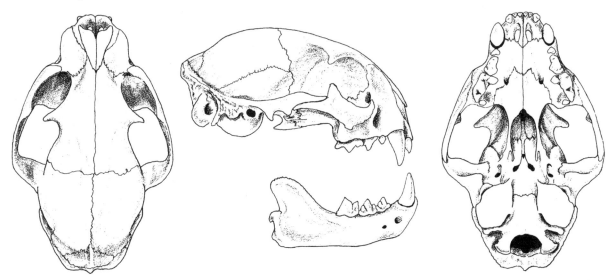

Figure 211 Skull of Ocelot *(Felis pardalis mearnsi),* ♂, 15 km. ESE Potrero Grande, Gromaco, Puntarenas
Prov., Costa Rica, LSUMZ 9289. Approx. ½ ×.

presence of the species in Louisiana do not add up to a positive record. Consequently, I have included the Ocelot in this work on a provisional basis, and, to indicate my doubts concerning its claim to a place on the state list, I have enclosed the name in brackets.

Subspecies

Felis pardalis albescens Pucheran

Felis albescens Pucheran, *in* I. Geoffroy Saint-Hilaire, Mammifères, in Petit-Thoars, Voyage autour du monde sur ... la *Venus* ..., Zoologie, 1855:149; type locality, Arkansas (exact locality unspecified).

Felis pardalis albescens, J. A. Allen, Bull. Amer. Mus. Nat. Hist., 22, 1906:219.

Populations of Ocelots in northeastern Mexico and Texas have been assigned to the race *albescens* (Goldman, 1943). Any of these cats that may have once wandered into Louisiana would surely also have been referable to this race.

Specimens examined from Louisiana. — None.

COUGAR   Plate XIII
(Felis concolor)

Vernacular and other names. — In Louisiana this species is most often called the panther, but in the Rocky Mountains and elsewhere it is frequently referred to as the mountain lion or puma. Calling it a mountain lion in this state does not seem quite appropriate, since our highest ground is Driskill Mountain with an elevation of only 535 feet above sea level. The word "cougar" is a corruption of a South American Tupi Indian name, and puma stems from the Quechua name for the big cat in Peru, by way of the Spanish conquistadores of the early 16th century. Other names include painter and catamount, the first being a colloquialism used in the eastern and southern United States, the second a contraction of cat-a-mountain. The early French settlers in southern Louisiana called it *chat-tigre* or *le tigre Américain*. The generic name *Felis* is the Latin word for cat. The specific name *concolor*

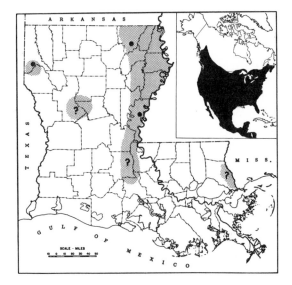

Map 65 Distribution of the Cougar *(Felis concolor). Shaded areas,* known or suspected sections where the species may still occur in the state; •, localities from which museum specimens have been examined; *inset,* overall range of the species in North America.

is the Latin prefix *con,* which can be translated as the same, attached to the Latin noun *color,* meaning hue or tint. Hence in combination with *Felis* it implies a cat of all one color, a reference to the adult Cougar's lack of spots. The subspecific name of the race that occurs in Louisiana, *coryi,* is a patronymic honoring Charles B. Cory, distinguished ornithologist and mammalogist of the early part of the 20th century.

Distribution. — The Cougar formerly occurred from northern British Columbia across southern Canada to New Brunswick and Nova Scotia and thence south all the way to Patagonia, at the southern extremity of South America, and had the largest range of any mammal in the Western Hemisphere. It has now been extirpated in many of its original haunts. In Louisiana it once occurred in most of the state's hardwood forest, but now only a few of these magnificent cats remain. (Map 65)

External appearance.—Among Western Hemisphere cats, the Cougar is rivaled in size only by the Jaguar. Adult males can measure seven feet or more in total length, of which the tail makes up more than 35 percent. The color of the upperparts is a uniform tawny cinnamon-buff, varying to a grayish buff. The throat, lower cheeks, lips, inside of ears, inside of legs, and belly are whitish. The back of the ears, moustache mark, and tip of the tail are dusky black.

Color phases.—No completely authenticated observation of a melanistic Cougar is known from the North American portion of the animal's range, but reports of so-called "black panthers" are irrepressible. I have heard of them over and over again in Louisiana in the last 30 years or so. Perhaps these reports stem from the verified accounts of melanism in South American populations of the species or from well-advertized cases of black Jaguars, including one now on display in the U.S. National Zoological Park in Washington, D.C., and another that appeared on TV in Walt Disney's *Wonderful World of Color.* The notion may also have been planted in the minds of the public by the writings of authors such as Rudyard Kipling, whose tale of the "black panther" Bagheera in *The Jungle Books* has delighted legions of readers. Bagheera was, however, a melanistic leopard, for true panthers do not occur in India.

Measurements and weights.—The adult male in the LSUMZ from Keithville measured as follows: total length, 2,176 (86 inches); tail, 787 (31 inches); hind foot, 288 (11 inches); ear, 77 (3 inches). It weighed 116 pounds. The adult male from 6 miles east of Hamburg, Ashley County, Arkansas (15.5 miles north of the Louisiana–Arkansas line), measured 83 inches in total length and weighed 152 pounds. The skulls of these two specimens plus an unsexed skull without a locality but probably that of a female from Louisiana, all in the LSUMZ, measured, respectively, as

follows: greatest length, 215.0, 204.5, 208.5; condylobasal length, 187.0, 177.0, 184.0; mastoid breadth, 90.7, 87.6, 85.1; zygomatic breadth, 143.7, 143.7, 144.0; postorbital breadth, 49.9, 45.9, 42.3; width of postorbital processes, 79.0, 80.3, 71.9; interorbital breadth, 43.5, 42.4, 41.1; width of nasals at anterior tips of frontals, 17.8, 19.6, 19.3; palatilar length, 80.3, 76.6, 79.2; postpalatal length, 96.6, 89.1, 88.7; maxillary toothrow, 65.6, 60.9, ——; cranial height (as defined by Goldman, *in* Young and Goldman, 1946), 78.3, 71.8, 75.1; height of coronoid, 69.1, 69.6, ——.

Cranial characters.—The skull is relatively broad and short, the greatest width being more than two-thirds the length; the rostrum is short, the nasals greatly expanded ante-

Plate XIII
1. Bobcat *(Lynx rufus).*
2. Cougar *(Felis concolor).*
3. American Black Bear *(Euarctos americanus).*

PRATT-1973

riorly; the postorbital processes, frontal processes of the zygoma, and the saggital crest are all highly developed; as in the Domestic Cat, the total number of teeth is 30, the first and last of the four cheek teeth being tiny with simple crowns. (Figure 212)

Sexual characters and age criteria.—Except for the primary and secondary sex organs and the slightly larger size of the male, the sexes of the Cougar are alike. Seasonal variation in color is not evident, but some individual variation occurs. In the kittens the ground color is near Pale Cinnamon or Pinkish Buff, the body and head are conspicuously marked with dark brown or blackish spots or blotches, and the tail shows dark bands, a pattern that is maintained until they are six months old.

Status and habits.—Cougars, or panthers as they are more often called in Louisiana, were once fairly numerous in the heavily forested areas of the state, especially where deer abounded. They doubtless occurred in greatest numbers in the bottomland swamps bordering the Mississippi, Tensas, Ouachita, Black, and Atchafalaya rivers, but they were also present in heavily wooded uplands. Unfortunately, these superb cats are now almost a thing of the past, although seldom does a year go by when at least one is not seen, especially by deer hunters in the northern part of the state. If only it were possible to persuade sportsmen merely to observe the panthers that they happen to see and not to shoot them, we might retain for some years to come a vestige of these wonderful creatures. What true sportsman could possibly take any delight in the knowledge that he killed the state's last surviving panther? Lest the readers of this passage get the impression that I am an overzealous conservationist opposed to hunting, let me hasten to make one thing clear. I love to hunt, and I have done my share of it—several thousand carefully preserved museum specimens bear witness! The harvesting of game by man for recreational

and other purposes is not only justifiable in my opinion but highly desirable. But rare and endangered species are not game. The only situation in which I would "shoot" a panther would be when one is in the viewfinder of my camera. Only by this kind of "hunting" and by the institution of rigid environmental protective measures can we hope to prevent this species and others in similar straits from passing from the American scene.

Some recent records of the Cougar in Louisiana and one from an adjacent part of Arkansas are given here in some detail. On 30 November 1965 two Caddo Parish deputies, Roy Lindsey and Frank Normand, obtained an adult male at Keithville, in Caddo Parish, only 13 miles south-southwest of downtown Shreveport. The two deputies generously donated the specimen to Louisiana State University, where it was mounted (Figure 213) and where it now reposes along with its skeleton, which is complete except for the terminal phalanges.

On 8 December 1969 another adult male was killed six miles east of Hamburg, Ashley County, Arkansas, by Harold Watts of Collingston, Louisiana (Noble, 1971). The locality where the animal was shot is only about 15 miles from the Louisiana–Arkansas state line.

On 9 January 1971, while in a deer stand in an open field surrounded by hardwood timber on the Spring Bayou Plantation in Madison Parish, L. A. Hamilton of Ruston and James Brown of Tallulah observed a large adult Cougar that Hamilton estimated to weigh close to 150 pounds. The cat was studied with binoculars and a 40-power spotting scope. Again, on 27 March 1971, in a stand located in a field about a mile from the point of his first sighting, Hamilton observed another large Cougar and watched it with binoculars as it slowly crossed the edge of the opening in front of his blind. This animal could have been, although not necessarily so, the same individual observed by Hamilton earlier in the same year. Hamilton and Brown are to be highly commended for hav-

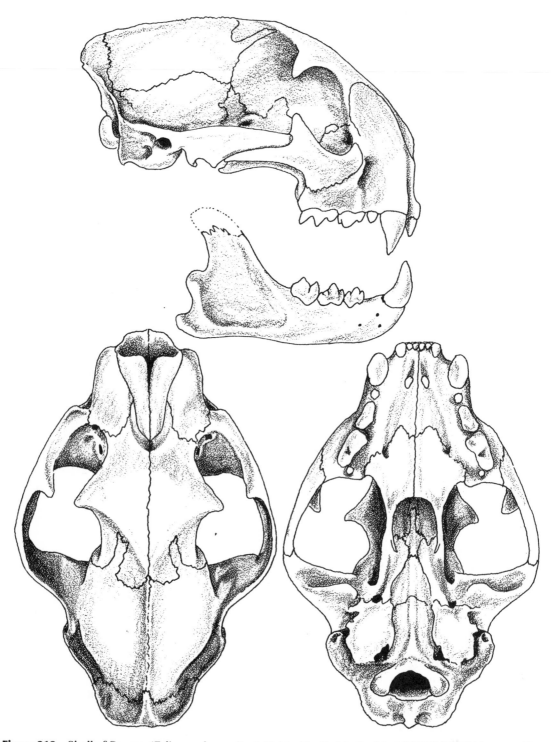

Figure 212   Skull of Cougar *(Felis concolor coryi)*, ♂, Keithville, Caddo Parish, LSUMZ 11363. Approx. ½ ×.

ing exercised restraint in spite of what must have been a strong temptation to shoot the animal. To my way of thinking they deserve a medal from some conservation agency for not having done so.

The most recent reports from the state are: a Cougar seen on several occasions by Joe H. Murphy in the spring of 1972 along Dorcheat Bayou near Sibley, in Webster Parish; another observed by Bob Landry crossing the road between Bush and Sun, in St. Tammany Parish, on 27 October 1972 (*St. Tammany Farmer,* 2 November 1972), and which was also heard to scream by Mrs. Margaret Stern (pers. comm.) back of her home near Bush a few nights earlier; and several sightings in 1972 and 1973 of a Cougar with offspring in the Glasscock chute-Lake Larto area in the southern part of Concordia and Catahoula parishes by George Barlow, Raymond Cowan, Leo Biles, and Bennet Landreneau.

Although other observations of Cougars have been reported in recent years, many of the sightings that have come to my attention leave something to be desired with regard to their authenticity. Of special interest is a series of events in the Baton Rouge area in the spring of 1958 that had their beginning in my

Figure 213   Photograph of the mounted specimen in the LSU Museum of the Cougar *(Felis concolor coryi)* that was obtained by Roy Lindsey and Frank Norman near Keithville, in Caddo Parish, on 30 November 1965. (Photograph by Joe L. Herring)

own backyard. My home is located a mile and a half southeast of the Louisiana State University campus on a 3.4-acre tract surrounded on three sides by a large wooded swamp. One evening just before twilight one of my daughters glanced out her bathroom window toward the lower end of the place and suddenly cried out for me to come at once. She had seen a large catlike animal on the edge of a garden some 120 feet from the house. When she shouted for me the animal took one leap into the nearby wood and disappeared. Of special significance was her insistence that the animal had a long tail and definitely was *not* a dog. The next morning my father found huge footprints in the soft earth of his freshly prepared garden. A plaster cast that he made of one of the prints caused some doubt to be expressed that the animal could have been a panther, for the impressions of the toes showed slight claw marks. I have since talked with numerous mountain lion experts, some of whom assert that the tracks of the big cat do occasionally show imprints of its retractile claws. The book entitled *A Field Guide To Animal Tracks* by Olaus J. Murie (1954) makes the same concession.

Little credence would have been placed in this incident had it not been for the events that followed. Shortly thereafter a workman on the Kleinpeter Dairy Farm, located an airline distance of only 11 miles from the site of the first observation, reported seeing a "huge long-tailed cat" cross the edge of a pasture. The next development took place on a farm less than two miles from the Kleinpeter Dairy, where a horse gave birth to a foal. Three nights later some unidentified predator killed the foal, leaving only some of the long bones as remains. Finally, at about the same time, a fisherman on the Amite River, some four miles distant, reported getting an excellent glimpse of a "large, long-tailed cat" on the opposite side of this small stream and asserted unequivocally that it was a panther.

Now, these observations, like so many of the reports of Cougars that come to light

from year to year, are highly suspect, to say the least, but they should not be dismissed summarily. The big cat is so furtive in its movements that it can easily escape detection, even where it is known to be present. The species has a reputation of wandering great distances in search of food—25 miles or more in a single night—and it can easily appear at rather widely separated places on successive days.

Although very few cases are on record of Cougars attacking man, most people residing in rural areas express abject fear, if not outright terror, at the thought of a panther being abroad in their area. As soon as one is said to have been seen the news travels fast; and, before hardly any time at all has elapsed, panthers are being reported from one source after another—usually with a few sightings of "black panthers" thrown in for good measure!

A track of a Cougar consists of four toe marks in a semicircle ahead of the imprint of the ball pad. The cat's fifth toe, the pollex of a front foot and the hallux of a hind foot, is located on the inside of the foot proper and somewhat above it, thus never takes part in the formation of the track. The imprints of the front feet are generally about three and one-half inches wide and three inches in length and are usually somewhat wider than those of the hind feet. In a slow, stalking gait the hind feet are often placed in the imprint made by the front paws. As previously noted, the claws are retractile. Even though an inch in length they are encased in a sheath. Consequently, they usually do not show in the track unless the animal slips and extends them in regaining its balance or in exploding into a sudden leap from soft ground. The tracks made by members of the dog family invariably show claw marks, and in many instances the imprints of the hind feet overstep the imprints of the forefeet. Cats almost invariably succeed in covering their scats by scratching dirt over them with the *front* feet, but in the case of dogs and their allies the parallel behavior is not much more than a ritual. The animal makes a few random scratches in the earth with its *hind* legs and only in the general direction of the fecal material.

Much controversy has been associated with sounds ascribed to the Cougar. Some writers allege, while others deny, that it sometimes gives voice to a blood-chilling scream likened to demoniacal laughter, to the wail of a child in terrible pain, or to a shriek of a woman in great distress. The fact that it does emit loud, terrifying, shrill sounds has been thoroughly

established (Young, *in* Young and Goldman, 1946). It also makes all the sounds of the Domestic Cat—caterwauls, mews, purrs, grunts, hisses, and whines—but increased manyfold in volume.

Just prior to the War Between the States, during the construction of the old Vicksburg, Shreveport & Pacific Railroad through the Tensas and Boeuf river sections of northeastern Louisiana, workmen allegedly could be kept on the job in the "head of the line" construction camps only with great difficulty because of the frequency with which panthers were actually seen or their sounds were heard at night (Lowery, 1936).

The Cougar does most of its hunting at twilight or under the cover of darkness, but it is sometimes abroad in full daylight. It walks with long strides, but when it is pursuing its prey or when it is being pursued itself by dogs, it bounds forward in long graceful leaps. Though mainly terrestrial it often ascends to the tops of tall trees by jumping from one limb to another (Figure 215).

The principal food of the Cougar is deer meat, but it also kills livestock, including colts, calves, sheep, goats, and pigs. Rabbits, armadillos, and rodents are also eaten. But of all these items, deer are by far the more important. The continued persistence of a few Cougars in some sections of the northern part of the state is, I am sure, attributable to the large population of deer that is found there. The Cougar obtains its prey by stealthily stalking it and, when it gets near enough, by pouncing on it. After partaking of a meal it drags the carcass to a thicket, where it conceals the remains under a mound of leaves and debris. Over a period of several days, or as long as the meat remains untainted, the cat continues to return to the cache for additional meals. The Cougar does not like putrid food, and rather than indulge in leftovers that have begun to spoil, it goes after a new kill.

Unprovoked attacks by Cougars on human beings are exceedingly rare, and those that have been known to occur may have been

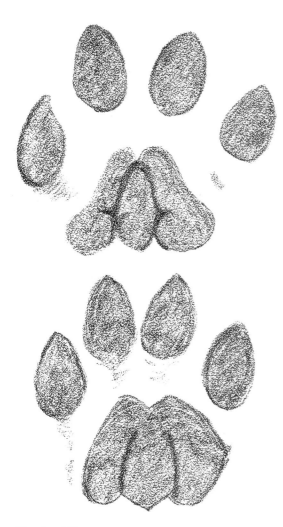

Tracks of Cougar

Figure 214 Sixteen-day-old baby Cougar, the eyes of which have only recently opened. Note that, unlike the adults, it is spotted. (Photographed at the Baton Rouge Zoo, August 1973, by H. Douglas Pratt)

executed by rabid animals. In one verified case, which occurred in California (Storer, 1923) a female Cougar set upon two boys and badly mutilated a young lady who came to their rescue. Seven weeks later rabies manifested itself in two of the victims, who soon died. In the opinion of Young (*in* Young and Goldman, 1946), livestock that appears to survive attacks by Cougars only to succumb later to what sometimes seems to be tetanus infections of their wounds may have died of rabies.

Cougars are extremely inquisitive animals and will not infrequently follow a man in the woods apparently for the sheer delight of doing so. Jack Kuhn, who was one of the finest woodsmen I have ever known and who helped me find Ivory-billed Woodpeckers in Madison Parish in the early 1930s, described to me an incident in which he was involved with a panther near Tallulah a few years earlier. He was returning home late one day through a section of virgin forest, when he sensed that he was being followed. Eventually he caught a glimpse of what was stalking him and saw that it was a panther. The animal

approached closer and closer and soon appeared to be ready to spring. Kuhn said that at the time he had a large hunting knife but no gun. So he backed himself against a tree and waited with knife ready for what he was sure would be an attack by the big cat. Just at that moment the old V.S.&P. railroad train, whose tracks happened to be not too far away, chanced to pass and the engineer fortunately blew his whistle, whereupon the panther bounded away and was not seen again.

Cougars are said to be at least two, probably more often three, years old when they begin to breed. The female is polyestrous, and she also comes into heat almost immediately after giving birth to young and remains in this state for approximately nine days. During the time a female is in heat, she is sometimes followed by two or more males, who fight for the first breeding privilege. This first copulation is often followed by union with other males. The duration of the gestation period is 96 days. The den or lair in mountainous areas is often in a rocky cave, but in Louisiana it is probably always in a dense thicket or canebrake. The number of young in a litter is usually three, but litters of six have been recorded. At birth the babies weigh eight to 16 ounces and measure eight to 12 inches in length, including the tail. By the end of the first week the eyes begin to open and at two weeks of age the process is completed. The young, unlike the adults, are spotted (Figure 214). By the time the kittens are eight weeks old they have attained a weight of approximately 10 pounds, and when six months of age they generally weigh 30 to 45 pounds, and they may double this amount by the time they become yearlings.

The flesh of the Cougar is edible, as I know from personal experience. When the Keithville cat was brought to LSU for preparation as a museum specimen, some of the graduate students who were present at the time could not bear to see all the fine flesh (that is, high-quality protein) going to waste. So they proceeded to cook and later to serve some of it at a gathering of museum personnel. I, for one, found it very tasty, as did also others in the group, particularly those who were not told what they were eating. I admit, though, that I would not order a serving of Cougar at Galatoire's or Antoine's, assuming it were to be found on their menus, in place of some more conventional delicacy.

Predators. — Aside from man, with his dogs, guns, and traps, the full-grown Cougar has few enemies against which it cannot adequately defend itself when the occasion arises. When it occurs side by side with the Jaguar, fierce encounters between them are

Figure 215 Cougar *(Felis concolor)* treed by dogs on the upper Hoh River, Olympic Peninsula, Washington. (Photograph courtesy of Thayer Collection, Western Historical Research Associates)

likely to take place, with the odds about even. But in the West, where it occupies the same habitat as the Grizzly Bear, it is said that the Cougar gives this powerful carnivore a wide berth.

Parasites and diseases.—External parasites known to infest the Cougar in North America include three ticks *(Dermacentor variabilis, Ixodes ricinus, I. cookei),* one louse *(Trichodectes felis),* and one flea *(Arctopsylla setosa).* Among the internal parasites are three tapeworms *(Hydatigera taeniaeformis, H. lyncis, Echinococcus granulosus)* and one nematode *(Physaloptera praeputialis).* Attention has already been called to the fact that the Cougar has been known to contract rabies.

Subspecies

Felis concolor coryi Bangs

Felis concolor floridana Cory, Hunting and fishing in Florida..., 1896:109; type locality, wilderness back of Sebastian, Brevard County, Florida. Not *F. floridana* Desmarest.
Felis coryi Bangs, Proc. Biol. Soc. Washington, 13, 1899:15; a renaming of *F. c. floridana* Cory.
Felis concolor coryi, Nelson and Goldman, Jour. Mamm., 10, 1929:347.
Felis arundivaga Hollister, Proc. Biol. Soc. Washington, 24, 1911:176; type locality, 12 mi. SW Vidalia, Concordia Parish, Louisiana.

This medium-sized subspecies is quite distinct from other named populations of *Felis concolor.* It occupies, or did so at one time, the Lower Mississippi Valley and the southeastern United States from Arkansas and Louisiana east through Mississippi, Alabama, Georgia, and Florida. It is distinguished by its short, rather stiff pelage and by several important cranial characters, including the broad, flat frontal region and the expanded, highly arched nasals. It intergrades with *F. c. stanleyana* Goldman in eastern Texas, Oklahoma, and Missouri, and with *F. c. couguar* Kerr in Tennessee and North and South Carolina. The specimen from Keithville might be expected to show tendencies toward *stanleyana,* but such is not the case. The skull of this specimen appears unequivocally assignable to *coryi.*

Specimens examined from Louisiana.—Total 4, as follows: *Caddo Par.:* Keithville, 1 (LSUMZ). *Concordia Par.:* 12 mi. SW Vidalia, 1 (USNM). *Morehouse Par.:* Prairie Mer Rouge, 2 (USNM).

Genus *Lynx* Kerr

Dental formula:

$$I\frac{3-3}{3-3} \ C\frac{1-1}{1-1} \ P\frac{2-2}{2-2} \ M\frac{1-1}{1-1} = 28$$

BOBCAT   Plate XIII
(*Lynx rufus*)

Vernacular and other names.—The name Bobcat calls attention to the short, bobbed tail that is so highly diagnostic of the cats of the genus *Lynx.* The animal is also frequently called the wildcat or bay lynx by woodsmen. I have heard three quite different explanations of the name bay lynx: that it is an allusion to the animal's basic reddish brown coat color, that it refers to the cat's tendency to come to bay by climbing a tree when pursued by hounds, and, finally, that it is based on the frequency with which the Bobcat is found in wet, swampy woodlands colloquially called "bays" in some sections of the state. The early French settlers termed it *pichou,* which may have come from the Mobilian Indian name *pishu,* "wild cat," or by way of Canadian travelers who borrowed the word from the Cree *pisiw* or the Nipissing *pishiu* (Read, 1963). The generic name is derived from a Greek word applied to this group of cats and may refer to its keen sight. The specific name *rufus* is a Latin word meaning reddish, applying to the general body color of some individuals.

Distribution.—The species occurs across southern Canada thence south over the entire United States, except for the midwestern corn belt, to southern Mexico. In Louisiana it is virtually statewide. (Map 66)

External appearance.—The Bobcat is medium-sized in comparison with a Domestic Cat on the one hand and a Cougar on the other. The total length is seldom greater than 35 inches, with the tail comprising six inches or less. It is rather long-legged and rangy. Its short face is set off by a ruff of fur from the ears down to and below the jowls that gives the effect of sideburns. The ears are prominent and pointed, often with tufts of black hairs, about one inch in length, at their tips. The general color is yellowish brown or reddish brown, streaked and spotted with black. The posterior surfaces of the ears are blackish, each with a central spot of white or gray. The underparts are white with black spots and several black bars on the inside of the legs. The tail is reddish brown above with several dusky bars, the last of which is the broadest and darkest. The tip and underside of the tail are whitish.

Color phases.—No true color phases occur in the Bobcat, but a case of melanism in a Florida specimen has been reported (Ulmer, 1941).

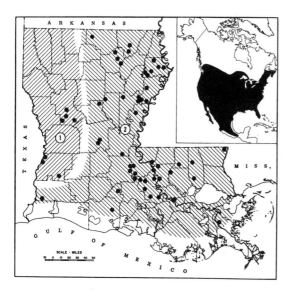

Map 66 Distribution of the Bobcat *(Lynx rufus)*. *Shaded areas,* known or presumed range of the species in the state; •, localities from which museum specimens have ben examined; *inset,* overall range of the species. 1. *L. r. texensis;* 2. *L. r. floridanus.*

Measurements and weights.—Adult males in the LSUMZ from various localities in the state (number of specimens in parentheses following the designation of measurement) averaged as follows: total length (3), 868 (834–900); tail (3), 147 (135–156); hind foot (3), 168 (160–175); ear (3), 68 (62–72); greatest length of skull (5), 122.4 (116.5–127.1); mastoid breadth (4), 52.5 (48.9–54.8); zygomatic breadth (6), 85.0 (79.4–88.5); interorbital breadth (7), 23.5 (22.0–26.2); palatilar length (6), 48.3 (46.4–49.4); postpalatal length (5), 55.0 (52.7–57.9); length of maxillary toothrow (7), 38.0 (36.0–40.3); cranial height (6), 45.8 (44.7–47.0). Females from the same source averaged as follows: total length (3), 786 (738–834); tail (3), 141 (130–154); hind foot (3), 141 (125–161); ear (3), 62 (53–70); greatest length of skull (5), 113.1 (108.7–114.8); mastoid breadth (5), 50.2 (48.9–53.8); zygomatic breadth (4), 79.2 (76.0–83.0); interorbital breadth (5), 21.3 (19.0–22.9); palatilar length (5), 43.0 (40.4–45.7); postpalatal length (5), 50.6 (49.1–56.4); length of maxillary toothrow (5), 34.8 (33.9–35.2); cranial height (4), 44.5 (40.5–47.5). Few weights of Bobcats from Louisiana are available, but males average about 20 pounds, females about 15 pounds. Harlan Hall (pers. comm.) furnished me with the weights of four females from East Baton Rouge Parish and one from Iberville Parish, and these averaged 15.5 (9.1–14.9) pounds. A single male obtained by him from East Baton Rouge Parish weighed 19.4 pounds. Edward Butler of St. Francisville sent me a photograph of a Bobcat killed near his home in April 1921 that weighed 45.25 pounds, despite being described as "very thin." It measured 1,118 mm (44 inches) in total length and was said to stand 21 inches at the shoulder. I would question the authenticity of this weight were it not for Mr. Butler's complete reliability in his lifetime as a naturalist in the Felicianas.

Cranial characters.—The Bobcat is the only carnivore of Louisiana that possesses only 28 teeth. The forehead is smoothly rounded and the eyes are set forward, producing a short face. The skull resembles that of the Domestic Cat except for its much larger size and two fewer teeth. (Figure 216)

Sexual characters and age criteria. — Males are identified by the penis and by the testes, which descend into a scrotum. Females have four teats, one pair abdominal, the other pair inguinal. The sexes are identical in color, but males average between 5 and 10 percent larger in some of their measurements. The kittens are heavily mottled, but as the fur develops they lose the spots on the upperparts. By the time they attain their first winter coat they are like the adults.

Status and habits. — The Bobcat is widely distributed over the state and still occurs in some numbers in nearly all heavily wooded regions. Even in densely populated East Baton Rouge Parish, it is not uncommon in Devil's Swamp, on Profit Island, and in the Faulkner's Lake area. All through the vast, hardwood bottomlands along the Mississippi River from the Arkansas line southward through the Atchafalaya Basin and the floodplain of

the lower Mississippi River, as well as in forested areas adjacent to other river systems in the state, the Bobcat maintains itself in appreciable numbers.

This species is one of the more handsome constituents of our state's mammal fauna (Figure 217). Its lithe, graceful body and its beautifully marked facial pattern combine to give it great physical character, and its skill as a hunter inspires admiration on the part of those who love the out-of-doors and the denizens of our forest, large and small. And one does not have to be a "cat lover" to admire *Lynx rufus.* "Quiet in all it does and possessed of a very keen sense of hearing, the bobcat will instantly crouch with all four feet under its body, remaining tense and motionless, listening and watching, when it detects any movement in the undergrowth. If it is an enemy, it will creep off without a sound betraying its retreat from the danger zone. If it proves to be some living thing it can make a meal of . . . it will creep on its victim with the utmost caution" (Arthur, 1928).

The Bobcat feeds on a wide variety of items, but rabbits comprise the most significant portion of its diet. These it supplements with squirrels, rats, mice, small birds, and

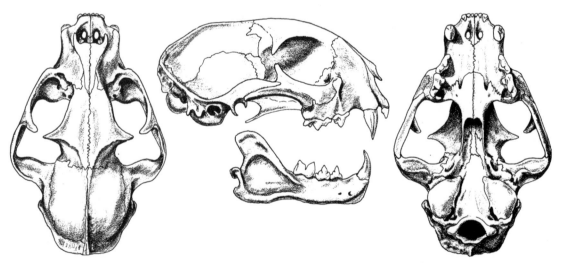

Figure 216 Skull of Bobcat *(Lynx rufus floridanus),* ♂, 3 mi. W Port Allen, West Baton Rouge Parish, LSUMZ 13747. Approx. ½ ×.

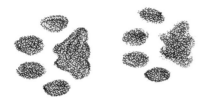

Tracks of Bobcat

occasionally even larger mammals such as fawns and young livestock. Some of the large animals are eaten as carrion. It often takes more than it needs and unless food is scarce it seldom returns to old kills even though it may have covered the remains with leaves. Its usual method of procuring its prey is to sneak up and then pounce upon it, but sometimes it lies in wait on the edge of a game trail until something comes along. After making a kill it consumes its prize voraciously, with growls and hisses if it is interrupted or disturbed in the process. Most of the sounds made by the Bobcat closely resemble those of the Domestic Cat but with amplification and an added quality of ferocity. In the breeding season it be-comes especially vociferous, its caterwauling consisting of howls, wails, screams, and snarls, some of which rend the air on an otherwise quiet night in the forest. Like most members of the cat family, it can see well in semidarkness. But it does not confine its hunting to the dark hours and is often on the prowl in broad daylight.

Breeding takes place in midwinter, and after a gestation period of approximately 62 days, the one to five (but usually two or three) blind, rather heavily furred young are born. At birth the kittens weigh about one pound, are spotted, and possess sharp claws. The den is usually located under an uprooted tree, or it may be in the middle of a cane thicket. In about 10 days the eyes have opened and the youngsters come out to frolic and play under the watchful eyes of the mother. By the time fall begins the kittens have learned to hunt to some extent on their own, but the family ties do not begin to dissolve until the middle of the winter, when again the mother becomes pregnant. (Young, 1958)

Figure 217   Bobcat *(Lynx rufus)*. (Photographs by Joe L. Herring)

Predators and other decimating factors.—Man is the only serious enemy of the Bobcat. Although it is considered a fur animal, few are taken for this purpose. Fur records of the Louisiana Wild Life and Fisheries Commission are somewhat ambiguous and do not give a clear picture of the number of Bobcats taken each year. More often than not all cats have been combined or else simply included under "Miscellaneous" pelts. Somewhat intriguing is the record of 1,430 "Miscellaneous Cats" marketed in the 1945–1946 season. I would hope that most of these were Domestic Cats, for one of the greatest enemies of wildlife in our state is the tabby that is allowed to run wild, fending for itself.

Parasites.—External parasites that have been found on *Lynx rufus* include two lice *(Felicola felis* and *F. subrostratus),* two tick *(Dermacentor variabilis* and *D. andersoni),* and six fleas *(Orchopeas wickhami, Cediopsylla simplex, Odontopsyllus multispinosus, Hoplopsyllus lynx, Echidnophaga gallinacea, Ctenocephalides felis).* Internal parasites are more numerous and include 15 cestodes, three trematodes, and 13 nematodes.

Subspecies

Lynx rufus floridanus Rafinesque

Lynx floridanus Rafinesque, Amer. Monthly Mag. 2 (1), 1817:46; type locality, Florida.
Lynx rufus var. *floridanus,* Baird, Mammals, *in* Repts. Expl. Serv...., 8 (1), 1858:91

This small, dark subspecies of *Lynx rufus* occupies the Lower Mississippi Valley and most of the southeastern United States. It is found throughout most of Louisiana, being replaced by *L. r. texensis* Allen in the central-western part of the state.

Specimens examined from Louisiana.—Total 71, as follows: *Avoyelles Par.:* 18 mi. NE Washington, 1 (USL). *Calcasieu Par.:* 25 mi. SW Lake Charles, 1 (USNM); locality unspecified, 2 (USNM). *Catahoula Par.:* 4 mi. NNE Foules, 1 (LSUMZ); 12 mi. E Harrisonberg, 1 (NLU); Ward 3, nw.

corner of parish, 2 (USNM). *Concordia Par.:* 15 mi. W Vidalia, 2 (USNM). *East Baton Rouge Par.:* Baton Rouge, 1 (LSUMZ); 15 mi. N Baton Rouge, 1 (LSUMZ); Profit Island, 4 (LSUMZ). *Grant Par.:* near Georgetown, 1 (LTU). *Iberville Par.:* 2 mi. S White Castle on hwy., 1 (LSUMZ); Bayou Blue, 1 (LSUMZ); vic. Grosse Tete, 1 (LSUMZ); 0.5 mi. E Bayou Sorrel, 1 (USL); 18 mi. NW Plaquemine, 1 (USNM). *Lafayette Par.:* 3 mi. N Lafayette, 1 (USL). *La Salle Par.:* 1 mi. E Tullos, 1 (NLU). *Livingston Par.:* Watson, Amite River swamp, 1 (LSUMZ). *Madison Par.:* Mississippi River swamp near Tallulah, 1 (LSUMZ); Tallulah, 3 (1 SLU; 2 USNM); 11 mi. SW Tallulah, 2 (NLU); 8 mi. NW Tallulah, 1 (USNM); 24 mi. SW Tallulah, 1 (USNM); 13 mi. SW Tallulah, 2 (USNM); 8 mi. SW Tallulah, 1 (USNM); 12 mi. S Delhi, 1 (LTU). *Morehouse Par.:* 3 mi. W Oak Ridge, 1 (NLU); 4 mi. W Oak Ridge, 1 (NLU); Mer Rouge, 3 (1 USNM; 2 MCZ). *Ouachita Par.:* Monroe, 1 (NLU); locality unspecified, 1 (NLU); 12 mi. SE Monroe, 1 (NLU); 9 mi. N of West Monroe, 1 (NLU). *Plaquemines Par.:* Burbridge, 1 (MCZ). *Pointe Coupee Par.:* Lottie, 4 (LSUMZ); 5 mi. E Melville, 2 (USL). *Rapides Par.:* Latanus swamp, 4.5 mi. NNE Lecompte, 1 (NSU); near Lecompte, 1 (USL). *St. Helena Par.:* 3 mi. E Pine Grove, 1 (LSUMZ). *St. James Par.:* 10 mi. NW Vacherie, 1 (LSUMZ). *St. Landry Par.:* Melville, 1 (LSUMZ); 3 mi. N Krotz Springs, 1 (LSUMZ). *St. Martin Par.:* Little Alabama Bayou, 1 (USL). *Tangipahoa Par.:* 2 mi. S Ponchatoula, 1 (SLU); Amite, 1 (USNM). *Tensas Par.:* locality unspecified, 1 (LTU); Tensas Bayou, 1 (USNM); mouth Fool River, 32 mi. SW Tallulah, 1 (USNM); 2 mi. S Newlight, 1 (NSU). *Union Par.:* near Spearsville, 1 (LTU). *Vermilion Par.:* Abbeville, 1 (USNM). *Washington Par.:* Angie, 1 (SLU). *West Baton Rouge Par.:* 3 mi. W Port Allen, 1 (LSUMZ). *West Feliciana Par.:* Thompson Creek at Hwy. 61, 1 (LSUMZ).

Lynx rufus texensis J. A. Allen

Lynx texensis J. A. Allen, Bull. Amer. Mus. Nat. Hist., 7, 1895:188; type locality, Castroville, Medina County, Texas.
Lynx rufus texensis, Mearns, Preliminary diagnoses of new mammals... from the Mexican boundary line, 1897:2 (preprint of Proc. U.S. Nat. Mus., 20, 1897:458).

This relatively large, more rufescent subspecies of the Bobcat occupies the central-western parishes and perhaps also the northwestern corner of the state, but from the latter area no specimens are yet available on which to base a judgment.

Specimens examined from Louisiana.—Total 10, as follows: *Beauregard Par.:* 7.5 mi. NW Merryville, 1 (MSU); below Ragley, 1 (MSU); 2 mi. W Merryville on Hwy. 190, 1 (MSU). *Natchitoches Par.:* Kisatchie, 1 (LSUMZ); Gorum, 3 (USNM); Ward 7, 2 (USNM); 1 mi. W Flora, 1 (NSU).

[SEALS AND WALRUS]
[Order Pinnipedia]

[The Eared Seal Family]
[Otariidae]

[CALIFORNIA SEA LION]
(*Zalophus californianus*)
Figure 218

Since 1965 several observations of eared seals on the Louisiana coast have been reported, and at least one has been substantiated by a photograph that permitted positive identification. Two seals, said to have been of this species, were present on Chandeleur Island in January 1966. One was probably the same individual as the seal seen on buoys in the Mobile Ship Channel in July of that year, when it was observed to have an identifying scar that showed it to be the seal found dead on Chandeleur Island on 11 August 1966 (Gunter, 1968). According to Gunter, the second seal may have been the one observed on a channel buoy off Pensacola in April 1967. Again, on 30 March 1968, a seal was observed at latitude 29°16′ N, longitude 88°45′ W, near the mouth of the Mississippi River, by a group of oil company employees. Finally, an eared seal, definitely identified by Dr. Carl L.

Hubbs from photographs as *Zalophus californianus,* visited an oil company barge in the Gulf 32 miles south of Cameron daily for nearly a month in August and September 1971 and sunned itself on the deck in spite of the presence of workmen a few yards away. It is said to have disappeared on 14 September, the day before the passage of Hurricane Edith.

The California Sea Lion occurs normally only on the Pacific Coast, from British Co-

Figure 218 California Sea Lion (*Zalophus californianus*) on bell buoy at mouth of Mobile Bay, Alabama, 1 July 1966. (Photograph by Fred Reese)

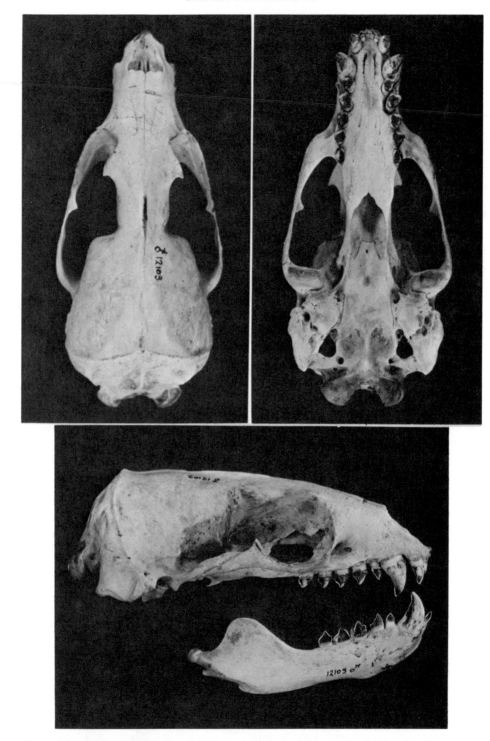

Figure 219 Skull of California Sea Lion *(Zalophus californianus)*, California, KU 12103. Total length of skull, 248 mm (9.7 in.). (Photograph by Joe L. Herring)

lumbia thence southward to the Tres Marias Islands off Nayarit. Its presence in the Gulf of Mexico must surely be attributable in some way to man, though inquiries directed to the various sea aquaria on the Gulf Coast have consistently failed to reveal any who reported having a California Sea Lion escape.

The skull of a sea lion is not unlike that of a typical carnivore and for this and other reasons some mammalogists give the pinnipeds only subordinal rank within the Carnivora rather than assigning them full ordinal status.

The skull of the sea lion is quite similar to the skull of an otter other than being much larger. It is decidedly flattened on top. (Figure 219)

The only seal native to the Gulf of Mexico is the West Indian Monk Seal, *Monachus tropicalis* (Gray), which is now considered extremely rare if not actually extinct. The species is, however, a member of the family Phocidae, known as the earless seals, since they lack pinnae and differ in other respects from the eared seals of the family Otariidae.

MANATEES, DUGONG, and SEA COW

Order Sirenia

THIS WHOLLY AQUATIC, bizarre group of mammals is presently found only in coastal waters bordering tropical or subtropical seas. One member of the order, the now extinct Steller's Sea Cow (*Hydrodamalis stelleri*), which was adapted for a life in the cold waters of the Bering Sea, was exterminated by man only 27 years after its original discovery in 1742 by Captain Vitus Bering and the naturalist on his expedition, Georg Steller. The family to which it belonged, Dugongidae, contains one other genus and species, the Dugong *(Dugong dugon)*, which occurs or has occurred in the Red Sea, in the waters along the east coast of Africa, around the edges of the Bay of Bengal, and in the waters surrounding southeastern Asia, the Philippines, New Guinea, Indonesia, and northern Australia. It has suffered so heavily at the hands of man that it has now been extirpated in some areas, and its ultimate survival as a species anywhere within its range is a matter of great concern to conservationists. Its hide is used for leather, its meat for food, and its oil and blubber for fuel and sundry other purposes.

A second family, Trichechidae, contains a single genus, *Trichechus*, with three species, one of which, *T. manatus*, occurs in Florida, casually north along the Atlantic Coast to North Carolina, and west in the Gulf of Mexico to southern Texas, as well as in the southern Gulf of Mexico, in the Caribbean, and in northern South America.

These massive, fusiform animals measure up to 15 feet in length and often weigh over half a ton (Steller's Sea Cow was estimated to weigh three tons). The forelimbs are modified as paddles, no dorsal fin nor hind limbs are present, and the tail is horizontally flattened as a flipper for propelling the animal through the water. The snout is abruptly rounded and the mouth is small. The nostrils are valvular and rather widely separated, being located on the upper surface of the muzzle with their openings directed posteriorly.

The members of this order are entirely vegetarians. The Dugong consumes prodigious quantities of algae and other marine plants. It never ventures far from shore nor into fresh water, but the manatees of American waters are often found considerable distances up freshwater streams, where they feed on various plants, including water hyacinths.

The Sirenia are without much doubt the basis of the Sirens of mythology and the mermaids of legendary tales of seafarers. Some of the stories told by European sailors on their return home described seeing mermaids that nursed their babies at the breast like human mothers. They probably had seen a docile sirenian, either a Dugong or a manatee, suckling its young at the two mammae on its chest, but their observations were faulty to say the least. In the manatees, and probably also in the Dugong, nursing takes place beneath the surface of the water with the mother in a horizontal position with her belly down. The calf is not clasped in her flippers, neither mother nor offspring has its head above water, and neither is in a vertical position.

Manatees and their relatives the Dugong and Steller's Sea Cow are considered close to the evolutionary line that also gave rise to the elephants and the hyraxes. This relationship is borne out by certain remarkable similarities in their skulls. The cheek teeth, for example, are replaced consecutively from the rear as the forward teeth are worn down and drop out, an arrangement also characteristic of the proboscidians.

The Manatee Family:
Trichechidae

THIS FAMILY contains only one genus, *Trichechus*, with three species, two of which occur in warm waters of North and South America, the other in equatorial West Africa. In the manatees the upper lip, which is deeply cleft with each half capable of being moved independently of the other, is used efficiently in browsing on aquatic plants. The tail is spatulate, whereas in the Dugong and Steller's Sea Cow it is notched. The large, drooping upper lip is provided with stiff bristles, and other bristlelike hairs are widely scattered over most of the body. Those of the lips assist the animal in its browsing. The forelimbs are used in forward propulsion, sometimes simulating crawling motions when a manatee is feeding on the bottom of a stream. They can also be used to a limited extent in directing food to the mouth. The pectoral appendages of the whales and dolphins of the order Cetacea lack any such mobility. The flippers of a manatee also differ from those of cetaceans in possessing vestigial nails on the tips of the terminal phalanges. Members of this family have only six neck vertebrae, whereas nearly all other mammals have seven. The nasal bones are present in manatees but absent in the Dugong and Steller's Sea Cow.

Another important peculiarity of the Sirenia is the density of their bones, including the skull and even the ribs. The latter are almost as solid as a piece of ivory. The increased specific gravity is probably of considerable advantage to a manatee in permitting it to sink with a heavy load of air and to stay submerged while feeding on underwater vegetation. Because of the virtual absence of bone marrow, the question arises as to where a manatee manufactures its hemoglobin. The spleen, of course, would be a logical possibility, but the one detailed account of the anatomy of the manatee of which I am aware (Quiring and Harlan, 1953) contains no reference at all to the spleen despite the fact that other organs are described in detail and their sizes and weights given.

Genus *Trichechus* Linnaeus

Dentition.—Incisors, which number two on each side of each jaw, are rudimentary and concealed beneath the horny plates and are lost before the animal reaches maturity. The molariform teeth are 11/11 but rarely more than 6/6 are in place at any one time; the anterior cheek teeth wear down and fall out, being replaced by new teeth at the posterior end of the toothrow.

WEST INDIAN MANATEE   Figure 220
(Trichechus manatus)

Vernacular and other names.—The word manatee has been said, probably erroneously, to be derived from two Spanish roots, *mano,* hand, and *tener,* to hold, supposedly alluding to the early seafarers' notion that these animals hold their offspring to their breast in human fashion while suckling them. Instead it is probably of Cariban origin, specifically from the Caribs of French Guiana, who employ for the species the word *manattouí* that came over into the Spanish as *manati* and *manaté.* The Carib word is said to mean "[a woman's] breast," thereby alluding to the superficial similarity of the mammae to those of the human species (Simpson, 1941c). The generic name *Trichechus* was formed from two Greek words (*trich,* hair, and *echus,* having) and was doubtless designed to call attention to the fact that these animals do indeed possess hairs even though they are stiff

bristles and few in number. The specific name *manatus* means having hands (the Latin word *manus,* hand, plus the Latin suffix *atus,* which, when added to a noun, means provided with). It refers to the somewhat mobile pectoral flippers. Since there are three species of manatees, a modifier is required in the vernacular name of each. The name "the Manatee" cannot be considered applicable solely to *Trichechus manatus.* Since *Trichechus senegalensis* occurs only in Africa, it can be called the African Manatee. And, since *Trichechus inunguis* is found only in the drainage of the Amazon and Orinoco rivers, it can be called the Amazon Manatee. Unfortunately, when one looks for a suitable modifier for our species, the problem is complicated by the fact that *Trichechus manatus* ranges from the United States south around the edges of the Gulf of Mexico and the Caribbean to northern South America. Since it extends, at least casually, as far north as 36°30′ N latitude, it could be called the Northern Manatee, but this name carries such a connotation of boreal distribution that I do not think it would be appropriate in this case. The name American Manatee, which has been used, is unsatisfactory because *Trichechus inunguis* of South America is also "American." Taking all factors into consideration, I believe the best name would be West Indian Manatee. No geographical name applied to a species can be expected to delineate the entire area occupied by that species. The West Indies are at least in the center of the overall range of the present species, and it was there that it was first seen by early explorers. Moreover, by restriction the West Indies were designated by Oldfield Thomas in 1911 as the type locality of the nominate subspecies.

Distribution.—The West Indian Manatee occurs along the coast and in coastal rivers of the southeastern United States from North Carolina southward to southern Florida and westward in the Gulf of Mexico to southern Texas and Veracruz, thence through most of the West Indies and Caribbean waters of Central America to northern South America. The inclusion of the species as part of the mammal fauna of Louisiana is based on two individuals found dead, one in Calcasieu Lake and the unconfirmed report of another in Sabine Lake, and on a single sight record from Lake St. Catherine, in Orleans Parish. (Map 67)

External appearance.—The general appearance of this large aquatic mammal is evident from the accompanying illustrations (Figures 220, 221, and 222). The rounded body along with small head, no visible neck, forelimbs modified as flippers, absence of hind limbs, spatulate tail, squarish snout with overhanging, centrally divided, bassethoundlike overhanging lips, absence of hair except for stiff bristles in the region of the mouth and a few scattered over the body, and lack of external ears all combine to make this species unique among the mammals of North America. The color is a dull gray to blackish and the skin is over two inches in thickness.

Figure 220   West Indian Manatee *(Trichechus manatus)*.

Color phases.—None.

Measurements and weights.—Adults measure 8 to 14 feet in length, but individuals that are over 12 feet are scarce. Weights usually range from 300 to 1,200 pounds, but manatees weighing 1,490 pounds are on record.

Cranial characters.—The skulls of *Trichechus manatus* and its close relative *T. inunguis* of South America are quite unlike that of any other mammal of the Western Hemisphere. The peculiarly-shaped rostrum, the massive lower jaw, and the manner in which the anterior molariform teeth drop out, being replaced by the addition of new teeth at the posterior end of the series, are, along with

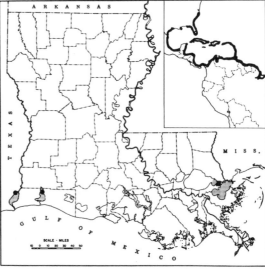

Map 67 Distribution of the West Indian Manatee (*Trichechus manatus*). *Shaded area,* known or presumed range within the state; •, localities from which manatees have been reported or collected in the state; *inset,* overall range of the species.

Figure 221 West Indian Manatee (*Trichechus manatus*), Kings Bay at head of Crystal River, Citrus County, Florida. (Photograph by James A. Sugar, courtesy of the photographer and the National Geographic Society)

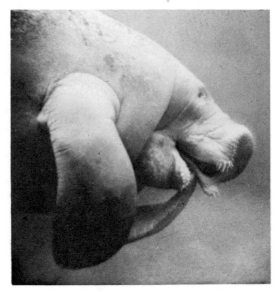

Figure 222 A female manatee using one of its flippers in a mouth-cleaning operation, illustrating the greater dexterity of the sirenian forelimb in contrast to that of a cetacean. Note the teat located on the posterior edge of her armpit. (Photograph by James A. Powell, Jr., courtesy of Daniel S. Hartman)

other distinctions, characters that are unique among aquatic mammals. (Figure 223)

Sexual characters and age criteria.—The sexes, except for the primary and secondary reproductive organs, are quite similar. Males average somewhat larger than females. The young (one, sometimes two) at birth weigh 40 to 60 pounds, measure slightly over three feet in length, and are pinkish in color. Fe-

males begin to breed at the age of three or four years. Any individual under eight feet in length is probably sexually immature.

Status and habits.—The West Indian Manatee is only of casual occurrence along the central northern Gulf Coast. Three instances of its presence in Louisiana waters are on record. First is the manatee that was found dead in Calcasieu Lake, below Lake Charles, in Janu-

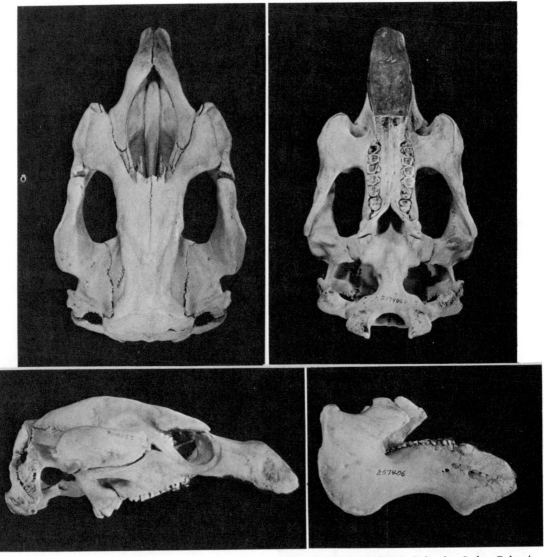

Figure 223 Skull of West Indian Manatee *(Trichechus manatus)*, USNM 257406, Calcasieu Lake, Calcasieu Parish, January 1929. Total length of skull 381 mm (15 in.). (Photograph by Joe L. Herring)

ary 1929 by Stanley C. Arthur, and its skull was preserved. It is now No. 257406 in the United States National Museum. Secondly, a dead adult male was reported by Gunter (1941) to have washed ashore at the mouth of Cow Bayou, where it empties into Sabine Lake on the boundary between Louisiana and Texas. According to Gunter an account of the incident appeared in the *Port Arthur News* in 1937, but I have been unable to locate the article, despite having perused each issue of the newspaper for the year in question. Likewise, no one in Port Arthur responded to my published appeal in the *News,* asking for anyone recalling the event to come forward with details that would confirm the record. The final report is that of a live manatee seen by two fishermen, Dominick Ciaccio and Joe Terranova, in Lake Saint Catherine near Grand Point, on 17 May 1943. The animal surfaced several times not far from their boat, once within 20 feet. An account of the incident appeared in the old and now defunct *New Orleans Item* in its issue of 18 May 1943.

Records of manatees from coastal Texas, especially in the lower Laguna Madre, are somewhat more numerous, but the species is casual even there. The source of manatees on the northern Gulf Coast is possibly by way of Mexico, since they are known from Veracruz and Tamaulipas. The species is intolerant of low temperatures, and even in Florida its numbers are often seriously reduced as a consequence of occasional spells of cold weather. Consequently, manatees that wander northward along the coast are probably decimated by the freezing weather associated with cold fronts that periodically push down into the Gulf coastal region.

The first recorded observation of the West Indian Manatee in the New World was by Christopher Columbus, who saw it along the coast of the Dominican Republic on his first voyage. On his return to Spain the Admiral presented his original journal to King Ferdinand and Queen Isabella, and a copy of it was made by court scribes, but now only subsequently revised and annotated versions are extant. The following excerpt is from a translation (1960) of the Bartolomé de las Casas version that was based on one of the original copies of the Admiral's logbook:

Wednesday, January 9 [1493]...The day before, when the admiral went to the Rio del Oro [probably the Río Yaque del Norte, Dominican Republic, *fide* Hartman, 1969], he said that he saw three sirens, who rose very high from the sea, but they were not as beautiful as they are depicted, for somehow their faces had the appearance of a man. He says that on other occasions he saw some in Guinea.

These huge, blimplike mammals, although somewhat brutish in appearance, are completely harmless. They are said to be about as dangerous as a contented cow. In Florida's warm springs, skin divers swim among the manatees with impunity. Some individuals are wary, others come up to the divers soliciting a back scratch. Those that avoid humans have every reason to do so, for the species has been severely persecuted, as have all members of the order. The flesh is said to be delicious, the hide makes excellent leather, and the oil derived from its blubber serves several purposes, including its use in cooking. Although now rigidly protected in Florida, with poachers and even molesters subject to a $500 fine, manatees are still killed occasionally with spearguns by unscrupulous skin divers. Even more serious is the great number that are fatally injured by the propellers of the host of motorboats that now churn the waters of the manatee's domain.

Manatees sometimes gather in small herds of 15 to 20 individuals. As many as 17 bulls have been seen following a cow in heat (Hartman, 1969). Manatees can remain submerged for as much as 16 minutes, but normally they surface for two or three breaths of air every 5 minutes or less. Among themselves they are not particularly sociable animals, but at times they carry on a muzzle-to-muzzle play. Courtship antics involve muzzling and "embracing" a prospective mate. Actual mating usually takes place in shallow water. The gestation period is said to last for 180 days. After parturition the calf is immediately escorted to the surface to breathe and it remains there for approximately 45 minutes. Afterwards the bull and the cow alternate in taking the calf to the surface. Each time it goes down it remains longer and longer, and in approximately four hours it is surfacing to breathe on its own.

Predators and other decimating factors.—Manatees have few enemies, if any, other than man. Sharks and crocodiles possibly pose some hazard to infant manatees in the tropical portions of the animal's range, but in United States waters these predators are probably of no great importance. Although the species is rigidly protected in Florida, a few are still said to be wantonly slaughtered from time to time by unconscionable individuals. As previously noted, the propellers of motorboats frequently cause lacerations of a manatee's thick hide and these injuries sometimes prove fatal. Perhaps the greatest decimating factor operating against the species in the northern part of its range is a night or two of freezing or near-freezing weather (Cahn, 1940), which is said to cause pneumonia leading to death. In Florida, manatees often migrate upstream to winter in the vicinity of warm springs.

Subspecies

Trichechus manatus subspecies?

[*Trichechus*] *Manatus* Linnaeus, Syst. nat., ed. 10, 1, 1758:34; type locality, restricted to West Indies by Thomas (Proc. Zool. Soc. London, 1911:132).

Hatt (1934) recognized two geographical races in this species: *T. m. manatus* Linnaeus (1758), occurring on the coasts and in coastal rivers of the West Indies and the adjacent mainland from southern Veracruz to northern South America, and *T. m. latirostris* (Harlan, 1824), occurring along the coast and in coastal rivers of the southeastern United States from North Carolina southward to southern Florida and westward along the northern Gulf Coast to Texas. The two alleged races are, however, of doubtful validity.

The one character that Hatt believed to be diagnostic was the ratio of the greatest vertical diameter of the foramen magnum to its greatest horizontal diameter. He stated that in Florida specimens examined by him the indices ranged from 0.54 to 0.61. In Guatemalan examples the corresponding indices ranged from 0.66 to 0.71, and in a Puerto Rican manatee the index was "something over 0.70." More recently, Moore (1951) has shown that eight manatees from one locality in Florida (Jupiter, Palm Beach County) have indices varying from 0.59 to 0.74, which seem to negate the usefulness of this character in separating *manatus* from *latirostris*. He further pointed out that two Texas specimens from which he was able to obtain measurements possessed indices of 0.515 and 0.765, again demonstrating a wide range of variability in this character. He gave the horizontal and vertical dimensions of the one specimen from Louisiana as 60.5 and 36.1, respectively, with the ratio being 0.597. My own measurement of the vertical diameter of the foramen magnum of this particular specimen is 32 mm (instead of 36.1), which makes the aforementioned ratio 0.529. Paradoxically, despite Moore's contention that the character is unreliable, he still referred the Texas material to *manatus* and argued that the occasional manatee found on the Texas coast represents an extension northward from the coasts of Central America and not an extension northward and westward from Florida. Since the species is extremely rare on the northern Gulf Coast anywhere from eastern Texas to the upper coast of the Florida Peninsula yet is regularly present on the coasts of Veracruz and Tamaulipas, I have no doubt that Texas and Louisiana occurrences represent individuals that have wandered northward from the Gulf and Caribbean coasts of southern Mexico and Central America. But my conclusion is based on zoogeographical probabilities and not on any subspecific differences that may or may not be displayed by the specimens available.

Specimens examined from Louisiana.—Total 1, as follows: *Cameron Par.:* Calcasieu Lake (USNM).

HOOFED MAMMALS WITH AN EVEN NUMBER OF TOES
Order Artiodactyla

THIS ORDER was once represented in Louisiana by three species, two of which, the Wapiti and the Bison, have been extirpated, leaving only the White-tailed Deer presently remaining. A specimen of a fourth species, the Collared Peccary, or javelina *(Tayassu angulatus)* was found dead in De Soto Parish on 3 March 1953 on the side of a highway three miles southwest of Mansfield on the road to Logansport. It had presumably been killed by an automobile and probably represented an individual brought into the state by man rather than one that found its way here by its own efforts.

In the artiodactyls, or the ungulates with an even number of toes, the main axis of the limbs passes down the center of the cannon bone and hence between the two median digits, the third and fourth. The first digit is always absent and the second and fifth are either vestigial or else in most forms not used in walking. The body weight is borne mainly or wholly by the third and fourth toes. In the perissodactyls, or ungulates with an odd number of toes, the weight is borne mainly by the third toe on each foot, through the center of which the main axis of the limb passes (Figure 224). The Perissodactyla include the horses, tapirs, and rhinoceroses, while the

Artiodactyla comprise the pigs, peccaries, hippopotamuses, camels, chevrotains, deer, giraffes, okapis, pronghorns, sheep, goats, cattle, and antelopes.

The order contains nine Recent families and some 75 genera and is worldwide except for Australia, New Zealand, Antarctica, and

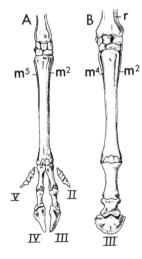

Figure 224 Bones of the left forefoot of (A) an artiodactyl (a deer, *Cervis* sp., ½ ×) and (B) a perissodactyl (a horse, *Equus caballus*, ⅛ ×): *r*. radius; m^2 to m^5, second to fifth metacarpals; *II–V*, second to fifth fingers. (Drawings adapted from Beddard, 1902)

oceanic and many other islands. The upper incisors are reduced or absent, as the canines also usually are, but the latter in some forms may be enlarged and tusklike. Frontal bone appendages, that is, antlers or horns, occur in many forms, including all three artiodactyls that have occurred in Louisiana. The stomach in the pigs, peccaries, and hippopotamuses is two- or three-chambered and nonruminating; in the camels and chevrotains it is three-chambered and ruminating; in the remaining families it is four-chambered and ruminating. The ruminants rapidly consume great amounts of grass and other vegetable matter without much chewing of it, and later regurgitate it into the mouth, where it is chewed again, this time more thoroughly, and again mixed with saliva before being reswallowed. It goes again into the first chamber, or rumen, before passing into the second chamber, or reticulum, thence into the third chamber, the omasum, and finally into the fourth chamber, the abomasum, the true digestive stomach. The course of the food through the four cham-

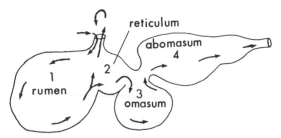

Figure 225 The complicated four-chambered stomach of ruminants. Arrows show the course of food from the time it is first swallowed until it is returned to the mouth for further mastication and then again swallowed to pass through the four chambers of the stomach and the remainder of the digestive tract. Deer chew their cuds as do domestic cattle and the Bison.

bers is shown in Figure 225. Cud-chewing by ruminants provides these animals with important advantages. They can quickly eat a great quantity of nutritionally low-grade food and then retire to a relatively safe place to rechew it at leisure and thereby allow their digestive juices the time required to convert it into usable nourishment.

The Deer Family
Cervidae

SIXTEEN GENERA and at least 37 Recent species make up this family, but some authorities recognize additional species. The group is composed of small to extremely large, long-legged artiodactyls. The smallest species is the Musk Deer (*Moschus moschiferus*), the largest, the Moose (*Alces alces*). Only two species lack antlers, the Musk Deer and the Chinese Water Deer (*Hydropotes inermis*), both of Asia. In the Caribou (*Rangifer tarandus*) both sexes acquire racks. Otherwise only male cervids possess antlers. The antlers arise from a bony pedicel, but during their growth and development they are well supplied with blood and are at first soft and tender, being covered with a thin skin and short hairs that resemble velvet. When the antlers have attained their full growth, the blood supply recedes, the velvety hair dries and loosens, and the animal rubs it off. The antlers are shed each year following the rut. The stomachs of cervids are four-chambered and ruminating, and the gall bladder is absent except in the Musk Deer. In three genera (*Moschus, Cervulus* [= *Muntiacus*], and *Elaphodus*), the upper canines are present and are saberlike tusks. In other cervids the upper canines are usually lacking. The lower canines resemble the incisors and are often confused with them. Nearly all deer have facial glands, located in a depression in front of the eyes, as well as tarsal, metatarsal, and interdigital glands.

Genus *Cervus* Linnaeus

Dental formula:

$$\text{I}\frac{0-0}{3-3}\ \text{C}\frac{1-1}{1-1}\ \text{P}\frac{3-3}{3-3}\ \text{M}\frac{3-3}{3-3}=34$$

WAPITI   Figure 226
(*Cervus canadensis*)

Vernacular and other names.—The name Wapiti comes from an Algonquian Indian word meaning white or whitish, and refers to the animal's white rump and tail. The animal is also called the American elk, but the name elk

Figure 226 Wapiti (*Cervus canadensis*), 30 miles south of Mammoth, Yellowstone National Park, Wyoming, September 1971. (Photograph by C. C. Lockwood)

in Europe is applied to what we call a Moose (*Alces alces*). The generic name *Cervus* is the Latin word for deer, and the specific name *canadensis* means belonging to Canada, the place from which the species first became known to zoologists.

Distribution.—The species formerly ranged from northern British Columbia across most of southern Canada to eastern Ontario and northern New York, thence south to northwestern California, southeastern Oregon, northeastern Nevada, the mountains of Arizona and New Mexico, southwestern Oklahoma, northern Louisiana, and Tennessee, and in the Allegheny Mountains to northern Georgia. The range is now greatly contracted, the species occupying only a fraction of the area it occupied at the time of early colonization. The basis for the inclusion of the Wapiti on the Louisiana list is detailed beyond. (Map 68)

External appearance.—This large deer is easily distinguished from the White-tailed Deer by

Map 68 Distribution of the Wapiti (*Cervus canadensis*). *Shaded areas,* and •, the two localities at which the species has been reliably recorded in the state; *inset,* overall range of the species.

its greater size, the immense rack of antlers in the male, and the large yellowish white rump patch. With the exception of the Moose *(Alces alces)*, the Wapiti is the largest of North American deer in weight. In total length, from the tip of the nose to the tip of the tail, it averages larger than the Moose. The neck is heavily maned, and a ruff extends posteriorly between the shoulders. The head and neck, the legs, and the underparts are dark brown, sometimes chestnut on the mane; these parts are much darker than the back and sides. Males possess large, widely branched antlers, the main beam of which sweeps backward and is sometimes four feet in length, having a well-developed brow tine and normally five other tines. The brow tine forms an obtuse angle with the main beam.

Measurements and weights. — The total length of adult males ranges from 6.7 to 9.8 feet; the tail is between 5 and 8 inches long, and the shoulder height is 4.5 to 4.8 feet. Weights of males average about 650 pounds, but Wapitis are on record that topped 1,000 pounds. Females are considerably smaller.

Cranial characters. — The skull of the Wapiti is easily distinguished from that of the White-tailed Deer by its larger size, and by the presence of a pair of prominent upper canines, which are normally absent in *Odocoileus* (Figure 227). In the male Moose *(Alces alces)* the antlers are broadly palmate, and in the Caribou *(Rangifer tarandus)* both sexes are antlered and racks of the male are immense with the brow tine on one side, sometimes both sides, somewhat palmate and projecting forward; the bez tine is rarely also semi-palmate distally.

Status. — I have been able to find only two references to the occurrence of the Wapiti, or American elk, in Louisiana. Murie (1951) quotes a letter received by him from Dr. Milton Dunn from Colfax, Grant Parish, dated 2 June 1918 that stated: "We had elk

here when this part of Louisiana was sectionized by surveyors in 1829." Murie also quotes a letter to the editor of the *Spirit of the Times* (H.S.D., 1843:555) in which the killing of a bull elk near Mounds, in Madison Parish, in December 1842 is recorded. The animal was shot between Roundaway Bayou and the Mississippi River. It weighed 704 pounds and measured 11 feet from the tip of the nose to the end of the hind foot. The height at the

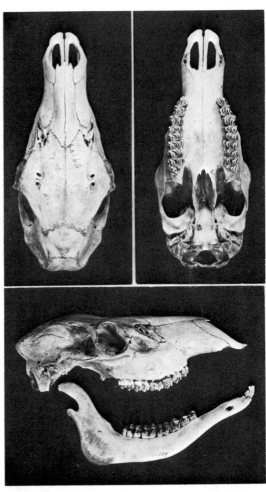

Figure 227   Skull of Wapiti *(Cervus canadensis)*, ♀, Univ. Michigan Mus. Zool. 59189, 30 mi. S Manila, Daggett County, Utah, 5 August 1928. Total length of skull, 435 mm (17.12 in.). (Photograph by Joe L. Herring)

withers was 5 feet 4 inches. The length of antlers was stated to be 4 feet 3 inches.

Although the Wapiti did occasionally extend as far south as Louisiana, it was probably never widespread nor common anywhere within our borders. Twenty were introduced near Urania, in La Salle Parish in February 1916, but the stock died out without establishing a herd.

Subspecies

Cervus canadensis canadensis Erxleben

[*Cervus elaphus*] *canadensis* Erxleben, Syst. regni animalis..., 1777:305; type locality, eastern Canada [= Quebec].

Assignment of Wapitis that wandered into Louisiana in days gone by to the race *canadensis* is done purely on the ground of geographical probability, since unfortunately no skeletal material is available on which to base an informed judgment. A rather remote possibility exists, however, that the now extinct *C. c. merriami*, which once occurred in the Wichita Mountains of Oklahoma, may have also extended still farther east into Arkansas and Louisiana.

Some mammalogists consider the Wapiti conspecific with the Red Deer (*Cervus elaphus* Linnaeus) of Europe. If their opinion should prevail, *Cervus elaphus* would take precedence over *C. canadensis* Erxleben as the name of the species.

Specimens examined from Louisiana. — None.

Genus *Odocoileus* Rafinesque

Dental formula:

$$I\frac{0-0}{3-3} \ C\frac{0-0}{1-1} \ P\frac{3-3}{3-3} \ M\frac{3-3}{3-3} = 32$$

White-tailed Deer   Plate XIV
(*Odocoileus virginianus*)

Vernacular and other names. — The word deer comes from the Anglo-Saxon *deor* or *dior* and originally meant any wild animal. The specific name White-tailed refers, of course, to the white undersurface of the tail, which the animal flashes when it bounds away. In Louisiana the species is often called simply the whitetail. Our French-speaking inhabitants refer to it as *le chevreuil*. The generic name *Odocoileus* is derived from two Greek words, *odon*, tooth, and *koilos*, hollowed. The cavernous depression in the crown of each unworn molar may have suggested the combination to Rafinesque, who named the genus. The specific name *virginianus* means of Virginia, the area from which the species first came to the attention of the zoologist who named it. The male deer is referred to as a buck, the female as a doe, and the young, while still in the spotted coat, as fawns.

Distribution. — The species is found in southern Canada and throughout most of the United States except for nearly all California, most of Nevada and Utah, and parts of western Colorado, northwestern New Mexico, and northern Arizona. It ranges southward over all Mexico, except Baja California, thence through Central America to South America. In Louisiana the species occurs statewide in suitable habitat. (Map 69)

External appearance. — The White-tailed Deer is so well known that it requires little description. Among Louisiana mammals it is immediately recognized by its long, spindly legs, hoofed toes, moderately short tail that is white beneath, and the presence of antlers in males during part of the year. The winter coat is grayish brown above. The underparts, the inside of the legs, and the chin and upper throat are white. There is also an area of white around the eyes and across the rostrum. In summer the upperparts are decidedly reddish.

Color phases. — No true color phases are known. Albinos, which have pink eyes, occur infrequently, as do also "partial albinos," ani-

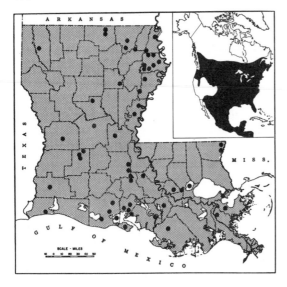

Map 69 Distribution of the White-tailed Deer *(Odocoileus virginianus). Shaded area,* known or presumed range within the state; •, localities from which museum specimens have been examined; *inset,* overall range of the species.

mals blotched with white. "Black-eyed whites" have been reported, but this mutation is rare. Cases of melanism are even rarer (Severinghaus and Cheatum, 1956). Cream-colored individuals with pink eyes are seen with surprising regularity in the coastal marshes, notably near Gum Cove, in Cameron Parish.

Measurements and weights. — The total length of the adult male whitetail in Louisiana ranges from 65 to 78 inches; the length of the tail is 9 to 10 inches; the height at the shoulders, 34 to 38 inches; the length of the hind foot, approximately 16 inches. The weight is usually in the neighborhood of 130 pounds but varies according to age and the condition of available food resources. The skulls of eight adult males in the LSUMZ from various localities in the state averaged as follows: condylobasal length, 261 (244–280); cranial breadth, 83.5 (78.3–97.0); zygomatic breadth, 107.0 (100.2–111.8); least interorbital breadth, 62.4 (52.4–72.7); width of orbit at frontojugal suture, 38.5 (35.7–41.0); length

of maxillary toothrow, 80.4 (75.3–84.7). The skulls of five adult females from the same source averaged as follows: condylobasal length, 238 (230–252); cranial breadth, 72.2 (69.7–76.5); zygomatic breadth, 98.5 (93.1–105.7); interorbital breadth, 54.7 (50.1–59.3); width of orbit at frontojugal suture, 37.4 (34.1–39.0); length of maxillary toothrow, 72.0 (67.7–76.5). Noble (1969) found that the mean weight of 152 adult does, 1.5 years or more in age, taken in November and December in batture habitat in Bolivar and Washington counties, Mississippi, was 108 pounds. The mean weight of 112 does of similar age taken in February, March, and April in the same region was 107 pounds, hence not significantly different from the earlier sample. The mean weight of 164 does, 1.5 years old or older, from the longleaf pine belt of Mississippi was only 76 pounds, clearly demonstrating the effect of low-grade habitat on deer weights.

Cranial characters. — The skull of the White-tailed Deer is easily identified by the presence of antlers in the male, by the shallow but prominent lacrimal fossae, by the absence of upper incisors and usually the absence of upper canines, and by the long diastema between the lower canines and the first molariform teeth of the lower toothrows (Figures 228 and 229). The antlers of the White-tailed Deer differ from those of the Mule Deer *(O. hemionus)* in having the main beam first directed backward then curved forward and outward, with the prongs usually unbranched and all appearing to rise vertically from the main beam. In the blacktail the main beam branches into two nearly equal parts, each with nearly equal dichotomous branches (Figure 230).

Plate XIV
Male and female White-tailed Deer *(Odocoileus virginianus)* in a moss-draped baldcypress swamp with Ivory-billed Woodpecker *(Campephilus principalis)* in background.

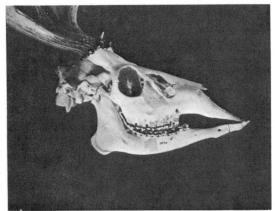

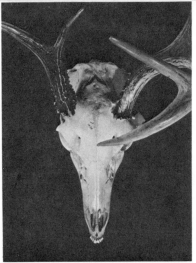

Figure 228   Skull of White-tailed Deer *(Odocoileus virginianus mcilhennyi)*, LSUMZ 1070, Belair, Plaquemines Parish. Note asymmetry of the rack. (Photograph by Joe L. Herring)

Figure 229   Skull of White-tailed Deer *(Odocoileus virginianus)*, LSUMZ 1980, showing supernumerary teeth, consisting of a pair of small canines. (Photograph by Joe L. Herring)

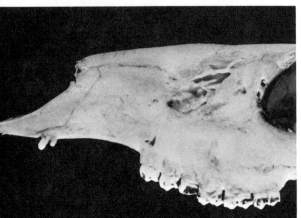

Sexual characters and age criteria. — In addition to the presence of antlers in the male and their usual absence in the female, considerable sexual dimorphism in size is evident in the White-tailed Deer. The female averages about two-thirds the size and weight of the male. During the rut the neck of the fully adult male, and to a lesser extent that of the spike buck, becomes swollen, but never does this occur in the female. The fawn is reddish brown above with numerous white spots. This spotted pelage is retained until an age of about four or five months, when the first winter pelage is assumed through the fall molt. One cannot tell how old an adult male is by the size of the antlers and the number of points, for nutritional factors drastically influence the growth and development of the rack. In other words, the quality and quantity of forage are factors of greater importance in determining the beam diameter and the number of points in a given rack than the animal's age. Old males frequently carry antlers that are considerably smaller than those of vigorous young males and have fewer points.

Age during the first 13 months can be estimated by the extent of dental development (Severinghaus, 1949; Schwartz and Schwartz, 1959). From birth to the end of the first week, the dentition is limited to four incisiform teeth on each side of the lower jaw.

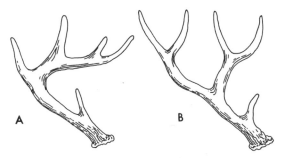

Figure 230   Antlers of (A) White-tailed Deer *(Odocoileus virginianus)* and (B) Mule Deer *(O. hemionus)*, showing the manner in which the branches of the latter fork dichotomously with the prongs about equal in size and length.

Additional teeth appear in the following or-der: two premolars in the second to fourth week; a third premolar in the fourth to tenth week; the first molar between the 10th week and the seventh month; another molar in the seventh to 13th months. A deer at 13 months of age or when a little older attains the full complement of four incisiform teeth (three incisors and one canine), three premolars, and three molars on each side of the lower jaw. The amount of wear, the progress of eruption, and certain structural features of the teeth themselves in the lower jaw provide excellent characters by which deer can be aged with a high degree of accuracy within the first 18 months of their lives. Between the fourth and seventh months the first perma-nent molar on each side erupts, and, if the temporary incisors have not by this time been replaced, they will be by the end of the ninth month. Of special importance is the third temporary lower premolar, which is retained until approximately the 18th month. The crown of this milk tooth has three cusps,

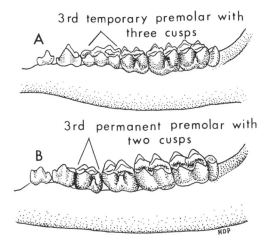

Figure 231 Cheek view of lower jaws of White-tailed Deer *(Odocoileus virginianus).* (A) Deer 18 months old or slightly less, showing the third pre-molar with three cusps on the crown in contrast with a deer (B) over 19 months old in which the third temporary premolar has been replaced by the permanent P$_3$ in which the crown has only two cusps.

while that of its permanent replacement has only two cusps (Figure 231).

By the end of the 19th month all six per-manent molariform teeth have erupted, but the third premolar and the third molar may not have attained their full height above the gum line. After the 20th month a deer can be aged by the differential amount of wear to the various cheek teeth. For instance, a deer 5½ years old will show the premolars only mod-erately worn and the lingual crests of the first and second molars worn away, while a deer that is 8½ years old will have the premolars and molars worn to within 1/16 inch of the gum on the cheek side and 3/16 inch on the lingual side (Schwartz and Schwartz, 1959). Another method of estimating age in older deer is by a histological examination or by gross inspection of the annular layers in the cementum of the teeth (Low and Cowan, 1963; Ransom, 1966; Gilbert, 1966; Lockard, 1972).

Status and habits. — The White-tailed Deer is both abundant and widespread in Louisiana. The present population in the state is esti-mated to exceed a quarter of a million indi-viduals, with the trend being steadily upward. According to information gathered by the Louisiana Wild Life and Fisheries Commis-sion, the number of deer killed by hunters in the state during the 1971–1972 season amounted to 60,923 (Roger Hunter, *in litt.*). Only one parish, Orleans, reported none taken. Parishes reporting the highest kills were Concordia (6,553), Tensas (5,481), Bi-enville (3,084), Union (2,798), Madison (2,556), Jackson (2,344), and Caldwell (2,020). No less than 12 additional parishes (Allen, Avoyelles, Bossier, Claiborne, De Soto, East Carroll, Iberville, Natchitoches, Pointe Coupee, Vermilion, Vernon, and Winn) recorded the taking of over 1,000. Some of the state's game management areas and some areas in certain parishes are so overpopulated with deer that the killing of does is not only allowed but encouraged as an absolutely essential device for keeping numbers in line with the carrying capacity of the habitat.

The present overall healthy condition of our deer population (Newsom, 1969) is attributable to the excellent management practices carried out by the state's corps of highly trained wildlife technicians. Under the wise supervision and attention that is being given our deer herds, they should continue to flourish and to promote indefinitely the kind of wholesome outdoor recreation for legions of hunters that only the whitetail can provide. But despite the apparent overall good condition of our deer population, the situation could change, and it could do so quite rapidly. Overpopulation leads to overbrowsed habitat, followed by food shortages, nutritional deficiencies, weight losses, reproductive failures, diseases, heavy parasite infestations, and other deleterious effects that often result in drastic die-offs. Serious crop depredations and damage to natural vegetation also occur when deer populations become excessive. The only way to prevent these consequences in most areas is to increase hunting pressure, including opening the season on does. The trapping of large numbers of the deer in overpopulated areas for use in restocking other sections of the state is not only an expensive operation but one for which there is no currently pressing need.

The antlers of the White-tailed Deer are formed and shed each year. Sometime in the period of late December to early April, but in any event following the rut, the shedding of antlers takes place. Within six or eight weeks the growth of new antlers begins, starting out from "buds" located on the frontal bones of the skull and covered with soft skin richly supplied with blood vessels that transport the calcium, phosphorus, and other building materials from which the antlers are made. From the time the antlers first begin to grow until they are fully developed, they are covered with skin that has a plushlike texture;

hence the name "velvet" has been applied to it. By late September the blood supply to the antlers is shut off and they begin to harden. The velvet begins to dry and to peel off rapidly, a process in which the buck assists by rubbing the antlers against limbs of trees and woody plants. Finally there remains only the hard, bony core, which the buck polishes by additional rubbing. The maturation of the antlers is stimulated by the male sex hormone called testosterone, which also effects other changes in the buck's reproductive physiology; so soon the rut is on. Because of hormonal imbalances, females have been known to develop spikes or small antlers, but this happens only rarely.

Figure 232 Native White-tailed Deer in a large fenced enclosure near Ferriday, Concordia Parish, fall of 1972. (Photograph by Lloyd Poissenot)

Breeding takes place in the whitetail from late September to early March. Roberson (1967) found considerable variation in the time when the most matings occurred in various herds in different parts of the state. Those on Avery Island peaked in October (38.5 percent between 15–31 October). Herds on the Evangeline Game Management Area in Rapides Parish and on the West Bay Game Management Area in Allen Parish also showed peak activity during the last two weeks of October, but those on the Red Dirt Game Management Area in Natchitoches Parish did not do so until the first two weeks of November, and those on the Jackson Parish–Bienville Parish Game Management Area not until the first two weeks of December. Paradoxically, the herd on the Delta National Wildlife Refuge, near the mouth of the Mississippi River, did not reach a breeding peak until the last two weeks of December, while the herd in Tensas Parish, at the opposite end of the state, peaked during the first two weeks of January, with some mating taking place considerably later. Roberson could offer no explanation that would account for this wide variation in the breeding seasons of the whitetail herds in the state. He did suggest, however, the possibility that the late breeding in the Delta Refuge and the Tensas Parish herds could be an adaptation to allow the fawns to be dropped when water levels are generally low.

Contrary to a general belief, the doe is not in heat at all times, merely awaiting the attention of an interested male. She is indeed polyestrous, but she comes into heat only four or five times a year, each for a period of about 24 hours (Severinghaus, 1955). In the event that she fails to encounter a male in this brief interval, she cannot be fertilized for another lunar month. Males are likewise not in breeding condition the year around. The shedding of the antlers coincides with a drastic drop in the testosterone levels, and at the same time spermatogenesis, or the production of sperm, ceases. Hunters are sometimes surprised when they observe spotted fawns in late January, but these are probably nothing more than fawns dropped by females that bred as late as January to March the previous year rather than over a year previous. The gestation period is 195 to 212 days in duration, the average being 202 days.

Young females generally produce only a single fawn, but most medium-aged females give birth to twins. Triplets and quadruplets are rare but do sometimes occur (Trodd, 1962). Where the range is poor or where overcrowding occurs, even old females tend to produce single offspring rather than twins.

The average age attained by male Louisiana deer in areas where they are heavily hunted is probably less than 2½ years. Although deer in captivity have been known to reach ages well in excess of 10 years, a deer in the wild that survives past 7 years is exceptional. Noble (1969), in his excellent study of reproduction in the whitetail in Mississippi, found only one female in 324 specimens collected that was over 10 years of age. He also found that does achieved their maximum weight, their best overall physical condition, and their greatest breeding potential at an average age of 4½ years.

The whitetail possesses three prominent

Tracks of White-tailed Deer

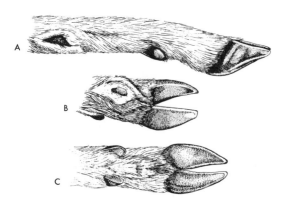

Figure 233 Foot glands of the White-tailed Deer: (a) metatarsal gland on the outside of the right hind foot; (b) top view of the right hind foot, showing the location of the pedal gland between toe 3 and toe 4; (c) ventral view of right hind foot. The tarsal gland, not illustrated, is located on the inside of the hind legs, slightly below the hock. (After Schwartz and Schwartz, 1959.)

sets of scent glands, the locations of which are easily discernible externally: the metatarsal glands, on the outside of the lower hind legs; the tarsal glands, on the inside of the hind legs at the hocks; and the interdigital glands, between the toes on all four feet (Figure 233). In addition, preorbital glands, which serve to lubricate the eyes, are to be found in the inner corners of the eyes. The scent glands exude an oily substance, which has a pungent, musky odor. Whitetails, and more frequently Mule Deer, urinate on the tarsal glands and thereby, it is said, increase their effectiveness. Hunters skinning a deer are cautioned to get rid of the metatarsal and hock glands as quickly as possible and not to rub their hands over that part of the limbs where the glands are located and then touch the meat. Should this happen, musky-smelling venison is likely to result. The scent glands are extremely important to whitetails in helping them find one another and retrace their steps as a doe might do in returning to a hidden fawn. Deer depend first on their sense of smell, secondly on hearing, and lastly on sight.

The color of antlers varies from almost immaculate ivory white to rich, mahogany brown. Some hunters believe that antlers are stained by blood during the velvet stage. Others contend that the color is acquired from plant juices when deer rub their antlers on certain trees and shrubs when the velvet is being shed. Both factors are probably involved to one extent or another, but I am inclined to think the first is the main agency responsible for the brown coloration. Pen-raised deer often have richly stained antlers even though they have no access to trees and shrubs on which to rub them while shedding the velvet.

Hunters are likely to be puzzled when they kill bucks during the hunting season with the antlers still in velvet. The shedding of the velvet follows the shutting off of the blood supply to the growing antlers and, according to Wislocki et al. (1947), is coincident with an abrupt decline in testosterone levels. When these levels remain high because of an endocrine malfunction, growth of the antlers continues. The proliferation of points is believed to be a consequence of this extra growth. The most frequently encountered number of points on a rack is 7 or 8, but antlers with a greater number are not uncommon. Severinghaus and Cheatum (1956) estimate that one buck in about 5,000 has 17 or more points.

The Boone and Crockett Club, with headquarters in the Carnegie Museum, Pittsburgh, Pennsylvania, has long maintained records of outstanding big game trophies. Employing rigid sets of rules and instructions (Figure 234) that govern a system of ratings, the club periodically publishes a list of the heads receiving the highest scores. The latest edition of this work, entitled *North American Big Game* and edited by a Committee on Records, appeared in 1971. Whitetails are categorized either as typical or nontypical, the latter having racks with a large number of points that are abnormally shaped and located in aberrant positions. Normal points are those arising from the main beam; abnor-

OFFICIAL SCORING SYSTEM FOR NORTH AMERICAN BIG GAME TROPHIES

RECORDS OF NORTH AMERICAN BIG GAME COMMITTEE

BOONE AND CROCKETT CLUB

Boone and Crockett Club
Records of North American Big Game Committee
c/o Carnegie Museum
4400 Forbes Ave. Pittsburgh, Pa. 15213

Minimum Score:	Deer
Whitetail:	Typical 170
Coues:	Typical 110

WHITETAIL and COUES DEER

KIND OF DEER

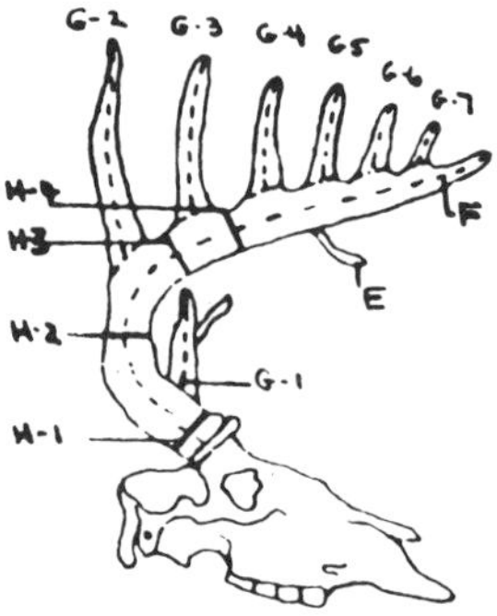

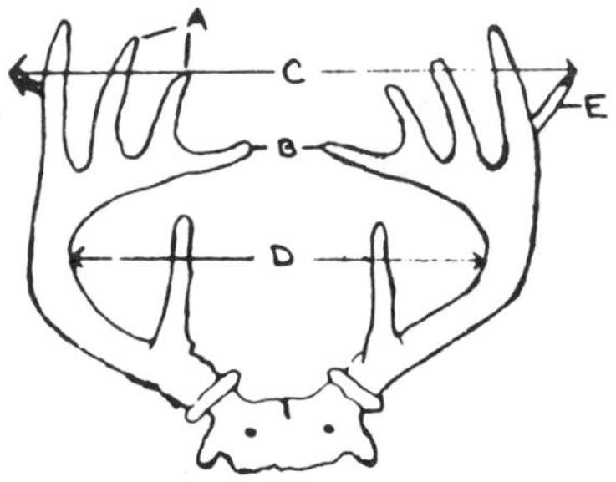

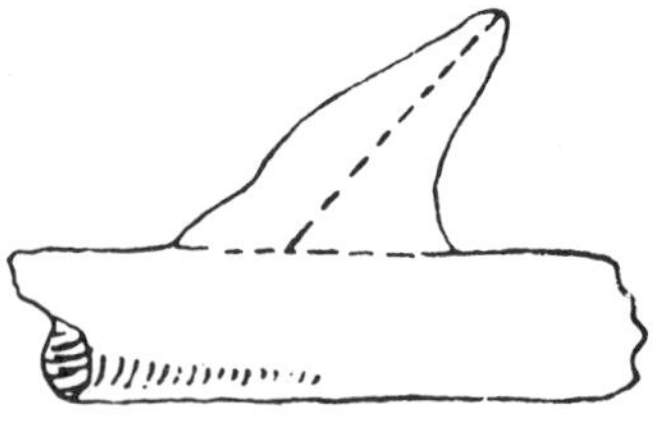

DETAIL OF POINT MEASUREMENT

SEE OTHER SIDE FOR INSTRUCTIONS	Supplementary Data R.	Supplementary Data L.	Column 1 Spread Credit	Column 2 Right Antler	Column 3 Left Antler	Column 4 Difference
A. Number of Points on Each Antler						
B. Tip to Tip Spread						
C. Greatest Spread						
D. Inside Spread of MAIN BEAMS — Spread credit may equal but not exceed length of longer antler						
IF Inside Spread of Main Beams exceeds longer antler length, enter difference						
E. Total of Lengths of all Abnormal Points						
F. Length of Main Beam						
G-1. Length of First Point, if present						
G-2. Length of Second Point						
G-3. Length of Third Point						
G-4. Length of Fourth Point, if present						
G-5. Length of Fifth Point, if present						
G-6. Length of Sixth Point, if present						
G-7. Length of Seventh Point, if present						
H-1. Circumference at Smallest Place Between Burr and First Point						
H-2. Circumference at Smallest Place Between First and Second Points						
H-3. Circumference at Smallest Place Between Second and Third Points						
H-4. Circumference at Smallest Place between Third and Fourth Points or half way between Third Point and Beam Tip if Fourth Point is missing						
TOTALS						

ADD	Column 1		Exact locality where killed
	Column 2		Date killed By whom killed
	Column 3		Present owner
	Total		Address
SUBTRACT Column 4			Guide's Name and Address
FINAL SCORE			Remarks: (Mention any abnormalities)

Figure 234 Facsimile reproduction of Boone and Crockett Club White-tailed Deer scoring sheet and instructions. (Reproduced courtesy of Boone and Crockett Club)

I certify that I have measured the above trophy on_____ 19 _____
at (address)_____ City _____ State _____
and that these measurements and data are, to the best of my knowledge and belief, made in accordance with the instructions given.

Witness: _____ __ _____ Signature: _____ _____

Boone and Crockett Official Measurer

INSTRUCTIONS

All measurements must be made with a flexible steel tape to the nearest one-eighth of an inch. Wherever it is necessary to change direction of measurement, mark a control point and swing tape at this point. To simplify addition, please enter fractional figures in eighths. Official measurements cannot be taken for at least sixty days after the animal was killed. Please submit photographs.

Supplementary Data measurements indicate conformation of the trophy, and none of the figures in Lines A, B and C are to be included in the score. Evaluation of conformation is a matter of personal preference. Excellent, but nontypical Whitetail Deer heads with many points shall be placed and judged in a separate class.

A. Number of Points on each Antler. To be counted a point, a projection must be at least one inch long AND its length must exceed the length of its base. All points are measured from tip of point to nearest edge of beam as illustrated. Beam tip is counted as a point but not measured as a point.

B. Tip to Tip Spread measured between tips of Main Beams.

C. Greatest Spread measured between perpendiculars at right angles to the center line of the skull at widest part whether across main beams or points.

D. Inside Spread of Main Beams measured at right angles to the center line of the skull at widest point between main beams. Enter this measurement again in "Spread Credit" column if it is less than or equal to the length of longer antler.

E. Total of lengths of all Abnormal Points. Abnormal points are generally considered to be those nontypical in shape or location.

F. Length of Main Beam measured from lowest outside edge of burr over outer curve to the most distant point of what is, or appears to be, the main beam. The point of beginning is that point on the burr where the center line along the outer curve of the beam intersects the burr.

G-1-2-3-4-5-6-7. Length of Normal Points. Normal points project from main beam. They are measured from nearest edge of main beam over outer curve to tip. To determine nearest edge (top edge) of beam, lay the tape along the outer curve of the beam so that the top edge of the tape coincides with the top edge of the beam on both sides of the point. Draw line along top edge of tape. This line will be base line from which point is measured.

H-1-2-3-4. Circumferences - If first point is missing, Take H-1 and H-2 at smallest place between burr and second point.

* * * * * * * * * *

TROPHIES OBTAINED ONLY BY FAIR CHASE MAY BE ENTERED IN ANY BOONE AND CROCKETT CLUB BIG GAME COMPETITION

To make use of the following methods shall be deemed UNFAIR CHASE and unsportsmanlike, and any trophy obtained by use of such means is disqualified from entry in any Boone and Crockett Club big game competition:

    I. Spotting or herding game from the air, followed by landing in its vicinity for pursuit;
    II. Herding or pursuing game with motor-powered vehicles;
    III. Use of electronic communications for attracting, locating or observing game, or guiding the hunter to such game.

* * * * * * * * * *

I certify that the trophy scored on this chart was not taken in UNFAIR CHASE as defined above by the Boone and Crockett Club.

I certify that it was not spotted or herded by guide or hunter from the air followed by landing in its vicinity for pursuit, nor herded or pursued on the ground by motor-powered vehicles.

I further certify that no electronic communications were used to attract, locate, observe, or guide the hunter to such game; and that it was taken in full compliance with the local game laws or regulations of the state, province or territory.

Date _____ Hunter _____

mal points are those branching from other points. The minimum acceptable score for a "typical" White-tailed Deer is 170. Because of the generally small size of our native deer, few of them qualify as record or near-record heads. Nevertheless, in the latest published tabulation, one typical deer, killed in 1960 by Jerry Loper at Cat Island, West Feliciana Parish, Louisiana, ties for the all-time 290th position with a score of 170. A deer with a nontypical rack, for which the minimum acceptable score is 195, ties for the 113th all-time position with a score of 206$\frac{6}{8}$. It was shot by Richard D. Ellison in Grant Parish in 1969. Another nontypical deer, shot near St. Francisville in 1970, occupies the all-time 141st position with a score of 202$\frac{3}{8}$. (Figure 235)

In those areas of the state where the hunt-ing of does is wisely permitted, sportsmen are sometimes distressed when they kill a female in late December with her udder filled with milk, which in the deer is thick and creamy. The immediate suspicion is that one or two fawns are then about to starve somewhere out in the forest, but such is not necessarily the case. After weaning its young, the female requires two to six weeks to "dry up."

During the rut, the buck moves about extensively in search of females, but at other times of the year his home range is on the average less than 600 acres (Lewis, 1968). The summer range is probably less than 300 acres (Robert E. Noble, pers. comm.). On poor deer range the animals perhaps move farther than in good habitat simply because they must travel about more in search of sufficient food. But in our rich bottomland

Figure 235   Record Louisiana White-tailed Deer heads in Boone and Crockett Club listings: *left,* typical category, score 170, rank 290th, shot by Jerry Loper, Cat Island, West Feliciana Parish, 27 November 1960; *right,* nontypical category, score 206$\frac{6}{8}$, rank 113th shot by Richard D. Ellison, Jr., 3 mi. ENE Fishville, Grant Parish, 23 November 1969. (Photographs by Joe L. Herring)

Figure 236 Two bucks with antlers locked, resulting in the ultimate death of both, Jackson Parish, 15 November 1967. (Photograph by Joe L. Herring)

swamps deer probably do not venture much more than a mile from the spot where they were dropped as fawns.

Bucks in rut often engage in severe competition with other males for mating privileges with a doe being courted. In the fights that take place, the bucks use both the antlers and the hoofs, and there is much pushing and shoving as each combatant awaits a momentary advantage in the struggle. The battles occasionally result in the death of one of the participants or more rarely, in the death of both fighters, when their antlers become inextricably locked together (Figure 236). A particularly aggressive and dominant male will often gather a harem of several females and service all of them until he is displaced by a more vigorous and competitively successful male.

The food plants utilized by deer in Louisiana vary widely according to the particular habitat and what is available. Although deer in the state have been found to eat over 112 kinds of plants, certain species are highly preferred when they are obtainable. Some of the important shrubs and vines eaten are: French mulberry (*Callicarpa americana*), trumpet creeper (*Campsis radicans*), titi (*Cyrilla racemiflora*), strawberry bush (*Euonymus americanus*), yellow jasmine (*Gelsemium sempervirens*), deciduous holly (*Ilex decidua*), yaupon (*Ilex vomitoria*), Virginia willow (*Itea virginia*), Japanese honeysuckle (*Lonicera japonica*), dewberry (*Rubus* spp.), and greenbriar (*Smilax* spp.). Of special importance to deer are oak mast (*Quercus* spp.) and the native wild pecan (*Carya illinoensis*). The absence of preferred species in some deer habitats is sufficient in itself to make these habitats inferior deer range. (Murphy, pers. comm.; Robert E. Noble, pers. comm.)

Predators.—One of the chief factors leading to the overpopulation of the White-tailed Deer in some sections of the state is the absence of effective predation. In primeval days numerous panthers and large packs of wolves took a heavy toll of deer, which tended to improve the vitality of the herds by eliminating substandard individuals. Now that panthers and wolves are scarce, the whitetail has few effective enemies. Bobcats doubtless account for some losses, as do also wild dogs, particularly among fawns and does, but otherwise the whitetail has only man and his hounds with which to contend. In hardwood bottomland swamps such as those in the parishes along the Mississippi River and in the Atchafalaya Basin, deer are able to cope quite effectively with wild dogs and hunting hounds because of the density of the cover and the great number of streams, sloughs, oxbow lakes, and other bodies of water in which they can take refuge and which they can use as avenues of escape. In the more open forests of our uplands, though, dogs are often able to run a deer until it dies of heat exhaustion, pneumonia, shock, or some other condition brought on by the long chase (Glasgow and Noble, 1971). Some wildlife experts believe that constant molestation by dogs often disrupts the rut, has a deleterious effect on breeding success, and results in considerable mortality. Recall also that the doe is in heat for only about 24 hours every 28 days over a span of only four or five months. She cannot afford to have one of her brief periods of estrus devoted to evading a pack of hounds.

Parasites and diseases.—Potentially pathogenic organisms known to occur in the White-tailed Deer include *Taxoplasma gondii, Anaplasma marginata, Brucella abortus, Leptospira* spp., and unidentified species of the genera *Babesia* and *Sarcocystis* (Shotts et al., 1958;

Hayes et al., 1960; Trainer, 1962). Metazoan parasites that have been found in or on the whitetail include 27 roundworms, 6 tapeworms, 5 flukes, 16 ticks, 4 mites, 4 nasal botflies, a screwworm, 1 flea, 2 biting lice, and 2 sucking lice. About 10 species of blackflies and tabanids have been found feeding on deer. Of the ticks, 3 *(Dermacentor nigrolineatus, Amblyomma maculatum,* and *A. cajennense)* are characteristic of southern deer herds (Anderson, 1962; Kellogg et al., 1971; Frank A. Hayes, Southeastern Cooperative Wildlife Disease Study, *in litt.,* 1972). The livers of 60 percent of the deer in the heavily populated bottomlands of Tensas Parish are said to be loaded with liver flukes and are, therefore, not fit to eat. But, deer liver is relished by hunters, and many a liver has doubtless been eaten that contained these parasites.

Subspecies

Odocoileus virginianus macrourus (Rafinesque)

Corvus [sic] *macrourus* Rafinesque, Amer. Monthly Mag., 1, 1817:436; type locality, "Plains of the Kangar [= Kansas] River [=plains near Wakarusa Creek, Douglas County, Kansas].

[*Odocoileus virginianus*] *macrurus* [sic], Trouessart, Catalogus mammalium . . . , Suppl., fasc. 3, 1905:704.

Odocoelus [sic] *virginianus louisianae* G. M. Allen, Amer. Nat., 35, no. 414, 1901:449; type locality, Mer Rouge, Morehouse Parish, Louisiana.

Odocoileus virginianus mcilhennyi F. W. Miller

Odocoileus virginianus mcilhennyi F. W. Miller, Jour. Mamm., 9, 1928:57; type locality, near Avery Island, Iberia Parish, Louisiana.

Because numerous introductions of northern whitetails have been made into Louisiana in the past 35 years, and because a great number of deer have been transplanted from one part of the state to another in restocking operations on the part of Wild Life and Fisheries Commission personnel, the present-day populations in the state cannot be expected to reflect the pattern of geographical variation that once prevailed under natural conditions. Consequently, nothing is to be gained by attempting to analyze the museum material that has been taken in the state in recent decades. Genetically it is a conglomerate of hybrids between several geographical races. For this reason I am following Hall and Kelson (1959) in their treatment of the subspecies, which is based on samples of natural populations collected prior to the extensive transplantations of deer from one part of the country to another.

Two fairly well-marked subspecies originally occurred in Louisiana. Deer in the northern half of the state were assigned to *O. v. macrourus,* whereas the deer of the southern tier of parishes and the coastal marshes were considered separable under the name *O. v. mcilhennyi.* The latter intergraded with *O. v. texanus* in southeastern Texas, with *O. v. macrourus* in southern Louisiana, and with *O. v. osceola* in southern Mississippi. The race *mcilhennyi* is said to have differed from both *texanus* and *macrourus* in being smaller and darker, in having slenderer antlers, and in possessing larger feet. The differences in body size and in the diameter of the beams of the antlers is possibly nothing more than a consequence of nutritional deficiencies in the food available to the marsh deer in coastal Louisiana. As to *mcilhennyi* having larger hooves, said to be an adaptation to its marsh habitat, John D. Newsom (pers. comm.) assures me that the differences between the hooves of marsh deer and those of deer from other parts of the state are indeed striking, the *mcilhennyi* hooves being longer and broader. He has also observed that the foot size is maintained in deer transplanted from the marshes of the Delta Refuge to hard ground and pen conditions near Baton Rouge, and that the offspring of the captive marsh deer also develop larger hooves.

Specimens examined from Louisiana but not identified to subspecies. — Total 118, as follows: *Allen Par.:* 5 mi. S Elizabeth, 2 (LSUMZ); 15 mi. N Oberlin, 1 (USL); West Bay Game Mgt. Area, 2 (USL). *Ascension Par.:* Sorrento, 3 (LSUMZ); locality unspecified, 2 (LSUMZ); 11 mi. SE Sorrento, 1 (LSUMZ). *Assumption Par.:* locality unspecified, 1 (LSUMZ). *Avoyelles Par.:* 4 mi. S Marksville, 1 (LTU). *Bossier Par.:* Benton, 1 (LTU). *Calcasieu Par.:* locality unspecified, 2 (USNM). *Caldwell Par.:* Caldwell Wildlife Mgt. Area, 1 (MSU). *Cameron Par.:* locality unspecified, 1 (USL); Peveto Beach Woods, 1 (USL). *Concordia Par.:* 10 mi. SW Vidalia, 1 (USL); Ferriday, 5 (USNM). *East Carroll Par.:* Lake Providence, 3 (USNM); 22 mi. SW Lake Providence, 1 (USNM). *Grant Par.:* 3 mi. N Williana (Catahoula Game Mgt. Area), 1 (USL). *Iberia Par.:* Marsh Island, 1 (LSUMZ); Weeks Island, 1 (USL); Avery Island, 6 (USNM); Lydia, 6 (USNM). *Iberville Par.:* 2 mi. W Rosedale, 1 (LSUMZ). *Madison Par.:* Tallulah, 2 (LSUMZ); s.w. corner, 1 (LSUMZ); 8 mi. NW Tallulah, 1 (USNM); 14 mi. SW Tallulah, 1 (USNM); 23 mi. SW Tallulah, 1 (USNM); 16 mi. NW Tallulah, 1 (USNM); 28 mi. SW Tallulah, 1 (USNM); 8 mi. SW Tallulah, 1 (USNM); 10 mi. NW Tallulah, 6 (USNM); 20 mi. SW Tallulah, 2 (USNM); Indian Lake, 1 (USNM). *Morehouse Par.:* 4 mi. W Oak Ridge, 1 (NLU); Mer Rouge, 4 (MCZ). *Orleans Par.:* Chef Menteur, 1 (USNM). *Plaquemines Par.:* Belair, 1 (LSUMZ). *Rapides Par.:* near Alexandria, 1 (LSUMZ). *Richland Par.:* 15 mi. N Rayville, 1 (USNM). *St. Bernard Par.:* locality unspecified, 4 (LSUMZ). *St. John the Baptist Par.:* Lake Maurepas, 2 (LSUMZ); locality unspecified, 1 (LSUMZ). *St. Landry Par.:* 11 mi. E Arnaudville, 1 (USL); 12 mi. E Port Barre, 1 (USL); 5 mi. N Krotz Springs, 2 (USL). *St. Martin Par.:* 6 mi. E Cecelia, 1 (USL). *St. Mary Par.:* locality unspecified, 2 (LSUMZ); south shore of Cote Blanche Island, 1 (USL); Morgan City, 7 (USNM). *Tensas Par.:* Newlight, 3 (LSUMZ); mouth Fool River, 30 mi. SW Tallulah, 3 (USNM). *Terrebonne Par.:* Billiot Shell Island, 1 (LSUMZ). *Union Par.:* Marion, 1 (LTU); 5 mi. N Marion, 1 (LTU). *Vermilion Par.:* Bayou Chene N Intracoastal Canal, 1 (USL); Red Fish Point, 1 (USL); 3 mi. N Intracoastal Canal, 1 (USL); Abbeville, 5 (USNM); 25 mi. SW Abbeville, 4 (USNM). *Vernon Par.:* Fort Polk Game Mgt. Area, 1 (USL). *Washington Par.:* 1 mi. S Angie, 1 (SLU); 10 mi. S Angie, 1 (SLU).

The Bovid Family
Bovidae

THIS LARGE family of 44 genera and approximately 111 Recent species includes some of the world's more important mammals. This group has given man his principal domestic animals and his main source of protein—cattle, sheep, and goats. They are a source of food in the form of meat, milk, cheese, and butter; they provide a wide variety of other commodities essential to man, such as leather, soap, glue, bone meal, fertilizer, and wool; and they serve as beasts of burden. In addition wild bovids are hunted for sport, hides, and trophies.

The body form of bovids differs widely. Some, such as the antelopes, are lithe and graceful. Others, like the oxen and the bisons, are lumbering and massive. The Water Buffalo *(Bubalus bubalis)* stands nearly six feet at the shoulders and weighs close to a ton, while the Pygmy Antelope *(Neotragus pygmaeus)* has a shoulder height of only 12 inches and a weight of less than 10 pounds.

Domestic cattle, sheep, and goats were all derived from Eurasian species. The domestication of the last two probably began in southwestern Asia between 8,000 and 9,000 years ago. Cattle *(Bos taurus)* were probably also domesticated by man before the beginning of written history. Although several wild stocks have unquestionably entered into the development of present-day cattle, one was certainly the Urus *(Bos primigenius)*. Native to Europe, North Africa, and southwestern Asia, it became extinct in about 1627. The Urus has sometimes been called the auroch, a name that has unfortunately been applied to the Wisent *(Bison bonasus)*.

Genus *Bison* Hamilton-Smith

Dental formula:

$$I\frac{0-0}{3-3} \ C\frac{0-0}{1-1} \ P\frac{3-3}{3-3} \ M\frac{3-3}{3-3} = 32$$

BISON   Figure 237
(Bison bison)

Vernacular and other names.—The Bison commonly goes by the name buffalo, but that name is properly applied only to certain bovids of Africa and southern Asia. The early French explorers and settlers in Louisiana called the Bison *boeuf sauvage* (French for wild ox) or simply *boeuf,* and they christened numerous physiographic features with the name—Boeuf River, Boeuf Lake, Bayou Boeuf (three separate bayous bear this name), and Bayou Terre aux Boeufs. Several old maps show "buffalo crossings" along small rivers and streams. The name bison serves as the generic, specific, and vernacular name and is the Latin cognate of the Old German Wisunt (modern German Wisent). Wisent is the English common name currently applied to the European form of the Bison.

Distribution. — This species originally ranged over much of North America and extended as far south as northern Florida, southern Louisiana, southern Texas, and north-central Mexico. It is now limited almost entirely to national parks, to private estates and ranches (a few in Louisiana), and to ranges set aside especially for the species in Alaska and Canada. The former status of the Bison in Louisiana is discussed beyond. See also Map 70.

External appearance. — The general features of the Bison are so familiar that the species requires no detailed description. It is the largest mammal to have inhabited Louisiana within historic times, sometimes standing six feet at the withers and weighing as much as a ton. A hump on the shoulders is produced by the extremely long spines on the anterior thoracic vertebrae. The head is carried low; the head, neck, and shoulders are massive and covered with long, dense, wooly hair; the tail is moderately long with a tuft of long hairs at the tip; true horns are present in both males and females, rising sidewise and up-

ward from bony cores that are part of the skull proper. The general coat color is dark brown, sometimes almost blackish brown, tending to be paler on the dorsum, palest on the shoulders.

Color phases. — Some individuals are quite dark in color but not melanistic in the strict sense. Albinos are extremely rare but do occur. In the days of the vast herds, albinos were occasionally seen, and Indians attached to them special religious significance.

Measurements and weights. — The total length of adult males ranges from 3,050 to 3,800 mm (10 to 12.5 ft.); tail, 550 to 815 (22 to 32 in.); hind foot, 585 to 660 (23 to 26 in.); height at withers, 1,675 to 1,830 (66 to 72 in.); weights, 1,600 to 2,000 pounds; greatest length of skull, 491 to 570; greatest width of skull, 271 to 343. The total body length of adult females ranges from 1,980 to 2,280 (6.5 to 7.5 ft.); tail, 430 to 545 (17 to 21 in.); hind foot, 460 to 555 (18 to 22 in.); height at withers, 1,420 to 1,625 (56 to 64 in.); weights, 900 to 1,200 pounds; greatest length of skull, 445 to 510; greatest width of skull, 245 to 310 (Jackson, 1961).

Figure 237   Bison *(Bison bison).*

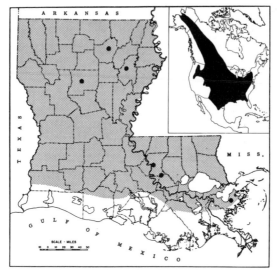

Map 70 Distribution of the Bison *(Bison bison)*. *Shaded area,* the portion of the state that was in all probability formerly occupied by the species; •, localities of occurrence specifically mentioned by early explorers and historians; *inset,* overall former range of the species.

Cranial characters.—The thick-set skull of the Bison (Figure 238), with its solid bone horn cores, which are permanent outgrowths of the skull proper, can be easily confused with the skull of the Domestic Cow. But it is sometimes separable by its larger size and more massive structure, by its heavier and longer upper molariform toothrows, and by the fact that the premaxillae do not extend back to touch the nasals as they do in a cow.

Sexual characters and age criteria.—Male Bison are considerably larger than females, averaging approximately 60 percent larger in both total length and weight. Females closely resemble males in color but are more uniformly dark brown. Newborn calves are decidedly reddish, and they retain this color until about three months old, when a postnatal molt puts them into adult pelage. In late summer, adults tend to become worn and faded.

Status and habits.—The Bison was present in considerable numbers throughout Louisiana during the period of French colonization. Many of the original French explorers, such as Bienville (Margry, 1878), Iberville (Butler, 1934), Dumont de Montigny (1753), Jolliet (Margry, 1878), and Bossu (1771), mentioned the Bison in that part of the Louisiana Territory now known as the state of Louisiana. They called the animal *boeuf sauvage* or sometimes simply *boeuf* and described its great abundance and its importance as a source of food, clothing, and other items. During the period embracing the last years of the 17th and the first years of the 18th centuries, vast numbers of Bison, which were part of the so-called southern herd (Hornaday, 1889), ranged extensively across North Louisiana into the southern part of the state. The Bayou Terre aux Boeufs section of Plaquemines and St. Bernard parishes, in southeastern Louisiana, was so named because Bison spent the winter months there (Arthur, 1928). And, as previously noted, the word *boeuf* was attached to other physiographic features by the French explorers. DeRémonville, writing on 10 December 1697, described the country around New Orleans and commented on the abundance of Bison in that area (French, 1869). He wrote: "We could also draw from thence a great quantity of buffalo hides every year, as the plains and forest are filled with the animals."

On 22 March 1700 the French explorer Bienville left the Tensas Indian villages on Lake St. Joseph, in what is now Tensas Parish, intending to visit the Natchitoches villages near the present-day town of that name. When about 15 leagues west of Lake St. Joseph, on 25 March, the party killed a Bison. They must have been in the vicinity of Winnsboro, in what is now Franklin Parish. Several days later, when somewhere in what is now Winn Parish, the party's hunters killed a bull Bison, a cow, and a calf. (Margry, 1878; Swanton, 1940)

That the Bison once occurred within the environs of Baton Rouge is evidenced by the interesting testimony of Pénicaut. Writing in 1704, this intrepid explorer and narrator had

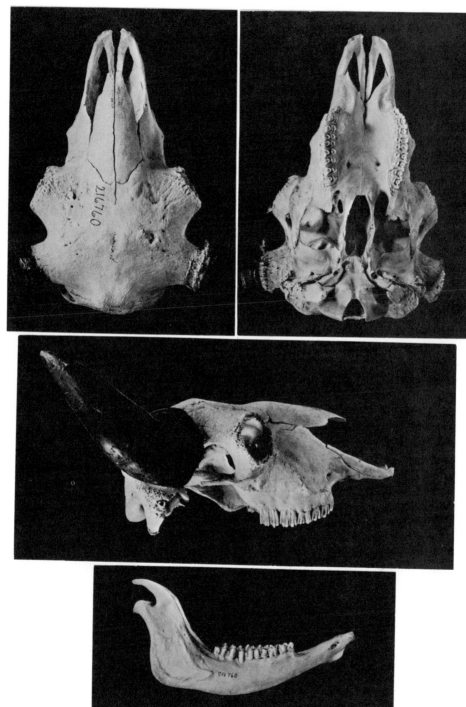

Figure 238 Skull of Bison *(Bison bison bison)*, ♂, USNM 216760, Yellowstone National Park, Wyoming, 1916. Total length of skull, 548 mm (21.6 in.). (Photograph by Joe L. Herring)

this to say (see McWilliams, 1953: 81–82 and 146):

When we got to Baton Rouge, we went ashore to hunt. We entered a forest, some ten of us who were together, the others staying with the boats to watch them and to keep fire burning. Beyond the forest into which we had entered we found a prairie. Never in my life have I seen such great numbers of buffalo, harts [male deer], and roes [female deer], as there were on that prairie. We killed five buffaloes, which we skinned and cut up in order to carry some to our comrades, who had stayed with the boats; and as there was fire burning we broiled some of it on spits and boiled some, too, in our kettle. Our comrades made some shelters on the bank of Missicipy while we went for the rest of our buffalos, which we transported in our boats. We felt so well off at that place that we remained more than ten days. Some of us went hunting every day, especially during the evening, in the woods where one commonly finds bustards and turkeys coming to roost in the trees; so we changed our menu from time to time.

Later, in 1712, probably in late spring or early summer, Pénicaut was on another journey up the Mississippi River when he wrote as follows: "On our way upstream we paused at the Manchacq [Bayou Manchac], where we killed about fifteen buffaloes. Again on the following day we went ashore to hunt. We killed eight buffaloes and just as many deer."

The Bison evidently became rare or absent in southern Louisiana during the last half of the 18th century. The last record of the presence of the animal anywhere in the state is apparently in the account of Bry (1847), in a geographical and historical treatise on the Ouachita River region. This author wrote as follows: "The last buffalo seen in the neighborhood of Fort Miro [the present site of Monroe] was killed in 1803."

The Bison, although probably never extremely abundant in Louisiana, was, however, exceedingly numerous in the Great Plains of the United States and the prairie provinces of Canada. There it occurred in vast herds of almost inestimable numbers. In early pioneer days, the population probably exceeded 60 million individuals. But, by the last years of the 19th century, wanton killing of the herds had reduced the species to fewer than 1,000 animals, and its total extermination seemed inevitable. Fortunately, through the efforts of various conservation groups, notably the American Bison Society, protective measures were initiated and refuges were established in the early 1900s that resulted not only in the species being saved but in substantial increases being achieved. Its total numbers in Canada and United States probably still do not exceed 50,000, including even those on private estates and ranches, but many of the present-day herds are made up of free-ranging animals that are living under natural conditions and not as the wards of man.

The Bison was a highly gregarious animal, and its immense herds, numbering millions of individuals, were absolutely incompatible with the building and operation of railroads, with the planting of vast acreages of wheat and corn, with the development of a huge livestock industry, and with the maturation of a multitude of other human enterprises associated with the growth and economic expansion of our nation's West. Fortunately, the annihilation was not permitted to continue until none of these magnificent creatures was left. That the Bison was not exterminated is to man's credit, although his record in the case of numerous other species is inexcusably bad.

Jackson (1961) had the following to say about the life history and habits of the Bison, and I consider the passages worthy of quotation:

It migrated in a slow movement of immense herds from the north toward the south in the autumn, and returned northward in the spring. These migrations or wanderings covered distances of two to four hundred miles. On its home range it left deeply worn paths or trails in the soil, and dropped huge flats of soft dung. These feces thoroughly dried became known as "buffalo chips," and were frequently used as fuel by Indians and white settlers. The buffalo is almost strictly diurnal in its general activities. It has great endurance and can run 40 to 50 miles without pause and can tire any horse. It is of uncertain temperament and may attack horse or man without provocation. A man on foot is almost a sure target for attack, and man on horseback sometimes may be attacked. Seldom if ever does a buffalo attempt an attack on a man in an automobile. There are many instances where a bison has charged and killed or seriously injured a man. Records in zoological parks and observations on animals on national wildlife refuges and private estates indicate that the normal longevity of the buffalo is about 12 to 15 years. The potential longevity, however, is at least 23 to 25 years, and there are records of individuals living 30 or more years.

Buffalo cows ordinarily mate when they are two years old and bear their first calves when three years old. Instances of two-year-old heifers producing calves have been recorded but are not common. Calves are usually born singly, cases of multiple births being rare. The cows are irregular breeders; they do not bear calves every year, but retain their fecundity to a rather advanced age. Cows 26 years old have been known to produce strong, well-developed calves. Bulls reach breeding age at about four years, but are not fully mature until eight years old. The breeding season normally occurs during July and August, but under artificial conditions induced by confinement and semidomestication, individuals may breed during all seasons. Mature bulls are dangerous during the rutting period. The gestation period is between 270 and 280 days. Calves are usually born during April, May, or June, although births have been recorded in all months. The calf weighs between 30 and 40 pounds at birth. It grows rapidly, and when one year old will weigh about three to four hundred pounds, and at two years old may weigh up to about six hundred pounds. The cow continues to grow until six or seven years old, and the bull until nine or ten years old. . . .

Aside from man and the wolf, the buffalo had other natural enemies such as confront many wild animals. In the case of the buffalo, prairie fires were no small hazard, yet sleet, snow, and ice probably killed even more. Many buffaloes were drowned in crossing rivers. Sometimes a rampaging herd would push many of its own members over a steep bluff or cliff to their destruction. The buffalo is a hardy animal, however, and is comparatively free from diseases and parasites. . . .

The buffalo is one of the most useful native quadrupeds that ever inhabited the United States. It supplied food, clothing, and shelter for hundreds of thousands of natives for centuries. Buffalo meat was utilized fresh and dried as pemmican. Dried dung was used as fuel and to smolder as an insect repellant. Hides were used for robes, both for clothing and bedding, and for making boats, tepees, shields, rawhide ropes, and sheets for winding the dead. Many ornamental articles were made from the bones, and the horns were made into spoons and other utensils. Sinews were utilized for bowstrings and for binding the arrow point into the arrow shaft. Tallow was molded into large balls and kept for waterproofing and other purposes, and buffalo marrow was preserved in buffalo bladders.

Subspecies

Bison bison bison (Linnaeus)

[Bos] *Bison* Linnaeus, Syst. nat., ed. 10, 1, 1758:72; type locality, ancient "Quivira"=central Kansas, *fide* Hershkovitz, 1957.

B[*ison*]. *bison*, Jordan, Manual of the vertebrate animals . . ., ed. 5, 1888:337.

The Bison that ranged into Louisiana in former days were surely referable to the nominate race, *B. b. bison*. The only other geographical race currently recognized is *B. b. athabascae* Rhoads (1898), which is distinguished from *bison* by its much larger size, darker color, and denser and more silky hair, by its slenderer, longer, and more incurved horns, and by other cranial characters. It occurred from Mackenzie, northwestern British Columbia, and western Alberta southward in the Rocky Mountains, particularly along the eastern slopes, to Colorado.

Specimens examined from Louisiana.—None. Although the late E. A. McIlhenny tells (*in* Gunter, 1943) of frequently finding bones and even the crania of Bison in Indian kitchen middens in southern and southwestern Louisiana before the turn of the century, his family at Avery Island can find no trace of any such items that he may have saved. I am informed, however, by Fred B. Kniffen and William G. Haag, distinguished anthropologists at Louisiana State University, that remains of Recent Bison are notoriously scarce or absent in Indian burial sites and kitchen middens, even in the center of the animals' main range.

DOMESTICATED MAMMALS

INCLUDED HERE is a résumé of the domesticated mammals inhabiting the state. Some are of immense economic importance, providing man with meat, various dairy products, and other useful items. Some are valued as pets and live with man under his own roof, serving him as cherished companions, partners on the hunt, and guardians of his property. Thus, no treatise on the mammals of the state would be complete without at least some reference to them. Although the following accounts are brief, I have tried to provide information on the probable origin of each species and to give details regarding identification, particularly with reference to the skulls. Since one often finds a skull of a horse, cow, goat, sheep, or pig in a pasture or even in a wooded area, questions regarding their identity are bound to arise. In fact if drawings of the skulls of these animals were not included one might, for example, confuse the skull of a goat with that of a deer. Even though these animals live mainly under domestication, all of them occur in a feral or semiferal state somewhere within our borders.

Order Carnivora
Family Canidae

DOMESTIC DOG
(Canis familiaris)

The Domestic Dog, through man's inventive genius in selective breeding, has been produced into so many diverse forms that the treatment of them all as a single species, *Canis familiaris,* would seem to be a paradox negating any biological definition of a species. The tiny Chihuahua, which is believed to antedate the Aztec civilization, could by some criteria be considered a good species, for it is certainly reproductively isolated from the Great Dane and certain other breeds. The fact remains that dogs present a special situation because of man's genetic manipulation of

507

them under domestication. The various types of dogs are not reacting to each other as do natural, wild populations, and, for this reason, a strict application of any biological definition of a species is probably not entirely applicable to them. Even the Dingo of Australia, which was long considered a distinct species, is now regarded by many systematists as nothing more than a subspecies of the dog *(Canis familiaris dingo),* because it is believed to have been brought to Australia by aboriginal immigrants. It differs, though, from a mere breed of dog in that it occurs now in a wild state and occupies a discrete geographical range, namely, Australia. Consequently, for these reasons and others, the Domestic Dog, rather than being a distinct species in the conventional biological sense, is doubtless instead a descendant from the various wild species of the genus *Canis* with which man has come in contact over the past several thousand years.

Pictures of what are clearly hunting dogs are found in the tomb of the Egyptian ruler Amten and have been dated about 3500 B.C. And drawings in the tomb of the monarch Antafee, who ruled in Egypt in 3000 B.C., depict him with four terrierlike mongrels at his feet. But dogs had been domesticated long before this period in history, for their bones have also been found in association with the remains of early man as far back as the Stone Age. Man doubtless domesticated wild canids to serve as his companion on hunts and as scavengers and sentinels around his abodes. The dog has certainly endeared itself to man and for the service it has rendered, the dog has earned the distinction of being called "man's best friend," an epithet bestowed on no other animal in the world.

The close relationship between the wild forms of the genus *Canis* and the Domestic Dog is borne out by the successful interbreeding that has taken place between the latter and its wild counterparts. And the resemblance of certain breeds of dogs to the Red Wolf *(Canis rufus),* as well as to the Coyote *(Canis latrans),* makes positive identification of Louisiana specimens sometimes difficult. Even the skulls are often quite similar (see Figure 239 and the account of the Coyote, page 380).

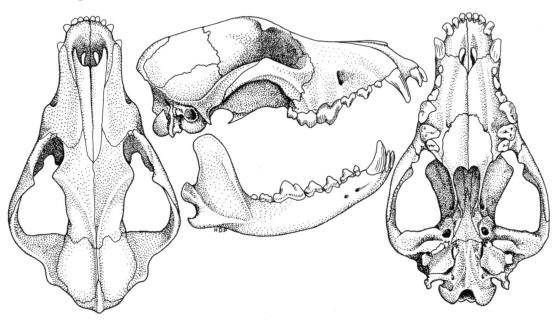

Figure 239   Skull of the Domestic Dog *(Canis familiaris).* Approx. ½ ×.

Feral and free-roaming dogs do great damage to wildlife and should be eliminated whenever possible. Particularly vulnerable are White-tailed Deer, which are sometimes run to exhaustion by packs of these hounds. A case in point is provided by events during the spring floods of 1973, when the opening of the Atchafalaya Spillway inundated vast acreage of deer range in the southern part of our state. Many of the deer in the Basin sought refuge on the spillway's containment levees only to be harassed or killed there by bands of marauding dogs.

Anyone especially interested in dogs should by all means read Konrad Lorenz's study of the canine personality entitled *Man Meets Dog*, published in 1955. This world famous authority on animal behavior writes about the dog, as one reviewer states, "with immense affection yet altogether without sentimentality." This book is unquestionably the best I have ever read about dogs, and along with his earlier work *King Solomon's Ring*, which first appeared in 1952, are "must" readings for anyone interested in the interpretation of animal behavior.

Canis familiaris **Linnaeus**

[*Canis*] *familiaris* Linnaeus, Syst. nat., ed. 10, 1, 1758:38.

Family Felidae

DOMESTIC CAT
(Felis catus)

Investigators of the history of domestication of the house cat once believed that it originated in Egypt from the African Wild Cat *(Felis libya)*, for innumerable mummies show that it was domesticated before the time of the oldest monuments of that civilization. The current consensus is, however, that early man elsewhere domesticated the species of different regions and allowed them to breed with tame cats of other species from other regions. For this reason the "tabby" as we know it today is a mixture that defies definite determination as to it exact origin. The wide array of breeds, ranging from the Siamese and the Manx to various long-haired forms, such as the Angoras and Persians, attests to the multiple origin of the animals we now lump under the name of *Felis catus*.

The relationship of the Domestic Cat with man is quite unlike that which exists between man and his dogs. The cat maintains its independent existence and seldom evinces any inclination to please its master. In taking up its abode in man's dwellings and his outhouses, it probably does so mainly because

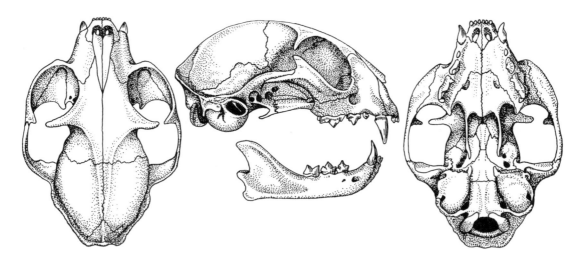

Figure 240 Skull of the Domestic Cat *(Felis catus)*. Approx. ⅔ ×.

there are more mice in such places than else-where. Lorenz (1955) puts the matter this way: "Only two animals have entered the human household otherwise than as prison-ers and become domesticated by other means than those of enforced servitude: the dog and the cat. Two things they have in common, namely, that both belong to the order of carnivores and both serve man in their capac-ity of hunters. In all other characteristics, and primarily in the manner of their association with man, they are as different as the night from the day. There is no domestic animal which has so radically altered its whole way of living, indeed its whole sphere of interest, that has become domestic in so true a sense as the dog; and there is no animal that . . . has altered so little as the cat. There is some truth in the assertion that the cat . . . is no domestic animal but a completely wild being."

One of the greatest decimators of wildlife,

particularly of song birds, are feral cats or "tabbies" that are allowed outside the house. People who take unwanted cats to the out-skirts of a town and release them to fend for themselves are doing irreparable harm to the bird populations in these areas. A more sen-sible course is to take such cats to the local S.P.C.A., which might be able to find a foster home for them where they would be wel-comed and properly cared for.

The skull of the Domestic Cat can be easily told from those of our native cats of the gen-era *Felis* and *Lynx* by its much smaller size, and it differs further from that of the Bobcat (*Lynx rufus*) in possessing an extra premolar on each side of the upper jaw (Figure 240). The dental formula is I 3/3, C 1/1, P 3/2, M 1/1 × 2=30.

Felis catus Linnaeus

[*Felis*] *Catus* Linnaeus, Syst. nat., ed. 10, 1, 1758:42.

Order Perissodactyla
Family Equidae

DOMESTIC HORSE
(*Equus caballus*)

The horse, as we know it today, was derived mainly from the Tarpan, *Equus caballus*, which no longer exists in a wild, unaltered state, but which formerly ranged widely over much of Europe and western Asia. It prob-ably contains also a mixture of other species as well, including the Mongolian Wild Horse (*E. przewalskii*), the only species that continues to survive even in limited numbers, except under man's domestication.

In addition to the horses, the family Equidae contains the zebras and asses. All are placed in the single genus *Equus*, and all have many features in common. Most possess a mane of hair on the neck, and many show a bang of hair, known as the forelock, on the front part of the head. The females have a

single pair of mammae located between the hind limbs. All Recent species have feet with a single functional toe, digit 3, which is termi-nally flattened, triangular, and hooved, on which the animal stands. Metacarpals 2 and 4 and metatarsals 2 and 4 are vestigial, the remainder are lost (Figure 224). The skull is long, massive, and hornless, and the dentition is complete (Figure 241). In males of Recent forms the dental formula is I 3/3, C 1/1, P 3/3 or 4/3, M 3/3 × 2 = 40 or 42. Rarely a small extra premolar is present in the lower jaw, increasing the total to 44. The diminutive extra premolars are called "wolf-teeth." The permanent canines do not erupt until the ani-mal is between four and five years of age (Sis-son and Grossman, 1953). In female equines the canines are vestigial or absent, and the incisors in both sexes are chisel shaped.

Most of the early evolution of horses took place in North America, beginning with the five-toed *Eohippus* in the Eocene, continuing through the three-toed *Mesohippus* of the Oligocene, and culminating in the one-toed *Equus* of the late Pliocene to the Recent. None of the Pleistocene species left any survivors in the Western Hemisphere, but some had moved across the Bering Bridge into Asia, where they completed their evolution into the forms that gave rise to the horse as we know it today. There in the Old World man domesticated it at least as early as 3000 B.C. It must have been an important source of food for humans long before the dawn of history, and the animal is surely in large measure responsible for man's accomplishments since that time. Its flesh is still a major item of food in many parts of the world, and the animal serves many other important functions. Although the horse disappeared from the Americas in the late Pleistocene, it came back with the arrival of the Spanish conquistadores led by De Soto in 1539 and Coronado in 1540. From the Spanish settlements in Mexico and in our own Southwest the American Indian acquired horses for the first time since the late Ice Age.

Mules are the hybrid progeny resulting from a cross between a female, or mare, of *E. caballus* and a jack, the male ass *(E. asinus)*. The reciprocal cross, between a stallion and a jennet or jenny, the she ass, is called a hinny and is comparatively rare. Although mules are generally thought always to be sterile, there are nevertheless notable exceptions such as the famous mule Old Bec at Texas A and M University, which produced several offspring (Growth, 1928; Anderson, 1939).

A mule is in many respects far superior to a horse. It has greater strength and endurance, is less susceptible to disease, is much easier to feed, possesses a more even temperament, and displays greater courage and sure-footedness on precipitous mountain trails when being employed as a mount or as a pack animal.

Equus caballus Linnaeus

[*Equus*] *Caballus* Linnaeus, Syst. nat., ed. 10, 1, 1758: 73.

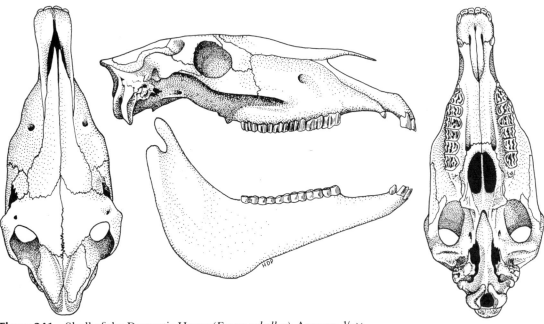

Figure 241   Skull of the Domestic Horse *(Equus caballus)*. Approx. ⅙ ×.

DOMESTIC ASS
(Equus asinus)

The skull of the Domestic Ass, or donkey, does not differ appreciably from that of a horse *(E. caballus)*. The animal differs, however, in being smaller and in having exceptionally long ears. Wild asses of this species now occur only in Somaliland and the Sudan, in northeastern Africa, but its domesticated version has been introduced almost worldwide. Domestication occurred at least 5,000 years ago, probably in southwestern Asia. The animal has been immensely important to man as a beast of burden and for the role it has played in producing mules (see previous account).

Equus asinus Linnaeus

[*Equus*] *Asinus,* Linnaeus, Syst. nat., ed. 10, 1, 1758:73.

Order Artiodactyla
Family Suidae

DOMESTIC HOG
(Sus scrofa)

The origin of the Domestic Hog, or pig, is shrouded in considerable mystery, although it probably originated in Europe or Asia from the wild boars *Sus scrofa* or *S. indicus,* or from a combination of the two. The first swine to reach the United States are said to have come ashore on the Gulf Coast in 1539 with Fernando De Soto (Lewis, 1907). The European wild boar occurs in a feral state in several parts of this country to which it has been introduced, notably in the mountains of Tennessee. The razorback, a small, rangy pig, roams wild in many sections of our state. It travels in bands of 6 to as many as 50. These feral pigs are omnivorous, eating fungi,

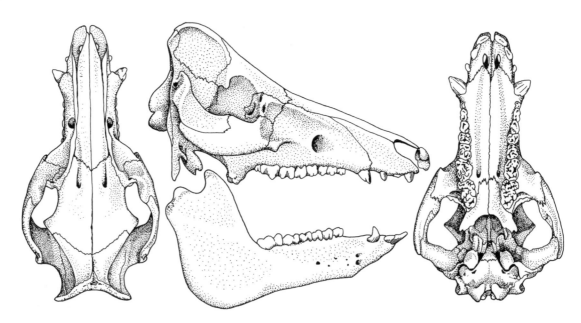

Figure 242   Skull of the Domestic Hog *(Sus scrofa)*. Approx. ⅕ ×.

leaves, acorns, fruits, roots, bulbs, and tubers, as well as snails, earthworms, frogs, reptiles, bird eggs, rats, mice, and even carrion. They are often hunted for sport and for their meat.

The numerous domestic breeds of hogs that have been developed usually fall into one of two categories, the bacon or the lard types. On farms in the United States alone there are some 62 million swine.

The Domestic Hog has no close affinities with the peccaries or javelinas of the New World, which are placed in the family Tayassuidae rather than Suidae. The combined ranges of the two species, the Collared and White-lipped Peccaries, extend from Texas to Patagonia. A Collared Peccary (*Tayassu tajacu*) was found dead on the side of a highway 3 miles southwest of Mansfield, in De Soto Parish, on 3 March 1953, but I do not believe that the record represents a natural occurrence of this species in the state. The Collared Peccary ranges widely in Texas but nowhere is known to approach Louisiana closer than 100 miles.

The skull of a Domestic Hog is easily recognized by the steeply elevated cranium, the absence of a bony ring around the eye socket, the presence of well-developed incisor teeth in the upper jaw, and upper canines that project outward and sometimes upward (Figure 242). The dental formula of the hog is I 3/3, C 1/1, P 4/4, M 3/3 × 2 = 44.

Sus scrofa Linnaeus

[*Sus*] *Scrofa* Linnaeus, Syst. nat., ed. 10, 1, 1758:49.

Family Bovidae

DOMESTIC COW
(Bos taurus)

The Domestic Cow dates back to the beginning of written history and doubtless even earlier. The genus *Bos* contains some seven Recent species, all of which have been domesticated to one degree or another and have con-

tributed in some measure to one or more of the various breeds of modern cattle to which the name *Bos taurus* is now generally applied. The seven nominal species are: *B. indicus*, the Zebu, or Brahma, of India; *B. grunniens*, the Yak of Tibet, China, and northern India; *B. gaurus*, the Gaur of India, Nepal, and Burma; *B. frontalis*, the Gayal of India, Assam, and northern Burma; *B. banteng*, the Banteng of Burma, Thailand, Indo-China, the Malay Peninsula, and Java; *B. sauveli*, the Kouprey of Cambodia; and *B. primigenius*, the Urus of North Africa, Europe, and southwestern Asia, which became extinct in about 1627 (Walker, *in* Walker et al., 1968).

Of all domesticated mammals, cattle are without much doubt most important to man. They provide him with his chief source of protein by their meat, milk, and other dairy products. Cattle are also useful in furnishing medicines such as insulin, in providing fat, fertilizers, glue, soap, leather, and other items of economic importance, and in serving as beasts of burden.

The skull of *Bos taurus* characteristically possesses in both sexes a pair of permanent horns that are not shed annually as are the antlers of a deer. The horns are formed around a bony core arising from the frontal bones. Some special, polled breeds are, of course, hornless. Upper incisors and canines are absent, but the roof of the mouth has a thick membrane called the dental pad that helps in chewing the cud. The dental formula is I 0/4, C 0/0, P 3/3, M 3/3 × 2=32. A drawing of the skull is shown in Figure 243.

Bos taurus Linnaeus

[*Bos*] *Taurus* Linnaeus, Syst. nat., ed. 10, 1, 1758:71.

DOMESTIC GOAT
(Capra hircus)

The earliest domestication of the goat may have taken place between eight and nine thousand years ago in southwestern Asia. Goats resemble sheep but differ in usually

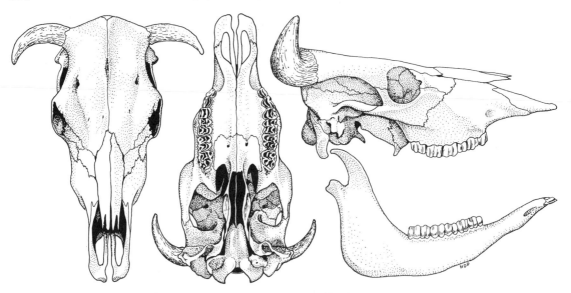

Figure 243   Skull of the Domestic Cow *(Bos taurus)*. Approx. $\frac{1}{7}\times$.

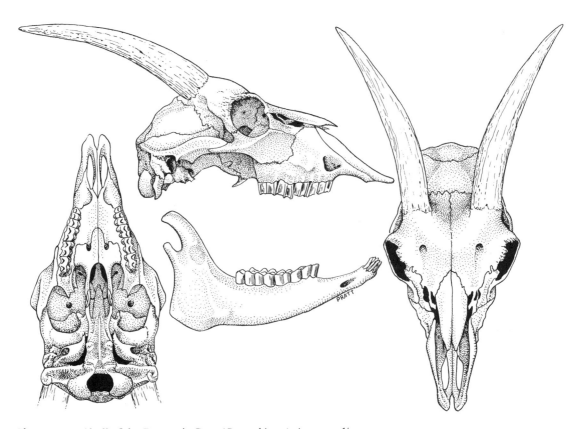

Figure 244   Skull of the Domestic Goat *(Capra hircus)*. Approx. $\frac{1}{3}\times$.

having long, almost parallel, recurved horns extending backward over the shoulders, in possessing a beard, and by the males having a gland at the base of the tail that makes them highly odiferous. The skull of a goat (Figure 244) lacks the depressions in front of the eye sockets characteristic of sheep, and the frontal bones possess a V- or U-shaped depression between the horns. Wild goats comprise some five species that generally, but not always, inhabit rugged mountain country. They range from Spain eastward to India, northward to Mongolia and Siberia, and southward to Egypt, the Sudan, Arabia, and the Greek Islands. The Domestic Goat is only somewhat distantly related to the North American Mountain Goat (Oreamos americanus) of the Rocky Mountains. The latter is actually a goat-antelope, not a true goat. The dental formula of the Domestic Goat is I 0/3, C 0/1, P 3/3, M 3/3 × 2=32.

Capra hircus Linnaeus

[Capra] Hircus Linnaeus, Syst. nat., ed. 10, 1, 1758:68.

DOMESTIC SHEEP
(Ovis aries)

The six species of sheep, including the two North American representatives, Ovis canadensis and O. dalli, are wide-ranging in mountainous areas of the world. The Domestic Sheep, to which the name O. aries is applied, is unknown to zoologists in a wild state. It has either become extinct in the wild or else was derived from one or more species still surviving and, through domestication, has become so modified that its ancestry cannot be traced. Its bones have been found in association with human settlements that date as far back as 5000 B.C. The skull of the Domestic Sheep (Figure 245) is readily recognized by the strongly convex roof to the cranium and the somewhat concave forehead, by the depres-

sions in front of the eye sockets, and by the spirally coiled hollow horns in the mature rams. The ewes and kids are either hornless or possess horns only a few inches in length. The dental formula is I 0/3, C 0/1, P 3/3, M 3/3 × 2=32.

Ovis aries Linnaeus

[Ovis] Aries Linnaeus, Syst. nat., ed. 10, 1, 1758:70.

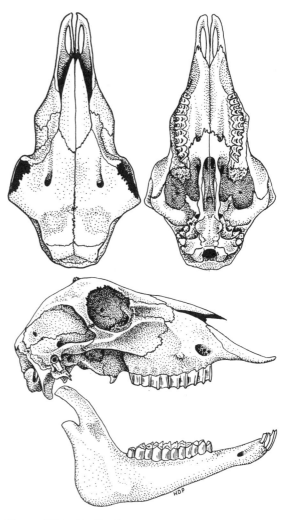

Figure 245 Skull of the Domestic Sheep (Ovis aries). Approx. ¼×.

FOSSIL MAMMALS

THE FOSSILIZED remains of mammals that have lived in Louisiana in the past are by no means plentiful in museum collections. This paucity of fossil material is attributable in some measure to neglect of Louisiana on the part of vertebrate paleontologists, who have tended to expend their energies excavating sites more noted for yielding important discoveries. Since in much of Louisiana the older formations are overlaid by relatively recent alluvial deposits, most of the fossil mammals that have been found are no older than Pleistocene in age. The few fossil localities in Louisiana that have received attention, notably Avery Island in Iberia Parish and the banks of streams cutting through the terraces of West Feliciana and East Baton Rouge parishes, have been shown to possess highly respectable Pleistocene faunas. None perhaps equals the richness of some of the limestone cave deposits in Florida or the fossil beds in many of our western states. But, on the other hand, the megafaunas of Avery Island and of numerous sites along the Mississippi River, in which both mastodon and mammoth bones frequently constitute significant elements, are impressive to say the least.

Only six kinds of fossil mammals have been found in the state in deposits older than Pleistocene. Among these was one of the most miraculous finds in the history of paleontology. During the drilling of an oil well in Caddo Parish in 1931 a short core was pulled from Paleocene beds at a depth of about 2,460 feet below the surface. By fortuitous circumstance the core bit had come down dead center on a skull with teeth intact of an ancient mammal, a member of the order Condylarthra, a primitive ungulate, the first vertebrate and only mammal of Paleocene age known from the state. It was described and appropriately named in 1932 by George Gaylord Simpson as *Anisonchus fortunatus.* But almost equally exciting was the discovery in Caldwell and Grant parishes of four Eocene-Miocene archaeocete whales *(Basilosaurus cetoides, Dorudon serratus, Pontogeneus brachyspondylus,* and *Zygorhiza kochii)* that had evidently stranded on the shore of the Gulf of Mexico in the long distant past when that body of water extended far inland from its present-day limits. Finally, fragments of a possibly Miocene extinct elephant of the family Gomphotheriidae were discovered in Vernon Parish and reported by Arata (1966).

The remaining 37 species so far represented by Louisiana fossil material date from the Pleistocene and are distributed among 11 orders. Many of the forms are of special interest because they persisted up to comparatively recent times, and some of them were with little or no doubt contemporary in Louisiana with Paleolithic man (Gagliano, 1964, 1967, and 1970). Mastodon remains are especially common and radiocarbon datings show that these huge pachyderms possibly survived to within 12,000 years or less of the present (Figures 246 and 247). At least two species of Pleistocene horses, *Equus complicatus* and *E.* cf. *scotti,* occurred at Avery Island and, in the case of *E. complicatus,* elsewhere within the present boundaries of Louisiana, and the remains of two or more kinds of ground sloths, a giant bison, a sabertoothed tiger, two kinds of tapirs, a giant beaver, and the Dire Wolf are found sometimes in associaton with bones that are essentially indistinguishable from those of our modern, present-day Virginia Opposum, White-tailed Deer, and Common Muskrat.

Domning (1969) has compiled a list of the vertebrate fossils known from Louisiana and Mississippi, along with a bibliography of published works citing references to fossil finds in these two states. The accompanying summary of Pleistocene mammals from the state is based primarily on information contained in the following sources: the paper by Domning (1969); papers by Arata (1964), Arata and Hutchison (1964), and Martin (1968); and an unpublished manuscript by Arata and Domning dealing with the late Pleistocene vertebrate fauna of West Feliciana Parish that Domning summarized in his 1969 work. Domning and Arata have kindly furnished me with supplementary West Feliciana Parish data and in addition I have included information on a few especially important specimens now in the LSUMZ that were not available at the time Domning made his compilation. One of these is a portion of the cranium of the extinct Pleistocene Woodland Musk-Ox *(Symbos cavifrons)* that was found recently on Bayou Sara, in West Feliciana Parish. It constitutes the first record for the species in the state and establishes by far the most southern extension of the animal's known range east of the Mississippi River. Where no reference at

Figure 246   American Mastodon *(Mammut americanum).*

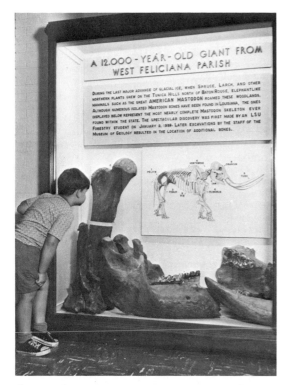

Figure 247 Portion of an exhibit in the LSU Museum of Natural Science showing remains of the most complete mastodon skeleton ever found in Louisiana. It was discovered on Tunica Bayou on 9 January 1959 by forestry students. The visible bones in sequence from left to right are two vertebrae, femur, rami of lower jaw, tusk, and floor of the cranium with three molariform teeth intact. (Photograph by Edgar J. Shore)

all is cited as the basis for the occurrence of a species in a given parish, the detailed record or the basic reference will be found in Domning's 1969 paper.

PLEISTOCENE MAMMALS OF LOUISIANA

Order Marsupialia (marsupials)
Didelphis virginiana Kerr [=*D. marsupialis* of Arata, 1964]. Virginia Opossum.
Iberia and West Feliciana parishes.

Order Insectivora (shrews and moles)
Blarina brevicauda (Say). Short-tailed Shrew.
West Feliciana Parish.

Scalopus aquaticus (Linnaeus). Eastern Mole.
West Feliciana Parish.

Order Chiroptera (bats)
identification indeterminate
Winn Parish (Harris, 1907).

Order Primates (monkeys, apes, and man)
Homo sapiens Linnaeus. Man.
Numerous references, notably Gagliano (1964, 1967, and 1970), in which basket fragments associated with mastodon, mammoth, horse, sloth, and bison bones are reported.

Order Edentata (sloths and armadillos)
Chlamytherium septentrionale (Leidy). Extinct giant armadillo.
West Feliciana Parish.
Dasypus bellus (Simpson). Extinct Pleistocene armadillo.
West Feliciana Parish (pers. comm., Domning, 1973).
Megalonyx jeffersoni (Desmarest). Extinct ground sloth.
Iberia and West Feliciana parishes.
Megalonyx sp. Extinct ground sloth.
Iberia and East Baton Rouge parishes.
Mylodon harlani Owen. Extinct ground sloth.
Iberia and West Feliciana parishes.
Mylodon sp. Extinct ground sloth.
Iberia Parish.

Order Lagomorpha (rabbits)
Sylvilagus floridanus (Allen). Cottontail.
West Feliciana Parish.
Sylvilagus sp.
Iberia Parish.

Order Rodentia (rodents)
Castor sp. Beaver.
West Feliciana Parish.
Castoroides ohioensis Foster. Extinct giant beaver.
"Louisiana" unspecified.
Glaucomys volans (Linnaeus). Southern Flying Squirrel.
West Feliciana Parish (Arata, pers. comm.).
Sigmodon hispidus Say and Ord. Hispid Cotton Rat.
West Feliciana Parish (Arata, pers. comm.).
Peromyscus gossypinus (Le Conte). Cotton Mouse.
West Feliciana Parish (Arata, pers. comm.).
Neotoma floridana (Ord). Eastern Wood Rat.
West Feliciana Parish (Arata, pers. comm.).
Microtus pennsylvanicus (Ord). Meadow Vole.
West Feliciana Parish.
Microtus pinetorum (Le Conte). Woodland Vole.
West Feliciana Parish (Arata, pers. comm.).
Synaptomys sp. Bog lemming.
West Feliciana Parish.
Hydrochreous sp. Extinct capybara.
Iberia Parish.

Order Carnivora (flesh-eating mammals)
Canis dirus Leidy. Extinct Dire Wolf.
Iberia Parish.
Euarctos americanus [=*Ursus americanus* of Arata, 1964 and of Domning 1969]. American Black Bear.
Iberia, West Feliciana, and Catahoula parishes.
Procyon lotor (Linnaeus). Northern Raccoon.
West Feliciana Parish.

Mustela cf. *vison* Schreber. North American Mink.
  West Feliciana Parish (Arata, pers. comm.).
Mustela cf. *frenata* Lichtenstein. Long-tailed Weasel.
  West Feliciana Parish (Arata, pers. comm.).
Smilodon cf. *floridanus* (Leidy). Extinct saber-tooth tiger.
  Iberia and West Feliciana parishes.
Lynx rufus (Schreber). Bobcat.
  West Feliciana Parish.

Order Proboscidea (elephants, mammoths, and mastodons)

Elephus sp. Extinct mammoth.
  Localities in Iberia, Lafayette, St. Landry, and West Feliciana parishes.
Mammut americanum Kerr. American Mastodon.
  Numerous localities in Ascension, Bienville, East Baton Rouge, East and West Feliciana, Iberia, St. Landry, St. Mary, Webster, and Winn parishes.

Order Perissodactyla (hooved mammals with odd-numbered toes)

Equus cf. *scotti* Gridley. Extinct Plains Horse.
  Iberia Parish.
Equus complicatus Leidy. Extinct Eastern Horse.
  Caddo, East Baton Rouge, Iberia, and West Feliciana parishes.
Equus sp. Extinct horse.
  East Carroll Parish (LSUMZ 9825).
Tapirus americanus Lacépède. Extinct tapir.
  St. Landry Parish.
Tapirus copei Simpson. Extinct tapir.
  West Feliciana Parish.
Tapirus cf. *veroensis* Sellards. Extinct tapir.
  St. Landry Parish; East Carroll Parish (LSUMZ 9824).

Order Artiodactyla (hooved mammals with even-numbered toes)

Odocoileus virginianus (Boddaert). White-tailed Deer.
  Iberia and West Feliciana parishes.
Bison latifrons Harlan. Extinct Giant Bison.
  Iberia Parish.
Bison sp. Bison.
  Iberia and West Feliciana parishes.
Symbos cavifrons (Leidy). Extinct Woodland Musk-Ox.
  West Feliciana Parish: near Hollywood, approx. 1,000 yds. south of benchmark 143, in Recent stream alluvium of Bayou Sara; found 16 April 1972 by Kenneth C. Corkum; identification by Clayton E. Ray, Curator of Vertebrate Paleontology, USNM; LSUMZ 17814.

LITERATURE CITED

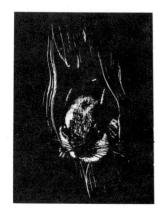

Adams, W. V., G. E. Sanford, E. E. Roth, and L. L. Glasgow
  1964. Nighttime capture of striped skunks in Louisiana. Jour. Wildl. Mgmt., 28:368–373.

Aguayo, C. G.
  1954. Notas sobre cetaceos de aguas Cubanas. Circ. Mus. y Bib. Zool. Havana, 13:1125–1126.

Allen, Elsa G.
  1938. The habits and life history of the eastern chipmunk, *Tamias striatus lysteri*. New York State Mus. Bull. No. 314:1–122.

Allen, G. M.
  1939. Bats. Harvard University Press, Cambridge. x + 368 p.

Allen, Harrison
  1894. A monograph of the bats of North America. Bull. U.S. Nat. Mus., 43:ix + 198 p.

Allison, V. C.
  1937. Evening bat flight from Carlsbad Caverns. Jour. Mamm., 18:80–82.

Alpers, Antony
  1961. Dolphins. The myth and the mammal. Houghton Mifflin Co., Boston. xviii + 268 p.

Anderson, R. C.
  1962. The parasites (helminths and arthropods) of white-tailed deer. Proc. First Nat. White-tailed Deer Disease Symp., Univ. Georgia, Athens: 162–173.

Anderson, Sydney, and J. K. Jones, eds.
  1967. Recent mammals of the world. Ronald Press, New York. viii + 453 p.

Anderson, W. S.
  1939. Fertile mare mules. Jour. Heredity, 30: 549–551.

Anonymous
  1965. Historic Claiborne '65. Claiborne Parish Historical Association. Homer, Louisiana. 126 p., illus.

Arata, A. A.
  1964. Fossil vertebrates from Avery Island. P. 69–76. *In* An archaeological survey of Avery Island by S. M. Gagliano, Coastal Studies Inst. ix + 76 p.
  1966. A Tertiary proboscidean from Louisiana. Tulane Stud. Geol., 4:73–74.

Arata, A. A., and J. H. Hutchison
  1964. The raccoon *(Procyon)* in the Pleistocene of North America. Tulane Stud. Geol., 2:21–27, 4 figs.

Arlton, A. V.
  1936. An ecological study of the mole. Jour. Mamm., 17:349–371.

Arthur, S. C.
  1928. The fur animals of Louisiana. Louisiana Dept. Conserv., Bull. 18:1–433.
  1931. The fur animals of Louisiana. Louisiana Dept. Conserv., Bull. 18 (revised):1–444.

Ash, Christopher
  1962. Whaler's eye. MacMillan Co., New York. viii + 245 p., illus.

Audubon, J. J., and John Bachman
  1846–1854. The viviparous quadrupeds of North America. 3 vols. J. J. and V. G. Audubon, New York.
  1851–1854. The quadrupeds of North America. 3 vols. V. G. Audubon, New York.

Babero, B. B.
  1960. A survey of parasitism in skunks *(Mephitis mephitis)* in Louisiana, with observations on pathological damages due to helminthiasis. Jour. Parasit., 46 (no. 5, sect. 2):26–27.

521

Babero, B. B., and J. W. Lee
  1961. Studies on the helminths of nutria, *Myocastor coypus* (Molina), in Louisiana with check-list of other worm parasites from this host. Jour. Parasit., 47:378–390.

Bachman, John
  1839. Monograph of the species of squirrel inhabiting North America. Proc. Zool. Soc. London, 1838, pt. 6:85–110.

Bailey, F. L.
  1951. Observations on the natural history of the free-tailed bat, *Tadarida cynocephala* (Le Conte). Unpublished Master's thesis, Louisiana State University, Baton Rouge. iv + 29 p.

Bailey, Vernon
  1900. Revision of the American voles of the genus Microtus. N. Amer. Fauna, 17:1–88.
  1905. Biological survey of Texas. N. Amer. Fauna, 25:1–222, 16 pls., 24 figs.
  1927. Beaver habits and experiments in beaver culture. U.S. Dept. Agric., Tech. Bull., 21:1–40.
  1937. Deodorizing skunks. Jour. Mamm., 18:481–482.

Baird, S. F.
  1859. Zoology of the boundary. Mammals. Rept. U.S. and Mexican Boundary Surv., 2, pt. 2:1–62, 27 pls.

Baker, R. J., and C. M. Ward
  1967. Distribution of bats in southeastern Arkansas. Jour. Mamm., 48:130–132.

Barbour, R. W.
  1942. Nests and habitat of the golden mouse in eastern Kentucky. Jour. Mamm., 23:90–91.
  1951. The mammals of Big Black Mountain, Harlan County, Kentucky. Jour. Mamm., 32:100–110.

Barbour, R. W., and W. H. Davis
  1969. Bats of America. The University Press of Kentucky, Lexington. 286 p., 24 pl., maps and other illus.

Barkalow, F. S.
  1967. A record gray squirrel litter. Jour. Mamm., 48:141.

Baughman, J. L.
  1946. On the occurrence of a rorqual whale on the Texas coast. Jour. Mamm., 27:392–393.

Baumgartner, L. L., and F. C. Bellrose, Jr.
  1943. Determination of sex and age in muskrats. Jour. Wildl. Mgmt., 7:77–81.

Baylis, H. A.
  1945. Helminths from the American cotton-rat *(Sigmodon hispidus)*. Ann. Mag. Nat. Hist., ser. 11, 12:189–195.

Beddard, F. E.
  1902. Mammalia. Macmillan and Co., New York. xii + 605 p.

Belisle, J. G.
  1912. History of Sabine Parish Louisiana from the first explorers and settlers to the present. Sabine Banner Press. 319 p.

Benton, A. H.
  1955. Observations on the life history of the northern pine mouse. Jour. Mamm., 36:52–62.

Best, P. B.
  1970. Records of the pygmy killer whale, *Feresa attenuata,* from southern Africa, with notes on behaviour in captivity. Ann. South African Mus., 57:1–14.

Blair, W. F.
  1937. The burrows and food of the prairie pocket mouse. Jour. Mamm., 18:188–191.
  1942. Systematic relationships of Peromyscus and several related genera as shown by the baculum. Jour. Mamm., 23:196–204.
  1950. Ecological factors in speciation of *Peromyscus.* Evolution, 4:253–275.

Bossu, [Jean Bernard, *fide* Lib. Cong. Cat.]
  1768. Nouveaux voyages aux Indes occidentales; contenant une relation des differens peuples qui habitent les environs du grand fleuve Saint-Louis, appelé vulgairement le Mississipi; leur religion; leur gouvernement; leurs moeurs; leurs guerres et leur commerce. 2nd ed. Le Jay, Paris. 244 + 263 p.
  1771. Nouveaux voyages dans l'Amerique Septentrionale, contenant une collection de lettres écrites sur les lieux, par l'auteur à son ami, M. Douin Chevalier, Capitaine dans les troupes du Roi, ci-devant son camarade dans le nouveau monde. Changuion, Amsterdam. xvi + 392 p.

Bradshaw, W. N.
  1965. Species discrimination in the *Peromyscus leucopus* group of mice. Texas Jour. Sci., 17:278–293.
  1968. Progeny from experimental mating tests with mice of the *Peromyscus leucopus* group. Jour. Mamm., 49:475–480.

Brass, Emil
  1911. Aus dem Reiche der Pelze. Im Verlage der Neuen Pelzwaren-Zeitung, Berlin. xxi + 709 p., illus., tables.

Breuer, J. P.
  1951. Gilchrist's whale. Texas Game and Fish, 9:24–25.

Brown, D. H., D. K. Caldwell, and Melba C. Caldwell
  1966. Observations on the behavior of wild and captive false killer whales, with notes on associated behavior of other genera of captive delphinids. Los Angeles County Mus., Contri. Sci., 95:1–32.

Brown, D. H., and K. S. Norris
  1956. Observations of captive and wild cetaceans. Jour. Mamm., 37:311–326.

Brown, L. G., and L. E. Yeager
  1945. Fox squirrels and gray squirrels in Illinois. Bull. Illinois Nat. Hist. Surv., 23:449–536.

Bry, H.
  1847. The Louisiana Ouachita region. De Bow's Review, 3:225–230.

Bryant, R. L.
1954. Population studies and food habits of cottontails and swamp rabbits in southeastern Louisiana. Unpublished Master's thesis, Louisiana State University, Baton Rouge. ix + 66 p., 6 maps, 12 tables, 9 pls.

Bullis, H. R., Jr., and J. C. Moore
1956. Two occurrences of false killer whales, and a summary of American records. Amer. Mus. Novit., no. 1756:1–5.

Burke, J. W., and R. C. Junge
1960. A new type of water dermatitis in Louisiana. Southern Med. Jour., 53:716–719.

Burns, R. K., and L. M. Burns
1956. Vie et reproduction de l'opossum Américain Didelphis marsupialis virginiana Kerr. Bull. Soc. Zool. France, 81:230–246.

Burt, W. H.
1936. A study of the baculum in the genera Perognathus and Dipodomys. Jour. Mamm., 17:145–156.
1940. Territorial behavior and populations of some small mammals in southern Michigan. Misc. Publs., Mus. Zool., Univ. Michigan, 45:7–58.
1960. Bacula of North American mammals. Misc. Publs. Mus. Zool., Univ. Michigan, 113:1–76, 25 pls.

Bushnell, D. I.
1909. The Choctaw of Bayou Lacombe, St. Tammany Parish. Bull. American Ethn., 48. Smithsonian Institution, Washington, D.C. ix + 37 p., 22 pls.

Butler, Ruth L., translator
1934. Journal of Paul du Ru [February 1 to May 8, 1700] The Caxton Club, Chicago. x + 74 p.

Cadenat, J.
1958. Notes sur les delphinids ouest-africaines. II. On specimen du genre Feresa capture sur les cotes du Senegal. Bull. Inst. Français Afrique Noire, (A), 20:1486–1491.

Cahn, A. R.
1940. Manatees and the Florida freeze. Jour. Mamm., 21:222–223.

Cain, G. D.
1966. Helminth parasites of bats from Carlsbad Caverns, New Mexico. Jour. Parasit., 52:351–357.

Caldwell, D. K.
1955. Notes on the spotted dolphin, Stenella plagiodon, and the first record of the common dolphin, Delphinus delphis, in the Gulf of Mexico. Jour. Mamm., 36:467–470.
1960. Notes on the spotted dolphin in the Gulf of Mexico. Jour. Mamm., 41:134–136.

Caldwell, D. K., and Melba C. Caldwell
1966. Observations on the distribution, coloration, behavior and audible sound production of the spotted dolphin, Stenella plagiodon (Cope). Los Angeles County Mus., Contri. Sci., 104:1–28.

1969. Gray's dolphin, Stenella styx, in the Gulf of Mexico. Jour. Mamm., 50:612–614.
1971 a. The pigmy killer whale, Feresa attenuata, in the western Atlantic, with a summary of world records. Jour. Mamm., 52:206–209.
1971 b. Underwater pulsed sounds produced by captive spotted dolphins, Stenella plagiodon. Cetology, 1:1–7.
1972. The world of the bottlenosed dolphin. J. B. Lippincott Company, Philadelphia and New York. 157 p.

Caldwell, D. K., Melba C. Caldwell, W. F. Rathjen, and J. R. Sullivan
1971. Cetaceans from the Lesser Antillean island of St. Vincent. Fishery Bull., 69:303–312.

Caldwell, D. K., Melba C. Caldwell, and C. M. Walker, Jr.
1970. Mass and individual strandings of false killer whales, Pseudorca crassidens, in Florida. Jour. Mamm., 51:634–636.

Caldwell, D. K., Melba C. Caldwell, and S. G. Zam
1971. A preliminary report on some ectoparasites and nasal-sac parasites from small odontocete cetaceans from Florida and Georgia. In Technical Report Number 5, Marineland Research Laboratory:1–7.

Caldwell, D. K., Anthony Inglis, and J. B. Siebenaler
1960. Sperm and pigmy sperm whales stranded in the Gulf of Mexico. Jour. Mamm., 41:136–138.

Caldwell, D. K., J. H. Layne, and J. B. Siebenaler
1956. Notes on a killer whale (Orcinus orca) from the northeastern Gulf of Mexico. Quart. Jour. Florida Acad. Sci., 19:189–196.

Caldwell, Melba C., D. K. Caldwell, and J. F. Miller
1973. Statistical evidence for individual signature whistles in the spotted dolphin, Stenella plagiodon. Cetology, 16:1–21.

Caldwell, Melba C., D. K. Caldwell, and J. B. Siebenaler
1965. Observations on captive wild Atlantic bottlenosed dolphins, Tursiops truncatus, in the northeastern Gulf of Mexico. Los Angeles County Mus., Contri. Sci., 91:1–10.

Caldwell, Melba C., D. K. Caldwell, and S. G. Zam
1968. Occurrence of the lung worm (Halocercus sp.) in Atlantic bottle-nosed dolphins (Tursiops truncatus) as a husbandry problem. Proc. Second Symposium on Diseases and Husbandry of Aquatic Mammals (Marineland Research Laboratory, St. Augustine, Florida): 11–15.

Calhoun, J. B.
1949. A method for self-control of population growth among mammals living in the wild. Science, 109:333–335.

Caslick, J. W.
1956. Color phases of the roof rat, Rattus rattus. Jour. Mamm., 37:255–257.

Catesby, Mark
1731–1743. The natural history of Carolina, Florida and the Bahama Islands, 2 vols. London.

Chabreck, R. H.
1957. A study of beaver in St. Tammany Parish, Louisiana. Unpublished Master's thesis, Louisiana State University, Baton Rouge. xi + 105 p., 10 tables, 31 figs.
1958. Beaver-forest relationships in St. Tammany Parish, Louisiana. Jour. Wildl. Mgmt., 22:179–183.

Chandler, A. C.
1941. Helminths of muskrats in southeast Texas. Jour. Parasit., 27:175–181.
1942 a. Helminths of tree squirrels in southeast Texas. Jour. Parasit., 28:135–140.
1942 b. The helminths of raccoons in east Texas. Jour. Parasit., 28:255–268.

Clark, G. M.
1959. Parasites of the gray squirrel. In Symposium on the gray squirrel. Vagn Flyger, ed. Contrib. 162, Maryland Dept. Research and Education. Extract from Proc. 13th Ann. Conf. Southeastern Assoc. Game and Fish Comms., 368–373.

Clark, G. M., and C. M. Herman
1959. Parasites of the raccoon, *Procyon lotor*. Jour. Parasit., 45 (4, sect. 2):58.

Cockrum, E. L.
1955. Reproduction in North American bats. Trans. Acad. Sci., 58:487–511.
1962. Introduction to mammalogy. The Ronald Press, New York. viii + 455 p.

Columbus, Christopher
1960. Journal. Translated by Cecil Jane (revised and annotated by L. A. Vigneras) with an appendix by R. A. Skelton. Hakluyt Society, London. xxiii + 227 p., 10 figs., illus.

Committee on Marine Mammals (Amer. Soc. Mamm.)
1961. Standardized methods for measuring and recording data on the smaller cetaceans. Edited by Kenneth S. Norris. Jour. Mamm., 42:471–476.

Committee on Records of North American Big Game
1971. North American big game. The Boone and Crockett Club, Pittsburgh, Pa. xvii + 403 p.

Conaway, C. H.
1958. Maintenance, reproduction and growth of the least shrew in captivity. Jour. Mamm., 39:507–512.
1959. The reproductive cycle of the eastern mole. Jour. Mamm., 40:180–194.

Constantine, D. G.
1958 a. Ecological observations on lasiurine bats in Georgia. Jour. Mamm., 39:64–70.
1958 b. Bleaching of hair pigment in bats by the atmosphere in caves. Jour. Mamm., 39:513–520.
1962. Rabies transmission by nonbite route. Pub. Health Repts., 77:287–289.
1966. Ecological observations on lasiurine bats in Iowa. Jour. Mamm., 47:34–41.

Cook, Fannye A.
1945. Fur resources of Mississippi. Survey Bull. Mississippi State Game and Fish Comm. 100 p. mimeographed.

Crabb, W. D.
1944. Growth, development and seasonal weights of spotted skunks. Jour. Mamm., 25:213–221.
1948. The ecology and management of the prairie spotted skunk in Iowa. Ecological Monographs, 18:201–232.

Crain, J. L. and R. L. Packard
1966. Notes on mammals from Washington Parish, Louisiana. Jour. Mamm., 47:323–325.

Croft, R. L.
1961. The value of food plots to cottontails and swamp rabbits in Louisiana. Unpublished Master's thesis, Louisiana State University, Baton Rouge. ix + 61 p., 21 tables, 14 figs.

D., H. S.
1943. Hunting extraordinary. Spirit of the Times (W. T. Porter, ed.), 12(47):555.

Dahlen, B. C.
1939. Beaver in Louisiana. Louisiana Cons. Rev., Summer:15–17.

Davis, D. E.
1948. The survival of wild brown rats on a Maryland farm. Ecology, 29:437–448.
1951 a. The relation between level of population and size and sex of Norway rats. Ecology, 32:462–464.
1951 b. The relation between the level of population and the prevalence of Leptospira, Salmonella, and Capillaria in Norway rats. Ecology, 32:465–468.
1951 c. The characteristics of global rat populations. Amer. Jour. Pub. Health, 41:158–163.
1951 d. The relation between level of population and pregnancy of Norway rats. Ecology, 32:459–461.

Davis, D. E., John Emlen, and A. W. Stokes
1948. Studies on home range in the brown rat. Jour. Mamm., 29:207–225.

Davis, D. E., and W. T. Fales
1950. The rat population of Baltimore, 1949. Amer. Jour. Hyg., 52:143–146.

Davis, R. B., C. F. Herreid, II, and H. L. Short
1962. Mexican free-tailed bats in Texas. Ecol. Monographs, 32:311–364.

Davis, W. B.
1939. A new *Peromyscus* from Texas. Occas. Papers Mus. Zool., Louisiana State Univ., 2:1–2.
1940. Distribution and variation of pocket gophers (genus Geomys) in the southwestern United States. Texas Agric. Exp. Stat., Bull. no. 590:1–38.
1966. The mammals of Texas (Rev. ed.) Texas Parks and Wildlife Dept., Bull., 41:1–267.

Davis, W. B., and Leonard Joeris
1945. Notes on the life-history of the little short-tailed shrew. Jour. Mamm., 26:136–138.

Davis, W. B., and G. H. Lowery, Jr.
1940. The systematic status of the Louisiana muskrat. Jour. Mamm., 21:212–213.

Davis, W. H., R. W. Barbour, and M. D. Hassell
1968. Colonial behavior of *Eptesicus fuscus*. Jour. Mamm., 49:44–50.

Davis, W. H., and W. Z. Lidicker, Jr.
1956. Winter range of the red bat, *Lasiurus borealis*. Jour. Mamm., 37:280–281.

Davis, W. H., and C. L. Rippy
1968. Distribution of *Myotis lucifugus* and *Myotis austroriparius* in the southeastern United States. Jour. Mamm., 49:113–117.

Dearborne, Ned
1946. Miscellaneous Notes. Jour. Mamm., 27:178.

Dice, L. R.
1937. Fertility relations in the *Peromyscus leucopus* group of mice. Contr. Lab. Vert. Gen., Univ. Michigan, 4:1–2.
1940. Relationships between the wood-mouse and the cotton-mouse in eastern Virginia. Jour. Mamm., 21:14–23.

Dingle, J. H.
1941. Infectious disease of mice. P. 349–474, *in* Biology of the laboratory mouse. G. D. Snell, ed. Blackston Company, Philadelphia. ix + 497 p., illus.

Domning, D. P.
1969. A list, bibliography, and index of the fossil vertebrates of Louisiana. Trans. Gulf Coast Assoc. Geol. Soc., 19:385–422.

Doran, D. J.
1954 a. A catalogue of the protozoa and helminths of North American rodents. I. Protozoa and Acanthocephala. Amer. Midl. Nat., 52:118–128.
1954 b. A catalogue of the protozoa and helminths of North American rodents. II. Cestoda. Amer. Midl. Nat., 52:469–480.
1955 a. A catalogue of the protozoa and helminths of North American rodents. III. Nematoda. Amer. Midl. Nat., 53:162–175.
1955 b. A catalogue of the protozoa and helminths of North American rodents. IV. Trematoda. Amer. Midl. Nat., 53:446–454.

Dozier, H. L.
1942. Identification of sex in live muskrats. Jour. Wildl. Mgmt., 6:292–293.

Dumont de Montigny, Butel
1753. Des animaux terrestres de la Louisianae. Chapitre 13, pages 74–86, *in* Memoires historiques sur Louisiane . . . Cl. J. B. Bauche, Paris, Tome Première, Première Partie. x + 261 p.

Dunaway, P. B.
1962. Litter-size record for eastern harvest mouse. Jour. Mamm., 43:428–429.

Dunford, Christopher
1972. Summer activity of eastern chipmunks. Jour. Mamm., 53:176–180.

Dunn, C. T.
1885. Historical and geographical description of Morehouse Parish. New Orleans. xxvi + 376 p., illus.

Eads, R. B.
1948. Ectoparasites from a series of Texas coyotes. Jour. Mamm., 29:268–271.

Eads, R. B., J. S. Wiseman, and G. C. Menzies
1957. Observations concerning the Mexican free-tailed bat, *Tadarida mexicana*, in Texas. Texas Jour. Sci., 9:227–242.

Edgerton, H. E., P. F. Spangle, and J. K. Baker
1966. Mexican free-tail bats: photography. Science, 153:201–203.

Enders, R. K.
1952. Reproduction in the mink (Mustela vison). Proc. Amer. Philos. Soc., 96:691–755.

English, P. F.
1932. Some habits of the pocket gopher, Geomys breviceps breviceps. Jour. Mamm., 13:126–132.

Erickson, A. B.
1944. Helminths of Minnesota Canidae in relation to food habits and a host list and key to the species reported from North America. Amer. Midl. Nat., 32:358–372.
1946. Incidence of worm parasites in Minnesota Mustelidae and host lists and keys to North American species. Amer. Midl. Nat., 36:494–509.

Essapian, F. S.
1962. An albino bottle-nosed dolphin, *Tursiops truncatus*, captured in the United States. Norsk Hvalfangst-tidende [Norwegian Whaling Gazette], 51:341–344.

Evans, James
1970. About nutria and their control. Resource Publ. 86, Bureau Sports Fisheries and Wildlife, Denver. vi + 65 p.

Fairchild, E. J., II
1950. Comparative studies of the baculum (os penis): Order Rodentia. Unpublished Master's thesis, Louisiana State University, Baton Rouge. xi + 124 p., 13 figs., 8 tables.

Findley, J. S., and Clyde Jones
1964. Seasonal distribution of the hoary bat. Jour. Mamm., 45:461–470.

Fitch, H. S., Phil Goodrum, and Coleman Newman
1952. The armadillo in the southeastern United States. Jour. Mamm., 33:21–37.

Fitzwater, W. D., and W. J. Frank
1955. Leaf nests of gray squirrels in Connecticut. Jour. Mamm., 25:160–170.

Flower, W. H.
1874. On Risso's dolphin, *Grampus griseus*. Trans. Zool. Soc. London, 8:22, 2 pls.

Frazer, F. C.
1949. Whales and dolphins. *In* J. R. Norman and F. C. Fraser, Field book of giant fishes. G. P. Put-

nam's Sons, New York. Pages 201–360, pls. 6–8, figs. 60–97.

1950. Description of a dolphin, *Stenella frontalis* (Cuvier), from the coast of French Equatorial Africa. Atlantide Rept., 1:61–84, pl. 6–9.

1953. Report on cetacea stranded on the British coast from 1938 to 1947. Brit. Mus. (Nat. Hist.), 13:1–48, illus., 9 maps.

1960. A specimen of the genus *Feresa* from Senegal. Bull. Inst. Français d'Afrique Noire, sér. A, 22:699–707.

1966. Guide for the identification and reporting of stranded whales, dolphins, and porpoises on the British coasts. Department of Zoology, Brit. Mus. (Nat. Hist.), London. ix + 34 p.

Fraser, F. C., and B. A. Noble
1970. Variation of pigmentation pattern in Meyen's dolphin, *Stenella coeruleoalba* (Meyen). Investigations on Cetacea, 2:147–163, 7 pls.

Fraser, F. C. and P. E. Purves
1960 a. Anatomy and functions of the cetacean ear. Proc. Roy. Soc., (B), 152:62–77.

1960 b. Hearing in cetaceans. Bull. Brit. Mus. (Nat. Hist.), Zool., 7:1–140.

Freeman, R. S.
1954. Studies on the biology of *Taenia crassiceps* (Zeder, 1800) Rudolphi, 1810. Jour Parasit., 40 (no. 5, sect. 2):41.

French, B. F.
1869. Historical collections of Louisiana and Florida. Pages 1–16, Memoir, addressed to Count de Pontchartrain, on the importance of establishing a colony in Louisiana, by M. de Rémonville, and Pages 35–166, Annals of Louisiana from 1698 to 1722, by M. Penicaut. New ser. J. Sabin and Sons, New York. ii–362 p.

Friley, C. E., Jr.
1949 a. Age determination, by use of the baculum, in the river otter, *Lutra c. canadensis* Schreber. Jour. Mamm., 80:102–110.

1949 b. Use of the baculum in age determination of Michigan beaver. Jour. Mamm., 30:261–267.

G., F. G.
1911. Wolves in the South. Forest and Stream, 76 (14):534.

Gagliano, S. M.
1964. An archaeological survey of Avery Island. Coastal Studies Inst., Louisiana State Univ. 76 p.

1967. Occupation sequence at Avery Island. Coastal Studies Series, 22:viii + 110.

1970. Progress Report: Archaeological and Geological Studies at Avery Island 1968–1970. Coastal Studies Inst., Louisiana State Univ. 21 p.

Gardner, A. L.
1973. The systematics of the genus Didelphis (Marsupialia: Didelphidae) in North and Middle America. Spec. Publ. Mus. Texas Tech Univ., 4:1–81, figs., maps.

Gardner, M. C.
1948. Albino cotton rats. Jour. Mamm., 29:185.

Gates, W. H.
1936. Keeping bats in captivity. Jour. Mamm., 17:268–273.

1937. Notes on the big brown bat. Jour. Mamm., 18:97–98.

1938. Raising the young of red bats on an artificial diet. Jour. Mamm., 19:461–464.

1941. A few notes on the evening bat, Nycticeius humeralis (Rafinesque). Jour. Mamm., 22: 53–56.

Genoways, H. H., and J. R. Choate
1972. A multivariate analysis of systematic relationships among populations of the short-tailed shrew (Genus *Blarina*) in Nebraska. Syst. Zool., 21:106–116.

George, J. E., and R. W. Strandtmann
1960. New records of ectoparasites of bats in west Texas. Southwestern Nat., 5:228–229.

Gilbert, F. F.
1966. Aging white-tailed deer by annuli in the cementum of the first incisor. Jour. Wildl. Mgmt., 30:200–202.

Glasgow, L. L., and R. E. Noble
1971. The importance of bottomland hardwood to wildlife. Proc. Symp. on Southeastern Hardwoods. U.S. Dept. Agri. (Forest Service). Mimeographed. 102 p.

Glass, B. P.
1947. Geographic variation in Perognathus hispidus. Jour. Mamm., 28:174–179.

1958. Returns of Mexican freetail bats banded in Oklahoma. Jour. Mamm., 39:435–437.

1959. Additional returns from free-tailed bats banded in Oklahoma. Jour. Mamm., 40:542–545.

1969. A key to the skulls of North American mammals. 9th printing. Privately printed by author. Stillwater, Oklahoma. ii + 53 p.

Goertz, J. W.
1965 a. Sex, age, and weight variation in cotton rats. Jour. Mamm., 46:471–477.

1965 b. Late summer breeding of flying squirrels. Jour. Mamm., 46:510.

Goldman, E. A.
1943. The races of the ocelot and margay in Middle America. Jour. Mamm., 24:372–385.

Goodpaster, W. W., and D. F. Hoffmeister
1954. Life history of the golden mouse, *Peromyscus nuttalli*, in Kentucky. Jour. Mamm., 35:16–27.

Goodrum, P. D.
1940. A population study of the gray squirrel in eastern Texas. Texas Agri. Exp. Stat., Bull. 591:1–34.

Gorham, J. R., and H. J. Griffiths
1952. Diseases and parasites of minks. U.S. Dept. Agric., Farmer's Bull. no. 2050. 41 p.

Gould, Edwin
1969. Communication in three genera of shrews (Soricidae): Suncus, Blarina, and Cryptotis. Communications in Behavioral Biology, Part A, 3:11–31.

Gould, Edwin, N. C. Negus, and Alvin Novick
1964. Evidence for echolocation in shrews. Jour. Exp. Zool., 156:19–38.

Gould, H. N., and Florence B. Kreeger
1948. The skull of the Louisiana muskrat (*Ondatra zibethica rivalicia* Bangs): 1. The skull in advanced age. Jour. Mamm., 29:138–149.

Grasse, J. E., and E. F. Putnam
1955. Beaver management and ecology in Wyoming. Wyoming Game and Fish Commission, Cheyenne. 74 p.

Gravier, Jacques
1900. Relation ou Journal du voyage du Père [Jacques] Gravier, de la Compagnie de Jesús en 1700 depuis le Pays des Illinois Jusqu'a l'Embouchure de Fleuve Mississippi. *In* The Jesuit Relations and Allied Documents. Travels and Explorations of the Jesuit Missionaries in New France 1610–1791. The original French, Latin, and Italian texts, with English translations and notes; illustrated by portraits, maps, and facsimiles. Vol. 65, pages 100–178. Reuben Gold Thwaites, ed. The Burrows Brothers Company, Cleveland. 273 p.

Greer, K. R.
1955. Yearly food habits of the river often in the Thompson Lake region, northwestern Montana, as indicated by scat analysis. Amer. Midl. Nat., 54:299–313.

Gregory, Tappan
1935. The black wolf of the Tensas. Program Activities Chicago Acad. Sci., 6:35–68, illus.

Growth, A. H.
1928. A fertile mare mule. Jour. Heredity, 19:413–416.

Gunter, Gordon
1941. Occurrence of the manatee in the United States, with records from Texas. Jour. Mamm., 22:60–64.
1942. Contributions to the natural history of the bottlenose dolphin, *Tursiops truncatus* (Montague), on the Texas coast, with particular reference to food habits. Jour. Mamm., 23:267–276.
1943. Remarks on American bison in Louisiana. Jour. Mamm., 24:398–399.
1951. Consumption of shrimp by the bottle-nosed dolphin. Jour. Mamm., 32:465–466.
1954. Mammals of the Gulf of Mexico. *In* Gulf of Mexico, its origin, waters, and marine life. Fishery Bull., Fish and Wildlife Serv., 55:543–551.
1955. Blainville's beaked whale, *Mesoplodon densirostris,* on the Texas coast. Jour. Mamm., 36:573–574.
1968. The status of seals in the Gulf of Mexico. Gulf Research Report, 2:301–308.

Gunter, Gordon, and J. Y. Christmas
1973. Stranding records of a finback whale, *Bataenoptera physalus,* from Mississippi and the goose-beaked whale, *Ziphius cavirostris,* from Louisiana. Gulf Research Reports, 4(2), in press.

Gunter, Gordon, C. L. Hubbs, and M. A. Beal
1955. Records of *Kogia breviceps* from Texas, with remarks on movements and distribution. Jour. Mamm., 36:263–270.

Hall, E. R.
1927. An outbreak of house mice in Kern County, California. Univ. California Publ. Zool., 30:189–203.
1936. Mustelid mammals from the Pleistocene of North America with systematic notes on some recent members of the genera Mustela, Taxidea and Mephitis. Pages 41–119, *in* Studies of Tertiary and Quaternary mammals of North America. Carnegie Inst. Washington, Publ. 473. 119 p., 5 pls.
1946. Mammals of Nevada. Univ. California Press, Berkeley and Los Angeles. xi + 710 p.
1951. American weasels. Univ. Kansas Publ., Mus. Nat. Hist., 4:1–466, 41 pls., 31 figs.
1965. The names of species of North American mammals north of Mexico. Univ. Kansas Mus. Nat. Hist., Misc. Publ., 43:1–16.

Hall, E. R., and E. L. Cockrum
1953. A synopsis of the North American microtine rodents. Univ. Kansas Publ., Mus. Nat. Hist., 5:373–498.

Hall, E. R., and W. W. Dalquest
1950. A synopsis of the American bats of the genus Pipistrellus. Univ. Kansas Publ., Mus. Nat. Hist., 1:591–602, 1 fig.

Hall, E. R., and J. K. Jones, Jr.
1961. North American yellow bats, "Dasypterus," and a list of the named kinds of the genus Lasiurus Gray. Univ. Kansas Publ., Mus. Nat. Hist., 14:73–98, 4 figs.

Hall, E. R., and K. R. Kelson
1952. Comments on the taxonomy and geographic distribution of some North American marsupials, insectivores and carnivores. Univ. Kansas Publ., Mus. Nat. Hist., 5:319–341.
1959. The mammals of North America. 2 vols. The Ronald Press, New York. xxx + 1083 p. + 158 p. index, illus.

Hall, J. S.
1963. Notes on *Plecotus rafinesquii* in central Kentucky. Jour. Mamm., 44:119–120.

Hamilton, R. B., and D. T. Stalling
1972. *Lasiurus borealis* with five young. Jour. Mamm., 53:190.

Hamilton, W. J., Jr.
1929. Breeding habits of the short-tailed shrew, Blarina brevicauda. Jour. Mamm., 10:125–134.
1930. The food of the Soricidae. Jour. Mamm., 11:26–39.

1933. The weasels of New York. Amer. Midl. Nat., 14:289–344.

1938. Life history notes on the northern pine mouse. Jour. Mamm., 19:163–170.

1939. Albino blarinas. Jour. Mamm., 20:252.

1944. The biology of the little short-tailed shrew, *Cryptotis parva*. Jour. Mamm., 25:1–7.

1946 a. A study of the baculum in some North American Microtinae. Jour. Mamm., 27:378–387.

1946 b. Habits of the swamp rice rat, *Oryzomys palustris palustris* (Harlan). Amer. Midl. Nat., 36:730–736.

1953. Reproduction and young of the Florida wood rat, *Neotoma f. floridana* (Ord). Jour. Mamm., 34:180–189.

Handley, C. O., Jr.
1959. A revision of the American bats of the genera Euderma and Plecotus. Proc. U.S. Nat. Mus., 110:95–246.

1960. Descriptions of new bats from Panama. Proc. U.S. Nat. Mus., 112, no. 3442:459–479.

1966. A synopsis of the genus *Kogia* (pygmy sperm whales). *In* Whales, dolphins, and porpoises. Kenneth S. Norris, ed. Univ. California Press, Berkeley and Los Angeles. xv + 789 p.

Hardberger, F. M.
1950. The nine-banded armadillo and the eastward expansion of its range. Turtox News, 28:74–177.

Harkema, Reinard
1936. The parasites of some North Carolina rodents [and rabbits]. Ecol. Monographs, 6:151–232.

1946. The metazoa parasitic in cotton rats of Wake County. Jour. Elisha Mitchell Sci. Soc., 62:142–143.

Harkema, Reinard, and Leo Kartman
1948. Observations on the helminths and ectoparasites of the cotton rat, *Sigmodon hispidus hispidus* Say and Ord, in Georgia and North Carolina. Jour. Elisha Mitchell Sci. Soc., 64:183–191.

Harlan, Richard
1823 [=1824]. On a species of Lamantin resembling the *Manatus senegalensis* (Cuvier) inhabiting the coast of east Florida. Jour. Acad. Nat. Sci. Philadelphia, ser. 1, 3 (pt. 2):390–394.

1825. Fauna Americana: being a description of the mammiferous animals inhabiting North America. Anthony Finely, Philadelphia. x + 11–318 + 2 p.

Harper, Francis
1942. The name of the Florida wolf. Jour. Mamm., 23:339.

Harris, D. W., and B. M. Hulse
1886. The history of Claiborne Parish, Louisiana. W. B. Stansbury & Co., New Orleans. 263 p.

Harris, G. D.
1907. Notes on the geology of the Winnfield sheet. La. Geol. Surv. Bull., 5:1–36.

Hartman, C. G.
1920. Studies in the development of the opossum Didelphys virginiana L. V. The phenomena of parturition. Anatomical Record, 19:251–261.

1922. Breeding habits, development, and birth of the opossum. Ann. Report, Smithsonian Inst., 1921:347–363, 10 pls.

1928. The breeding season of the opossum (Didelphis virginiana) and the rate of intra-uterine and post-natal development. Jour. Morph. and Physiol., 46:142–215.

1952. Possums. Univ. Texas Press, Austin. xiii + 174 p.

Hartman, D. S.
1969. Florida's manatees, mermaids in peril. Nat. Geog., 136:342–353.

Harwood, P. D., and Virgil Cooke
1949. The helminths from a heavily parasitized fox squirrel, *Sciurus niger*. Ohio Jour. Sci., 49:146–148.

Hastings, E. F.
1954. Life history studies of cottontail and swamp rabbits in Louisiana. Unpublished Master's thesis, Louisiana State University, Baton Rouge. xi + 85 p., 11 tables, 23 pls.

Hatt, R. T.
1934. A manatee collected by the American Museum Congo Expedition, with observations on the recent manatees. Bull. Amer. Mus. Nat. Hist., 66:533–566, 1 pl.

1938. Feeding habits of the least shrew. Jour. Mamm., 19:247–248.

1944. A large beaver-felled tree. Jour. Mamm., 25:313.

Hayes, F. A., W. T. Gerard, E. B. Shotts, and Gloria J. Dills
1960. Further serologic studies of brucellosis in white-tailed deer of the southeastern United States. Jour. Amer. Vet. Med. Assoc., 137:190–191.

Hayne, D. W.
1949. Two methods for estimating population from trapping records. Jour. Mamm., 30:399–411.

Hays, H. A.
1941. The mammals of Natchitoches Parish [Louisiana]. Unpublished Master's thesis, Louisiana State University, Baton Rouge. viii + 62 p., 4 pls.

Henisch, B. A., and H. K. Henisch
1970. Chipmunk portrait. Carnation Press, State College, Pa. 98 p.

Hershkovitz, Philip
1957. The type locality of *Bison bison* Linnaeus. Proc. Biol. Soc. Washington, 70:31–32.

1961. On the nomenclature of certain whales. Fieldiana, Zool., 39:547–565.

1966. Catalog of living whales. Bull. U.S. Nat. Mus., 246. viii + 259 p.

Hisaw, F. L.
1923. Observations on the burrowing habits of moles (Scalopus aquaticus machrinoides). Jour. Mamm., 4:79–88.

Holloway, H. L.
1956. The acanthocephala of Mountain Lake mammals. Virginia Jour. Sco., 7:285.

Hooper, E. T.
1943. Geographic variation in harvest mice of the species *Reithrodontomys humulis*. Occas. Papers Mus. Zool., Univ. Michigan, 477:1–19.
1958. The male phallus in mice of the genus *Peromyscus*. Misc. Publ., Mus. Zool. Univ. Michigan, 105:1–24, i–xiv pls.
1960. The glans penis in *Neotoma* (Rodentia) and allied genera. Occas. Papers Mus. Zool., Univ. Michigan, 618:1–20, 1 table, 11 pls.
1962. The glans penis in *Sigmodon, Sigmomys,* and *Reithrodon* (Rodentia, Cricetinae). Occas. Papers Mus. Zool., Univ. Michigan, 625:1–11.

Hoover, R. L.
1954. Seven fetuses in western fox squirrel *(Sciurus niger rufiventer)*. Jour. Mamm., 35:447–448.

Hornaday, W. T.
1889. The extermination of the American bison. Rept. U.S. Nat. Mus., for 1887:367–548, 1 map.

Howell, A. H.
1908. Notes on diurnal migrations of bats. Proc. Biol. Soc. Washington, 21:35–38.
1921. A biological survey of Alabama. I. Physiography and life zones. II. The mammals. N. Amer. Fauna, 45:1–88, illus.
1929. Revision of the American chipmunks (genera Tamias and Eutamias). N. Amer. Fauna, 52:1–57, illus.

Howell, R. B., and W. U. Richardson
1885. Bienville Parish, Louisiana. F. F. Hansell, New Orleans. 26 p.

Hubbs, C. L.
1953. Dolphin protecting dead young. Jour. Mamm., 34:498.

Huckaby, D. G.
1967. A systematic study of the *Peromyscus leucopus* species group in the Lower Mississippi Valley. Unpublished Master's thesis. Louisiana State University, Baton Rouge. vii + 32 p.

Hunt, T. P.
1959. Breeding habits of the swamp rabbit with notes on its life history. Jour. Mamm., 40:82–91.

Huxley, J. S. (editor)
1940. The new systematics. Clarendon Press, Oxford. viii + 583 p.

International Commission on Zoological Nomenclature
1964. International code of zoological nomenclature. Second edition. Published by International Trust for Zoological Nomenclature, London. xx + 176 p.

Jackson, H. H. T.
1961. Mammals of Wisconsin. Univ. Wisconsin Press, Madison. xiii + 504 p.

James, Pauline, F. W. Judd, and J. C. Moore
1970. First western Atlantic occurrence of the pigmy killer whale. Fieldiana, Zool., 58:1–3.

Jameson, D. K.
1959. A survey of the parasites of five species of bats. Southwestern Nat., 4:61–65.

Jameson, E. W., Jr.
1947. Natural history of the prairie vole (mammalian genus Microtus). Univ. Kansas Publ., Mus. Nat. Hist., 1:125–151.
1950. The external parasites of the short-tailed shrew, *Blarina brevicauda* (Say). Jour. Mamm., 31:138–145.

Jellison, W. L.
1945. A suggested homolog of the os penis or baculum of mammals. Jour. Mamm., 26:146–147.

Jemison, E. S., and R. H. Chabreck
1962. Winter barn owl foods in a Louisiana coastal marsh. Wilson Bull., 74:95–96.

Jenkins, Winchester
1933. Wild life of Mississippi from forty-five years experience. The Reporter Printing Co., Natchez. 155 p., illus.

Johnson, C. E.
1921. The "hand-stand" habit of the spotted skunk. Jour. Mamm., 2:87–89.

Jones, Clyde
1967. Growth, development, and wing loading in the evening bat, *Nycticeius humeralis* (Rafinesque). Jour. Mamm., 48:1–19.

Jones, Clyde, and John Pagels
1968. Notes on a population of *Pipistrellus subflavus* in southern Louisiana. Jour. Mamm., 49:134–139.

Jones, Sarah V. H.
1923. Color variations in wild animals. Jour. Mamm., 4:172–177.

Jordan, H. E., and F. A. Hayes
1959. Gastrointestinal helminthes of raccoons *(Procyon lotor)* from Ossabaw Island, Georgia. Jour. Parasit., 45:249–252.

Kale, H. W., II
1965. Ecology and bioenergetics of the Long-billed Marsh Wren in Georgia salt marshes. Publs. Nuttall Ornith. Club, 5. xiii + p., 22 figs., 61 tables.

Kalmbach, E. R.
1944. The armadillo: its relation to agriculture and game. Game, Fish and Oyster Comm., Austin, Tex. iv + 60 p.

Kaye, S. V.
1959. A study of the eastern harvest mouse, *Reithrodontomys h. humulis* (Audubon and Bachman). Unpublished Master's thesis, North Carolina State College. 93 p.

Kays, C. E.
1956. An ecological study with emphasis on nutria

(Myocastor coypus) in the vicinity of Price Lake, Rockefeller Refuge, Cameron Parish, Louisiana. Unpublished Master's thesis, Louisiana State University, Baton Rouge. xiv + 145 p.

Kellogg, F. E., T. P. Kistner, R. K. Strickland, and R. R. Gerrish
1971. Arthropod parasites collected from white-tailed deer. Jour. Med. Ent., 8:495–498.

Kellogg, Remington
1940. Whales, giants of the sea. Nat. Geog., 77:35–90.

Kellogg, W. N.
1958. Echo-ranging in the porpoise. Science, 128:982–988.
1961. Porpoises and sonar. Univ. Chicago Press, Chicago. xiv + 177 p.

Kennerly, T. E.
1958. The baculum in the pocket gopher. Jour. Mamm., 39:445–446.

Kennicott, Robert
1857. The quadrupeds of Illinois, injurious and beneficial to the farmer. U.S. Patent Office Rept. (Agric.) for 1856:52–110, pls. 5–14.

Kent, George C.
1973. Comparative anatomy of the vertebrates. 3rd ed. The C. V. Mosby Co., St. Louis. 414 p.

Kidd, J. B.
1954. The fox and gray squirrels of Louisiana. Louisiana Conserv., 7(1):2–5.

Kilpatrick, A. R.
1852. Historical and statistical collections of Louisiana. The parish of Catahoula. De Bow's Review, 12:256–275, 631–646.

Kinsella, J. M.
1971. *Angiostrongylus schmidti* sp. n. (Nematoda: Metastrongyloidea) from the rice rat, *Oryzomys palustris*, in Florida, with a key to the species of *Angiostrongylus* Kamensky, 1905. Jour. Parasit., 57:494–497.
In Press. Helminth parasites of the cotton rat in Florida. Quart. Jour. Florida Acad. Sci.

Kirkpatrick, C. M.
1956. Coprophagy in the cottontail. Jour. Mamm., 37:300.

Korschgen, L. J.
1957. Food habits of the coyote in Missouri. Jour. Wildl. Mgmt., 21:424–435.

Kurten, Bjorn
1965. The Pleistocene Felidae of Florida. Bull. Florida State Mus., 9:215–273.

Lapham, V. T.
1950. A study of the genus Sylvilagus in the vicinity of Baton Rouge, Louisiana. Unpublished Master's thesis, Louisiana State University, Baton Rouge. x + 74 p., 2 tables, 7 figs., 24 pls.

Latham, R. M.
1951. The ecology and economics of predator management. Final Report, Pittman-Robertson Proj. 36-R, Report 2. Pennsylvania Game Comm., Harrisburg. 96 p., illus.

LaVal, R. K.
1970. Infraspecific relationships of bats of the species *Myotis austroriparius*. Jour. Mamm., 51:542–552.

Lay, D. W.
1942. Ecology of the opossum in eastern Texas. Jour. Mamm., 23:147–159.

Layne, J. N.
1959. Growth and development of the eastern harvest mouse, *Reithrodontomys humulis*. Bull. Florida State Mus., 4:61–82.
1965. Observations on marine mammals in Florida waters. Bull. Florida State Mus., Biol. Sci., 9:131–181.
1968. Host and ecological relationships of the parasitic helminth *Capallaria hepatica* in Florida mammals. Zoologica, 53:107–123.

Layne, J. N., D. E. Birkenholz, and J. V. Griffo, Jr.
1960. Records of *Dracunculus insignis* (Leidy, 1858) from raccoons in Florida. Jour. Parasit., 46:685.

Lee, Hong-Fang
1962. Susceptibility of mammalian host to experimental infection with *Heterobilharzia americana*. Jour. Parasit., 48:740–745.

Leidy, Joseph
1870. [*Ovibos cavifrons* from Natchez, Mississippi.] Proc. Acad, Nat. Sci. Philadelphia, 22:73.

Leopold, A. S.
1959. Wildlife of Mexico. Univ. California Press, Berkeley and Los Angeles. xiii + 568 p., illus.

Le Page du Pratz, A. S.
1758. Histoire de la Louisiane. de Bure, La Veuve Delaguette, et Lambert, Paris. 3 vols.

Lewis, D. M.
1968. Telemetry studies of white-tailed deer on Red Dirt Game Management Area, Louisiana. Unpublished Master's thesis, Louisiana State University, Baton Rouge. viii + 65 p., 11 illus.

Lewis, T. H.
1907. Spanish explorers in the southern United States 1528–1543. Charles Scribner's Sons, New York. 411 p.

Lilly, J. C.
1958. Some considerations regarding basic mechanisms of positive and negative types of motivations. Amer. Jour. Psychiatry, 115:498–504.
1961. Man and dolphin. Doubleday and Company, Inc., Garden City, N. Y. 312 p.
1966. Sonic-ultrasonic emissions of the bottlenose dolphin. Pages 503–509, *in* Whales, dolphins, and porpoises. Kenneth S. Norris, ed. Univ. California Press, Berkeley and Los Angeles. xv + 789 p.

Lilly, J. C., and Alice M. Miller
1961. Vocal exchanges between dolphins. Science, 134:1873–1876.

Link, V. B.
1955. A history of plague in the United States. Public Health Monograph No. 26, 70. vii + 120 p.

Linnaeus, Carolus
1758. Systema naturae, secundum, classes, ordines, genera, species cum characteribus differentiis, synonymis, locis. Ed. 10, tomus 1. Regnum animale. 824 p.

Linné, Carl von (see Carolus Linnaeus)

Little, M. D.
1965. Dermatitis in a human volunteer infected with *Strongyloides* of nutria and raccoon. Amer. Jour. Trop. Med. & Hgy., 14:1007–1009.
1966. Seven new species of *Strongyloides* (Nematoda) from Louisiana. Jour. Parasit., 52:85–97.

Llewellyn, L. M., and F. M. Uhler
1952. The foods of fur animals of the Patuxent Research Refuge, Maryland. Amer. Midl. Nat., 48:193–203.

Lockard, G. R.
1972. Further studies of dental annuli for aging white-tailed deer. Jour. Wildl. Mgmt., 36:46–55.

Lockett, S. H.
1873. Louisiana as it is. Original manuscript. LSU Archives. xv + 232 p.

Lorenz, K. Z.
1952. King Solomon's ring. Translated by Marjorie Kerr Wilson. Thomas Y. Crowell Company, New York. 202 p.
1955. Man meets dog. Translated by Marjorie Kerr Wilson. Houghton Mifflin Company, Boston. 211 p.
1963. On aggression. Translated by Marjorie Kerr Wilson. Harcourt, Brace & World, Inc., New York. xiv + 306 p.

Low, W. A., and I. McT.Cowan
1963. Age determination of deer by annular structure of dental cementum. Jour. Wildl. Mgmt., 27:466–471.

Lowery, G. H., Jr.
1936. A preliminary report on the distribution of the mammals of Louisiana. Proc. Louisiana Acad. Sci., 3:11–39.
1943. Check-list of the mammals of Louisiana and adjacent waters. Occas. Papers Mus. Zool., Louisiana State Univ., 13:213–257.

Lucker, J. T.
1943. A new trichostrongylid nematode from the stomachs of American squirrels. Jour. Washington Acad. Sci., 33:75–79.

Lumsden, R. D., and Carol Ann Winkler
1962. The opossum, *Didelphis virginiana* (Kerr), a host for the cyathocotylid trematode *Linstowiella szidati* (Anderson, 1944) in Louisiana. Jour. Parasit., 48:503.

Lumsden, R. D., and J. A. Zischke
1961. Seven trematodes from small mammals in Louisiana. Tulane Studies Zool., 9:89–98.

McBride, A. F.
1940. Meet Mister Porpoise *(Tursiops truncatus)*, Nat. Hist., 45(1):16–29.

McBride, A. F., and D. O. Hebb
1948. Behavior of the captive bottle-nose dolphin, *Tursiops truncatus*. Jour. Comp. and Physio. Psychol., 41:111–123.

McBride, A. F., and Henry Kritzler
1951. Observations on pregnancy, parturition, and postnatal behavior in the bottlenose dolphin. Jour. Mamm., 32:251–266.

McCarley, W. H.
1954 a. The ecological distribution of the *Peromyscus leucopus* species group in eastern Texas. Ecology, 35:375–379.
1954 b. Natural hybridization in the *Peromyscus leucopus* species group of mice. Evolution, 8:314–323.
1954 c. Fluctuations and structure of *Peromyscus gossypinus* populations in eastern Texas. Jour. Mamm., 85:526–532.
1958. Ecology, behavior and population dynamics of *Peromyscus nuttalli* in eastern Texas. Texas Jour. Sci., 10:147–171.
1959. An unusually large nest of *Cryptotis parva*. Jour. Mamm., 40:243.
1964. Ethological isolation in the cenospecies *Peromyscus leucopus*. Evolution, 18:331–332.

McClure, H. E.
1942. Summer activities of bats (Genus *Lasiurus*) in Iowa. Jour. Mamm., 23:430–434.

McCoy, C. J.
1960. Albinism in *Tadarida*. Jour. Mamm., 41:119.

McKeever, Sturgis
1958. Reproduction in the opossum in southwestern Georgia and northwestern Florida. Jour. Wildl. Mgmt., 22:303.
1971. *Zonorchis komareki* (McIntosh, 1939) (Trematoda:Dicrocoeliidae) from *Reithrodontomys humulis* (Audubon and Bachman). Jour. Parasit., 57:865.

McWilliams, R. G. (translator and editor)
1953. Fleur de Lys and Calumet, being the Pénicaut narrative of French adventure in Louisiana. Louisiana State Univ. Press, Baton Rouge. xxxiii + 282 p.

Mahken, Thomas, and R. M. Gilmore
1960. Suckerfish on porpoise. Jour. Mamm., 41:134.

Malek, E. A., L. R. Ash, H. F. Lee, and M. D. Little
1960. Schistosomiasis in Louisiana mammals. Jour. Parasit., 46(no. 5, sect. 2):34

Margry, Pierre
1878. Découvertes et éstablissements des Français dans l'ouest et dans le sud de l'Amerique Septentrionale (1614–1754). Quatrième partie. Découverte par mer des bouches du Mississippi et éstablissements de Lemoyne d'Iberville sur le golfe du Mexique (1694–1703). Maisonneuve et Cie, Paris. 653 p.

Martin, E. P.
1956. A population study of the prairie vole (Microtus ochrogaster). Univ. Kansas Publ., Mus. Nat. Hist., 8:361–416, 19 figs.

Martin, R. A.
1968. Late Pleistocene distribution of *Microtus pennsylvanicus*. Jour. Mamm., 49:265–271.

Matson, J. R.
1952. Litter size in the black bear. Jour. Mamm., 33:246–247.

Maynard, C. J.
1889. Singular effects produced by the bite of a short-tailed shres, Blarina brevicauda. Contributions to Science, 1:57–59.

Mayr, Ernst
1940. Speciation phenomena in birds. Amer. Nat., 74:249–278.
1942. Systematics and the origin of species, from the viewpoint of a zoologist. Columbia Univ. Press, New York. xiv + 334 p., num. illus.
1965. Animal species and evolution. Belknap Press of Harvard Univ., Cambridge, xiv + 797 p., num. illus.

Mease, J. A.
1929. Tularemia from opossums. Jour. Amer. Med. Assoc., 92:1042.

Melville, Herman
1930. Moby-Dick. Random House, New York. xxxi + 825 p.

Mengel, R. M.
1971. A study of dog-coyote hybrids and implications concerning hybridization in *Canis*. Jour. Mamm., 52:316–336.

Merriam, C. H.
1884. The vertebrates of the Adirondack region, northeastern New York. (Mammalia, concluded.) Trans. Linnaean Soc. New York, 2:5–214.

Miller, G. C., and Reinard Harkema
1964. Studies on helminths of North Carolina vertebrates. V. Parasites of the mink, *Mustela vison* Schreber. Jour. Parasit., 50:717–720.

Miller, G. S., Jr.
1924. List of North American recent mammals, 1923. Bull. U.S. Nat. Mus., 128. xvi + 673 p.
1928. The pollack whale in the Gulf of Campeche. Proc. Biol. Soc. Washington, 41:171.

Miller, G. S., Jr., and G. M. Allen
1928. The American bats of the genera Myotis and Pizonyx. Bull. U.S. Nat. Mus., 144. viii + 218 p.

Miller, G. S., Jr., and Remington Kellogg
1955. List of North American recent mammals. Bull. U.S. Nat. Mus., 205. xii + 954 p.

Moore, C. R., and David Bodian
1940. Opossum pouch young as experimental material. Anat. Rec., 76:319–327.

Moore, J. C.
1951. The range of the Florida manatee. Quart. Jour. Florida Acad. Sci., 14:1–19.
1953. Distribution of marine mammals to Florida waters. Amer. Midl. Nat., 49:117–158.
1957. The natural history of the fox squirrel, *Sciurus*

niger shermani. Bull. Amer. Mus. Nat. Hist., 113:1–71.
1960. New records of the Gulf-stream beaked whale, *Mesoplodon gervaisi*, and some taxonomic considerations. Amer. Mus. Novit., 1993:1–35.
1966. Diagnoses and distributions of beaked whales of the genus *Mesoplodon* known from North American waters. Pages 32–61, *in* Whales, dolphins, and porpoises. Kenneth S. Norris, ed. Univ. California Press, Berkeley and Los Angeles. xv + 789 p.
1968. Relationships among the living genera of beaked whales with classifications, diagnoses and keys. Fieldiana, Zool., 53:iv + 209–298.

Moore, J. C., and Eugenie Clark
1963. Discovery of right whales in the Gulf of Mexico. Science, 141, no. 3577:269.

Moore, J. C., and F. G. Wood, Jr.
1957. Differences between the beaked whales, *Mesoplodon mirus* and *Mesoplodon gervaisi*. Amer. Mus. Novit., 1831:1–25.

Morlan, H. B.
1952. Host relationships and seasonal abundance of some southwest Georgia ectoparasites. Amer. Midl. Nat., 48:74–93.

Murie, O. J.
1951. The elk of North America. Stackpole Company, Harrisburg, Pa. and Wildlife Mgmt. Inst., Washington, D.C. [x +] 376 p., illus.
1954. A field guide to animal tracks. Houghton Mifflin Co., Boston. xxii + 375 p.

National Livestock Historical Association
1905. Prose and poetry of the livestock industry of the United States. 3 vols., illus.

Neal, W. A.
1967. A study of the ecology of the woodrat in the hardwood forests of the lower Mississippi River basin. Unpublished Master's thesis, Louisiana State University, Baton Rouge. x + 116 p., 18 tables, 13 figs., 8 pls.

Negus, N. C., and R. K. Chipman
1956. A record of the piked whale, *Balaenoptera acutorostrata*, off the Louisiana coast. Proc. Louisiana Acad. Sci., 19:41–42.

Negus, N. C., and H. A. Dundee
1965. The nest of *Sorex longirostris*. Jour. Mamm., 46:495.

Negus, N. C., Edwin Gould, and R. K. Chipman
1961. Ecology of the rice rat, *Oryzomys palustris* (Harlan), on Breton Island, Gulf of Mexico, with a critique of social stress theory. Tulane Studies Zool., 8:93–123.

Newman, H. H.
1910. A large sperm whale captured in Texas waters. Science, 31, no. 799:631.

Newsom, J. D.
1969. History of deer and their habitat in the South. Proc. Symposium, white-tailed deer in southern forest habitat. Southern Forest Exp. Stat., Nacogdoches, Texas: 1–4.

Nishiwaki, Masaharu
1966. A discussion of varieties among the small cetaceans caught in Japanese waters. Pages 192–204, *in* Whales, dolphins, and porpoises. Kenneth S. Norris, ed. Univ. California Press, Berkeley and Los Angeles. xv + 789 p.
1972. General biology. Pages 3–204, *in* Mammals of the sea. Sam H. Ridgway, ed. Charles C. Thomas, Springfield, Illinois. xiii + 812 p.

Nishiwaki, Masaharu, T. Kasuya, T. Kamiya, T. Tobavama, and M. Nakajima
1965. Feresa attenuata captured on the Pacific coast of Japan in 1963. Sci. Repts., Whales Res. Inst., 19:65–90.

Noble, R. E.
1958. The survival, adaptation, distribution and ecology of transplanted beaver *(Castor canadensis)* in Louisiana. Unpublished Master's thesis, Louisiana State University, Baton Rouge. xviii + 215 p.
1969. Reproductive characteristics of the Mississippi white-tailed deer with notes on history, weights and age-class composition. Unpublished Doctoral dissertation, Michigan State University, East Lansing. 136 p.
1971. A recent record of the puma *(Felis concolor)* in Arkansas. Southwestern Nat., 16:209.

Norris, K. S.
1964. Some problems of echolocation in cetaceans. Vol. 1, pages 317–336, *in* Marine Bio-Acoustics. William N. Tavolga, ed. Pergamon Press, New York. xii + 413 p.

Norris, K. S., J. H. Prescott, P. V. Asa-Dorian, and Paul Perkins
1961. An experimental demonstration of echo-location behavior in the porpoise, *Tursiops truncatus* (Montague). Biol. Bull., 120:163–176.

Nowak, R. M.
1967. The red wolf in Louisiana. Defenders of Wildlife News, 42:60–70.
1970. Report on the red wolf. Defenders of Wildlife News, 45:82–94.
1972. The mysterious wolf of the South. Nat. Hist., 81:50–53, 74–77.
1973. A possible occurrence of the jaguar in Louisiana. Southwestern Nat., 17:430–432.

Odum, Eugene
1955. An eleven year history of a *Sigmodon* population. Jour. Mamm., 36:368–378.

Oesch, R. D.
1969. Fossil Felidae and Machairondontidae from two Missouri caves. Jour. Mamm., 50:367–368.

Olive, J. R.
1950. Some parasites of the prairie mole, Scalopus aquaticus machrinus (Rafinesque). Ohio Jour. Sci., 50:263–266.

Olsen, Ørjan
1913. On the external characters and biology of Bryde's whale, a new rorqual from the coast of South Africa. Proc. Zool. Soc. London, 1913:1073–1090, 5 pls., 14 figs.

Olson, R. W.
1969. Agonistic behavior of the short-tailed shrew *(Blarina brevicauda).* Jour. Mamm., 50:494–500.

Omura, Hideo
1959. Bryde's whale from the coast of Japan. Whales Research Inst., 14:1–33, 6 pls., 10 figs.

O'Neil, Ted
1949. The muskrat in the Louisiana coastal marshes. Louisiana Dept. Wild Life and Fisheries. 152 p., num. illus.
1968. The fur industry in retrospect. Louisiana Conser., 20 (9 and 10): 9–14.

Orr, R. T.
1966. Risso's dolphin on the Pacific Coast of North America. Jour. Mamm., 47:341–343.

Osgood, W. H.
1909. Revision of the mice of the American genus Peromyscus. N. Amer. Fauna, 28:1–280, illus.

Oswald, V. H.
1958. Helminth parasites of the short-tailed shrew in central Ohio. Ohio Jour. Sci., 58:325–334.

Owen, R. D., and R. M. Shackelford
1942. Color aberrations in Microtus and Pitymys. Jour. Mamm., 23:306–314.

Packard, R. L.
1961. Additional records of mammals from eastern Texas. Southwestern Nat., 6:193–195.
1969. Taxonomic review of the golden mouse, Ochrotomys nuttalli. Misc. Publ., Univ. Kansas Mus. Nat. Hist., 51:373–406.

Packard, R. L., and Herschel Garner
1964. Arboreal nests of the golden mouse in eastern Texas. Jour. Mamm., 45:369–374.

Palmisano, A. W.
1967. Ecology of *Scirpus olneyi* and *Scirpus robustus* in Louisiana coastal marshes. Unpublished Master's thesis, Louisiana State University, Baton Rouge. xii + 145 p.
1970. Plant community-soil relationships in Louisiana coastal marshes. Unpublished Doctoral dissertation, Louisiana State University, Baton Rouge. xii + 98 p.
1971. Louisiana's fur industry. Commercial Wildlife Work Unit Report of Louisiana Wild Life and Fisheries Commission to U. S. Army Corps of Engineers. New Orleans District. Mimeographed.

Paradiso, J. L.
1958. The common blackfish in Virginia coastal waters. Jour. Mamm., 39:440–441.
1968. Canids recently collected in east Texas, with comments on the taxonomy of the red wolf. Amer. Midl. Nat., 80:529–534.

Paradiso, J. L., and R. M. Nowak
1971. A report on the taxonomic status and distribution of the red wolf. Special Scientific Report–Wildlife no. 145. ii + 36 p.
1972. Canis rufus. Mammalian Species No. 22:1–4.

Parker, J. C.
1968. Parasites of the gray squirrel in Virginia. Jour. Parasit., 54:633–634.

Parmalee, P. W.
1954. Food of the Great Horned Owl and Barn Owl in East Texas. Auk, 71:469–470.

Paul, J. R.
1968. Risso's dolphin, *Grampus griseus*, in the Gulf of Mexico. Jour. Mamm., 49:746–748.

Pearson, O. P.
1942. On the cause and nature of a poisonous action produced by the bite of a shrew *(Blarina brevicauda)*. Jour. Mamm., 23:159–166.
1950. The submaxillary glands of shrews. Anat. Record, 107:161–169, 1 pl.

Pearson, O. P., and Anita K. Pearson
1947. Owl predation in Pennsylvania, with notes on the small mammals of Delaware County. Jour. Mamm., 28:137–147.

Pearson, P. G.
1952. Observations concerning the life history and ecology of the woodrat, *Neotoma floridana floridana* (Ord). Jour. Mamm., 33:459–463.
1953. A field study of Peromyscus populations in Gulf Hammock, Florida. Ecology, 34:199–207.

Penn, G. H., Jr.
1942. Parasitological survey of Louisiana muskrats. Jour. Parasit., 28:348–349.

Penn, G. H., Jr., and E. C. Martin
1941. The occurrence of porocephaliasis in the Louisiana muskrat. Jour. Wildl. Mgmt., 5:13–14.

Perrin, W. F.
1972. Color patterns of spinner porpoises *(Stenella* cf. *S. longirostris)* of the eastern Pacific and Hawaii, with comments on delphinid pigmentation. Fishery Bull., 70(3):983–1003.

Perrin, W. F., and C. L. Hubbs
1969. Observations on a young pygmy killer whale *(Feresa attenuata* Gray) from the eastern tropical Pacific Ocean. Trans. San Diego Soc. Nat. Hist., 15:297–308.

Petrides, G. A.
1949. Sex and age determination in the opossum. Jour. Mamm., 30:364–378.
1950. The determination of sex and age ratios in fur animals. Amer. Midl. Nat., 43:355–382.

Philippi-B., R. A.
1893. Los delfines de la punta austral de la America del sur. Annales del Museo Nacional Chile, Zool., 6:p. 14, pl. 5, fig. 3.

Phillips, G. L.
1966. Ecology of the big brown bat (Chiroptera:Vespertilionidae) in northeastern Kansas. Amer. Midl. Nat., 75:168–198.

Polderboer, E. B., L. W. Kuhn, and G. O. Hendrickson
1941. Winter and spring habits of weasels in central Iowa. Jour. Wildl. Mgmt., 5:115–119.

Pournelle, G. H.
1952. Reproduction and early post-natal development

of the cotton mouse, *Peromyscus gossypinus gossypinus*. Jour. Mamm., 33:1–20.

Pryor, Taylor, Karen Pryor, and K. S. Norris
1965. Observations on a pygmy killer whale *(Feresa attenuata* Gray) from Hawaii. Jour. Mamm., 46:450–461.

Quiring, D. P., and C. F. Harlan
1953. On the anatomy of the manatee. Jour. Mamm., 34:192–203.

R., W. H.
1898. Game in the South. Southern Sportsman, 1:10–12.

Rainey, D. G.
1956. Eastern wood rat, Neotoma floridana: Life history and ecology, Univ. Kansas Mus. Nat. Hist., 8:535–645, 12 pls., 13 figs.

andolph, N. M., and R. B. Eads
1946. An ectoparasite survey of mammals from Lavaca County, Texas. Ann. Ento. Soc., Amer., 39:597–601.

Ransom, A. B.
1966. Determining age of white-tailed deer from layers in cementum of molars. Jour. Wildl. Mgmt., 30:197–199.

Raun, G. G., H. D. Hoese, and Frank Mosely
1970. Pygmy sperm whales, genus *Kogia*, on the Texas coast. Texas Jour. Sci., 21:269–274.

Raynor, G. S.
1960. Three litters in a pine mouse nest. Jour. Mamm., 41:275.

Read, W. A.
1931. Louisiana-French. University Studies No. 5. Louisiana State University Press, Baton Rouge. xxiv + 253 p.
1963. Louisiana-French, (Revised Edition.) Louisiana State University Press, Baton Rouge. xxiv + 263 p.

Reynolds, H. C.
1945. A contribution to the life history and ecology of the opossum in central Missouri. Jour. Mamm., 26:361–379.

Reysenbach de Haan, F. W.
1960. Some aspects of mammalian hearing under water. Proc. Roy. Soc., (B) 152:54–62.
1966. Listening underwater:thoughts on sound and cetacean hearing. Pages 583–596, *in* Whales, dolphins, and porpoises. Kenneth S. Norris, ed. Univ. California Press. xv + 789 p.

Rhoads, S. N.
1897. [=19 January 1898]. Notes on the living and extinct species of North American Bovidae. Proc. Acad. Nat. Sci., Philadelphia, 49:483–502.

Rice, D. W.
1957. Life history and ecology of *Myotis austroriparius* in Florida. Jour. Mamm., 38:15–32.
1965. Bryde's whale in the Gulf of Mexico. Norsk Hvalfangst-Tidende, 54(5):114–115.
1967. Cetaceans. *In* Recent mammals of the world, a synopsis of the families. Sydney Anderson and J. K. Jones, eds. Ronald Press, New York. 453 p.

Rice, D. W., and D. K. Caldwell
1961. Observations on the habits of the whalesucker *(Remilegia australis)*. Norsk Hvalfangst-Tidende, 50(5):181–182, 185–186, 189.

Rice, D. W., and V. B. Scheffer
1968. A list of the marine mammals of the world. U.S. Fish and Wildl. Serv., Special Sci. Rept.—Fisheries, 579:1–12.

Richard, J.
1936. Notes sur les Cétacés et les Pinnipedès. *In* Résultats campagnes scientifiques du Prince du Monaco, 94:34–71, 8 pls.

Richardson, L. V.
1963. Food preferences and nutritive content of selected plants fed to cottontail and swamp rabbits. Unpublished Master's thesis, Louisiana State University, Baton Rouge. x + 77 p., 23 tables, 9 figs., 2 pls.

Ridgway, Robert
1912. Color standards and color nomenclature. Privately published, Washington, D.C. iii + 43 p., 53 pls.

Riley, G. A., and R. T. McBride
1972. A survey of the red wolf *(Canis rufus)*. Spec. Sci. Rept.-Wildlife No. 162. iii + 15 p.

Roberson, J. H., Jr.
1967. A survey of breeding seasons of white-tailed deer *(Odocoileus virginianus)* in Louisiana. Unpublished Master's thesis, Louisiana State University, Baton Rouge. xii + 89 p., 10 tables, 10 figs.

Roberts, H. A., and R. C. Early
1952. Mammal survey of southeastern Pennsylvania. Final Report Pittman-Robertson Proj. 43-R. Pennsylvania Game Comm., Harrisburg. 70 p.

Roosevelt, Theodore
1908. In the Louisiana canebrakes. Scribner's Magazine, 43:47–60.

Roth, E. C., W. V. Adams, G. E. Sanford, Jr., Betty Greer, Kay Newman, Mary Moore, Patricia Mayeax, and Donna Linder
1963. The bacteriologic and serologic incidence of leptospirosis among striped skunks in Louisiana. Zoonoses Research, 2:13–39.

Rue, L. L., III
1965. Cottontail. Thomas Y. Crowell, New York. xii + 112 p.

Sawyer, T. K.
1961. The American otter, *Lutra canadensis vaga*, as a host for two species of trematodes previously unreported from North America. Proc. Helmint. Soc. Washington, 28:175–176.

Schacher, J. F., and E. C. Faust
1956. Occurrence of *Dioctophyma renale* in Louisiana, with remarks on the size of infertile eggs of this species. Jour. Parasit., 42:533–535.

Scheffer, V. B.
1969. The year of the whale. Charles Scribners' Sons, New York. ix + 214 p.

Scheffer, V. B., and D. W. Rice
1963. A list of the marine mammals of the world. U.S. Fish and Wildl. Serv., Special Sci. Rept.–Fisheries, 431:1–12.

Schevill, W. E.
1964. Underwater sounds of cetaceans. Pages 307–316, *in* Marine Bio-Acoustics, Vol. 2, William N. Tavolga, ed. Pergamon Press, New York. 413 p.

Schevill, W. E., and Barbara Lawrence
1956. Food-finding by a captive porpoise *(Tursiops truncatus)*. Brevoria, 53:1–15.

Schiller, E. L.
1952. Studies on the helminth fauna of Alaska. V. Notes on Adak rats *(Rattus norvegicus* Berkenhout) with special reference to helminth parasites. Jour. Mamm., 33:38–49.

Schmidly, D. J., M. H. Beleau, and Henry Hildebran [sic = Hildebrand]
1972. First record of Cuvier's dolphin from the Gulf of Mexico with comments on the taxonomic status of *Stenella frontalis*. Jour. Mamm., 53:625–628.

Schmidly, D. J., C. O. Martin, and G. F. Collins
1972. First occurrence of a black right whale *(Balaena glacialis)* along the Texas coast. Southwestern Nat., 17:214–215.

Schmidly, D. J., and Betty A. Reeves
In Press. Annotated checklist and key to the cetaceans of Texas waters. Southwestern Nat.

Schmidt, G. D.
1965. *Molineus mustelae* sp. n (Nematoda:Trichostrongylidae) from the long-tailed weasel in Montana and M. *chabaudi* nom. n., with a key to the species of *Molineus*. Jour. Parasit., 51:164–168.

Schofield, R. D.
1955. Analysis of muskrat age determination methods and their application in Michigan. Jour. Wildl. Mgmt., 19:463–466.

Scholander, P. F., Laurence Irving, and S. W. Grinnell
1943. Respiration of the armadillo, with possible implications as to its burrowing. Jour. Cell. Comp. Physiol., 21:53.

Schwartz, Albert, and E. P. Odum
1957. The woodrats of the eastern United States. Jour. Mamm., 38:197–206.

Schwartz, C. W., and Elizabeth R. Schwartz
1959. The wild mammals of Missouri. Univ. Missouri Press and Missouri Conserv. Comm. xvi + 341 p.

Sealander, J. A., Jr.
1956. A provisional check-list and key to the mammals of Arkansas (with annotations). Amer. Midl. Nat., 56:257–296.

Self, J. T., and J. H. Esslinger
1955. A new species of bothriocephalid cestode from the fox squirrel *(Sciurus niger* Linn.). Jour. Parasit., 41:256–258.

Sergeant, D. E.
  1962. On the external characters of the blackfish or pilot whales (genus *Globicephala*). Jour. Mamm., 43:395–413.

Seton, E. T.
  1929. Lives of game animals. Vols. 1–4. Doubleday, Doran and Co., Inc., Garden City, New York.

Severinghaus, C. W.
  1949. Tooth development and wear as criteria of age in white-tailed deer. Jour. Wildl. Mgmt., 13:195–241.
  1955. Some observations on the breeding behavior of deer. New York Fish and Game Jour., 2:239–241.

Severinghaus, C. W., and E. L. Cheatum
  1956. Life and times of the white-tailed deer. Pages 57–186, *in* The deer of North America. W. P. Taylor, ed. The Stackpole Co., Harrisburg, Pa., and The Wildl. Mgmt. Inst., Washington, D.C. xvii + 668 p., illus.

Shadowen, H. E.
  1956. Rodent population dynamics in uncultivated fields of Louisiana. Unpublished Doctoral dissertation, Louisiana State University, Baton Rouge. xi + 86 p.
  1963. A live-trap study of small mammals in Louisiana. Jour. Mamm., 44:103–108.

Shanks, C. E.
  1948. The pelt-primeness method of aging muskrats. Amer. Midl. Nat., 39:179–187.

Sharp, H. F., Jr.
  1967. Food ecology of the rice rat, *Oryzomys palustris* (Harlan), in a Georgia salt marsh. Jour. Mamm., 48:557–563.

Sherman, H. B.
  1930. Birth of the young of Myotis austroriparius. Jour. Mamm., 11:495–503.
  1937. Breeding habits of the free-tailed bat. Jour. Mamm., 18:176–187.
  1951. Aberrant color phases of the cotton rat *Sigmodon*. Jour. Mamm., 32:217.

Short, H. L., R. B. Davis, and C. F. Herreid, Jr.
  1960. Movements of the Mexican free-tailed bat in Texas. Southwestern Nat., 5:208–216.

Shotts, E. B., William Greer, and F. A. Hayes
  1958. A preliminary survey of the incidence of brucellosis and leptospirosis among white-tailed deer (*Odocoileus virginianus*) of the Southeast. Jour. Amer. Vet. Med. Assoc., 133:359–361.

Siebenaler, J. B., and D. K. Caldwell
  1956. Cooperation among adult dolphins. Jour. Mamm., 37:126–128.

Simpson, George G.
  1941 a. Discovery of jaguar bones and footprints in a cave in Tennessee. Amer. Mus. Novit., no. 1131:1–12.
  1941 b. Large Pleistocene felines of North America. Amer. Mus. Novit., no. 1136:1–27.

1941 c. Vernacular names of South American mammals. Jour. Mamm., 22:1–17.
  1971. Concluding remarks: Mesozoic mammals revisited. Pages 181–198, *in* Early mammals. D. M. and K. A. Kermack, editors. Academic Press, London.

Sisson, Septimus, and J. D. Grossman
  1953. The anatomy of the domestic animals. Fourth ed. W. B. Saunders Co., Philadelphia. 972 p.

Skipwith, H.
  1892. East Feliciana, Louisiana: past and present. Hopkins Printing Office, New Orleans. 61 p.

Smith, John
  1608. A true relation. London.
  1624. The general historie of Virginia. Second Book. Pages 21–40, Michael Sparkes, London. 14 unnumbered pages + 248 p., 3 maps, 2 pls.

Smith, K. C.
  1964. The beaver in Louisiana. [Louisiana] Wildl. Educ. Bull., 83:1–10.

Sogandares-Bernal, Franklin
  1961. *Sellacotyle vitellosa*, a new troglotrematid trematode from the mink in Louisiana. Jour. Parasit., 47:911–912.

Sollberger, D. E.
  1943. Notes on the breeding habits of the eastern flying squirrel (*Glaucomys volans volans*). Jour. Mamm., 24:163–173.

Southern, H. N.
  1942. Periodicity of refection in the wild rabbit. Nature, 149:553–554.

Springer, J. W.
  1973. The prehistory and cultural geography of coastal Louisiana. Unpublished Doctoral dissertation, Yale University. 184 p.

St. Amant, L. S.
  1959. Louisiana wildlife inventory and management plan. Louisiana Wildl. and Fish. Comm. xx + 329 p.

Standley, P. C.
  1921. Albinism in the black bear. Science (n.s.), 54, no. 1386:74.

Stevenson, H. M.
  1962. Occurrence and habits of the eastern chipmunk in Florida. Jour. Mamm., 43:110–111.

Stiles, C. W., and Clara Edith Baker
  1935. Key-catalogue of parasites reported for Carnivora (cats, dogs, bears, etc.) with their possible health importance. Nat. Inst. Health Bull. no. 163:913–1223.

Stiles, C. W., and S. F. Stanley
  1932. Key-catalogue of parasites reported for Insectivora (moles, shrews, etc.) with their possible public health importance. Nat. Inst. Health Bull. no. 159:791–911.

Stoddard, H. L.
  1920. The flying squirrel as a bird killer. Jour. Mamm., 1:95–96.

1936. The bobwhite quail. Charles Scribner's Sons, New York. xxix + 599 p.

Stoner, Dayton
1938. New York state records for the common dolphin, *Delphinus delphis*. New York State Mus. Circ., 21:1–16.

Storer, T. I.
1923. Rabies in a mountain lion. California Fish and Game, 9(2):1–4.

Strandtmann, R. W.
1962. *Nycteriglyphus bifolium* n. sp., a new cavernicolous mite associated with bats (Chiroptera) (Acarina: Glycyphagidae). Acarologia, 4:623–631.

Strecker, J. K.
1926. The extension of range of the nine-banded armadillo. Jour. Mamm., 7:206–210.

Stuewer, F. W.
1943. Raccoons: their habits and management in Michigan. Ecol. Monog., 13:203–258.

Sullivan, E. G., and A. O. Haugen
1956. Age determination of foxes by x-ray of forefeet. Jour. Wildl. Mgmt., 20:210–212.

Svihla, Arthur
1929. Life history notes on Sigmodon hispidus hispidus. Jour. Mamm., 10:352–353.
1931. Life history of the Texas rice rat (Oryzomys palustris texensis). Jour. Mamm., 12:238–242.

Svihla, Arthur, and Ruth D. Svihla
1933. Notes on the life history of the woodrat, Neotoma floridana rubida Bangs. Jour. Mamm., 14:73–75.

Svihla, Ruth D.
1930a. Notes on the golden harvest mouse. Jour. Mamm., 11:53–54.
1930b. A family of flying squirrels. Jour. Mamm., 11:211–213.

Swanton, J. R.
1940. American bison in northern Louisiana. Jour. Mamm., 21:222.

Taber, F. W.
1945. Contribution on the life history and ecology of the nine-banded armadillo. Jour. Mamm., 26:211–226.

Talmage, R. V., and G. D. Buchanan
1954. The armadillo *(Dasypus novemcinctus)*, a review of its natural history, ecology, anatomy and reproductive physiology. The Rice Inst. Pamphlet, 41. viii + 135 p.

Tavolga, Margaret C.
1966. Behavior of the bottle-nosed dolphin *(Tursiops truncatus)*:social interactions in a captive colony. Pages 718–730, *in* Whales, dolphins, and porpoises. Kenneth S. Norris, ed. Univ. California Press, Berkeley and Los Angeles. xv + 789 p.

Tavolga, Margaret C., and F. S. Essapian
1957. The behavior of the bottle-nosed dolphin *(Tursiops truncatus)*:mating, pregnancy, parturition, and mother-infant behavior. Zoologica: New York Zool. Soc., 42:11–31.

Taylor, D. F.
1971. A radio-telemetry study of the black bear *(Euarctos americanus)* with notes on its history and present status in Louisiana. Unpublished Master's thesis, Louisiana State University, Baton Rouge. x + 87 p., 19 figs., 7 tables.

Taylor, W. P.
1947. Recent record of the jaguar in Texas. Jour. Mamm., 28:66.

Thigpen, T. D.
1950. A study of beaver in Louisiana. Unpublished Master's thesis, Louisiana State University, Baton Rouge. xii + 168 p., 15 tables, 6 figs., 55 pls.

Thomas, Kim R.
1973. The activity variations and burrow systems of the eastern chipmunk *(Tamias striatus)* in Louisiana, with notes on other aspects of ecology and behavior. Unpublished Master's thesis, Louisiana State University, Baton Rouge. vi + 46 p., illus.

Thomas, Oldfield
1921. Bats on migration. Jour. Mamm., 2:167.

Tomilin, A. G.
1957. Animals of USSR and adjacent lands. Tom. 9. Cetaceans. Akademiya Nauk SSSR, Moscow. 756 p., 12 pls.

Townsend, C. H.
1935. The distribution of certain whales as shown by logbook records of American whaleships. Zoologica, 19(1):3–50, 4 maps.

Trainer, D. O.
1962. Protozoan diseases of white-tailed deer *(Odocoileus virginianus)*. Proc. First Nat. White-tailed Deer Disease Symp., Univ. Georgia, Athens:155–161.

Trippensee, R. E.
1948. Wildlife management. McGraw-Hill, New York. x + 479 p.

Trodd, L. L.
1962. Quadruplet fetuses in a white-tailed deer from Espanola, Ontario. Jour. Mamm., 43:414.

True, F. W.
1885. On a spotted dolphin apparently identical with Prodelphinus doris of Gray. Rept. U.S. Nat. Mus. for 1884:317–324.
1889. Contributions to the natural history of the cetaceans, a review of the family Delphinidae. Bull. U.S. Nat. Mus., 36:1–192, 47 pls.
1904. Whalebone whales of the western North Atlantic compared with those occurring in European waters with some observations on the species of the North Pacific. Smithson. Contrib. Knowledge, 33. vii + 332 p., 50 pls., 97 figs.

Tuttle, M. D.
1964. Additional record of *Sorex longirostris* in Tennessee. Jour. Mamm., 45:146–147.

Ubelaker, J. E., and J. F. Downhower
1965. Parasites recovered from *Geomys bursarius* in Douglas County. Trans. Kansas Acad. Sci., 68:206–208.

Uhlig, H. G.
1955. The gray squirrel—its life history, ecology, and population characteristics in West Virginia. Conserv. Comm. West Virginia. 75 p.

Ulmer, F. A., Jr.
1940. Albinism in Blarina. Jour. Mamm., 21:89–90.
1941. Melanism in the Felidae, with special reference to the genus Lynx. Jour. Mamm., 22:285–288.

Van Bree, P. J. H.
1971. On the skull of *Stenella longirostris* (Gray, 1828) from the eastern Atlantic. Notes on Cetacea, Delphinoidea IV. Beaufortia, 19, no. 251: 99–106.

Van Gelder, R. G.
1953. The egg-opening technique of a spotted skunk. Jour. Mamm., 34:255–256.
1959. A taxonomic revision of the spotted skunks (genus *Spilogale*). Bull. Amer. Mus. Nat. Hist., 117:233–392, 47 figs., 32 tables.

van Zyll de Jong, C. G.
1972. A systematic review of the Nearctic and Neotropical river otters (genus *Lutra,* Mustelidae, Carnivora). Life Sci. Contri. R Ontario Mus., 80:1–104.

Vaughan, T. A.
1962. Reproduction in the plains pocket gopher in Colorado. Jour. Mamm., 43:1–13.

Verts, B. J.
1967. The biology of the striped skunk. Univ. Illinois Press, Urbana and Chicago. vii + 218 p.

Vigoureaux, P.
1960. Underwater sound. Proc. Roy. Soc., (B) 152: 49–51.

Villa R., Bernardo
1956. *Tadarida brasiliensis mexicana* (Saussure), el murcielago guanero, es una subespecie migratoria. Acta Zool. Mex., 1:1–11.
1969. La ballena rorcual o ballena de aleta, *Balaenoptera borealis* Lesson, 1828, en la costa de Veracruz, México. Ann. Inst. Biol., Univ. Nat. Auton. México, 40, Ser. Zool. (1):129–138.

Villa R., Bernardo, and E. L. Cockrum
1962. Migration of the guano bat *Tadarida brasiliensis mexicana* (Saussure). Jour. Mamm., 43:43–64.

Voge, Marietta
1955. A list of cestode parasites from California mammals. Amer. Midl. Nat., 54:413–417.

Wade, Otis, and P. T. Gilbert
1940. The baculum of some Sciuridae and its significance in determining relationships. Jour. Mamm., 21:52–63.

Waldo, Ednard
1957. Whales in Gulf of Mexico. Louisiana Conserv., 9(4):13–15.

Walker, Alex.
1930. The "hand-stand" and some other habits of the Oregon spotted skunk. Jour. Mamm., 11:227–229.

Walker, E. P., Florence Warnick, Sybil E. Hamlet, K. I. Lange, Mary A. Davis, H. E. Uible, and P. F. Wright
1968. Mammals of the world. Second edition, with revisions by J. L. Paradiso. Johns Hopkins Press, Baltimore. 2 vols. xlviii + 1500 p.

Watson, J. S., and R. H. Taylor
1955. Reingestion in the hare *Lepus europaeus* Pal. Science, 121:314.

Wenk, P.
1969. Investigations of population dynamics in cotton rat filariasis (*Litomosoides carninii*: Filariidac). [Abstract.] Parasitology, 59:11–12.

Werner, H. J., W. W. Dalquest, and J. H. Roberts
1950. Histological aspects of the glands of the bat, *Tadarida cynocephala* (Le Conte). Jour. Mamm., 31:395–399.

Whitaker, J. O., Jr.
1968. Parasites. Pages 254–311, *in* Biology of *Peromyscus* (Rodentia). J. A. King, ed. Spec. Publ. No. 2, American Society of Mammalogy. xiii + 593 p.

Wilks, B. J., and H. E. Laughlin
1961. Roadrunner preys on a bat. Jour. Mamm., 42:98.

Williamson, F. W., and G. T. Goodman
1939. Eastern Louisiana: a history of the watershed of the Ouachita River and the Florida Parishes. The Historical Record Assoc., Louisville, Monroe, and Shreveport. 3 vols.

Wimsatt, W. A.
1945. Notes on breeding behavior, pregnancy, and parturition in some vespertilionid bats of the eastern United States. Jour. Mamm., 26:23–33.

Wislocki, G. B.
1933. Location of the testes and body temperature in mammals. Quant. Rev. Biol, 8:385–396.

Wislocki, G. B., J. C. Aub, and C. M. Waldo
1947. The effects of gonadectomy and the administration of testosterone propionate on the growth of antlers in male and female deer. Endocrinology, 40:202–224.

Wood, A. E.
1955. A revised classification of the rodents. Jour. Mamm., 36:165–187.

Wood, F. G., Jr.
1953. Underwater sound production and concurrent behavior of captive porpoises, *Tursiops truncatus* and *Stenella plagiodon*. Bull. Mar. Sci. Gulf and Caribbean, 3:120–133.

Wood, F. G., Jr., D. K. Caldwell, and Melba Caldwell
1970. Behavioral interactions between porpoises and sharks. Pages 264–277, Vol. 2, *in* Investigations on Cetacea. G. Pilleri ed. Hirnanatomisches Institute der Universitat. Berne, Switz.

Wood, J. E.
  1955. Notes on young pocket gophers. Jour. Mamm., 36:143–144.

Worth, C. B.
  1950 a. Field and laboratory observations on roof rats, *Rattus rattus* (Linnaeus), in Florida. Jour. Mamm., 31:293–304.
  1950 b. Observations on the behavior and breeding of captive rice rats and woodrats. Jour. Mamm., 31:421–426.

Wright, P. L.
  1942. Delayed implantation in the long-tailed weasel (Mustela frenata), and short-tailed weasel (Mustela cicognani), and the marten (Martes americana). Anat. Rec., 83:341–353, 2 pls.
  1947. The sexual cycle of the male long-tailed weasel (Mustela frenata). Jour. Mamm., 28:343–352.
  1948 a. Breeding habits of captive long-tailed weasels (Mustela frenata). Amer. Midl. Nat., 39:388–344.
  1948 b. Preimplantation stages in the long-tailed weasel (Mustela frenata). Anat. Rec., 100:593–607, 2 pls.

Yamada, Munesato
  1954. An account of a rare porpoise, *Feresa* Gray from Japan. Sci. Repts., Whales Res. Inst., 9:59–88.

Yeh, James, and D. E. Davis
  1950. Seasonal changes in abundance of fleas on rats at Baltimore, Md. Public Health Reports, 65:337–342.

Young, S. P.
  1958. The bobcat of North America. Stackpole Company, Harrisburg, Pa., and Wildl. Mgmt. Inst., Washington, D.C. xiv + 193 p., 128 pls., 10 figs.

Young, S. P., and E. A. Goldman
  1944. The wolves of North America. Part I, by Young. Their history, life habits, economic status, and control. Part II, by Goldman. Classification of wolves. The American Wildl. Inst., Washington, D.C. xxi + 636 p.
  1946. The puma, mysterious American cat. Part I, by Young. History, life habits, economic status, and control. Part II, by Goldman. Classification of the races of the puma. American Wildl. Inst., Washington, D.C. xiv + 358 p., 93 pls., 6 figs.

Young, S. P., and H. H. T. Jackson
  1951. The clever coyote. Part I, by Young. Its history, life habits, economic status, and control. Part II, by Jackson. Classification of the races of the coyote. Stackpole Co., Harrisburg, Pa., and Wildl. Mgmt. Inst., Washington, D.C. xvi + 411 p., 81 pls., 28 figs.

Zam, S. G., D. K. Caldwell, and Melba C. Caldwell
  1971. Some endoparasites from small odontocete cetaceans collected in Florida and Georgia. Cetology, 2:1–11.

Zippin, Calvin
  1958. The removal method of population estimation. Jour. Wildl. Mgmt., 22:82–90.

INDEX

A **boldface** number refers to the page on which a given species account begins.

sexual characters and age
criteria of, 192; status and
habits of, 193; subspecies of,
193; underground burrows of,
192
Wood, F. G., 305
Woodland Vole, **264**–268

X

Xenobalanus globicipitus, 347

Y

Yak, 513

Z

Zalophus, dental formula of, 15
   californianus, **471**–473, Fig. 218;
      cranial characters of, 473, Fig.
      219; distribution of, 471, 473
Zebu, 513
zibethicus, Ondatra, **268**–279
Ziphiidae, characters of, 309
Ziphius, 309, 313; dentition of, 309
   cavirostris, **309**–312, Fig. 133;
      cranial and other skeletal
      characters of, 310, Fig. 134;
      distribution of, 309, Map 49;
      external appearance of, 309,
      Fig. 133; food of, 312;
      measurements and weights of,
      310; other names of, 309; status
      and habits of, 310
Zonorchis komareki, 229, 233
zygomatic breadth, definition of, 51
Zygorhiza kochii, 517